Linear Algebra
Basics, Practice, and Theory

Linear Algebra
Basics, Practice, and Theory

Bernard Gelbaum

North-Holland
New York • Amsterdam • London

No responsibility is assumed by the publisher for any injury and/or damage to persons or property as a matter of products liability, negligence, or otherwise, or from any use or operation of any methods, products, instructions, or ideas contained in the material herein.

Elsevier Science Publishing Co., Inc.
52 Vanderbilt Avenue, New York, New York 10017

Sole distributors outside the United States and Canada:

Elsevier Science Publishers B.V.
P.O. Box 211, 1000 AE Amsterdam, the Netherlands

© 1989 by Elsevier Science Publishing Co., Inc.

This book has been registered with the Copyright Clearance Center, Inc.
For further information, please contact the Copyright Clearance Center, Inc.,
Salem, Massachusetts.

Library of Congress Cataloging-in-Publication Data

Gelbaum, Bernard R.
   Linear algebra : basics, practice, and theory / Bernard R. Gelbaum
      p. cm.
   Bibliography: p.
   Includes index.
   ISBN 0-444-01460-8
   1. Algebras, Linear. I. Title
QA184.G43 1989
512'.5--dc19   88-26003
                                                           CIP

Current printing (last digit)
10 9 8 7 6 5 4 3 2 1

Manufactured in the United States of America

CONTENTS

PART I BASICS

Chapter 1 ORIENTATION

Chapter 2 The *PROCESS* (Simple elimination)

Chapter 3 DETERMINANTS (A direct approach)

Contents

Chapter 7 QUADRATIC FORMS

Chapter 8 MISCELLANY

PREFACE

This book has three parts: **Basics, Practice,** and **Theory.**

The study of linear transformations between finite-dimensional vector spaces is carried out ultimately, both in practice and in theory, in terms of matrices. This is true because every finite-dimensional vector space V is, for some natural number n, essentially the set of all n-tuples of elements of a field K, e.g., real or complex numbers. Thus much of this book is concerned with the study of n-dimensional real space $\mathbf{R}^n$ or n-dimensional complex space $\mathbf{C}^n$ and linear transformations among and within them.

The identification of V with one of those spaces can be made in an infinite number of ways and, so viewed, V can be studied abstractly. The consequence is that a linear transformation T can be represented as a matrix in infinitely many ways and the search for a useful matrix representation of T is regarded as the search for a useful identification of V with $\mathbf{C}^n$, in effect a search for a useful basis for V. This approach to linear algebra is described in **Part III** but is *not* the one used in **Part I.**

In **Part I** the emphasis is on matrices themselves, and on their algebraic relations. The fundamentals of linear algebra are presented in the context of matrices as rectangular arrays and of vectors as special kinds of matrices, i.e., $n \times 1$ or $1 \times n$ matrices. The more abstract development in **Part III** serves, in particular, to illuminate the concrete discussion of **Part I.**

Part I. Basics.

> **Chapter 1.** *Orientation.* Here the motivations for the study of linear algebra are given in terms of five problem classes.

> **Chapter 2.** *The PROCESS* (**Simple elimination**). Several versions are given of the elimination method for analyzing a system of finitely many equations in finitely many unknowns. The experience with this kind of analysis is made to lead naturally to the study of matrix algebra and the language of linear algebra. The notions of invertible matrices, rank, dimension, solvability of systems of equations, etc., are repeatedly brought back to their roots in the different versions of elimination.

> **Chapter 3.** *Determinants* (**A direct approach**). The standard approaches to determinant theory are replaced by a simpler description, again based firmly in the elimination methods discussed in **Chapter 2.** Fortunately the simpler definition of a determinant is not merely equivalent to the standard definitions. It provides also for an easier proof of the basic multiplication theorem for determinants $(det(AB) = det(A)det(B))$ and for a direct and algorithmic method for the evaluation of a determinant.

> **Chapter 4.** *Useful Forms For Matrices.* From the standpoint of matrix algebra and the functional calculus for matrices, there is given a motivated study of a SQUARE matrix and the possibilities for reducing it to more easily manipulable, e.g., diagonal and Jordan normal, forms. The presentation here, like the presentation of determinant theory, is novel

and makes more accessible the whole topic of useful forms for matrices. The possibility of doing most of functional calculus with *polynomial* functions of a matrix is emphasized and illustrated.

Part II. Practice.

Chapter 5. *Applying Linear Algebra.*Applications of the developments in **Part I** are given to a wide variety of "real world" problems, particularly those sketched in **Chapter 1** but others as well. The computational aspects, considered today to be central in current studies of linear algebra, are discussed at some length. As well there are treatments of linear programming (including the Bland rules and the Karmarkar algorithm), 0-sum 2-person game theory, and Markov chains.

Part III. Theory.

Chapter 6. *General Vector Spaces.*The contents of **Part I** are given a broader setting via a reformulation in terms of abstract vector spaces and their endo-, epi-, homo-, iso-, and monomorphisms. Generalizations to infinite-dimensional spaces, in particular to Hilbert space, are indicated for the basic results derived in **Part I**. The limitations on these generalizations are exemplified.

Chapter 7. *Quadratic Forms.*Sylvester's Law of Inertia is proved. Definite quadratic forms are examined and characterized by determinantal criteria. Quadratic forms are applied to the study of extrema in multivariable calculus.

Chapter 8.*Miscellany.*The elements of vector space duality, of multilinear forms (with some reference to determinants), and of tensor algebra conclude the book.

Numerical **Exercises** and **Examples** abound throughout the text. Some of the simpler facts about linear algebra are given as **Exercises** in the body of the text.

References such as [**MW**] refer to items in the **Bibliography**.

There are occasional notes on etymologies, particularly of those words that are peculiar and intrinsic to the subject of linear algebra.

Thanks are extended to Roberto Leggiero, who, at a crucial time for the writer, was a consultant at the SUNY/Buffalo University Computing Services. Mr. Leggiero's patience and assistance in showing the author how to use TEX for typesetting the contents of this book were most helpful.

SUGGESTIONS FOR COURSE OUTLINES

A one-semester introductory course can be based on **Sections 1.1 - 3.1** and **4.1 - 4.4**. The level of the course determines the thoroughness with which the material is covered. For example, the treatment in **Section 3.1** provides a simple, direct, and easily understood definition of the determinant function.

The study and derivation of the properties of determinants can be passed over lightly so long as the **WORKING RULES** are explained.

A second semester permits coverage of **Sections 4.5 - 4.8** and **Part II**.

In a third semester the material in **Section 3.1** and in **Part III** provides the foundation for the study of abstract vector spaces and their elaborations.

Note on terminology. In the mathematical community, the words "Gauss," "Jordan," "elimination,", "reduction," "row-reduction," and "echelon form" in appropriate combinations are used to describe the procedures and results arising in the analysis of systems of linear equations. The essential ideas are embodied in what is here called the *PROCESS*. Successively refined variants of the *PROCESS* are dubbed the GEM (Gauss Elimination Method) that yields the echelon form, the GJM (Gauss Jordan Method) that produces via extended row-reduction an echelon form in which each pivot is 1 and each pivot column is a canonical basis vector.

SUNY/Buffalo
June 1988

PART I

BASICS

CHAPTER 1

ORIENTATION

1.1. Where linear algebra is useful and used

The subject of linear algebra crops up in modern applied mathematics and in many ways. Linear algebra is used to address and solve practical problems for which the methods of differential and integral calculus are of little value. It also provides the means for getting concrete numerical answers to a vast array of problems that arise in the applications of calculus itself, e.g., in ordinary and partial differential equations, in integral equations, and, most spectacularly, in quantum mechanics. In the last instance certain infinite-dimensional generalizations of finite-dimensional linear algebra seem to provide the most successful mathematical and practical solutions to the problems in atomic physics (quantum mechanics).

In the next few paragraphs there are posed some very simple and yet very practical problems. Linear algebra plays a central role in the solution of each of these problems. They are readily grouped into classes characterized by the kinds of *matrix manipulations* used to solve the problems.

The problem classes are listed below. They are illustrated in this section. In later parts of the book there are thorough treatments of the most effective methods of linear algebra that can be used to solve the problems in a given class.

The mathematics — linear algebra — stands on its own feet. It is sometimes motivated and even to some extent assisted by an understanding of the technical details of the physics, biology, ecology, economics, etc., from which the problems in the various classes come. However to understand the mathematics of linear algebra one need not be at all familiar with any of applied fields mentioned.

THE PROBLEM CLASSES

    I. Systems of finitely many linear equations in finitely many unknowns.

    II. Stability of populations of living organisms, of mechanical systems, and

---

of biological and chemical processes, etc.

III. Approximation of solutions to equations in classical analysis.

IV. Linear programming and game theory.

V. Functional equations and functional calculus.

There follow illustrative examples for each of the classes listed above.

The problem classes I and IV.

In the field of economics much use is made of so-called "input-output" matrices or arrays. The next **Example** in the logistics of nutrition and food supply gives the basic idea.

Example 1.1.1. Suppose that the nutrients: PROTEINS (P), FATS (F), and CARBOHYDRATES (C) for a population are to be derived from the eating of MEAT PRODUCTS (M), DAIRY PRODUCTS (D), and GRAIN PRODUCTS (G). Suppose further that the following table represents the proportions of the nutrients that the three kinds of foods provide. Slanted capital letters such as P, F, ..., G are symbols representing *both* the items PROTEINS, FATS, ..., GRAIN PRODUCTS *and* the amounts of these items as well. Similar remarks apply later as the example is elaborated. (Simple numbers have been chosen so as not to obscure the essential structure. Since only their proportions are of consequence these are chosen to be reasonable in what follows. Units of measurement are not identified.)

	M	D	G
P	7	1	2
F	4	1	0
C	0	2	23

Table 1.1.1.

The array of numbers in **Table 1.1.1** above is the input-output *matrix* relating the "inputs" MEAT PRODUCTS, DAIRY PRODUCTS, and GRAIN PRODUCTS to the "outputs" PROTEINS, FATS, and CARBOHYDRATES. Thus the input-output matrix

$$\mathcal{A} \stackrel{\mathrm{def}}{=} \begin{pmatrix} 7 & 1 & 2 \\ 4 & 1 & 0 \\ 0 & 2 & 23 \end{pmatrix}$$

is the array of coefficients in the left members of the following system of equations:

$$7M + D + 2G = P$$
$$4M + D + 0G = F \qquad (1.1.1)$$
$$0M + 2D + 23G = C.$$

These equations show how the inputs are combined to produce the desired outputs.

The horizontal strips of numbers in a matrix are its *rows*. The vertical strips of numbers are its *columns*. The numbers are called the *entries*. The *entry* in the ith row and the jth column is called the (i, j) entry. Thus in $\mathcal{A}$ the $(2,1)$ entry is 4, the $(3,3)$ entry is 23. A matrix having m *rows* and n *columns* is called an $m \times n$ matrix and its *size* is $m \times n$. Thus $\mathcal{A}$ is a 3×3 matrix of size 3×3.

[*Matrix* is Latin for — womb, pregnant animal — and is derived from *mater* — mother. The Indo-European root is *ma* — mother — and appears in — mama, mamma, mammal, maieutic, etc. *Matrix* occurs in many forms of discourse in which the word is used to describe a mold or a medium that *holds* something or a group of things in place, e.g., the womb holding the fetus, the mold for casting type faces in printing, etc.

The use of *matrix* in linear algebra stems, perhaps, from the way in which matrices are depicted, namely as arrays of numbers bracketed by parentheses that hold or contain them.]

Suppose next that MEAT PRODUCTS, DAIRY PRODUCTS, and GRAIN PRODUCTS may be supplied from CATTLE (B), SHEEP (S), HOGS (H), RICE (R), WHEAT (W), and CORN (K).

(To avoid a collision between "C" for "CARBOHYDRATES" and "C" for "CATTLE" or "C" for "CORN," "B" serves for "CATTLE" (Latin "BOS" for "CATTLE" or English "BEEF") and "K" serves for "CORN" ("KERNEL").

	B	S	H	R	W	K
M	2	1	3	0	0	0
D	10	1	0	0	0	2
G	0	0	0	4	6	10

Table 1.1.2.

The array of numbers in **Table 1.1.2** is the 3×6 input-output matrix

$$\mathcal{B} \overset{\text{def}}{=} \begin{pmatrix} 2 & 1 & 3 & 0 & 0 & 0 \\ 10 & 1 & 0 & 0 & 0 & 2 \\ 0 & 0 & 0 & 4 & 6 & 10 \end{pmatrix}$$

relating the "inputs" CATTLE, SHEEP, HOGS, RICE, WHEAT, and CORN, to

---

the "outputs" MEAT PRODUCTS, DAIRY PRODUCTS, and GRAIN PROD-UCTS.

If specific numbers are given for the demands P, F, and C then solving the first system (1.1.1) for the three unknowns M, D, and G reveals the amounts of MEAT PRODUCTS, DAIRY PRODUCTS, and GRAIN PRODUCTS needed to satisfy the demands. The resulting values for M, D, and G can then be entered into the system of equations derived from **Table 1.1.2** relating M, D, and G to B, S, H, R, W, and K, namely

$$2B + 1S + 3H + 0R + 0W + 0K = M$$
$$10B + 1S + 0H + 0R + 0W + 2K = D \qquad (1.1.2)$$
$$0B + 0S + 0H + 4R + 6W + 10K = G.$$

Replacing M, D, and G in (1.1.1) by the expressions in (1.1.2) leads to a third system, namely

$$24B + 8S + 21H + 8R + 12W + 22K = P$$
$$18B + 5S + 12H + 0R + 0W + 2K = F \qquad (1.1.3)$$
$$20B + 2S + 0H + 92R + 138W + 234K = C.$$

The coefficients in the left members (1.1.3) form the 3×6 input-output matrix

$$C \stackrel{\text{def}}{=} \begin{pmatrix} 24 & 8 & 21 & 8 & 12 & 22 \\ 18 & 5 & 12 & 0 & 0 & 2 \\ 20 & 2 & 0 & 92 & 138 & 234 \end{pmatrix}$$

relating the raw materials CATTLE, SHEEP, HOGS, RICE, WHEAT, and CORN to the desired nutrients PROTEINS, FATS, and CARBOHYDRATES.

The coefficient 24 of B in the first equation of (1.1.3)) is the $(1,1)$ entry in the matrix above. How that coefficient is calculated exemplifies how all the coefficients in (1.1.3) are found:

$$P = 7M + D + 2G$$
$$= 7(2B + 1S + 3H + 0R + 0W + 0K) + 1(10B + 1S + 0H + 0R + 0W + 2K)$$
$$\qquad + 2(0B + 0S + 0H + 4R + 6W + 10K)$$
$$= (7 \cdot 2 + 1 \cdot 10 + 2 \cdot 0)B + \cdots \qquad (1.1.4)$$
$$= 24B + \cdots .$$

The coefficients, 7, 1, and 2, in the formula preceding B in (1.1.4) come from *row* 1 of A. The coefficients, 2, 10, 0, come from *column* 1 of of B. So that *row* 1 of A combined in a special way with *column* 1 of B provides the $(1,1)$ entry of C:

$$\begin{bmatrix} 7 \times 2 \\ + \\ 1 \times 10 \\ + \\ 2 \times 0 \end{bmatrix} = 24.$$

Row 1 of $\mathcal{A}$ is rotated $90°$ clockwise, its entries are matched against the entries in *column* 1 of $\mathcal{B}$, the matched entries are multiplied together, and the results are added together.

A careful examination of the way in which $\mathcal{C}$ above is derived from $\mathcal{A}$ and $\mathcal{B}$ shows that the illustration above is a model for all the calculations of the entries in $\mathcal{C}$.

The array of coefficients in the left members of $(1.1.3)$ is related to the arrays in the left members of $(1.1.1)$ and $(1.1.2)$ in a manner that is the genesis of the idea for *multiplying* matrices. In other words

$$\underbrace{\begin{pmatrix} 7 & 1 & 2 \\ 4 & 1 & 0 \\ 0 & 2 & 23 \end{pmatrix}}_{\mathcal{A}} \underbrace{\begin{pmatrix} 2 & 1 & 3 & 0 & 0 & 0 \\ 10 & 1 & 0 & 0 & 0 & 2 \\ 0 & 0 & 0 & 4 & 6 & 10 \end{pmatrix}}_{\mathcal{B}} = \underbrace{\begin{pmatrix} 24 & 8 & 21 & 8 & 12 & 22 \\ 18 & 5 & 12 & 0 & 0 & 2 \\ 20 & 2 & 092 & 138 & 234 & \end{pmatrix}}_{\mathcal{C}}$$

$$\mathcal{AB} = \mathcal{C}.$$

Note that the **3 × 6** matrix $\mathcal{C}$ arises from the particular combination of the **3 × 3** matrix $\mathcal{A}$ and the **3 × 6** matrix $\mathcal{B}$. Note also that 3, the number of *columns* of $\mathcal{A}$, is the same as 3, the number of *rows* of $\mathcal{B}$.

More generally if

$$A = \begin{pmatrix} a_{11} & \cdots & a_{1n} \\ \vdots & \ddots & \vdots \\ a_{m1} & \cdots & a_{mn} \end{pmatrix} \quad \text{and} \quad B = \begin{pmatrix} b_{11} & \cdots & b_{1q} \\ \vdots & \ddots & \vdots \\ b_{n1} & \cdots & b_{nq} \end{pmatrix}$$

then A is an **m × n** matrix, B is a $n \times q$ matrix, and their *matrix product* or simply *product AB in the order just written* is defined to be the **m × q** matrix

$$P \stackrel{\text{def}}{=} \begin{pmatrix} p_{11} & \cdots & p_{1q} \\ \vdots & \ddots & \vdots \\ p_{m1} & \cdots & p_{mq} \end{pmatrix}.$$

in which the (i, j) entry p_{ij} in AB is given by the formula:

$$p_{ij} = a_{i1}b_{1j} + a_{i2}b_{2j} + \cdots + a_{in}b_{nj}.$$

In tighter form the preceding formula may be written:

$$p_{ij} = \sum_{\mathbf{K}=1}^{n} a_{i\mathbf{K}} b_{\mathbf{K}j}, \quad 1 \le i \le m, \quad 1 \le j \le q.$$

The match-up of the indices as shown above is helpful in remembering the formula. The index of summation $(\mathbf{K})$ is emphasized in **boldface**. The equation above says that the entry in the *ith row* and *jth column* of the matrix AB arises

---

by summing the products of the matched corresponding elements of the *ith row of A* and the *jth column* of B. Thus, with **boldface** letters marking the relevant entries, and with other entries suppressed, the formula above can be visualized as

$$\begin{pmatrix} \vdots & \ddots & \vdots \\ \mathbf{a_{i1}} & \cdots & \mathbf{a_{in}} \\ \vdots & \ddots & \vdots \end{pmatrix} \begin{pmatrix} \cdots & \mathbf{b_{1j}} & \cdots \\ \ddots & \vdots & \ddots \\ \cdots & \mathbf{b_{nj}} & \cdots \end{pmatrix} = \begin{pmatrix} \vdots & \vdots & \vdots \\ \cdots & \mathbf{p_{ij}} & \cdots \\ \vdots & \vdots & \vdots \end{pmatrix}.$$

$$\underbrace{m \times n}\quad\quad \underbrace{n \times q}\quad\quad \underbrace{m \times q}$$

The *i*th row of A is rotated $90°$ clockwise, aligned with the *j*th column of B so that corresponding entries match up, the matched entries are multiplied together and the products so formed are added together. The result is the (i, j) entry in AB:

$$\begin{bmatrix} \mathbf{a_{i1}} \times \mathbf{b_{1j}} \\ + \\ \mathbf{a_{i2}} \times \mathbf{b_{2j}} \\ + \\ \vdots \\ + \\ \mathbf{a_{in}} \times \mathbf{b_{nj}} \end{bmatrix} = \mathbf{p_{ij}}.$$

The matching of the entries in the *j*th column of B with the entries in the *i*th row of A reflects the need that the number n of *columns* of A be the same as the number n of *rows* of B. No restrictions are placed on the number of rows of A or on the number of columns of B.

Only because n, the *number of columns of A*, is the same as n, the *number of rows of B*, can the product AB as defined be formed. Conversely, if the number of columns of A is the same as the number of rows of B, then the product AB as defined can be formed.

If A is an $m \times n$ matrix and B is a $p \times q$ matrix then AB as defined can be formed if and only if ("iff") $n = p$. The matrices A and B are called *compatible for multiplication* iff $n = p$. Their product AB is an $m \times q$ matrix.

Two different populations in widely separated parts of the world may have different agricultural techniques, different breeds of farm animals, different kinds of soil, etc. In that event the numbers in array **Table 1.1.2** might well be different for the second population, e.g.,

	B	S	H	R	W	K
M	1	2	5	0	0	0
D	6	2	0	0	0	5
G	0	0	0	8	5	6

Table 1.1.3.

If these two populations provide equal amounts of B, S, H, R, W, and K then the total production of the two populations of M, D, and G is governed by the array below:

	B	S	H	R	W	K
M	3	3	8	0	0	0
D	16	3	0	0	0	7
G	0	0	0	12	11	16

Table 1.1.4.

in which the entry in each box is the *sum* of the entries in the counterpart boxes of **Table 1.1.2** and **Table 1.1.3**.

The array in **Table 1.1.4** might be regarded as the *sum* of the arrays **Table 1.1.2** and **Table 1.1.3**. The idea of *adding* entries in corresponding boxes of two matched arrays is the genesis of the idea of *addition* of matrices. In mathematical terms addition of matrices may be illustrated as follows:

$$\begin{pmatrix} 2 & 1 & 3 & 0 & 0 & 0 \\ 10 & 1 & 0 & 0 & 0 & 2 \\ 0 & 0 & 0 & 4 & 6 & 10 \end{pmatrix} + \begin{pmatrix} 1 & 2 & 5 & 0 & 0 & 0 \\ 6 & 2 & 0 & 0 & 0 & 5 \\ 0 & 0 & 0 & 8 & 5 & 6 \end{pmatrix}$$
$$\mathcal{B} \qquad\qquad \mathcal{D}$$

$$= \begin{pmatrix} 3 & 3 & 8 & 0 & 0 & 0 \\ 16 & 3 & 0 & 0 & 0 & 7 \\ 0 & 0 & 0 & 12 & 11 & 16 \end{pmatrix}$$
$$\mathcal{G}$$

$$\mathcal{B} + \mathcal{D} = \mathcal{G}.$$

Each entry in $\mathcal{G}$ is the sum of the corresponding entries in $\mathcal{B}$ and $\mathcal{D}$. Two matrices A and B may be added together iff they are of the same *size*, e.g., each matrix should be a 3×6 matrix, like the matrices $\mathcal{B}$, $\mathcal{D}$. The sum of two matrices of equal size is another matrix of the same size, e.g., $\mathcal{G}$ is, like $\mathcal{B}$ and $\mathcal{D}$, a 3×6 matrix. Two matrices of the same size are called *compatible for addition*.

The term "compatible" alone is to be interpreted in the context where it is used, e.g., in matrix multiplication or in matrix addition.

Matrix addition and matrix multiplication are the basic operations of matrix algebra. In the preceding survey and examples these operations crop up naturally in the course of organizing the analysis of the question of providing nutrition.

THE ELEMENTS OF MATRIX ALGEBRA

The following paragraphs constitute a short excursion into *matrix algebra*, the subject of combining matrices in various ways. For the most part each *entry* a_{pq} in the matrices to be considered is a *complex number*. Thus there is a symbol i such that $i^2 = -1 = (-i)^2$ and there are two real numbers r and s such that

$$a_{pq} \stackrel{\text{def}}{=} r + si \in \mathbf{C}, \ r, s \in \mathbf{R}$$

When geometric interpretations of linear algebra are desirable each entry is taken to be real. Following the excursion the discussion of the applications resumes.

In formal terms the addition of two matrices can be described by the following equation:

$$\begin{pmatrix} a_{11} & \cdot & a_{1n} \\ \cdot & \cdot & \cdot \\ a_{m1} & \cdot & a_{mn} \end{pmatrix} + \begin{pmatrix} b_{11} & \cdot & b_{1n} \\ \cdot & \cdot & \cdot \\ b_{m1} & \cdot & b_{mn} \end{pmatrix} \stackrel{\text{def}}{=} \begin{pmatrix} a_{11} + b_{11} & \cdot & a_{1n} + b_{1n} \\ \cdot & \cdot & \cdot \\ a_{m1} + b_{m1} & \cdot & a_{mn} + b_{mn} \end{pmatrix}.$$

Thus if the matrix

$$\begin{pmatrix} a_{11} & \cdot & a_{1n} \\ \cdot & \cdot & \cdot \\ a_{m1} & \cdot & a_{mn} \end{pmatrix}$$

is denoted $(a_{ij})_{i,j=1}^{m,n}$ then the formula for the (r,c) entry of

$$(c_{ij})_{i,j=1}^{m,n} \stackrel{\text{def}}{=} (a_{ij})_{i,j=1}^{m,n} + (b_{ij})_{i,j=1}^{m,n}$$

is

$$c_{rc} = a_{rc} + b_{rc}.$$

It is natural to write

$$(a_{ij})_{i,j=1}^{m,n} + (a_{ij})_{i,j=1}^{m,n} \stackrel{\text{def}}{=} 2(a_{ij})_{i,j=1}^{m,n} \stackrel{\text{def}}{=} (2a_{ij})_{i,j=1}^{m,n}$$

and more generally, for any number t, $t(a_{ij})_{i,j=1}^{m,n} \stackrel{\text{def}}{=} (ta_{ij})_{i,j=1}^{m,n}$.

Exercise 1.1.1. If the two populations provide not *equal* amounts of B, S, H, R, W, and K but amounts in the ratio $2 : 3$, what is the counterpart of **Table 1.1.4**?

Exercise 1.1.2. What is the counterpart *matrix* if the ratio is $4 : 7$?

Exercise 1.1.3. If s and t are natural numbers what is the counterpart matrix if the ratio is $s : t$?

Exercise 1.1.4. What are the entries c_{ij} if s and t are numbers and

$$
\begin{pmatrix} c_{11} & \cdots & c_{1n} \\ \vdots & \ddots & \vdots \\ c_{m1} & \cdots & c_{mn} \end{pmatrix} \overset{\text{def}}{=} s \begin{pmatrix} a_{11} & \cdots & a_{1n} \\ \vdots & \ddots & \vdots \\ a_{m1} & \cdots & a_{mn} \end{pmatrix} + t \begin{pmatrix} b_{11} & \cdots & b_{1n} \\ \vdots & \ddots & \vdots \\ b_{m1} & \cdots & b_{mn} \end{pmatrix} \; ?
$$

Exercise 1.1.5. Let $\mathcal{A}$, $\mathcal{B}$, and $\mathcal{D}$ be the numerical matrices introduced earlier. Write $\mathcal{A}(\mathcal{B} + \mathcal{D})$ as a single matrix. Compare the result with $\mathcal{A}\mathcal{B} + \mathcal{A}\mathcal{D}$ written as a single matrix.

Exercise 1.1.6. Let A, B, C, and D be defined as follows:

$$
A \overset{\text{def}}{=} \begin{pmatrix} 1 & -2 & 3+2i & 0 \\ 2 & -i & 3 & 5 \\ 3 & 0 & -2 & 1 \end{pmatrix} \qquad B \overset{\text{def}}{=} \begin{pmatrix} -2+5i & 6 & -3 \\ 4 & 0 & 8 \end{pmatrix},
$$

$$
C \overset{\text{def}}{=} \begin{pmatrix} 1-i & 3+4i \end{pmatrix} \qquad D \overset{\text{def}}{=} \begin{pmatrix} 1 & -1 & 1 & -1 \\ \cos\alpha & \cos\beta & \cos\gamma & \cos\delta \\ -1 & i & -i & i \end{pmatrix}.
$$

Let A^t, B^t, C^t, and D^t denote the result of replacing A, B, C, and D by matrices with the same entries but *rows and columns interchanged*. For example

$$
A^t \overset{\text{def}}{=} \begin{pmatrix} 1 & 2 & 3 \\ -2 & -i & 0 \\ 3+2i & 3 & -2 \\ 0 & 5 & 1 \end{pmatrix}.
$$

The matrix A^t, the *transpose* of A, is called "A-transpose." Let R, S, and T be any one of A, A^t, $\ldots$, D, and D^t and let r, s, and t be numbers. There are many situations in which the matrices listed above are compatible for the given operations. For each of the following choose a pair or a trio of compatible matrices and calculate the indicated combination.

$$
\begin{aligned}
& R + S, \ R + (S + T), \ (R + S) + T, \ rR + sS, \ (r + s)T, \ rT + sT, \ T + (-1)T; \\
& \qquad RS, \ R(ST), \ (RS)T, \ r(sT), \ (rs)T; \\
& \qquad R(S + T), \ RS + RT, \ (S + T)R, \ SR + TR.
\end{aligned}
$$

Exercise 1.1.7. What are the entries η_{ij} if

$$
\begin{pmatrix}
\eta_{11} & \cdots & \eta_{1n} \\
\vdots & \ddots & \vdots \\
\eta_{m1} & \cdots & \eta_{mn}
\end{pmatrix}
\overset{\text{def}}{=}
$$

$$
\begin{pmatrix}
\alpha_{11} & \cdots & \alpha_{1p} \\
\vdots & \ddots & \vdots \\
\alpha_{m1} & \cdots & \alpha_{mp}
\end{pmatrix}
\left[
\begin{pmatrix}
\beta_{11} & \cdots & \beta_{1n} \\
\vdots & \ddots & \vdots \\
\beta_{p1} & \cdots & \beta_{pn}
\end{pmatrix}
+
\begin{pmatrix}
\gamma_{11} & \cdots & \gamma_{1n} \\
\vdots & \ddots & \vdots \\
\gamma_{p1} & \cdots & \gamma_{pn}
\end{pmatrix}
\right] ?
$$

Exercise 1.1.8. Calculate

$$
\left[
\begin{pmatrix}
a & b & c \\
d & e & f \\
g & h & k
\end{pmatrix}
\begin{pmatrix}
\alpha & \beta & \gamma \\
\delta & \epsilon & \zeta \\
\eta & \theta & \kappa
\end{pmatrix}
\right]
\begin{pmatrix}
A & B & C \\
D & E & F \\
G & H & K
\end{pmatrix}
$$

and compare the result with

$$
\begin{pmatrix}
a & b & c \\
d & e & f \\
g & h & k
\end{pmatrix}
\left[
\begin{pmatrix}
\alpha & \beta & \gamma \\
\delta & \epsilon & \zeta \\
\eta & \theta & \kappa
\end{pmatrix}
\begin{pmatrix}
A & B & C \\
D & E & F \\
G & H & K
\end{pmatrix}
\right] .
$$

Exercise 1.1.9. Let A, B, and C be

$$
\begin{pmatrix}
a_{11} & \cdots & a_{1n} \\
\vdots & \ddots & \vdots \\
a_{m1} & \cdots & a_{mn}
\end{pmatrix},
\quad
\begin{pmatrix}
b_{11} & \cdots & b_{1p} \\
\vdots & \ddots & \vdots \\
b_{n1} & \cdots & b_{np}
\end{pmatrix},
\quad \text{and} \quad
\begin{pmatrix}
c_{11} & \cdots & c_{1q} \\
\vdots & \ddots & \vdots \\
c_{p1} & \cdots & c_{pq}
\end{pmatrix}.
$$

Show that if $m = 3$, $n = 5$, $p = 4$, and $q = 7$ then

$$
(AB)C = A(BC).
$$

More generally, show that for arbitrary m, n, p, and q

$$
\sum_{k=1}^{p} \left(\sum_{j=1}^{n} a_{rj} b_{jk} \right) c_{kc} = \sum_{j=1}^{n} a_{rj} \left(\sum_{k=1}^{p} b_{jk} c_{kc} \right).
$$

Show also that if s and t are numbers then

$$
t(AB) = (tA)B
$$
$$
(s+t)A = sA + sA
$$
$$
(st)A = s(tA)
$$
$$
t(A + C) = tA + tC \quad \text{if} \quad m = p \quad \text{and} \quad n = q.
$$

Exercise 1.1.10. Show that if A, B, and C are $m \times n$ matrices then

$$A + B = B + A$$
$$A + (B + C) = (A + B) + C.$$

Exercise 1.1.11. Compare

$$\begin{pmatrix} 0 & 1 \\ 0 & 0 \end{pmatrix} \begin{pmatrix} 0 & 0 \\ 1 & 0 \end{pmatrix} \quad \text{with} \quad \begin{pmatrix} 0 & 0 \\ 1 & 0 \end{pmatrix} \begin{pmatrix} 0 & 1 \\ 0 & 0 \end{pmatrix}.$$

(If A and B are SQUARE matrices of the same size, AB and BA *can* be different.)

Exercise 1.1.12. Let I denote the $m \times m$ SQUARE matrix

$$\begin{pmatrix} 1 & 0 & 0 & \dots & 0 \\ 0 & 1 & 0 & \dots & 0 \\ 0 & 0 & 1 & \dots & 0 \\ \vdots & \vdots & \vdots & \ddots & \vdots \\ 0 & 0 & 0 & \dots & 1 \end{pmatrix}.$$

Show that if $m = 5$ and M is a 5×7 matrix then $IM = M$ and that if R is an 10×5 matrix then $RI = R$.

Alternatively show that, for any natural numbers p and q, if M is an $m \times p$ matrix and if R is an $r \times m$ matrix then $IM = M$ and $RI = R$.

Because I behaves in the manner described in **Exercise 1.1.12** I is called the *identity matrix* or simply the *identity*. It behaves in matrix multiplication just as the number 1 behaves in ordinary multiplication of numbers.

Exercise 1.1.13. Let O denote an $m \times n$ matrix in which all entries are 0's. Show that if M is an $m \times n$ matrix then $M + O = O + M = M$. Show also that if T is an $n \times q$ matrix and if S is a $p \times m$ then OT and SO are also matrices in which all entries are 0's.

The matrix O interacts with matrices just as the number 0 interacts with numbers.

The results in **Exercises 1.1.9 - 1.1.13** are parts of the system of laws of matrix algebra. These are listed below and are examined at some length in **Section 2.6**.

THE LAWS OF MATRIX ALGEBRA

It is assumed that whenever matrices are combined they are compatible for the combinations indicated. Symbols such as A, B, ... represent matrices; symbols such as r, s, ... stand for real or complex numbers.

Group i. Addition.

$A + B = B + A$ (addition is commutative);

$A + (B + C) = (A + B) + C$ (addition is associative);

$O + A = A + O = A$;

$(A + B)^t = A^t + B^t$.

Group ii. Multiplication.

$A(BC) = (AB)C$ (multiplication is associative);

$IA = A$, $BI = B$;

$(AB)^t = B^t A^t$;

$OA = O$, $BO = O$.

No mention is made of the commutativity of multiplication. As **Exercise 1.1.11** illustrates, AB and BA can be different, even if both are SQUARE and of the same size.

Group iii. Multiplication, addition, and multiplication by numbers — also called *scalars*.

a. $A(B + C) = AB + AC$, $(B + C)A = BA + CA$ (the distributive laws);

b. $A(rB) = r(AB)$;

c. $r(A + B) = rA + rB$;

d. $r(AB) = (rA)B$;

e. $r(sA) = (rs)A$;

f. $(r + s)A = rA + sA$;

g. $(rA)^t = rA^t$.

When A is an $m \times n$ matrix and B and C are $n \times 1$ matrices (*column vectors*) then the first equation in *iiia* and *iiib* are the properties that give rise to the term *linear transformation* describing the transforming action of a matrix A on (column) vectors.

The function f such that for some real number a

$$f(x) = ax$$

enjoys the two properties: $f(x + y) = f(x) + f(y)$, $f(rx) = rf(x)$. The graph of f is a straight *line* (through the origin). The analogy of the behavior of f and the behavior of the matrix A in its transforming action on vectors is the root of the term "*line*ar transformation."

On the basis of experience thus far these "laws" hold. Proofs that they hold are direct consequences of the definitions of matrices, matrix addition, matrix multiplication, multiplication of matrices by scalars, and the laws of arithmetic.

For example, the proof of the associative law of multiplication, requires the most calculation. However with the use of "sigma-notation" even that proof is straightforward. Thus if

$$A \stackrel{\text{def}}{=} \begin{pmatrix} a_{11} & \cdots & a_{1n} \\ \vdots & \ddots & \vdots \\ a_{m1} & \cdots & a_{mn} \end{pmatrix}, \quad B \stackrel{\text{def}}{=} \begin{pmatrix} b_{11} & \cdots & b_{1p} \\ \vdots & \ddots & \vdots \\ b_{n1} & \cdots & b_{np} \end{pmatrix},$$

and

$$C \stackrel{\text{def}}{=} \begin{pmatrix} c_{11} & \cdots & c_{1q} \\ \vdots & \ddots & \vdots \\ c_{p1} & \cdots & c_{pq} \end{pmatrix}$$

then the (s,t) entry of $(AB)C$ is

$$\sum_{k=1}^{p} \left(\sum_{i=1}^{n} a_{si} b_{ik} \right) c_{kt} \tag{1.1.5}$$

while the (s,t) entry of $A(BC)$ is

$$\sum_{i=1}^{n} a_{si} \left(\sum_{k=1}^{p} b_{ik} c_{kt} \right). \tag{1.1.6}$$

Reversing the order of summation in (1.1.6), i.e., summing first on i and then on k yields (1.1.5) and proves the associative law of multiplication for matrices.

Exercise 1.1.14. Let A be the 3×5 matrix

$$\begin{pmatrix} 1 & 2 & 3 & 4 & 5 \\ -1 & 2 & -3 & 4 & -5 \\ 1 & 3 & 5 & 7 & 9 \end{pmatrix}$$

and let B be the 5×6 matrix

$$\begin{pmatrix} 2 & 4 & 6 & 8 & -6 & -4 \\ 1 & -5 & 4 & -7 & 9 & 2 \\ 0 & 1 & -1 & 0 & 1 & 0 \\ i & 4 & 6+2i & 4 & 0 & -2 \\ 1 & -3 & 4 & 5 & 6 & 7 \end{pmatrix}.$$

Calculate the third row, the fourth column, and the $(2,5)$ entry of AB.

Some easily remembered rules of matrix multiplication.

If A is an $m \times n$ matrix and B is an $n \times p$ matrix let

$$\mathbf{r}_1, \mathbf{r}_2, \ldots, \mathbf{r}_m \text{ resp. } \mathbf{c}_1, \mathbf{c}_2, \ldots, \mathbf{c}_p$$

---

denote the rows ($1 \times n$ matrices or *row vectors*) of A resp. the columns ($n \times 1$ matrices or *column vectors*) of B. Then:

i. the *j*th *column* of AB is the matrix product of the $m \times n$ matrix A and the *j*th *column* $\mathbf{c}_j$ ($n \times 1$ matrix) of B:

$$j\text{th column of } AB = A \cdot \mathbf{c}_j;$$

ii. the *i*th *row* of AB is the matrix product of the *i*th *row* $\mathbf{r}_i$ ($1 \times n$ matrix) of A and the $n \times p$ matrix B:

$$i\text{th row of } AB = \mathbf{r}_i \cdot B;$$

iii. The (i, j) entry in the *i*th *row* and the *j*th *column* of AB is the matrix product of the *i*th *row* $\mathbf{r}_i$ ($1 \times n$ matrix) of A and the *j*th *column* $\mathbf{c}_j$ ($n \times 1$ matrix) of B:

$$(i, j) \text{ entry of } AB = \mathbf{r}_i \cdot \mathbf{c}_j.$$

[The root of *vector* is traced back to the Latin root *vehere* — to carry — and then to the Indo-European root *wegh* — to go, transport in a *vehicle.*

In physics vectors represent directed quantities of displacement, velocity, acceleration, force, etc., generally associated with movement or cause of movement. Hence vectors are pictured as arrows pointing in the direction of the movement (or its cause). In a rectangular coordinate system an arrow can be specified by its x-, y-, and z-components and hence as a triple of numbers. The notion of a vector as an n-tuple of numbers is a simple generalization.]

In the subsequent material, the shorter notation ij rather than (i, j) is occasionally used in designating the position of an entry in a matrix. Since 213 may be interpreted as $(21,3)$ or $(2,13)$, the notation (i, j) is used when numbers rather than letters are the indices and there is some possibility for confusion.

For a SQUARE $n \times n$ matrix A let A^k be the k-fold product of A with itself:

$$A^k = \underbrace{A \cdot A \cdots A}_{k \text{ factors}}.$$

(The associative law of multiplication makes the definition unambiguous, e.g.,

$$(A \cdot A)A = A(A \cdot A),$$
$$A^2 \cdot A = A \cdot A^2,$$
$$A^3 = A^3, \text{ etc.})$$

Conventionally A^0 is *defined* to be the identity matrix I: $A^0 \stackrel{\text{def}}{=} I$.

Thus polynomial functions of SQUARE matrices such as $3A^2 - 4A + I$ may be formed. Since

$$3z^2 - 4z + 1 = (3z - 1)(z - 1) = (z - 1)(3z - 1),$$

it follows also that

$$3A^2 - 4A + I = (3A - I)(A - I) = (A - I)(3A - I).$$

More generally, despite the absence of a general commutative law of matrix multiplication, matrix multiplication *is* commutative if the factors are polynomial functions of a fixed SQUARE matrix.

If A is a SQUARE matrix and if p and q are polynomials, then

$$p(A)q(A) = q(A)p(A)$$

(restricted commutative law of matrix multiplication).

For example,

$$(5A^3 + 3A^2 - 7A + 6I)(A^2 + A + 4I) = (A^2 + A + 4I)(5A^3 + 3A^2 - 7A + 6I)$$
$$= 5A^5 + 8A^4 + 16A^3 + 11A^2 - 22A + 24I.$$

Exercise 1.1.15. For any natural numbers m and n and any pair (i, j) such that $1 \le i \le m$, $1 \le j \le n$ let U_{ij} be the $m \times n$ matrix in which the (i, j) entry is 1 and all other entries are 0's. Thus

$$U_{ij} \overset{\text{def}}{=} \text{ith row} \rightarrow \begin{pmatrix} 0 & \cdots & 0 & \cdots & \overset{\text{jth column}}{\underset{\downarrow}{0}} & \cdots & 0 \\ \vdots & \ddots & \vdots & \ddots & \vdots & \ddots & \vdots \\ 0 & \cdots & 0 & \cdots & 1 & \cdots & 0 \\ \vdots & \ddots & \vdots & \ddots & \vdots & \ddots & \vdots \\ 0 & \cdots & 0 & \cdots & 0 & \cdots & 0 \end{pmatrix}.$$

The U_{ij} are known as *matrix units*. Write all (four) 2×2 matrix units.

Exercise 1.1.16. Show that for any matrix $A \overset{\text{def}}{=} (a_{ij})_{i,j=1}^{m,n}$

$$A = \sum_{i,j=1}^{m,n} a_{ij} U_{ij}.$$

Exercise 1.1.17. Assume $m = n$ and $1 \leq i, j, k, l \leq n$. Calculate $U_{ij}U_{kl}$.

Exercise 1.1.18. Assume $m = n$ and $1 \leq i, j, k, l \leq n$. Show there is precisely one pair xy such that $U_{ij}U_{xy}U_{kl} \neq O$. Identify U_{xy} and calculate $U_{ij}U_{xy}U_{kl}$ (cf. [**GeS**]).

THE PROBLEM CLASSES (cont.)

The system (1.1.3) deserves attention from another point of view. In a real society there is usually a desire to do things efficiently, cheaply, economically. Since there are costs associated with

$$B, \; S, \; H, \; R, \; W, \text{ and } K$$

it is desirable to meet the requirements for P, F, and C at least cost.
Thus let

$$c_B, \; c_S, \; c_H, \; c_R, \; c_W, \text{ and } c_K$$

be the respective unit costs of providing B, S, H, R, W, and K. The total cost of providing the amounts required is then

$$T \stackrel{\text{def}}{=} c_B B + c_S S + c_H H + c_R R + c_W W + c_K K.$$

If the unit costs are the following

$$c_B = 20, \; c_S = 15, \; c_H = 12, \; c_R = 5, \; c_W = 8, \text{ and } c_K = 6, \tag{1.1.7}$$

and if

$$P = 5, \; F = 3, \text{ and } C = 10, \tag{1.1.8}$$

the problem of providing economical nutrition can be stated as follows:
Find nonnegative values of B, S, H, , R, W, and K so that

$$24B + 8S + 21H + 8R + 12W + 22K \geq 5$$
$$18B + 5S + 12H + 0R + 0W + 2K \geq 3 \tag{1.1.9}$$
$$20B + 2S + 0H + 92R + 138W + 234K \geq 10$$

and so that

$$T \stackrel{\text{def}}{=} 20B + 15S + 12H + 5R + 8W + 6K \tag{1.1.10}$$

is *least*.

Note that in (1.1.9) there are no *equations*, only *inequalities*. The methods of differential calculus, so effective in solving "extremal problems with constraints" are of no avail in the situation above. The problem is in the field of *linear programming* and a technique that *is* used is one of several variants of the so-called *simplex method* to be discussed later in this book.

There is a one-one correspondence between problems in linear programming and problems in so-called *zero-sum two-person games.* Solutions of a problem in linear programming are solutions of the corresponding problem in game theory and vice versa. This situation is explored in **Part II.**

The problem classes **I** and **IV** are characterized by the requirement that systems of linear equations or linear inequalities be solved.

Let m and n be natural numbers, i.e.,

$$m, \ n \in \mathbf{N} \overset{\text{def}}{=} \{1, 2, \ldots\}.$$

Then the typical system of linear equations is

$$a_{11}x_1 + \cdots + a_{1n}x_n = b_1$$

$$\vdots \qquad \ddots \qquad \vdots \qquad \vdots$$

$$a_{m1}x_1 + \cdots + a_{mn}x_n = b_m.$$

The a_{ij} and the b_j are known, given (data). The system is to be analyzed to discover whether there are numbers x_i ("unknowns") that satisfy the system. If there are such numbers they are to be found or described.

If

$$A \overset{\text{def}}{=} \begin{pmatrix} a_{11} & \cdots & a_{1n} \\ \vdots & \ddots & \vdots \\ a_{m1} & \cdots & a_{mn} \end{pmatrix}, \quad \mathbf{x} \overset{\text{def}}{=} \begin{pmatrix} x_1 \\ \vdots \\ x_n \end{pmatrix}, \quad \text{and} \quad \mathbf{b} \overset{\text{def}}{=} \begin{pmatrix} b_1 \\ \vdots \\ b_m \end{pmatrix}$$

then the system may be written in terms of matrix multiplication:

$$A\mathbf{x} = \mathbf{b}.$$

By abuse of notation the typical linear programming problem may be described as that of finding a solution to the problem:

Let

$$\mathbf{c} \overset{\text{def}}{=} (c_1, \ldots, c_n), \quad A \overset{\text{def}}{=} \begin{pmatrix} a_{11} & \cdots & a_{1n} \\ \vdots & \ddots & \vdots \\ a_{m1} & \cdots & a_{mn} \end{pmatrix}, \quad \text{and} \quad \mathbf{b} \overset{\text{def}}{=} \begin{pmatrix} b_1 \\ \vdots \\ b_m \end{pmatrix}$$

be given. Then find

$$\mathbf{x} \overset{\text{def}}{=} \begin{pmatrix} x_1 \\ \vdots \\ x_n \end{pmatrix}$$

so that $Ax \geq b$ and the number cx is minimized. Instead of the abused notation there is available the notation $Ax \succeq b$ signifying that each component of Ax is greater than or equal to the corresponding component of b. Analogously the notation $Ax \succ b$ signifies that each component of Ax is greater than the corresponding component of b. Thus if

$$ \mathbf{x} \overset{\text{def}}{=} \begin{pmatrix} -1 \\ -2 \end{pmatrix} \text{ and } \mathbf{y} \overset{\text{def}}{=} \begin{pmatrix} -3 \\ 0 \end{pmatrix} $$

then $O \succeq y$, $O \succ x$, $y \preceq O$, and $x \prec O$. The reader is urged to decipher the symbolic statements $Ax \preceq b$ and $Ax \prec b$ and to exemplify further the use of each of the symbols $\succeq, \succ, \preceq$, and $\prec$.

The problem classes II and V.

Example 1.1.2. A simplified description of life on Earth, of the *biosphere*, is the following.

There are two forms of life, plant and animal, living together by living *off* each other and *on* two fundamental inorganic materials, air and water. In a fixed period of time the amounts of air, water, plant mass, and animal mass change from their values at the beginning of the period to their values at the end of the period.

Animals consume plants and other animals, air, and water. Animals die and decompose and some of their products of decomposition restore some of the air and water the animals consumed while they were alive. Some of their products of decomposition are sources of sustenance for plants and for other animals. While animals live they contribute air and water in various forms (in exhaled breath, in perspiration, in urine, etc.) and also various forms of plant and animal matter in feces.

Plants consume air (the carbon dioxide in it), water, nitrogenous materials produced by bacteria that "fix" nitrogen in the air for use by plants, decayed animal matter, and, in some instances, other plants and animals (usually insects, ants, spiders, etc.). At certain times when they are not photosynthesizing, e.g., when it is dark, plants actually consume oxygen. Living plants add oxygen and water transpired from their surfaces. Their seeds, fruits, leaves, and most of their other organic structures are consumed by various forms of animal and plant life while the plants live and by other plants and animals after the plants die.

These phenomena can be described by what might be called the *life matrix* or the *symbiosis* matrix as derived from **Table 1.1.5** below.

End Beginning →	Plant	Animal	Air	Water
↓				
Plant	L_{11}	L_{12}	L_{13}	L_{14}
Animal	L_{21}	L_{22}	L_{23}	L_{24}
Air	L_{31}	L_{32}	L_{33}	L_{34}
Water	L_{41}	L_{42}	L_{43}	L_{44}

Table 1.1.5.

The symbiosis matrix is thus

$$\mathcal{L} \overset{\text{def}}{=} \begin{pmatrix} L_{11} & L_{12} & L_{13} & L_{14} \\ L_{21} & L_{22} & L_{23} & L_{24} \\ L_{31} & L_{32} & L_{33} & L_{34} \\ L_{41} & L_{42} & L_{43} & L_{44} \end{pmatrix}.$$

[The word *symbiosis* comes from the Indo-European root *sem* — one and *gwei* — to live. Hence *symbiosis* — living as one or living together. The cognates of these roots make surprising appearances in English, e.g., for *sem*: *single, seemly, symphony, synthesis,* etc., and for *gwei*: *biology, vivid, whiskey, quick.*]

For example, **Table 1.1.5**, in which the entries are $L_{...}$ (L for *Life*), says that L_{21} units of plant matter, L_{22} units of animal matter, L_{23} units of air, and L_{24} units of water at the *beginning* of a time period will produce *one* unit of animal matter at the *end* of the same time period. Thus assume

$$X(n) \overset{\text{def}}{=} \begin{pmatrix} x_1(n) \\ x_2(n) \\ x_3(n) \\ x_4(n) \end{pmatrix} \overset{\text{def}}{=} \begin{pmatrix} \text{amount of plant matter} \\ \text{amount of animal matter} \\ \text{amount of air} \\ \text{amount of water} \end{pmatrix}$$

is the 4×1 matrix (*column vector*) representing the amounts of plant matter, animal matter, air, and water available at the *beginning* of time period n. At the *end* of the same time period there is a new column vector $X(n+1)$. The two column vectors are related by the (recursion) equation:

$$x_1(n+1) = L_{11}x_1(n) + L_{12}x_2(n) + L_{13}x_3(n) + L_{14}x_4(n)$$
$$x_2(n+1) = L_{21}x_1(n) + L_{22}x_2(n) + L_{23}x_3(n) + L_{24}x_4(n)$$
$$x_3(n+1) = L_{31}x_1(n) + L_{32}x_2(n) + L_{33}x_3(n) + L_{34}x_4(4)$$
$$x_4(n+1) = L_{41}x_1(n) + L_{42}x_2(n) + L_{43}x_3(n) + L_{44}x_4(n).$$

Hence in terms of matrix multiplication

$$X(n+1) = \mathcal{L}X(n) = \mathcal{L}\mathcal{L}X(n-1) = \ldots . \qquad (1.1.11)$$

As suggested earlier, $\mathcal{L}\mathcal{L}$ may be denoted $\mathcal{L}^2$, $\mathcal{L}\mathcal{L}\mathcal{L}$ may be denoted $\mathcal{L}^3, \ldots$, and

$$\underbrace{\mathcal{L} \cdot \mathcal{L} \cdots \mathcal{L}}_{n \text{ factors}}$$

may be denoted $\mathcal{L}^n$. Then (1.1.11) may be written as follows:

$$X(n+1) = \mathcal{L}X(n)$$
$$X(n) = \mathcal{L}^n X(0). \qquad (1.1.12)$$

Let $X(0)$ be interpreted as the vector describing the situation "now." The formula in (1.1.12) shows how to predict what the proportions of plants, animals, air, and water will be in the future (n years from now) on the basis of what they are now. To do this effectively and within a reasonable length of time one must be able to calculate $\mathcal{L}^n$ for large values of n.

If the matrix $\mathcal{L}$ is *sparse* in that most of its entries are 0's then the calculations are relatively easy. Indeed, e.g., if $L_{ij} = 0$ unless $i = j$, then (blanks representing 0's)

$$\mathcal{L} = \begin{pmatrix} L_{11} & & & \\ & L_{22} & & \\ & & L_{33} & \\ & & & L_{44} \end{pmatrix} .$$

The rules of matrix multiplication show that

$$\mathcal{L}^n = \begin{pmatrix} L_{11}^n & & & \\ & L_{22}^n & & \\ & & L_{33}^n & \\ & & & L_{44}^n \end{pmatrix} ,$$

for which the calculation is straightforward and simple.

On the other hand if the matrix $\mathcal{L}$ is not sparse the calculations can become physically impossible. Alternatively, if a computer is used, it *must* round off all numbers to a fixed finite number of decimal places. The result may be rather inaccurate. For example suppose

$$\mathcal{L} = \begin{pmatrix} 1 & 2 & 1 & 1 \\ 2 & 1 & 1 & 1 \\ 1 & 1 & 2 & 1 \\ 1 & 1 & 1 & 2 \end{pmatrix} .$$

This matrix is *not* sparse. A prediction for 20 years hence calls for a calculation of $\mathcal{L}^{20}$. The reader is encouraged to calculate

$$\mathcal{L}^2$$
$$\mathcal{L}^4 = (\mathcal{L}^2)^2$$
$$\mathcal{L}^8 = (\mathcal{L}^4)^2$$
$$\mathcal{L}^{16} = (\mathcal{L}^8)^2$$

and then find the result according to the formula

$$\mathcal{L}^{20} = \mathcal{L}^4 \mathcal{L}^{16}$$

to get a notion of the difficulties involved. If $n = 50$ or $n = 100$, corresponding to a short length of time as life on earth goes, the calculation without some simplifying technique may well be worthless.

One of the objects of linear algebra is the development of techniques for avoiding these difficulties. There is an extensive treatment of the problem in **Chapter 4.**

A more dynamic version of the problem is that in which change is viewed as taking place essentially continuously. The approximate version of continuous change is change that takes place in short intervals of time (as in the quarterly, monthly, daily, hourly, etc., rather than the yearly compounding of interest). Correspondingly there is a matrix C in the following *differential equation* for the vector X, now regarded no longer as a function of n but as a function of *time t*:

$$\frac{dX(t)}{dt} = CX(t). \tag{1.1.13}$$

If the number of variables in the life model is m, and if $C \overset{\text{def}}{=} (c_{ij})_{i,j=1}^{m,m}$ then an expanded form of (1.1.13) may be written as

$$\frac{dx_i(t)}{dt} = \sum_{j=1}^{m} c_{ij} x_j(t), \ 1 \leq i \leq m. \tag{1.1.14}$$

In even more explicit form (1.1.13)) may be written as

$$\frac{dx_1(t)}{dt} = c_{11} x_1(t) + \cdots + c_{1m} x_m(t)$$
$$\vdots \qquad \vdots \qquad \ddots \qquad \vdots \tag{1.1.15}$$
$$\frac{dx_m(t)}{dt} = c_{m1} x_1(t) + \cdots + c_{mm} x_m(t).$$

If $X(t)$ is simply a numerical rather than a vector function of t and if C is simply a number rather than a matrix, then (1.1.13) is the familiar differential equation of growth (or decay). For some constant K the solution to (1.1.13) is

$$X(t) = e^{Ct} K. \tag{1.1.16}$$

Remarkably, (1.1.16) is the the solution to (1.1.13) even in the "vector-matrix" situation. However K is no longer a constant *number* but a constant $m \times 1$ *matrix*, i.e., a column vector.

The expression $e^{Ct} K$ is easily interpretable as

$$(I + Ct + \frac{C^2 t^2}{2!} + \cdots + \frac{C^k t^k}{k!} + \cdots)K.$$

How power series involving matrices, more generally how functions of matrices, are handled and simplified is a topic ("the functional calculus") that is treated in **Chapter 4**. At this point **Example 1.1.2** serves to emphasize the effectiveness of matrices in modelling "real world" situations.

Exercise 1.1.19. Let D be the matrix

$$\begin{pmatrix} 1 & 0 & 0 \\ 0 & -2 & 0 \\ 0 & 0 & 3 \end{pmatrix}.$$

i. Evaluate D^2 and D^3.

ii. Give a (closed) formula for the matrix D^n.

iii. Give a simple closed formula for the matrix

$$e^D \stackrel{\text{def}}{=} I + D + \frac{D^2}{2!} + \cdots + \frac{D^k}{k!} + \cdots = \sum_{k=0}^{\infty} \frac{D^k}{k!}.$$

Example 1.1.3. There is a chemistry matrix T consisting of m rows, one for each compound, and n columns, one for each chemical element. In each row the entries indicate how many atoms of each element are needed to make the compound corresponding to that row.

Since most compounds are made of relatively few elements, this matrix is *sparse* in that most of the entries in any row are 0's. Thus the row corresponding to the simple compound water (H_2O) consists of the entry 2 in the hydrogen column, the number 1 in the oxygen column and 0's in all the $n - 2$ remaining columns.

If $\mathbf{x}$ is an $n \times 1$ matrix, i.e., a column of n nonnegative numbers representing how much of each element is available for an experiment, then $T\mathbf{x}$ is an $m \times 1$ matrix, i.e., a column $\mathbf{y}$ of m numbers. The jth entry in $\mathbf{y}$ may be interpreted as the number of units of compound j that can be produced from the quantities listed in $\mathbf{x}$ *if no elements tagged by the nonzero entries in row j are used for other compounds.*

Example 1.1.4. To solve systems of linear differential equations that occur throughout applied mathematics, approximative methods are used. These methods essentially reverse the procedure used above to pass in the "life" problem from a discrete variable (n) to a continuous variable (t). Thus a system

$$\frac{dX_i(t)}{dt} = \sum_{j=1}^{n} c_{ij} X_j(t),$$

$$X_i(0) = a_0, \quad 1 \le i \le m \tag{1.1.17}$$

is, for a small nonzero value of h, replaced by a system

$$X_i(n+1) = \sum_{j=1}^{m} c_{ij} h X_j(n),$$

$$X_i(0) = a_0, \quad 1 \le i \le m. \tag{1.1.18}$$

If

$$\mathcal{X}(n+1) \stackrel{\text{def}}{=} \begin{pmatrix} X_1(n+1) \\ \vdots \\ X_m(n+1) \end{pmatrix}$$

then (1.1.18) may be written in abbreviated form as follows:

$$\mathcal{X}(n+1) = h^n C^n \mathcal{X}(1). \tag{1.1.19}$$

The values of all the $X_i(t)$ can be calculated for all reasonable values t of the form nh and if h is small then these values provide a fairly accurate notion of the behavior of the $X_i(t)$ for all reasonable values of t.

The problem class **II** shows the need to analyze the behavior of *powers*, e.g., $\mathcal{L}^n$, of matrices. The problem class **V** shows the need to interpret and find meaningful formulae for *functions* of matrices, e.g., $\sin \mathcal{L}$. The solution of problems in classes **II** and **V** lies in the development of a *functional calculus* for matrices. For example, if A is a SQUARE matrix, a useful meaning can be given to A^n and thereby, e.g., to $6A^2 + 2A + 6I$ and, indeed, for any polynomial p, to $p(A)$. Beyond such relatively simple functions as polynomials one may consider others such as:

$$\frac{3A^2 - 4A + 7I}{5A^3 - A^2 + 8A - 5I} \text{ (rational functions)},$$

$$\sin A \text{ (trigonometric functions)},$$

$$e^A \text{ (exponential functions)},$$

etc. Using power series and other devices from analysis one can provide even such functions of matrices with coherent meanings. The results are useful in applications of linear algebra to the solution of systems of differential, to the study of integral equations (cf. below), etc. Most of the developments in **Chapter 4** are the basis for achieving the generality in question. The reader interested in pursuing this topic can do so by absorbing the contents of **Chapter 2** and then going on directly to **Chapter 4**.

The problem class III.

In many parts of applied mathematics there arise problems of the following kind:

Let K be a given function of two variables: $K : (x, y) \mapsto K(x, y)$ and let g be a given function of one variable: $g : x \mapsto g(x)$. Find a function f such that

$$\int_a^b K(x, y)f(y)dy = g(x). \tag{1.1.20}$$

Equations like (1.1.20) are called *integral equations*.

An approximate solution to (1.1.20) may be found by rewriting the equation in approximate form. Let h be a small positive quantity so that for large n $nh = b - a$. Let y_j be $a + jh$ and let $x_k = a + kh$. Then owing to the definition of the definite integral it is approximately correct to write instead of (1.1.20)

$$\sum_{j=1}^{n} K(x_k, y_j)f(y_j)h = g(x_k), \quad 1 \leq k \leq n. \tag{1.1.21}$$

Equation (1.1.21) may be rewritten in the following notations:

$$\begin{aligned}
g_k &\stackrel{\text{def}}{=} g(x_k) \\
K_{kj} &\stackrel{\text{def}}{=} hK(x_k, y_j) \\
f_j &\stackrel{\text{def}}{=} f(y_j)
\end{aligned} \tag{1.1.22}$$

as

$$\begin{pmatrix} K_{11} & \cdots & K_{1n} \\ \vdots & \ddots & \vdots \\ K_{n1} & \cdots & K_{nn} \end{pmatrix} \begin{pmatrix} f_1 \\ \vdots \\ f_n \end{pmatrix} = \begin{pmatrix} g_1 \\ \vdots \\ g_n \end{pmatrix}. \tag{1.1.23}$$

If

$$\begin{aligned}
\mathbf{g} &\stackrel{\text{def}}{=} \begin{pmatrix} g_1 \\ \vdots \\ g_n \end{pmatrix} \\
\mathcal{K} &\stackrel{\text{def}}{=} \begin{pmatrix} K_{11} & \cdots & K_{1n} \\ \vdots & \ddots & \vdots \\ K_{n1} & \cdots & K_{nn} \end{pmatrix} \\
\mathbf{f} &\stackrel{\text{def}}{=} \begin{pmatrix} f_1 \\ \vdots \\ f_n \end{pmatrix}
\end{aligned} \tag{1.1.24}$$

then (1.1.21) and (1.1.23) may be written

$$\mathcal{K}\mathbf{f} = \mathbf{g}. \qquad (1.1.25)$$

Thus the approximate solution to (1.1.20) can be found as a solution of (1.1.21) or of (1.1.25).

Exercise 1.1.20. Write a system of linear equations that corresponds to solving approximately the problem of finding f so that

$$\int_a^b e^{xy} f(y)\,dy = e^x.$$

Exercise 1.1.21. Find an *exact* formula for f if

$$\int_0^1 e^{x-y} f(y)\,dy = e^x.$$

The problem class **III** is characterized by the use of matrices to find approximate solutions for a variety of systems of ordinary and partial differential equations and integral equations.

CHAPTER 2

THE PROCESS $\left(\text{\small Simple elimination}\right)$

2.1. Introduction: The system $A\mathbf{x} = \mathbf{b}$

The simplest example of the system $A\mathbf{x} = \mathbf{b}$ is that in which A, $\mathbf{x}$, and $\mathbf{b}$ are 1×1 matrices, denoted for simplicity a, x, and b. Thus the elementary version of the general equation is

$$ax = b. \tag{2.1.1}$$

Three examples of such an equation are

$$3x = 6, \qquad\qquad 0x = 6, \qquad\qquad \text{and} \qquad\qquad 0x = 0.$$

The first has one and only one solution: $x = 2$; the second has no solution; the third has infinitely many solutions since any value of x serves.

The three examples above are instances of the following general situations:

$$
\begin{array}{lllcc}
a \neq 0 & \text{and} & b \text{ arbitrary}: & \text{1 solution} & \mathbf{1}; \\
a = 0 & \text{and} & b \neq 0: & \text{no solution} & \mathbf{0}; \\
a = 0 & \text{and} & b = 0: & \text{infinitely many solutions} & \infty.
\end{array}
$$

Hence associated with the equation (2.1.1) is one and only one of the symbols $\mathbf{1}$, $\mathbf{0}$, ∞.

Equation (2.1.1) may be viewed as declaring that the given number a acts on the unknown x and (perhaps) produces the given number b. To solve (2.1.1) one should analyze it to find whether it has any solutions at all, and, if it does have solutions, find all the solutions or at least a description of all the solutions.

Similarly the system of equations

$$
\begin{aligned}
ax + by &= c \\
fx + gy &= h
\end{aligned} \tag{2.1.2}
$$

gives rise to three possibilities. For example the three systems

$$
\begin{aligned}
2x + 3y &= 2 \\
5x - 4y &= 6
\end{aligned}, \qquad
\begin{aligned}
2x + 3y &= 2 \\
4x + 6y &= 5
\end{aligned}, \qquad \text{and} \qquad
\begin{aligned}
2x + 3y &= 2 \\
4x + 6y &= 4
\end{aligned} \tag{2.1.3}
$$

---

have respectively one and only one solution, no solution, and infinitely many solutions.

The first system in (2.1.3) is

$$2x + 3y = 2$$
$$5x - 4y = 6$$

and may be analyzed as follows:

Step 1. Subtract $\frac{5}{2}$ of each member of the first equation from the corresponding members of the second equation. The second equation is now $-\frac{23}{2}y = 1$ whence $y = -\frac{2}{23}$.

Step 2. Replace y in the first equation by $-\frac{2}{23}$ and solve the resulting equation for x: $x = \frac{26}{23}$.

Step 3. Substitute in each equation the values found for x and y to determine whether they constitute a solution of the system. (They do.)

The second system in (2.1.3) is

$$2x + 3y = 2$$
$$4x + 6y = 5$$

and it may be analyzed similarly:

Step 1. Subtract twice each member of the first equation from the corresponding member of the second equation. The second equation is now $0 = 1$.

Step 2. The last equation is contradictory and so the system has no solution.

The third system in (2.1.3) is

$$2x + 3y = 2$$
$$4x + 6y = 4$$

and it may be analyzed as follows:

Step 1. Subtract twice each member of the first equation from the corresponding member of the second equation. The second equation is now $0 = 0$.

Step 2. This time the last equation in no way constrains the values of x and y. The first equation requires that x be $\frac{1}{2}(2 - 3y)$. Whatever value is given to y the corresponding value of x is determined. The system has infinitely many solutions.

It should be noted that there are systems such as

$$2x + 3y = 2 \qquad \text{(fewer equations than unknowns)},$$

and

$$2x + 3y = 2$$
$$4x - 3y = 5$$
$$-7x + y = -9 \qquad \text{(more equations than unknowns)}.$$

Section 2.1. Introduction: The system $A\mathbf{x} = \mathbf{b}$

In the former case any value of y determines a value of x and the system has infinitely many solutions.

In the latter case the analysis proceeds as follows:

Step 1. To *eliminate* x from the second equation subtract twice each member of the first equation from the corresponding member of the second equation. To eliminate x from the third equation add $\frac{7}{2}$ of each member of the first equation to the corresponding member of the third equation. There emerge the following equations from which x has been eliminated.

$$-12y = -1, \quad y = -\frac{1}{12}$$
$$\frac{23}{2}y = -2, \quad y = -\frac{4}{23}$$

These equations are mutually contradictory, and so the system has no solution.

However the system

$$2x + 3y = 2$$
$$4x - 3y = 5$$
$$2x + 12y = 1$$

yields, after a similar elimination of x from the second and third equations, $y = -\frac{1}{9}$ and $y = -\frac{1}{9}$. If $x = \frac{7}{6}$ then the x and y values found satisfy each equation of the system.

For a complete discussion of all the possibilities, the treatment that follows concerns itself with systems of m equations in n unknowns. All the cases: $m < n$, $m = n$, and $m > n$, are handled in one stroke. The theme and purpose of the method is the successive *elimination* of unknowns and, along the way, successive reduction of the number of equations. Throughout m and n stand for natural numbers, i.e., m and n belong to $\mathbf{N} \overset{\text{def}}{=} \{1, 2, \ldots\}$.

The task is to examine in great detail the problem of determining for the typical system displayed below whether solutions exist and, if they exist, what they are. The system looks like this:

$$a_{11}x_1 + \cdots + a_{1n}x_n = b_1$$
$$\vdots \qquad \ddots \qquad \vdots \qquad \vdots$$
$$a_{m1}x_1 + \cdots + a_{mn}x_n = b_m. \tag{2.1.4}$$

The a_{ij} and the b_i are known or given complex numbers and the x_j are unknowns, to be found if a solution exists at all. To say that no solution exists is to say that the equations cannot be satisfied by any choice of values for the unknowns. Note that the system consists of m equations in n unknowns and that m and n are arbitrary natural numbers ($m < n$, $m = n$, or $m > n$).

    Chapter 2. THE PROCESS (Simple elimination)

From the start equation (2.1.4) should be regarded as saying that the array
or *matrix*

$$\begin{pmatrix} a_{11} & \cdots & a_{1n} \\ \vdots & \ddots & \vdots \\ a_{m1} & \cdots & a_{mn} \end{pmatrix} \qquad (2.1.5)$$

consisting of given numbers a_{ij} (data) acts on the array or *matrix*

$$\begin{pmatrix} x_1 \\ \vdots \\ x_n \end{pmatrix}$$

of unknowns and produces the array or matrix

$$\begin{pmatrix} b_1 \\ \vdots \\ b_m \end{pmatrix}$$

of given numbers, (data). The matrix in (2.1.5) is denoted $(a_{ij})_{i,j=1}^{m,n}$.

The analysis is carried out through use of what is called in this book the
PROCESS and variations on it. The classical name for one of its variations is
Gauss(ian) elimination.

C. F. Gauss, at the age of nine years, was recognized as a mathematical
prodigy . His early promise was fulfilled in his long and productive life.
Many scientists regard Gauss as the greatest mathematician of the nineteenth
century. Some regard him as the greatest mathematician since Archimedes.

Before the plunge into the application of the simple steps of the *PROCESS*,
some comments are in order.

i. Owing to the step-by-step *("algorithmic")* nature of the *PROCESS*, it is in
all its versions readily programmable for use by computers.

ii. The subject of this text is called "linear" algebra for the following reasons.
An equation of the form $ax + by = c$ is usually interpreted as an equation
of a straight *line*. Since the equation is of the first degree, all equations
of the first degree, no matter how many unknowns are involved, are called
*"line*ar" equations. Linear algebra is the study of systems of finitely many
linear equations in finitely many unknowns. As the study progresses and
deepens it turns to the study of matrices and matrix algebra. Matrices
in turn are inextricably connected with *linear transformations* from *vector
spaces* to vector spaces.

These subjects and the relationships among them are treated in this book.

[The words *line(ar)*, *algebra*, *algorithm(ic)* are of diverse origins. The Indo-European root *lino* — flax — gives rise to *linen*, *line*, (the nautical term for a rope) and *line* the geometric object. The Semitic roots of *algebra* are *al* — the — and *jabr* — re-uniting, composing, bone-setting. The 9th century mathematician Abu Jafar Mohammed ibn Musa al-Khowarizmi wrote a treatise *Al jabr w'al muqabala* — Reunion and comparison — introduced Arabic numerals to Europe, and dealt at some length with the solution of equations. An inexact version of his surname is the source of *algorithm(ic)*.]

Until the study turns from the concrete to the abstract, in **Part III**, the term *vector* means, for some natural number n, either a $1 \times n$ matrix, a *row vector*, or an $n \times 1$ matrix, a *column vector*. For example

$$\mathbf{x} \stackrel{\text{def}}{=} (x_1, \ldots, x_n) \text{ or } \mathbf{y} \stackrel{\text{def}}{=} \begin{pmatrix} y_1 \\ \vdots \\ y_n \end{pmatrix} \tag{2.1.6}$$

are vectors: $\mathbf{x}$ is a row vector and $\mathbf{y}$ is a column vector. Thus a vector is an n-tuple of complex numbers written in a row or in a column. If the context makes the meaning clear *vector* is used instead of *row vector* or *column vector*. The entries in matrices of these special kinds are called *components*. Thus in (2.1.6) the x_i are the components of $\mathbf{x}$ and the y_j are the components of $\mathbf{y}$. The numbers $1, e + i\pi, -2, 0$, and $-\frac{2}{81}$ are the components of the row vector

$$\left(1, e + i\pi, -2, 0, -\tfrac{2}{81}\right)$$

and the numbers $i, 16$, and $\sqrt{2}$ are the components of the column vector

$$\begin{pmatrix} i \\ 16 \\ \sqrt{2} \end{pmatrix}.$$

Exercise 2.1.1. Analyze each of the following systems of equations. Determine in each case whether the system has ONE solution, NO solution, or INFINITELY MANY solutions. If there are solutions describe them.

$$i: \begin{array}{l} 2x - 3y = -1 \\ x + y = 4 \end{array} ; \quad ii: \begin{array}{l} -x + 2y = 1 \\ 3x - 6y = 3 \end{array} ; \quad iii: \begin{array}{l} -5x + 2y + z = 1 \\ 3x - y + 4z = 2 \end{array}.$$

Exercise 2.1.2. Analyze each of the following systems of equations and repeat the analysis specified in **Exercise 2.1.1.**

$$i: \begin{array}{l} -2x_1 + 5x_2 + 6x_3 = 2 \\ x_1 - 3x_3 = -1 \end{array} ; \quad ii: \begin{array}{l} u - v + w - z = 0 \\ 2u + v - 4w + z = -7 \end{array} ; \quad iii: \begin{array}{l} a + b + c = 0 \\ a + b - c = 0 . \\ a - b - c = 0 \end{array}$$

Exercise 2.1.3. For each system below determine the values of λ for which the system has a solution (x, y) in which at least one of x and y is *not* 0.

$$i : \begin{array}{l} -5x + 3y = \lambda x \\ 12x + 4y = \lambda y \end{array} ; \quad ii : \begin{array}{l} 2x + 2y = \lambda x \\ 5x - 7y = \lambda y \end{array} .$$

For each system and for each λ meeting the requirements for that system describe the set of all solutions.

2.2. Examples

Assume that in (1.1.1) the values of P, F, and C are respectively 5, 3, and 10. The values of M, D, and G are found by analysis of the system.

$$7M + D + 2G = 5$$
$$4M + D + 0G = 3$$
$$0M + 2D + 23G = 10. \tag{2.2.1}$$

Step 1. Subtract $\frac{4}{7}$ of the the left and right members of the first equation from the corresponding members of the second equation. Leave the third equation unaltered since there the coefficient of M is already 0. The result is two new equations (one fewer than in the original system) in which M is absent — it has been *eliminated*.

$$\frac{3}{7}D - \frac{8}{7}G = \frac{1}{7}$$
$$2D + 23G = 10. \tag{2.2.2}$$

Step 2. In (2.2.2) subtract $\frac{14}{3}$ of each member of the first equation from the corresponding members of the second equation. Thereby produce one equation from which D has been eliminated: $\frac{85}{3}G = \frac{28}{3}$. It follows that $G = \frac{28}{85}$.

Step 3. Back substitution for G in (2.2.2) permits D to be found: $D = \frac{103}{85}$. Back substitution in (2.2.1) yields the value of M: $M = \frac{38}{85}$.

The *Steps* used in analyzing (1.1.1) may be used in (1.1.3). Just the outline of the results is given below.

Step 1. Elimination of B from the second and third equations of (1.1.3) leads to

$$-S - \frac{15}{4}H - 6R - 9W - \frac{29}{2}K = -\frac{3}{4}$$
$$-\frac{14}{3}S - \frac{35}{2}H + \frac{256}{3}R + 128W + \frac{647}{3}K = \frac{35}{6}. \tag{2.2.3}$$

Elimination of S from the equations in (2.2.3) leads to the next equation:

$$\frac{340}{3}R + 170W + \frac{850}{3}K = \frac{28}{3}. \tag{2.2.4}$$

Coincidentally H has been eliminated. This favorable outcome, (the elimination of two unknowns in one *Step*) crops up only occasionally in the course of solving large systems.

From (2.2.4) it follows that

$$R = -\frac{3}{2}W - \frac{5}{2}K + \frac{7}{85}. \tag{2.2.5}$$

Hence W and K may be chosen at will. Their values determine R, and the values of R, W, and K determine S via the first equation in (2.2.3):

$$S = -\frac{15}{4}H + \frac{1}{2}K + \frac{87}{340}. \tag{2.2.6}$$

The value of H (because it was coincidentally eliminated when S was eliminated) may be chosen at will. The values of the "free variables" H, K, W, and the values they determine of R and S determine B via the first equation of (1.1.3):

$$B = \frac{3}{8}H - \frac{1}{4}K + \frac{13}{136}. \tag{2.2.7}$$

Back substitution of the values found for R, S, and B in (2.2.5) - (2.2.7) confirms that for any values of the "free variables" H, K, and W and the corresponding values of R, S, and B all equations of (1.1.3) are satisfied.

If the reality of the situation is respected, only nonnegative values of H, W, and K may be used and furthermore only values that cause R, S, and B to be nonnegative are admissible. Hence the following conditions must be satisfied:

$$H \geq 0$$
$$W \geq 0$$
$$K \geq 0$$
$$R = -\frac{3}{2}W - \frac{5}{2}K + \frac{7}{85} \geq 0 \tag{2.2.8}$$
$$S = -\frac{15}{4}H + \frac{1}{2}K + \frac{87}{340} \geq 0 \tag{2.2.9}$$
$$B = \frac{3}{8}H - \frac{1}{4}K + \frac{13}{136} \geq 0. \tag{2.2.10}$$

The total cost T, in (1.1.10), of providing the requirements may now be written in terms of the free variables:

$$\begin{aligned}
T &= 20B + 15S + 12H + 5R + 8W + 6K \\
&= -\frac{147}{4}H - 10K + \frac{1}{2}W + \frac{419}{68}
\end{aligned} \tag{2.2.11}$$

Thus the right member of (2.2.11) is to be minimized while the conditions in (2.2.8)-(2.2.10) are to be observed.

In this instance "common sense" provides a solution. Indeed from (2.2.9) and (2.2.10) it follows that

$$\frac{1}{2}S + B = -\frac{3}{2}H + \frac{19}{85} \geq 0 \qquad (2.2.12)$$

$$S + 10B = -2K + \frac{103}{85} \geq 0. \qquad (2.2.13)$$

whence $H \leq \frac{38}{255}$ and $K \leq \frac{103}{170}$. From (2.2.8) and the nonnegativity of W it follows that $K \leq \frac{14}{425} - \frac{3}{5}W \leq \frac{14}{425}$. Hence if $H = \frac{38}{255}$, $K = \frac{14}{425}$ and $W = 0$ then

$$T = -\frac{147}{4}\frac{38}{255} - \frac{28}{81} + \frac{419}{68} = 0.339615105. \qquad (2.2.14)$$

Since the maximum allowable values are used for H and K and since W must be nonnegative it follows that the value given for T in (2.2.14) is the minimum that T can assume in the presence of the given constraints.

Geometrically the equation $T = constant \overset{\text{def}}{=} t$ is interpretable as a plane Π_t in 3-dimensional space $\mathbf{R}^3$. The inequalities (2.2.8)- (2.2.10) are interpretable as half-spaces (each is one "side" of a plane together with the plane itself). The simultaneous satisfaction of the inequalities (2.2.8)-(2.2.10) places the point with coordinates (H, K, W) in the intersection Q of these half-spaces. Each of these half-spaces is *convex*, i.e., a set C such that if P and P' are in C then the whole line segment $\overline{PP'}$ lies in C. As the intersection of convex sets Q is itself a convex set.

Minimizing T while satisfying the nutritional requirements means minimizing t while the plane Π_t passes through some point of Q. Since Q is a convex polyhedron lying in the the first octant of $\mathbf{R}^3$ the minimizing value of t will cause the plane Π_t to pass through a *vertex* of Q.

One very *inefficient* way to solve the problem is to find all the vertices of Q and then test the coordinates of each vertex in the formula for T. In relatively simple cases this kind of calculation is, despite its inefficiency, quite feasible, if the number of vertices of Q is small. However in realistic problems the number of variables is in the hundreds or even thousands, the number of inequalities is equally large and the number of vertices of Q would outstrip the capacity of a world-wide consortium of all existing computers working in unison.

A more realistic approach is via the *simplex method*. The essence of the *simplex method* and its variants is a technique for starting at *some* vertex of Q and moving along some judiciously chosen edge issuing from that vertex to a neighboring vertex where t is smaller than it is at the original vertex.

The simplex method involves the finding of a starting vertex of Q. Finding such a vertex entails the solution of a system $Ax = \mathbf{b}$ of equations and so at least one part of the simplex method requires a systematic procedure for finding a solution to such a system. In **Section 2.3** the simplex method is illustrated in finding the complete solution to a sample problem **Example 2.3.7**.

2.3. The *PROCESS*

The system to be analyzed is $A\mathbf{x} = \mathbf{b}$, in more explicit form,

$$\begin{pmatrix} a_{11} & \cdots & a_{1n} \\ \vdots & \ddots & \vdots \\ a_{m1} & \cdots & a_{mn} \end{pmatrix} \begin{pmatrix} x_1 \\ \vdots \\ x_n \end{pmatrix} = \begin{pmatrix} b_1 \\ \vdots \\ b_m \end{pmatrix}, \tag{2.3.1}$$

or in even more explicit form,

$$a_{11}x_1 + \cdots + a_{1n}x_n = b_1$$
$$\vdots \qquad \ddots \qquad \vdots \qquad \vdots$$
$$a_{m1}x_1 + \cdots + a_{mn}x_n = b_m. \tag{2.3.2}$$

The *augmented* matrix

$$\left(\begin{array}{ccc|c} a_{11} & \cdots & a_{1n} & b_1 \\ \vdots & \ddots & \vdots & \vdots \\ a_{m1} & \cdots & a_{mn} & b_m \end{array} \right) \stackrel{\text{def}}{=} A\,|\,\mathbf{b} \tag{2.3.3}$$

contains all the data with which the *PROCESS* is concerned. The vertical line
in (2.3.3) represents the column of equal-signs (='s) in (2.3.2). The matrix to
the left of the vertical line is A. The column to the right of the vertical line is $\mathbf{b}$,
the column of right members of the equations of the system. The "unknowns" x_i
are simply placeholders and are ignored. By manipulating (2.3.3) one can arrive
at a new matrix that is readily interpreted as belonging to a system logically
equivalent to (2.3.2). The new system is one for which an analysis is very simple
and from which solutions (if they exist) can be read off essentially by inspection.
The analysis and the solutions are valid for the original system (2.3.2) and thus
constitute a complete answer to the questions about (2.3.2).

The next **Example** provides a motivation for the *PROCESS*.

Example 2.3.1. Let the system to be analyzed be

$$-5y + 2z + 5w = -1$$
$$2x + 3y - 4z + 6w = 8$$
$$9z - 2w = 5.$$

The last equation implies that $z = \frac{(5+2w)}{9}$. Substitution in the first equation
shows $y = \frac{(-1-2z-5w)}{-5} = \frac{(-19-49w)}{9}$. Substitution in the second equation leads
to $x = \frac{(8-3y+4z-6w)}{2} = \frac{(149+155w)}{18}$. The solution is

$$x = \frac{(149 + 155w)}{18}$$
$$y = \frac{(-19 - 49w)}{9}$$
$$z = \frac{(5 + 2w)}{9}.$$

The value of w is unrestricted and, whatever its value, it determines the values of x, y, and z.

The ease in analyzing the solution in the **Example** above is based on the form of the system. One of its equations (the last) involves only z and w, the first equation involves only y, z, and w, and the second equation involves x, y, z, and w. A sequence of back-substitutions leads to the complete solution.

The *PROCESS* offers a systematic method for converting a general system of linear equations into an *equivalent* system in which back-substitution is feasible and readily performed.

The *PROCESS* consists of two steps applied in sequence again and again. These steps are:

SEARCH.

Search the first *column*

$$\begin{pmatrix} a_{11} \\ \vdots \\ a_{m1} \end{pmatrix}$$

(consisting of the *coefficients* of x_1). Moving from top to bottom of the column, choose the first *nonzero* number in the column. If *all* numbers in the first column are zero, search the second column

$$\begin{pmatrix} a_{12} \\ \vdots \\ a_{m2} \end{pmatrix}$$

for *its* first nonzero entry.

If the SEARCH continues and *all* entries a_{ij} in the first k columns are found to be zero, then the unknowns $x_1, \ldots, x_k$ may assume any values so long as $k < n$.

(Such a curious situation can arise. For example, in an application of linear algebra there can be a model that gives rise to a system such as the one under consideration. Each time the model is used, actual data, *numbers*, are substituted for the a_{ij} and the b_j. In some cases the numbers a_{ij} in the first k columns can indeed be 0's or can be so small that the need for roundoff in computation causes the computer program to treat these numbers as 0's. In either circumstance, the *PROCESS* as realized in a computer program must provide for these contingencies.

For example, in the study of systems of differential equations there occasionally appear systems of linear equations that look like this:

$$(1 - \lambda)x_1 + 2x_2 + x_3 = 1$$
$$(5 - \lambda)x_2 + 2x_3 = 3$$
$$-2x_2 + (4 - \lambda)x_3 = -1.$$

If $\lambda = 1$ the system becomes one in which all coefficients of x_1 are 0's.)

If *all* the entries in *all* the first n columns are found to be zero, then the left member of each equation is zero. In that case

> EITHER: SOME $b_i \neq 0$ and then there is *NO* solution.
>
> OR: EACH $b_i = 0$ and then there are
>
> *INFINITELY MANY* solutions.

Indeed any n-tuple

$$\begin{pmatrix} x_1 \\ \vdots \\ x_n \end{pmatrix}$$

serves as a solution.

For the sake of interest it may be assumed that some (first)

$$a_{rc} \neq 0$$

(subscript "$_r$" for "row" r; subscript "$_c$" for "column" c).

Hence in the most general case the system looks like this:

$$0x_1 + \cdots + 0x_c + \cdots + a_{1n}x_n = b_1$$

$$\vdots \qquad \ddots \qquad \vdots \qquad \ddots \qquad \vdots \qquad \vdots$$

$$0x_1 + \cdots + a_{rc}x_c + \cdots + a_{rn}x_n = b_r$$

$$\vdots \qquad \ddots \qquad \vdots \qquad \ddots \qquad \vdots \qquad \vdots$$

$$0x_1 + \cdots + a_{mc}x_c + \cdots + a_{mn}x_n = b_m.$$

The array is designed to suggest that all coefficients "north," "west," "north-west," or "southwest" of a_{rc} are zero and that a_{rc} itself is not zero. All other coefficients are unexplored.

At the present stage of the *PROCESS* the augmented matrix has the following form:

$$\left(\begin{array}{ccccc|c} 0 & \cdots & 0 & \cdots & a_{1n} & b_1 \\ \vdots & \ddots & \vdots & \ddots & \vdots & \vdots \\ 0 & \cdots & a_{rc} & \cdots & a_{rn} & b_r \\ \vdots & \ddots & \vdots & \ddots & \vdots & \vdots \\ 0 & \cdots & a_{mc} & \cdots & a_{mn} & b_m \end{array} \right).$$

[**Remark 2.3.1:** Any nonzero entry to the left of the vertical line is called a *pivot candidate* of the matrix A. Thus the entry a_{rc} is a pivot candidate of A. SEARCH leads to a_{rc} as the *first* pivot candidate. A variant, MAXSEARCH, may be used to find not the *first* pivot candidate but the pivot candidate of largest absolute value.

Another variant, PARTIALMAXSEARCH, of SEARCH may be used. Then, in the first column where there is some nonzero entry, the SEARCH is pursued until a the pivot candidate of largest absolute value in that first column is found. In any procedure, the pivot candidate that is actually used is called simply the *pivot*.]

ELIMINATION.

There are two special cases:

a. $m = 1$: There is only one row. The *PROCESS* stops since x_c can be written in terms of $x_{c+1}, \ldots, x_n$:

$$x_c = b_1 - \frac{(a_{1,c+1} x_{c+1} + \cdots + a_{1n} x_n)}{a_{1c}}. \tag{2.3.4}$$

The values of $x_1, \ldots, x_{c-1}, x_{c+1}, \ldots, x_n$ are unrestricted and any choice for them determines the value of x_c. (If there is only one unknown, i.e., if $n = 1$, the "system" is really: $ax = b$, and nothing more need be said.)

b. $r = m > 1$: SEARCH leads to the bottom row. The bottom row, column c, and the columns to the left of column c are marked as items to be ignored in the next SEARCH/ELIMINATION cycle.

The special cases aside, if $r < m$ regard the rows as $1 \times (n + 1)$ matrices so that they may be added, multiplied by constants, etc.

Leave row r unchanged. Row r is called the *active* row, i.e., the row that is used to *act* on the rows below it. The rows $r + 1$, ..., m are called the *affected* rows, i.e., the rows that are to be *affected* by adding to (or subtracting from) them multiples of the *active* row.

Use the pivot a_{rc}, which is not 0, and the rules of matrix algebra (cf. **Section 1.1**) to *act* on row $r + 1$ by replacing it with

$$\text{row } r + 1 - \frac{a_{r+1,c}}{a_{rc}} \times \text{row } r.$$

Then use the pivot a_{rc} and the laws of matrix algebra to *act* on row $r + 2$ by replacing it with

$$\text{row } r + 2 - \frac{a_{r+2,c}}{a_{rc}} \times \text{row } r,$$

etc.

Finally use the pivot a_{rc} to *act* on row m by replacing it with

$$\text{row } m - \frac{a_{mc}}{a_{rc}} \times \text{row } r.$$

All these operations do nothing more than replace column c by a column in which the only nonzero entry is a_{rc}: all entries (if there are any) above

---

a_{rc} are 0's before the operations begin; all entries below a_{rc} are 0's after the operations end. All entries to the left of column c are 0's before the operations begin and they remain 0's after the operations end. Entries to the right of column c and below row r are *affected*. All other entries to the right of column c remain unchanged.

Then *mark* row r (the *active* row), column c (the *affected* column), and all the columns to the left of column c as items to be ignored in the next SEARCH/ELIMINATION cycle.

The effects of completing the first SEARCH/ELIMINATION cycle are:

i. All entries except one (a_{rc}) in column c are 0's.

ii. The marked columns correspond in the system to *eliminated* unknowns.

iii. The marked row corresponds in the system to a kind of (*elimination*), i.e., a reduction by 1 of the number of equations.

Repeat the SEARCH/ELIMINATION cycle (hereafter called the S/E cycle) on the matrix of *un*marked rows and columns. Compared to $A|\mathbf{b}$ this *un*marked matrix is smaller by one row and, since $c \geq 1$, by at least one column. The effect of the S/E cycle in the *un*marked matrix is analogous to the result in the original matrix at the end of the first S/E cycle .

There are only finitely many rows and at the end of each S/E cycle one row is marked and subsequently ignored. Hence, at the end of finitely many S/E cycles the *un*marked matrix consists of one row and the *PROCESS* stops. Then

EITHER

To the left of the vertical line in the *un*marked matrix all the entries are 0's.

OR

There is some (first) nonzero entry to the left of the vertical line.

In the first instance if the entry to the right of the vertical line is *not* 0, there is *NO* solution. Otherwise there are *INFINITELY MANY* solutions.

In the second instance an equation like (2.3.4) is available. If there is only one unknown left ("$c = n$") there is only *ONE* solution. If several unknowns are left ("$c < n$") there are *INFINITELY MANY* solutions.

If the *PROCESS* is applied to a system $A\mathbf{x} = \mathbf{b}$ then at the end of the operation the augmented matrix $A|\mathbf{b}$ is replaced by a new augmented matrix to be denoted $A_P|\mathbf{c}$. The column

$$
\mathbf{c} \overset{\text{def}}{=} \begin{pmatrix} c_1 \\ \vdots \\ c_m \end{pmatrix}
$$

is the result of the operation of the *PROCESS* on the original column

$$\mathbf{b} \overset{\text{def}}{=} \begin{pmatrix} b_1 \\ \vdots \\ b_m \end{pmatrix}.$$

[**Remark 2.3.2:** Although the operations of the *PROCESS* are applied to the augmented matrix $A|\mathbf{b}$, the *PROCESS* stops if there are no *un*marked columns in A itself. In this sense the *PROCESS* never crosses the vertical line.]

The logic behind the *PROCESS* is very simple. At the end of each cycle if the *un*marked matrix is not empty it corresponds to a system that must have a solution if the original system has a solution. Furthermore each row marked *after* an S/E cycle can be converted to its form *before* the cycle. For example, after the first cycle, old row $r + k$ can be recovered according to the formula

$$\text{old row } r + k = \text{new row } (r + k) + \frac{a_{r+k,c}}{a_{rc}}(\text{row } r).$$

Thus one can roll up from the end of the *PROCESS* to the original system by a sequence of "row-recoveries." Hence any solution to the system corresponding to the *PROCESSED* augmented matrix is a solution to the original system and vice versa. The systems are logically equivalent and whatever is true of the unknowns in one is true of the unknowns in the other.

The logic of the *PROCESS* is revalidated when the *PROCESS* is described in the language of matrix algebra. The discussion of this aspect of the *PROCESS* appears in **Sections 2.6** and **2.8**.

The **Examples** and **Exercises** that follow are useful in clarifying the possibilities, twists, and turns that are encountered in dealing with systems of the form: $A\mathbf{x} = \mathbf{b}$.

The three kinds of examples to be considered are those:

$$\text{for which } m < n;$$
$$\text{for which } m = n;$$
$$\text{and for which } m > n.$$

What follows illustrates each of these kinds and presents opportunities to observe a variety of interesting phenomena associated with the application of the *PROCESS*.

Example 2.3.2. Analyze:

$$3x + 4y + 2z = 4$$
$$2x - z = -2. \tag{2.3.5}$$

Here $m = 2 < 3 = n$ and the augmented matrix to be considered is

$$\left(\begin{array}{ccc|c} 3 & 4 & 2 & 4 \\ 2 & 0 & -1 & -2 \end{array}\right). \tag{2.3.6}$$

In the first S/E cycle row 1 is the *active* row used to *affect* row 2 by replacing it with

$$\text{row } 2 - \frac{2}{3} \times \text{ row } 1$$

and after row 1 and column 1 are marked the partly *PROCESS*ed matrix is

$$\left(\begin{array}{ccc|c} 3 & 4 & 2 & 4 \\ 0 & -\frac{8}{3} & -\frac{7}{3} & -\frac{14}{3} \end{array}\right). \tag{2.3.7}$$

In (2.3.7) only one unmarked row remains. The *PROCESS* ends. Removing the marks from the *PROCESS*ed matrix leads to the *PROCESS*ed matrix

$$\left(\begin{array}{ccc|c} 3 & 4 & 2 & 4 \\ 0 & -\frac{8}{3} & -\frac{7}{3} & -\frac{14}{3} \end{array}\right).$$

It corresponds to the system

$$\begin{aligned} 3x + 4y + 2z &= 4 \\ -\frac{8}{3}y - \frac{7}{3}z &= -\frac{14}{3} \end{aligned} \tag{2.3.8}$$

from which it follows that

$$y = \frac{14}{8} - \frac{7}{8}z = \frac{7}{4} - \frac{7}{8}z.$$

Back-substitution in (2.3.8) shows that

$$\begin{aligned} x &= \frac{(4 - 4y - 2z)}{3} \\ &= -1 + \frac{1}{2}z. \end{aligned} \tag{2.3.9}$$

(It is in the very nature of *substitution* that the values of the unknowns are found in reverse order.)

Since the *PROCESS* is, as noted earlier, reversible, a solution of the system has been found. The value of z may be assigned arbitrarily and then the corresponding values of x and y together with z satisfy the equations of the original system. Thus there are INFINITELY MANY SOLUTIONS for the original system.

Substitution in (2.3.5) confirms that for *any* value of z: $x = -1 + \frac{1}{2}z$, $y = \frac{7}{4} - \frac{7}{8}z$, and $z = z$ is in fact a solution of (2.3.5). Although the theory as developed shows that such a check is unnecessary, it is nevertheless good

---

practice since in the many arithmetical operations performed mechanical errors can easily occur. When the *PROCESS* is carried out by a computer, even if no gross errors are committed, *roundoff* errors sneak in and make checking the solutions desirable.

Exercise 2.3.1. Assume that the number -2 in the second member of the second equation of (2.3.5) is replaced by $\frac{8}{3}$. What does the analysis of the resulting system show?

Exercise 2.3.2. Show how to recover row 2 of (2.3.6) from (2.3.7).

Example 2.3.3. Analyze:

$$x + y - 2z = 3$$
$$3x + z = -1$$
$$4x - 7y + z = 0.$$

Here $m = n = 3$.

The augmented matrix to be considered is

$$\left(\begin{array}{ccc|c} 1 & 1 & -2 & 3 \\ 3 & 0 & 1 & -1 \\ 4 & -7 & 1 & 0 \end{array} \right).$$

After one S/E cycle there emerges

$$\left(\begin{array}{ccc|c} \not{1} & \not{1} & \not{-2} & \not{3} \\ \not{0} & -3 & 7 & -10 \\ \not{0} & -11 & 9 & -12 \end{array} \right).$$

The system corresponding to the unmarked entries consists of two equations and two unknowns (y and z).

The S/E cycle is repeated and produces

$$\left(\begin{array}{ccc|c} \not{1} & \not{1} & \not{-2} & \not{3} \\ \not{0} & \not{-3} & 7 & \not{-10} \\ \not{0} & \not{-0} & -\frac{50}{3} & \frac{74}{3} \end{array} \right)$$

which corresponds to one equation for one unknown (z). The *PROCESS* stops. Now the one and only, i.e., unique, solution can be read off as follows:

$$z = -\frac{74}{50} = -\frac{37}{25},$$
$$y = \frac{10}{3} + \frac{7}{3}z = -\frac{3}{25},$$
$$x = 3 - y + 2z = \frac{4}{75}.$$

Notice how again the solution results from *back-substitution*: the last shall be first.

Example 2.3.4. Analyze:

$$\begin{aligned} 2x + 3y - 5z + 8u - v &= 1 \\ x - y + z + u + v &= 0 \\ 7x + 8y - 14z + 25u - 2v &= 3 \\ -x - y + 6z - 7u + v &= -2. \end{aligned}$$

Here $m = 4 < 5 = n$.

For practice the analysis below is abbreviated and offers only bare suggestions of the S/E cycles that achieve the matrices displayed.

$$\left(\begin{array}{rrrrr|r} 2 & 3 & -5 & 8 & -1 & 1 \\ 1 & -1 & 1 & 1 & 1 & 0 \\ 7 & 8 & -14 & 25 & -2 & 3 \\ -1 & -1 & 6 & -7 & 1 & -2 \end{array}\right) \quad \text{(original matrix)}.$$

After the first S/E cycle the matrix looks like this:

$$\left(\begin{array}{rrrrr|r} 2 & 3 & \cancel{-5} & 8 & \cancel{-1} & 1 \\ \cancel{0} & -\frac{5}{2} & \frac{7}{2} & -3 & \frac{3}{2} & -\frac{1}{2} \\ \cancel{0} & -\frac{5}{2} & \frac{7}{2} & -3 & \frac{3}{2} & -\frac{1}{2} \\ \cancel{0} & \frac{1}{2} & \frac{7}{2} & -3 & \frac{1}{2} & -\frac{3}{2} \end{array}\right).$$

After the second S/E cycle the matrix looks like this:

$$\left(\begin{array}{rrrrr|r} 2 & 3 & \cancel{-5} & 8 & \cancel{-1} & 1 \\ \cancel{0} & \cancel{-\frac{5}{2}} & \frac{7}{2} & \cancel{-3} & \frac{3}{2} & \cancel{-\frac{1}{2}} \\ \cancel{0} & \cancel{0} & 0 & 0 & 0 & 0 \\ \cancel{0} & \cancel{0} & \frac{21}{5} & -\frac{18}{5} & \frac{4}{5} & -\frac{8}{5} \end{array}\right). \qquad (2.3.10)$$

After the next cycle there emerges the matrix

$$\left(\begin{array}{rrrrr|r} 2 & 3 & \cancel{-5} & 8 & \cancel{-1} & 1 \\ \cancel{0} & \cancel{\frac{5}{2}} & \frac{7}{2} & \cancel{3} & \frac{3}{2} & \cancel{\frac{1}{2}} \\ \cancel{0} & \cancel{0} & 0 & 0 & 0 & 0 \\ \cancel{0} & \cancel{0} & \cancel{\frac{21}{5}} & \cancel{\frac{18}{5}} & \cancel{\frac{4}{5}} & \cancel{\frac{8}{5}} \end{array}\right)$$

in which the *un*marked zero 1×3 matrix is one in which the solid row of 0's to the left of the vertical line is matched by a 0 to the right of the vertical line. The *PROCESS* ends. There are *INFINITELY MANY* solutions. The matrix in (2.3.10) shows that $z = -\frac{8}{21} + \frac{6}{7}u - \frac{4}{21}v$ and from back-substitution it follows that

$$y = \frac{1}{5} + \frac{7}{5}z - \frac{6}{5}u + \frac{3}{5}v = -\frac{1}{3} + \frac{1}{3}v$$

$$x = \frac{1}{2} - \frac{3}{2}y + \frac{5}{2}z - 4u + \frac{1}{2}v = \frac{1}{21} - \frac{13}{7}u - \frac{10}{21}v.$$

     Chapter 2. THE PROCESS (**Simple elimination**)

There are INFINITELY MANY solutions since u and v may be given arbitrary values, which then determine x, y, and z.

Not always found in a problem like this one is an interesting feature of the solutions in that y depends only on v whereas x and z depend on *both* u and v.

Exercise 2.3.3. What does the analysis of the problem above reveal if the right member of the third equation of the (original) system is 2 rather than 3?

Example 2.3.5. Analyze the following system:

$$
\begin{aligned}
2x + 3y - 5z &= 1 \\
x - y + z &= 0 \\
7x + 8y - 14z &= 4 \\
-x - y + 6z &= 2.
\end{aligned}
\tag{2.3.11}
$$

Here $m = 4 > 3 = n$.

This time the analysis is restricted to the discussion of the system after the *PROCESS* has been completed, i.e., when the system looks like this:

$$
\begin{aligned}
2x + 3y - 5z &= 1 \\
-\frac{5}{2}y + \frac{7}{2}z &= -\frac{1}{2} \\
0z &= -1 \\
-\frac{21}{5}z &= -\frac{12}{5}.
\end{aligned}
$$

The third equation above allows no solution and so the original system has no solution.

Exercise 2.3.4. What does the analysis of the system (2.3.11) reveal if 4 in the third equation is replaced by 3?

Example 2.3.6. Analyze the system

$$
\begin{aligned}
0w + 0x + 0y + 0z + u - v &= 1 \\
0w + 0x + 0y - z - u + v &= -1 \\
0w + 2x - 3y - z + 3u - 4v &= 0 \\
0w + 5x - y + z + 2u - 3v &= 2.
\end{aligned}
$$

Here $m = 4 < 6 = n$. Note the anomaly: all the coefficients of w are 0's.

After the first S/E cycle the modified augmented matrix looks like this:

$$
\left(
\begin{array}{cccccc|c}
\cancel{0} & 0 & 0 & 0 & 1 & -1 & 1 \\
\cancel{0} & 0 & 0 & -1 & -1 & 1 & -1 \\
\cancel{0} & 2 & \cancel{-3} & \cancel{-1} & 3 & \cancel{-4} & \cancel{0} \\
\cancel{0} & \cancel{0} & \frac{13}{2} & \frac{7}{2} & -\frac{11}{2} & 7 & \frac{9}{2}
\end{array}
\right).
$$

(After the cycle just completed there occurs the situation in which SEARCH leads to the bottom row.)

After the next cycle the matrix is

$$\left(\begin{array}{cccccc|c} \cancel{0} & 0 & 0 & 0 & 1 & -1 & 1 \\ \cancel{0} & \cancel{0} & \cancel{0} & \cancel{1} & \cancel{1} & 1 & \cancel{1} \\ \cancel{0} & 2 & \cancel{3} & \cancel{1} & 3 & \cancel{1} & \cancel{0} \\ \cancel{0} & \cancel{0} & \frac{13}{2} & \frac{7}{2} & \cancel{\frac{11}{2}} & 7 & \cancel{2} \end{array}\right). \qquad (2.3.12)$$

There is only one row in the *un*marked matrix. The *PROCESS* stops and from the *PROCESS*ed matrix there can be constructed the corresponding system:

$$u - v = 1 \qquad (2.3.13)$$

$$-z - u + v = -1 \qquad (2.3.14)$$

$$2x - 3y - z + 3u - 4v = 0 \qquad (2.3.15)$$

$$\frac{13}{2}y + \frac{7}{2}z - \frac{11}{2}u + 7v = 2. \qquad (2.3.16)$$

It follows that the solution is

$$u = 1 + v \quad \text{from } (2.3.13) \qquad (2.3.17)$$

$$z = 0 \quad \text{from } (2.3.14),\ (2.3.17) \qquad (2.3.18)$$

$$y = \frac{(15 - 3v)}{13} \quad \text{from } (2.3.16),\ (2.3.17),\ (2.3.18) \qquad (2.3.19)$$

$$x = \frac{(3 + 2v)}{13} \quad \text{from } (2.3.15),\ (2.3.17),\ (2.3.18),\ (2.3.19). \qquad (2.3.20)$$

Note that v and w may be given any values.

Exercise 2.3.5. There are 19,683 3×3 matrices in which each entry is one of the numbers 0, ± 1. Choose any five significantly different matrices from this set of matrices. Regard each choice as representing the *augmented* matrix corresponding to a system: $Ax = b$. Analyze each system.

Exercise 2.3.6. There are 244,140,625 3×4 matrices in which each entry is one of the numbers 0, ± 1, ± 2. Choose any five significantly different matrices from this set of matrices. Regard each choice as representing the *augmented* matrix of a system: $Ax = b$. Analyze each system.

Exercise 2.3.7. There are 390,625 2×4 matrices in which each entry is one of the numbers 0, ± 1, ± 2. Choose any five significantly different matrices in which each entry in the fourth column is 0. Let $\mathbf{O}$ denote the vector

$$\begin{pmatrix} 0 \\ \vdots \\ 0 \end{pmatrix}.$$

Regard each choice as representing the *augmented* matrix of a system: $A\mathbf{x} = \mathbf{O}$. Analyze each system and show that it has a solution.

Exercise 2.3.8. Generalize the result in **Exercise 2.3.7** to the following:

THEOREM. IF A IS ANY $m \times n$ MATRIX THEN THERE IS ALWAYS A SOLUTION FOR THE SYSTEM:

$$\begin{pmatrix} a_{11} & \cdots & a_{1n} \\ \vdots & \ddots & \vdots \\ a_{m1} & \cdots & a_{mn} \end{pmatrix} \begin{pmatrix} x_1 \\ \vdots \\ x_n \end{pmatrix} = \begin{pmatrix} 0 \\ \vdots \\ 0 \end{pmatrix}.$$

Exercise 2.3.9.

i. Analyze the system

$$\begin{aligned} -3x + 2y - 4z + 6u &= -7 \\ x + 5z - 8u &= -1 \\ 4x - y + 2u &= 13 \\ -2x + 9y + 12z - 30u &= -83. \end{aligned} \tag{2.3.21}$$

ii. Replace the right members by 0's and analyze the resulting *homogeneous* system.

iii. Show that the difference of any two solutions of (2.3.21) is a solution of the *homogeneous* system.

iv. Show that if a is a solution of (2.3.21) and if h is a solution of the homogeneous system then $\mathbf{a} + \mathbf{h}$ is also a solution of (2.3.21).

v. Show that if

$$\mathbf{a} \stackrel{\text{def}}{=} \begin{pmatrix} a_1 \\ a_2 \\ a_3 \\ a_4 \end{pmatrix}$$

is one solution of (2.3.21) then for any other solution

$$\mathbf{c} \stackrel{\text{def}}{=} \begin{pmatrix} c_1 \\ c_2 \\ c_3 \\ c_4 \end{pmatrix}$$

of (2.3.21) there is a solution

$$\mathbf{h} \stackrel{\text{def}}{=} \begin{pmatrix} h_1 \\ h_2 \\ h_3 \\ h_4 \end{pmatrix}$$

of the homogeneous system and $\mathbf{c} = \mathbf{a} + \mathbf{h}$.

In other words, there are infinitely many solutions of (2.3.21), any two differing by a solution of the homogeneous version of (2.3.21).

Exercise 2.3.10. Generalize the result in **Exercise 2.3.9** to the following:

THEOREM. IF $\mathbf{a}$ AND $\mathbf{c}$ ARE SOLUTIONS OF $A\mathbf{x} = \mathbf{b}$ THEN $\mathbf{a} - \mathbf{c}$ IS A SOLUTION OF $A\mathbf{x} = \mathbf{O}$. FURTHERMORE IF $\mathbf{a}$ IS ANY SOLUTION OF $A\mathbf{x} = \mathbf{b}$ AND IF $\mathbf{c}$ IS ANY OTHER SOLUTION OF $A\mathbf{x} = \mathbf{b}$ THEN FOR SOME SOLUTION $\mathbf{h}$ OF $A\mathbf{x} = \mathbf{O}$, $\mathbf{c}=\mathbf{a}+\mathbf{h}$.

Example 2.3.7. Find the solution of the following linear programming problem:

Maximize: $G \overset{\text{def}}{=} 5x + 2y$ under the constraints

$$(1): \ x - 2y \leq 2 \ (L_1); \quad (2): \ x - y \leq 3 \ (L_2); \quad (3): \ 2x + y \leq 12 \ (L_3);$$
$$(4): \ -x + 2y \leq 9 \ (L_4); \quad (5): \ -x + y \leq 4 \ (L_5);$$
$$(6): \ x \geq 0; \quad (7): \ y \geq 0. \tag{2.3.22}$$

The inequalities (2.3.22) determine, as shown in **Figure 2.3.1**, a *convex* set $\mathcal{Q}$ in the first quadrant of the xy-plane ($\mathbf{R}^2$).

The lines l_k for which the equations are

$$5x + 2y = k, \quad -\infty < k < \infty, \tag{2.3.23}$$

are parallel to each other. Hence one seeks a value of k that will cause the line l_k to meet $\mathcal{Q}$. If l_k meets $\mathcal{Q}$ the coordinates (x, y) of any point on l_k and in $\mathcal{Q}$, i.e., the coordinates of any point P in

$$l_k \cap \mathcal{Q},$$

satisfy the inequalities that determine $\mathcal{Q}$. If and only if $0 \leq k \leq 29$ does l_k meet $\mathcal{Q}$. From **Figure 2.3.1** it follows that the maximizing value 29 of k corresponds to a line l_k that passes through a vertex of $\mathcal{Q}$. Thus the search for a maximizing vertex of $\mathcal{Q}$ solves the problem.

The vertices of $\mathcal{Q}$ can be found by converting each of the seven inequalities

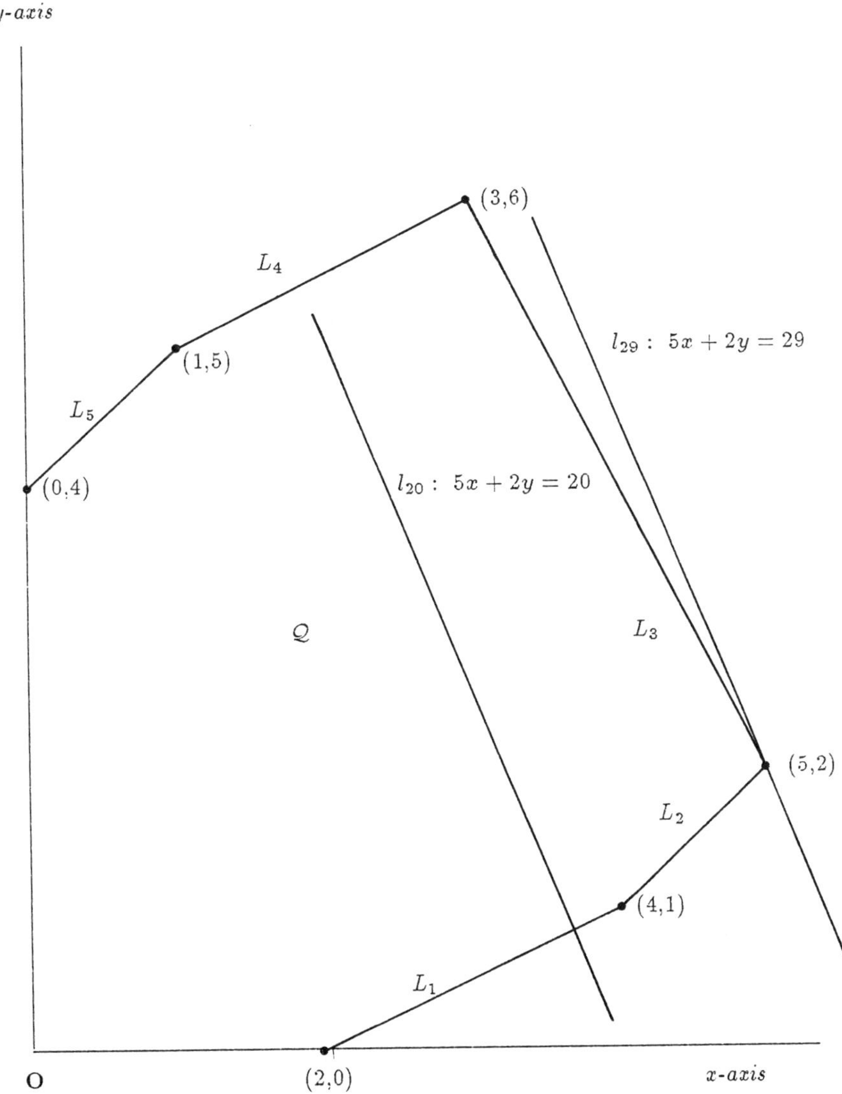

Figure 2.3.1.

(1) - (7) in (2.3.22) to equations, solving each of the 21 systems corresponding to each pair of those equations, and then choosing those vertices that lie in the first quadrant. If each of the coordinate pairs for these vertices is substituted in the expression $5x + 2y$ for G, then the largest of those G values is the maximum sought.

Such a procedure is clearly inefficient. For example a relatively simple problem in economics might involve 10,000 inequalities, each involving 1,000 variables. The corresponding convex set Q has a vertex corresponding to each of the

$$\binom{10,000}{1,000} \approx 10^{2,000}$$

systems of 1,000 equations (in 1,000 unknowns) that can be drawn from the 10,000 inequalities. The number of these "vertices" exceeds the number of elementary particles in the universe.

However, if just one vertex is found, a move to a neighboring "best" vertex can be determined, and the operation can be repeated until there is no "better" vertex is available. This is the essence of the *simplex method*. It and its improved variants are far more practical than the method of examining all vertices.

For example, (0,0) is a vertex of Q. The two edges of Q that pass through (0,0) are the x- and y- axes.

Move out along the x-axis where $y = 0$. On the x-axis the vertex *nearest* (0,0) corresponds to the *least* value of x that satisfies all the inequalities in (2.3.22) when $y = 0$. This value is 2 which is required by (1) and vertex (2,0) is a neighbor of the vertex (0,0). At (2,0) the value of G is 10.

Go back to (0,0) and move out along the y-axis where $x = 0$. By virtue of (5) the nearest vertex this time is (0,4) where the value of G is 8. Hence (2,0) is a better vertex than (0,4). (In defiance of the rules of grammar (2,0) is the "best" of the two neighbors to (0,0).)

The new edge passing through (2,0) is on the line L_1 for which the equation is $x = 2 + 2y$. Move along this edge by allowing y to increase from 0 so long as the corresponding value $2 + 2y$ of x violates none of the inequalities (1) - (7). The inequality (2) forces y not to exceed 1, when $x = 4$. The neighboring vertex is (4,1) where the value of G is 22.

The new edge passing through (4,1) is on the line L_2 for which the equation is $x = 3 + y$. Unless y *increases* from 1 the inequality (1) corresponding to the equation of L_1 will be violated. However y may not increase beyond 2 since then x exceeds 5 and then $2x + y$ exceeds 12 in violation of (3). Hence the neighbor is (5,2) where $G = 29$.

The new edge passing through (5,2) is on the line L_3 for which the equation is $x = \frac{(12-y)}{2}$. Move along this edge by allowing y to increase from 1 so long as the corresponding value $\frac{(12-y)}{2}$ of x violates none of the inequalities (1) - (7). (Note that moving on L_3 as y *decreases* from 1 causes a violation of (2). That is why the move is made by allowing y to *increase*.) This time (4) requires that y not exceed 6. The neighboring vertex is (3,6) where the value of G is 27.

Since the value of G decreases in passing from (5,2) to (3,6), it follows that (5,2) is the maximizing (optimizing) vertex for G. The maximum value for G is 29.

It should be noted that the optimal vertex is the fourth vertex tested. On

the other hand there are 21 pairs of lines to be examined if the method of finding
all vertices is used.

The simplex method, in the version shown above, achieves the optimum,
but not necessarily in an optimal manner.

Example 2.3.8. In the following problem the simplex method is not optimal. Assume $G = 2x + y$ and that Q is defined by the following conditions:

$$(1): \ y \leq 5x \ (2): \ 2y - x \leq 9 \ (3): \ y \leq 5.5$$
$$(4): \ 2y + 2x \leq 17 \ (5): \ 2y - 7x \geq -19 \ (6): \ 2y \geq x.$$

Find the maximum value of G if (x, y) must satisfy conditions (1)-(6).

One vertex of Q is $(0,0)$. The two neighboring vertices of Q are $(2,1)$ and
$(1,5)$. At $(1,5)$ the value of G is 7; at $(2,1)$ the value of G is 5. The simplex
method suggests that one seek the "best" vertex neighboring $(1,5)$. That vertex
is $(2,5.5)$ where the value of G is 9.5. The next "best" vertex is $(3,5.5)$ where
the value of G is 11.5. The next "best" vertex is $(4,4.5)$ where the value of G is
12.5. The only other vertex is $(2,1)$ where the value of G is 5. Hence $(4,4.5)$ is
the maximizing vertex and the maximum value of G in Q is 12.5. The simplex
method leads to the maximizing vertex in *four* steps.

On the other hand, despite the fact that $(2,1)$ is not the "best" vertex to
test after $(0,0)$, the "best" neighboring vertex to $(2,1)$ is $(4,4.5)$, the maximizing
vertex, reached in only *two* steps rather than in the *four* steps prescribed by the
simplex method when the first vertex chosen is $(0,0)$.

Nevertheless the simplex method *is* quicker than the method in which all
vertices are found and tested. In **Example 2.3.7** only five vertices are tested
whereas the number of line intersections is 21 and only six of those line intersections are vertices of Q.

Exercise 2.3.11. For the constraints (1) - (7) in **Example 2.3.7** find:

i. the maximum of $H \stackrel{\text{def}}{=} 5x + y$;

ii. the maximum of $J \stackrel{\text{def}}{=} 6x + y$.

Exercise 2.3.12. Let the constraints on (x, y) be:

$$(1): \ x - 2y \leq 0; \ (2): \ x - y \leq 6; \ (3): \ x + y \leq 9;$$
$$(4): \ x + 2y \leq 14; \ (5): \ -3x + y \leq 0; \ (6): \ 2x + y \geq 5. \qquad (2.3.24)$$

For the constraints (1) - (6) in (2.3.24) find:

i. the maximum and minimum of $R \stackrel{\text{def}}{=} 5x + 3y$;

ii. the maximum and minimum of $T \stackrel{\text{def}}{=} 5x + 4y$;

iii. the maximum and minimum of $M \stackrel{\text{def}}{=} y - x$;

---

iv. the minimum of $F \overset{\text{def}}{=} 2y - x$.

Exercise 2.3.13. Assume $L = 6x + y$. Maximize L subject to the conditions (1)-(6) of **Example 2.3.8**. Use the simplex method and start at the vertex $(0,0)$. Is the simplex method optimal in this case?

Exercise 2.3.14. If, in **Exercise 2.3.11, 2.3.12,** or **2.3.13**, there appears to be no unique best vertex, interpret the result.

[**Remark 2.3.3:**   The heart of the simplex method when only two variables (x and y) are involved is an algorithm that in effect moves a "line" parallel to itself and through a sequence of better and better vertices of a convex polygon.

In the general case there are n variables and the operation takes place in $\mathbf{R}^n$. If $n = 3$ the "line" is a plane; if $n > 3$ the "line" is called a *hyperplane*. The equation of such a hyperplane in $\mathbf{R}^n$ is a *linear* equation:

$$a_1 x_1 + \cdots + a_n x_n = k.$$

The coefficients $a_1, \ldots, a_n$ are held fixed, and the "parameter" k is varied. Hyperplanes corresponding to different values of k are called "parallel." Varying k causes the hyperplane to move parallel to itself.

The domain through which k is varied is chosen so that each value of k corresponds to a hyperplane intersecting a convex "polyhedron" Q, which is defined by a finite set of *linear* inequalities, e.g.,

$$\sum_{j=1}^{n} a_{ij} x_j \leq K_i, \ 1 \leq i \leq N$$

or

$$\sum_{j=1}^{n} a_{ij} x_j \geq K_i, \ 1 \leq i \leq N.$$

The number N counts the number of constraints or inequalities to which the variables $x_1, \ldots, x_n$ are subjected. The numbers n and N are independent of each other, e.g., n might be 6 while N is 20, or n might be 200 while N is 13.

A "face" of Q is the intersection of Q with one of the hyperplanes that define Q; an $n - 2$-dimensional "face" of Q is defined as consisting of the points of Q in the intersection of *two* intersecting hyperplanes defining Q; ...; a "vertex " of Q is a point lying in Q and in the intersection of n of the hyperplanes defining Q.]

2.4. Complements

I.

SHAPES

The matrices considered next are the matrices consisting of the coefficients of the *unknowns* in the system. For the moment the *augmented* matrices

$$A|\mathbf{b} \overset{\text{def}}{=} \left(\begin{array}{ccc} a_{11} & \cdots & a_{1n} \\ \vdots & \ddots & \vdots \\ a_{m1} & \cdots & a_{mn} \end{array} \middle| \begin{array}{c} b_1 \\ \vdots \\ b_m \end{array} \right)$$

are in abeyance. Only the matrices

$$A \overset{\text{def}}{=} \left(\begin{array}{ccc} a_{11} & \cdots & a_{1n} \\ \vdots & \ddots & \vdots \\ a_{m1} & \cdots & a_{mn} \end{array} \right)$$

appearing to the *left* of the long vertical lines are of current interest.

The three possibilities $m > n$, $m < n$, and $m = n$ for the system

$$a_{11}x_1 + \cdots + a_{1n}x_n = b_1$$
$$\vdots \qquad \ddots \qquad \vdots \qquad \vdots$$
$$a_{m1}x_1 + \cdots + a_{mn}x_n = b_m \qquad\qquad (*)$$

may be remembered by the *shape* (specified by the numbers m and n) of the matrix

$$A \overset{\text{def}}{=} \left(\begin{array}{ccc} a_{11} & \cdots & a_{1n} \\ \vdots & \ddots & \vdots \\ a_{m1} & \cdots & a_{mn} \end{array} \right).$$

The matrix A is:

HIGH iff $m > n$:
In A there are more rows than there are columns. In $(*)$ there are more equations than there are unknowns — few unknowns are asked to satisfy many equations. The chance that the system $(*)$ has a solution is poor.

WIDE iff $m < n$:
In A there are fewer rows than there are columns. In $(*)$ there are fewer equations than there are unknowns — many unknowns are asked to satisfy few equations. The chance that the system $(*)$ has a solution is excellent. (*In particular, if all* $b_j = 0$, *i.e., if the system* $(*)$ *is* homogeneous, *there are always* INFINITELY MANY *solutions.*) Hence if the system is WIDE and homogeneous there is a *nonzero* solution.

SQUARE iff $m = n$:

In A there are as many rows as there are columns. In $(*)$ there are as many equations as there are unknowns — the unknowns are asked to do what is reasonable. The chance that the system $(*)$ has a solution is fair.

Only a careful analysis, e.g., via the *PROCESS*, can provide assurances, statements that are untainted by chance, about the existence of solutions for the system $(*)$. Nevertheless the intuition provided by the *shape* of A should not be ignored. The *shapes* of the matrix A in its different forms look like this:

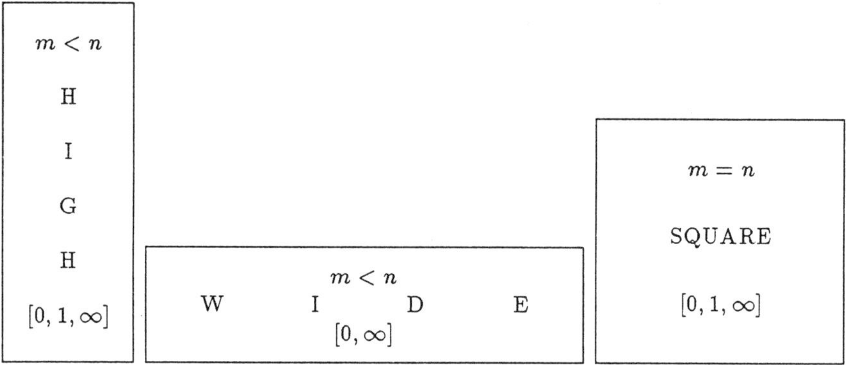

Figure 2.4.1.

Systems of equations with HIGH, WIDE, or SQUARE matrices (*not* the augmented matrices, just the matrices consisting of the coefficients of the unknowns) are called HIGH, WIDE, or SQUARE systems.

From the *shape* of a system one can see in a helpful but somewhat imprecise way what to expect as the *PROCESS* is performed.

[**Remark 2.4.1:**

In a HIGH system there are fewer columns than rows and so the marking of columns will stop before all rows are marked.

Hence, in the *PROCESSING* of a HIGH system there will inevitably appear at least one row of 0's to the left of the vertical line.

In a WIDE system there are fewer rows than columns and so the marking of rows will stop before all columns are marked.

Hence, in the *PROCESSING* of a WIDE system there will inevitably appear a $1 \times n$ *un*marked matrix.

For the *PROCESS*ing of a SQUARE system no general prediction can be made.]

II.

CONSTRAINTS

Any equation (or inequality) involving unknowns may be regarded as a *constraint* imposed on the unknowns, in that they are no longer free to assume any values but only values that satisfy the equation or inequality. A system of m equations or inequalities in n unknowns imposes m constraints on n unknowns.

In particular, a HIGH system imposes on the unknowns more constraints than there are unknowns since in a HIGH system there are more equations than there are unknowns, i.e., $m > n$. Superficially the unknowns are overburdened with requirements, are tightly constrained, and may not be able to meet them all.

On the other hand a WIDE system imposes fewer constraints than there are unknowns, i.e., $m < n$. Superficially the unknowns are underburdened with requirements, are loosely constrained, and should be able to meet, perhaps in many different ways, all the requirements of the system.

A SQUARE system seems to be just right, imposing just as many constraints as there are unknowns, i.e., $m = n$. There are just as many constraints as the unknowns should be able to handle.

This highly oversimplifying way of regarding systems of equations is helpful in providing at least a hint of what to expect in dealing with them. However only more detailed analysis, e.g., via the *PROCESS*, can tell the whole story. Alternatives or modifications of the *PROCESS* are given below in **V**.

One may regard a system of equations or inequalities as a means of *determining* the value(s) of the unknowns in the system. A HIGH system appears to be *over-determining*; a WIDE system appears to be *under-determining*; a SQUARE system appears to be just plain *determining*. Not surprisingly, appearances can deceive and only further analysis can pierce the deceptions.

III.

LOOSENESS

Regard the quantity LOOSENESS defined by the formula

LOOSENESS = (NUMBER OF UNKNOWNS)
− (NUMBER OF EQUATIONS)

as (crudely) measuring the chances of finding a solution for a system. The greater the looseness, the greater are the chances that the system has at least

---

one solution. In particular, a negative looseness bodes poorly for the existence of any solution at all. Of course, the detailed analysis provided by the *PROCESS* permits a proper sorting out of what really goes on.

IV.

CONSISTENCY

Call a system INCONSISTENT if it has no solution; otherwise call it CONSISTENT. The sign $[1, \mathbf{0}, \infty]$ can be modified so that $[0]$ stands for an INCONSISTENT system, $[1, \infty]$ for a CONSISTENT system, $[1]$ for a (CONSISTENT) system having exactly one solution, $[\infty]$ for a (CONSISTENT) system having infinitely many solutions, etc.

Exercise 2.4.1. Can a HIGH system have the sign $[1]$?

Exercise 2.4.2. Can a WIDE system have the sign $[1]$?

Exercise 2.4.3. How many signs $[\ldots]$ can be valid descriptions of solution possibilities for a system? What are these valid signs?

The systems in **Example 2.4.1** point up the the need to exercise caution in jumping to conclusions based on the crude hints.

Example 2.4.1. The HIGH systems

$$
\begin{array}{ll}
2x + 3y = 3 & 2x + 3y = 3 \\
5x - y = 1 \quad \text{resp.} & 4x + 6y = 6 \\
7x + 2y4 & 6x + 9y = 9
\end{array}
$$

are determined resp. underdetermined; the WIDE system

$$
\begin{array}{l}
2x + 3y + 5z = 1 \\
4x + 6y + 10z = 3
\end{array}
$$

is over-determined (a WIDE system is never determining); the SQUARE systems

$$
\begin{array}{lll}
2x + 3y = 3 & & 2x + 3y = 3 \\
4x + 6y = 6 & \text{resp.} & 4x + 6y = 7
\end{array}
$$

are under-determined resp. over-determined. The following display reveals all that can be said (short of carrying out a detailed analysis, e.g., by the *PROCESS*) about the various types of systems.

   Chapter 2. THE PROCESS (**Simple elimination**)

TYPE	LOOSENESS	SIGN
HIGH	NEGATIVE	$[1, 0, \infty]$
WIDE	POSITIVE	$[0, \infty]$
SQUARE	ZERO	$[1, 0, \infty]$

Table 2.4.1

V.

OTHER FORMS OF THE *PROCESS*

In the *PROCESS* as described at no point are rows exchanged.

Some prefer modifications of the *PROCESS* that permit reading off more easily whether there are solution(s) and reading off more easily the solution(s) themselves, if there are any, from the final form of the augmented matrix. The description and discussion that follow take up methods that facilitate this kind of improvement in the visible form of the matrix after it has been *PROCESS*ed.

Thus let

$$a_{11}x_1 + \cdots + a_{1n}x_n = b_1$$

$$\vdots \qquad \ddots \qquad \vdots \qquad \vdots$$

$$a_{m1}x_1 + \cdots + a_{mn}x_n = b_m \tag{2.4.1}$$

be the system to be analyzed. Four steps constitute one method, to be denoted hereafter as the GEM (Gauss Elimination Method).

SEARCH

This step is exactly the same as SEARCH in the *PROCESS*.

EXCHANGE (as needed)

In this step the GEM arranges to place at the *top* of column c the *nonzero* entry found in the SEARCH. Hence the operation is the exchange of rows r and 1 to produce the matrix:

$$\left(\begin{array}{ccccc|c}
0 & \cdots & a_{rc} & \cdots & a_{rn} & b_r \\
0 & \cdots & 0 & \cdots & a_{2n} & b_2 \\
\vdots & \ddots & \vdots & \ddots & \vdots & \vdots \\
0 & \cdots & 0 & \cdots & a_{1n} & b_1 \\
0 & \cdots & a_{r+1,c} & \cdots & a_{r+1,n} & b_{r+1} \\
\vdots & \ddots & \vdots & \ddots & \vdots & \vdots \\
0 & \cdots & a_{mc} & \cdots & a_{mn} & b_m
\end{array}\right).$$

Note that the nonzero entry a_{rc} in the new matrix is now the *first* nonzero entry that is found if SEARCH is applied to the new matrix.

[EXCHANGE is omitted if $a_{1c} \neq 0$. In that case $r = 1$.]

ELIMINATION

This step is the same as ELIMINATION in the *PROCESS*. The result of ELIMINATION on rows $r + 1$ through m is a matrix

$$\left(\begin{array}{ccccccc|c}
0 & \cdot & a_{rc} & a_{r,c+1} & \cdot & a_{rn} & & b_r \\
0 & \cdot & 0 & a_{2,c+1} & \cdot & a_{2n} & & b_2 \\
\cdot & \cdot & \cdot & \cdot & \cdot & \cdot & & \cdot \\
0 & \cdot & 0 & a_{1,c+1} & \cdot & a_{1n} & & b_1 \\
0 & \cdot & 0 & a_{r+1,c+1} - \frac{a_{r+1,c}}{a_{rc}} a_{r,c+1} & \cdot & a_{r+1,n} - \frac{a_{r+1,c}}{a_{rc}} a_{rn} & b_{r+1} - \frac{a_{r+1,c+1}}{a_{rc}} b_r \\
\cdot & \cdot & \cdot & \cdot & \cdot & \cdot & & \cdot \\
0 & \cdot & 0 & a_{m,c+1} - \frac{a_{mc}}{a_{rc}} a_{r,c+1} & \cdot & a_{mn} - \frac{a_{mc}}{a_{rc}} a_{rn} & b_m - \frac{a_{mc}}{a_{rc}} b_r
\end{array}\right)$$

in which the *affected* rows are prominently displayed to show the modifications stemming from the *active* row.

The original matrix has been altered so that in the new matrix the first row is simpler. Below the first row all that needs attention is the smaller matrix consisting of *un*marked rows numbered $2, 3, \ldots, m$, and *un*marked columns numbered $c + 1, \ldots, n + 1$. These rows and columns are outlined by the $*$ pattern in the next display.

$$\left(\begin{array}{cccccc|c}
0 & \cdots & a_{rc} & a_{r,c+1} & \cdots & a_{rn} & b_r \\
0 & \cdots & 0 & * & \cdots & * & * \\
\vdots & \ddots & \vdots & \vdots & \ddots & \vdots & \vdots \\
0 & \cdots & 0 & * & \cdots & * & * \\
\vdots & \ddots & \vdots & \vdots & \ddots & \vdots & \vdots \\
0 & \cdots & 0 & * & \cdots & * & *
\end{array}\right)$$

The corresponding system of equations is smaller in two ways: it contains fewer unknowns and involves fewer equations because the number of unknowns in the original system is n and in the reduced system the number of unknowns is $n - c$. The number of equations is diminished and is now at most $m - 1$.

Quite possibly there are p solid rows of 0's to the left of the vertical line in the smaller matrix now under consideration. Then the number of equations requiring further attention is only $m - p - 1$. Thus for some nonnegative integer p the new system consists of $m - p - 1$ equations in $n - c$ unknowns. This reduction in the number of equations and in the number of unknowns is the whole point of the *PROCESS* and of the GEM. Each successively eliminates equations and unknowns until the system is reduced to one equation in one or more unknowns.

The rest of the GEM operation consists of repeating the cycle

SEARCH

EXCHANGE (as needed)

ELIMINATION on the new and smaller matrix produced in the previous cycle of SEARCH, EXCHANGE, and (as needed) ELIMINATION.

Just as in the *PROCESS* the GEM operation must stop after a finite number of cycles. When the GEM operation stops:

EITHER *i.*:
At the conclusion of some performance of ELIMINATION below some row (r) all entries are 0's — they form a *block* of 0's — to the *left* of the long vertical line. Then the GEM operation stops and the system can be analyzed by inspection. Whether there are solutions is decided by whether each row of 0's to the left of the vertical line is matched by a 0 in the same row and to the right of the vertical line.

OR *ii.*:
The matrix is WIDE or SQUARE $(m \leq n)$. The GEM operation reaches the bottom row, and, ** denoting a nonzero number, converts the bottom row to the form

$$0 \quad \ldots \quad 0 \quad \ldots \quad ** \quad * \quad \ldots \quad * \quad | \quad *.$$

The system can be analyzed by inspection.

In the first instance i, which is certainly the case if the system is HIGH (more equations than unknowns), the original system is replaced by a system that looks like this: (Each $*$ represents some number (possibly 0) that has been produced in the course of the GEM; ** represents a nonzero number.)

$$**x_c + *x_{c+1} + \cdots\cdots\cdots\cdots\cdots\cdots\cdots\cdots\cdots\cdots + *x_n = *$$
$$**x_{c+k} + *x_{c+k+1} + \cdots\cdots\cdots\cdots\cdots\cdots + *x_n = *$$
$$\cdots\cdots\cdots\cdots\cdots\cdots\cdots\cdots\cdots\cdots\cdots \quad \cdots$$
$$**x_{c+k+\cdots+l} + *x_{c+k+\cdots+l+1} + \cdots + *x_n = *$$
$$0x_n = *$$
$$\vdots \quad \vdots$$
$$0x_n = *.$$

The first nonzero coefficient in each equation is **. The system has no solution if any left member for which all coefficients are 0's is not matched by a right member $*$ that is also zero.

On the other hand if each left member for which all coefficients are 0's is matched by a right member $*$ that is also zero, the system has a solution. To find it start with the last *nonzero* left member and solve for its $x_{c+k+\cdots+l}$ (which has a nonzero coefficient $**$). Then use back substitution to solve the preceding equation for *its* first $x_{c+k+\cdots+?}$ having nonzero coefficient $**$, etc.

(Some liken back-substitution in the GEM to rolling up a window shade or a rug. The **Examples** below illustrate the situation.)

In the second instance *ii* there is no obstacle and one proceeds with back-substitution (as in the preceding case, when the left members that are zero are properly matched by zeros in the right members.) The **Examples** below illustrate this situation as well.

If *i* obtains and if there is one solution there are infinitely many. The sign for systems in this instance is thus $[0, \infty]$.

If *ii* obtains there is always at least one solution and there can be infinitely many. The sign for such systems is thus $[1, \infty]$.

It is important to note that each step of the GEM operation is *reversible*. Indeed if, *before* a cycle at any one stage of the GEM operation is begun, the numbers $x_1, x_2, \ldots, x_n$ satisfy the equations of the system then those same numbers $x_1, x_2, \ldots, x_n$ satisfy the equations of the modified system *after* the same cycle is finished. Conversely, if, *after* some cycle is finished, the numbers $x_1, x_2, \ldots, x_n$ satisfy the equations of a modified system then those same numbers $x_1, x_2, \ldots, x_n$ satisfy the equations of the system *before* that cycle is begun. In other words, the numbers $x_1, x_2, \ldots, x_n$ satisfy the original system of equations iff the numbers satisfy each of the modified systems that emerge as the GEM operation goes through its cycles.

Thus the analysis that is carried out on the system at the end reveals precisely what is going on in the original system. Later, when *matrix algebra* is taken up, this state of affairs is illuminated from a new perspective in the proof of THEOREM 2.9.2.

Note also that the GEM may be applied equally well to the system of equations or to the augmented matrix that corresponds to the system. Dealing with the augmented matrix rather than with the equations makes the work much easier. The needless and repeated writing of $x_1, x_2, \ldots, x_n$ vanishes.

The following indicates the form of the augmented matrix derived from (2.4.1) after the GEM is applied to it:

$$
\left(
\begin{array}{ccccccccccccc|c}
0 & \ldots & 0 & ** & * & \ldots & * & \ldots & * & * & * & \ldots & * & * \\
0 & \ldots & 0 & 0 & ** & \ldots & * & \ldots & * & * & * & \ldots & * & * \\
0 & \ldots & 0 & 0 & 0 & \ldots & 0 & \ldots & ** & * & * & \ldots & * & * \\
0 & \ldots & 0 & 0 & 0 & \ldots & 0 & \ldots & 0 & 0 & 0 & \ldots & 0 & * \\
\ldots & \ldots & \ldots & \ldots & \ldots & \ldots & \ldots & \ldots & \ldots & \ldots & & & \vdots \\
0 & \ldots & 0 & 0 & 0 & \ldots & 0 & 0 & 0 & 0 & 0 & \ldots & 0 & * \\
\end{array}
\right).
$$

The solid bottom rows of 0's (to the left of the vertical line) occur in *i* and not necessarily in *ii*.

The GEM applied to a matrix A produces a matrix denoted A_E and called the *echelon form* of A. Thus if c stands for the column of *'s to the right of the vertical line when the GEM operation is over, the final augmented matrix is $A_E|$c. Note how the blocks of 0's form *steps* and in the corner of each step there stands a nonzero number **.

If

$$A_E \overset{\text{def}}{=} \begin{pmatrix} \alpha_{11} & \cdots & \alpha_{1n} \\ \vdots & \ddots & \vdots \\ \alpha_{m1} & \cdots & \alpha_{mn} \end{pmatrix}$$

the first nonzero entry of each nonzero row of A_E is a *pivot* of the matrix A.

[The word *echelon* comes from an Indo-European root *skand* — to climb. It is related to a*scend*, *scal*e and to the Latin *scalae* — ladder or steps. The blocks of 0's in *steps* consisting of the blocks of 0's against which and on which the pattern of 1's and *'s is placed account for the term *echelon* used for matrices with this step-like structure.]

The description and discussion above are offered for completeness. The GEM is easy to understand when it is applied to explicit numerical situations where subscripts and indices and letters are absent, do not obscure what is really happening, and can cause no confusion. In the next **Section** there are several **Examples** designed to illustrate the different possibilities that arise and to show how the *PROCESS* handles them all with equal ease and directness.

Exercise 2.4.4. Give mathematical descriptions of the geographical terms "north," "west," etc., used above in describing the locations of some of the 0's with respect to the entry a_{rc}.

[*Hint:* The entries "north" of a_{rc} are those a_{ij} for which $i < r$ and $j = c$.]

Exercise 2.4.5. Use the GEM to analyze each of the systems corresponding to the input-output matrices for the logistics of nutrition as in **Section 2.2**.

Akin to GEM is another version, the GJM (<u>G</u>auss-<u>J</u>ordan <u>M</u>ethod), where DIVISION (by a_{rc}) precedes ELIMINATION. The steps of the GJM are:

SEARCH;

EXCHANGE (as needed);

DIVISION: Divide all elements of the (new) first row by the nonzero (!) number a_{rc}.

ELIMINATION The topmost entry in column c is 1. All entries to the left of a_{rc} are now 0 and the matrix looks like this:

$$\left(\begin{array}{ccccc|c}
0 & \cdots & 1 & \cdots & \frac{a_{rn}}{a_{rc}} & \frac{b_r}{a_{rc}} \\
0 & \cdots & 0 & \cdots & a_{2n} & \vdots \\
\vdots & \ddots & \vdots & \ddots & \vdots & b_1 \\
0 & \cdots & 0 & \cdots & a_{1n} & b_2 \\
0 & \cdots & a_{r+1,c} & \cdots & a_{r+1,n} & b_{r+1} \\
\vdots & \ddots & \vdots & \ddots & \vdots & \vdots \\
0 & \cdots & a_{mc} & \cdots & a_{mn} & b_m
\end{array}\right) .$$

Owing to the the fact that the (1,c) entry is 1, ELIMINATION in the GJM reduces to a simpler operation: Regard the rows of the matrix as $1 \times n + 1$ matrices (vectors). Then subtract from row $r + 1$ the matrix $a_{r+1,c} \times$ row 1. Repeat the corresponding operation for rows $r + 2, r + 3, \ldots, m$.

Hence the typical cycle in the GJM is S/(E)/D/E.

Some prefer the GEM or the GJM to the *PROCESS* because when a GEM or GJM operation stops the new but equivalent system has a visually appealing (echelon) form. Although DIVISION is not carried out if the GEM is used, when the back-substitution is applied after the GEM cycles are finished, DIVISION may be needed to solve for the unknowns as back-substitution proceeds.

The GEM leads from A to the echelon form A_E. The GJM leads also from A to an echelon form, denoted A_J. In both echelon forms the steps are visible. In A_J the numbers in the corners are all 1's. In A_E the numbers in the corners are *not* 0's but need not be 1's, i.e., the numbers in the corners are simply nonzero numbers. The matrix A_J looks like this:

$$\left(\begin{array}{cccccccccccc}
0 & \cdots & 0 & 1 & * & \cdots & * & \cdots & * & * & * & \cdots & * \\
0 & \cdots & 0 & 0 & 1 & \cdots & * & \cdots & * & * & * & \cdots & * \\
0 & \cdots & 0 & 0 & 0 & \cdots & 0 & \cdots & 1 & * & * & \cdots & * \\
0 & \cdots & 0 & 0 & 0 & \cdots & 0 & \cdots & 0 & 0 & 0 & \cdots & 0 \\
\cdots & \cdots & \cdots & \cdots & \cdots & \cdots & \cdots & \cdots & \cdots & \cdots & & & \\
0 & \cdots & 0 & 0 & 0 & \cdots & 0 & 0 & 0 & 0 & 0 & \cdots & 0
\end{array}\right) .$$

The following display is useful in sorting out the differences among the three versions.

Methods→	*PROCESS*	GEM	GJM
Operations	SEARCH ELIMINATION	SEARCH EXCHANGE ELIMINATION	SEARCH EXCHANGE DIVISION ELIMINATION
Final form of A	A_P	A_E	A_J

Table 2.4.2

The echelon forms A_E and A_J come about because of the use of EX-CHANGE in the GEM and the GJM. The sequence of EXCHANGEs has the effect of placing the *first* nonzero entry in each row "northwest" of all the nonzero entries in the rows below. Consequently back-substitution *appears* to be systematic because it starts at the bottom and works its way up, row-by-row. From the standpoint of computer operation this "systematization" is, at best, no improvement on what is done during back-substitution at the end of the *PROCESS*. At worst, the EXCHANGEs are simply extra computer operations, admittedly trivial, but an unnecessary nuisance.

Section 2.6 relates the *PROCESS*, the GEM, and the GJM to matrix multiplication. When that discussion is over there is an opportunity for a second look at the *PROCESS*, the GEM, and the GJM. Viewed from the standpoint of matrix algebra these procedures show their relative advantages and disadvantages.

The essential operations in the *PROCESS*, in the GEM, and in the GJM are SEARCH and ELIMINATION. EXCHANGE alone in the GEM and EX-CHANGE/DIVISION in the GJM are "embellishments of convenience" that add little clarification to what is going on.

Exercise 2.4.6. For each numerical **Example** and **Exercise** preceding this **Exercise** count the number p of EXCHANGES used in bringing the matrix to echelon form. (If the matrix is SQUARE the *parity*, i.e., the oddness or evenness, of this number p is of value in the study and use of determinants (cf. **Chapter 3**). Only the parity of p is of interest because only the number $(-1)^p$ figures in studying determinants.)

Exercise 2.4.7. Let A be a $p \times q$ matrix and let A_E be its echelon form: $(\eta_{ij})_{i,j=1}^{p,q}$, $\eta_{ij} = 0$ if $i > j$. Denote the diagonal entries $\eta_{11}, \eta_{22}, \ldots$ by $\delta_1, \delta_2, \ldots$. Show that:

   i. if $\delta_k = 0$ then $\delta_{k+1} = \delta_{k+2} = \cdots = 0$;
   ii. if $\delta_k \neq 0$ then $\delta_1 \delta_2 \cdots \delta_k \neq 0$;
   iii. if $p > q$ and if $r > q$ then for all j: $\eta_{rj} = 0$.

State and prove analogous results for the entries of the Jordan elimination form A_J of A.

2.5. More examples

The examples treated in **Section 2.3** by the *PROCESS* are treated below by the GJM.

Example 2.5.1. Analyze:

$$3x + 4y + 2z = 4$$
$$2x - z = -2.$$

Here $m = 2 < 3 = n$ and the matrix to be considered is

$$\left(\begin{array}{ccc|c} 3 & 4 & 2 & 4 \\ 2 & 0 & -1 & -2 \end{array} \right).$$

SEARCH produces 3 (the topmost entry in the first column of the matrix). EXCHANGE is skipped because SEARCH ended at the top of a column. DIVISION leads to

$$\left(\begin{array}{ccc|c} 1 & \frac{4}{3} & \frac{2}{3} & \frac{4}{3} \\ 2 & 0 & -1 & -2 \end{array} \right).$$

ELIMINATION produces

$$\left(\begin{array}{ccc|c} 1 & \frac{4}{3} & \frac{2}{3} & \frac{4}{3} \\ 0 & -\frac{8}{3} & -\frac{7}{3} & -\frac{14}{3} \end{array} \right).$$

The cycle is repeated by invoking SEARCH. Since this means that SEARCH is applied to the (single-rowed matrix)

$$\left(\begin{array}{ccc|c} 0 & -\frac{8}{3} & -\frac{7}{3} & -\frac{14}{3} \end{array} \right)$$

the number $-\frac{8}{3}$ is found.

Then because no exchange of rows is possible EXCHANGE is skipped. DIVISION leads to

$$\left(\begin{array}{ccc|c} 0 & 1 & \frac{7}{8} & \frac{7}{4} \end{array} \right)$$

and the GJM operation stops since no further cycle is possible when the bottom row of the matrix is reached.

Starting from the original matrix the GJM leads to the final matrix

$$\left(\begin{array}{ccc|c} 1 & \frac{4}{3} & \frac{2}{3} & \frac{4}{3} \\ 0 & 1 & \frac{7}{8} & \frac{7}{4} \end{array} \right).$$

which in turn corresponds to the system

$$x + \tfrac{4}{3}y + \tfrac{2}{3}z = \tfrac{4}{3}$$
$$y + \tfrac{7}{8}z = \tfrac{7}{4}.$$

Thus solving for y and using *back-substitution* leads to the equations

$$y = \tfrac{7}{4} - \tfrac{7}{8}z$$
$$x = \tfrac{4}{3} - \tfrac{4}{3}y - \tfrac{2}{3}z$$
$$= -1 + \tfrac{1}{2}z.$$

(It is in the very nature of *back-substitution* that the values of the unknowns are found in reverse order.)

Since the *Steps* of the GJM are, as noted earlier, reversible, a solution of the system has been found. The value of z may be assigned arbitrarily, the corresponding values of x and y are found from the last set of equations above. Thus there are INFINITELY MANY SOLUTIONS for the original system.

Direct substitution in the original system of the values of of x and y (which depend on the arbitrary value assigned to z) can reassure the doubting reader that a solution has, indeed, been found. The *reversibility* of each *Step* of the GJM makes this direct substitution unnecessary in *theory*. Good *practice* suggests direct substitution as a helpful way to catch calculational errors since they easily creep in during the operation of the the GJM (or, for that matter, in any operation requiring many numerical calculations).

Example 2.5.2. Analyze:

$$x + y - 2z = 3$$
$$3x + z = -1$$
$$4x - 7y + z = 0.$$

Here $m = n = 3$.

The *augmented* matrix to be considered is

$$\left(\begin{array}{ccc|c} 1 & 1 & -2 & 3 \\ 3 & 0 & 1 & -1 \\ 4 & -7 & 1 & 0 \end{array} \right)$$

which, after SEARCH, yields 1, the coefficient of x in the first equation. Hence EXCHANGE and DIVISION are omitted. After ELIMINATION is applied to rows 2 and 3 there emerges

$$\left(\begin{array}{ccc|c} 1 & 1 & -2 & 3 \\ 0 & -3 & 7 & -10 \\ 0 & -11 & 9 & -12 \end{array} \right).$$

The stage where there are only two equations in two unknowns has been reached.

The next cycle of the GJM operation is begun and at the end of the cycle the result is

$$\begin{pmatrix} 1 & 1 & -2 & \Big| & 3 \\ 0 & 1 & -\frac{7}{3} & \Big| & \frac{10}{3} \\ 0 & 0 & 1 & \Big| & -\frac{74}{50} \end{pmatrix}.$$

Now the one and only , i.e., unique, solution can be read off as follows:

$$z = -\tfrac{74}{50} = -\tfrac{37}{25},$$
$$y = \tfrac{10}{3} + \tfrac{7}{3}z = -\tfrac{3}{25},$$
$$x = 3 - y + 2z = \tfrac{4}{25}.$$

Notice how again the solution results from *back-substitution*: the last shall be first.

Example 2.5.3. Analyze:

$$\begin{aligned} 2x + 3y - 5z + 8u - v &= 1 \\ x - y + z + u + v &= 0 \\ 7x + 8y - 14z + 25u - 2v &= 3 \\ -x - y + 6z - 7u + v &= -2. \end{aligned}$$

Here $m = 4 < 5 = n$.

For practice the analysis below is abbreviated and offers only bare suggestions of what *Steps* of the GJM have been used to achieve the matrices displayed.

$$\begin{pmatrix} 2 & 3 & -5 & 8 & -1 & \Big| & 1 \\ 1 & -1 & 1 & 1 & 1 & \Big| & 0 \\ 7 & 8 & -14 & 25 & -2 & \Big| & 3 \\ -1 & -1 & 6 & -7 & 1 & \Big| & -2 \end{pmatrix} \quad \text{(original matrix)};$$

$$\begin{pmatrix} 1 & \frac{3}{2} & -\frac{5}{2} & 4 & -\frac{1}{2} & \Big| & \frac{1}{2} \\ 0 & -\frac{5}{2} & \frac{7}{2} & -3 & \frac{3}{2} & \Big| & -\frac{1}{2} \\ 0 & -\frac{5}{2} & \frac{7}{2} & -3 & \frac{3}{2} & \Big| & -\frac{1}{2} \\ 0 & \frac{1}{2} & \frac{7}{2} & -3 & \frac{1}{2} & \Big| & -\frac{3}{2} \end{pmatrix} \quad \begin{array}{l} \text{SEARCH,} \\ \text{no EXCHANGE,} \\ \text{DIVISION,} \\ \text{and} \\ \text{ELIMINATION;} \end{array}$$

$$\begin{pmatrix} 1 & \frac{3}{2} & -\frac{5}{2} & 4 & \frac{1}{2} & \Big| & \frac{1}{2} \\ 0 & 1 & -\frac{7}{5} & \frac{6}{5} & -\frac{3}{5} & \Big| & \frac{1}{5} \\ 0 & 0 & 0 & 0 & 0 & \Big| & 0 \\ 0 & 0 & \frac{21}{5} & -\frac{18}{5} & \frac{4}{5} & \Big| & -\frac{8}{5} \end{pmatrix} \quad \begin{array}{l} \text{SEARCH,} \\ \text{no EXCHANGE,} \\ \text{DIVISION,} \\ \text{and} \\ \text{ELIMINATION;} \end{array}$$

$$\begin{pmatrix} 1 & \frac{3}{2} & -\frac{5}{2} & 4 & -\frac{1}{2} & \Big| & \frac{1}{2} \\ 0 & 1 & -\frac{7}{5} & \frac{6}{5} & -\frac{3}{5} & \Big| & \frac{1}{5} \\ 0 & 0 & 1 & -\frac{6}{7} & \frac{4}{21} & \Big| & -\frac{8}{21} \\ 0 & 0 & 0 & 0 & 0 & \Big| & 0 \end{pmatrix} \quad \begin{array}{l} \text{EXCHANGE,} \\ \text{resume cycle,} \\ \text{SEARCH,} \\ \text{and} \\ \text{DIVISION.} \end{array}$$

Hence

$$z = -\tfrac{8}{21} + \tfrac{6}{7}u - \tfrac{4}{21}v$$
$$y = \tfrac{1}{5} + \tfrac{7}{5}z - \tfrac{6}{5}u + \tfrac{3}{5}v$$
$$= -\tfrac{1}{3} + \tfrac{1}{3}v$$
$$x = \tfrac{1}{2} - \tfrac{3}{2}y + \tfrac{5}{2}z - 4u + \tfrac{1}{2}v$$
$$= \tfrac{1}{21} - \tfrac{13}{7}u - \tfrac{10}{21}v.$$

There are INFINITELY MANY SOLUTIONS since u and v may be given arbitrary values, which then determine x, y, and z.

Example 2.5.4. Analyze

$$2x + 3y - 5z = 1$$
$$x - y + z = 0$$
$$7x + 8y - 14z = 3$$
$$-x - y + 6z = 2.$$

Here $m = 4 > 3 = n$.

This time the analysis is restricted to the discussion of the system after the GJM operation has been completed, i.e., when the system looks like this:

$$x + \tfrac{3}{2}y - \tfrac{5}{2}z = \tfrac{1}{2}$$
$$y - \tfrac{7}{5}z = \tfrac{1}{5}$$
$$z = \tfrac{4}{7}.$$

Back-substitution reveals that $y = 1$ and $x = \tfrac{3}{7}$ and substitution in the equations of the original system confirms that the values found for $x, y,$ and z satisfy all four equations of the system.

Exercise 2.5.1. Analyze the systems in **Examples 2.5.1-2.5.4** by:

i. the GEM;
ii. the GJM in which MAXSEARCH rather than simple SEARCH is used;
iii. the *PROCESS* in which PARTIALMAXSEARCH is used.

Exercise 2.5.2. Consider the system

$$a_{11}x_1 + \cdots + a_{1n}x_n = b_1$$
$$\vdots \qquad \ddots \qquad \vdots \qquad \vdots$$
$$a_{m1}x_1 + \cdots + a_{mn}x_n = b_m$$

for the following "parameter" values:

i. $m = n = 4$, a_{ij}, $b_j = 0$ or ± 1 ad lib.;

---

ii. $m = 5$, $n = 3$, a_{ij}, $b_j = 0$ or ± 1, ± 2, or ± 3 ad lib.;

iii. $m = 3$, $n = 5$, a_{ij}, $b_j = 0$ or ± 2, ± 5, or ± 0.2 ad lib.

For each of *i* - *iii* choose a sample system and use the GJM or the GEM to analyze the system. In each case according as the sign of the resulting system is **1, 0,** or ∞ determine how to adjust some of the a_{ij} or some of the b_j so as to change the sign to at least one of the remaining signs. Check each solution found by the GJM or by the GEM against the other and also by substitution in the original system.

Exercise 2.5.3. In **Example 2.5.3** let the right member of the third equation be 2 rather than 3. Repeat the analysis of the system.

Exercise 2.5.4. In **Example 2.5.4** let the right member of the third equation be 4 rather than 3. Repeat the analysis of the system.

Exercise 2.5.5. For what value(s) of λ does the system

$$(1 - \lambda)x + 2y + 3z = 0$$
$$4x + (5 - \lambda)y + 6z = 0$$
$$7x + 8y + (9 - \lambda)z = 0$$

have a nonzero solution, i.e., a solution (x, y, z) such that at least one of x, y, and z is not 0?

[*Hint:* Let A be the matrix

$$\begin{pmatrix} 1 & 2 & 3 \\ 4 & 5 & 6 \\ 7 & 8 & 9 \end{pmatrix}.$$

Use the GEM (no divisions) on $A - \lambda I$. After the last cycle the (3,3) entry of $(A - \lambda I)_E$ indicates which values of λ *might* lead to a nonzero solution. Test each such value of λ.]

2.6. The *PROCESS*, the **GEM**, the **GJM**, and matrix algebra

The *PROCESS*, the GEM, or the GJM applied to any system

$$a_{11}x_1 + \cdots + a_{1n}x_n = b_1$$
$$\vdots \qquad \ddots \qquad \vdots \qquad \vdots$$
$$a_{m1}x_1 + \cdots + a_{mn}x_n = b_m \qquad (2.6.1)$$

provides an analysis of the system and, if the system has a solution or solutions, each method provides the solution or solutions in full generality. Frequently the same matrix $A \overset{\text{def}}{=} (a_{ij})_{i,j=1}^{m,n}$ occurs again and again while

$$\mathbf{b} \overset{\text{def}}{=} \begin{pmatrix} b_1 \\ \vdots \\ b_m \end{pmatrix}$$

varies from one time to the next. In **Example 1.1.1** the vector

$$\begin{pmatrix} P \\ F \\ C \end{pmatrix}$$

can vary from population to population while the input-output matrix corresponding to **Table 1.1.1** remains the same.

In the development below it is shown that once any of the procedures is performed on A then each time the system (2.6.1) is analyzed for a different vector $\mathbf{b}$, the number of arithmetical operations (principally multiplications) to be performed is considerably lower than the number performed in bringing A to one of its forms, A_P, A_E, or A_J. Each analysis after the first costs about 2% of what the first analysis costs if the matrix is of size 50×50.

Matrix algebra helps reveal how these economies arise, because the *PROCESS*, the GEM, and the GJM can equally well be described by means of matrix operations, i.e., in the language of matrix algebra.

ELEMENTARY MATRICES

In what follows it is shown that the results of the operations ELIMINATION, EXCHANGE, and DIVISION applied to an $m \times n$ matrix A are the same as the results of multiplying A by appropriate SQUARE $m \times m$ matrices.

ELIMINATION:
This operation subtracts a multiple of one row, say row i of A, from another row, say row j of A.
Let the multiplier of row i be p_i. Subtract $p_i \times$ row i from row j in the $m \times m$ identity matrix I and call the result $E_{ij}(-p_i)$. For example if $m = 5$ then

$$I = \begin{pmatrix} 1 & 0 & 0 & 0 & 0 \\ 0 & 1 & 0 & 0 & 0 \\ 0 & 0 & 1 & 0 & 0 \\ 0 & 0 & 0 & 1 & 0 \\ 0 & 0 & 0 & 0 & 1 \end{pmatrix}.$$

If $i = 2$, $j = 4$, $p_i = 7$, then

$$E_{24}(-7) = \begin{pmatrix} 1 & 0 & 0 & 0 & 0 \\ 0 & 1 & 0 & 0 & 0 \\ 0 & 0 & 1 & 0 & 0 \\ 0 & -7 & 0 & 1 & 0 \\ 0 & 0 & 0 & 0 & 1 \end{pmatrix} \quad \leftarrow \text{ row } \mathbf{4} - 7 \times \text{ row } \mathbf{2}.$$

Exercise 2.6.1. Let A be the 5×6 matrix

$$\begin{pmatrix} 0 & 2 & 3 & 4 & 5 & 6 \\ 9 & 1 & 2 & 3 & 4 & 5 \\ 0 & 6 & 1 & 2 & 3 & 4 \\ 7 & 4 & 6 & 1 & 2 & 3 \\ 3 & 4 & 5 & 6 & 1 & 2 \end{pmatrix}.$$

i. Use ELIMINATION to replace the 7 by 0 in row 4.

ii. Calculate $A_1 \stackrel{\text{def}}{=} E_{24}(-\frac{7}{9}))A$ and compare A_1 with the result in *i.*

iii. Calculate

$$A_2 \stackrel{\text{def}}{=} E_{25}(-\tfrac{1}{3})A_1 = E_{25}(-\tfrac{1}{3})E_{24}(-\tfrac{7}{9}))A$$

and see that A_2 is the intermediate result, after the first S/E cycle, of *PROCESS*ing A.

The notation $E_{ij}(r)$ is easy to remember. For example in **Exercise 2.6.1**:

i. The pair **24** in $E_{24}(-\frac{7}{9})$ is precisely the index pair for row **2** and row **4**, the rows involved in this particular ELIMINATION.

ii. The number $-\frac{7}{9}$ is linked to the first index 2 of the index pair 24 and thus $E_{24}(-\frac{7}{9})$ means that row **2** is multiplied by $-\frac{7}{9}$ and the result is added to row **4**.

It is helpful to read $E_{24}(-\frac{7}{9}))$ as follows:

Add $-\frac{7}{9} \times$ row2 to row 4 and leave row 2 unchanged.

Note that row **2** is left unchanged during the operation.

EXCHANGE:

This operation EXCHANGEs two rows, say rows i and j, of A. Let E_{ij} be the result of interchanging rows i and j in I.

When EXCHANGE is applied to I the result is a simple matrix. For example if rows 2 and 4 are EXCHANGED in the 5×5 identity matrix I the result is E_{24} given below:

$$\begin{pmatrix} 1 & 0 & 0 & 0 & 0 \\ 0 & 0 & 0 & 1 & 0 \\ 0 & 0 & 1 & 0 & 0 \\ 0 & 1 & 0 & 0 & 0 \\ 0 & 0 & 0 & 0 & 1 \end{pmatrix}.$$

On the other hand

$$E_{24}A = \begin{pmatrix} 0 & 2 & 3 & 4 & 5 & 6 \\ 7 & 4 & 6 & 1 & 2 & 3 \\ 0 & 6 & 1 & 2 & 3 & 4 \\ 9 & 1 & 2 & 3 & 4 & 5 \\ 3 & 4 & 5 & 6 & 1 & 2 \end{pmatrix}$$

and it appears that EXCHANGE can be implemented by matrix multiplication.

Exercise 2.6.2. Exchange rows 1 and 2 of A above by means of matrix multiplication.

Exercise 2.6.3. Calculate $E_{12}E_{34}$.

DIVISION:
In the GJM DIVISION consists of *multiplying* some row by a *nonzero* number, namely the reciprocal of some a_{rc}. Let $E_r(t)$ be the result of multiplying the rth row of I by t. If the third row of I is multiplied by the number t the result is $E_3(t)$ and looks like this:

$$\begin{pmatrix} 1 & 0 & 0 & 0 & 0 \\ 0 & 1 & 0 & 0 & 0 \\ 0 & 0 & t & 0 & 0 \\ 0 & 0 & 0 & 1 & 0 \\ 0 & 0 & 0 & 0 & 1 \end{pmatrix}.$$

Unsurprisingly

$$E_3(t)A = \begin{pmatrix} 0 & 2 & 3 & 4 & 5 & 6 \\ 9 & 1 & 2 & 3 & 4 & 5 \\ 0 & 6t & t & 2t & 3t & 4t \\ 7 & 4 & 6 & 1 & 2 & 3 \\ 3 & 4 & 5 & 6 & 1 & 2 \end{pmatrix}.$$

Only exceptionally $t = 0$.

Exercise 2.6.4. Find a 5×5 matrix E such that EA is the matrix A except that its fifth row is divided by 6.

A careful study of the rules for multiplying matrices shows that the examples above are not anomalies. Rather they are models for the following general rules:

ROW COMBINATION:
Let A be an $m \times n$ matrix and, in the SQUARE $m \times m$ identity matrix I, let $E_{ij}(-p_i)$ be the result of subtracting from row **j** the $1 \times n$ matrix $p_i \times$ row **i** *and leaving row* **i** *unchanged.* Then

$$B \stackrel{\text{def}}{=} E_{ij}(-p_i)A$$

is the matrix A in which row **j** is replaced by

$$\text{row } \mathbf{j} - p_i \times \text{row } \mathbf{i}$$

---

and row i *remains unchanged.* In other words,

Multiplication by $E_{ij}(-p_i)$ *adds* $-p_i \times$ row i to row j and leaves row i unchanged.

ROW COMBINATION is the means used for ELIMINATION in the *PROCESS*, in the GEM, and in the GJM.

ROW EXCHANGE:

Let E_{ij} be the $m \times m$ identity matrix I with its rows i and j exchanged. Then $E_{ij}A$ is the matrix A with *its* rows i and j exchanged.

ROW EXCHANGE is the means used for EXCHANGE in the in the GEM and in the GJM.

ROW SCALING:

Let $E_i(s)$ be the $m \times m$ identity matrix I with its row i replaced by $s \times$ row i. Then $E_i(s)A$ is the matrix A with *its* row i multiplied (*scaled*) by s.

ROW SCALING is the means used for DIVISION in the GJM. The scaling factor s in DIVISION is the *reciprocal of the first nonzero entry* a_{rc} found by SEARCH at the beginning of each cycle of the GJM.

All the SQUARE matrices $E_{ij}(-p_i)$ (more generally $E_{ij}(r)$), $E_i(s)$, and E_{ij}, and are called *elementary* matrices.

Typical forms for these elementary matrices are the following:

ROW COMBINATION matrices: if $i < j$

$$E_{ij}(r) = \begin{array}{r} \\ \\ \\ \\ i\text{th row } \rightarrow \\ \\ j\text{th row } \rightarrow \\ \\ \\ \end{array} \begin{pmatrix} & & \text{column } i & & \text{column } j & & \\ & & \downarrow & & \downarrow & & \\ 1 & \cdots & 0 & \cdots & 0 & \cdots & 0 \\ \vdots & \ddots & \vdots & \ddots & \vdots & \ddots & \vdots \\ 0 & \cdots & 1 & \cdots & 0 & \cdots & 0 \\ \vdots & \ddots & \vdots & \ddots & \vdots & \ddots & \vdots \\ 0 & \cdots & r & \cdots & 1 & \cdots & 0 \\ \vdots & \ddots & \vdots & \ddots & \vdots & \ddots & \vdots \\ 0 & \cdots & 0 & \cdots & 0 & \cdots & 1 \end{pmatrix};$$

if $i > j$

$$E_{ij}(r) = \begin{array}{c} \\ \\ \\ \\ \\ j\text{th row} \;\longrightarrow \\ \\ \\ i\text{th row} \;\longrightarrow \\ \\ \\ \\ \end{array}
\left(\begin{array}{ccccccccc}
 & & \overset{\textstyle\overset{\text{column}}{j}}{\downarrow} & & \overset{\textstyle\overset{\text{column}}{i}}{\downarrow} & & \\
1 & \cdots & 0 & \cdots & 0 & \cdots & 0 \\
\vdots & \ddots & \vdots & \ddots & \vdots & \ddots & \vdots \\
0 & \cdots & 1 & \cdots & r & \cdots & 0 \\
\vdots & \ddots & \vdots & \ddots & \vdots & \ddots & \vdots \\
0 & \cdots & 0 & \cdots & 1 & \cdots & 0 \\
\vdots & \ddots & \vdots & \ddots & \vdots & \ddots & \vdots \\
0 & \cdots & 0 & \cdots & 0 & \cdots & 1
\end{array}\right) ;$$

ROW EXCHANGE matrices: if $i > j$

$$E_{ij} = \begin{array}{c} \\ \\ \\ \\ \\ i\text{th row} \;\longrightarrow \\ \\ \\ j\text{th row} \;\longrightarrow \\ \\ \\ \\ \end{array}
\left(\begin{array}{ccccccccc}
 & & \overset{\textstyle\overset{\text{column}}{i}}{\downarrow} & & \overset{\textstyle\overset{\text{column}}{j}}{\downarrow} & & \\
1 & \cdots & 0 & \cdots & 0 & \cdots & 0 \\
\vdots & \ddots & \vdots & \ddots & \vdots & \ddots & \vdots \\
0 & \cdots & 0 & \cdots & 1 & \cdots & 0 \\
\vdots & \ddots & \vdots & \ddots & \vdots & \ddots & \vdots \\
0 & \cdots & 1 & \cdots & 0 & \cdots & 0 \\
\vdots & \ddots & \vdots & \ddots & \vdots & \ddots & \vdots \\
0 & \cdots & 0 & \cdots & 0 & \cdots & 1
\end{array}\right) .$$

[**Note 2.6.1:** Since $E_{ij} = E_{ji}$, the representation above is valid if $i < j$ or if $i > j$.]

ROW SCALING matrices:

$$E_i(s) = \begin{array}{c} \\ \\ \\ \\ i\text{th row} \;\longrightarrow \\ \\ \\ \end{array}
\left(\begin{array}{ccccc}
 & & \overset{\textstyle\overset{\text{column}}{i}}{\downarrow} & & \\
1 & \cdots & 0 & \cdots & 0 \\
\vdots & \ddots & \vdots & \ddots & \vdots \\
0 & \cdots & s & \cdots & 0 \\
\vdots & \ddots & \vdots & \ddots & \vdots \\
0 & \cdots & 0 & \cdots & 1
\end{array}\right) .$$

Any product Π of matrices of the type E_{ij} is called a *permutation matrix* because it effects a sequence of EXCHANGEs, i.e., a rearrangement or *permutation* (of the rows of a matrix). Furthermore any permutation of the rows of a

---

matrix is achievable by a sequence of ROW EXCHANGEs. Indeed let

$$\pi \overset{\text{def}}{=} \begin{pmatrix} 1 & 2 & 3 & \ldots & n \\ i_1 & i_2 & i_3 & \ldots & i_n \end{pmatrix}$$

depict the permutation π in which row 1 is replaced by row i_1, row 2 is replaced by row i_2, row 3 is replaced by row i_3, ..., and row n is replaced by row i_n.

The permutation π is certainly achievable by ROW EXCHANGEs if $n = 1$ or $n = 2$. If $n = 1$ there is no genuine *rearrangement* possible. Then π is the *identity* permutation that interchanges nothing; π results from multiplication by $E_{11} = I$. If $n = 2$ only one genuine rearrangement is possible and is achievable by multiplying by E_{12}.

If every permutation of $n - 1$ rows can be achieved by a sequence of ROW EXCHANGEs, then the permutation π above of n rows is achievable by first multiplying by E_{1i_1} and then by a sequence of ROW EXCHANGEs among the $n - 1$ rows numbered by all the numbers except the number i_1. Hence 3 rows can be rearranged arbitrarily by the use of a sequence of ROW EXCHANGEs, hence 4 rows can be rearranged by a sequence of ROW EXCHANGEs, etc.

[**Remark 2.6.1:** The rows here are irrelevant. What is relevant is that the numbers $1, 2, 3, \ldots, n$ have been rearranged in the order $i_1, i_2, i_3, \ldots, i_n$. Thus π is conveniently and generally viewed as a permutation of the numbers $1, 2, 3, \ldots, n$ rather than as a permutation of the rows numbered $1, 2, 3, \ldots, n$.]

Exercise 2.6.5. Repeat the analyses in **Examples 2.5.1 - 2.5.4** and this time indicate how each operation is produced by matrix multiplication.

Exercise 2.6.6. Let n be 3.

i. Write out the following matrices: E_{23}; $E_3(-2)$; $E_{13}(\frac{2}{5})$; $E_{23}(4)$.
ii. Show that $E_{ij} = E_{ji}$ and that $E_{ij}^2 = I$.
iii. What is E_{ii}?

Example 2.6.1. Let A be the 3×3 matrix

$$\begin{pmatrix} 1 & 2 & 3 \\ 0 & -1 & 5 \\ 3 & 1 & 2 \end{pmatrix}.$$

Then:

$$E_{23}A = \begin{pmatrix} 1 & 2 & 3 \\ 3 & 1 & 2 \\ 0 & -1 & 5 \end{pmatrix}$$

$$E_3(-2)A = \begin{pmatrix} 1 & 2 & 3 \\ 0 & -1 & 5 \\ -6 & -2 & -4 \end{pmatrix}$$

$$E_{23}(\tfrac{2}{5})A = \begin{pmatrix} 1 & 2 & 3 \\ 0 & -1 & 5 \\ 3 & \frac{3}{5} & 4 \end{pmatrix}.$$

On the other hand

$$AE_{23} = \begin{pmatrix} 1 & 3 & 2 \\ 0 & 5 & -1 \\ 3 & 2 & 1 \end{pmatrix}$$

$$AE_3(-2) = \begin{pmatrix} 1 & 9 & -6 \\ 0 & 15 & -10 \\ 3 & 6 & -4 \end{pmatrix}$$

$$AE_{23}(\tfrac{2}{5}) = \begin{pmatrix} 1 & \frac{16}{5} & 3 \\ 0 & 1 & 5 \\ 3 & \frac{9}{5} & 2 \end{pmatrix}.$$

Matrix transposes (cf. **Section 1.1**) are quite helpful in understanding what *post-multiplication* by elementary matrices achieves. Thus if $E_{\ldots}$ is an elementary matrix then

$$AE_{\ldots} = (E_{\ldots}^t A^t)^t.$$

Since $E_{ij}^t = E_{ij}$ and $E_i^t(s) = E_i(s)$ and since the *rows* of A^t are the *columns* of A it follows that *post-multiplication* by E_{ij} or $E_i(s)$ does to *columns* what *premultiplication* does to *rows*.

In **Example 2.6.1** multiplication by $E_{23}(\tfrac{2}{5})$ results in the alteration of the *third row* by the addition of two-fifths of *second row*. On the other hand, $AE_{23}(\tfrac{2}{5})$ results in the alteration of the *second column* by the addition of two-fifths of the *third column*. The rôles of the second and third rows are switched.

The switch is explained by the fact that if $i \neq j$ then

$$E_{ij}^t(r) = E_{ji}(r),$$

i.e., i and j are switched. Thus, in general, whereas $E_{ij}(r)A$ is just like A except that its jth *row* is altered by the addition of r times the ith *row*, $AE_{ij}(r)$ is just like A except that the ith *column* of A is changed by the addition of r times the jth *column*. The rôles of i and j are switched.

Exercise 2.6.7. Let A be a $p \times q$ matrix

$$\begin{pmatrix} a_{11} & \cdots & a_{1q} \\ \vdots & \ddots & \vdots \\ a_{p1} & \cdots & a_{pq} \end{pmatrix}.$$

i. Describe in words the matrices

$$E_{ij}A, \ E_i(s)A, \text{ and } E_{ij}(r)A.$$

ii. For the $n \times n$ matrices $E \ldots$ in *i* what is n?

iii. Describe in words the matrices

$$AE_{ij}, \ AE_i(s), \text{ and } AE_{ij}(r).$$

iv. For the $n \times n$ matrices $E \ldots$ in *iii* what is n?

Regarded as means of performing operations on matrices, the elementary matrices have *inverses*, i.e., operations that undo, reverse, or invert whatever the original operations do. Thus

The inverse of the ROW COMBINATION resulting from multiplication by $E_{ij}(-p_i)$ is the operation that

$$\text{replaces (row } \mathbf{j} - p_i \times \text{row } \mathbf{i}) \text{ by row } \mathbf{j}.$$

In other words the inverse of a ROW COMBINATION adds back what the ROW COMBINATION takes away. Hence the inverse of the ROW COMBINATION resulting from multiplication by $E_{ij}(-p_i)$ results from multiplication by $E_{ij}(p_i)$.

The inverse of a ROW EXCHANGE resulting from multiplication by E_{ij} is the same ROW EXCHANGE performed again. Hence the inverse of the ROW EXCHANGE resulting from multiplication by E_{ij} results from multiplication by E_{ij} again.

The inverse of a ROW SCALING resulting from multiplication by $E_i(s)$ results, if $s \neq 0$, from multiplication by $E_i(\frac{1}{s})$.

Hence whatever an elementary matrix $E_{ij}(s)$ or E_{ij} *does* is *undone* by an elementary matrix of the same type. Similarly if $t \neq 0$ whatever $E_r(t)$ *does* is *undone* by an elementary matrix of the same type. (Call any $E_r(0)$ the *trivial* elementary matrix.) Since each elementary matrix $E\ldots$ may be viewed as "doing" something to I, i.e., $E\ldots = E\ldots(I) = E\ldots I$, it appears that for each nontrivial elementary matrix $E\ldots$ there is another elementary matrix, called the *inverse* of $E\ldots$ and denoted $E\ldots^{-1}$, such that $E\ldots^{-1}E\ldots = I$. Since $E\ldots$ undoes what $E\ldots^{-1}$ does, it is also true that $E\ldots E\ldots^{-1} = I$.

Exercise 2.6.8. For each type of nontrivial 3×3 *elementary matrix* $E\ldots$ find matrices A and B such that $A(E\ldots) = (E\ldots)B = I$. Show that in each instance that A is again an elementary matrix and $A = B$.

---

 *Chapter 2. THE PROCESS (***Simple elimination***)*

Exercise 2.6.9. Write the general formula for the inverse of each type of nontrivial elementary matrix. For example, $E_{ij}^{-1} = E_{ij} = E_{ji}$.

More generally, if A is a SQUARE $n \times n$ matrix and if C is a SQUARE $n \times n$ matrix such that $AC = CA = I$ then C is called the *inverse* of A and is denoted A^{-1}. Furthermore, in these circumstances, A is called *invertible*.

Example 2.6.2. If $A^{-1}A = AA^{-1} = B^{-1}B = BB^{-1} = I$ then

$$(B^{-1}A^{-1})(AB) = B^{-1}(A^{-1}A)B = B^{-1}IB = I,$$
$$(AB)(B^{-1}A^{-1}) = A(BB^{-1})A^{-1} = AIA^{-1} = I.$$

Thus $(AB)^{-1} = B^{-1}A^{-1}$,

"the inverse of a product is the product of the inverses in opposite order."

In **Section 2.8** the subject of inverses of SQUARE matrices is explored in some detail.

Exercise 2.6.10. For each of the following assume that the matrices are of size $n \times n$.

i. Show that

$$E_{ii}(r) = E_i(r+1), \quad E_{ij}(r)E_{ij}(t) = E_{ij}(r+t), \quad E_i(s)E_i(t) = E_i(st)$$
$$E_{ji}E_{ij}(r)E_{ji} = E_{ji}(r)$$

$$E_{ij}E_i(s)E_{ij} = E_j(s), \quad E_{ij}E_{jk} = E_{ik}.$$

ii. For $1 \le i, j, p, q \le n$ develop general formulae for x and y so that

$$E_{ij}E_{pq}(r)E_{ij} = E_{xy}(r).$$

iii. Show that if the pairs ij and pq have nothing in common i.e., if $i \ne p, q$ and $j \ne p, q$ $(\{i,j\} \cap \{p,q\} = \emptyset)$, then $E_{ij}E_{pq} = E_{pq}E_{ij}$.

Exercise 2.6.11. Show that if A is an $m \times n$ matrix then there is an invertible $m \times m$ matrix $\mathcal{P}$ (the "*PROCESS*" or "*PROCESS*ing" matrix), a product of elementary matrices $(E_{...})$ and such that $\mathcal{P}A = A_P$. For the 2×3 matrix

$$\begin{pmatrix} 1 & -2+3i & 5 \\ -3 & i & 1-4i \end{pmatrix}$$

calculate the *PROCESS* matrix $\mathcal{P}$ and describe it as the product of elementary matrices used in the *PROCESS*.

Exercise 2.6.12. Repeat **Exercise 2.6.11** for $\mathcal{E}$ (the "GEM" matrix) and $\mathcal{J}$ (the "GJM" matrix).

[**Note 2.6.2:** In view of the conclusions in **Exercises 2.6.10** and **2.6.11** it follows that if $\mathbf{x}$ is a solution of $A\mathbf{x} = \mathbf{b}$ iff it is also a solution of the systems

$$A_P\mathbf{x} = \mathcal{P}\mathbf{b}$$
$$A_E\mathbf{x} = \mathcal{E}\mathbf{b}$$
$$A_J\mathbf{x} = \mathcal{J}\mathbf{b}$$

Since these systems arise as a result of applying the *PROCESS*, the GEM, and the GJM, to the original system $A\mathbf{x} = \mathbf{b}$ it follows that the *PROCESS*, the GEM, and the GJM are reversible, as remarked earlier.]

Exercise 2.6.13. Let E be an arbitrary $n \times n$ elementary matrix. What are E_P, E_E, and E_J?

In a matrix $A \stackrel{\text{def}}{=} (a_{ij})_{i,j=1}^{m,n}$ the entries a_{ii} are called the *diagonal* entries. The "line" on which those entries lie is called the *diagonal*, even when A is not SQUARE. The nomenclature is particularly apt if $m = n$, i.e., if A is SQUARE.

If A is a matrix then the EXCHANGE operations in the GEM and in the GJM are performed so that in the echelon form A_E and in the Jordan elimination version A_J all entries "west" of the diagonal are 0's. If A is SQUARE it is natural to call such matrices *upper triangular*. The nomenclature is applied equally well to an arbitrary matrix in which all entries "west" of the diagonal are 0's. Such matrices are very handy in computations. They account for the imagery "rolling up a window shade or a rug" in back-substitution associated with the GEM or the GJM (cf. **Section 2.4**).

The counterpart of an upper triangular matrix is a *lower triangular* matrix in which all entries "east" of the diagonal are 0's. Upper and lower triangular matrices play an important rôle in study of solutions for systems. Thus they deserve a special discussion.

A SQUARE upper triangular matrix looks like this:

$$\begin{pmatrix} a_{11} & a_{12} & \cdots & a_{1n} \\ & a_{22} & \cdots & a_{2n} \\ & & \ddots & \vdots \\ & & & a_{nn} \end{pmatrix}.$$

    Chapter 2. THE PROCESS (Simple elimination)

A SQUARE lower triangular matrix looks like this:

$$\begin{pmatrix} a_{11} & & & \\ a_{21} & a_{22} & & \\ \vdots & \vdots & \ddots & \\ a_{n1} & a_{n2} & \ldots & a_{nn} \end{pmatrix}.$$

In general a matrix $(a_{ij})_{i,j=1}^{m,n}$ is upper triangular iff $a_{ij} = 0$ whenever $j < i$. A matrix $(a_{ij})_{i,j=1}^{m,n}$ is lower triangular iff $a_{ij} = 0$ whenever $j > i$.

A matrix D that is both upper triangular and lower triangular is called a *diagonal* matrix. Thus if $D = (d_{ij})_{i,j=1}^{m,n}$ then D is a diagonal matrix iff $d_{ij} = 0$ whenever $i \neq j$. A SQUARE diagonal matrix looks like this:

$$\begin{pmatrix} d_{11} & & & \\ & d_{22} & & \\ & & \ddots & \\ & & & d_{nn} \end{pmatrix}$$

and is very often presented merely as

$$\begin{pmatrix} d_1 & & & \\ & d_2 & & \\ & & \ddots & \\ & & & d_n \end{pmatrix}$$

in which double indices are avoided.

Exercise 2.6.14. Give examples of upper triangular 3×5 and 5×3 matrices. Give examples of 5×7 and 7×5 lower triangular matrices. Give examples of 6×9 and 9×6 diagonal matrices.

In the *PROCESS* every ELIMINATION operation treats a row *below* the active row. Thus whenever ELIMINATION takes place it involves multiplication by an elementary matrix $E_{ij}(r)$ in which $i < j$. Such an elementary matrix is lower triangular.

[**Remark 2.6.2:**

The *only* elementary matrices used in the *PROCESS* are of the form $E_{ij}(r)$, i.e., ROW COMBINATION matrices in which $i < j$.

The *only* elementary matrices used in the GEM are of the form $E_{ij}(r)$ in which $i < j$ or of the form E_{ij}, i.e., either ROW COMBINATION matrices in which $i < j$ or ROW EXCHANGE matrices.

In the GJM all the different kinds of elementary matrices can crop up although in particular instances some are not used.]

Example 2.6.3.

If the GJM is applied to I *no* elementary matrices are used.

If the GJM is applied to an upper triangular matrix *only* ROW SCALING matrices $E_i(s)$ are used.

If the GJM is applied to a permutation matrix only ROW EXCHANGE matrices are used.

Whenever an elementary matrix of the form $E_{ij}(r)$ is used, $i < j$.

For similar reasons the following rule obtains:

If $E_{...}$ is an elementary matrix, if $E_{...}$ is *not* of the type E_{ij}, and if $E_{...}$ is used in the GEM or in the GJM then $E_{...}$ is lower triangular.

Exercise 2.6.15. Give an example of a matrix $E_{ij}(r)$ that is *not* lower-triangular. Explain why such a matrix is never used in the *PROCESS*, in the *GEM*, or in the GJM.

Exercise 2.6.16.

$i.$ Assume that A is a lower triangular $m \times n$ matrix and let B be a lower triangular $n \times p$ matrix. Show that the product $C \stackrel{\text{def}}{=} AB$ is also lower triangular.

[*Hint:* If $A = (a_{ij})_{i,j=1}^{m,n}$ and $B = (b_{rs})_{r,s=1}^{n,p}$ and both are lower triangular matrices and if $i < j$ then the ij entry of AB is found by matching the ith row of A against the jth column of B and then adding up products of matched entries. Thus if $C = (c_{is})_{i,s=1}^{m,p} = AB$ it follows that

$$
c_{ij} = \begin{array}{c}
a_{i1} \times b_{1j} \\
+ \\
\vdots \\
+ \\
a_{ii} \times b_{ij} \\
+ \\
a_{i,i+1} \times b_{i+1,j} \\
+ \\
\vdots \\
+ \\
a_{in} \times b_{np}
\end{array} \ .
$$

Now use the lower triangularity of A and B.]

ii. Show that the product of compatible upper triangular matrices is upper triangular.

iii. Show that the product of compatible diagonal matrices is diagonal.

iv. Show that if A and B are $n \times n$ (SQUARE) diagonal matrices

$$A = \begin{pmatrix} a_1 & & & \\ & a_2 & & \\ & & \ddots & \\ & & & a_n \end{pmatrix}, \text{ and } B = \begin{pmatrix} b_1 & & & \\ & b_2 & & \\ & & \ddots & \\ & & & b_n \end{pmatrix}$$

then

$$AB = \begin{pmatrix} a_1 b_1 & & & \\ & a_2 b_2 & & \\ & & \ddots & \\ & & & a_n b_n \end{pmatrix} = BA.$$

[**Remark 2.6.3:** Two $n \times n$ (SQUARE) matrices A and B can be such that

$$AB \neq BA.$$

Nevertheless, for compatible, SQUARE, *diagonal* matrices D_1 and D_2 matrix multiplication *is* commutative: $D_1 D_2 = D_2 D_1$.]

It follows that if A is an $m \times n$ matrix $(a_{ij})_{i,j=1}^{m,n}$ then the product $\widetilde{L}$ of all the $m \times m$ elementary matrices used to *PROCESS* the system

$$a_{11} x_1 + \cdots + a_{1n} x_n = b_1$$
$$\vdots \quad \ddots \quad \vdots \quad \vdots \quad (A\mathbf{x} = \mathbf{b})$$
$$a_{m1} x_1 + \cdots + a_{mn} x_n = b_m \tag{2.6.2}$$

is itself a lower triangular matrix, whence the PROCESSed version A_P of A may be written as $\widetilde{L}A$: $A_P = \widetilde{L}A$.

Even more is true. Each elementary matrix used in the *PROCESS* is not merely lower triangular, but *all* diagonal *entries are 1's*. In the next **Example** it is shown that the lower triangular matrix having only nonzero diagonal entries has an inverse. The following **Exercise** provides the general result.

Example 2.6.4. Let A be the lower triangular matrix

$$\begin{pmatrix} 1 & 0 & 0 \\ 2 & 3 & 0 \\ 4 & 5 & 6 \end{pmatrix}.$$

To find a 3×3 matrix F such that $AF = I$, let F be

$$\begin{pmatrix} a & b & c \\ d & e & f \\ g & h & k \end{pmatrix}.$$

The rules of matrix multiplication show that

$$AF = \begin{pmatrix} a & b & c \\ 2a + 3d & 2b + 3e & 2c + 3f \\ 4a + 5d + 6g & 4b + 5e + 6h & 4c + 5f + 6k \end{pmatrix}.$$

If AF is to be I it follows that:

$a = 1$, $b = 0$, $c = 0$, from the comparison of entries in the first rows;

$d = -\dfrac{2}{3}$, $e = \dfrac{1}{3}$, $f = 0$, from comparison of entries in the second rows;

and

$g = -\dfrac{1}{9}$, $h = -\dfrac{5}{18}$, $k = \dfrac{1}{6}$, from comparison of the entries in the third rows.

A direct check shows that, with the values found for a, b, c, d, e, and f, indeed, $AF = I$. Furthermore a direct check shows also that $FA = I$. Note that F, like A, is lower triangular.

Exercise 2.6.17. Assume that A is an $n \times n$ SQUARE lower triangular matrix and furthermore that all diagonal entries are nonzero, i.e., $a_{ii} \neq 0$, $1 \leq i \leq n$.

i. Show how to find a matrix F such that $AF = I$ and a matrix G such that $GA = I$.

ii. Show that F and G are lower triangular and that $F = G$.

iii. Show that there is a lower triangular matrix L such that each diagonal element l_{ii} of L is 1 and

$$A = L \begin{pmatrix} d_1 & & & \\ & d_2 & & \\ & & \ddots & \\ & & & d_n \end{pmatrix} \overset{\text{def}}{=} LD.$$

(Is the last assertion valid if the hypothesis: $a_{ii} \neq 0$, $1 \leq i \leq n$, is dropped?)

iv. Repeat *i - iii* for SQUARE upper triangular matrices.

v. Assume A is a SQUARE triangular matrix and that one of its diagonal entries a_{ii} is 0. Find a nonzero column vector $\mathbf{x}$ such that $A\mathbf{x} = \mathbf{O}$.

vi. Use *v* to show that if a diagonal entry of a SQUARE triangular matrix A is 0 then there is no F such that $AF = I$ and there is no G such that $GA = I$.

Of the equation $A_P = \tilde{L}A$ it can now be said that $\tilde{L}$ has an inverse L. Thus $L\tilde{L} = I$, $L\tilde{L}A = IA$, i.e.,

$$A = LA_P. \qquad\qquad (2.6.3)$$

Note that (2.6.3) is valid for HIGH, WIDE, or SQUARE matrices.

In particular matrix algebra shows why the PROCESSed system and the original system have the same solutions if there are any at all. Indeed,

$$A\mathbf{x} = \mathbf{b} \;\Leftrightarrow\; LA_P\mathbf{x} = \mathbf{b} \;\Leftrightarrow\; A_P\mathbf{x} = L^{-1}\mathbf{b}.$$

Thus the PROCESSed system $A_P\mathbf{x} = L^{-1}\mathbf{b}$ has a solution iff the original system $A\mathbf{x} = \mathbf{b}$ has a solution, and any solution for either system is a solution for the other.

The form of A_P can be somewhat ragged, although its form in no way influences the ease and speed of back-substitution that wraps up the task of finding a solution to the system. For example the following matrix might be the result of PROCESSing some augmented matrix $A|\mathbf{b}$:

$$\left(\begin{array}{ccccc|c}
0 & 0 & 0 & 7 & 2 & -6 \\
0 & 2 & 5 & -6 & 3 & 0 \\
4 & 8 & 1 & 2 & 1 & 2 \\
0 & 0 & 6 & -5 & 1 & -3
\end{array}\right).$$

In the display above

$$A_P = \begin{pmatrix}
0 & 0 & 0 & 7 & 2 \\
0 & 2 & 5 & -6 & 3 \\
4 & 8 & 1 & 2 & 1 \\
0 & 0 & 6 & -5 & 1
\end{pmatrix}.$$

The first pivot candidate 4 lies to the left of all nonzero entries in A_P. The second pivot candidate 2 lies to the left of all nonzero entries in the unmarked 3×4 matrix remaining after row 3 and column 1 are marked. The third pivot candidate 6 lies to the left of all nonzero entries in the unmarked matrix remaining after row 2 and column 2 of the original matrix are marked, etc.

An EXCHANGE of rows 1 and 3 and, in the resulting matrix, an EXCHANGE of rows 3 and 4, bring about

$$\begin{pmatrix}
4 & 8 & 1 & 2 & 1 \\
0 & 2 & 5 & -6 & 3 \\
0 & 0 & 6 & -5 & 1 \\
0 & 0 & 0 & 7 & 2
\end{pmatrix}$$

which is the echelon form A_E of A. Furthermore the EXCHANGEs applied to the PROCESSed augmented matrix yield for the echelon form of the augmented matrix

$$\left(\begin{array}{ccccc|c}
4 & 8 & 1 & 2 & 1 & 2 \\
0 & 2 & 5 & -6 & 3 & 0 \\
0 & 0 & 6 & -5 & 1 & -3 \\
0 & 0 & 0 & 7 & 2 & -6
\end{array}\right).$$

The EXCHANGEs above are produced by multiplying $A|\mathbf{b}$ by

$$E_{34}E_{13}.$$

In general, if $A \neq O$ then in A_P one of the rows has a nonzero entry, the first pivot candidate, to the left of all other nonzero entries in the matrix A. Once the first pivot's column and row are marked, either the unmarked matrix remaining consists entirely of 0's or there is a second pivot candidate, and it is to the left of all other nonzero entries in the unmarked matrix, etc. Hence by a rearrangement or permutation of the rows of A_P there can be produced an upper triangular matrix in which the first pivot candidate occurs in the first row, the second pivot candidate in the second row, etc. The permutation is achievable by a sequence of ROW EXCHANGEs, hence by a sequence of multiplications by elementary permutation matrices E_{ij}.

Exercise 2.6.18. A product Π of elementary permutation matrices is called a *permutation matrix*. Prove that if Π is a permutation matrix then:

i. in each column of Π there is one and only nonzero entry and it is a 1;
ii. in each row of Π there is one and only one nonzero entry and it is a 1.

Show also that Π is lower triangular or upper triangular iff $\Pi = I$.

Exercise 2.6.19. Let A be an $m \times n$ matrix and let A_P, A_E, and A_J be its *PROCESS*ed, echelon, and Jordan elimination versions. Show that there is a permutation matrix Π and a lower triangular matrix Λ, a product of lower triangular elementary matrices of the type $E_i(r)$, such that:

i. $\Pi A_P = A_E$;
ii. $\Lambda A_E = A_J$;
iii. $\Lambda \Pi A_P = A_J$.

It is advantageous to do all the rearranging of the rows of the system or of the relevant matrices *before* the SEARCHes and ELIMINATIONs are carried out. In fact, if this kind of preparatory step is taken appropriately, each SEARCH leads either to an entry in the row immediately below the last active row used in ROW COMBINATION or to the conclusion that there are no further nonzero entries.

Example 2.6.5. Let A be the 3×3 matrix

$$\begin{pmatrix} 1 & 3 & 1 \\ 2 & 6 & -1 \\ 4 & -1 & 3 \end{pmatrix}.$$

Then the product of the elementary matrices used to simplify A by the GEM is

$$E_{23}E_{13}(-4)E_{12}(-2)$$

     Chapter 2. THE PROCESS (Simple elimination)

and

$$A_E = E_{23}E_{13}(-4)E_{12}(-2)A = \begin{pmatrix} 1 & 3 & 1 \\ 0 & -13 & -1 \\ 0 & 0 & -3 \end{pmatrix}.$$

The permutation matrix E_{23} can be slid past the factors $E_{13}(-4)$ and $E_{12}(-2)$ according to the following procedure:
Since

$$L_1 \stackrel{\text{def}}{=} E_{13}(-4)E_{12}(-2) = \begin{pmatrix} 1 & 0 & 0 \\ -2 & 1 & 0 \\ -4 & 0 & 1 \end{pmatrix}$$

L_1 is a lower triangular matrix. Since $E_{23}^2 = I$ it follows that

$$E_{23}L_1 = E_{23}L_1 I = E_{23}L_1 E_{23} E_{23}$$
$$= \begin{pmatrix} 1 & 0 & 0 \\ 0 & 0 & 1 \\ 0 & 1 & 0 \end{pmatrix} \begin{pmatrix} 1 & 0 & 0 \\ -2 & 1 & 0 \\ -4 & 0 & 1 \end{pmatrix} \begin{pmatrix} 1 & 0 & 0 \\ 0 & 0 & 1 \\ 0 & 1 & 0 \end{pmatrix} \begin{pmatrix} 1 & 0 & 0 \\ 0 & 0 & 1 \\ 0 & 1 & 0 \end{pmatrix}.$$

When the first three matrices are multiplied together the result is a new lower triangular matrix

$$L_2 \stackrel{\text{def}}{=} \begin{pmatrix} 1 & 0 & 0 \\ -4 & 1 & 0 \\ -2 & 0 & 1 \end{pmatrix}$$

whence

$$E_{23}L_1 = \begin{pmatrix} 1 & 0 & 0 \\ -4 & 1 & 0 \\ -2 & 0 & 1 \end{pmatrix} \begin{pmatrix} 1 & 0 & 0 \\ 0 & 0 & 1 \\ 0 & 1 & 0 \end{pmatrix} = L_2 E_{23}$$

which is the product of the new lower triangular matrix L_2 and the *same* permutation matrix E_{23}, but on the *right* of L_2, whereas E_{23} is on the *left* of L_1 in the original formula for A_E.
Hence

$$A_E = L_2 E_{23} A,$$

i.e., by a preliminary rearrangement (via E_{23}) of the rows of A, one can produce A_E by a sequence of ELIMINATIONs only, hence by multiplying $E_{23}A$ by a lower triangular matrix L_2. Since the factors that make up L_2 are lower triangular matrices having only 1's on their diagonals it follows that L_2 also has only 1's on its diagonal. By means of inverses, it follows that

$$E_{23}A = L_2^{-1} A_E.$$

A careful calculation shows that

$$E_{23}A = \begin{pmatrix} 1 & 0 & 0 \\ 4 & 1 & 0 \\ 2 & 0 & 1 \end{pmatrix} \begin{pmatrix} 1 & 3 & 1 \\ 0 & -13 & -1 \\ 0 & 0 & -3 \end{pmatrix},$$

---

the product of a lower triangular matrix L and an upper triangular matrix $\tilde{U}$. It is also true that

$$E_{23}A = \begin{pmatrix} 1 & & \\ 4 & 1 & \\ 2 & 0 & 1 \end{pmatrix} \begin{pmatrix} 1 & & \\ & -1 & \\ & & -3 \end{pmatrix} \begin{pmatrix} 1 & 3 & 1 \\ & 1 & 1 \\ & & 1 \end{pmatrix},$$

the product of (the same) lower triangular matrix L, a diagonal matrix D and an upper triangular matrix U. The diagonal entries of L and of U are all 1's. Thus

$$E_{23}A = L\tilde{U} = LDU.$$

In **Section 2.7** next it is shown that if A is any matrix then there can be found

$i.$ a permutation matrix Π,
$ii.$ a lower triangular matrix L in which all diagonal entries are 1's,
$iii.$ a diagonal matrix D,

and

$iv.$ an upper triangular matrix U in which all diagonal entries are 1's

so that

$$\Pi A = LDU. \tag{2.6.4}$$

Exercise 2.6.20. Write each of the following matrices according to the pattern in (2.6.4).

$$i. \begin{pmatrix} 1 & 2 & 3 \\ 4 & 5 & 6 \end{pmatrix}, \quad ii. \begin{pmatrix} -1 & 2 & -3 \\ 4 & -5 & 6 \\ -7 & 8 & -9 \\ 8 & -7 & 6 \end{pmatrix}, \quad iii. \begin{pmatrix} 1 & 2 & -3 \\ 4 & 8 & -6 \\ -7 & -8 & -9 \end{pmatrix}.$$

[*Hint:* In iii one can slide E_{23} past $E_{12}(-4)$ according to the following scheme:

$$E_{23}E_{12}(-4) = E_{23}E_{12}(-4)E_{23}E_{23}$$
$$= E_{13}(-4)E_{23}$$

and thus place the permutation matrix E_{23} next to A.]

Exercise 2.6.21. In each row of a permutation matrix there is precisely one 1 and in each column there is precisely one 1, cf. **Exercise 2.6.18**. Use the GEM to show that every matrix in which there is precisely one 1 in each row and precisely one 1 in each column is the product of elementary permutation matrices. Use the conclusion to prove again that every permutation of rows is achievable via a sequence of *transpositions*, i.e., permutations in which all but two rows are left fixed and the remaining two are interchanged or *transposed*.

2.7. Details about the elementary matrices

Up to this point the fine details of the GEM and the GJM remain unexplored. To get a better understanding of the GEM and the GJM it is desirable to examine some of their nooks and crannies.

If A is an $m \times n$ matrix then, after SEARCH, the first cycle of the GEM begins with multiplying A first by some E_{1j_1}. If $a_{1c} = 0$, then $j_1 > 1$. If $a_{1c} \neq 0$ then $j_1 = 1$ (in which case the first multiplication is by E_{11} $(= I)$). The next part of the cycle consists of ROW COMBINATIONs, hence multiplication by $E_{12}(r_{12})$, then multiplication by $E_{13}(r_{13})$, ..., and finally multiplication by $E_{1m}(r_{1m})$. At the end of the first cycle the matrix A is converted to

$$A^{(1)} \stackrel{\text{def}}{=} E_{1m}(r_{1m}) \cdots E_{12}(r_{12})E_{1j_1} A.$$

Similarly by the end of the second cycle A is converted to

$$A^{(2)} \stackrel{\text{def}}{=} E_{2m}(r_{2m}) \cdots E_{23}(r_{23})E_{2j_2} A^{(1)}$$
$$= E_{2m}(r_{2m}) \cdots E_{23}(r_{23})E_{2j_2} E_{1m}(r_{1m}) \cdots E_{12}(r_{12})E_{1j_1} A.$$

At the end of the last $((m-1)\text{st})$ cycle A is converted to

$$A^{(m-1)} \stackrel{\text{def}}{=} A_E = E_{m-1,m}(r_{m-1,m}) \cdots E_{m-1,j_{m-1}} A^{(m-2)}$$
$$= A_E = E_{m-1,m}(r_{m-1,m}) \cdots E_{m-1,j_{m-1}} \cdots$$
$$E_{2m}(r_{2m}) \cdots E_{23}(r_{23})E_{2j_2} E_{1m}(r_{1m}) \cdots$$
$$E_{12}(r_{12})E_{1j_1} A. \tag{2.7.1}$$

The numbers r_{pq} are at this stage of no interest or consequence. The numbers $j_{\ldots}$ are of interest only to the extent that each $j_k > k$. What *is* of great interest is that to the right of each elementary permutation (ROW EXCHANGE) matrix E_{ij_i} are ROW COMBINATION matrices $E_{pq}(r_{pq})$ for which $p > i$ and other ROW EXCHANGE matrices E_{kj_k}. (It is true that $k > i$, but this fact cuts no ice in the current considerations.)

In the GJM the expression in (2.7.1) is still further complicated by the presence at the beginning of each cycle of a ROW SCALING matrix $E_i(s_i)$. For completeness, but hardly for clarity, the GJM counterpart of (2.7.1) is given below where the ROW SCALING matrices $E_i(s_i)$ appear in boldface:

$$A_J = \mathbf{E_m}(\mathbf{s_m})E_{m-1,m}(r_{m-1,m}) \cdots \mathbf{E_{m-1}}(\mathbf{s_{m-1}})E_{m-1,j_{m-1}} \cdots$$
$$E_{2m}(r_{2m}) \cdots E_{23}(r_{23})\mathbf{E_2}(\mathbf{s_2})E_{2j_2} E_{1m}(r_{1m}) \cdots$$
$$E_{12}(r_{12})\mathbf{E_1}(\mathbf{s_1})E_{1j_1} A. \tag{2.7.2}$$

In (2.7.2) to the right of each ROW EXCHANGE matrix E_{ij_i} are, again, ROW COMBINATION matrices $E_{pq}(r_{pq})$ for which $p > i$, other ROW EXCHANGE matrices E_{kj_k}, ROW SCALING matrices $E_l(s_l)$, and A. The numbers r_{pq} and s_l and the fact that $k > i$ and $l > i$ are again of no interest or consequence.

[**Remark 2.7.1:** Although one picture is often worth a thousand words, in the current situation it is simpler to describe the important relations displayed above in a few words, viz.:

A ROW EXCHANGE matrix is never to the left of a ROW COMBINATION matrix for which the active row used is *at or below* those in the ROW EXCHANGE.]

The object of the analysis that follows is to show that any *elementary permutation matrix* E_{pq} that crops up during the application of the GEM or the GJM can be slid to the right as often as needed until there are no matrices save A itself or other elementary permutation matrices to the right of E_{pq}. Furthermore, after all the sliding is over there is a cluster of elementary permutation matrices immediately preceding (to the left of) A. To the left of the cluster of elementary permutation matrices there are only lower triangular elementary matrices $E_{pq}(r)$ for ROW COMBINATION and, in the GJM, $E_i(s)$ for ROW SCALING.

[**Remark 2.7.2:** The preceding discussion is rather formal. An informal approach to the subject of rearranging the elementary matrices is the following: If ROW EXCHANGEs are carried out in the GEM or in the GJM then they might as well be carried out first, before any of the SEARCH or ELIMINATION or ROW SCALING operations are performed. If this is done, the matrix to be treated is multiplied first by a product of ROW EXCHANGE matrices and then the other kinds of operations are applied.

The catch is that the sequence of other kinds of operations performed *after* the rearrangement of the rows can be different from the sequence used in the standard application of the GEM or the GJM. The argument given next deals with this catch.]

If one admits the use of matrices E_{ii} and $E_{ij}(0)$ each of which is the identity I, then the GEM may be viewed as the formation of A_E by means of the following product:

$$A_E = E_{n-1,n}(r_{n-1,n})E_{n-1,i_{n-1}}E_{n-2,n}(r_{n-2,n}) \cdots$$
$$E_{2n}(r_{2n}) \cdots E_{23}(r_{23})E_{2i_2}E_{1n}(r_{1n}) \cdots E_{12}(r_{12})E_{1i_1}A.$$

In other words, one always starts with some, possibly trivial, ROW EXCHANGE (via E_{1i_1}, which might be $E_{11} = I$), then $n - 1$ ROW COMBINATIONs, (via $E_{1k}(r_{1k})$, $2 \leq k \leq n$, some of which might be $E_{1k}(0) = I$), then some ROW EXCHANGE (via E_{2i_2}, which might be $E_{22} = I$), then $n - 2$ ROW COMBINATIONs (via $E_{2k}(r_{2k})$, $3 \leq k \leq n$, some of which might be $E_{2k}(0) = I$), etc. It follows that whenever a ROW EXCHANGE is performed

---

via some E_{ij} in which $i \leq j$ then i is greater than any p in some $E_{pq}(r_{pq})$, in which $p < q$ and that stands to the *right* of E_{ij}. Since $E_{ii} = I$ it suffices to consider only the situation in which $p < i < j$ and $p < q$.

It is shown below that:

i. if $p < i < j$ and $p < q$ then for some q', such that $p < q'$,

$$E_{ij} E_{pq}(r) = E_{pq'}(r) E_{ij};$$

ii. for some k'

$$E_{ij} E_k(s) = E_{k'}(s) E_{ij}.$$

The argument makes the values of q', and k' clear. If the assertions i and ii above are accepted one may indeed conclude that E_{2i_2} can be slid right until it reaches $E_{1i_1} A$, then E_{3i_3} can be slid right until it reaches $E_{2i_2} E_{1i_1} A$, then, $\ldots$, then $E_{n-1,i_{n-1}}$ can be slid right until it reaches $E_{n-2,i_{n-2}} \cdots E_{2i_2} E_{1i1} A$. When all is said and done, all the E_{ij} cluster just to the left of A and to *their* left are:

for the GEM: only lower triangular elementary matrices used for ROW COMBINATION;

for the GJM: only lower triangular matrices used for ROW COMBINATION and diagonal (hence lower triangular) matrices used for ROW SCALING.

The product of all the E_{ij} clustered before A is a permutation matrix Π.

The heart of the "sliding" argument is based on the simple observation that $E_{ij}^2 = I$. Hence for any compatible SQUARE matrix M and any pair i, j

$$E_{ij} M = (E_{ij} M E_{ij}) E_{ij} \overset{\text{def}}{=} \widetilde{M} E_{ij}. \tag{2.7.3}$$

Since $E_{ij} = E_{ji}$ one may assume always that $i \leq j$. Since $E_{ii} = I$ the only interesting situation is that in which $i < j$. Hence the only cases requiring discussion are:

i. if M is any ROW SCALING matrix $E_k(s)$,
OR
ii. if M is a ROW COMBINATION matrix $E_{pq}(r)$ in which the active row is above row i, i.e., if

$$M = E_{pq}(r), \ p < q \text{ and } p < i < j.$$

A direct calculation, given below, shows that M and $\widetilde{M}$ are matrices of the same kind, and, in particular, $\widetilde{M}$ is also lower triangular.

If M is a ROW SCALING matrix $E_k(s)$ then there are three cases to be analyzed: $i, j \neq k$, $i = k$, and $j = k$.

If $i, j \neq k$ then

$$E_{ij} E_k(s) E_{ij} = E_k(s), \ k' = k.$$

<hr>

If $i = k$ then

$$E_{ij} E_k(s) E_{ij} = E_j(s), \ k' = j.$$

If $j = k$ then

$$E_{ij} E_k(s) E_{ij} = E_i(s), \ k' = i.$$

Since all ROW SCALING matrices are diagonal they are all lower triangular.

If M is a ROW COMBINATION matrix $E_{pq}(r)$ there are three cases to be analyzed: $i, \ j \neq q$, $i = q < j$, and $i < j = q$.

If $i, \ j \neq q$ then

$$E_{ij} E_{pq}(r) E_{ij} = E_{pq}(r), \ q' = q.$$

If $p < i = q < j$ then

$$E_{ij} E_{pq}(r) E_{ij} = E_{pj}(r), \ q' = j.$$

If $i < j = q$ then

$$E_{ij} E_{pq}(r) E_{ij} = E_{pi}(r), \ q' = i.$$

Owing to the relations $p < i < j$ the new ROW COMBINATION matrices are again lower triangular.

Example 2.7.1. The following illustrates the necessity of the (weaker) requirement $p < i \leq j$. If $n = 3$, $i = 1$, $j = 2$, $p = 2$, $q = 3$ then $i = 1 < 2 = p$ and

$$E_{13}M \overset{\text{def}}{=} E_{13} \begin{pmatrix} 1 & 0 & 0 \\ 0 & 1 & 0 \\ 0 & 2 & 1 \end{pmatrix} = \begin{pmatrix} 0 & 2 & 1 \\ 0 & 1 & 0 \\ 1 & 0 & 0 \end{pmatrix} \overset{\text{def}}{=} M'.$$

Since there are only three columns, a simple count of all possible column-switches shows that no column-switch can convert M' to a lower triangular matrix. Hence there is no lower triangular matrix L and no elementary permutation matrix E_{ij} such that $E_{13}M = LE_{ij}$. In other words the sliding operation can fail to produce another lower triangular matrix if $i < p$.

Exercise 2.7.1. Show that if $i = p < q = j$ then E_{ij} can be slid from the right side of $E_{pq}(r)$ to its left side and thereby leaves on the right of E_{ij} an upper triangular ROW COMBINATION matrix.

Exercise 2.7.2. Assume each elementary matrix below is of size 3×3.

$i.$ Find $E_{pq}(r)$ and $E_i(s)$ so that

$$E_{13}E_{23}(-5) = E_{pq}(r)E_{13}$$
$$E_{12}E_3(4) = E_i(s)E_{12}.$$

$ii.$ Calculate $\Pi \overset{\text{def}}{=} E_{12}E_{32}E_{13}$.

    Chapter 2. THE PROCESS (**Simple elimination**)

iii. For Π in *ii* calculate Π^{-1}.

Exercise 2.7.3. Let A be an $m \times n$ matrix. If A_P is the *PROCESS*ed version of a matrix A then for some permutation matrix Π, $\Pi A_P = A_E$, the echelon form of A, cf. **Exercise 2.6.19**.

Show the following is also true:

for a permutation matrix $\widetilde{\Pi}$ the *PROCESS* applied to $\widetilde{\Pi} A$ yields A_E.

The echelon form A_E has a variety of diagonal entries and some of them may be 0's.

Example 2.7.2. Assume A_E looks like this:

$$\begin{pmatrix} 2 & -1 & 6 & 3 & 8 \\ 0 & -3 & 2 & 0 & 4 \\ 0 & 0 & 6 & -4 & -1 \\ 0 & 0 & 0 & 0 & 0 \end{pmatrix}.$$

For a proper choice of a, b, c, d, e, f, and g, A_E may be written

$$\begin{pmatrix} 2 & 0 & 0 & 0 \\ 0 & -3 & 0 & 0 \\ 0 & 0 & 6 & 0 \\ 0 & 0 & 0 & 0 \end{pmatrix} \begin{pmatrix} 1 & a & b & c & d \\ 0 & 1 & e & f & g \\ 0 & 0 & 1 & h & k \\ 0 & 0 & 0 & 1 & r \end{pmatrix}$$

the product of a matrix that is both lower triangular and upper triangular, i.e., a *diagonal* matrix, and an upper triangular matrix in which all diagonal entries are 1's. Indeed multiplying the two matrices above reveals that if

$$2a = -1, \ 2b = 6, \ 2c = 3, \ 2d = 8,$$
$$-3e = 2, \ -3f = 0, \ -3g = 4,$$
$$6h = -4, \ 6k = -1,$$
$$r = \text{ any number}$$

then

$$A_E = \begin{pmatrix} 2 & 0 & 0 & 0 \\ 0 & -3 & 0 & 0 \\ 0 & 0 & 6 & 0 \\ 0 & 0 & 0 & 0 \end{pmatrix} \begin{pmatrix} 1 & -\frac{1}{2} & 3 & \frac{3}{2} & 4 \\ 0 & 1 & -\frac{2}{3} & 0 & -\frac{4}{3} \\ 0 & 0 & 1 & -\frac{2}{3} & -\frac{1}{6} \\ 0 & 0 & 0 & 1 & r \end{pmatrix} \tag{2.7.4}$$

Example 2.7.3. The matrices I and $E_i(s)$ are diagonal matrices. A diagonal matrix D is a matrix that is both upper and lower triangular. Conversely, a matrix that is both lower triangular and upper triangular is a diagonal matrix.

At this point the work above suggests that the following might be true:

If A is an $m \times n$ matrix then there are four matrices:

an $m \times m$ permutation matrix Π;

a lower triangular $m \times m$ matrix L in which all the diagonal entries are 1's, i.e., a *unit* lower triangular matrix;

a diagonal $m \times m$ matrix D;

an upper triangular $m \times n$ matrix U in which all diagonal entries are 1's, i.e., a *unit* upper triangular matrix.

The matrix A and the four matrices Π, L, D, and U are related by the equation $\Pi A = LDU$.

The matrix Π is a SQUARE $m \times m$ matrix, — a permutation matrix, a product of elementary permutation matrices, — in which there is precisely one nonzero entry in each row and in each column and that nonzero entry is 1.

The matrix $L \overset{\text{def}}{=} (l_{ij})_{i,j=1}^{m,m}$ is a lower triangular SQUARE matrix in which all entries l_{ii} on the diagonal are 1's. Thus L looks like this:

$$\begin{pmatrix} 1 & & & \\ l_{21} & 1 & & \\ \vdots & \vdots & \ddots & \\ l_{m1} & l_{m2} & \cdots & 1 \end{pmatrix}.$$

The matrix D is a diagonal matrix and looks like this:

$$\begin{pmatrix} d_1 & & & \\ & d_2 & & \\ & & \ddots & \\ & & & d_n \end{pmatrix}.$$

The matrix U is an upper triangular matrix $(u_{ij})_{i,j=1}^{m,n}$ in which all diagonal entries are 1's. Since the $m \times n$ matrix U is not necessarily SQUARE no single example illustrates all its possibilities. If $m = n$ the matrix U looks like this:

$$\begin{pmatrix} 1 & u_{12} & \cdots & u_{1n} \\ & 1 & \cdots & u_{2n} \\ & & \ddots & \vdots \\ & & & 1 \end{pmatrix}.$$

The matrix DU is precisely A_E, the upper triangular matrix $\tilde{U}$ mentioned at the end of **Section 2.6**: $A_E = \tilde{U}$, $A = L\tilde{U}$.

The matrices Π, L, D, U, and $\tilde{U}$ $(=A_E)$ associated with A play important rôles in most of the study of the matrix A. These rôles are listed next and are treated more fully in appropriate paragraphs below and in **Chapter 3**.

i. In the study of the *determinant* of A, when A is SQUARE, the matrix Π fixes the *sign* $\pm$ in the determinant formula (cf. **Chapter 3**).

ii. The matrix L is a record of the *negatives* of the numbers r_i that appear in the matrices $E_{ij}(r_i)$.

iii. The matrices Π, L, D, and U are determined uniquely by A.

iv. If A is SQUARE then the absence of 0's on the diagonal of the (diagonal) matrix D is necessary and sufficient for the existence of a matrix B such that $AB = BA = I$. Equivalently, there is a matrix B such that $AB = BA = I$ iff each diagonal entry in D is *not* 0.

Example 2.7.4 below illustrates how, starting with a SQUARE matrix A, one can find the matrices Π, L, D, and U so that $\Pi A = LDU$.

Example 2.7.4. Let A be the matrix

$$\begin{pmatrix} 1 & 2 & 3 \\ 4 & 5 & 6 \\ 7 & 8 & 9 \end{pmatrix} .$$

Then the product of the elementary matrices used to *PROCESS* A is

$$\tilde{L} \stackrel{\text{def}}{=} E_{23}(-2)E_{13}(-7)E_{12}(-4) = \begin{pmatrix} 1 & 0 & 0 \\ -4 & 1 & 0 \\ 1 & -2 & 1 \end{pmatrix} ,$$

a lower triangular matrix, (no EXCHANGEs are needed) and

$$\tilde{L}A = A_P = \begin{pmatrix} 1 & 2 & 3 \\ 0 & -3 & -6 \\ 0 & 0 & 0 \end{pmatrix} .$$

Owing to the upper triangular form of A_P the rows 1, 2, and 3 are marked and ignored in normal sequence. Hence $A_P = A_E$. In other words, the *PROCESS* and the GEM are indistinguishable when they are applied to this particular matrix A for which the GEM requires *no* ROW EXCHANGEs.

The pivot candidates encountered in the course of the *PROCESS* are 1 and -3, the nonzero numbers appearing on the diagonal of A_P. Despite the fact that there is a 0 on the diagonal of A_P, it is nevertheless possible to factor A_P as follows:

$$A_P = \begin{pmatrix} 1 & 0 & 0 \\ 0 & -3 & 0 \\ 0 & 0 & 0 \end{pmatrix} \begin{pmatrix} 1 & 2 & 3 \\ 0 & 1 & 2 \\ 0 & 0 & 1 \end{pmatrix} \stackrel{\text{def}}{=} DU.$$

In the display above, D is the diagonal matrix and U is the upper triangular matrix with only 1's on its diagonal. Since

$$\widetilde{L}A = A_P = DU$$

and since $\widetilde{L}$ is a lower triangular matrix in which all diagonal entries are 1's, there is a lower triangular matrix L such that $L\widetilde{L} = I$ whence $A = LA_P = LDU$.

The generalization of the conclusions in the **Example 2.7.4** are outlined next.

Exercise 2.7.4. Let A be an $m \times n$ matrix and let its echelon form be A_E.

i. Show that A_E is upper triangular.

ii. Show that every elementary matrix $E_{ij}(t)$ or $E_i(t)$ used in applying the GEM to A is lower triangular.

iii. Assume no row exchanges are involved in applying the GEM to A, and let $\mathcal{E}$ be the product of the elementary matrices used in the GEM. Show $\mathcal{E}$ is lower triangular.

iv. Again assume no row exchanges are involved in applying the GEM to A. Show that for some lower triangular matrix L_1 and some upper triangular matrix U_1,
$$L_1 A = U_1.$$

Show also that there is a lower triangular matrix L_2 such that

$$L_2 L_1 = L_1 L_2 = I \text{ and}$$
$$A = L_2 U_1.$$

v. Once more assume no row exchanges occur in applying the GEM to A. Show that for some lower triangular matrix L_1, the diagonal matrix D the entries of which are the k *pivots* of A and $n - k$ 0's, and some upper triangular matrix U in which each diagonal entry is 1,

$$L_1 A = DU \text{ or } A = L_2 DU \stackrel{\text{def}}{=} LDU.$$

If A is lower (upper) triangular what are L, D and U?

vi. Use v to show that in any event there is a matrix $\widetilde{\Pi}$, the product of elementary *permutation* matrices, such that for some lower triangular matrix $\widetilde{L}$, some upper triangular matrix $\widetilde{U}$, and some diagonal matrix D,

$$\widetilde{L}\widetilde{\Pi}A = \widetilde{U} = DU$$
$$\widetilde{\Pi}A = LDU.$$

[*Hint:* The discussion following **Remark 2.7.2** shows how to slide all the E_{ij} used in the GEM or in the GJM so that they cluster just to the left of A and to the left of *them* only lower triangular elementary matrices appear.]

Exercise 2.7.5. If A is the matrix

$$\begin{pmatrix} a_{11} & & \cdots & & \\ a_{21} & a_{22} & & & \\ \vdots & \vdots & \ddots & & \\ a_{n-1,1} & a_{n-1,2} & \cdots & a_{n-1,n-1} & \\ a_{n1} & a_{n2} & \cdots & a_{n,n-1} & a_{nn} \end{pmatrix},$$

if B is the matrix

$$\begin{pmatrix} b_{11} & & \cdots & & \\ b_{21} & b_{22} & & & \\ \vdots & \vdots & \ddots & & \\ b_{n-1,1} & b_{n-1,2} & \cdots & b_{n-1,n-1} & \\ b_{n1} & b_{n2} & \cdots & b_{n,n-1} & b_{nn} \end{pmatrix},$$

and if $AB \overset{\text{def}}{=} C$ is the matrix

$$\begin{pmatrix} c_{11} & \cdots & c_{1n} \\ \vdots & \ddots & \vdots \\ c_{n1} & \cdots & c_{nn} \end{pmatrix},$$

then (**Exercise 2.6.16**) C is also lower triangular. Show that $c_{ii} = a_{ii}b_{ii}, i = 1, \ldots, n$. Repeat for upper triangular matrices.

[*Hint:* Try the problem for the case of 3×3 matrices.]

Exercise 2.7.6. Assume that the SQUARE matrix A is lower triangular and that some $a_{ii} = 0$.

i. Show that then neither F nor G as required in **Exercise 2.6.17** can be found.
ii. Repeat *i* for upper triangular matrices.
iii. Combine the results of *i* in this **Exercise** and in **Exercise 2.7.5** into a single "iff" statement.

Exercise 2.7.7. Show that if A is a lower triangular $n \times n$ matrix such that all its diagonal entries are 0 then $A^n = O$. Repeat for upper triangular matrices.

[**Remark 2.7.3:** A matrix A such that for some natural number k $A^k = O$ is called a *nilpotent* matrix since some *power* of the matrix is O (nil).]

Exercise 2.7.8.

i. Show that there is a matrix Π as in **Exercise 2.7.4***vi*, a lower triangular matrix L, the diagonal matrix D, and an upper triangular matrix U, all as in **Exercise 2.7.4***v* and such that $\Pi A = LDU$.

ii. For each of the SQUARE matrices appearing in the numerical **Exercises** and **Examples** of this **Chapter** write the matrix in the forms $\widetilde{L}\widetilde{U}$ and LDU if those forms are achievable. Otherwise write the matrix so that $\Pi(\text{given matrix}) = LDU$.

iii. Show that
$$E_{ij} = E_{ji}(-1)E_j(-1)E_{ij}(-1)E_{ji}(1).$$

In other words, show that the GEM or the GJM can be carried out without recourse to EXCHANGE, (but at some cost in the number of calculations).

iv. Let A be SQUARE and assume

$$L_1,\ L_2,\ D_1,\ D_2,\ U_1,\ U_2$$

are SQUARE matrices of the types considered in *i*. Show that if

$$L_1 D_1 U_1 = L_2 D_2 U_2$$

then

$$L_1 = L_2,\ D_1 = D_2,\ U_1 = U_2.$$

Why can uniqueness fail if A is not SQUARE?

[*Hint:* There is a lower triangular matrix L such that $LL_2 = I$ and there is an upper triangular matrix U such that $U_1 U = I$. Hence

$$LL_1 D_1 = D_2 U_2 U.$$

The left member above is lower triangular and the right member above is upper triangular. Hence both sides are diagonal.]

It is now possible to show how the factorization $A = LDU$ or the factorization $\Pi A = LDU$ speeds the analysis of the typical system: $A\mathbf{x} = \mathbf{b}$.

Example 2.7.5. If A as in **Example 2.6.5** is part of a system $A\mathbf{x} = \mathbf{b}$ then the system may be written $\Pi A\mathbf{x} = LDU\mathbf{x} = \Pi\mathbf{b} \overset{\text{def}}{=} \mathbf{c}$. The permutation matrix Π indicates how the rows of the system are rearranged. In other words, row 1, row 2, and row 3 are replaced by rows 1, 3, and 2. Hence instead of working with the system $A\mathbf{x} = \mathbf{b}$ one may work with the system $LDU\mathbf{x} = \mathbf{c}$. For example if

$$\mathbf{b} \overset{\text{def}}{=} \begin{pmatrix} b_1 \\ b_2 \\ b_3 \end{pmatrix} = \begin{pmatrix} 6 \\ -2 \\ 5 \end{pmatrix} \text{ then } \mathbf{c} = \begin{pmatrix} 6 \\ 5 \\ -2 \end{pmatrix}$$

---

and

$$LDU\mathbf{x} = \begin{pmatrix} 6 \\ 5 \\ -2 \end{pmatrix}.$$

Let $\widetilde{U}$ denote DU. Then

$$\widetilde{U}\mathbf{x} \stackrel{\text{def}}{=} \mathbf{y} \stackrel{\text{def}}{=} \begin{pmatrix} y_1 \\ y_2 \\ y_3 \end{pmatrix}.$$

Because L is a lower triangular matrix that is the product of elementary matrices the system $Ly = \mathbf{c}$ can be analyzed and then its solution can be read off by *front-substitution*, viz.,

$$y_1 = 6, \ y_2 = 5 - 4y_1 = 5 - 24 = -19, \text{ and } y_3 = -2 - 2y_1 = -2 - 12 = -14.$$

Then

$$\widetilde{U}\mathbf{x} = \begin{pmatrix} 1 & 3 & 1 \\ 0 & -13 & -1 \\ 0 & 0 & -3 \end{pmatrix} \begin{pmatrix} x_1 \\ x_2 \\ x_3 \end{pmatrix} = \begin{pmatrix} 6 \\ -19 \\ -14 \end{pmatrix}.$$

Back-substitution then leads to

$$x_3 = \frac{14}{3},$$

$$x_2 = \frac{\left(-19 + \frac{14}{3}\right)}{-13} = \frac{43}{39},$$

$$x_1 = 6 - \frac{14}{3} - 3\frac{43}{39} = -\frac{77}{39},$$

which is indeed the solution of the system.

When an $n \times n$ triangular matrix permits finding a solution via back- (or, as in the case of a lower triangular matrix, front-) substitution, there is one multiplication performed in finding the first unknown, there are two multiplications performed in finding the second unknown, etc., hence the number of multiplications performed is at most

$$1 + 2 + \cdots + n = \frac{n(n+1)}{2}.$$

When the LDU approach is used two sets of substitutions are carried out, a front-substitution when L is used and a back-substitution when DU is used. In sum not more than $n^2 + n$ multiplications are performed.

The number of multiplications performed in carrying out the GEM or the GJM on an $m \times n$ matrix is at most

$$\mu(m,n) = \begin{cases} \sum_{k=0}^{m-2}(n-k)(m-k-1) = \frac{m(m-1)(3n-m+2)}{6}, & \text{if } m \leq n; \\ \sum_{k=0}^{n-1}(n-k)(m-k-1) = \frac{n(3mn+3(m-n)-n^2-2)}{6}, & \text{otherwise.} \end{cases}$$

When $m = n$ both formulae reduce to

$$\frac{n(n-1)(n+1)}{3} = \mu(n,n).$$

The guidelines above reveal that n^3 is the measure of the number of operations in the GEM or in the GJM and n^2 is the number of operations in back-substitution. Thus if the same system is analyzed with $M + 1$ different right members $\mathbf{b}$, then without LDU methods, about $(M+1)n^3$ operations are carried out. On the other hand if LDU methods are used, about $n^3 + Mn^2$ operations are performed. The relative amount of work in the shorter to the longer method is, for large M, measured by the ratio

$$\frac{(n+M)}{(M+1)n} \approx \frac{1}{n}.$$

For $n = 50$ and large M the LDU work done is about 2% of the work done if the straightforward repetition of the GEM or of the GJM is used.

Generally to solve $A\mathbf{x} = \mathbf{b}$ repeatedly for varying $\mathbf{b}$, the matrices L and $\widetilde{U}$ are used as follows:

First solve the equation $L\mathbf{y} = \mathbf{b}$ for $\mathbf{y}$. Then solve $\widetilde{U}\mathbf{x} = \mathbf{y}$ for $\mathbf{x}$. It follows that $\Pi A\mathbf{x} = \Pi L\widetilde{U}\mathbf{x} = \Pi L\mathbf{y} = \Pi\mathbf{b} = \mathbf{c}$ and so $A\mathbf{x} = \mathbf{b}$, i.e., the $\mathbf{x}$ found in the second step is a solution of the problem.

In **Part II** there is an extensive discussion of applications and of the practical methods employed in applications of linear algebra.

Example 2.7.6. Let the matrix displayed below be the Jordan elimination version A_J of some matrix A:

$$A_J \stackrel{\text{def}}{=} \begin{pmatrix} 0 & 0 & 0 & 1 & 6 & -1 & 4 \\ 0 & 0 & 0 & 0 & 0 & 1 & 3 \\ 0 & 0 & 0 & 0 & 0 & 0 & 1 \\ 0 & 0 & 0 & 0 & 0 & 0 & 0 \end{pmatrix}.$$

A direct calculation shows that

$$E_{21}(1)E_{31}(-4)E_{32}(-3)A_J = \begin{pmatrix} 0 & 0 & 0 & 1 & 6 & 0 & 0 \\ 0 & 0 & 0 & 0 & 0 & 1 & 0 \\ 0 & 0 & 0 & 0 & 0 & 0 & 1 \\ 0 & 0 & 0 & 0 & 0 & 0 & 0 \end{pmatrix}.$$

Notice that $i > j$ for each $E_{ij}(r)$ used above whereas $i < j$ for each $E_{ij}(r)$ used in the $PROCESS$, the GEM, or the GJM.

In each column containing a 1 sitting on a step all other entries are 0's.

Exercise 2.7.9. Let A_J be the Jordan elimination version of a matrix. Show that there is a sequence of ROW COMBINATION matrices of the form

$$E_{ij}(r)$$

in which $i > j$ and such that if $\mathcal{R}$ is the product of those ROW COMBINATION matrices then $\mathcal{R}A_J$ is

$$\begin{pmatrix} 0 & \dots & 0 & \underline{1} & 0 & \dots & * & \dots & 0 & * & * & \dots & * \\ 0 & \dots & 0 & 0 & \underline{1} & \dots & * & \dots & 0 & * & * & \dots & * \\ 0 & \dots & 0 & 0 & 0 & \dots & 0 & \dots & \underline{1} & * & * & \dots & * \\ 0 & \dots & 0 & 0 & 0 & \dots & 0 & \dots & 0 & 0 & 0 & \dots & 0 \\ \dots & \dots & \dots & \dots & \dots & \dots & \dots & \dots & \dots & \dots & \dots & \dots & \dots \\ 0 & \dots & 0 & 0 & 0 & \dots & 0 & 0 & 0 & 0 & 0 & \dots & 0 \end{pmatrix}. \qquad (2.7.5)$$

Again, in each column containing a 1 sitting on a step, all other entries are 0's.

The matrix in the right member of (2.7.5) is called the *row-reduced* form of A and is denoted A_R.

Exercise 2.7.10. For each numerical matrix A found in the preceding **Exercises** and **Examples** find the row-reduced form A_R.

Matrix algebra permits a succinct statement to replace the long-winded statement about the circumstances in which $A\mathbf{x} = \mathbf{b}$ does or not have a solution.

Indeed, let $\mathcal{E}$ be the product of the elementary matrices used in the GEM. If A is an $m \times n$ matrix, then $\mathcal{E}$ is an $m \times m$ SQUARE matrix and $\mathbf{b}$ is an $m \times 1$ matrix (*column vector*). Assume that the last k rows of A_E consist entirely of 0's. Then $A\mathbf{x} = \mathbf{b}$ has a solution iff the last k rows (entries) of $\mathcal{E}\mathbf{b}$ are also 0's.

In particular, if $m = n$, i.e., if A is SQUARE, then $A\mathbf{x} = \mathbf{b}$ has a *unique* solution for *every* $\mathbf{b}$ iff *no* row of A_E consists entirely of 0's, i.e., iff every *diagonal* entry of A_E is nonzero.

The last criterion can be rephrased in terms of the zero vector

$$\mathbf{O} \overset{\text{def}}{=} \begin{pmatrix} 0 \\ \vdots \\ 0 \end{pmatrix}.$$

EITHER
$A\mathbf{x} = \mathbf{O}$ has a solution that is *not* the zero vector $\mathbf{O}$
OR
$A\mathbf{x} = \mathbf{b}$ has a *unique* solution for every $\mathbf{b}$.

Example 2.7.7. The echelon form of

$$A \overset{\text{def}}{=} \begin{pmatrix} 1 & 2 & 3 \\ 4 & 5 & 6 \\ 7 & 8 & 9 \end{pmatrix}$$

---

is

$$A_E \overset{\text{def}}{=} \begin{pmatrix} 1 & 2 & 3 \\ 0 & -3 & -6 \\ 0 & 0 & 0 \end{pmatrix}.$$

Furthermore the GEM matrix $\mathcal{E}$ is the lower triangular matrix

$$E_{13}(-7)E_{12}(-4) = \begin{pmatrix} 1 & 0 & 0 \\ -4 & 1 & 0 \\ -7 & 0 & 1 \end{pmatrix}.$$

Let b be

$$\begin{pmatrix} b_1 \\ b_2 \\ b_3 \end{pmatrix}.$$

Then $\mathcal{E}$b has 0 as its last entry iff $-7b_1 + b_3 = 0$. For this **Example** Ax = b has a solution iff $-7b_1 + b_3 = 0$.

Exercise 2.7.11. To this point there are among the **Exercises** and **Examples** several SQUARE matrices with numerical entries. For each such matrix A determine the vectors b such that Ax = b has a solution.

Exercise 2.7.12. Describe the circumstances in which the product of two elementary matrices is again an elementary matrix.

Exercise 2.7.13. Prove or disprove that if E is an elementary matrix then E^t is also an elementary matrix.

Exercise 2.7.14. Assume

$$E^{(1)}, \ldots, E^{(K)} \text{ and } F^{(1)}, \ldots, F^{(L)}$$

are 2×2 *elementary permutation matrices*, none is I, and that

$$\prod_{k=1}^{K} E^{(k)} = \prod_{l=1}^{L} F^{(l)}.$$

Show that $K - L$ is an even integer: $K - L \in \{0, \pm 2, \pm 4, \ldots, \pm 2p, \ldots\}$.

2.8. Inverses of SQUARE matrices

SQUARE matrices, e.g., elementary matrices and the matrices of SQUARE systems, play a special rôle in linear algebra. Some SQUARE matrices have a special character — they are *invertible* — as described next.

DEFINITION **2.8.1.** A SQUARE MATRIX A IS CALLED INVERTIBLE IFF THERE IS A MATRIX B, CALLED AN INVERSE OF A, SUCH THAT $AB = BA = I$.

If A, B, and C are $n \times n$ matrices such that both B and C are inverses of A then

$$B(AC) = BI = B$$
$$B(AC) = (BA)C = IC = C,$$

i.e.,

$$B = C.$$

[**Remark 2.8.1:** In other words, if *an* inverse of A exists then only one inverse exists. The unique inverse of A is called "A-inverse" and is denoted A^{-1}.]

Example 2.8.1. Here is a list of the inverses of all invertible elementary matrices:

$$I^{-1} = I;$$
$$E_{ij}^{-1} = E_{ij};$$
$$E_{ij}^{-1}(r) = E_{ij}(-r), \ r \in \mathbf{C};$$
$$E_i^{-1}(s) = E_i(\frac{1}{s}), \ s \neq 0;$$

The only elementary matrix having no inverse is $E_i(0)$.

If A is invertible so is A^{-1} and

$$(A^{-1})^{-1} = A.$$

Indeed, $A^{-1}A = AA^{-1} = I$.

Not every matrix has an inverse. For example, O has no inverse since for any compatible matrix C, $OC = O$. Any triangular matrix in which at least one diagonal entry is 0 is not invertible.

If A and B are invertible $n \times n$ matrices then

$$(B^{-1}A^{-1})AB = B^{-1}(A^{-1}A)B = B^{-1}IB = I$$
$$AB(B^{-1}A^{-1}) = A(BB^{-1})A^{-1} = AIA^{-1} = I.$$

[**Remark 2.8.2:**   In other words, if A and B are invertible $n \times n$ matrices then AB is invertible and $(AB)^{-1} = B^{-1}A^{-1}$.

The product of invertible matrices is invertible and the inverse of the product is the product of the inverses in opposite order.]

Hence the matrices $\mathcal{P}$, $\mathcal{E}$, $\mathcal{J}$, and, $\mathcal{R}$ such that

$$\mathcal{P}A = A_P, \text{ the PROCESSed form of } A$$
$$\mathcal{E}A = A_E, \text{ the echelon form of } A$$
$$\mathcal{J}A = A_J, \text{ the Jordan elimination form of } A$$
$$\mathcal{R}A = A_R, \text{ the row-reduced form of } A$$

---

are invertible — *even if A itself is not invertible or even if A is not* SQUARE.

Exercise 2.8.1. Show that if A is a SQUARE matrix then A is invertible iff the transpose A^t is invertible. Show that if A is invertible then

$$(A^t)^{-1} = (A^{-1})^t.$$

Example 2.8.2. Each of the following matrices is *not* invertible:

$$A \overset{\text{def}}{=} \begin{pmatrix} 1 & 2 \\ 2 & 4 \end{pmatrix};$$

$$B \overset{\text{def}}{=} \begin{pmatrix} -4 & 1 & 2 & 3 \\ 0 & 0 & -6 & \pi \\ 0 & 0 & 6 & -3 \\ 0 & 0 & 5 & 2 \end{pmatrix};$$

$$C \overset{\text{def}}{=} \begin{pmatrix} a & b & c \\ d & e & f \\ xa + yd & xb + ye & xc + yf \end{pmatrix}.$$

Indeed, if

$$\begin{pmatrix} p & q \\ r & s \end{pmatrix} \begin{pmatrix} 1 & 2 \\ 2 & 4 \end{pmatrix} = \begin{pmatrix} 1 & 0 \\ 0 & 1 \end{pmatrix}$$

then comparison of the (1,1) and (1,2) entries in the left and right members above leads to the system of equations

$$p + 2q = 1$$
$$2p + 4q = 0.$$

The *PROCESS* applied to the system leads to the contradiction: 2=0.

If $M \overset{\text{def}}{=} (m_{ij})_{i,j=1}^{4,4}$ is such that $MB = I$ then

$$m_{11} \cdot (-4) = 1 \text{ and } m_{11} \cdot 1 = 0,$$

an impossibility.

Direct calculation shows that

$$F \overset{\text{def}}{=} E_{23}(-y)E_{13}(-x)C = \begin{pmatrix} a & b & c \\ d & e & f \\ 0 & 0 & 0 \end{pmatrix}.$$

Hence if $QC = CQ = I$ then

$$FQ = (E_{23}(-y)E_{13}(-x)C)Q = E_{23}(-y)E_{13}(-x)(CQ) = E_{23}(-y)E_{13}(-x).$$

 Chapter 2. THE PROCESS (Simple elimination)

As a product of elementary, hence invertible, matrices FQ is invertible. On the other hand, since the last row of F consists entirely of 0's, the last row of FQ consists entirely of 0's. Since FQ is invertible, for some G, $(FQ)G = I$, and so the last row of I consists entirely of 0's, a contradiction.

If A is a SQUARE invertible matrix then the system $A\mathbf{x} = \mathbf{b}$ has as its unique solution $A^{-1}\mathbf{b}$ since:

   i. $A(A^{-1}\mathbf{b}) = (AA^{-1})\mathbf{b} = I\mathbf{b} = \mathbf{b}$

     and

   ii. if $A\mathbf{x} = \mathbf{b}$ then

$$A^{-1}(A\mathbf{x}) = (A^{-1}A)\mathbf{x} = I\mathbf{x} = \mathbf{x}$$
$$= A^{-1}\mathbf{b}$$
$$\mathbf{x} = A^{-1}\mathbf{b}.$$

However, inverses are rarely if ever computed for the purpose of solving SQUARE systems. The utility of inverses shows up in the search for *useful forms of matrices* (cf. **Chapter 4**). The next lines indicate what a useful form for a matrix might be and how such a useful form can be exploited.

In **Example 1.1.2** there naturally arises the need to calculate powers of a matrix. For an $n \times n$ matrix A the simple formula for A^2 requires n^3 multiplications and $n^2(n-1)$ additions. To calculate A^2, A^3, ..., A^k or more generally, for numbers $a_0, \ldots a_k$, to calculate

$$a_0 I + a_1 A + \cdots + a_k A^k$$

requires about kn^3 arithmetical operations, a great deal of daunting work.

On the other hand, *if* there is an invertible $n \times n$ matrix P such that

$$P^{-1}AP \stackrel{\text{def}}{=} D = \begin{pmatrix} \lambda_1 & & & \\ & \lambda_2 & & \\ & & \ddots & \\ & & & \lambda_n \end{pmatrix}$$

is a diagonal matrix then

$$A = PDP^{-1}, \quad A^2 = PDP^{-1} \cdot PDP^{-1} = PD^2P^{-1}, \quad \ldots, \quad A^k = PD^kP^{-1}.$$

Since

$$D^k = \begin{pmatrix} \lambda_1^k & & & \\ & \lambda_2^k & & \\ & & \ddots & \\ & & & \lambda_n^k \end{pmatrix}$$

---

it follows that once P and P^{-1} are known the calculation of A^k requires only about $2n^2 + n(k - 1)$ arithmetic operations. If $n = 100$ the savings amount to about 99%.

Furthermore for the diagonal matrix D and any reasonable function f the formula

$$f(D) \overset{\text{def}}{=} \begin{pmatrix} f(\lambda_1) & & & \\ & f(\lambda_2) & & \\ & & \ddots & \\ & & & f(\lambda_n) \end{pmatrix}$$

suggests itself and then, by natural definition,

$$f(A) \overset{\text{def}}{=} P f(D) P^{-1}.$$

This *functional calculus* for matrices is an important tool in finding the solutions of systems of ordinary linear differential equations.

Thus the diagonal matrix D, when it exists, is *a* useful form of the matrix A (there are others). Difficult calculations involving A are enormously simplified when they are performed with D and the results, in terms of D, are easily translated into the desired results in terms of A.

The details about the circumstances in which such a P can be found and the means for determining such a P when it can be found are in **Chapter 4.** There are matrices, e.g.,

$$A \overset{\text{def}}{=} \begin{pmatrix} 0 & 1 \\ 0 & 0 \end{pmatrix},$$

such that for no invertible P is it true that $P^{-1}AP$ is diagonal. Indeed, if $P^{-1}AP \overset{\text{def}}{=} D$ is diagonal, then

$$A = PDP^{-1}, \ A^2 = O = P^{-1}D^2P, \ D^2 = O, \ D = O, \ A = O,$$

a contradiction.

There is a simple criterion for establishing whether a SQUARE matrix A has an inverse.

THEOREM 2.8.1. LET A BE AN $n \times n$ (SQUARE) MATRIX AND LET α_{nn} BE THE "SOUTHEAST" ENTRY OF THE ECHELON FORM A_E OF A:

$$A_E = \begin{pmatrix} \alpha_{11} & \cdots & \alpha_{1n} \\ & \ddots & \vdots \\ & & \alpha_{nn} \end{pmatrix}.$$

THEN A IS INVERTIBLE IFF $\alpha_{nn} \neq 0$ AND IF $\alpha_{nn} \neq 0$ THEN $A^{-1} = \mathcal{R}$.

[**Remark 2.8.3:**   A *number* z has a multiplicative inverse iff $z \neq 0$. The criterion above says that a matrix A has a multiplicative inverse iff a very particular and easily computed number α_{nn} associated with A is not 0. In **Chapter 3** the study of determinants leads to a different but related number — $det(A)$ — associated with a SQUARE matrix A. It is also true that A is invertible iff $det(A) \neq 0$. However, $det(A)$ is somewhat off the direct line of development of linear algebra whereas α_{nn} in A_E is one of the first significant items encountered as the subject is studied.]

PROOF. If $\alpha_{nn} \neq 0$ all diagonal entries α_{ii}, $1 \leq i \leq n$, are nonzero and back-substitution in the system $A_E \mathbf{x} = \mathcal{E}\mathbf{b}$ yields a solution of any system

$$A\mathbf{x} = \mathbf{b}. \tag{2.8.1}$$

Hence, in particular, (2.8.1) can be solved if

$$\begin{pmatrix} b_1 \\ \vdots \\ b_n \end{pmatrix} = \begin{pmatrix} 0 \\ \vdots \\ 1 \\ \vdots \\ 0 \end{pmatrix} \quad \leftarrow \text{ row } j$$

$$\overset{\text{def}}{=} \mathbf{e}_j, \ 1 \leq j \leq n.$$

If the corresponding solution is

$$\left(y_{1j}, \ldots, y_{nj} \right)^t$$

then the matrix $Y \overset{\text{def}}{=} (y_{ij})_{i,j=1}^{n,n}$ is such that

$$AY = I. \tag{2.8.2}$$

On the other hand, since all diagonal entries of A_E are nonzero it follows that the row-reduced form A_R of A is I, i.e.,

$$\mathcal{R}A = A_R = I. \tag{2.8.3}$$

From (2.8.2) and (2.8.3) it follows that

$$(\mathcal{R}A)Y = Y = \mathcal{R}(AY) = \mathcal{R} = A^{-1}.$$

Conversely, if A is invertible then $A_E = \mathcal{E}A$, as the product of invertible matrices, is also invertible. If $\alpha_{nn} = 0$ the bottom row of A_E is a row of 0's and so the bottom row of $A_E(A_E)^{-1}$ is also a row of 0's and cannot be I, a contradiction.

$$\Omega$$

The following crucial facts about inverses are directly derivable from THE-
OREM **2.8.1**.

Exercise 2.8.2. Let $A \stackrel{\text{def}}{=} (a_{ij})_{i,j=1}^{n,n}$ be an $n \times n$ (SQUARE) matrix and
let $A_E \stackrel{\text{def}}{=} (\alpha_{ij})_{i,j=1}^{n,n}$ be its echelon form. Then the following are equivalent:

i. $\alpha_{nn} \neq 0$.

ii. The matrix A is invertible.

iii. The matrix A^t is invertible.

iv. There is a matrix B such that $AB = I$.

v. There is a matrix C such that $CA = I$.

vi. FOR ANY VECTOR **b** THE EQUATION $A\mathbf{x} = \mathbf{b}$ HAS A UNIQUE SO-
LUTION **x**. (This statement is sometimes abbreviated as: UNISOL.)

vii. $A\mathbf{x} = \mathbf{O}$ iff **x**=**O**.

viii. Each diagonal entry of the echelon form A_E of A is nonzero.

ix. Each diagonal entry of the Jordan elimination version A_J of A is 1.

x. Each row of the *PROCESS*ed version A_P of A has at least one nonzero
entry.

xi. $\mathcal{R}A = I$.

xii. $A_R = I$.

[**Note 2.8.1:**   Item *viii* implies A has no inverse iff some $\alpha_{ii} = 0$.]

The following is a skeleton key to the complete solution of **Exercise 2.8.2**.
Start with *i*. Use the scheme

$$i \Rightarrow ii \Rightarrow iii \Rightarrow iv \Rightarrow v \Rightarrow vi \Rightarrow vii$$
$$\Rightarrow viii \Rightarrow ix \Rightarrow x \Rightarrow xi \Rightarrow xii \Rightarrow i$$

to prove everything. In the outline below the *PROCESS* and its variants, the
GEM and the GJM are central in the argument. It is worth noting that the
associative law for matrix multiplication $(A(BC) = (AB)C)$ is used in some of
the demonstrations.

$i \Rightarrow ii$: THEOREM **2.8.1**.

$ii \Rightarrow iii$: $(AB)^t = B^t A^t$.

$iii \Rightarrow iv$: $(AB)^t = B^t A^t$.

$iv \Rightarrow v$: A^t is invertible.

$v \Rightarrow vi$: $C\mathbf{b}$ is a solution.

$vi \Rightarrow vii$: **O** is a solution.

$vii \Rightarrow viii$: If any diagonal entry of A_E is 0 then so is α_{nn}.

$viii \Rightarrow ix$: The diagonal entries of A_J and A_E are proportional.

$ix \Rightarrow x$: There is a permutation matrix Π such that $\Pi A_P = A_E$.

$x \Rightarrow xi$: Each diagonal entry of A_E is not 0.

$xi \Rightarrow xii$: $A_R = I$.

$xii \Rightarrow i$: If $\alpha_{nn} = 0$ each entry in the bottom row of A_E is 0.

[**Remark 2.8.4:** A SQUARE matrix A is invertible iff there is a matrix B such that $AB = I$ and then there is only one such B, and $BA = I$ as well.]

Statement vi deserves particular attention. If $c_1, \ldots, c_n$ are the column vectors that make up the $n \times n$ matrix A and if

$$\mathbf{x} \stackrel{\text{def}}{=} \begin{pmatrix} x_1 \\ \vdots \\ x_n \end{pmatrix}$$

then the column vector $A\mathbf{x}$ is, according to the rules for multiplying matrices,

$$x_1 c_1 + \cdots + x_n c_n,$$

a *linear combination* (cf. below) of the the column vectors c_i. Thus UNISOL (vi) says that A is invertible iff any column vector

$$\mathbf{b} \stackrel{\text{def}}{=} \begin{pmatrix} b_1 \\ \vdots \\ b_n \end{pmatrix}$$

may be written as a linear combination of the column vectors c_i:

$$\mathbf{b} = \sum_{i=1}^{n} x_i c_i.$$

Furthermore the linear combination is unique since if also

$$\mathbf{b} = \sum_{i=1}^{n} y_i c_i$$

then

$$\sum_{i=1}^{n} (x_i - y_i) c_i = \mathbf{O}$$

or

$$\begin{pmatrix} a_{11} & \cdots & a_{1n} \\ \vdots & \ddots & \vdots \\ a_{n1} & \cdots & a_{nn} \end{pmatrix} \begin{pmatrix} x_1 - y_1 \\ \vdots \\ x_n - y_n \end{pmatrix} = \begin{pmatrix} 0 \\ \vdots \\ 0 \end{pmatrix}$$

whence $x_i = y_i$, $1 \leq i \leq n$. In other words, not only may any vector $\mathbf{b}$ be written as a linear combination of the column vectors making up the columns of A but the linear combination is *unique*. For this reason the set $\{c_1, \ldots, c_n\}$ is called a *basis* (cf. **Section 2.9**) for $\mathbf{C}^n$.

Conversely, if the k $n \times 1$ column vectors $c_1, \ldots, c_k$ constitute a basis for $\mathbf{C}^n$ then $k = n$ and the matrix A in which the jth column is c_j is invertible.

Indeed, if $k < n$, then A is HIGH and the bottom row of A_E consists of 0's. Let e_n be the $n \times 1$ vector

$$\begin{pmatrix} 0 \\ 0 \\ \vdots \\ 0 \\ 1 \end{pmatrix}$$

and if $\mathcal{E}A = A_E$ let b be the vector $\mathcal{E}^{-1}e_n$. If

$$\mathbf{x} \overset{\text{def}}{=} \begin{pmatrix} x_1 \\ \vdots \\ x_k \end{pmatrix} \quad \text{and} \quad x_1 c_1 + \cdots + x_k c_k = \mathbf{b}$$

then

$$\begin{pmatrix} a_{11} & \cdots & a_{1k} \\ \vdots & \ddots & \vdots \\ a_{n1} & \cdots & a_{nk} \end{pmatrix} \begin{pmatrix} x_1 \\ \vdots \\ x_k \end{pmatrix} = \begin{pmatrix} b_1 \\ \vdots \\ b_n \end{pmatrix},$$

i.e.,

$$A\mathbf{x} = \mathbf{b}$$

$$\mathcal{E}A\mathbf{x} = A_E\mathbf{x} = \mathcal{E}\mathbf{b} = \mathcal{E}\mathcal{E}^{-1}e_n = e_n.$$

The nth entry of $A_E\mathbf{x}$ must be 0 because the bottom row of A_E consists of 0's. But the nth entry of e_n is 1, and the contradiction implies $k \geq n$.

If $k > n$ the matrix A is WIDE and so there are infinitely many solutions of

$$A\mathbf{x} = \mathbf{O}.$$

If

$$\mathbf{x} \overset{\text{def}}{=} \begin{pmatrix} x_1 \\ \vdots \\ x_k \end{pmatrix}$$

is a nonzero solution then

$$\sum_{i=1}^{k} x_i c_i = \mathbf{O}.$$

Since $0c_1 + \cdots + 0c_k = \mathbf{O}$ there are two linear combinations of the c_i that yield $\mathbf{O}$, a contradiction of the assumption that the c_i form a basis for $\mathbf{C}^n$.

Hence $k = n$, A is SQUARE and UNISOL is true for A whence A is invertible.

IF A IS AN INVERTIBLE $n \times n$ (SQUARE) MATRIX ITS COLUMNS CONSTITUTE A BASIS FOR $\mathbf{C}^n$.

IF k $n \times 1$ VECTORS $\mathbf{c}_i$ CONSTITUTE A BASIS FOR $\mathbf{C}^n$ THEN $k = n$ AND THE MATRIX A IN WHICH THE COLUMNS ARE THE $\mathbf{c}_i$ IS AN INVERTIBLE SQUARE MATRIX.

SQUARE matrices are divided into two classes, the invertible SQUARE matrices and all others. The "others" are unusual and so they are called *singular*. An invertible SQUARE matrix is thus called *nonsingular* (cf. **Remark 2.8.5** following **Example 2.8.3** below).

Exercise 2.8.3. The following statements are consequences, corollaries, or restatements of THEOREM 2.8.1 above. Prove each one.

i. If k is a natural number, i.e., if $k \in \mathbf{N} \overset{\text{def}}{=} \{1, 2, \ldots, n, \ldots\}$, and if A is a SQUARE matrix then A^k is invertible iff A is invertible and then

$$(A^k)^{-1} = (A^{-1})^k \overset{\text{def}}{=} A^{-k}.$$

ii. If A and B are $n \times n$ matrices then AB is invertible iff both A and B are invertible and then
$$(AB)^{-1} = B^{-1}A^{-1}.$$

(In particular, if A is invertible and $t \neq 0$ then tA is also invertible and

$$(tA)^{-1} = A^{-1}t^{-1} = \frac{1}{t}A^{-1}.)$$

iii. If A_i, $1 \leq i \leq k$ are $n \times n$ matrices then

$$\prod_{i=1}^{n} A_i$$

is invertible iff each A_i is invertible and then

$$\left(\prod_{i=1}^{n} A_i\right)^{-1} = \prod_{i=1}^{n} A_{n-i+1}^{-1}.$$

[Sometimes it is helpful to distinguish between vectors written as columns,

$$\begin{pmatrix} x_1 \\ \vdots \\ x_n \end{pmatrix},$$

i.e., column vectors, and vectors written as rows: $(y_1, \ldots, y_n)$, i.e., row vectors. When the context makes the distinction obvious, no special notation is used. Occasionally, for clarity, $\mathbf{O}^c$ denotes

$$\begin{pmatrix} 0 \\ \vdots \\ 0 \end{pmatrix},$$

the column vector in which all entries (*components*) are 0's, and $\mathbf{O}_r$ denotes $(0,\ldots,0)$, the *row* vector in which all components are 0's.]

iv. The $n \times n$ matrix A is singular iff there is some nonzero column vector

$$\mathbf{x} \stackrel{\text{def}}{=} \begin{pmatrix} x_1 \\ \vdots \\ x_n \end{pmatrix}$$

such that

$$A\mathbf{x} = \mathbf{O}^c \stackrel{\text{def}}{=} \begin{pmatrix} 0 \\ \vdots \\ 0 \end{pmatrix}.$$

v. The $n \times n$ matrix A is singular iff there is some nonzero row vector

$$\mathbf{y} \stackrel{\text{def}}{=} (y_1, \ldots, y_n)$$

such that

$$\mathbf{y}A = \mathbf{O}_r \stackrel{\text{def}}{=} (0, \ldots, 0).$$

vi. Let A be an $n \times n$ matrix and let its columns be denoted $\mathbf{c}_1, \ldots, \mathbf{c}_n$. Then A is singular iff there is a nonzero column vector

$$\begin{pmatrix} t_1 \\ \vdots \\ t_n \end{pmatrix}$$

such that

$$t_1 \mathbf{c}_1 + \cdots + t_n \mathbf{c}_n = \sum_{j=1}^{n} t_j \mathbf{c}_j = \mathbf{O}^c. \tag{$*$}$$

vii. Let A be an $n \times n$ matrix and let its rows be denoted $\mathbf{r}_1, \ldots, \mathbf{r}_n$. Then A is singular iff there is a nonzero row vector $(s_1, \ldots, s_n)$ such that

$$s_1 \mathbf{r}_1 + \cdots + s_n \mathbf{r}_n = \sum_{i=1}^{n} s_i \mathbf{r}_i = \mathbf{O}_r. \tag{$**$}$$

viii. Every WIDE $m \times n$ matrix W is a *submatrix* of a singular SQUARE $n \times n$ matrix A in which the first m rows are the rows of W and the last $n - m$ rows consist entirely of 0's.

ix. Every HIGH $m \times n$ matrix H is submatrix of a singular SQUARE $m \times m$ matrix H in which the first n columns are the columns of H and the last $m - n$ columns consist entirely of 0's.

---

x. If k is a natural number and W is an $n \times (n + k)$ (WIDE) matrix with columns denoted $c_1, \ldots, c_{n+k}$, there is a nonzero column vector

$$\begin{pmatrix} t_1 \\ \vdots \\ t_{n+k} \end{pmatrix}$$

such that

$$\sum_{j=1}^{n+k} t_j c_j = O^c.$$

xi. If k is a natural number and H is an $(n + k) \times n$ (HIGH) matrix with rows denoted $r_1, \ldots, r_{n+k}$, there is a nonzero row vector $(s_1, \ldots, s_{n+k})$ such that

$$\sum_{i=1}^{n+k} s_i r_i = O_r.$$

LINEAR COMBINATIONS AND LINEAR INDEPENDENCE

The leftmost members

$$t_1 c_1 + \cdots + t_n c_n$$
$$s_1 r_1 + \cdots + s_n r_n$$

in $(*)$ and $(**)$ are called *linear combinations* with coefficients

$$t_1, \ldots, t_n \text{ resp. } s_1, \ldots, s_n$$

of the column vectors $c_1, \ldots, c_n$ resp. the row vectors $r_1, \ldots, r_n$.

Since *at least one* t_k resp. *at least one* s_l is *not* 0 and since the linear combinations *are* O^c resp. O_r, the sets

$$C \overset{\text{def}}{=} \{c_1, \ldots, c_n\}, \quad R \overset{\text{def}}{=} \{r_1, \ldots, r_n\}$$

are called *linearly dependent*. Occasionally, in these circumstances, one says that the vectors themselves are linearly dependent.

A set $X \overset{\text{def}}{=} \{x_1, \ldots, x_m\}$ of column vectors is called *linearly independent* iff X is *not* linearly dependent, in other words, iff there is *no* nonzero column vector

$$\begin{pmatrix} \alpha_1 \\ \vdots \\ \alpha_m \end{pmatrix}$$

such that

$$\alpha_1 \mathbf{x}_1 + \cdots + a_m \mathbf{x}_m = \mathbf{O}^c.$$

Similarly, a set $Y \stackrel{\text{def}}{=} \{\mathbf{y}_1, \ldots, \mathbf{y}_n\}$ of row vectors is called linearly independent iff Y is *not* linearly dependent, in other words, iff there is *no* nonzero row vector $(\beta_1, \ldots, \beta_n)$ such that

$$\beta_1 \mathbf{y}_1 + \cdots + \beta_n \mathbf{y}_n = \mathbf{O}_r.$$

From the definitions of linear independence and linear dependence it follows that if S is a set of vectors, say in $\mathbf{C}^n$, then:

i. if $\mathbf{O} \in S$ then S is linearly dependent;
ii. if S is linearly independent and $T \subset S$ then T is linearly independent;
iii. if S is linearly dependent and $S \subset T$ then T is linearly dependent;
iv. if S is linearly independent, if $\{\mathbf{x}_1, \ldots, \mathbf{x}_k\} \subset S$, and if

$$\sum_{i=1}^{k} t_i \mathbf{x}_i = \mathbf{O}$$

then $t_1 = \cdots = t_k = 0$.

Furthermore:

Exercise 2.8.3x says that any set consisting of more than n $n \times 1$ (column) vectors is linearly dependent.
Exercise 2.8.3xi says that any set consisting of more than n $1 \times n$ (row) vectors is linearly dependent

Exercise 2.8.4. Show that $X \stackrel{\text{def}}{=} \{\mathbf{x}_1, \ldots, \mathbf{x}_m\}$ consisting of column vectors is linearly dependent iff the matrix $\mathcal{X}$, in which the columns are the column vectors making up X, is such that for some nonzero column vector $\mathbf{w}$, $\mathcal{X}\mathbf{w} = \mathbf{O}^c$.

Exercise 2.8.5. Show that $Y \stackrel{\text{def}}{=} \{\mathbf{y}_1, \ldots, \mathbf{y}_n\}$ is linearly dependent iff the matrix $\mathcal{Y}$, in which the rows are the row vectors making up Y, is such that for some nonzero row vector $\mathbf{v}$, $\mathbf{v}\mathcal{Y} = \mathbf{O}_r$.

Exercise 2.8.6. Show that a SQUARE matrix is nonsingular iff either its rows are linearly independent or its columns are linearly independent.

Exercise 2.8.7. Let W and H be defined as follows:

$$W \stackrel{\text{def}}{=} \begin{pmatrix} 1 & 2 & -3 \\ -5 & 7 & -8 \end{pmatrix}, \quad H \stackrel{\text{def}}{=} \begin{pmatrix} 2 & 4 \\ 3 & -1 \\ 7 & -9 \end{pmatrix}.$$

i. Make each of W resp. H a part of a singular SQUARE matrix $\widetilde{W}$ resp. $\widetilde{H}$.

---

ii. For W resp. H find

$$\mathbf{t} \overset{\text{def}}{=} \begin{pmatrix} t_1 \\ t_2 \\ t_3 \end{pmatrix} \quad \text{resp. } \mathbf{s} \overset{\text{def}}{=} (s_1, s_2, s_3)$$

so that, $\mathbf{c}_1, \mathbf{c}_2, \mathbf{c}_3$ denoting the columns of W and $\mathbf{r}_1, \mathbf{r}_2, \mathbf{r}_3$ denoting the rows of H,

$$t_1 \mathbf{c}_1 + t_2 \mathbf{c}_2 + t_2 \mathbf{c}_3 = \mathbf{O}^c,$$
$$s_1 \mathbf{r}_1 + \sigma_2 \mathbf{r}_2 + s_3 \mathbf{r}_3 = \mathbf{O}_r.$$

[*Hint:* The first requirement is equivalent to the requirement:

$$W\mathbf{t} = \mathbf{O}^c;$$

Similarly there is an equivalent version of the second requirement.]

Example 2.8.3. There are some general kinds of vectors that always constitute linearly independent sets:

i. the $1 \times n$ vectors

$$\mathbf{f}_i \overset{\text{def}}{=} (0, \quad \ldots, \quad \underset{\underset{\text{column } i}{\uparrow}}{1,} \quad \ldots, \quad 0) \quad , \ 1 \le i \le n,$$

are linearly independent;

ii. the $n \times 1$ vectors

$$\mathbf{e}_j \overset{\text{def}}{=} \begin{pmatrix} 0 \\ \vdots \\ 1 \\ \vdots \\ 0 \end{pmatrix} \quad \leftarrow \text{ row } j \ ,$$

$1 \le j \le n$ are linearly independent;

iii. if $\mathbf{x}_1, \mathbf{x}_2, \ldots$ are vectors such that jth component of $\mathbf{x}_j$ is its first nonzero component, then the vectors $\mathbf{x}_1, \mathbf{x}_2, \ldots$ are linearly independent. For example, the $*$'s denoting arbitrary numbers, the vectors

$$\mathbf{x}_1 \overset{\text{def}}{=} (1, *, *, *, *, *, *)$$
$$\mathbf{x}_2 \overset{\text{def}}{=} (0, \pi, *, *, *, *, *)$$
$$\mathbf{x}_3 \overset{\text{def}}{=} (0, 0, i, *, *, *, *)$$
$$\vdots$$

---

are linearly independent vectors.

iv. more generally if $1 \le j_1 < j_2 < \cdots < j_k \le n$ and if the $1 \times n$ vectors $\mathbf{x}_i$ are such that only $\mathbf{x}_i$ has a nonzero j_ith component then the vectors $\mathbf{x}_1, \mathbf{x}_2, \ldots$ are linearly independent.

For example, if $n = 9$, if $j_1 = 3, j_2 = 7$, and $j_3 = 9$ then, the $*$'s denoting arbitrary numbers, the vectors

$$\mathbf{x}_1 \overset{\text{def}}{=} (*, *, 1, *, *, *, 0, *, 0)$$
$$\mathbf{x}_2 \overset{\text{def}}{=} (*, *, 0, *, *, *, 1, *, 0)$$
$$\mathbf{x}_3 \overset{\text{def}}{=} (*, *, 0, *, *, *, 0, *, 1)$$

are linearly independent. In other words if an $m \times n$ matrix A is such that among its columns there are the m pairwise different $m \times 1$ column vectors

$$\mathbf{e}_i \overset{\text{def}}{=} \begin{pmatrix} 0 \\ \vdots \\ 1 \\ \vdots \\ 0 \end{pmatrix} \quad \leftarrow \text{ row } i\,, \ 1 \le i \le m,$$

then the rows of A regarded as row vectors are linearly independent.

[**Remark 2.8.5:** Are invertible matrices rare? As the next lines show they are in no reasonable sense rare. Indeed they seem to be everywhere, and in the precise notions of measure theory they are "almost everywhere." Hence SQUARE matrices having no inverses are the rarities, the unusual ones, the *singular* ones.

If

$$A \overset{\text{def}}{=} \begin{pmatrix} a_{11} & \cdots & a_{1n} \\ \vdots & \ddots & \vdots \\ a_{n1} & \cdots & a_{nn} \end{pmatrix}$$

is invertible there is some positive δ such that any matrix

$$C \overset{\text{def}}{=} \begin{pmatrix} c_{11} & \cdots & c_{1n} \\ \vdots & \ddots & \vdots \\ c_{n1} & \cdots & c_{nn} \end{pmatrix}$$

for which

$$|a_{ij} - c_{ij}| < \delta, \ 1 \le i, j \le n \tag{2.8.4}$$

is also invertible.

A sketch of the proof of the last statement above follows.

Assume $0 < \delta < \frac{1}{n^2}$ and assume that $B \overset{\text{def}}{=} (b_{ij})_{i,j=1}^{n,n}$ is such that

$$|b_{ij}| < \delta, \ 1 \le i, j \le n.$$

Then each entry $b_{ij}^{(k)}$ in B^k is such that

$$b_{ij}^{(k)} < n^{k-1}\delta^k < n^{-k-1}$$

whence

$$S_K \overset{\text{def}}{=} \sum_{k=1}^{K} B^k, \ 1 \le K < \infty$$

are matrices in which the entries are the partial sums of convergent series. Thus

$$S \overset{\text{def}}{=} \sum_{k=1}^{\infty} B^k$$

is meaningful. The direct calculation

$$(1 - x)(1 + x + x^2 + \cdots) = 1, \ \text{if } |x| < 1$$

used to derive the formula for the sum of an infinite geometric progression shows that

$$(I - B)(I + S) = (I - B)(I + B + B^2 + \cdots) = I.$$

Hence if B is near O then $I - B$ is invertible.

If A is invertible and B is near enough to O then $A^{-1}B$ is also near O and thus

$$(A - B) = A(I - A^{-1}B)$$

is invertible. Hence if the entries of $A - C \overset{\text{def}}{=} B$ are sufficiently small then $C \ (= A - B)$ is invertible.

It is also possible to show that arbitrarily near any matrix there is an invertible matrix. The proof, depending as it does upon results in the theory of functions of several complex variables and in measure theory, is beyond the scope of this book, cf. [**GeL, Co**].

From the standpoint of measure theory the set of singular matrices is a set of measure zero, a "null" set — the nonsingular matrices are "almost everywhere." Interpreted in probabilistic terms, the last sentence says that there is no "chance" that a "randomly" chosen SQUARE matrix is singular. Yet in **Chapter 4** the search for singular matrices is the central topic!

<hr>

The statements above need serious modification if the context is not that of the system **C** of complex numbers or the system **R** of real numbers. For example if the context is floating point arithmetic, i.e., the arithmetic of a *finite* set of numbers as found in a computer, then there are no arbitrarily small positive numbers and the δ in the discussion above may not exist. Furthermore since there are only finitely many floating point numbers in any computer operation, the only "null" set is the empty set and there is indeed a positive probability (some "chance") that a "randomly" chosen SQUARE matrix is singular.

Indeed, if all numbers are restricted to consist of 1 binary digit (0 or ± 1) then there are only 3^4 ($= 81$) 2×2 matrices that can be formed with such numbers. Of these 81 matrices 48 are nonsingular, 33 are singular. The "chance" of finding a singular matrix among them is substantial:

$$\frac{33}{81} = .407 + .$$

There are large classes of matrices for which diagonal forms can be found. In the same senses of "nearness" and "chance" employed above in the discussion of singular and nonsingular matrices the following obtain:

    i. arbitrarily "near" any matrix is a matrix for which a diagonal form can be found;

    ii. if A is a matrix for which a diagonal form can be found, then all matrices sufficiently "near" A enjoy the same property;

    iii. there is "no chance" of finding a matrix for which there is no diagonal form, although such matrices exist, e.g.,

$$M \stackrel{\text{def}}{=} \begin{pmatrix} 1 & 1 \\ 0 & 1 \end{pmatrix};$$

    iv. in the floating point arithmetic used in computers all the statements just made are false unless they are modified.

From the standpoint of practice and practicality the theoretical conclusions above are not of great significance.]

Exercise 2.8.8. Show that there is no invertible 2×2 matrix P such that

$$P^{-1} \begin{pmatrix} 1 & 1 \\ 0 & 1 \end{pmatrix} P \stackrel{\text{def}}{=} P^{-1} M P$$

is a diagonal matrix D.

    [*Hint:* $P^{-1} M P = D$ iff $M P = P D$.]

To find whether an $n \times n$ SQUARE matrix $A \overset{\text{def}}{=} (a_{ij})_{i,j=1}^{n,n}$ has an inverse, i.e., whether A is invertible, the following procedure may be used:

i. Construct the $n \times 2n$ matrix K in which the first n columns are those of A and the last n columns are those of I:

$$K = \begin{pmatrix} a_{11} & \cdots & a_{1n} & 1 & \cdots & 0 \\ \vdots & \ddots & \vdots & \vdots & \ddots & \vdots \\ a_{n1} & \cdots & a_{nn} & 0 & \cdots & 1 \end{pmatrix}.$$

ii. Use the GJM to convert K to *row-reduced* form K_R.

iii. If the first n columns of K_R constitute the matrix I then A is invertible and A^{-1} is the matrix consisting of the last n columns of K_R.

iv. If the first n columns of K_R do not constitute the matrix I then A is not invertible.

The method is validated by the fact that, $\mathcal{R}$ denoting the product of the elementary matrices used to bring A to row-reduced form, the last n columns of K_R constitute $\mathcal{R}$. If the first n columns of K_R constitute the columns of I then $\mathcal{R}A = I$, i.e., A is invertible and $A^{-1} = \mathcal{R}$.

Example 2.8.4. Let A be the matrix

$$\begin{pmatrix} 1 & 2 \\ 3 & 4 \end{pmatrix}.$$

Then

$$K = \begin{pmatrix} 1 & 2 & 1 & 0 \\ 3 & 4 & 0 & 1 \end{pmatrix}$$

and, $\mathcal{R}$ denoting the product of elementary matrices that bring K to row-reduced form,

$$K_R = E_{21}(-2)E_2(-\tfrac{1}{2})E_{12}(-3)K = \mathcal{R}K$$
$$= \begin{pmatrix} 1 & 0 & -2 & 1 \\ 0 & 1 & \frac{3}{2} & -\frac{1}{2} \end{pmatrix}.$$

Hence A is invertible and

$$A^{-1} = \mathcal{R} = \begin{pmatrix} -2 & 1 \\ \frac{3}{2} & -\frac{1}{2} \end{pmatrix}.$$

Example 2.8.5. Let A be the matrix

$$\begin{pmatrix} 1 & 2 & 3 \\ 4 & 5 & 6 \\ 7 & 8 & 9 \end{pmatrix}.$$

Then

$$K = \begin{pmatrix} 1 & 2 & 3 & 1 & 0 & 0 \\ 4 & 5 & 6 & 0 & 1 & 0 \\ 7 & 8 & 9 & 0 & 0 & 1 \end{pmatrix},$$

and

$$K_J = E_{23}E_2\left(-\frac{1}{3}\right)E_{13}(-7)E_{12}(-4)K = JK$$

$$= \begin{pmatrix} 1 & 2 & 3 & 1 & 0 & 0 \\ 0 & 1 & 2 & \frac{4}{3} & -\frac{1}{3} & 0 \\ 0 & 0 & 0 & 1 & -2 & 1 \end{pmatrix}.$$

Since the first three entries in the bottom row of K_J are 0's the first three columns of K_R cannot possibly constitute the columns of I. Hence A is not invertible.

Exercise 2.8.9. Examine each of the following SQUARE matrices to determine whether each has an inverse and, if it is invertible, what the inverse is:

$$i. \begin{pmatrix} 1 & 2 & 3 \\ 6 & 5 & 4 \\ 7 & 8 & 9 \end{pmatrix}; \quad ii. \begin{pmatrix} 1 & 1 & 1 \\ 2 & 3 & 4 \\ 4 & 9 & 16 \end{pmatrix}; \quad iii. \begin{pmatrix} 1 & -1 & 1 \\ 3 & 4 & 5 \\ 5 & 2 & 7 \end{pmatrix}.$$

Exercise 2.8.10. Assume $1 \le m < n$ and that

$$\mathbf{a}_i \overset{\text{def}}{=} (a_{i1}, \ldots, a_{in}), \quad 1 \le i \le m,$$

are linearly independent $1 \times n$ (row) vectors. Let A be the $m \times n$ matrix

$$\begin{pmatrix} a_{11} & \cdots & a_{1n} \\ \vdots & \ddots & \vdots \\ a_{m1} & \cdots & a_{mn} \end{pmatrix}.$$

$i.$ In row k of A_E, the echelon form of A, the first nonzero entry appears in some column, say column j_k, $1 \le k \le m$. Since $m < n$ (A is WIDE) there remain in A $n - m$ columns other than those just defined. Let the other columns be numbered $r_1, \ldots, r_{n-m}$ and let the $1 \times n$ (row) vector

$$\mathbf{b}_l, \quad 1 \le l \le n - m \tag{2.8.5}$$

consist of a 1 in column r_l and 0's elsewhere. Form the $n \times n$ matrix B in which the first m rows are those of A_E and the last $n - m$ rows are those in (2.8.5). Show that B is nonsingular (invertible). (For example, if

$$A_E = \begin{pmatrix} 6 & * & * & * & * & * & * \\ 0 & 0 & 0 & 2 & * & * & * \\ 0 & 0 & 0 & 0 & 0 & 1 & * \end{pmatrix}$$

then

$$m = 3 < 7 = n,$$
$$n - m = 4,$$
$$j_1 = 1, j_2 = 4, j_3 = 6,$$
$$r_1 = 2, r_2 = 3, r_3 = 5, r_4 = 7,$$
$$\mathbf{b}_1 = (0, 1, 0, 0, 0, 0, 0),$$
$$\mathbf{b}_2 = (0, 0, 1, 0, 0, 0, 0),$$
$$\mathbf{b}_3 = (0, 0, 0, 0, 1, 0, 0),$$
$$\mathbf{b}_4 = (0, 0, 0, 0, 0, 0, 1),$$

$$B = \begin{pmatrix} 6 & * & * & * & * & * & * \\ 0 & 0 & 0 & 2 & * & * & * \\ 0 & 0 & 0 & 0 & 0 & 1 & * \\ 0 & 1 & 0 & 0 & 0 & 0 & 0 \\ 0 & 0 & 1 & 0 & 0 & 0 & 0 \\ 0 & 0 & 0 & 0 & 1 & 0 & 0 \\ 0 & 0 & 0 & 0 & 0 & 0 & 1 \end{pmatrix} .)$$

ii. Let $\mathcal{E}$ be an invertible $m \times m$ matrix such that

$$\mathcal{E}A = A_E.$$

Show $\mathcal{E}$ is unique, i.e., that if $\mathcal{E}'$ is invertible and

$$\mathcal{E}'A = A_E$$

then

$$\mathcal{E} = \mathcal{E}'.$$

[*Hint:* The rows of A are linearly independent.]

iii. Let $\mathcal{F}$ be the $n \times n$ matrix in which:
   a. the $m \times m$ submatrix (block) in the "northwest" corner is the *inverse* of $\mathcal{E}$;
   b. the $m \times (n - m)$ submatrix (block) in the "northeast" corner is O;
   c. the $(n - m) \times (n - m)$ submatrix (block) in the "southeast" corner is the $(n - m) \times (n - m)$ identity matrix I_{n-m};
   d. the $(n - m) \times m$ submatrix (block) in the "southwest" corner is O.
Thus $\mathcal{F}$ in block matrix form looks like this:

$$\begin{array}{cc} & \begin{array}{cc} m & n - m \end{array} \\ \begin{array}{c} m \\ n - m \end{array} & \begin{pmatrix} \mathcal{E}^{-1} & O \\ O & I \end{pmatrix} \end{array}.$$

Show that $\mathcal{F}$ is invertible and that, for B as in *i*,

$$\mathcal{F}B \stackrel{\text{def}}{=} C$$

is an (invertible) matrix in which the first m rows are those of A.

The result above shows how to *fill out* a set of m linearly independent $1 \times n$ (row) vectors (the first m rows of A) with $n - m$ other $1 \times n$ (row) vectors (the last $n - m$ rows of C) so that the resulting set S of n $1 \times n$ (row) vectors, say $\{\mathbf{x}_1, \ldots, \mathbf{x}_n\}$, is linearly independent.

Since no linearly independent set of $1 \times n$ vectors can contain more than n members, the set S constructed is a *maximally linearly independent* set. That is to say, S is a linearly independent set and if any $1 \times n$ vector $\mathbf{x}$ not in S is adjoined to S then the resulting set $\widetilde{S} \overset{\text{def}}{=} S \cup \{\mathbf{x}\}$ is linearly *dependent*. Thus, any $1 \times n$ vector $\mathbf{x}$ (in S or not in S) is a linear combination of members of S:

$$\mathbf{x} = \sum_{i=1}^{n} t_i \mathbf{x}_i. \tag{2.8.6}$$

If $\mathbf{x} \in S$, say $\mathbf{x} = \mathbf{x}_j$, then in (2.8.6) all t_i except t_j are 0's and $t_j = 1$. If $\mathbf{x} \notin S$ then $S \cup \{\mathbf{x}\}$ is linearly dependent, and so for some coefficients $a, a_1, \ldots, a_n$, not all 0's,

$$a\mathbf{x} + \sum_{i=1}^{n} a_i \mathbf{x}_i = \mathbf{O}. \tag{2.8.7}$$

If $a = 0$ then, since *some* coefficient in (2.8.7) is *not* 0, S is linearly dependent, a contradiction. Hence $a \neq 0$ and (2.8.6) is valid if

$$t_i = -\frac{a_i}{a}.$$

Furthermore the coefficients t_i in (2.8.6) are unique since if

$$\mathbf{x} = \sum_{i=1}^{n} s_i \mathbf{x}_i$$

then

$$\sum_{i=1}^{n} (t_i - s_i)\mathbf{x}_i = \mathbf{x} - \mathbf{x} = \mathbf{O}. \tag{2.8.8}$$

Because S is linearly independent the coefficients in the left member of (2.8.8) must be 0's.

Exercise 2.8.11. Fill out $M \overset{\text{def}}{=} \{(1, 2, 3, 4, 5), (5, 4, 3, 2, 1)\}$ to a maximally linearly independent set S containing M and consisting of 5 vectors.

Exercise 2.8.12. Fill out

$$N \overset{\text{def}}{=} \{(1, 2, 3, 4, 5, 6), (5, 10, 2, 20, 1, 4), (3, 6, 9, 12, 14, 13)\}$$

     Chapter 2. THE PROCESS (**Simple elimination**)

to a maximally linearly independent set T containing N and consisting of 6 vectors.

No set of linearly independent $n \times 1$ vectors can consist of more than n vectors. Any linearly independent set of $n \times 1$ vectors can be filled out to a maximally linearly independent set of $n \times 1$ vectors. It follows that:

Every maximally linearly independent set of $n \times 1$ vectors contains precisely n vectors. Similar remarks apply to sets of linearly independent $1 \times n$ vectors.

A maximally linearly independent set of $n \times 1$ vectors is a *basis* (for the set $\mathbf{C}^n$ of all $n \times 1$ vectors). Similarly a maximally linearly independent set of $1 \times n$ vectors is a *basis* (for the set $\mathbf{C}_n$ of all $1 \times n$ vectors).

Thus the discussion above may be rephrased and summarized as follows:

i. A set S of vectors in $\mathbf{C}_n$ resp. $\mathbf{C}^n$ is:
  a. *linearly independent* iff the zero vector $\mathbf{O}$ of $\mathbf{C}_n$ resp. $\mathbf{C}^n$ can be expressed *uniquely* as a linear combination of vectors in S;
  b. a *basis* iff *every* vector in $\mathbf{C}_n$ resp. $\mathbf{C}^n$ can be expressed uniquely as a linear combination of vectors in S.
ii. Every basis for $\mathbf{C}_n$ consists of precisely n vectors. Every basis for $\mathbf{C}^n$ consists of precisely n vectors.
iii. A SQUARE $n \times n$ matrix is invertible iff its rows constitute a basis for $\mathbf{C}_n$ iff its columns constitute a basis for $\mathbf{C}^n$.
iv. If $m < n$ and if A is an $m \times n$ (WIDE) matrix in which the rows are linearly independent then A can be enlarged to an *invertible* $n \times n$ matrix.
v. If $m > n$ and if A is an $m \times n$ (HIGH) matrix in which the columns are linearly independent then A can be enlarged to an *invertible* $m \times m$ matrix.

Exercise 2.8.13.

i. Fill out $S \stackrel{\text{def}}{=} \{(1, -2, 3, -4, 5)\}$ to a basis for $\mathbf{C}_5$.
ii. Fill out

$$S \stackrel{\text{def}}{=} \left\{ \begin{pmatrix} 1 \\ 0 \\ -1, \\ 2 \end{pmatrix}, \begin{pmatrix} 2 \\ 4 \\ 6 \\ -3 \end{pmatrix} \right\}$$

to a basis for $\mathbf{C}^4$.
iii. Let $S \stackrel{\text{def}}{=} \{(x_1, \ldots, x_n)\}$ be a set consisting of one vector in which at least one component is not 0. Fill S out to a basis for $\mathbf{C}_n$.

Example 2.8.6. Let A be a SQUARE $(n \times n)$ matrix:

$$\begin{pmatrix} a_{11} & \cdots & a_{1n} \\ \vdots & \ddots & \vdots \\ a_{n1} & \cdots & a_{nn} \end{pmatrix}.$$

Suppose that B is the matrix:

$$\begin{pmatrix} b_{11} & \cdots & b_{1n} \\ \vdots & \ddots & \vdots \\ b_{n1} & \cdots & b_{nn} \end{pmatrix}$$

and that $AB = I_n$. Then the n^2 numbers b_{rc} that are the entries in the matrix B satisfy n^2 linear equations. These equations are grouped into n systems, each consisting of n equations in n unknowns. They are most simply expressed in terms of the column vectors that constitute the columns of B. Thus let $b_1, \ldots, b_n$ be the column vectors that are the columns of B. Then e_i designating the column vector

$$\begin{pmatrix} 0 \\ \vdots \\ 1 \\ \vdots \\ 0 \end{pmatrix} \quad \leftarrow \text{ row } i \, ,$$

the systems of equations are

$$Ab_i = e_i, \ 1 \le i \le n.$$

In other words, if B exists it can be found by solving a system consisting of n^2 equations in n^2 unknowns. Moreover this system of equations can be solved by solving n systems, each consisting of n equations in n unknowns.

One may regard B as a solution of the single *matrix* equation $AX = I$ in which the unknown is the $n \times n$ *matrix* X.

Now it is clearly appropriate to consider *matrix* equations of the following general form: $AX = C$. Such an equation makes sense so long as A is an $m \times n$ matrix, X is an $n \times q$ matrix, and C is an $m \times q$ matrix.

Example 2.8.7. Let A be

$$\begin{pmatrix} 1 & 2 \\ -1 & 3 \\ 2 & -5 \end{pmatrix},$$

let X be

$$\begin{pmatrix} x_{11} & x_{12} & x_{13} \\ x_{21} & x_{22} & x_{23} \end{pmatrix},$$

and let C be the matrix

$$\begin{pmatrix} 9 & 0 & -2 \\ 1 & -7 & 4 \\ 0 & 2 & 5 \end{pmatrix}.$$

Then the equation $AX = C$ leads to three systems of equations. The first of these is

$$A \begin{pmatrix} x_{11} \\ x_{21} \end{pmatrix} = \begin{pmatrix} 9 \\ 1 \\ 0 \end{pmatrix}$$

or

$$x_{11} + 2x_{21} = 9$$
$$-x_{11} + 3x_{21} = 1$$
$$2x_{11} - 5x_{21} = 0.$$

The unique solution of this system is $x_{11} = 5$, $x_{21} = 2$. On the other hand since the echelon form A_E of A is

$$\begin{pmatrix} 1 & 2 \\ 0 & 1 \\ 0 & 0 \end{pmatrix},$$

the corresponding system

$$A \begin{pmatrix} x_{12} \\ x_{22} \end{pmatrix} = \begin{pmatrix} 0 \\ -7 \\ 2 \end{pmatrix}$$

for the second column of X uses the same matrix A and the same A_E. Hence

$$x_{12} + 2x_{22} = 0$$
$$x_{22} = -\tfrac{7}{5}$$
$$0x_{22} = -\tfrac{53}{5}$$

for which there is no solution. Thus the equation $AX = B$ has no solution.

In other instances $AX = B$ does have solutions, e.g., simply choose for X an arbitrary 2×3 matrix and then let B be AX.

Exercise 2.8.14. Let A be the 3×1 HIGH matrix

$$\begin{pmatrix} 1 \\ 2 \\ 3 \end{pmatrix}.$$

Show there is *no* 1×3 WIDE matrix $B \overset{\text{def}}{=} (x_1, x_2, x_3)$ such that

$$AB = \begin{pmatrix} 1 & 0 & 0 \\ 0 & 1 & 0 \\ 0 & 0 & 1 \end{pmatrix} \quad \text{(the } 3 \times 3 \text{ SQUARE identity matrix).}$$

On the other hand, show that if $A = (1, 2, 3)$ (a 1×3 WIDE matrix) there is a 3×1 HIGH matrix

$$\stackrel{\text{def}}{=} \begin{pmatrix} x_1 \\ x_2 \\ x_3 \end{pmatrix}$$

such that $AB = 1$ (the 1×1 SQUARE identity matrix).

Exercise 2.8.15. Assume $m > n$. Show that if A is an $m \times n$ (HIGH) matrix there is *no* $n \times m$ (WIDE) matrix B such that AB is the $m \times m$ SQUARE identity matrix I_m.

On the other hand, show that if $m < n$ it is *sometimes* possible for a given $m \times n$ matrix A to find an $n \times m$ matrix B such that AB is the $m \times m$ SQUARE identity matrix I_m.

In short:

$$\text{Never: HIGH·WIDE} = \text{identity};$$

$$\text{Sometimes: WIDE·HIGH} = \text{identity}.$$

Exercise 2.8.16. Let A be the 3×4 WIDE matrix

$$\begin{pmatrix} 1 & 2 & 3 & -1 \\ 3 & 4 & 5 & 0 \\ -1 & 2 & -3 & 4 \end{pmatrix}.$$

Show that there is a 4×3 HIGH matrix

$$B \stackrel{\text{def}}{=} \begin{pmatrix} a_{11} & a_{12} & a_{13} \\ a_{21} & a_{22} & a_{23} \\ a_{31} & a_{32} & a_{33} \\ a_{41} & a_{42} & a_{43} \end{pmatrix}$$

such that AB is the 3×3 SQUARE identity matrix I_3.

Example 2.8.8. Consider the following 8 matrices:

$$A \stackrel{\text{def}}{=} (a_{ij})_{i,j=1}^{3,4}, \quad B \stackrel{\text{def}}{=} (b_{ij})_{i,j=1}^{3,2}, \quad C \stackrel{\text{def}}{=} (c_{ij})_{i,j=1}^{6,4}, \quad D \stackrel{\text{def}}{=} (d_{ij})_{i,j=1}^{6,2};$$
$$M \stackrel{\text{def}}{=} (m_{ij})_{i,j=1}^{4,8}, \quad N \stackrel{\text{def}}{=} (n_{ij})_{i,j=1}^{4,7}, \quad P \stackrel{\text{def}}{=} (p_{ij})_{i,j=1}^{2,8}, \quad Q \stackrel{\text{def}}{=} (q_{ij})_{i,j=1}^{2,7}.$$

Form the block matrices

$$\mathcal{A} \stackrel{\text{def}}{=} \begin{pmatrix} a_{11} & \cdots & a_{14} & b_{11} & b_{12} \\ \vdots & \ddots & \vdots & \vdots & \vdots \\ a_{31} & \cdots & a_{34} & b_{31} & b_{32} \\ c_{11} & \cdots & c_{14} & d_{11} & d_{12} \\ \vdots & \ddots & \vdots & \vdots & \vdots \\ c_{61} & \cdots & c_{64} & d_{61} & d_{62} \end{pmatrix} \stackrel{\text{def}}{=} \begin{pmatrix} A & B \\ C & D \end{pmatrix}$$

and

$$\mathcal{B} \overset{\text{def}}{=} \left(\begin{array}{cccccc} m_{11} & \cdots & m_{18} & n_{11} & \cdots & n_{17} \\ \vdots & \ddots & \vdots & \vdots & \ddots & \vdots \\ m_{41} & \cdots & m_{48} & n_{41} & \cdots & n_{47} \\ & & & & & \\ p_{11} & \cdots & p_{18} & q_{11} & \cdots & q_{17} \\ p_{21} & \cdots & p_{28} & q_{21} & \cdots & q_{27} \end{array} \right) \overset{\text{def}}{=} \left(\begin{array}{cc} M & N \\ P & Q \end{array} \right).$$

Thus $\mathcal{A}$ is a 9×6 matrix and $\mathcal{B}$ is a 6×15 matrix. The *matrices* A, B, C, and D, are submatrices or *blocks* in $\mathcal{A}$. The matrices M, N, P, and Q are submatrices or *blocks* in $\mathcal{B}$. The sizes of the blocks have been chosen so that the following block matrix

$$\mathcal{C} \overset{\text{def}}{=} \left(\begin{array}{cc} AM + BP & AN + BQ \\ CM + DP & CN + DQ \end{array} \right) \tag{2.8.9}$$

is meaningful, i.e., the matrix products appearing in $\mathcal{C}$ can be formed. Direct calculation shows that

$$\mathcal{C} = \mathcal{A}\mathcal{B}.$$

The equation (2.8.9) illustrates the idea of *block multiplication* of *block matrices*.

Example 2.8.9. Let A be

$$\left(\begin{array}{ccc} a_{11} & \cdots & a_{16} \\ \vdots & \ddots & \vdots \\ a_{41} & \cdots & a_{46} \end{array} \right)$$

and let B be

$$\left(\begin{array}{ccc} b_{11} & \cdots & b_{19} \\ \vdots & \ddots & \vdots \\ b_{61} & \cdots & b_{69} \end{array} \right).$$

Let A be decomposed into *blocks* to form a block matrix

$$\mathcal{A} \overset{\text{def}}{=} \left(\begin{array}{cc} A_{11}(2 \times 2) & A_{12}(2 \times 4) \\ A_{21}(2 \times 2) & A_{22}(2 \times 4) \end{array} \right)$$

and let B be decomposed into blocks to form a block matrix

$$\mathcal{B} \overset{\text{def}}{=} \left(\begin{array}{cc} B_{11}(2 \times 7) & B_{12}(2 \times 2) \\ B_{21}(4 \times 7) & B_{22}(4 \times 2) \end{array} \right).$$

Then

$$AB = \left(\begin{array}{cc} A_{11}B_{11} + A_{12}B_{21} & A_{11}B_{12} + A_{12}B_{22} \\ A_{21}B_{11} + A_{22}B_{21} & A_{21}B_{12} + A_{22}B_{22} \end{array} \right)$$

$$= \mathcal{A}\mathcal{B}.$$

In analogy with elementary matrices there is the concept of *elementary block matrices*. The analogs are block matrices each block of which is either a constant (matrix) multiple of an appropriately sized I or an appropriately sized O. Thus for the matrix A above there are the following basic types of elementary block matrices:

i.

$$\mathcal{E}_{I,II} \stackrel{\text{def}}{=} \begin{pmatrix} O & I_2 \\ I_2 & O \end{pmatrix};$$

ii. If T is a 2×2 matrix then

$$\mathcal{E}_{II}(T) \stackrel{\text{def}}{=} \begin{pmatrix} I_2 & O \\ O & TI_2 \end{pmatrix} = \begin{pmatrix} I_2 & O \\ O & T \end{pmatrix};$$

iii. If T is a 2×2 matrix then

$$\mathcal{E}_{II,I}(T) \stackrel{\text{def}}{=} \begin{pmatrix} I_2 & T \\ O & I_2 \end{pmatrix}.$$

Exercise 2.8.17. Define the three types of elementary block matrices that can operate on the block matrix $\mathcal{B}$ in **Example 2.8.9**.

Exercise 2.8.18. Use block multiplication to calculate the product of

$$\mathcal{A} \stackrel{\text{def}}{=} \begin{pmatrix} 1 & 2 & 3 & 3 & 2 \\ -2 & 3 & -1 & -2 & 3 \\ & & & & \\ a & b & c & d & e \\ -c & b & -a & -e & d \\ b & -c & a & -d & -e \end{pmatrix}$$

and

$$\mathcal{B} \stackrel{\text{def}}{=} \begin{pmatrix} \alpha & \beta & \gamma \\ \pi & \sigma & \tau \\ \xi & \eta & \zeta \\ & & \\ 1 & 2 & 3 \\ -1 & 4 & 5 \end{pmatrix}.$$

Exercise 2.8.19. Generalize the idea of block multiplication of two block matrices

$$\mathcal{A} \stackrel{\text{def}}{=} (A_{ij})_{i,j=1}^{m_i,n_j} \quad \text{and} \quad \mathcal{B} \stackrel{\text{def}}{=} (B_{rs})_{r,s=1}^{p_r,q_s}.$$

The sizes $m_i \times n_j$ and $p_r \times q_s$ of the blocks $(A_{ij})_{i,j=1}^{m_i,n_j}$ in $\mathcal{A}$ and the blocks $(B_{rs})_{r,s=1}^{p_r,q_s}$ in $\mathcal{B}$ must be chosen so that matrix multiplication of the blocks is possible. Thus part of the problem is to relate the block sizes properly.

     Chapter 2. THE PROCESS (Simple elimination)

Then show that $\mathcal{AB}$ can be calculated by block multiplication.

Exercise 2.8.20. Assume $m < n$ and let A be an $m \times n$ matrix. Denote by A_1 the $m \times m$ matrix made up of the first m columns of A. Show that there is an $n \times m$ matrix B such that AB is the $m \times m$ identity matrix I_m iff A_1 is invertible.

[*Hint:* Use block multiplication of matrices.]

Exercise 2.8.21. For each matrix below find the value of s such that the given matrix has no inverse.

$$
i. \begin{pmatrix} 2 & -3 \\ s & 4 \end{pmatrix}, \quad
ii. \begin{pmatrix} 9 & 8 & 7 \\ -6 & -5 & s \\ 3 & 2 & 1 \end{pmatrix}, \quad
iii., \begin{pmatrix} 1 & 1 & 1 & 1 \\ 0 & 1 & 0 & 1 \\ -1 & 0 & s & 0 \\ 1 & -1 & 1 & 3 \end{pmatrix}.
$$

Exercise 2.8.22. Show that the matrix

$$
\begin{pmatrix} \cos\theta & \sin\theta \\ -\sin\theta & \cos\theta \end{pmatrix}
$$

always has an inverse and then find it.

Exercise 2.8.23. For what values of p, q, and r does the matrix

$$
\begin{pmatrix} p & -4 \\ q & r \end{pmatrix}
$$

have an inverse?

Exercise 2.8.24. Show that if E is a SQUARE matrix such that

$$
E^2 = E \neq I
$$

then E is singular. Is the result valid for a SQUARE matrix F such that

$$
F^k = F \neq I, \ 3 \leq k \in \mathbf{N}?
$$

Repeat the discussion for a SQUARE matrix G such that

$$
G^k = G^l, \ G \neq I, \ k,l \in \mathbf{N}, \ k \neq l.
$$

Exercise 2.8.25. Find the matrix A such that

$$A^{-1} = \begin{pmatrix} 1 & 2 & 3 \\ 4 & 5 & 6 \\ 7 & 8 & 10 \end{pmatrix}.$$

Exercise 2.8.26. Rephrase the criterion in THEOREM **2.8.1** in terms of a special entry in A_P rather than a special entry in A_E.

Exercise 2.8.27. Assume A is a SQUARE $n \times n$ matrix such that for every $n \times n$ matrix B, $AB = BA$. Prove that there is a number a such that $A = aI$.

Exercise 2.8.28. Assume A is an $n \times n$ matrix such that for every $n \times n$ diagonal matrix D, $AD = DA$. Prove that A is itself a diagonal matrix.

Exercise 2.8.29. Assume A is an $n \times n$ matrix such that for every upper triangular matrix U, $AU = UA$. Show that for some a in $\mathbf{C}$: $A = aI$.

Exercise 2.8.30. Assume S is a set of matrices in Mat_{nn} and that:

   i. if A and B are in S then $A + B$ is in S;
  ii. if $A \in S$ and $B \in Mat_{nn}$ then $AB \in S$ and $BA \in S$.

Prove $S = \{O\}$ or $S = Mat_{nn}$.

Exercise 2.8.31. Consider the following statements about the SQUARE matrix A:

   i. there is a matrix X such that $AX = I$;
  ii. $\mathcal{R}A = I$;
 iii. $X = \mathcal{R}$;
 iv. there is a matrix Y such that $YA = I$;
   v. A is invertible.

Without reference to previous **Exercises** establish the equivalence of $i - v$ by showing that

$$i \Rightarrow ii \Rightarrow iii \Rightarrow iv \Rightarrow v \Rightarrow i.$$

2.9. Using and extending the language of linear algebra

The sets $\mathbf{C}^n$ resp. $\mathbf{C}_n$ consisting of all n-component column vectors resp. row vectors are examples of *vector spaces*. The elements of $\mathbf{C}^n$ and $\mathbf{C}_n$ are special matrices, and so may be added together and multiplied by complex numbers according to the laws governing such operations with matrices (cf. **Section 1.1**). One says that $\mathbf{C}^n$ and $\mathbf{C}_n$ are *closed* with respect to (vector) addition and to multiplication by complex numbers.

The sets $\mathbf{R}^n$ resp. $\mathbf{R}_n$ of all n-component column resp. row vectors in which the components are exclusively real numbers are also vector spaces. Vectors in these spaces may be multiplied only by real numbers since multiplication of, e.g.,

$$\begin{pmatrix} 1 \\ 2 \\ -6 \end{pmatrix}$$

in $\mathbf{R}^3$ by, e.g., $2 - 3i$, leads to

$$\begin{pmatrix} 2 - 3i \\ 4 - 6i \\ -12 + 18i \end{pmatrix}$$

which is no longer in $\mathbf{R}^3$. Thus $\mathbf{R}^n$ resp. $\mathbf{R}_n$ are closed with respect to addition and to multiplication by *real* numbers and are *not* closed with respect to multiplication by complex numbers. The vector spaces $\mathbf{R}^1$, $\mathbf{R}^2$, and $\mathbf{R}^3$ are particularly useful in illustrating the geometric interpretations of what goes on in linear algebra.

Example 2.9.1. Let A be the matrix

$$\begin{pmatrix} 2 & -3 \\ -6 & 9 \end{pmatrix}$$

and let W in $\mathbf{C}^2$ be the set of all vectors

$$\mathbf{x} \overset{\text{def}}{=} \begin{pmatrix} x_1 \\ x_2 \end{pmatrix}$$

such that $A\mathbf{x} = \mathbf{O}$. Then the *PROCESS* reveals that W consists of all column vectors that are complex multiples of

$$\begin{pmatrix} 3 \\ 2 \end{pmatrix}.$$

The set W is *closed* with respect to vector addition and to multiplication by complex numbers, i.e.,

if a and b are numbers and $\mathbf{x}$ and $\mathbf{y}$ are vectors in W then $a\mathbf{x} + b\mathbf{y}$ is again a vector in W.

Thus W may be regarded as a vector space, a *(vector) subspace* of $\mathbf{C}^2$. Traditionally the subspace W is called the *kernel* of A:

$$W = ker(A) \overset{\text{def}}{=} \{\, \mathbf{x} \ : \ A\mathbf{x} = \mathbf{O} \,\}$$
$$= \left\{\, \mathbf{x} \ : \ \text{for some } t, \ \mathbf{x} = t \begin{pmatrix} 3 \\ 2 \end{pmatrix} \,\right\}.$$

On the other hand let V in $\mathbf{C}^2$ be the set of all vectors

$$\mathbf{y} \stackrel{\text{def}}{=} \begin{pmatrix} y_1 \\ y_2 \end{pmatrix}$$

such that there is some vector $\mathbf{x}$ for which $A\mathbf{x} = \mathbf{y}$. A direct check shows that V is also a subspace of $\mathbf{C}^2$ and indeed that V is the set of all vectors that are multiples of the vector

$$\begin{pmatrix} 1 \\ 3 \end{pmatrix} .$$

It is reasonable to call V the *image* of A:

$$V = im(A) \stackrel{\text{def}}{=} \left\{ \mathbf{y} \ : \ \text{for some } \mathbf{x}, \ \mathbf{y} = A\mathbf{x} \right\}$$
$$= \left\{ \mathbf{y} \ : \ \text{for some } t, \ \mathbf{y} = t \begin{pmatrix} 1 \\ 3 \end{pmatrix} \right\} .$$

DEFINITION 2.9.1. IF V IS A SUBSET OF $\mathbf{C}^n$ OR $\mathbf{C}_n$ AND IF V IS CLOSED WITH RESPECT TO BOTH VECTOR ADDITION AND TO MULTIPLICATION BY COMPLEX NUMBERS THEN V IS CALLED A *subspace* OF $\mathbf{C}^n$ OR $\mathbf{C}_n$.

Example 2.9.2. More generally if $A \stackrel{\text{def}}{=} (a_{ij})_{i,j=1}^{m,n}$ then each vector

$$\mathbf{x} \stackrel{\text{def}}{=} \begin{pmatrix} x_1 \\ \vdots \\ x_n \end{pmatrix}$$

in $\mathbf{C}^n$ is *transformed* by A into a vector $\mathbf{y}$ in $\mathbf{C}^m$:

$$A : \mathbf{x} = \begin{pmatrix} x_1 \\ \vdots \\ x_n \end{pmatrix} \mapsto \mathbf{y} \stackrel{\text{def}}{=} \begin{pmatrix} y_1 \\ \vdots \\ y_m \end{pmatrix} = A\mathbf{x} = \begin{pmatrix} a_{11}x_1 + \cdots + a_{1n}x_n \\ \vdots \\ a_{m1}x_1 + \cdots + a_{mn}x_n \end{pmatrix} .$$

Furthermore, according to the laws for combining matrices, A acts like a *linear transformation* in that (cf. **Section 1.1**): if $p, q \in \mathbf{C}$ and $\mathbf{x}, \mathbf{y} \in \mathbf{C}^n$ then

$$A(p\mathbf{x} + q\mathbf{y}) = pA\mathbf{x} + qA\mathbf{y}.$$

It follows that

$$W \stackrel{\text{def}}{=} \left\{ \mathbf{x} \ : \ \mathbf{x} \in \mathbf{C}^n, \ A\mathbf{x} = \mathbf{O} \right\}$$

resp.

$$V \stackrel{\text{def}}{=} \left\{ \mathbf{y} \ : \ \mathbf{y} \in \mathbf{C}^m, \ \text{for some } \mathbf{x} \text{ in } \mathbf{C}^n, \ \mathbf{y} = A\mathbf{x} \right\}$$

     Chapter 2. THE PROCESS (Simple elimination)

are (vector) subspaces of $\mathbf{C}^n$ resp. $\mathbf{C}^m$. These subspaces have special importance and have special names:

$$W \text{ is called the } kernel \text{ or the } nullspace \text{ of } A,$$
$$W \text{ is denoted } ker(A) \text{ or } \mathcal{N}_A;$$
$$V \text{ is called the } image \text{ or the } range \text{ of } A,$$
$$V \text{ is denoted } im(A) \text{ or } \mathcal{R}_A.$$

Exercise 2.9.1. Show that $ker(A)$ resp. $im(A)$ are subspaces of $\mathbf{C}^n$ resp. $\mathbf{C}^m$.

The definition of the vector $A\mathbf{x}$ shows that it is a linear combination of the *columns* of A. (The coefficients in this linear combination are the *components* of $\mathbf{x}$.) Hence $im(A)$ is called the *column space* of A.

DEFINITION 2.9.2. IF S IS A SET OF VECTORS IN $\mathbf{C}^n$ AND IF W IS THE SET OF ALL LINEAR COMBINATIONS

$$\left\{ \mathbf{x} \; : \; \mathbf{x} = \sum_{i=1}^{m} a_i \mathbf{y}_i, \; \mathbf{y}_i \in S, \; m \in \mathbf{N} \right\}$$

THEN W IS CALLED THE *span* OF S AND IS DENOTED $span(S)$.

Example 2.9.3. Let $\mathbf{R}^3$ be visualized as 3-dimensional (Euclidean) space. Thus a 3×1 matrix, i.e., a three-component column vector

$$\mathbf{x} \stackrel{\text{def}}{=} \begin{pmatrix} x_1 \\ x_2 \\ x_3 \end{pmatrix}$$

is viewed as an "arrow" having its tail at the origin and its head at the point with coordinates (x_1, x_2, x_3). If

$$\mathbf{u} \stackrel{\text{def}}{=} \begin{pmatrix} u_1 \\ u_2 \\ u_3 \end{pmatrix} \text{ and } \mathbf{v} \stackrel{\text{def}}{=} \begin{pmatrix} v_1 \\ v_2 \\ v_3 \end{pmatrix}$$

then $\{ \mathbf{z} \; : \; \text{for real numbers } a \text{ and } b, \; \mathbf{z} = a\mathbf{u} + b\mathbf{v} \}$ is the (real) span of the set $\{\mathbf{u}, \mathbf{v}\}$ and corresponds in Euclidean space to:

 i. the origin iff $\mathbf{u} = \mathbf{v} = \mathbf{O}$;
 ii. a straight line L through the origin and containing both $\mathbf{u}$ and $\mathbf{v}$ iff at least one of $\mathbf{u}$ and $\mathbf{v}$ is not $\mathbf{O}$ and and $\mathbf{u}$ and $\mathbf{v}$ are linearly dependent;

---

iii. a plane P through the origin and containing both $\mathbf{u}$ and $\mathbf{v}$ iff $\mathbf{u}$ and $\mathbf{v}$ are linearly independent.

Exercise 2.9.2. Show that for any set S of vectors in $\mathbf{C}^n$, $span(S)$ is a subspace of $\mathbf{C}^n$.

Thus, in the language of linear algebra,

$$im(A) = \big\{\, \mathbf{y} \ : \ \text{for some } \mathbf{x} \text{ in } \mathbf{C}^n, \ \mathbf{y} = A\mathbf{x} \,\big\}$$
$$= \text{column space of } A$$
$$= span(\text{columns of } A).$$

If the subspace V of $\mathbf{C}^n$ or of $\mathbf{C}_n$ is not $\{\mathbf{O}\}$ then V contains a maximally linearly independent subset, i.e., a subset S that is linearly independent and is not a proper subset of any other linearly independent set in V. Such a maximally linearly independent subset of V is called a *basis for V*.

[**Note 2.9.1:** The set $\{\mathbf{O}\}$ consisting of only the vector $\mathbf{O}$ is a subspace V of $\mathbf{C}^n$ and of $\mathbf{C}_n$. Yet, according to the definition of a basis, $\{\mathbf{O}\}$ has no basis! Explain.]

THEOREM 2.9.1. IF X AND Y ARE TWO BASES FOR THE SUBSPACE W THEN THE NUMBER OF VECTORS IN X IS THE SAME AS THE NUMBER OF VECTORS IN Y: THE VECTORS IN TWO BASES CAN BE PUT IN ONE-ONE CORRESPONDENCE, i.e., EACH OF X AND Y CONSISTS OF AS MANY VECTORS AS DOES THE OTHER.

PROOF. Assume $X \stackrel{\text{def}}{=} \{\mathbf{x}_1, \ldots, \mathbf{x}_p\}$, $Y \stackrel{\text{def}}{=} \{\mathbf{y}_1, \ldots, \mathbf{y}_q\}$, and that $p < q$. Then each $\mathbf{y}_i$ is uniquely expressible as a linear combination of the $\mathbf{x}_j$:

$$\mathbf{y}_i = \sum_{j=i}^{p} a_{ij}\mathbf{x}_j, \ 1 \le i \le q.$$

The matrix $(a_{ij})_{i,j=1}^{q,p}$ consists of q rows, each a $1 \times p$ vector. Since there can never be more than p linearly independent $1 \times p$ vectors and since $q > p$ it follows that the rows of $(a_{ij})_{i,j=1}^{q,p}$ are linearly dependent: for constants t_i, not all 0's,

$$\sum_{i=1}^{q} t_i(a_{i1}, \ldots a_{ip}) = (0, \ldots, 0), \ \text{i.e.,}$$

$$\sum_{i=1}^{q} t_i a_{ij} = 0, \ 1 \le j \le p.$$

Hence

$$\sum_{i=1}^{q} t_i \mathbf{y}_i = \sum_{i=1}^{q} t_i \left(\sum_{j=1}^{p} a_{ij} \mathbf{x}_j \right)$$

$$= \sum_{j=1}^{p} \left(\sum_{i=1}^{q} t_i a_{ij} \right) \mathbf{x}_j = \sum_{j=1}^{p} 0 \mathbf{x}_j = \mathbf{O}$$

in contradiction of the linear independence of Y. Thus $p \geq q$. If the argument is repeated with the rôles of X and Y reversed the conclusion is that $q \geq p$ and so $p = q$.

$$\Omega$$

DEFINITION 2.9.3. IF W IS A SUBSPACE OF $\mathbf{C}^n$ (OR OF $\mathbf{C}_n$) THE NUMBER d OF VECTORS IN ANY AND (HENCE EVERY) MAXIMALLY LINEARLY INDEPENDENT SUBSET (*basis*) OF W IS CALLED THE *dimension* OF W AND IT IS DENOTED $dim(W)$.

Thus the number of vectors in a(ny) basis of W is the dimension of W.

[**Note 2.9.2:** If V and W are subspaces of $\mathbf{C}^n$ and if $V \subset W$ then $dim(V) \leq dim(W)$ and $dim(V) = dim(W)$ iff $V = W$.]

As subspaces, $ker(A)$ and $im(A)$ have maximally linearly independent subsets, i.e., bases. Any two bases of $ker(A)$ contain equally many vectors; any two bases of $im(A)$ contain equally many vectors. The number of vectors in any basis for $ker(A)$ is the *dimension* of $ker(A)$. The number r of vectors in any basis for $im(A)$ is the *dimension* of $im(A)$; r is called the *rank* of A: $r = rank(A)$. It follows that

$$dim(\mathbf{C}^n) = n, \ dim(\mathbf{C}_n) = n;$$

$$0 \leq dim(ker(A)) \stackrel{\mathrm{def}}{=} \text{the dimension of } ker(A) \ \leq n;$$

$$0 \leq dim(im(A)) \stackrel{\mathrm{def}}{=} \text{the dimension of } im(A) \stackrel{\mathrm{def}}{=} r \leq m.$$

Let k be the dimension of $ker(A)$: $k = dim(ker(A))$. Let

$$X' \stackrel{\mathrm{def}}{=} \{\mathbf{x}_1, \ldots, \mathbf{x}_k\}$$

be a basis for $ker(A)$. Then fill X' out to a basis X for $\mathbf{C}^n$. Thus there are $n - k$ vectors $\mathbf{x}_{k+1}, \ldots, \mathbf{x}_n$ such that

$$X = \{\mathbf{x}_1, \ldots, \mathbf{x}_k, \mathbf{x}_{k+1}, \ldots, \mathbf{x}_n\}.$$

The vectors $A\mathbf{x}_{k+1}, \ldots, A\mathbf{x}_n$, are linearly independent. Otherwise there are constants t_j not all 0's and such that

$$\sum_{j=k+1}^{n} t_j A\mathbf{x}_j = A\left(\sum_{j=k+1}^{n} t_j \mathbf{x}_j \right) = \mathbf{O}.$$

Hence

$$\sum_{j=k+1}^{n} t_j \mathbf{x}_j \in ker(A)$$

and thus for some numbers $s_1, \ldots, s_k$

$$\sum_{i=1}^{k} s_i \mathbf{x}_i = \sum_{j=k+1}^{n} t_j \mathbf{x}_j, \quad \sum_{i=1}^{k} s_i \mathbf{x}_i - \sum_{j=k+1}^{n} t_j \mathbf{x}_j = \mathbf{O},$$

contradicting the linear independence of X.

Furthermore if $\mathbf{y}$ is a vector in $im(A)$ then for some $\mathbf{x}$ in $\mathbf{C}^n$ $A\mathbf{x} = \mathbf{y}$ and for a unique n-tuple $\{t_1, \ldots, t_n\}$ of complex numbers

$$\mathbf{x} = \sum_{i=1}^{n} t_i \mathbf{x}_i = \sum_{i=1}^{k} t_i \mathbf{x}_i + \sum_{i=k+1}^{n} t_i \mathbf{x}_i.$$

However, since $\{\mathbf{x}_1, \ldots, \mathbf{x}_k\}$ is a basis for $ker(A)$ it follows that

$$A\mathbf{x} = \sum_{i=1}^{k} t_i \mathbf{O} + \sum_{i=k+1}^{n} t_i A\mathbf{x}_i = \sum_{i=k+1}^{n} t_i A\mathbf{x}_i,$$

i.e., the set $\{A\mathbf{x}_{k+1}, \ldots, A\mathbf{x}_n\}$ is a basis for $im(A)$. Hence

$$dim(im(A)) = r = rank(A) = n - k = n - dim(ker(A)),$$
$$dim(ker(A)) + dim(im(A)) = n.$$

In accordance with standard practice $\mathbf{C}^n$ is called the *domain* of the linear transformation (*function, mapping*) A: $\mathbf{C}^n \stackrel{\text{def}}{=} dom(A)$. Thus

$$dim(ker(A)) + dim(im(A)) = dim(dom(A)).$$

If M is an *invertible* SQUARE $m \times m$ matrix then the $m \times 1$ column vectors $\mathbf{z}_1, \ldots, \mathbf{z}_p$ are linearly dependent iff the $m \times 1$ column vectors $M\mathbf{z}_1, \ldots, M\mathbf{z}_p$ are linearly dependent. Indeed,

$$\sum_{j=1}^{p} a_j \mathbf{z}_j = \mathbf{O} \Rightarrow \sum_{j=i}^{p} a_j M\mathbf{z}_j = \mathbf{O};$$
$$\sum_{j=1}^{p} a_j M\mathbf{z}_j = \mathbf{O} \Rightarrow M^{-1}(\sum_{j=1}^{p} a_j M\mathbf{z}_j) = \mathbf{O};$$
$$M^{-1}(\sum_{j=1}^{p} a_j M\mathbf{z}_j) = \mathbf{O} \Rightarrow \sum_{j=1}^{p} a_j \mathbf{z}_j = \mathbf{O},$$

from which the assertion follows. (Hence it is also true that $z_1, \ldots, z_p$ are linearly independent iff $Mz_1, \ldots, Mz_p$ are linearly independent.)

Let $\mathcal{E}$ be the unique invertible $m \times m$ matrix arising from the application of the GEM (algorithm). Then

$$\mathcal{E}A = A_E = \text{the echelon form of } A.$$

Thus a set of columns of A is linearly independent iff the corresponding columns of A_E are linearly independent. Let the number of nonzero rows in A_E be s. Then rows numbered 1 through s are the nonzero rows of A_E. Let the first nonzero entry in row 1 be in column numbered $c_1, \ldots$, the first nonzero entry in row s be in column numbered c_s. Then

$$c_1 < c_2 < \cdots < c_s.$$

The columns numbered $c_1, \ldots, c_s$ are linearly independent and form a basis for the span of the set of columns in A_E and the first s rows of A_E form a basis for the span of the set of rows of A_E (cf. **Example 2.8.3** *iii, iv*). Hence the rank r of A can be read off from A_E:

$$r = rank(A) = dim(im(A)) = dim(\text{column space of } A)$$
$$= \text{the number of nonzero rows in } A_E$$
$$= s = dim(\text{row space of } A_E).$$

The span of the rows of A is the *row space* of A and the dimension of the row space of A is the *row rank* of A. The dimension of the column space is correspondingly the *column rank* of A and according to the convention adopted above: column rank of A = rank of A. More importantly it is shown next that

$$\text{row rank of } A = \text{column rank of } A \ (= rank(A)).$$

Indeed, the equation $A = \mathcal{E}^{-1}A_E$ shows that every row of A is in the row space of A_E. Hence

$$dim(\text{row space of } A) \leq dim(\text{row space of } A_E)$$
$$dim(\text{row space of } A_E) = dim(\text{column space of } A_E)$$
$$= dim(\text{column space of } A)$$

whence for any matrix A

$$dim(\text{row space of } A) \leq dim(\text{column space of } A) \tag{$*$}$$

Owing to the properties of matrix transposes,

$$dim(\text{row space of } A^t) = dim(\text{column space of } A)$$
$$\leq dim(\text{column space of } A^t)$$
$$dim(\text{column space of } A^t) = dim(\text{row space of } A)$$

whence for any matrix A

$$dim(\text{column space of } A) \leq dim(\text{row space of } A) \qquad (**)$$

and thus from $(*)$ and $(**)$ it follows that

$$rank(A) = \text{column rank of } A = \text{row rank of } A.$$

Exercise 2.9.3. Show that the row space of A is $im(A^t)$.

In the discussion just concluded the following relations are the important ones:

$$im(A) = \text{column space of } A \subset \mathbf{C}^m;$$
$$ker(A) \subset \mathbf{C}^n;$$
$$im(A^t) = \text{row space of } A \subset \mathbf{C}^n;$$
$$ker(A^t) \subset \mathbf{C}^m;$$
$$dim(ker(A)) + dim(im(A)) = dim(dom(A)); \qquad (2.9.1)$$
$$dim(\text{column space of } A) = dim(\text{row space of } A) = rank(A); \qquad (2.9.2)$$
$$= \text{the number of nonzero rows in } A_E;$$
$$\text{the row rank of } A = \text{the column rank of } A;$$
$$\text{the row space of } A \text{ is the column space of } A^t.$$

[**Note 2.9.3:** If $m \neq n$ then $im(A)$ and $ker(A)$ lie in different vector spaces: $im(A) \subset \mathbf{C}^m$ and $ker(A) \subset \mathbf{C}^n$. Nevertheless the dimensions of $im(A)$ and $ker(A)$ are related as in (2.9.1).

Similarly if $m \neq n$ the row space and the column space $(im(A))$ lie in different vector spaces:

$$\text{column space of } A = im(A) \subset \mathbf{C}^m$$
$$\text{row space of } A \subset \mathbf{C}^n.$$

Nevertheless the dimensions of the column space and of the row space are related as in (2.9.2): they are equal.

A distinguishing feature of a SQUARE $n \times n$ matrix A is the fact that both $im(A)$ and $ker(A)$ are in the *same* vector space: both are in $\mathbf{C}^n$.]

Example 2.9.4. Let A be the matrix

$$\begin{pmatrix} 0 & 1 \\ 0 & 0 \end{pmatrix}.$$

then

$$im(A) = \left\{ \begin{pmatrix} a \\ 0 \end{pmatrix} \, : \, a \in \mathbf{C} \right\} ;$$

$$ker(A) = \left\{ \begin{pmatrix} a \\ 0 \end{pmatrix} \, : \, a \in \mathbf{C} \right\} .$$

In this case $im(A)$ and $ker(A)$ are not only in the same vector space $\mathbf{C}^2$ but they are the same subspace of $\mathbf{C}^2$.

On the other hand, if A is the $n \times n$ matrix I then

$$im(A) = \mathbf{C}^n;$$

$$ker(A) = \{\mathbf{O}\}.$$

In this case although $im(A)$ and $ker(A)$ are both in $\mathbf{C}^n$, they are as different as they can be.

Example 2.9.5. Let A be the matrix

$$\begin{pmatrix} 1 & 2 & 3 & 4 & 5 \\ -5 & -3 & -1 & 1 & 17 \end{pmatrix} .$$

The GJM applied to A in the context of the system $A\mathbf{x} = \mathbf{O}$ of homogeneous equations leads to the system:

$$x_1 + 2x_2 + 3x_3 + 4x_4 + 5x_5 = 0$$
$$x_2 + 2x_3 + 3x_4 + 6x_5 = 0.$$

Back substitution shows

$$x_1 = -2(-2x_3 - 3x_4 - 6x_5) - 3x_3 - 4x_4 - 5x_5 = x_3 + 2x_4 + 7x_5$$
$$x_2 = -2(-2x_3 - 3x_4 - 6x_5) = 4x_3 + 6x_4 + 12x_5.$$

The values of x_3, x_4, x_5 may be assigned arbitrarily and they determine the values of x_1 and x_2.

If each of x_3, x_4, x_5 is given the value 1 while the other two are given the value 0 there result the following three solutions of the system:

$$\mathbf{x_1} \overset{\text{def}}{=} \begin{pmatrix} 1 \\ 4 \\ 1 \\ 0 \\ 0 \end{pmatrix} , \quad \mathbf{x_2} \overset{\text{def}}{=} \begin{pmatrix} 2 \\ 6 \\ 0 \\ 1 \\ 0 \end{pmatrix} , \quad \text{and} \quad \mathbf{x_3} \overset{\text{def}}{=} \begin{pmatrix} 7 \\ 12 \\ 0 \\ 0 \\ 1 \end{pmatrix} .$$

Again **Example 2.8.3** *iii, iv* shows that the three vectors above constitute a basis for $ker(A)$: $dim(ker(A)) = 3$.

Since the rows of A_J are nonzero multiples of the rows of A_E the rank of A is 2. Furthermore the first two columns

$$\mathbf{y}_1 \overset{\text{def}}{=} \begin{pmatrix} 1 \\ -5 \end{pmatrix} \quad \text{and} \quad \mathbf{y}_2 \overset{\text{def}}{=} \begin{pmatrix} 2 \\ -3 \end{pmatrix}$$

constitute a basis for $im(A)$: $dim(im(A)) = 2$. In particular,

$$dim(ker(A)) + dim(im(A)) = 3 + 2 = 5 = dim(dom(A)) = dim(\mathbf{C}^5).$$

Exercise 2.9.4. For each of the matrices, say A, below find a basis for its row space and a basis for its column space. Also determine:

$$rank(A), \ dim(im(A)), \ dim(ker(A)), \ \text{and} \ dim(dom(A)).$$

$$i. \begin{pmatrix} 1 & 2 & 3 & 4 & 5 \\ 5 & 1 & 2 & 3 & 4 \\ 4 & 5 & 1 & 2 & 3 \\ 3 & 4 & 5 & 1 & 2 \\ 2 & 3 & 4 & 5 & 1 \end{pmatrix}; \quad ii. \begin{pmatrix} 1 & -1 & 1 \\ -1 & 1 & -1 \end{pmatrix}; \quad iii. \begin{pmatrix} 1 & 2 \\ 3 & 4 \\ 5 & 6 \\ 7 & -8 \end{pmatrix}.$$

Exercise 2.9.5.

i. Construct a 2×3 matrix of rank 1 and in which *every* entry is not 0.

ii. Show that an $m \times n$ matrix A is of rank 1 iff there are two nonzero vectors

$$\mathbf{x} \overset{\text{def}}{=} \begin{pmatrix} x_1 \\ \vdots \\ x_m \end{pmatrix} \quad \text{and} \quad \mathbf{y} \overset{\text{def}}{=} (y_1, \ldots, y_n)$$

such that $A = \mathbf{x} \cdot \mathbf{y}$.

[*Hint:* A matrix is of rank 1 iff at least one of its entries is not 0 and, furthermore, iff any two rows (columns) are linearly dependent.]

Exercise 2.9.6. Show that $rank(A) = rank(A^t)$.

Exercise 2.9.7. Show that if A is an $m \times n$ matrix then $rank(A) \leq minimum(m, n)$.

Exercise 2.9.8. Show

$$rank(A + B) \leq rank(A) + rank(B),$$
$$rank(AB) \leq min(rank(A), rank(B)).$$

For each of the inequalities above give an example for which "$<$" holds and an example for which "$=$" holds.

Exercise 2.9.9. Show that if P and Q are invertible then:

i. $rank(PA) = rank(A) = rank(AQ)$;
ii. if A is SQUARE then $rank(A) = rank(P^{-1}AP)$.

Exercise 2.9.10. Show that if A is an $n \times n$ matrix then A^{-1} exists iff $rank(A) = n$.

Exercise 2.9.11. Let A be and $m \times n$ matrix and let B be a $p \times q$ matrix and let C be the $r \times s$ matrix described by the following display:

$$\begin{pmatrix} A & O \\ O & B \end{pmatrix}.$$

i. What are r and s?
ii. Show that:
$$rank(C) = rank(A) + rank(B).$$

iii. Generalize the problem so that k matrices $A_1, \ldots, A_k$ are arranged in the pattern below
$$\begin{pmatrix} A_1 & O & \ldots & O \\ O & A_2 & \ldots & O \\ \vdots & \vdots & \ddots & \vdots \\ O & O & \ldots & A_k \end{pmatrix}$$

to form a block matrix A. Again determine the size of A in terms of the sizes of the A_i and prove:

$$rank(A) = \sum_{i=1}^{k} rank(A_i).$$

The presence (or absence) of a row of 0's in the echelon form A_E of the $m \times n$ matrix A is related to $rank(A)$. If row k of A_E consists of 0's and row $k - 1$ contains at least nonzero entry then, $rank(A) = k - 1$. Furthermore, $\eta_{k1}, \ldots, \eta_{km}$ denoting the entries in the kth row of $\mathcal{E}$ and $\mathbf{a}_1, \ldots, \mathbf{a}_m$ denoting the rows of A,

$$\sum_{j=1}^{m} \eta_{kj} \mathbf{a}_j = \mathbf{O}. \tag{2.9.3}$$

Since no row of the invertible matrix $\mathcal{E}$ can consist entirely of 0's (otherwise I has a row of 0's), (2.9.3) implies the rows of A are linearly dependent.

Thus the presence of a row of 0's in some row r of A_E has a simple interpretation. It indicates that the linear combination of the unknowns in some row k of the system

$$a_{11}x_1 + \cdots + a_{1n}x_n = b_1$$
$$\vdots \qquad \ddots \qquad \vdots \qquad \vdots$$
$$a_{m1}x_1 + \cdots + a_{mn}x_n = b_m$$

of equations is a *consequence* of the linear combinations of unknowns in *other* rows. Hence the left member of row k *depends* in a particular way on the left members of other rows. Thus, if there is a solution, the number in the right member of row k must depend in the same way on the numbers in the right members of the other rows.

For example the system

$$2x + 3y = 2$$
$$4x + 6y = 5$$

(Section 1.1) has no solution since the left member of the second equation is twice the left member of the first equation, but the right member of the second equation is *not* twice the right member of the first equation.

The system

$$2x + 3y = 2$$
$$4x + 6y = 4$$

has infinitely many solutions since the second equation says nothing more than the first equation and the first equation has infinitely many solutions.

A rôle of *rank* is illustrated in the next result.

THEOREM 2.9.2. LET A BE AN $m \times n$ MATRIX, LET $\mathbf{b}$ BE AN $m \times 1$ MATRIX (A COLUMN VECTOR), AND LET $A|\mathbf{b}$ DENOTE THE *augmented* MATRIX

$$\left(\begin{array}{ccc|c} a_{11} & \cdots & a_{1n} & b_1 \\ \vdots & \ddots & \vdots & \vdots \\ a_{m1} & \cdots & a_{mn} & b_m \end{array} \right).$$

THEN THE EQUATION $A\mathbf{x} = \mathbf{b}$ HAS A SOLUTION IFF $rank(A) = rank(A|\mathbf{b})$. IN OTHER WORDS,

$$a_{11}x_1 + \cdots + a_{1n}x_n = b_1$$
$$\vdots \qquad \ddots \qquad \vdots \qquad \vdots$$
$$a_{m1}x_1 + \cdots + a_{mn}x_n = b_m$$

HAS A SOLUTION IFF THE RANK OF A AND THE RANK OF THE AUGMENTED MATRIX $A|\mathbf{b}$ ARE EQUAL.

PROOF. When the GEM is applied to the system $A\mathbf{x} = \mathbf{b}$ the matrix A is multiplied on the left by a sequence of elementary (invertible!) $m \times m$ matrices. If their product is $\mathcal{E}$ then $\mathcal{E}A\mathbf{x} = A_E\mathbf{x} = \mathcal{E}\mathbf{b} \overset{\text{def}}{=} \mathbf{c}$ and since $\mathcal{E}$ is invertible, $A\mathbf{x} = \mathbf{b}$ has a solution iff $A_E\mathbf{x} = \mathbf{c}$ has a solution.

The equation $A_E\mathbf{x} = \mathbf{c}$ has a solution iff every solid row of zeros in A_E, say the ith row, is matched by a zero as the ith component of $\mathbf{c}$. Re-interpreted in terms of *rank* the previous sentence says that $A_E\mathbf{x} = \mathbf{c}$ has a solution iff $rank(A_E) = rank(A_E|\mathbf{c})$. Since $\mathcal{E}$ is invertible the result follows from **Exercise 2.9.9**.

$$\Omega$$

Exercise 2.9.12. For the system

$$x - y + 2z - u = b_1$$
$$3x + y - 3z = b_2$$
$$5x - y + z - 2u = b_3$$
$$x + y + 4z - 2z + u = b_4$$

determine the conditions that the numbers b_1, b_2, b_3, and b_4 must satisfy if the system is to have a solution.

Exercise 2.9.13. Show that if A is a WIDE $m \times n$ matrix and $rank(A) = m$ then for any $\mathbf{b}$ the system $A\mathbf{x} = \mathbf{b}$ has infinitely many solutions.

Exercise 2.9.14. Show that if A is an $m \times n$ matrix then:

i. the equation $A\mathbf{x} = \mathbf{O}$ has a nonzero solution if $m < n$;
ii. if $m \geq n$ the equation $A\mathbf{x} = \mathbf{O}$ has a nonzero solution iff $rank(A) < n$.

In **Exercise 2.9.15** there are several *equivalent* definitions of a basis for a vector space V. Each is useful in one or another context of linear algebra.

Exercise 2.9.15. Show that a subset $X \overset{\text{def}}{=} \{\mathbf{x}_1, \ldots, \mathbf{x}_m\}$ of a vector space V is a basis for V iff

i. X is a maximal linearly independent subset of V;
ii. X is linearly independent and every vector in V is in the span of X:

$$V \subset span(X);$$

iii. $V = span(X)$ and if $X' \subsetneq X$ then $span(X') \subsetneq V$, i.e., iff X is a *minimal spanning* subset of V;
iv. every vector in V may be written *uniquely* (in one and only one way) as a linear combination of vectors in X;
v. for every invertible matrix P the set

$$PX \overset{\text{def}}{=} \{P\mathbf{x}_1, \ldots, P\mathbf{x}_m\}$$

---

is a basis for V;

$vi.$ when $V = \mathbf{C}^m$,

    $a.$ the $m \times m$ matrix C, in which the kth column of C is the (column) vector $\mathbf{x}_k$, is invertible;

    $b.$ the $m \times m$ matrix R, in which the kth row of R is the (row) vector $\mathbf{x}_k^t$, is invertible;

    $c.$ (cf. via) $CE \stackrel{\text{def}}{=} \{Ce_1, \ldots, Ce_m\}$ is a basis for $\mathbf{C}^m$;

    $d.$ (cf. vib) $RE \stackrel{\text{def}}{=} \{Re_1, \ldots, Re_m\}$ is a basis of $\mathbf{C}^m$;

$vii.$ whenever $Y \stackrel{\text{def}}{=} \{\mathbf{y}_1, \ldots, \mathbf{y}_m\}$ is a basis for V there is a unique invertible $m \times m$ matrix R_{XY}, depending on both X and Y, and such that

$$R_{XY}\mathbf{y}_k = \mathbf{x}_k, \quad 1 \le k \le m.$$

[**Remark 2.9.1:** As noted earlier (cf. **Section 2.8**) in connection with the vector spaces $\mathbf{C}_n$ resp. $\mathbf{C}^n$, a basis in a vector space V is a set S of vectors in V and such that every vector in V is uniquely expressible as a linear combination of vectors in S. A linearly independent set $\mathcal{I}$ is a set of vectors such that $\mathbf{O}$ is uniquely expressible as a linear combination of vectors in $\mathcal{I}$. Since $\mathbf{O}$ is always expressible in at least one way as a linear combinations of any set of vectors, a linearly dependent set $\mathcal{D}$ of vectors is one that permits at least two representations of $\mathbf{O}$ as linear combinations of vectors in $\mathcal{D}$. In one of these representations at least one coefficient is not 0. Thus every basis in V is linearly independent but not every linearly independent set S in V is a basis.]

In the next **Exercise** there are several equivalent definitions of an invertible SQUARE matrix.

Exercise 2.9.16. Show that each of the following definitions is equivalent to each of the others as a definition of an invertible matrix. These alternative definitions are helpful in varying contexts of linear algebra.

$o.$ An $n \times n$ matrix A is invertible iff there is an $n \times n$ matrix B such that $AB = BA = I$.

$i.$ An $n \times n$ matrix A is invertible iff for some basis

$$Z \stackrel{\text{def}}{=} \{\mathbf{z}_1, \ldots, \mathbf{z}_n\}$$

of $\mathbf{C}^n$

$$AZ \stackrel{\text{def}}{=} \{A\mathbf{z}_1, \ldots, A\mathbf{z}_n\}$$

is also a basis for $\mathbf{C}^n$.

$ii.$ An $n \times n$ matrix A is invertible iff the rows of A constitute a basis of $\mathbf{C}_n$.

iii. An $n \times n$ matrix A is invertible iff the columns of A constitute a basis for $\mathbf{C}^n$.

iv. An $n \times n$ matrix A is invertible iff its rows constitute a linearly independent set of vectors in $\mathbf{C}_n$.

v. An $n \times n$ matrix A is invertible iff its columns constitute a linearly independent set of vectors in $\mathbf{C}^n$.

vi. An $n \times n$ matrix A is invertible iff $im(A) = \mathbf{C}^n$, i.e., iff "$A\mathbf{C}^n = \mathbf{C}^n$," i.e., iff for every column vector

$$\mathbf{b} \stackrel{\text{def}}{=} \begin{pmatrix} b_1 \\ \vdots \\ b_n \end{pmatrix}$$

the system

$$a_{11}x_1 + \cdots + a_{1n}x_n = b_1$$
$$\vdots \quad\quad \ddots \quad\quad \vdots \quad\quad \vdots \quad\quad (Ax = b)$$
$$a_{n1}x_1 + \cdots + a_{nn}x_n = b_n$$

has a solution.

vii. An $n \times n$ matrix A is invertible iff for each row vector $\mathbf{y}$ in $\mathbf{C}_n$ there is in $\mathbf{C}_n$ a row vector $\mathbf{x}$ such that $\mathbf{x}A = \mathbf{y}$, i.e., iff "$\mathbf{C}_n A = \mathbf{C}_n$."

If X is a set $\{\mathbf{x}_1, \ldots, \mathbf{x}_m\}$ of m vectors in $\mathbf{C}^n$ or in $\mathbf{C}_n$ one may form the "column vector"

$$\mathcal{X}^c \stackrel{\text{def}}{=} \begin{pmatrix} \mathbf{x}_1 \\ \vdots \\ \mathbf{x}_m \end{pmatrix} \tag{2.9.4}$$

and the "row vector"

$$\mathcal{X}_r \stackrel{\text{def}}{=} (\mathbf{x}_1, \ldots, \mathbf{x}_m). \tag{2.9.5}$$

These "vectors" look like vectors in $\mathbf{C}^m$ and $\mathbf{C}_m$. However their *components* are themselves vectors rather than numbers. If $\mathbf{x}_i \stackrel{\text{def}}{=} (x_{i1}, \ldots, x_{in}) \in \mathbf{C}_n$, $1 \le i \le m$, i.e., each $\mathbf{x}_i$ is a row vector in $\mathbf{C}_n$, then $\mathcal{X}^c$ may be viewed as an $m \times n$ *matrix* in which the entries are numbers. Similarly if

$$\mathbf{x}_j \stackrel{\text{def}}{=} \begin{pmatrix} x_{1j} \\ \vdots \\ x_{nj} \end{pmatrix},$$

i.e., each $\mathbf{x}_j$ is a column vector in $\mathbf{C}^n$, then $\mathcal{X}_r$ may be viewed as an $n \times m$ *matrix* in which the entries are numbers. If the superscript ^{c} is dropped, as in $\mathcal{X}$, then

$$\begin{pmatrix} \mathbf{x}_1 \\ \vdots \\ \mathbf{x}_n \end{pmatrix}$$

is intended.

Exercise 2.9.17. Show that if A is a SQUARE matrix then $rank(AA^t) = n$ iff $rank(A) = n$.

Exercise 2.9.18. Let the following matrix be the *augmented* matrix of a system of linear equations.

$$\left(\begin{array}{ccc|c} x & 1 & 2 & 1 \\ 3 & 4 & y & x \\ x & 0 & y & y \end{array} \right).$$

Find the values of x and y for which the system has

   i. exactly ONE solution;
  ii. NO solution;
 iii. INFINITELY MANY solutions.

Exercise 2.9.19. Find a basis for the span of the set

$$\{(1, 2, 3, 4), (-1, 2 - 3, 4), (1, 9, 7, 5), (0, 3, 4, -7)\}$$

in $\mathbf{C}^4$.

Exercise 2.9.20. Let A be the matrix

$$\left(\begin{array}{ccccc} 1 & 2 & 3 & 4 & 5 \\ 3 & -2 & 3 & -6 & 1 \\ 0 & 8 & 6 & 18 & 14 \end{array} \right).$$

Find:

    i. a basis for $im(A)$;
   ii. a basis for $ker(A)$;
  iii. $dim(im(A))$;
  iv. $dim(ker(A))$;
   v. $dim(ker(A)) + dim(im(A))$;
  vi. in $\mathbf{C}^5$ a subspace W such that
        $dim(W) = dim(im(A))$;
        $A(W) = im(A)$;
 vii. a basis for W.

Exercise 2.9.21. Let $X \overset{\text{def}}{=} \{\mathbf{x}_1, \ldots, \mathbf{x}_n\}$ be a basis for $\mathbf{C}^n$. Hence for each vector $\mathbf{y}$ in $\mathbf{C}^n$ there is a unique set $S \overset{\text{def}}{=} \{a_1, \ldots, a_n\}$ of numbers such that

$$\mathbf{y} = a_1 \mathbf{x}_1 + \cdots + a_n \mathbf{x}_n \equiv \sum_{i=1}^{n} a_i \mathbf{x}_i.$$

For X fixed the vector $\mathbf{y}$ determines the numbers a_i, i.e., each a_i *depends* on $\mathbf{y}$, is a *function* of $\mathbf{y}$:

$$a_i \stackrel{\text{def}}{=} \mathbf{f}_i(\mathbf{y}), \ 1 \leq i \leq n.$$

i. Show that if $\mathbf{y}$ and $\mathbf{z}$ are in $\mathbf{C}^n$ and p and q are numbers then

$$\mathbf{f}_i(p\mathbf{y} + q\mathbf{z}) = p\mathbf{f}_i(\mathbf{y}) + q\mathbf{f}_i(\mathbf{z}), \ 1 \leq i \leq n. \tag{2.9.6}$$

ii. Show that

$$\mathbf{f}_i(\mathbf{x}_j) \stackrel{\text{def}}{=} \delta_{ij} = \begin{cases} 1, & \text{if } i = j; \\ 0, & \text{if } i \neq j. \end{cases} \tag{2.9.7}$$

[**Remark 2.9.2:** Because each $\mathbf{f}_i$ behaves in the manner described in (2.9.6) each $\mathbf{f}_i$ is called a *linear functional*. The terminology goes back to the fact that if

$$y = f(x)$$

is the equation of a straight *line* through the origin then

$$\text{for some number } b \ f(x) = bx$$
$$\text{and}$$
$$f(pw + qz) = pf(w) + qf(z).$$

The relations described in equation (2.9.7) are called *biorthogonality* relations and the sets

$$X \stackrel{\text{def}}{=} \{\mathbf{x}_1, \ldots, \mathbf{x}_n\} \text{ and } F \stackrel{\text{def}}{=} \{\mathbf{f}_1, \ldots, \mathbf{f}_n\}$$

are constitute a *biorthogonal* pair (X, F). The function δ_{ij} is the *Kronecker delta (function)*. Its value is 1 if $i = j$ and 0 if $i \neq j$.]

Example 2.9.6. Let $X \stackrel{\text{def}}{=} \{\mathbf{x}_1, \ldots, \mathbf{x}_n\}$ be a basis for $\mathbf{C}^n$. Let b be the column vector

$$\begin{pmatrix} b_1 \\ \vdots \\ b_n \end{pmatrix}.$$

If $\mathbf{a} = (a_1, \ldots, a_n)$ and

$$\mathbf{C}^n \ni \mathbf{x} \stackrel{\text{def}}{=} \sum_{i=1}^{n} a_i \mathbf{x}_i$$

define $\mathbf{f}(\mathbf{x})$ by the equation

$$\mathbf{f}(\mathbf{x}) \stackrel{\text{def}}{=} \mathbf{ba} = \sum_{i=1}^{n} b_i a_i.$$

Then direct calculation shows that **f** is a linear functional. The conclusion of the following **Exercise** is that the converse is true.

Exercise 2.9.22. Let **f** be a linear functional defined on $\mathbf{C}^n$:

$$\mathbf{f} : \mathbf{C}^n \ni \mathbf{x} \mapsto \mathbf{f}(\mathbf{x}) \in \mathbf{C}.$$

Assume $X \overset{\text{def}}{=} \{\mathbf{x}_1, \dots, \mathbf{x}_n\}$ is a basis for $\mathbf{C}^n$. Show that if

$$\mathbf{f}(\mathbf{x}_i) \overset{\text{def}}{=} b_i$$

and

$$\mathbf{x} \overset{\text{def}}{=} \sum_{i=1}^{n} a_i \mathbf{x}_i$$

then

$$\mathbf{f}(\mathbf{x}) = \sum_{i=1}^{n} b_i a_i = \begin{pmatrix} b_1 \\ \vdots \\ b_n \end{pmatrix} (a_1, \dots, a_n) \overset{\text{def}}{=} \mathbf{ba}.$$

Exercise 2.9.23. Show that if A is a SQUARE nonsingular $n \times n$ matrix then the set X of row vectors constituting the rows of A and the set F of column vectors constituting the columns of A^{-1} are a biorthogonal pair (X, F).

Exercise 2.9.24. If $A \overset{\text{def}}{=} (a_{ij})_{i,j=1}^{m,n}$, $\overline{A} \overset{\text{def}}{=} (\overline{a}_{ij})_{i,j=1}^{m,n}$, and $A^* \overset{\text{def}}{=} \overline{A}^t$ then associated with A are four sets:

$$im(A), \ ker(A), \ im(A^*), \ \text{and} \ ker(A^*).$$

i. Which of the sets above are in $\mathbf{C}^n$ and which are in $\mathbf{C}^m$?
ii. Which of the sets above are subspaces?
iii. Show that $ker(A) \cap im(A^*) = \{\mathbf{O}\}$.
iv. Show that $ker(A^*) \cap im(A) = \{\mathbf{O}\}$.
v. Show that each vector $\mathbf{x}$ in $\mathbf{C}^n$ may be written uniquely as a sum of a vector in $ker(A)$ and a vector in $im(A^*)$:

$$\mathbf{x} = \mathbf{y} + \mathbf{z}, \ \mathbf{y} \in ker(A), \ \mathbf{z} \in im(A^*).$$

vi. Show that each vector $\mathbf{u}$ in $\mathbf{C}^m$ may be written uniquely as a sum of a vector in $ker(A^*)$ and a vector in $im(A)$:

$$\mathbf{u} = \mathbf{v} + \mathbf{w}, \ \mathbf{v} \in ker(A^*), \ \mathbf{w} \in im(A).$$

vii. Show
$$dim(im(A)) + dim(ker(A^*)) = m$$
$$dim(im(A^*)) + dim(ker(A)) = n$$

viii. Show that the system $A\mathbf{x} = \mathbf{b}$ has a solution iff

$$\mathbf{b} \in im(A),$$

i.e.,

$$\mathbf{b} \in \text{ column space of } A,$$

i.e.,

$$\mathbf{b} \in span(\text{columns of } A).$$

[**Remark 2.9.3:** The four sets,

$$im(A), \ ker(A), \ im(A^*), \ \text{and} \ ker(A^*)$$

are sometimes called the *four fundamental subspaces* associated with an $m \times n$ matrix A.]

Exercise 2.9.25. Prove: $dim(ker(A)) - dim(ker(A^*)) = n - m$.

Example 2.9.7. Let A be the 3×3 matrix

$$\begin{pmatrix} 1 & 2 & 3 \\ 4 & 5 & 6 \\ 7 & 8 & 9 \end{pmatrix}.$$

Then the row-reduced form of A is

$$A_R = \begin{pmatrix} 1 & 0 & -1 \\ 0 & 1 & 2 \\ 0 & 0 & 0 \end{pmatrix}$$

and the rank of A is 2. Let C be the matrix

$$\begin{pmatrix} 1 & 2 \\ 4 & 5 \\ 7 & 8 \end{pmatrix}$$

consisting of the columns of A that correspond to the first 2 nonzero columns of A_R. These are column vectors that constitute a *basis* of the *column space* of A. Correspondingly let R be the matrix

$$\begin{pmatrix} 1 & 0 & -1 \\ 0 & 1 & 2 \end{pmatrix}$$

---

consisting of the first two rows of A_R. These are the row vectors that constitute a *basis* of the *row space* of A_R. Direct calculation shows $A = CR$:

$$\begin{pmatrix} 1 & 2 \\ 4 & 5 \\ 7 & 8 \end{pmatrix} \begin{pmatrix} 1 & 0 & -1 \\ 0 & 1 & 2 \end{pmatrix} = \begin{pmatrix} 1 & 2 & 3 \\ 4 & 5 & 6 \\ 7 & 8 & 9 \end{pmatrix}.$$

It is generally true that if an $n \times n$ SQUARE matrix A has rank r a similar "CR" decomposition is available.

Indeed let A_R be the row reduced form of A and let columns $j_1, \ldots, j_r$ be the first r nonzero columns of A_R. (These first r nonzero columns of A_R are the column vectors $e_1, \ldots, e_r$.) Let C be the $n \times r$ matrix consisting of the columns $j_1, \ldots, j_r$ of A. Because $A_R = \mathcal{R}A$, because $\mathcal{R}$, as the product of invertible elementary matrices, is invertible, and because

$$\text{column } j_k \text{ of } A_R = \mathcal{R}(\text{column } j_k \text{ of } A), \tag{2.9.8}$$

it follows that the rank of C is also r and that the columns of C constitute a basis for the column space of A.

Similarly let R be the $r \times n$ matrix consisting of the first r rows of A_R. Then the rank of R is r and the rows of R constitute a basis for the row space of A_R.

Finally let J be the $n \times r$ matrix

$$\begin{pmatrix} 1 & 0 & \ldots & 0 \\ 0 & 1 & \ldots & 0 \\ \vdots & \vdots & \ddots & \vdots \\ 0 & 0 & \ldots & 1 \\ 0 & 0 & \ldots & 0 \\ 0 & 0 & \ldots & 0 \\ \vdots & \vdots & \ddots & \vdots \\ 0 & 0 & \ldots & 0 \end{pmatrix},$$

a block matrix consisting of the $r \times r$ identity matrix I_r above the $(n - r) \times r$ zero matrix $O_{n-r,r}$:

$$J = \begin{pmatrix} I_r \\ O_{n-r,r} \end{pmatrix}.$$

Then

$$JR = A_R$$
$$C = \mathcal{R}^{-1}J \quad (\text{see } (2.9.8))$$
$$CR = \mathcal{R}^{-1}JR = \mathcal{R}^{-1}A_R = A.$$

Exercise 2.9.26. Find the "CR" decomposition of the matrix

$$A \overset{\text{def}}{=} \begin{pmatrix} -2 & 1 & 4 & 3 + 2i \\ 3 & 6 & -1 & 0 \\ -i & 2 - i & 5 & 2 \\ 4 - 2i & 17 - 2i & 12 & 7 + 2i \end{pmatrix}.$$

   Chapter 2. THE PROCESS (Simple elimination)

Exercise 2.9.27. Show that if C and R are used for the CR decomposition of an $n \times n$ matrix A then $CR = RC$ iff $rank(A) = n$ iff $RC = A$.

Exercise 2.9.28. If a matrix A is not SQUARE is there a CR decomposition for A?

Example 2.9.8. Let X be the basis consisting of the vectors

$$\mathbf{x}_1 \overset{\text{def}}{=} \begin{pmatrix} 1 \\ 2 \\ 3 \end{pmatrix}, \quad \mathbf{x}_2 \overset{\text{def}}{=} \begin{pmatrix} 4 \\ 5 \\ 6 \end{pmatrix}, \quad \text{and } \mathbf{x}_3 \overset{\text{def}}{=} \begin{pmatrix} 7 \\ 8 \\ 10 \end{pmatrix}$$

and let Y be the basis consisting of the vectors

$$\mathbf{y}_1 \overset{\text{def}}{=} \begin{pmatrix} 1 \\ 1 \\ 0 \end{pmatrix}, \quad \mathbf{x} \overset{\text{def}}{=} \begin{pmatrix} 1 \\ 0 \\ 1 \end{pmatrix}, \quad \text{and } \mathbf{y}_3 \overset{\text{def}}{=} \begin{pmatrix} 0 \\ 1 \\ 1 \end{pmatrix}.$$

Then

$$\mathcal{X} \overset{\text{def}}{=} \begin{pmatrix} \mathbf{x}_1 \\ \mathbf{x}_2 \\ \mathbf{x}_3 \end{pmatrix} = \begin{pmatrix} 0 & 1 & 1 \\ \frac{3}{2} & \frac{5}{2} & \frac{7}{2} \\ -\frac{3}{2} & \frac{17}{2} & \frac{3}{2} \end{pmatrix} \begin{pmatrix} \mathbf{y}_1 \\ \mathbf{y}_2 \\ \mathbf{y}_3 \end{pmatrix}$$

$$\overset{\text{def}}{=} R_{XY}\,\mathcal{Y}$$

$$\mathcal{Y} \overset{\text{def}}{=} \begin{pmatrix} \mathbf{y}_1 \\ \mathbf{y}_2 \\ \mathbf{y}_3 \end{pmatrix} = \begin{pmatrix} -\frac{26}{9} & \frac{7}{9} & \frac{1}{9} \\ -\frac{5}{6} & \frac{1}{6} & \frac{1}{6} \\ \frac{11}{6} & -\frac{1}{6} & -\frac{1}{6} \end{pmatrix} \begin{pmatrix} \mathbf{x}_1 \\ \mathbf{x}_2 \\ \mathbf{x}_3 \end{pmatrix}$$

$$\overset{\text{def}}{=} R_{YX}\,\mathcal{X}.$$

Direct calculation shows that

$$\begin{pmatrix} 0 & 1 & 1 \\ \frac{3}{2} & \frac{5}{2} & \frac{7}{2} \\ -\frac{3}{2} & \frac{17}{2} & \frac{3}{2} \end{pmatrix} \begin{pmatrix} -\frac{26}{9} & \frac{7}{9} & \frac{1}{9} \\ -\frac{5}{6} & \frac{1}{6} & \frac{1}{6} \\ \frac{11}{6} & -\frac{1}{6} & -\frac{1}{6} \end{pmatrix} = \begin{pmatrix} 1 & 0 & 0 \\ 0 & 1 & 0 \\ 0 & 0 & 1 \end{pmatrix}$$

$$R_{XY}\,R_{YX} = I.$$

Exercise 2.9.29. Let $X \overset{\text{def}}{=} \{\mathbf{x}_1, \ldots, \mathbf{x}_n\}$ and $Y \overset{\text{def}}{=} \{\mathbf{y}_1, \ldots, \mathbf{y}_n\}$ be two bases for $\mathbf{C}^n$. Show that there are $n \times n$ matrices R_{XY} and R_{YX} such that

$$\mathcal{X} = R_{XY}\,\mathcal{Y}$$
$$\mathcal{Y} = R_{YX}\,\mathcal{X}$$
$$R_{XY}\,R_{YX} = I.$$

If $z \stackrel{\text{def}}{=} a + ib \in \mathbf{C}$ it is customary to denote a and b as follows:

$$a \stackrel{\text{def}}{=} \Re(z) \text{ and } b \stackrel{\text{def}}{=} \Im(z).$$

Exercise 2.9.30. Consider the following subsets of $\mathbf{C}^2$. For each determine whether it is a subspace and, if it is a subspace, a basis for it.

  $i.$ $S_1 \stackrel{\text{def}}{=} \left\{ \mathbf{x} \stackrel{\text{def}}{=} (x_1, x_2) \; : \; x_1 = 3 \right\}$;

  $ii.$ $S_2 \stackrel{\text{def}}{=} \left\{ \mathbf{x} \stackrel{\text{def}}{=} (x_1, x_2) \; : \; x_1 = 3x_2 \right\}$;

  $iii.$ $S_3 \stackrel{\text{def}}{=} \left\{ \mathbf{x} \stackrel{\text{def}}{=} (x_1, x_2) \; : \; x_1 + x_2 = 1 \right\}$;

  $iv.$ $S_4 \stackrel{\text{def}}{=} \left\{ \mathbf{x} \stackrel{\text{def}}{=} (x_1, x_2) \; : \; x_2 = 0 \right\}$;

  $v.$ $S_5 \stackrel{\text{def}}{=} \left\{ \mathbf{x} \stackrel{\text{def}}{=} (x_1, x_2) \; : \; \Re(x_1) = -2\Im(x_2) \right\}$;

  $vi.$ $S_{ij} \stackrel{\text{def}}{=} S_i \cap S_j$, for any pair i, j such that $1 \leq i, j \leq 5$.

Exercise 2.9.31. Consider the following subsets of $\mathbf{C}^3$. For each determine whether it is a subspace and, if it is a subspace, a basis for it.

  $i.$ $S_1 \stackrel{\text{def}}{=} \left\{ \mathbf{x} \stackrel{\text{def}}{=} (x_1, x_2, x_3) \; : \; x_1^2 + x_2^2 + x_3^2 = 1 \right\}$;

  $ii.$ $S_2 \stackrel{\text{def}}{=} \left\{ \mathbf{x} \stackrel{\text{def}}{=} (x_1, x_2, x_3) \; : \; a_1 x_1 + a_2 x_2 + a_3 x_3 = k \right\}$;

  $iii.$ $S_3 \stackrel{\text{def}}{=} \left\{ \mathbf{x} \stackrel{\text{def}}{=} (x_1, x_2, x_3) \; : \; 2\Re(x_1) + 3\Im(x_2) - x_3 = 0 \right\}$;

  $iv.$ $S_4 \stackrel{\text{def}}{=} \left\{ \mathbf{x} \stackrel{\text{def}}{=} (x_1, x_2, x_3) \; : \; \Re(x_1) \leq \Im(x_2) \right\}$;

  $v.$ $S_5 \stackrel{\text{def}}{=} \left\{ \mathbf{x} \stackrel{\text{def}}{=} (x_1, x_2, x_3) \; : \; \mathbf{x} = t(1, 2, 3) \right\}$;

  $vi.$ $S_{ij} \stackrel{\text{def}}{=} S_i \cap S_j$, for any pair i, j such that $1 \leq i, j \leq 5$.

Exercise 2.9.32. The following statements constitute another proof of the equation

$$\text{row rank of } A = \text{column rank of } A$$

(cf. (2.9.2)). Prove each statement.

  $i.$ the row rank of A is the same as the row rank of A_E;
  $ii.$ the row rank of A_E is the same as the column rank of A_E;
  $iii.$ the column rank of A_E is not less than the column rank of A;
  $iv.$ the row rank of A is not less than the column rank of A;
  $v.$ the row rank of A^t is the same as the column rank of A;
  $vi.$ the row rank of A^t is not less than the column rank of A^t;
  $vii.$ the column rank of A is not less than the row rank of A.

     Chapter 2. THE PROCESS (**Simple elimination**)

Exercise 2.9.33. Show that any n linearly independent vectors in $\mathbf{C}^n$ form a basis for $\mathbf{C}^n$.

Exercise 2.9.34. Let A be a SQUARE matrix. Show:

i. $ker(A^k) \subset ker(A^{k+1})$;

ii. $im(A^k) \supset im(A^{k+1})$;

iii. $ker(A^k) = ker(A^{k+1})$ iff for all natural numbers n

$$ker(A^k) = ker(A^{k+n});$$

iv. $im(A^k) = im(A^{k+1})$ iff for all natural numbers n

$$im(A^k) = im(A^{k+n});$$

v. there is in $\mathbf{N}$ a number k_A such that for all n in $\mathbf{N}$

$$ker(A) \subsetneq ker(A^2) \subsetneq \cdots \subsetneq ker(A^{k_A}) = ker(A^{k_A+n});$$
$$im(A) \supsetneq im(A^2) \supsetneq \cdots \supsetneq im(A^{k_A}) = im(A^{k_A+n}).$$

[*Hint:* For *iii* note that if $ker(A^k) = ker(A^{k+1})$ and $A^{k+2}\mathbf{x} = \mathbf{O}$ then $A\mathbf{x} \in ker(A^{k+1}) = ker(A^k)$. Use mathematical induction. For *iv* use *iii* and the symbolic equation $dim[ker] + dim[im] = dim[dom]$.]

The burden of **Exercise 2.9.34** is the following statement.

For any SQUARE matrix A there is a natural number k_A and a subspace $K_A \stackrel{\text{def}}{=} ker(A^{k_A})$ such that for all vectors $\mathbf{x}$ in K_A and all natural numbers m $A^m\mathbf{x} = \mathbf{O}$.

Exercise 2.9.35. Let A be an $m \times n$ matrix and let Y be a set of linearly independent vectors $\mathbf{y}_i$, $1 \leq i \leq k$ in $\mathbf{C}^m$. Assume that for each $\mathbf{y}_i$ there is in $\mathbf{C}^n$ a vector $\mathbf{x}_i$ such that $A\mathbf{x}_i = \mathbf{y}_i$, $1 \leq i \leq k$. Show that $X \stackrel{\text{def}}{=} \{\mathbf{x}_i,\ 1 \leq i \leq k\}$ is a linearly independent set.

Exercise 2.9.36. Let n be a natural number greater than 2. Show that the rank of the matrix

$$\begin{pmatrix} 1 & 2 & \ldots & n \\ n+1 & n+2 & \ldots & 2n \\ \vdots & \vdots & \ddots & \vdots \\ (n-1)n+1 & (n-1)n+2 & \ldots & n^2 \end{pmatrix}$$

is 2.

Exercise 2.9.37. Let $a, a+d, a+2d, \ldots$ be an arithmetic progression. Show that if $n > 2$ the rank of the $n \times n$ matrix

$$\begin{pmatrix} a & a+d & \cdots & a+(n-1)d \\ a+nd & a+(n+1)d & \cdots & a+(2n-1)d \\ \vdots & \vdots & \ddots & \vdots \\ a+(n-1)nd & a+[(n-1)n+1]d & \cdots & a+(n^2-1)d \end{pmatrix}$$

is not more than 2. Show that the rank is 0 iff $a = d = 0$, the rank is 1 iff $a \neq 0$ and $d = 0$, and that the rank is 2 iff $d \neq 0$.

Exercise 2.9.38. Let $a, ar, ar^2, \ldots$ be a geometric progression. Show that if $n > 1$ the rank of the $n \times n$ matrix

$$\begin{pmatrix} a & ar & \cdots & ar^{n-1} \\ ar^n & ar^{n+1} & \cdots & ar^{2n-1} \\ \vdots & \vdots & \ddots & \vdots \\ ar^{(n-1)n} & ar^{(n-1)n+1} & \cdots & ar^{n^2-1} \end{pmatrix}$$

is not more than 1. Show that the rank is 1 iff $ar \neq 0$.

Exercise 2.9.39. Assume that $S \overset{\text{def}}{=} \{\mathbf{x}_1, \ldots, \mathbf{x}_k\}$ is a linearly dependent set of vectors.

i. Show that there is a set $\{\alpha_1, \ldots, \alpha_k\}$ of numbers such that at least one is *positive* and $\sum_{i=1}^k \alpha_i \mathbf{x}_i = \mathbf{O}$.

ii. Show that at least one of the vectors, say $\mathbf{x}_{i_0}$, may be written as a linear combination of the others in S.

Exercise 2.9.40. Let A be an $n \times n$ matrix and let $A|\mathbf{b}$ be a corresponding augmented matrix. Show that

EITHER

$$rank(A) = rank(A|\mathbf{b})$$

OR

$$rank(A) = rank(A|\mathbf{b}) - 1.$$

2.10. Matrices and linear transformations

If $A \overset{\text{def}}{=} (a_{ij})_{i,j=1}^{m,n}$ is an $m \times n$ matrix then the transpose A^t of A is the $n \times m$ matrix $(\alpha_{ji})_{j,i=1}^{n,m}$ in which $\alpha_{ji} = a_{ij}$. Thus if

$$A = \begin{pmatrix} 1 & 2 & 3 \\ 4 & 5 & 6 \end{pmatrix}$$

then

$$A^t = \begin{pmatrix} 1 & 4 \\ 2 & 5 \\ 3 & 6 \end{pmatrix}.$$

In words:

The transpose of an $m \times n$ matrix A is an $n \times m$ matrix A^t in which the rows are the columns of A and the columns are the rows of A.

In **Exercise 2.8.2** transposes are handy in some of the proofs. Moreover, the ideas and language of **Section 2.9** lead to a natural rôle that transposes play in the subject of linear algebra. For simplicity the transposes of SQUARE matrices are studied first.

Let

$$A \stackrel{\text{def}}{=} \begin{pmatrix} a_{11} & \cdots & a_{1n} \\ \vdots & \ddots & \vdots \\ a_{n1} & \cdots & a_{nn} \end{pmatrix}$$

be an $n \times n$ SQUARE matrix.

For each column vector

$$\mathbf{x} \stackrel{\text{def}}{=} \begin{pmatrix} x_1 \\ \vdots \\ x_n \end{pmatrix}$$

there is another column vector

$$\mathbf{y} \stackrel{\text{def}}{=} \begin{pmatrix} y_1 \\ \vdots \\ y_n \end{pmatrix} \stackrel{\text{def}}{=} \begin{pmatrix} a_{11} & \cdots & a_{1n} \\ \vdots & \ddots & \vdots \\ a_{n1} & \cdots & a_{nn} \end{pmatrix} \begin{pmatrix} x_1 \\ \vdots \\ x_n \end{pmatrix} = A\mathbf{x}.$$

The vector $\mathbf{y}$ depends on $\mathbf{x}$, is a function of $\mathbf{x}$: $\mathbf{y} = T(\mathbf{x})$ and the laws of matrix algebra show that if p and q are numbers and $\mathbf{u}$ and $\mathbf{v}$ are vectors then

$$T(p\mathbf{u} + q\mathbf{v}) = pT(\mathbf{u}) + qT(\mathbf{v}). \tag{2.10.1}$$

The equation (2.10.1) is reminiscent of the equation (2.9.6) and so T is called a *linear transformation*, i.e., a function that *transforms* the vector $\mathbf{x}$ into the vector $\mathbf{y}$ and does so in the linear fashion described in (2.10.1).

As usual let $\mathbf{e}_i$ be

$$\begin{pmatrix} 0 \\ \vdots \\ 1 \\ \vdots \\ 0 \end{pmatrix} \quad \leftarrow \text{ row } i$$

and let $E \stackrel{\text{def}}{=} \{\mathbf{e}_1, \ldots, \mathbf{e}_n\}$ be the standard basis for $\mathbf{C}^n$. Then each vector $T(\mathbf{e}_i)$, as a vector in $\mathbf{C}^n$ is a linear combination of the vectors in the basis E:

$$T(\mathbf{e}_i) = \alpha_{i1}\mathbf{e}_1 + \cdots + \alpha_{in}\mathbf{e}_n, \ 1 \le i \le n. \tag{2.10.2}$$

The striking fact is that the matrix

$$T_{EE} \stackrel{\text{def}}{=} \begin{pmatrix} \alpha_{11} & \cdots & \alpha_{1n} \\ \vdots & \ddots & \vdots \\ \alpha_{n1} & \cdots & \alpha_{nn} \end{pmatrix}$$

is the transpose A^t of A:

$$T_{EE} = A^t. \tag{2.10.3}$$

The next calculation shows why (2.10.3) is true. By definition

$$T(\mathbf{e}_i) = A\mathbf{e}_i = \begin{pmatrix} a_{1i} \\ \vdots \\ a_{ni} \end{pmatrix}$$

$$= a_{1i}\mathbf{e}_1 + \cdots + a_{ni}\mathbf{e}_n. \tag{2.10.4}$$

Comparison of (2.10.2) and (2.10.4) confirms the validity of (2.10.3).

Example 2.10.1. Let A be the matrix

$$\begin{pmatrix} 1 & 2 \\ 3 & 4 \end{pmatrix}.$$

If

$$\mathbf{x} = \begin{pmatrix} x_1 \\ x_2 \end{pmatrix}.$$

then

$$\mathbf{y} = T(\mathbf{x}) = \begin{pmatrix} x_1 + 2x_2 \\ 3x_1 + 4x_2 \end{pmatrix}.$$

In particular

$$T(\mathbf{e}_1) = \begin{pmatrix} 1 \\ 3 \end{pmatrix}$$

$$T(\mathbf{e}_2) = \begin{pmatrix} 2 \\ 4 \end{pmatrix}$$

or

$$T(\mathbf{e}_1) = 1\mathbf{e}_1 + 3\mathbf{e}_2$$

$$T(\mathbf{e}_2) = 2\mathbf{e}_1 + 4\mathbf{e}_2. \tag{2.10.5}$$

Let

$$\mathcal{E} \stackrel{\text{def}}{=} \begin{pmatrix} \mathbf{e}_1 \\ \mathbf{e}_2 \end{pmatrix}$$

be the *vectorial vector* in which the components are themselves the vectors $\mathbf{e}_1$ and $\mathbf{e}_2$. Then the last two equations in (2.10.5) may be written $T(\mathcal{E}) = A^t\mathcal{E}$, which illustrates the natural rôle played by the transpose.

Exercise 2.10.1. Let A be the matrix

$$\begin{pmatrix} -6 & 4 & 5 \\ i & 1 & 0 \\ 3 & -2 & 9 \end{pmatrix}$$

and let it define a linear transformation T according to the pattern set out above. Find

$$T\left[\begin{pmatrix} a \\ b \\ c \end{pmatrix}\right].$$

More generally, let

$$X \overset{\text{def}}{=} \{\mathbf{x}_1, \ldots, \mathbf{x}_n\}$$

be a basis for $\mathbf{C}^n$ and let

$$A \overset{\text{def}}{=} \begin{pmatrix} a_{11} & \cdots & a_{1n} \\ \vdots & \ddots & \vdots \\ a_{n1} & \cdots & a_{nn} \end{pmatrix}$$

be an $n \times n$ matrix. Then the formula

$$T : \mathbf{C}^n \ni \mathbf{x} \overset{\text{def}}{=} \sum_{i=1}^{n} x_i \mathbf{x}_i \mapsto \sum_{i=1}^{n} \left(\sum_{j=1}^{n} a_{ij} x_j \right) \mathbf{x}_i \in \mathbf{C}^n$$

$$\overset{\text{def}}{=} \sum_{i=1}^{n} y_i \mathbf{x}_i \overset{\text{def}}{=} \mathbf{y}$$

defines a linear transformation T and $T(\mathbf{x}) = \mathbf{y}$. Let

$$\begin{pmatrix} \mathbf{x}_1 \\ \vdots \\ \mathbf{x}_n \end{pmatrix}$$

be the vectorial vector $\mathcal{X}$ the components of which are the vectors in the basis X. Then straightforward calculation shows that with abuse of notation

$$T(\mathcal{X}) = A^t \mathcal{X} \overset{\text{def}}{=} T_{XX} \mathcal{X}.$$

(The linear transformation T transforms the vectorial vector $\mathcal{X}$ into the vectorial vector $T_{XX}\mathcal{X}$.)

If a vector $\mathbf{x}$ is given by the equation

$$\mathbf{x} \overset{\text{def}}{=} \sum_{i=1}^{n} x_i \mathbf{x}_i$$

---

call the numbers $x_1, \ldots, x_n$ the X-*components* of the vector $\mathbf{x}$. The *same* vector $\mathbf{x}$ may be written in terms of the basis E:

$$\mathbf{x} = \sum_{i=1}^{n} \eta_i \mathbf{e}_i.$$

The numbers $\eta_1, \ldots, \eta_n$ are the E-components of $\mathbf{x}$. Thus it is convenient to write

$$\mathbf{x}_X \overset{\text{def}}{=} \begin{pmatrix} x_1 \\ \vdots \\ x_n \end{pmatrix}$$

$$\mathbf{x}_E \overset{\text{def}}{=} \begin{pmatrix} \eta_1 \\ \vdots \\ \eta_n \end{pmatrix}.$$

However it is not generally true that $\mathbf{x}_X = \mathbf{x}_E$ unless each $\mathbf{x}_i$ is $\mathbf{e}_i$. The column vector

$$\mathbf{y}_X \overset{\text{def}}{=} A\,\mathbf{x}_X \overset{\text{def}}{=} \begin{pmatrix} y_1 \\ \vdots \\ y_n \end{pmatrix}$$

gives rise to the vector $\mathbf{y}$ according to the formula

$$\mathbf{y} = \sum_{i=1}^{n} y_i \mathbf{x}_i,$$

i.e., $\mathbf{y} = T(\mathbf{x})$.

    In sum,

$$\mathbf{x} = \mathbf{x}_X^t\, \mathcal{X}$$
$$\mathbf{y} = \mathbf{y}_X^t\, \mathcal{X}$$
$$\mathbf{y}_X = A\,\mathbf{x}_X$$
$$\mathbf{y} = T(\mathbf{x})$$
$$\mathbf{y} = \mathbf{y}_X^t\, \mathcal{X}$$
$$T(\mathcal{X}) = A^t \mathcal{X} \overset{\text{def}}{=} T_{XX}\, \mathcal{X}. \tag{2.10.6}$$

Exercise 2.10.2. Let

$$A \overset{\text{def}}{=} \begin{pmatrix} 1 & 2 & 3 & 4 & 5 \\ 6 & 7 & 8 & 9 & 8 \\ 7 & 6 & 5 & 4 & 3 \\ 2 & 1 & 2 & 3 & 4 \\ 5 & 6 & 7 & 8 & 9 \end{pmatrix}$$

be a 5×5 matrix and let A define a linear transformation T in accordance with the pattern set out above. Let

$$\{e_1 + e_2, e_3 - e_4, e_2 + e_5, e_1 + e_3 + e_5, e_1 + 2e_2 + 4e_4\}$$
$$\stackrel{\text{def}}{=} \{\mathbf{x}_1, \mathbf{x}_2, \mathbf{x}_3, \mathbf{x}_4, \mathbf{x}_5\}$$

be a set X of five vectors in $\mathbf{C}^5$.

   i. Show that X is a basis for $\mathbf{C}^5$.

  ii. Let

$$\mathbf{x} \stackrel{\text{def}}{=} \sum_{k=1}^{n} (-1)^k k \mathbf{x}_k$$

be a vector in $\mathbf{C}^5$. Write out the column vectors $\mathbf{x}_X$ and $\mathbf{x}_E$.

iii. Assume

$$\begin{pmatrix} y_1 \\ \vdots \\ y_n \end{pmatrix} \stackrel{\text{def}}{=} \mathbf{y}_X \stackrel{\text{def}}{=} A\mathbf{x}_X$$

and $\mathbf{y} \stackrel{\text{def}}{=} \sum_{i=1}^{n} y_n \mathbf{x}_i$. Find the column vector $\mathbf{y}_E$.

 iv. Verify the relations in (2.10.6).

  v. Describe the vectorial vector $\mathcal{X}$.

 vi. Describe the vectorial vector $A\mathcal{X}$.

vii. Write out the matrix T_{XX}.

Example 2.10.2. Let X be the basis

$$\{e_1, e_1 + e_2, e_1 + e_2 + e_3\}$$

for $\mathbf{C}^3$. The vectorial vector $\mathcal{X}$ is

$$\begin{pmatrix} e_1 \\ e_1 + e_2 \\ e_1 + e_2 + e_3 \end{pmatrix}.$$

The matrix in which the columns are the vectorial *components* of $\mathcal{X}$ is

$$Q \stackrel{\text{def}}{=} \begin{pmatrix} 1 & 1 & 1 \\ 0 & 1 & 1 \\ 0 & 0 & 1 \end{pmatrix} \quad \text{and} \quad Q^{-1} = \begin{pmatrix} 1 & -1 & 0 \\ 0 & 1 & -1 \\ 0 & 0 & 1 \end{pmatrix}.$$

Denote the transpose Q^t of Q by R_{XE}. It is the matrix that shows how the vectors in the basis X are expressed in terms of the vectors in the basis E.

$$R_{XE} \stackrel{\text{def}}{=} Q^t = \begin{pmatrix} 1 & 0 & 0 \\ 1 & 1 & 0 \\ 1 & 0 & 1 \end{pmatrix};$$

$$\mathcal{X} = R_{XE}\mathcal{E}.$$

Since

$$(Q^{-1})^t = (Q^t)^{-1} = \begin{pmatrix} 1 & 0 & 0 \\ -1 & 1 & 0 \\ 0 & -1 & 1 \end{pmatrix}$$

and since direct calculation shows that $\mathcal{E} = (Q^t)^{-1}\mathcal{X}$ it is natural to denote $(Q^t)^{-1}$ by R_{EX} and to write $\mathcal{E} = R_{EX}\mathcal{X}$.

Exercise 2.10.3. Let E be the standard basis for $\mathbf{C}^n$ and let $\{\mathbf{x}_1,\ldots,\mathbf{x}_n\}$ be another basis X for $\mathbf{C}^n$. Let the typical column vector $\mathbf{x}_j$ have E-components

$$\begin{pmatrix} x_{1j} \\ \vdots \\ x_{nj} \end{pmatrix}.$$

Let $\mathcal{E}$ and $\mathcal{X}$ be the corresponding vectorial vectors and let Q be the matrix in which the jth column is the column vector $\mathbf{x}_j$:

$$Q \stackrel{\text{def}}{=} \begin{array}{cc} \begin{array}{ccc} \mathbf{x}_1 & \cdots & \mathbf{x}_n \end{array} & \\ \begin{pmatrix} x_{11} & \cdots & x_{1n} \\ \vdots & \ddots & \vdots \\ x_{n1} & \cdots & x_{nn} \end{pmatrix} & . \end{array}$$

Define R_{XE} and R_{EX} according to the pattern set out above.

   i. Show that:

$$\mathcal{X} = Q^t\mathcal{E} = R_{XE}\mathcal{E};$$
$$\mathcal{E} = (Q^t)^{-1}\mathcal{X} = R_{EX}\mathcal{X}.$$

   ii. Let $Y \stackrel{\text{def}}{=} \{\mathbf{y}_1,\ldots,\mathbf{y}_n\}$ be yet another basis for $\mathbf{C}^n$. Define R_{YE} and R_{EY} according to the pattern set out above. Show that

$$\mathcal{Y} = R_{YE}R_{EX}\mathcal{X};$$
$$\mathcal{X} = R_{XE}R_{EY}\mathcal{Y}.$$

   iii. Let R_{XY} and R_{YX} be the matrices relating the bases X and Y: $\mathcal{X} = R_{XY}\mathcal{Y}$ and $\mathcal{Y} = R_{YX}\mathcal{X}$. Show that:

$$R_{XY} = R_{XE}R_{EY};$$
$$R_{YX} = R_{YE}R_{EX};$$
$$I = R_{XY}R_{YX} = R_{XE}R_{EX} = R_{YE}R_{EY}.$$

One may start afresh with an $n \times n$ matrix $A \stackrel{\text{def}}{=} (a_{ij})_{i,j=1}^{n,n}$ and a basis $X \stackrel{\text{def}}{=} \{\mathbf{x}_1,\ldots,\mathbf{x}_n\}$ but this time one may define a transformation $\tilde{T}$ by the

---

    *Chapter 2. THE PROCESS (**Simple elimination**)*

formulae:

$$\widetilde{T}(\mathbf{x}_i) \overset{\text{def}}{=} \sum_{j=1}^{n} a_{ij}\mathbf{x}_j, \ 1 \le i \le n,$$

$$\widetilde{T}(\sum_{i=1}^{n} x_i\mathbf{x}_i) \overset{\text{def}}{=} \sum_{i=1}^{n} x_i\widetilde{T}\mathbf{x}_i.$$

A direct calculation shows that $\widetilde{T}$ is a linear transformation. Furthermore, since X is a basis for $\mathbf{C}^n$ it follows that there are unique numbers y_i such that

$$\widetilde{T}(\sum_{i=1}^{n} x_i\mathbf{x}_i) = \sum_{i=1}^{n} y_i\mathbf{x}_i.$$

How are

$$\mathbf{x}_X \overset{\text{def}}{=} \begin{pmatrix} x_1 \\ \vdots \\ x_n \end{pmatrix} \ \text{and} \ \mathbf{y}_X \overset{\text{def}}{=} \begin{pmatrix} y_1 \\ \vdots \\ y_n \end{pmatrix}$$

related? As one might expect the answer is found in terms of the transpose of A:

$$\mathbf{y}_X = A^t\,\mathbf{x}_X.$$

Indeed

$$
\begin{aligned}
\widetilde{T}(\sum_{i=1}^{n} x_i\mathbf{x}_i) &= \sum_{i=1}^{n} x_i\widetilde{T}(\mathbf{x}_i) \\
&= \sum_{i=1}^{n} x_i(\sum_{j=1}^{n} a_{ij}\mathbf{x}_j) \\
&= \sum_{j=1}^{n}(\sum_{i=1}^{n} x_i a_{ij})\mathbf{x}_j \\
&= \sum_{j=1}^{n} y_j\mathbf{x}_j.
\end{aligned}
\tag{2.10.7}
$$

Hence

$$y_j = \sum_{i=1}^{n} x_i a_{ij}$$

i.e., $\mathbf{y} = A^t\mathbf{x}$.

In sum:

$$
\begin{aligned}
\mathbf{x} &= \mathbf{x}_X^t \, \mathcal{X} \\
\mathbf{y} &= \mathbf{y}_X^t \, \mathcal{X} \\
\mathbf{y} &= \widetilde{T}(\mathbf{x}) = A^t \mathbf{x}_X \, \mathcal{X} \\
\mathbf{y} &= \mathbf{y}_X^t \, \mathcal{X} = \mathbf{x}_X^t \, A \, \mathcal{X} \\
\widetilde{T}(\mathcal{X}) &= A \mathcal{X}
\end{aligned}
\qquad\qquad
\begin{aligned}
\mathbf{y} &= \mathbf{y}_X^t \, \mathcal{X} \\
\mathbf{x} &= \mathbf{x}_X^t \, \mathcal{X} \\
\mathbf{y}_X &= A \, \mathbf{x}_X \\
\mathbf{y} &= T(\mathbf{x}) \\
\mathbf{y} &= \mathbf{y}_X^t \, \mathcal{X} \\
T(\mathcal{X}) &= A^t \mathcal{X} \stackrel{\text{def}}{=} T_{XX}\,\mathcal{X}.
\end{aligned}
$$

$$(2.10.8)$$

A careful comparison between the displays (2.10.6) and (2.10.8) is instructive.

Exercise 2.10.4. Use the matrix A and the basis X of **Exercise 2.10.2** to define a linear transformation $\widetilde{T}$ in accordance with the second pattern just discussed. Check the relations (2.10.8).

Finally, assume that

$$
T : \mathbf{C}^n \ni \mathbf{x} \mapsto \mathbf{y} \stackrel{\text{def}}{=} T(\mathbf{x}) \in \mathbf{C}^n
$$

is a linear transformation and that X as above is a basis for $\mathbf{C}^n$. Then because X is a basis for $\mathbf{C}^n$ each $T(\mathbf{x}_i)$ is a linear combination of the $\mathbf{x}_j$:

$$
T(\mathbf{x}_i) = \sum_{j=1}^n a_{ij}\mathbf{x}_j.
$$

On the other hand, for $\mathbf{x}$ in $\mathbf{C}^n$

$$
\begin{aligned}
\mathbf{x} &= \sum_{i=1}^n x_i \mathbf{x}_i \\
T(\mathbf{x}) &= \sum_{i=1}^n x_i T(\mathbf{x}_i) \\
&= \sum_{i=1}^n x_i \Big(\sum_{j=1}^n a_{ij}\mathbf{x}_j \Big) \\
&= \sum_{j=1}^n \Big(\sum_{i=1}^n x_i a_{ij} \Big)\mathbf{x}_j.
\end{aligned}
$$

It follows that if

$$
\mathbf{y} \stackrel{\text{def}}{=} T(\mathbf{x}) \stackrel{\text{def}}{=} \sum_{i=1}^n y_i \mathbf{x}_i
$$

$$
= \mathbf{y}_X^t \, \mathcal{X}
$$

---

then, T_{XX} designating the matrix,

$$\begin{pmatrix} a_{11} & \cdots & a_{1n} \\ \vdots & \ddots & \vdots \\ a_{n1} & \cdots & a_{nn} \end{pmatrix},$$

i.e., $T(\mathcal{X}) = T_{XX}\mathcal{X}$, (which shows how each $T(\mathbf{x}_i)$ is written in terms of the basis X), there obtains

$$\mathbf{y}_X \overset{\text{def}}{=} \begin{pmatrix} y_1 \\ \vdots \\ y_n \end{pmatrix}$$

$$= \begin{pmatrix} a_{11} & \cdots & a_{n1} \\ \vdots & \ddots & \vdots \\ a_{1n} & \cdots & a_{nn} \end{pmatrix} \begin{pmatrix} x_1 \\ \vdots \\ x_n \end{pmatrix}$$

$$= T_{XX}^t \mathbf{x}_X.$$

In sum this time:

i. When A is given and T is defined by the action of A on the X-components of a vector $\mathbf{x}$ then T is a linear transformation. The X-components of $T(\mathbf{x}_i)$ constitute the ith row of A^t.

ii. When A is given and T is defined by the action of A on the vectorial vector $\mathcal{X}$ then T is a linear transformation. The X-components y_i of $T(\mathbf{x})$ are related to the X-components x_i of $\mathbf{x}$ by the matrix A^t:

$$\mathbf{y}_X = A^t \mathbf{x}_X = (\mathbf{x}_X^t A)^t.$$

iii. If T is given as a linear transformation of $\mathbf{C}^n$ into itself then T naturally generates a matrix

$$T_{XX} \overset{\text{def}}{=} \begin{pmatrix} a_{11} & \cdots & a_{1n} \\ \vdots & \ddots & \vdots \\ a_{n1} & \cdots & a_{nn} \end{pmatrix}$$

when each $T(\mathbf{x}_i)$ is written in terms of the basis X:

$$T(\mathbf{x}_i) = \sum_{j=1}^{n} a_{ij} \mathbf{x}_j.$$

If

$$\mathbf{x} = \sum_{i=1}^{n} x_i \mathbf{x}_i$$

and if

$$\mathbf{y}_X \stackrel{\text{def}}{=} \begin{pmatrix} y_1 \\ \vdots \\ y_n \end{pmatrix} = T^t_{XX}\mathbf{x}_X = \begin{pmatrix} a_{11} & \cdots & a_{n1} \\ \vdots & \ddots & \vdots \\ a_{1n} & \cdots & a_{nn} \end{pmatrix}\begin{pmatrix} x_1 \\ \vdots \\ x_n \end{pmatrix}$$

then

$$\mathbf{y} = T(\mathbf{x}) = \sum_{i=1}^{n} y_i\mathbf{x}_i$$
$$= \mathbf{y}^t_X \mathcal{X}$$
$$= \mathbf{x}^t_X T_{XX}\mathcal{X}.$$

The preceding discussion should make clearer the fact that matrix transposes arise naturally in the course of the development of linear algebra.

[**Note 2.10.1:** Once a matrix A and a basis X have been chosen the corresponding linear transformation T is defined once and for all. It finds its expression or *representation* as a matrix T_{XX} with respect to the basis X and as another matrix T_{YY} in terms of another basis Y. As the bases change so *may* the matrix representations change but the linear transformation remains the same.

The matrix representation sometimes does not change. For example, the linear transformation that leaves all vectors fixed is called the *identity linear transformation* and, for convenience, is also denoted I. No matter what the basis X, the matrix representation I_{XX} of I is always the same, namely the identity matrix I: $I_{XX} = I$. More generally, if T is a linear transformation and T_{XX} is its representation with respect to a basis X and if Y is a new basis then the representation T_{YY} of T with respect to Y is given by the formula:

$$T_{YY} = R_{YX}T_{XX}R_{XY}.$$

If $R_{YX}T_{XX} = T_{XX}R_{YX}$ then, since $R_{YX}R_{XY} = I$ it follows that the matrix representations T_{XX} and T_{YY} of T are the same.]

Linear transformations and arbitrary matrices

One may ring the changes on the theme above by considering arbitrary, not necessarily SQUARE, matrices A and linear transformations T from some $\mathbf{C}^n$ to some $\mathbf{C}^m$. The results analogous to those above are given below and are followed by **Exercise 2.10.5** in which the details can be worked out.

The situation just described may be summarized as follows.

Let $A \stackrel{\text{def}}{=} (a_{ij})_{i,j=1}^{m,n}$ be an $m \times n$ matrix and let

$$X \stackrel{\text{def}}{=} \{\mathbf{x}_i, \ldots, \mathbf{x}_n\}$$
resp.
$$Y \stackrel{\text{def}}{=} \{\mathbf{y}_1, \ldots, \mathbf{y}_m\}$$

be bases for $\mathbf{C}^n$ resp. $\mathbf{C}^m$.

Then:

i. The rule

$$\begin{pmatrix} x_1 \\ \vdots \\ x_n \end{pmatrix} \mapsto \begin{pmatrix} y_1 \\ \vdots \\ y_m \end{pmatrix} \overset{\text{def}}{=} \begin{pmatrix} a_{11} & \cdots & a_{1n} \\ \vdots & \ddots & \vdots \\ a_{m1} & \cdots & a_{mn} \end{pmatrix} \begin{pmatrix} x_1 \\ \vdots \\ x_n \end{pmatrix}$$

defines a linear transformation by the formula

$$T(\mathbf{x}) \overset{\text{def}}{=} T(\sum_{i=1}^{n} x_i \mathbf{x}_i) = \sum_{j=1}^{m} y_j \mathbf{y}_j \overset{\text{def}}{=} \mathbf{y}.$$

In these circumstances

$$T(\mathbf{x}_i) = \sum_{j=1}^{m} t_{ij} \mathbf{y}_j, \ 1 \le i \le n.$$

The $n \times m$ matrix $(t_{ij})_{i,j=1}^{n,m}$ is naturally denoted T_{XY} and

$$T_{XY} = \begin{pmatrix} t_{11} & \cdots & t_{1n} \\ \vdots & \ddots & \vdots \\ t_{n1} & \cdots & t_{nn} \end{pmatrix} = \begin{pmatrix} a_{11} & \cdots & a_{n1} \\ \vdots & \ddots & \vdots \\ a_{1n} & \cdots & a_{nn} \end{pmatrix} = A^t.$$

ii. The rule

$$\mathbf{x}_i \mapsto \sum_{i=1}^{m} a_{ji} \mathbf{y}_j$$

$$\mathcal{X} \mapsto A^t \mathcal{Y}$$

defines a linear transformation T such that $T(\mathcal{X}) = A^t \mathcal{Y}$ and it is natural to denote A^t by T_{XY}. If

$$\mathbf{x} \overset{\text{def}}{=} \sum_{i=1}^{n} x_i \mathbf{x}_i$$

$$\mathbf{y} \overset{\text{def}}{=} T(\mathbf{x}) \overset{\text{def}}{=} \sum_{j=1}^{n} y_j \mathbf{y}_j$$

then

$$y_j = \sum_{i=1}^{n} a_{ji} x_i, \ 1 \le j \le m$$

$$\mathbf{y}_X = T_{XY}^t \mathbf{x}_X.$$

---

iii. If

$$T(\mathbf{x}_i) \overset{\text{def}}{=} \sum_{j=1}^{m} a_{ij}\mathbf{y}_j$$

$$T : \mathbf{C}^n \ni \mathbf{x} \overset{\text{def}}{=} \sum_{i=1}^{n} x_i\mathbf{x}_i \mapsto T(\mathbf{x}) \overset{\text{def}}{=} \mathbf{y} \overset{\text{def}}{=} \sum_{i=1}^{n} x_i T(\mathbf{x}_i)$$

then T is a linear transformation. If

$$\mathbf{x} \overset{\text{def}}{=} \sum_{i=1}^{n} x_i\mathbf{x}_i$$

and

$$T(\mathbf{x}) \overset{\text{def}}{=} \mathbf{y} = \sum_{j=1}^{m} y_j\mathbf{y}_j$$

then

$$\mathbf{y}_Y \overset{\text{def}}{=} \begin{pmatrix} y_1 \\ \vdots \\ y_m \end{pmatrix} = \begin{pmatrix} a_{11} & \cdots & a_{m1} \\ \vdots & \ddots & \vdots \\ a_{1n} & \cdots & a_{mn} \end{pmatrix} \begin{pmatrix} x_1 \\ \vdots \\ x_n \end{pmatrix} = A^t \mathbf{x}_X$$
$$= (\mathbf{x}_X^t A)^t.$$

It is natural to denote A by T_{XY}.

Let

$$X \overset{\text{def}}{=} \{\mathbf{x}_1, \ldots, \mathbf{x}_n\} \ \text{resp.} \ Y \overset{\text{def}}{=} \{\mathbf{y}_1, \ldots, \mathbf{y}_m\}$$

be bases for $\mathbf{C}^n$ resp. $\mathbf{C}^m$. The following relationships hold.

i. When $A \overset{\text{def}}{=} (a_{ij})_{i,j=1}^{m,n}$ is given and $T : \mathbf{C}^n \mapsto \mathbf{C}^m$ is defined by the action of A on the X-components of a vector $\mathbf{x}$ then T is a linear transformation. The Y-components of $T(\mathbf{x}_i)$ constitute the ith row of A^t:

$$T_{XY} = A^t.$$

ii. When A is given and T is defined by the action of A on the vectorial vector $\mathcal{X}$ then T is a linear transformation and $T_{XY} = A$. The Y-components y_i of $T(\mathbf{x})$ are related to the X-components x_i of $\mathbf{x}$ by the matrix A^t:

$$\mathbf{y}_Y = A^t \mathbf{x}_X = (\mathbf{x}_X^t A)^t.$$

iii. If T is given as a linear transformation of $\mathbf{C}^n$ into $\mathbf{C}^m$ then T naturally generates an $n \times m$ matrix $T_{XY} \overset{\text{def}}{=} (a_{ij})_{i,j=1}^{n,m}$ when each $T(\mathbf{x}_i)$ is written in terms of the basis Y:

$$T(\mathbf{x}_i) = \sum_{j=1}^{m} a_{ij}\mathbf{y}_j, \ 1 \le i \le n.$$

    Chapter 2. THE PROCESS (Simple elimination)

If

$$\mathbf{x} = \sum_{i=1}^{n} x_i \mathbf{x}_i$$

and if

$$\begin{pmatrix} x_1 \\ \vdots \\ x_n \end{pmatrix} \overset{\text{def}}{=} \mathbf{x}_X$$

$$\begin{pmatrix} y_1 \\ \vdots \\ y_n \end{pmatrix} \overset{\text{def}}{=} \begin{pmatrix} a_{11} & \cdots & a_{n1} \\ \vdots & \ddots & \vdots \\ a_{1m} & \cdots & a_{nm} \end{pmatrix} \begin{pmatrix} x_1 \\ \vdots \\ x_n \end{pmatrix}$$

i.e.,

$$\mathbf{y}_Y = T_{XY}^t \mathbf{x}_X = (\mathbf{x}_X^t T_{XY})^t$$

then

$$T(\mathbf{x}) = \sum_{j=1}^{m} y_j \mathbf{y}_j$$
$$= \mathbf{y}_Y^t \mathcal{Y}$$
$$= \mathbf{x}_X^t T_{XY} \mathcal{Y}.$$

To solidify the understanding of what is going on **Exercise 2.10.5** is helpful.

Exercise 2.10.5. Let

$$A \overset{\text{def}}{=} \begin{pmatrix} 1 & 2 & 3 & 4 \\ 5 & 6 & 7 & 8 \end{pmatrix}$$

be regarded as transforming each column vector

$$\mathbf{x} \overset{\text{def}}{=} \begin{pmatrix} x_1 \\ x_2 \\ x_3 \\ x_4 \end{pmatrix}$$

to the vector

$$\mathbf{y} \overset{\text{def}}{=} \begin{pmatrix} 1 & 2 & 3 & 4 \\ 5 & 6 & 7 & 8 \end{pmatrix} \begin{pmatrix} x_1 \\ x_2 \\ x_3 \\ x_4 \end{pmatrix} \overset{\text{def}}{=} \begin{pmatrix} y_1 \\ y_2 \end{pmatrix}.$$

Let

$$X \overset{\text{def}}{=} \{\mathbf{e}_1, \mathbf{e}_1 + \mathbf{e}_2, \mathbf{e}_1 + \mathbf{e}_2 + \mathbf{e}_3, \mathbf{e}_1 + \mathbf{e}_2 + \mathbf{e}_3 + \mathbf{e}_4\}$$

be chosen as a basis for $\mathbf{C}^4$ and let

$$Y \overset{\text{def}}{=} \{\mathbf{e}_1 + \mathbf{e}_2, \mathbf{e}_1 - 2\mathbf{e}_2\}$$

be chosen as a basis for $\mathbf{C}^2$.

Then A defines a transformation T according to the rule

$$T : \mathbf{C}^4 \ni \mathbf{x} \overset{\text{def}}{=} \mathbf{x}_X^t \mathcal{X} \mapsto \mathbf{y} \overset{\text{def}}{=} \mathbf{y}_Y^t \mathcal{Y}$$
$$= \mathbf{x}_X^t A^t \mathcal{Y} \in \mathbf{C}^2. \tag{2.10.9}$$

   i. Show that T is a linear transformation.

  ii. Show that

$$T(\mathcal{X}) = A^t \mathcal{Y}. \tag{2.10.10}$$

[**Note 2.10.2:** The transpose is natural in this context. In particular when $m \neq n$ and A is an $m \times n$ matrix the adjoint A^t is an $n \times m$ matrix. It just fits the sizes of the vectors to which it is applied.]

Exercise 2.10.6. Let

$$T : \mathbf{C}^n \mapsto \mathbf{C}^m \text{ and } S : \mathbf{C}^m \mapsto \mathbf{C}^l$$

be linear transformations. Their *composition* denoted ST is a transformation

$$R \overset{\text{def}}{=} ST : \mathbf{C}^n \mapsto \mathbf{C}^l$$

that takes the vector $\mathbf{x}$ in $\mathbf{C}^n$ into the vector $S(T(\mathbf{x}))$ in $\mathbf{C}^l$.

   i. Show R is a linear transformation.

  ii. Show that if $U : \mathbf{C}^l \mapsto \mathbf{C}^k$ is a linear transformation then $U(ST) = (US)T$ and the result is again a linear transformation. (Hence the result is designated unambiguously UST.)

 iii. Assume

$$X \overset{\text{def}}{=} \{\mathbf{x}_1, \ldots, \mathbf{x}_n\} \text{ is a basis for } \mathbf{C}^n;$$
$$Y \overset{\text{def}}{=} \{\mathbf{y}_1, \ldots, \mathbf{y}_m\} \text{ is a basis for } \mathbf{C}^m;$$
$$Z \overset{\text{def}}{=} \{\mathbf{z}_1, \ldots, \mathbf{z}_l\} \text{ is a basis for } \mathbf{C}^l.$$

Let T resp. S resp. R engender matrices T_{XY} resp. S_{YZ} resp. R_{XZ} acting on the bases X resp. Y resp. X.

 iv. Show: T_{XY} is an $n \times m$ matrix; S_{YZ} is an $l \times m$ matrix; R_{XZ} is an $n \times l$ matrix.

  v. Show that $R_{XZ} = T_{XY}S_{YZ}$. Note the order reversal:

$$(ST)_{XZ} = T_{XY}S_{YZ}$$

and the easily remembered match-up of subscripts:

$$R_{XZ} = T_{XY}S_{YZ}.$$

Example 2.10.3. Let the context of the situation above be the following:

Vector spaces	Bases	Matrices	Linear transformations
$\mathbf{C}^3$	$\mathcal{X} \stackrel{\text{def}}{=} \begin{pmatrix} e_1 - e_2 \\ e_1 + e_2 \\ 2e_2 - e_3 \end{pmatrix}$	$A \stackrel{\text{def}}{=} \begin{pmatrix} 1 & 2 & 3 & 4 \\ 5 & 6 & 7 & 8 \\ 9 & 1 & 2 & 3 \end{pmatrix}$	$T(\mathcal{X}) = A\mathcal{Y}$
$\mathbf{C}^4$	$\mathcal{Y} \stackrel{\text{def}}{=} \begin{pmatrix} e_1 \\ e_1 + e_2 \\ e_1 + e_2 + e_3 \\ e_2 + e_3 - e_4 \end{pmatrix}$	$A \stackrel{\text{def}}{=} \begin{pmatrix} 1 & 2 \\ 3 & 4 \\ 5 & 6 \\ 7 & 8 \end{pmatrix}$	$S(\mathcal{Y}) = B\mathcal{Z}$
$\mathbf{C}^2$	$\mathcal{Z} \stackrel{\text{def}}{=} \begin{pmatrix} e_1 \\ e_1 - 3e_2 \end{pmatrix}$	AB	$ST(\mathcal{X}) = AB\mathcal{Z}$

Then

$$A = T_{XY}, \quad B = S_{YZ}, \quad R = ST, \quad R_{XZ} = T_{XY}S_{YZ}.$$

Exercise 2.10.7. In the situation of **Example 2.10.3** let $\mathbf{x}$ be the vector

$$\begin{pmatrix} 1 \\ -2 \\ 3 \end{pmatrix},$$

i.e.,

$$\mathbf{x}_E = \begin{pmatrix} 1 \\ -2 \\ 3 \end{pmatrix}.$$

What is $\mathbf{x}_X$? Calculate $T(\mathbf{x})$ and $ST(\mathbf{x})$.

An application of these ideas yields the next interesting result.

THEOREM 2.10.1. IF $A \stackrel{\text{def}}{=} (a_{ij})_{i,j=1}^{m,n}$ IS AN $m \times n$ MATRIX THERE IS AN $n \times m$ MATRIX B SUCH THAT

$$A = ABA. \tag{2.10.11}$$

[**Remark 2.10.1:** If A is SQUARE and invertible then (2.10.11) is true if $B = A^{-1}$: $A = AA^{-1}A$. The impact of the THEOREM is that it is true even if A is SQUARE and *not* invertible, and, indeed, even if A is *not* SQUARE.]

PROOF. The preceding developments make the derivation of (2.10.11) relatively simple.

The matrix A, acting on the E-components of a vector in $\mathbf{C}^n$ defines a linear transformation

$$T : \mathbf{C}^n \mapsto \mathbf{C}^m$$

according to the formula:

$$T : \mathbf{C}^n \ni \begin{pmatrix} x_1 \\ \vdots \\ x_n \end{pmatrix} \mapsto \begin{pmatrix} y_1 \\ \vdots \\ y_m \end{pmatrix} \overset{\text{def}}{=} \begin{pmatrix} a_{11} & \cdots & a_{1n} \\ \vdots & \ddots & \vdots \\ a_{m1} & \cdots & a_{mn} \end{pmatrix} \begin{pmatrix} x_1 \\ \vdots \\ x_n \end{pmatrix} \in \mathbf{C}^m.$$

A basis

$$\widetilde{Y} \overset{\text{def}}{=} \{\mathbf{y}_1, \ldots, \mathbf{y}_k\}$$

for $im(A)$ can be filled out to a basis

$$Y \overset{\text{def}}{=} \{\mathbf{y}_1, \ldots, \mathbf{y}_k, \mathbf{y}_{k+1}, \ldots, \mathbf{y}_m\}$$

for $\mathbf{C}^m$.

For each vector $\mathbf{y}_p$ in $\widetilde{Y}$ there is in $\mathbf{C}^n$ a vector $\mathbf{x}_p$ such that

$$T(\mathbf{x}_p) = \mathbf{y}_p, \ 1 \le p \le k.$$

Since $\widetilde{Y}$ is a basis (for $im(A)$), the vectors $\mathbf{y}_p$ in $\widetilde{Y}$ are linearly independent and the corresponding vectors $\mathbf{x}_p$ are also linearly independent. (If

$$\sum_{p=1}^{k} t_p \mathbf{x}_p = \mathbf{O}$$

then

$$\mathbf{O} = T(\sum_{p=1}^{k} t_p \mathbf{x}_p) = \sum_{p=1}^{k} t_b T(\mathbf{x}_p)$$

$$= \sum_{p=1}^{k} t_p \mathbf{y}_p$$

whence each t_p is 0.)

Fill out the set $\{\mathbf{x}_1, \ldots, \mathbf{x}_k\}$ to a basis

$$X \overset{\text{def}}{=} \{\mathbf{x}_1, \ldots, \mathbf{x}_k, \mathbf{x}_{k+1}, \ldots, \mathbf{x}_n\}$$

      Chapter 2. THE PROCESS (Simple elimination)

for $\mathbf{C}^n$.

Define a transformation $S : \mathbf{C}^m \mapsto \mathbf{C}^n$ according to the formulae

$$S(\mathbf{y}_p) = \mathbf{x}_p, \ 1 \leq p \leq k$$
$$S(\mathbf{y}_q) = \mathbf{O}, \ p+1 \leq q \leq m$$
$$S\left(\sum_{j=1}^{m} c_j \mathbf{y}_j\right) = \sum_{j=1}^{m} c_j S(\mathbf{y}_j).$$

Since Y is a basis for $\mathbf{C}^m$ a direct check shows that S is a linear transformation. Furthermore an equally direct check shows that $T = TST$. Hence if B corresponds to S as A corresponds to T it follows that $A = ABA$ as promised.
$$\Omega$$

Exercise 2.10.8. Let $A \overset{\text{def}}{=} (a_{ij})_{i,j=1}^{3,5}$ be

$$\begin{pmatrix} 1 & 2 & 3 & 4 & 5 \\ 4 & 5 & 6 & 7 & 8 \\ -3 & 4 & -5 & 6 & -7 \end{pmatrix}.$$

Find B such that $A = ABA$.

Exercise 2.10.9. Let A be the matrix

$$\begin{pmatrix} 1 & 2 & 3 \\ 4 & 5 & 6 \\ 7 & 8 & 9 \\ 8 & 7 & 6 \\ 5 & 4 & 3 \end{pmatrix}.$$

Find a matrix B such that $ABA = A$.

Exercise 2.10.10. Let A be the matrix

$$\begin{pmatrix} 1 & 2 & 3 \\ 4 & -5 & 6 \\ -7 & 8 & 10 \end{pmatrix}.$$

i. Is A invertible?
ii. Find a matrix B such that $ABA = A$.
iii. What is AB?

Exercise 2.10.11. Let A be the matrix

$$\begin{pmatrix} 1 & 2 & 3 \\ 4 & 5 & 6 \\ 7 & 8 & 9 \end{pmatrix}.$$

Find a matrix B such that $ABA = A$. What is AB?

Exercise 2.10.12. Let A be the matrix

$$\begin{pmatrix} 0 & 1 \\ 0 & 0 \end{pmatrix}.$$

Find two *different* matrices B_1 and B_2 such that $AB_1A = AB_2A = A$. Show that there are infinitely many different matrices B_λ such that $AB_\lambda A = A$.

Let A be an $n \times n$ matrix of rank r and consider its "CR" decomposition (cf. **Example 2.9.7** and its sequel). The matrix $B \stackrel{\text{def}}{=} RC \stackrel{\text{def}}{=} (b_{pq})_{p,q=1}^{r,r}$ is an $r \times r$ matrix for which there is a simple and interesting interpretation. The dimension of the subspace $W \stackrel{\text{def}}{=} im(A)$ is r: $dim[im(A)] = r$. The columns $c_{j_1}, \ldots, c_{j_r}$ of C constitute a basis, again denoted C for W. If $y \in W$ there are numbers η_p such that

$$y = \sum_{p=1}^{r} \eta_p c_{j_p}. \tag{2.10.12}$$

Let η be the column vector

$$\begin{pmatrix} \eta_1 \\ \vdots \\ \eta_r \end{pmatrix}$$

and let ξ be $B\eta$. Then (2.10.12) may be written

$$y = C\eta$$

and so

$$Ay = AC\eta = CRC\eta$$

$$= CB\eta = C\xi = \sum_{p=1}^{r} \xi_p c_{j_p}.$$

In other words B is the matrix that converts the C-components η_p of a vector y in W into the C-components ξ_p of the vector Ay (also in W): $\xi = B\eta$, i.e.,

$$\begin{pmatrix} \xi_1 \\ \vdots \\ \xi_r \end{pmatrix} = \begin{pmatrix} r_{11} & \cdots & r_{1r} \\ \vdots & \ddots & \vdots \\ r_{b1} & \cdots & r_{br} \end{pmatrix} \begin{pmatrix} \eta_1 \\ \vdots \\ \eta_r \end{pmatrix}.$$

If A is an $n \times n$ matrix and W is a subspace of $\mathbf{C}^n$ then W is an *invariant subspace* for A iff for every vector x in W, $Ax \in W$.

Example 2.10.4. For any $n \times n$ matrix A the subspaces $\{O\}$ and $\mathbf{C}^n$ are invariant subspaces for A. Of greater interest is the subspace $im(A)$ which is also invariant for A.

Exercise 2.10.13. Let A be a SQUARE matrix and let W be a subspace for A. If W is an invariant subspace and $1 \leq dim(W) \stackrel{\text{def}}{=} r < n$ then W is a *reducing* subspace for A. Assume W is a reducing subspace for A and that $\mathbf{x}_1, \ldots, \mathbf{x}_r$ is a basis for W. Fill out this basis to a basis

$$X \stackrel{\text{def}}{=} \{\mathbf{x}_1, \ldots, \mathbf{x}_r, \mathbf{x}_{r+1}, \ldots, \mathbf{x}_n\}$$

for $\mathbf{C}^n$. Assume $\mathbf{x}_j \stackrel{\text{def}}{=} (x_{1j}, \ldots, x_{nj})^t$.

i. Show that the matrix $P \stackrel{\text{def}}{=} (x_{ij})_{i,j=1}^{n,n}$ is invertible.

ii. Show that $P^{-1}AP$ has the block matrix form

$$
\begin{array}{cc}
 & \begin{array}{cc} r & n-r \end{array} \\
\begin{array}{c} r \\ n-r \end{array} & \begin{pmatrix} B & C \\ O & D \end{pmatrix}
\end{array}.
\tag{2.10.13}
$$

iii. Show that if Q is an invertible matrix such that $Q^{-1}AQ$ has the form (2.10.13) then the first r columns of Q constitute a basis of a reducing subspace for A.

[**Note 2.10.3:** A SQUARE matrix A for which there is a reducing subspace is *reducible*. Hence a SQUARE matrix is reducible iff for some invertible matrix P the form of $P^{-1}AP$ is that shown in (2.10.13) and is achievable with r positive and less than n. If $A = O$ then every subspace is invariant and A is reducible. If A is invertible then $im(A)$ is invariant and $dim[im(A)] = n$ and so $im(A)$ is *not* a reducing subspace for A. Nevertheless, it is shown in **Chapter 4** that every SQUARE matrix *is* reducible.]

In the notation used above let W be a reducing subspace for the SQUARE matrix A. If $\{\mathbf{x}_{r+1}, \ldots, \mathbf{x}_n\}$ can be found to fill out $\{\mathbf{x}_1, \ldots, \mathbf{x}_r\}$ to a basis in such a manner that $V \stackrel{\text{def}}{=} span(\{\mathbf{x}_{r+1}, \ldots, \mathbf{x}_n\})$ is also invariant (hence also reducing) then A is *completely reducible*. The pair (W, V) of subspaces is *completely reducing*.

The subspaces

$$W \stackrel{\text{def}}{=} span(\{\mathbf{x}_1, \ldots, \mathbf{x}_r\})$$

$$V \stackrel{\text{def}}{=} span(\{\mathbf{x}_{r+1}, \ldots, \mathbf{x}_n\})$$

are always *complementary*: for each vector $\mathbf{x}$ there is a unique pair $\{\mathbf{w}, \mathbf{v}\}$ of vectors such that

$$\mathbf{w} \in W, \ \mathbf{v} \in V, \ \mathbf{x} = \mathbf{w} + \mathbf{v}.$$

Hence one writes $\mathbf{C}^n = W \oplus V$: $\mathbf{C}^n$ is the *direct sum* of W and V. Every completely reducing pair is automatically complementary. The converse is false (cf. **Exercise 2.10.15**).

Exercise 2.10.14. Show that a SQUARE matrix A is completely reducible iff there is a positive r less than n and such that for some invertible matrix R

$$R^{-1}AR = \begin{matrix} & r n-r \\ \begin{matrix} r \\ n-r \end{matrix} & \begin{pmatrix} B & O \\ O & C \end{pmatrix} \end{matrix}.$$

Exercise 2.10.15. Show that the reducible, indeed reduced, matrix

$$A \overset{\text{def}}{=} \begin{pmatrix} 1 & 1 \\ 0 & 1 \end{pmatrix}$$

is not completely reducible. Identify a reducing subspace W and a pair (W, V) of complementary subspaces (necessarily not completely reducing).

[*Hint:* Let $R \overset{\text{def}}{=} (\rho_{ij})_{i,j=1}^{2,2}$ be an invertible matrix such that

$$R^{-1}AR = \begin{pmatrix} B & O \\ O & C \end{pmatrix} \overset{\text{def}}{=} D. \tag{2.10.14}$$

Recognize the form of D and compare entries in the left and right members of the equation $AR = RD$.]

Exercise 2.10.16. Suppose Π is a permutation matrix and the SQUARE matrix A is such that

$$\Pi^{-1}A\Pi$$

has the form displayed in (2.10.13) resp. (2.10.14). Identify a reducing subspace W resp. a pair of completely reducing subspaces W and V.

Exercise 2.10.17. Let V be a finite-dimensional vector space. If $S \subset V$ let $\overline{S}$ denote the span of S: $\overline{S} = span(S)$. Show:

i. $S \subset \overline{S}$;
ii. $\overline{\overline{S}} = \overline{S}$;
iii. $S \subset T \Rightarrow \overline{S} \subset \overline{T}$;
iv. if $\mathbf{x} \in \overline{S}$ then there is in S a finite subset $S_{\mathbf{x}}$ such that $x \in \overline{S_{\mathbf{x}}}$;
v. if $\mathbf{x} \in \overline{S \cup \mathbf{y}}$ and $\mathbf{x} \notin \overline{S}$ then $\mathbf{y} \in \overline{S \cup \mathbf{x}}$.

[**Remark 2.10.2:** The statements *i–iv* were given by Steinitz. As axioms they characterize a large class of set operations of which "span" is one example. In particular *iv* is the *exchange axiom* which, as a valid statement in linear algebra, underlies simple elimination.]

CHAPTER 3

DETERMINANTS (A direct approach)

3.1. Introduction

The rôle played by determinants in the modern study of linear algebra is a minor one. For instructional purposes, determinants present some modest advantages. In applied worked they are used in only the smallest and most elementary problems. Nevertheless determinants occupy a respected niche in the history of linear algebra. In this **Chapter** there are presented the essentials for working with them. Since the GEM figures so significantly in the analysis of matrices, the GEM is used as the simplest device for defining determinants and for deriving their various properties.

Linear algebra deals with matrices of all shapes, HIGH, WIDE, or SQUARE. Determinants are defined *only* for SQUARE matrices. In **Section 2.8** the discussion of inverses of $n \times n$ SQUARE matrices shows that the existence of the inverse of $A \overset{\text{def}}{=} (a_{ij})_{i,j=1}^{n,n}$ is intimately bound up with the last diagonal entry α_{nn} of the echelon form A_E of A: A is invertible iff $\alpha_{nn} \neq 0$ iff each $\alpha_{ii} \neq 0$. The echelon form A_E of A is the result of applying the GEM to A. In the course of applying the GEM there arises an algorithmically defined number p of ROW EXCHANGEs.

It turns out that the determinant $det(A)$ is simply $(-1)^p \alpha_{11} \alpha_{22} \cdots \alpha_{nn}$.

DEFINITION 3.1.1. ASSUME $A \overset{\text{def}}{=} (a_{ij})_{i,j=1}^{n,n}$. IF THE GEM PRODUCES $A_E \overset{\text{def}}{=} (\alpha_{ij})_{i,j=1}^{n,n}$ AND INVOLVES THE USE OF p ROW EXCHANGEs THEN

$$det(A) = (-1)^p \prod_{i=1}^{n} \alpha_{ii}. \tag{3.1.1}$$

If q is an integer then $2q$ is an even integer and $(-1)^{p+2q} = (-1)^p$ and since p is even or odd iff $p + 2q$ is even or odd only the *parity* – the

---

evenness or the oddness – of p, is of consequence in the definition of $det(A)$.

[**Remark 3.1.1:** If one permits the abuse of language whereby "pivot" means any diagonal entry α_{ii} of A_E, then $det(A)$ is simply

$$(-1)^p \times (\text{product of the pivots of } A),$$

even if some $\alpha_{ii} = 0$.]

If $1 \le i < n$ and $\alpha_{ii} = 0$ then $\alpha_{jj} = 0$ whenever $j > i$. Hence:

A is singular iff $det(A) = 0$;
A is nonsingular iff $det(A) \neq 0$.

Example 3.1.1. Let each matrix below be an $n \times n$ matrix. **Definition 3.1.1** implies directly that:

i.

$$det(E_{ij}) = \begin{cases} -1, & \text{if } i \neq j; \\ 1, & \text{otherwise}; \end{cases}$$

ii.

$$det[E_i(s)] = s \ (s \text{ may be } any \text{ number, even } 0);$$

iii.

$$det[E_{ij}(r)] = \begin{cases} 1 & \text{if } i \neq j; \\ r+1 & \text{if } i = j; \end{cases}$$

iv. if $n > 1$ and U_{ij} is an $n \times n$ matrix unit then $det(U_{ij}) = 0$;

v.

$$det\left[\begin{pmatrix} 3 & 0 \\ 0 & -6 \end{pmatrix}\right] = 3 \cdot (-6) = -18;$$

vi. if

$$A = \begin{pmatrix} 0 & 2 \\ 3 & 4 \end{pmatrix}$$

then $p = 1$, $\alpha_{11} = 3$, $\alpha_{22} = 2$ and

$$det(A) = (-1)^1 \cdot 3 \cdot 2 = -6.$$

vii.

$$det\left[\begin{pmatrix} d_1 & & & \\ & d_2 & & \\ & & \ddots & \\ & & & d_n \end{pmatrix}\right] = d_1 d_2 \cdots d_n;$$

(The determinant of a diagonal matrix is the product of its diagonal entries.)

$$det\left[\begin{pmatrix} d_1 & & & \\ & d_2 & & \\ & & \ddots & \\ & & & d_n \end{pmatrix}\begin{pmatrix} \delta_1 & & & \\ & \delta_2 & & \\ & & \ddots & \\ & & & \delta_n \end{pmatrix}\right]$$

$$= det\left[\begin{pmatrix} d_1\delta_1 & & & \\ & d_2\delta_2 & & \\ & & \ddots & \\ & & & d_n\delta_n \end{pmatrix}\right]$$

$$= (d_1\delta_1)\cdots(d_n\delta_n)$$

$$= det\left[\begin{pmatrix} d_1 & & & \\ & d_2 & & \\ & & \ddots & \\ & & & d_n \end{pmatrix}\right] det\left[\begin{pmatrix} \delta_1 & & & \\ & \delta_2 & & \\ & & \ddots & \\ & & & \delta_n \end{pmatrix}\right];$$

(The determinant of the product of diagonal matrices is the product of their determinants.)

ix. if at least one of A_1 and A_2, say A_1, is singular then $A_1 A_2$ is also singular and

$$det(A_1 A_2) = 0 = 0 \cdot det(A_2) = det(A_1)det(A_2);$$

(The determinant of the product of matrices at least one of which is singular is the product of their determinants.)

x. if $A \stackrel{\text{def}}{=} (a_{ij})_{i,j=1}^{n,n}$ is an upper triangular SQUARE matrix then either:

    a. A is invertible in which case each diagonal entry $\alpha_{ii} = a_{ii} \neq 0$ and $p = 0$ whence

$$det(A) = a_{11} \cdot \cdots \cdot a_{nn};$$

    or

    b. A is singular, some $a_{ii} = 0$, and hence $\alpha_{nn} = 0$, whence, again

$$det(A) = 0 = a_{11} \cdots a_{nn};$$

i.e., whether or not A is invertible

$$det(A) = a_{11} \cdot \cdots \cdot a_{nn};$$

(The determinant of an upper triangular matrix is the product of its diagonal entries.)

xi. if A and B are upper triangular matrices then $det(AB) = det(A)det(B)$;
(The determinant of the product of upper triangular matrices is the product of their determinants.)

xii. if $A \stackrel{\text{def}}{=} (a_{ij})_{i,j=1}^{n,n}$ is lower triangular and if each diagonal entry is nonzero then $\alpha_{ii} = a_{ii}$, $1 \leq i \leq n$, $p = 0$ and

$$det(A) = (-1)^p \prod_{i=1}^{n} a_{ii} = \prod_{i=1}^{n} a_{ii};$$

---

if some diagonal entry is 0 then some $\alpha_{ii} = 0$ and so

$$0 = det(A) = (-1)^p \prod_{i=1}^{n} a_{ii};$$

hence if A and B are lower triangular matrices then

$$det(AB) = det(A)det(B).$$

(The determinant of the product of lower triangular matrices is the product of their determinants.)

Example 3.1.2. Let A be the 2×2 matrix

$$\begin{pmatrix} a_{11} & a_{12} \\ a_{21} & a_{22} \end{pmatrix}.$$

To calculate $det(A)$ according to its definition the task is broken down into cases.

Case i, $a_{11} \neq 0$.

The first pivot is a_{11} and after the first cycle of the GEM there emerges the matrix

$$\begin{pmatrix} a_{11} & a_{12} \\ 0 & a_{22} - \frac{a_{21}a_{12}}{a_{11}} \end{pmatrix}.$$

If A is singular, then, by definition, $det(A) = 0$ and also

$$\alpha_{22} = a_{22} - \frac{a_{21}a_{12}}{a_{11}} = 0$$
$$a_{11}\left(a_{22} - \frac{a_{21}a_{12}}{a_{11}}\right) = a_{11}a_{22} - a_{12}a_{21}$$
$$a_{11}\left(a_{22} - \frac{a_{21}a_{12}}{a_{11}}\right) = a_{11}0 = 0.$$

Hence

$$det(A) = a_{11}a_{22} - a_{12}a_{21} = 0.$$

If A is *not* singular then the second pivot

$$\alpha_{22} = a_{22} - \frac{a_{21}a_{12}}{a_{11}} \neq 0$$

and so

$$det(A) = \text{product of the pivots} = a_{11}\left(a_{22} - \frac{a_{21}a_{12}}{a_{11}}\right)$$
$$= a_{11}a_{22} - a_{12}a_{21}.$$

Case ii, $a_{11} = 0$.

If $a_{21} \neq 0$ the GEM leads to

$$\begin{pmatrix} a_{21} & a_{22} \\ 0 & a_{12} \end{pmatrix}.$$

If A is singular then $a_{12} = 0$, $det(A) = 0$, and again

$$det(A) = 0 = a_{11}a_{22} - a_{12}a_{21}.$$

Finally, if A is not singular the nonzero pivots are a_{21} and a_{12} and, by definition, since *one* row interchange occurs, yet again, because $a_{11} = 0$,

$$det(A) = (-1)^1 a_{21}a_{12} = a_{11}a_{22} - a_{12}a_{21}.$$

Hence

$$det\left[\begin{pmatrix} a_{11} & a_{12} \\ a_{21} & a_{22} \end{pmatrix}\right] = a_{11}a_{22} - a_{12}a_{21}.$$

Exercise 3.1.1. Let A be the 3×3 matrix

$$\begin{pmatrix} a_{11} & a_{12} & a_{13} \\ a_{21} & a_{22} & a_{23} \\ a_{31} & a_{32} & a_{33} \end{pmatrix}.$$

Imitate the analysis in **Example 3.1.2** and thereby show that

$$det(A) = a_{11}a_{22}a_{33} + a_{12}a_{23}a_{31} + a_{13}a_{21}a_{32}$$
$$- a_{13}a_{22}a_{31} - a_{12}a_{21}a_{33} - a_{11}a_{23}a_{32}.$$

[*Hint:* Assume first that $a_{11} \neq 0$ and after one cycle of the GEM use the formula already established for the determinant of a 2×2 matrix. If $a_{11} = 0$ first perform a ROW EXCHANGE.]

[**Note 3.1.1:** In **Example 3.1.2** and in **Exercise 3.1.1** the formulae for the determinant are the *sums* of 2! resp. 3! terms, each of which is a product of 2 resp. 3 entries. The *first* indices of the factors in each product appear in their natural sequence. The *second* indices of the factors are different from one product to the next and indeed the second indices appear in every possible permutation of the natural order of the indices. Since there are 2! $= 2$ resp. 3! $= 6$ permutations of 2 resp. 3 indices there are 2! resp. 3! terms in the formulae for the determinants.]

The following may be remarked in the results of **Example 3.1.2** and **Exercise 3.1.1**:

d1. If $\widetilde{A}$ is the matrix resulting from exchanging two rows of A then $det(\widetilde{A}) = -det(A)$.

d2. If the first row (a row vector) varies and the other rows are held fixed then the function *det* is a *linear* function F of the row vector that is the first row of A. For example, if $\mathbf{a}_1 \overset{\text{def}}{=} (a_{11}, a_{12})$ and $F(\mathbf{a}_1) \overset{\text{def}}{=} det(A)$ then a direct check shows that if $r, s \in \mathbf{C}$ then

$$F(r\mathbf{a}_1 + s\mathbf{b}_1) = rF(\mathbf{a}_1) + sF(\mathbf{b}_1).$$

d3. If $A = I$ then $det(A) = det(I) = 1$.

The formulae for the determinants of 2×2 and 3×3 matrices can be interpreted by amusing and helpful mnemonic diagrams for "basket-weaving." Thus following the arrows in the diagrams below and affixing "+" to products corresponding to each $\searrow$ and "−" to products corresponding to each $\nearrow$ leads to the value of the determinant when the products with signs affixed are added together.

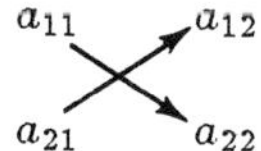

Figure 3.1.1.

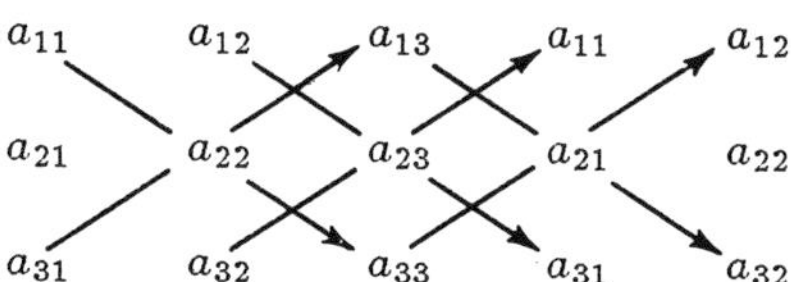

Figure 3.1.2.

[**Remark 3.1.2:** Regrettably, if $n > 3$, no simple "basket-weaving" mnemonic is available for the calculation of determinants of $n \times n$ matrices.]

Exercise 3.1.2. Calculate the determinants of each of the matrices below. Confirm that each matrix is invertible iff its determinant is not 0.

$$i.\ \begin{pmatrix} 1 & 2 \\ 3 & 4 \end{pmatrix};\ ii.\ \begin{pmatrix} 1 & 2 \\ 3 & 6 \end{pmatrix};\ iii.\ \begin{pmatrix} 1 & 2 & 3 \\ 3 & 4 & 5 \\ 4 & 5 & 6 \end{pmatrix};\ iv.\ \begin{pmatrix} 1 & 2 & 3 \\ 4 & 5 & 6 \\ 7 & 8 & 9 \end{pmatrix}.$$

In what follows there is a proof of the n-dimensional generalizations of $d1 - d3$.

Let Mat_{mn} denote the set of all $m \times n$ matrices. The generalizations of $d1 - d3$ are:

D1. If two rows of a matrix A are exchanged the determinant of the new matrix is the negative of the determinant of the original matrix. Thus if $i \neq j$ then $det(E_{ij}A) = -det(A) = det(E_{ij})det(A)$.

D2. The function

$$det : Mat_{nn} \ni A \mapsto det(A) \in \mathbf{C}$$

is a linear function of the first row $\mathbf{a}_1$ of A if the other rows are held fixed and the first row $\mathbf{a}_1$ varies.

[**Note 3.1.2:** If $A = (a_{ij})_{i,j=1}^{n,n}$ and

$$\mathbf{a}_i \overset{\text{def}}{=} (a_{i1}, \ldots, a_{in}) = i\text{th row of } A$$

then A may be written as

$$\begin{pmatrix} \mathbf{a}_1 \\ \vdots \\ \mathbf{a}_n \end{pmatrix}.$$

To say that det is a linear function of the first row of A is to say that if r and s are numbers and $\mathbf{b}_1 = (b_{11}, \ldots, b_{1n})$ then

$$det\left[\begin{pmatrix} r\mathbf{a}_1 + s\mathbf{b}_1 \\ \mathbf{a}_2 \\ \vdots \\ \mathbf{a}_n \end{pmatrix}\right] = r\,det\left[\begin{pmatrix} \mathbf{a}_1 \\ \mathbf{a}_2 \\ \vdots \\ \mathbf{a}_n \end{pmatrix}\right] + s\,det\left[\begin{pmatrix} \mathbf{b}_1 \\ \mathbf{a}_2 \\ \vdots \\ \mathbf{a}_n \end{pmatrix}\right].$$

The consequences of this formula are extensive.]

D3. $det(I) = 1$.

Before the proof that det satisfies D1-D3 the following **Exercises** and **Example** offer some insight and background for the derivation.

Exercise 3.1.3. Verify that D1-D3 are valid for 3×3 matrices.

Example 3.1.3. Let δ be a function mapping Mat_{22} (the set of all 2×2 matrices) into $\mathbf{C}$ and satisfying D1-D3. Then from D1 and D3 follows that

$$\delta\left[\begin{pmatrix} 0 & 1 \\ 1 & 0 \end{pmatrix}\right] = -1. \tag{3.1.2}$$

It follows from D1-D3 that

$$\delta\left[\begin{pmatrix} a_{11} & a_{12} \\ a_{21} & a_{22} \end{pmatrix}\right] = a_{11}\delta\left[\begin{pmatrix} 1 & 0 \\ a_{21} & a_{22} \end{pmatrix}\right] + a_{12}\delta\left[\begin{pmatrix} 0 & 1 \\ a_{21} & a_{22} \end{pmatrix}\right]$$

$$= -a_{11}\delta\left[\begin{pmatrix} a_{21} & a_{22} \\ 1 & 0 \end{pmatrix}\right] - a_{12}\delta\left[\begin{pmatrix} a_{21} & a_{22} \\ 0 & 1 \end{pmatrix}\right]$$

$$= -a_{11}a_{21}\delta\left[\begin{pmatrix} 1 & 0 \\ 1 & 0 \end{pmatrix}\right] - a_{11}a_{22}\delta\left[\begin{pmatrix} 0 & 1 \\ 1 & 0 \end{pmatrix}\right]$$

$$- a_{12}a_{21}\delta\left[\begin{pmatrix} 1 & 0 \\ 0 & 1 \end{pmatrix}\right] - a_{12}a_{22}\delta\left[\begin{pmatrix} 0 & 1 \\ 0 & 1 \end{pmatrix}\right]. \quad (3.1.3)$$

However D1 and D3 imply

$$\delta\left[\begin{pmatrix} 0 & 1 \\ 1 & 0 \end{pmatrix}\right] = -\delta\left[\begin{pmatrix} 1 & 0 \\ 0 & 1 \end{pmatrix}\right] = -1$$

while the equation

$$\begin{pmatrix} 1 & 0 \\ 1 & 0 \end{pmatrix} = E_{12}\begin{pmatrix} 1 & 0 \\ 1 & 0 \end{pmatrix}$$

and D1 imply that

$$\delta\left[\begin{pmatrix} 1 & 0 \\ 1 & 0 \end{pmatrix}\right] = \delta\left[E_{12}\begin{pmatrix} 1 & 0 \\ 1 & 0 \end{pmatrix}\right] = -\delta\left[\begin{pmatrix} 1 & 0 \\ 1 & 0 \end{pmatrix}\right],$$

i.e., that

$$\delta\left[\begin{pmatrix} 1 & 0 \\ 1 & 0 \end{pmatrix}\right] = 0.$$

Similarly there obtains the equation:

$$\delta\left[\begin{pmatrix} 0 & 1 \\ 0 & 1 \end{pmatrix}\right] = 0.$$

Hence

$$\delta\left[\begin{pmatrix} a_{11} & a_{12} \\ a_{21} & a_{22} \end{pmatrix}\right] = a_{11}a_{22} - a_{12}a_{21} = \det\begin{pmatrix} a_{11} & a_{12} \\ a_{21} & a_{22} \end{pmatrix}$$

and so δ and $\det$ are the *same* function on Mat_{22}.

Exercise 3.1.4. Verify *Cramer's rule* for systems of 2 equations in 2 unknowns, viz.,

if $a_{11}a_{22} - a_{12}a_{21} \neq 0$ and if

$$a_{11}x_1 + a_{12}x_2 = b_1$$
$$a_{21}x_1 + a_{22}x_2 = b_2$$

    Chapter 3. DETERMINANTS (**A direct approach**)

then

$$x_1 = \frac{det\left[\begin{pmatrix} b_1 & a_{12} \\ b_2 & a_{22} \end{pmatrix}\right]}{det\left[\begin{pmatrix} a_{11} & a_{12} \\ a_{21} & a_{22} \end{pmatrix}\right]}$$

$$x_2 = \frac{det\left[\begin{pmatrix} a_{11} & b_1 \\ a_{21} & b_2 \end{pmatrix}\right]}{det\left[\begin{pmatrix} a_{11} & a_{12} \\ a_{21} & a_{22} \end{pmatrix}\right]}$$

Exercise 3.1.5. Formulate in words Cramer's rule as illustrated above.

Exercise 3.1.6. Show that if δ is a mapping of Mat_{nn} into $\mathbf{C}$ and if δ satisfies D1-D3 then $\delta(A)$ depends linearly on *each* row (not only row 1) of the $n \times n$ matrix A if the other rows are held fixed: if r and s are numbers and $\mathbf{b}_k = (b_{k1}, \ldots, b_{kn})$ then

$$\delta\left[\begin{pmatrix} \mathbf{a}_1 \\ \vdots \\ r\mathbf{a}_k + s\mathbf{b}_k \\ \vdots \\ \mathbf{a}_n \end{pmatrix}\right] = r\delta\left[\begin{pmatrix} \mathbf{a}_1 \\ \vdots \\ \mathbf{a}_k \\ \vdots \\ \mathbf{a}_n \end{pmatrix}\right] + s\delta\left[\begin{pmatrix} \mathbf{a}_1 \\ \vdots \\ \mathbf{b}_k \\ \vdots \\ \mathbf{a}_n \end{pmatrix}\right].$$

[*Hint:* Perform a ROW EXCHANGE of rows 1 and k and apply D1 and D2.]

Exercise 3.1.7. Verify that if A and B are two 2×2 matrices then:

i.

$$det(A) = det(A^t);$$

ii. $det(AB) = det(A)det(B)$, the *product rule* for 2×2 determinants.

Exercise 3.1.8. Show that if δ is a mapping of Mat_{nn} into $\mathbf{C}$, if δ satisfies D1-D3, and if $\delta(B) \neq 0$ then Δ given by

$$\Delta(A) \overset{\text{def}}{=} \frac{\delta(AB)}{\delta(B)} \tag{3.1.4}$$

also satisfies D1-D3.

The one formula that is missing thus far from what has been shown about determinants is the *product rule*:

$$det(AB) = det(A)det(B) \qquad (3.1.5)$$

when neither A nor B is singular. In what follows there are established three results:

   i. only one function can satisfy D1-D3;
  ii. *det* is a function satisfying D1-D3;
 iii. the product rule (3.1.5).

If $i - ii$ above are accepted then the result in **Exercise 3.1.8** implies that if $det(B) \neq 0$ then

$$\Delta(A) \stackrel{\text{def}}{=} \frac{det(AB)}{det(B)} = det(A)$$

i.e., $det(AB) = det(A)det(B)$ $(= det(B)det(A))$, and so *even if neither matrix is singular* the product rule holds. Once the general validity of the product rule is proved most of the familiar and useful properties of the function *det* can be derived with some ease.

Exercise 3.1.9. Show that if $\delta : Mat_{nn} \mapsto \mathbf{C}$ satisfies D2 then δ satisfies D1 iff $\delta(A) = 0$ whenever two rows of A are the same.

If δ satisfies D2 alone then repeated application of D2 shows that

$$\delta\left[\begin{pmatrix} t_1\mathbf{v}_1 + t_2\mathbf{v}_2 + \cdots + t_m\mathbf{v}_m \\ \mathbf{a}_2 \\ \vdots \\ \mathbf{a}_n \end{pmatrix}\right]$$

$$= t_1\delta\left[\begin{pmatrix} \mathbf{v}_1 \\ \mathbf{a}_2 \\ \vdots \\ \mathbf{a}_n \end{pmatrix}\right] + \delta\left[\begin{pmatrix} t_2\mathbf{v}_2 + \cdots + t_m\mathbf{v}_m \\ \mathbf{a}_2 \\ \vdots \\ \mathbf{a}_n \end{pmatrix}\right]$$

$$= \sum_{j=1}^{m} t_j\delta\left[\begin{pmatrix} \mathbf{v}_j \\ \mathbf{a}_2 \\ \vdots \\ \mathbf{a}_n \end{pmatrix}\right].$$

More generally, if each row of A is a linear combination of row vectors $\mathbf{v}_1, \ldots, \mathbf{v}_m$ then

$$\delta\left[\begin{pmatrix} t_{11}\mathbf{v}_1 + \cdots + t_{1m}\mathbf{v}_m \\ t_{21}\mathbf{v}_1 + \cdots + t_{2m}\mathbf{v}_m \\ \vdots \\ t_{n1}\mathbf{v}_1 + \cdots + t_{nm}\mathbf{v}_m \end{pmatrix}\right] =$$

$$= \sum_{j_1=1}^{m} t_{1j_1} \delta\left[\begin{pmatrix} \mathbf{v}_{j_1} \\ t_{21}\mathbf{v}_1 + \cdots + t_{2m}\mathbf{v}_m \\ \vdots \\ t_{n1}\mathbf{v}_1 + \cdots + t_{nm}\mathbf{v}_m \end{pmatrix}\right]$$

$$= \sum_{j_1=1}^{m} t_{1j_1} \sum_{j_2=1}^{m} t_{2j_2} \delta\left[\begin{pmatrix} \mathbf{v}_{j_1} \\ \mathbf{v}_{j_2} \\ \vdots \\ t_{n1}\mathbf{v}_1 + \cdots + t_{nm}\mathbf{v}_m \end{pmatrix}\right]$$

$$\vdots$$

$$= \sum_{j_1=1}^{m} \sum_{j_2=1}^{m} \cdots \sum_{j_n=1}^{m} t_{1j_1} t_{2j_2} \cdots t_{nj_n} \delta\left[\begin{pmatrix} \mathbf{v}_{j_1} \\ \mathbf{v}_{j_2} \\ \vdots \\ \mathbf{v}_{j_n} \end{pmatrix}\right]$$

$$(3.1.6)$$

describes the situation.

The calculation in (3.1.6) may be viewed as a "break-out" operation that expands a very tight formula into an elaborate sum. The number

$$\delta\left[\begin{pmatrix} \mathbf{v}_{j_1} \\ \mathbf{v}_{j_2} \\ \vdots \\ \mathbf{v}_{j_n} \end{pmatrix}\right]$$

is 0 if any two of the rows $\mathbf{v}_{j_i}$ are the same. If $m < n$ then at least two of the rows $\mathbf{v}_{j_i}$ are the same and *every* term in (3.1.6) is 0. If $m \geq n$ then although (3.1.6) consists of m^n terms, at most

$$_mP_n \stackrel{\text{def}}{=} m(m-1)\cdots(m-n+1)$$

of those terms are not 0's. If $m = n$ then at most $n!$ terms rather than n^n terms are not 0. If $n = 10$, then $n^n = 10,000,000,000$ whereas $n! = 3,628,800$, which is 0.036288% of 10,000,000,000.

In particular, if

$$\mathbf{v}_j = \begin{cases} \mathbf{f}_j \overset{\mathrm{def}}{=} (0 \quad \cdots \quad \underset{\underset{\text{column } j}{\uparrow}}{1} \quad \cdots \quad 0) \end{cases}, \ 1 \le j \le n$$

then

$$\delta(A) = \sum_{j_1=1}^{m} \sum_{j_2=1}^{m} \cdots \sum_{j_n=1}^{m} t_{1j_1} t_{2j_2} \cdots t_{nj_n} \delta \left[\begin{pmatrix} \mathbf{f}_{j_1} \\ \mathbf{f}_{j_2} \\ \vdots \\ \mathbf{f}_{j_n} \end{pmatrix} \right]. \tag{3.1.7}$$

Each matrix

$$\begin{pmatrix} \mathbf{f}_{j_1} \\ \mathbf{f}_{j_2} \\ \vdots \\ \mathbf{f}_{j_n} \end{pmatrix} \tag{3.1.8}$$

in which no two rows are the same is the result of rearranging, i.e., permuting, the rows of the $n \times n$ identity matrix I. Since the GEM applied to any matrix (3.1.8) involves *only* ROW EXCHANGEs it follows that

$$det \left[\begin{pmatrix} \mathbf{f}_{j_1} \\ \mathbf{f}_{j_2} \\ \vdots \\ \mathbf{f}_{j_n} \end{pmatrix} \right] = \begin{cases} \pm 1, & \text{if no two rows are equal} \\ 0, & \text{otherwise.} \end{cases}$$

Hence if *det* is shown to satisfy D1-D3 then

$$det \left[\begin{pmatrix} a_{11} & \cdots & a_{1n} \\ \vdots & \ddots & \vdots \\ a_{n1} & \cdots & a_{nn} \end{pmatrix} \right] = \sum_{1 \le j_1, \ldots, j_n \le n} \pm a_{1j_1} \cdots a_{nj_n}$$

$$= \sum_{\substack{\text{no two } j_k \text{ the same} \\ 1 \le j_1, \ldots, j_n \le n}} \pm a_{1j_1} \cdots a_{nj_n}.$$

The "$\pm$" in each term is calculated according to the number of ROW EX-CHANGEs used when the GEM is applied to the matrix (3.1.8).

It is important to note that the formula (3.1.7) is derived on the basis of D1 and D2 alone.

Exercise 3.1.10. Show that if δ_1 and δ_2 are maps of Mat_{nn} into **C**, if t_1 and t_2 are numbers, and if δ_1 and δ_2 satisfy D1 and D2 then

$$\delta_3 \overset{\mathrm{def}}{=} t_1 \delta_1 + t_2 \delta_2$$

also satisfies D1 and D2. In particular $\delta_1 - \delta_2$ satisfies D1 and D2.

Exercise 3.1.11. Show that if δ_1 and δ_2 are as in **Exercise 3.1.10** and if, furthermore, both satisfy D3 then $\delta_1 = \delta_2$.

[*Hint:* Consider the difference $\delta_1 - \delta_2 \overset{\text{def}}{=} \delta_3$. Show $\delta_3(I) = 0$. Hence, by virtue of D1 if Π is a permutation matrix $\delta_3(\Pi) = 0$. Use the break-out.]

Exercise 3.1.11 implies that at most one function satisfies D1-D3. What remains to be shown is that *det* satisfies D1-D3.

The next paragraphs contain a proof that if n is an arbitrary natural number then

$$det : Mat_{nn} \ni A \mapsto det(A) \in \mathbf{C}$$

satisfies D1-D3.

The basic tool in most of the arguments is mathematical induction. The fact that *det* applied to 2×2 matrices satisfies D1-D3 is the very motivation for considering D1-D3. Hence the first step in mathematical induction is done: the assertion is true if $n = 2$. (The assertion is trivially true if $n = 1$ and that fact is sufficient to begin the induction. It is instructive and reassuring to know that the assertion is true if $n = 2$.)

The property D3 is disposed of at once because the GEM applied to I shows that *each* α_{ii} is 1 and no ROW EXCHANGEs are involved, whence $det(I) = 1$.

Thus in all that follows the underlying assumption is that D1-D3 (and thus the product rule) hold for *det* applied to $(n - 1) \times (n - 1)$ matrices. On this premise the proof is given that D1-D3 hold as well for *det* applied to $n \times n$ matrices. The proof of the next LEMMA uses mathematical induction and a device that reduces the proof that *det* satisfies D1 to the case in which $i = 1$ and $j = 2$.

LEMMA **3.1.1.** IF A IS AN $n \times n$ MATRIX AND IF $i \neq j$ THEN $det(E_{ij} A) = -det(A)$.

PROOF. The case in which the first column of A is the zero (column) vector may be disregarded since then $\alpha_{11} = 0$ and $det(A) = det(E_{12} A) = 0$.

The line of the proof is the following:

i. Show $det(E_{12} A) = -det(A)$.
ii. If $\{i, j\} \neq \{1, 2\}$ reduce the argument to that in *i*.

The proof that $det(E_{12} A) = -det(A)$ is handled by cases.

Case *i.* $a_{11} a_{21} \neq 0$ (neither a_{11} nor a_{21} is 0).

Let $B \overset{\text{def}}{=} (b_{ij})_{i,j=1}^{n,n}$ be $E_{12} A$. Thus $b_{11} = a_{21}$, $b_{21} = a_{11}$ and $b_{k1} = a_{k1}$ if $k > 2$. Let $\mathbf{a}'_1, \ldots, \mathbf{a}'_n$ be the $n - 1$-component row vectors consisting of the last $n - 1$ components of the rows of A:

$$\mathbf{a}'_i \overset{\text{def}}{=} (a_{i2}, \ldots, a_{in}).$$

Let B' be the $(n-1) \times (n-1)$ matrix in the "southeast" corner after the completion of the first cycle of the GEM applied to B. Then

$$B' = \begin{pmatrix} \mathbf{a}_1' - \frac{a_{11}}{a_{21}} \mathbf{a}_2' \\ \mathbf{a}_3' - \frac{a_{31}}{a_{21}} \mathbf{a}_2' \\ \vdots \\ \mathbf{a}_n' - \frac{a_{n1}}{a_{21}} \mathbf{a}_2' \end{pmatrix}.$$

If the echelon form B_E of B is $(\beta_{ij})_{i,j=1}^{n,n}$ then $\beta_{11} = a_{21}$ and $\beta_{22}, \ldots, \beta_{nn}$ are the diagonal entries of of B_E'. Hence

$$det(B) = a_{21} det(B') = a_{21} det \left[\begin{pmatrix} \mathbf{a}_1' - \frac{a_{11}}{a_{21}} \mathbf{a}_2' \\ \mathbf{a}_3' - \frac{a_{31}}{a_{21}} \mathbf{a}_2' \\ \vdots \\ \mathbf{a}_n' - \frac{a_{n1}}{a_{21}} \mathbf{a}_2' \end{pmatrix} \right]. \qquad (3.1.9)$$

By inductive assumption, D1-D3 hold for all $n-1 \times n-1$ matrices and so does the product rule.

For notational convenience let

$$\begin{pmatrix} \mathbf{x}_1 \\ \mathbf{x}_2 \\ \vdots \\ \mathbf{x}_n \end{pmatrix}_{k,\mathbf{y}} , \; 1 \le k \le n$$

be the matrix

$$\begin{pmatrix} \mathbf{x}_1 \\ \mathbf{x}_2 \\ \vdots \\ \mathbf{x}_n \end{pmatrix}$$

with row k replaced by $\mathbf{y}$. Then the multilinearity of det for $n-1 \times n-1$ matrices implies:

$$det(B) = a_{21} det \left[\begin{pmatrix} \mathbf{a}_1' \\ \mathbf{a}_3' \\ \vdots \\ \mathbf{a}_n' \end{pmatrix} \right] - a_{31} det \left[\begin{pmatrix} \mathbf{a}_1' \\ \mathbf{a}_3' \\ \vdots \\ \mathbf{a}_n' \end{pmatrix}_{2,\mathbf{a}_2'} \right] - \cdots$$

$$- a_{n1} det \left[\begin{pmatrix} \mathbf{a}_1' \\ \mathbf{a}_3' \\ \vdots \\ \mathbf{a}_n' \end{pmatrix}_{n-1,\mathbf{a}_2'} \right] - a_{11} det \left[\begin{pmatrix} \mathbf{a}_1' \\ \mathbf{a}_3' \\ \vdots \\ \mathbf{a}_n' \end{pmatrix}_{1,\mathbf{a}_2'} \right]. \qquad (3.1.10)$$

Similarly,

$$det(A) = a_{11}det\left[\begin{pmatrix}\mathbf{a}'_2\\\mathbf{a}'_3\\\vdots\\\mathbf{a}'_n\end{pmatrix}\right] - a_{31}det\left[\begin{pmatrix}\mathbf{a}'_2\\\mathbf{a}'_3\\\vdots\\\mathbf{a}'_n\end{pmatrix}_{2,\mathbf{a}'_1}\right] - \cdots$$

$$- a_{n1}det\left[\begin{pmatrix}\mathbf{a}'_2\\\mathbf{a}'_3\\\vdots\\\mathbf{a}'_n\end{pmatrix}_{n-1,\mathbf{a}'_1}\right] - a_{21}det\left[\begin{pmatrix}\mathbf{a}'_2\\\mathbf{a}'_3\\\vdots\\\mathbf{a}'_n\end{pmatrix}_{1,\mathbf{a}'_1}\right]. \quad (3.1.11)$$

Comparison of (3.1.10) and (3.1.11) shows that if $a_{11} \neq 0$ then $det(E_{12}A) = -det(A) = det(E_{12})det(A)$.

Case *ii*. $a_{11} = 0$, $a_{21} \neq 0$.

The GEM applied to A begins with the ROW EXCHANGE of rows 1 and 2. The GEM applied to $E_{12}A$ does not involve the ROW EXCHANGE of rows 1 and 2. All other operations of GEM are the same for A and for $E_{12}A$. The GEM applied to $E_{12}A$ involves 1 ROW EXCHANGE fewer than does the GEM applied to A, whence $det(E_{12}A) = -det(A)$.

Case *iii*. $a_{11} \neq 0$, $a_{21} = 0$.

This time the GEM applied to $E_{12}A$ involves 1 ROW EXCHANGE more than does the GEM applied to A, whence again $det(E_{12}A) = -det(A)$.

Case *iv*. $a_{11} = a_{21} = 0$.

Here, there is an r greater than 2 and such that the GEM applied to A begins with a ROW EXCHANGE of rows 1 and r. Let $(E_{1r}A)'$ resp. $(E_{1r}B)'$ denote the $n-1 \times n-1$ matrix in the "southeast" corner of $E_{1r}A$ resp. $E_{1r}B$. Then the GEM applied to A produces the equality:

$$det(A) = a_{1r}det[(E_{1r}A)'].$$

The GEM applied to B produces the equality:

$$det(B) = a_{1r}det[(E_{1r}B)'].$$

However the $n-1 \times n-1$ matrices $(E_{1r}B)'$ and $E_{1,r-1}(E_{1r}A)'$ are the same: $(E_{1r}B)' = E_{1,r-1}(E_{1r}A)'$. By inductive assumption

$$det[(E_{1r}B)'] = det[E_{1,r-1}(E_{1r}A)'] = -det[(E_{1r}A)']$$

whence $det(E_{12}A) = -det(A) = det(E_{12})det(A)$.

If both i and j are greater than 2 then the argument given in Case *iv* above implies that

$$det(E_{ij}A) = -det(A).$$

If $i = 1$ and $j > 2$ then

$$E_{ij}A = E_{1j}A = E_{12}E_{2j}E_{12}A.$$

The result established above for the effect of E_{12} implies that

$$det(E_{12}E_{2j}E_{12}A) = -det(E_{2j}E_{12}A).$$

Since E_{2j} produces a ROW EXCHANGE among the last $n - 1$ rows of A, the argument given for Case *iv* above implies

$$det(E_{2j}E_{12}A) = -det(E_{12}A).$$

A second appeal to the result established for the effect of E_{12} shows that

$$det(E_{12}A) = -det(A)$$

and the *three* sign changes uncovered show $det(E_{ij}A) = -det(A)$.
   If $i > 2$ and $j = 1$ apply the argument to $E_{ji}\ (= E_{ij})$.

$$\Omega$$

LEMMA **3.1.2.** ON Mat_{nn} THE DETERMINANT FUNCTION det IS A LINEAR FUNCTION OF THE FIRST ROW OF EACH MATRIX, i.e., IF

$$A \overset{\text{def}}{=} \begin{pmatrix} a_{11} & \cdots & a_{1n} \\ \vdots & \ddots & \vdots \\ a_{n1} & \cdots & a_{nn} \end{pmatrix},$$

IF THE LAST $n - 1$ ROWS OF A ARE HELD FIXED, AND THE FIRST ROW VARIES THEN AS A FUNCTION OF THE FIRST ROW, det VARIES LINEARLY AS THE FIRST ROW VARIES.

PROOF. If $a_{i1} = 0$, $2 \leq i \leq n$ then, in the notation used earlier,

$$det\left[\begin{pmatrix} r\mathbf{a}_1 + s\mathbf{b}_1 \\ \mathbf{a}_2 \\ \vdots \\ \mathbf{a}_n \end{pmatrix}\right] = (ra_{11} + sb_{11})det\left[\begin{pmatrix} \mathbf{a}_2' \\ \vdots \\ \mathbf{a}_n' \end{pmatrix}\right]$$

$$= r\,det\left[\begin{pmatrix} \mathbf{a}_1 \\ \mathbf{a}_2 \\ \vdots \\ \mathbf{a}_n \end{pmatrix}\right] + s\,det\left[\begin{pmatrix} \mathbf{b}_1 \\ \mathbf{a}_2 \\ \vdots \\ \mathbf{a}_n \end{pmatrix}\right].$$

If $1 < i \le n$ and $a_{i1} \ne 0$ then

$$det\left[\begin{pmatrix} r\mathbf{a}_1 + s\mathbf{b}_1 \\ \mathbf{a}_2 \\ \vdots \\ \mathbf{a}_n \end{pmatrix}\right] = -det\left[\begin{pmatrix} \mathbf{a}_i \\ \mathbf{a}_2 \\ \vdots \\ r\mathbf{a}_1 + s\mathbf{b}_1 \\ \vdots \end{pmatrix}\right]$$

$$= -a_{i1}det\left[\begin{pmatrix} \mathbf{a}_2' - \frac{a_{21}}{a_{i1}}\mathbf{a}_i' \\ \vdots \\ r\mathbf{a}_1' + s\mathbf{b}_1' - \frac{ra_{11}+sb_{11}}{a_{i1}}\mathbf{a}_i' \\ \vdots \end{pmatrix}\right]$$

$$= -a_{i1}r\,det\left[\begin{pmatrix} \mathbf{a}_2' - \frac{a_{21}}{a_{i1}}\mathbf{a}_i' \\ \vdots \\ \mathbf{a}_1' - \frac{a_{11}}{a_{i1}}\mathbf{a}_i' \\ \vdots \end{pmatrix}\right] - a_{i1}s\,det\left[\begin{pmatrix} \mathbf{a}_2' - \frac{a_{21}}{a_{i1}}\mathbf{a}_i' \\ \vdots \\ \mathbf{b}_1' - \frac{b_{11}}{a_{i1}}\mathbf{a}_i' \\ \vdots \end{pmatrix}\right]$$

$$= -r\,det\left[\begin{pmatrix} \mathbf{a}_i \\ \vdots \\ \mathbf{a}_1 \\ \vdots \end{pmatrix}\right] - s\,det\left[\begin{pmatrix} \mathbf{a}_i \\ \vdots \\ \mathbf{b}_1 \\ \vdots \end{pmatrix}\right]$$

$$= r\,det\left[\begin{pmatrix} \mathbf{a}_1 \\ \mathbf{a}_2 \\ \vdots \\ \mathbf{a}_i \\ \vdots \end{pmatrix}\right] + s\,det\left[\begin{pmatrix} \mathbf{b}_1 \\ \mathbf{a}_2 \\ \vdots \\ \mathbf{a}_i \\ \vdots \end{pmatrix}\right]$$

as required.

$$\Omega$$

The two preceding results, LEMMA **3.1.1** and LEMMA **3.1.2**, together with the equation $det(I) = 1$ show that det satisfies D1-D3.

Exercise 3.1.12. Let A be an $n \times n$ matrix. Show that:

i. if $\mathcal{E}$ is the GEM matrix such that $\mathcal{E}A = A_E$ then $det(\mathcal{E}) = \pm 1$;

ii. if $\mathcal{R}$ is the matrix such that $\mathcal{R}A = A_R$, the row-reduced form of A, and if A is invertible then $A_R = I$ and $det(\mathcal{R}) = det(A)^{-1}$.

Exercise 3.1.13. Show that if A is an upper or lower triangular matrix then $det(A) = det(A^t)$.

Exercise 3.1.14. Show that if A is a singular matrix then $det(A) = det(A^t)$.

Exercise 3.1.15. Show that if Π is a permutation matrix, i.e., if Π is the product of elementary matrices of the form E_{ij} then $det(\Pi) = \pm 1$.

Exercise 3.1.16. Use the factorization $\Pi A = LU$, the product rule for determinants, and the result in **Exercise 3.1.15** to show that for *any* matrix A $det(A) = det(A^t)$.

Exercise 3.1.17. Show that if two rows or two columns of A are the same then $det(A) = 0$.

Exercise 3.1.18. Show that if any column or any row of the SQUARE matrix $A \stackrel{\text{def}}{=} (a_{ij})_{i,j=1}^{n,n}$ consists of 0's then $det(A) = 0$.

A SQUARE matrix A is called *skew-symmetric* iff $A = -A^t$.

Exercise 3.1.19. Assume n is odd and that A is skew-symmetric. Show $det(A) = 0$.

Exercise 3.1.20. Show that if A is nonsingular then

$$det(A^{-1}) = (det(A))^{-1}.$$

Exercise 3.1.21. Show that if B is nonsingular then $det(B^{-1}AB) = det(A)$.

[**Remark 3.1.3:** Thus the determinant of a matrix is a *similarity invariant* (cf. Section 4.1).]

Exercise 3.1.22. Show that the determinant of a matrix is a linear function of each row when the other rows are held fixed. Show that the determinant of a matrix is a linear function of each column when the other columns are held fixed.

3.2. Details about determinants

If A is the 2×2 matrix $(a_{ij})_{i,j=1}^{2,2}$ then $det(A)$ may be (re)written as

$$+a_{11}a_{22} - a_{12}a_{21}, \tag{3.2.1}$$

to emphasize the way in which "+" and "−" are affixed to the different products in the formula.

The entries a_{ij} all bear double subscripts. The first subscript is i and the second subscript is j.

---

     *Chapter 3. DETERMINANTS (**A direct approach**)*

Each product in (3.2.1) is written so that the *first* subscripts appear in their natural order:

$$+a_{11}a_{22} - a_{12}a_{21}.$$

In the first product the second subscripts are also in natural order. The number of order reversals of the second subscripts is 0, an *even* number. The term is preceded by a positive sign ("+").

In the second product the second subscripts are in *reverse* order. The number of order reversals of the second subscripts is 1, an *odd* number. The term is preceded by a negative sign ("−"):

$$+a_{11}a_{22} - a_{12}a_{21}.$$

The associating of a positive sign ("+") resp. a negative sign ("−") with an *even* resp. an *odd* number of order reversals is not accidental. In the following paragraphs there is a development that has among its consequences the generalization of the "sign" phenomenon just described.

Again the row vector in which the jth component is 1 and all other components are 0's is denoted $\mathbf{f}_j$. Thus the matrix $A \overset{\text{def}}{=} (a_{ij})_{i,j=1}^{n,n}$ may be written as follows:

$$A = \begin{pmatrix} \sum_{j=1}^{n} a_{1j}\mathbf{f}_j \\ \vdots \\ \sum_{j=1}^{n} a_{ij}\mathbf{f}_j \\ \vdots \\ \sum_{j=1}^{n} a_{nj}\mathbf{f}_j \end{pmatrix} \qquad \leftarrow \ i\text{th row} .$$

The break-out operation, based upon properties D1 and D2, shows that

$$det(A) = det\left[\begin{pmatrix} \sum_{j=1}^{n} a_{1j}\mathbf{f}_j \\ \vdots \\ \sum_{j=1}^{n} a_{ij}\mathbf{f}_j \\ \vdots \\ \sum_{j=1}^{n} a_{nj}\mathbf{f}_j \end{pmatrix}\right] = \sum_{j_1=1}^{n} a_{1j_1} det\left[\begin{pmatrix} \mathbf{f}_{j_1} \\ \vdots \\ \sum_{j=1}^{n} a_{ij}\mathbf{f}_j \\ \vdots \\ \sum_{j=1}^{n} a_{nj}\mathbf{f}_j \end{pmatrix}\right]$$

$$\vdots$$

$$= \sum_{j_1=1}^{n}\sum_{j_2=1}^{n}\cdots\sum_{j_n=1}^{n} a_{1j_1}a_{2j_2}\cdots a_{nj_n} det\left[\begin{pmatrix} \mathbf{f}_{j_1} \\ \mathbf{f}_{j_2} \\ \vdots \\ \mathbf{f}_{j_n} \end{pmatrix}\right]. \qquad (3.2.2)$$

As in **Section 3.1** each matrix

$$\begin{pmatrix} \mathbf{f}_{j_1} \\ \mathbf{f}_{j_2} \\ \vdots \\ \mathbf{f}_{j_n} \end{pmatrix}$$

is either a permutation matrix Π or a matrix in which at least two rows are the same. Hence each nonzero factor

$$det\left[\begin{pmatrix} \mathbf{f}_{j_1} \\ \mathbf{f}_{j_2} \\ \vdots \\ \mathbf{f}_{j_n} \end{pmatrix}\right]$$

in (3.2.2) is ± 1. Hence the value of each nonzero term in (3.2.2) depends on, among other things, the particular arrangement, i.e., permutation π,

$$j_1, j_2, \ldots, j_n$$

of the numbers $1, 2, \ldots, n$.

Then π determines whether "$+$" or "$-$" occurs. Hence one writes $sgn(\pi)$ (read "sign of π") for

$$det\left[\begin{pmatrix} \mathbf{f}_{j_1} \\ \mathbf{f}_{j_2} \\ \vdots \\ \mathbf{f}_{j_n} \end{pmatrix}\right] \tag{3.2.3}$$

which is ± 1 whenever it is not 0.

The following notations,

$$\pi \stackrel{\text{def}}{=} \begin{pmatrix} 1 & 2 & \ldots & n \\ j_1 & j_2 & \ldots & j_n \end{pmatrix}$$

or, alternatively,

$$\pi \stackrel{\text{def}}{=} \begin{pmatrix} 1 & 2 & \ldots & n \\ \pi(1) & \pi(2) & \ldots & \pi(n) \end{pmatrix},$$

i.e., $\pi(1) = j_1, \ldots, \pi(n) = j_n$, for an arbitrary permutation of the numbers $1, 2, \ldots, n$ are useful in what follows.

If $x, y, \ldots$ is an ordered arrangement of pairwise distinguishable objects, let the symbol

$$S_{x,y,\ldots}$$

denote the set of all ordered arrangements (*permutations*) of the objects. If the number of objects is a finite number n then $S_{x,y,\ldots}$ consists

---

   *Chapter 3. DETERMINANTS (**A** direct approach)*

of $n!$ permutations. Thus if $\#(S)$ denotes the number of elements in a set S it follows that

$$\#(S_{x,y,\dots}) = n!.$$

The symbol

$$\sum_{\pi \in S_{1,2,\dots,n}}$$

signifies summation taken over all $n!$ possible permutations π of the numbers $1, 2, \dots, n$.

Then (3.2.2) may be written

$$det\left[\begin{pmatrix} a_{11} & \cdots & a_{1n} \\ \vdots & \ddots & \vdots \\ a_{n1} & \cdots & a_{nn} \end{pmatrix}\right] = \sum_{\pi \in S_{1,2,\dots,n}} sgn(\pi)a_{1\pi(1)}a_{2\pi(2)} \cdots a_{n\pi(n)}, \qquad (3.2.4)$$

a sum consisting of $n!$ terms some of which can be 0's.

Exercise 3.2.1. Calculate the sign of each of the following permutations:

$$i. \begin{pmatrix} 1 & 2 & 3 & 4 \\ 2 & 4 & 3 & 1 \end{pmatrix}, \quad ii. \begin{pmatrix} 1 & 2 & 3 & 4 & 5 \\ 3 & 5 & 4 & 1 & 2 \end{pmatrix}.$$

If π_1 and π_2 are two permutations of $1, 2, \dots, n$ then

$$\pi_3 : j \mapsto \pi_2(\pi_1(j)), \ 1 \le j \le n,$$

is another permutation: $\pi_3 \overset{\text{def}}{=} \pi_2\pi_1$.

The *inverse* of a permutation π is the permutation

$$\begin{pmatrix} \pi(1) & \pi(2) & \cdots & \pi(n) \\ 1 & 2 & \cdots & n \end{pmatrix}$$

and it is denoted π^{-1}.

Example 3.2.1. Assume

$$\pi_1 = \begin{pmatrix} 1 & 2 & 3 & 4 \\ 3 & 1 & 4 & 2 \end{pmatrix}, \pi_2 = \begin{pmatrix} 1 & 2 & 3 & 4 \\ 4 & 1 & 2 & 3 \end{pmatrix}.$$

Then

$$\pi_2\pi_1 = \begin{pmatrix} 1 & 2 & 3 & 4 \\ 2 & 4 & 3 & 1 \end{pmatrix}$$

and

$$\pi_1^{-1} = \begin{pmatrix} 3 & 1 & 4 & 2 \\ 1 & 2 & 3 & 4 \end{pmatrix} = \begin{pmatrix} 1 & 2 & 3 & 4 \\ 2 & 4 & 1 & 3 \end{pmatrix}.$$

Furthermore

$$sgn(\pi_1) = (-1)^3 = -1$$
$$sgn(\pi_2) = (-1)^3 = -1$$
$$sgn(\pi_2\pi_1) = (-1)^4 = 1 = sgn(\pi_2)sgn(\pi_1)$$
$$sgn(\pi_1^{-1}) = (-1)^3 = -1 = sgn(\pi_1)$$
$$sgn(\pi_1) = \prod_{1 \le i < j \le 4} \frac{(i-j)}{(\pi(i) - \pi(j))} = \prod_{1 \le i < j \le 4} \frac{(x_i - x_j)}{(x_{\pi(i)} - x_{\pi(j)})}.$$

Exercise 3.2.2. Prove the following general formulae for π_1 and π_2 in $S_{1,2,\dots,n}$:

i.

$$sgn(\pi_2\pi_1) = sgn(\pi_2)sgn(\pi_1);$$

ii.

$$sgn(\pi^{-1}) = sgn(\pi);$$

iii.

$$sgn(\pi) = \prod_{1 \le i < j \le n} \frac{(i-j)}{(\pi(i) - \pi(j))} = \prod_{1 \le i < j \le n} \frac{(x_i - x_j)}{(x_{\pi(i)} - x_{\pi(j)})}.$$

[*Hint:* Use the product rule for $i - ii$. For iii note that to each denominator $(\pi(i) - \pi(j))$ there corresponds exactly one numerator $(i' - j')$ such that:

$$(i' - j') = \begin{cases} (\pi(i) - \pi(j)) & \text{if } \pi(i) < \pi(j) \text{ -- no order reversal;} \\ -(\pi(i) - \pi(j)) & \text{if } \pi(i) > \pi(j) \text{ -- an order reversal.} \end{cases}$$

The correspondence is one-one between the set of *all* numerators and the set of *all* denominators.]

[**Remark 3.2.1:** If π is a permutation of $1, 2, \dots, n$ then an order reversal in the sequence

$$\pi(1), \dots, \pi(n)$$

is a pair $\{\pi(i), \pi(j)\}$ for which $i < j$ and $\pi(i) > \pi(j)$. The number $sgn(\pi)$ may be calculated in terms of the number ρ of order reversals in the sequence $\pi(1), \pi(2), \dots, \pi(n)$.

Indeed, the permutation matrix

$$\Pi \overset{\text{def}}{=} \begin{pmatrix} \mathbf{f}_{\pi(1)} \\ \vdots \\ \mathbf{f}_{\pi(n)} \end{pmatrix}$$

may be converted to I by ROW EXCHANGEs as follows:

   i. The row of Π in which $\mathbf{f}_1$ appears is switched successively with each row above it. The result is a possibly new permutation matrix Π_1 in which the first row is $\mathbf{f}_1$.

  ii. The row of Π_1 in which $\mathbf{f}_2$ appears is switched successively with each row, except row 1, above it. The result is a possibly new permutation matrix Π_2 in which the first row is $\mathbf{f}_1$ and the second row is $\mathbf{f}_2$.

iii. etc.

The ROW EXCHANGEs described above are achieved by a number of multiplications by elementary permutation matrices. That number is the total number of order reversals in the original permutation π.

Indeed, the first batch of ROW EXCHANGEs in i counts the number of order reversals involving 1. The second batch of ROW EXCHANGEs in ii counts the number of order reversals involving 2 and not involving 1. The third batch ..., etc.

Hence $sgn(\pi) = (-1)^{\rho}$ since both the GEM applied to Π and the procedure above applied to Π reduce Π to I.]

The formula (3.2.4) may be described verbally as follows:

The determinant of A is a sum of products of entries in A. Each product consists of n factors, each an entry, and there is one and only one entry from each row and one and only one entry from each column. Every possible product of that kind appears in the sum and there is affixed to each product a "+" or a "−" in accordance with rules that link the order of the indices to the sign.

From (3.2.4) and the equation $det(A) = det(A^t)$ it follows that $det(A)$ may be written as a linear expression involving the entries in any row or in any column, viz.,

$$det(A) = \sum_{i=1}^{n} a_{ij} A_{ij}, \tag{3.2.5}$$

$$= \sum_{j=1}^{n} a_{ij} B_{ij}. \tag{3.2.6}$$

Although (3.2.5) and (3.2.6) appear to be very similar, they are significantly different.

In (3.2.5) the index of summation is i while j is held fixed. Since j is held fixed, all the a_{ij} are in the same *column* and as i varies during the summation, a pass is made down all the entries in column j. Hence (3.2.5) is called *the expansion of $det(A)$ according to the entries of the jth column.*

In (3.2.6) the index of summation is j while i is held fixed. Since i is held fixed, all the a_{ij} are in the same *row* and as j varies during the summation, a pass is made across all the entries in row i. Hence (3.2.6) is called *the expansion of $det(A)$ according to the entries of the the ith row.*

Expansions according to entries of a row or of a column go under the general name of *cofactor* expansions.

Example 3.2.2. If

$$det(A) \stackrel{\text{def}}{=} det \left[\begin{pmatrix} a_{11} & a_{12} & a_{13} \\ a_{21} & a_{22} & a_{23} \\ a_{31} & a_{32} & a_{33} \end{pmatrix} \right]$$

is expanded according to the entries of the second column the result is

$$det(A) = a_{12}(-(a_{21}a_{33} - a_{23}a_{31})) + a_{22}(a_{11}a_{33} - a_{13}a_{31})$$
$$+ a_{32}(-(a_{11}a_{23} - a_{13}a_{31}))$$
$$\stackrel{\text{def}}{=} a_{12}A_{12} + a_{22}A_{22} + a_{32}A_{32}$$

If $det(A)$ is expanded according to the entries of the third row the result is

$$det(A) = a_{31}(a_{12}a_{23} - a_{13}a_{22}) + a_{32}(-(a_{11}a_{23} - a_{13}a_{21}))$$
$$+ a_{33}(a_{11}a_{22} - a_{12}a_{21})$$
$$= a_{31}B_{31} + a_{32}B_{32} + a_{33}B_{33}$$

Note that in the expansions above, the numbers A_{ij} and B_{ij} are themselves determinants of 2×2 submatrices. More specifically, if M_{ij} is the 2×2 submatrix that results from deleting the ith row and the jth column from A, then

$$A_{ij} = B_{ij} = (-1)^{i+j} det(M_{ij}).$$

The expansion according to entries in a row or the entries in a column is helpful particularly when some column has some 0's. In fact a helpful strategy is to expand according to a row or column that has the largest number of 0's.

Exercise 3.2.3. Expand the following determinant according to the entries in the third column and also according to the entries in the first row:

$$det \left[\begin{pmatrix} 3 & -2 & 1 \\ 9 & 4 & -2 \\ -5 & 1 & 7 \end{pmatrix} \right].$$

Exercise 3.2.4. Use the most convenient row or column expansion to calculate

$$det \left[\begin{pmatrix} 2 & 4 & 5 \\ 0 & 1 & 4 \\ 3 & 6 & 9 \end{pmatrix} \right].$$

The formulae (3.2.5) and (3.2.6) may be related to the formula (3.2.4). For example, in (3.2.4) each entry a_{ij} appears in $(n-1)!$ terms.

Hence in $det(A)$ a_{11} is the coefficient of

$$A_{11} \overset{\text{def}}{=} \sum_{\pi \in S_{1,\ldots,n}, \pi(1)=1} sgn(\pi) a_{2\pi(2)} \cdots a_{n\pi(n)}$$

and the right member above is precisely the determinant of the $(n-1) \times (n-1)$ matrix M_{11} that arises when row 1 and column 1 of A are deleted.

The coefficient of a_{21} is

$$A_{21} \overset{\text{def}}{=} \sum_{\pi \in S_{1,\ldots,n}, \pi(2)=1} sgn(\pi) a_{1\pi(1)} a_{3\pi(3)} \cdots a_{n\pi(n)}.$$

If M_{21} is the matrix arising from A by deleting row 2 and column 1 then

$$det(M_{21}) = \sum_{\pi' \in S_{1,3,\ldots,n}} sgn(\pi') a_{1\pi'(1)} a_{3\pi'(3)} \cdots a_{n\pi'(n)}.$$

If $\pi(2) = 1$ then π looks like this:

$$\begin{pmatrix} 1 & 2 & 3 & \cdots & n \\ j_1 & 1 & j_3 & \cdots & j_n \end{pmatrix}. \tag{3.2.7}$$

The permutation π' in $S_{1,3,\ldots,n}$ that looks like this

$$\begin{pmatrix} 1 & 3 & \cdots & n \\ j_1 & j_3 & \cdots & j_n \end{pmatrix} \tag{3.2.8}$$

is the one and only π' such that $\pi'(k) = \pi(k)$ if $k \neq 2$. The presence of 1 in the second position in (3.2.7) gives rise to exactly 1 order reversal produced by π and *not* produced by π'. Hence

$$sgn(\pi') = (-1)^1 sgn(\pi).$$

The coefficient of a_{i1} is

$$A_{i1} \overset{\text{def}}{=} \sum_{\pi \in S_{1,\ldots,n}, \pi(i)=1} sgn(\pi) a_{1\pi(1)} \cdots a_{i-1,\pi(i-1)} a_{i+1,\pi(i+1)} \cdots a_{n\pi(n)}.$$

In general, if $\pi \in S_{1,\ldots,n}$ and π looks like this

$$\begin{pmatrix} 1 & \ldots & i & \ldots & n \\ j_1 & \ldots & 1 & \ldots & j_n \end{pmatrix} \qquad (3.2.9)$$

then there is in $S_{1,\ldots,i-1,i+1,\ldots,n}$ exactly one π' such that $\pi'(k) = \pi(k)$ if $k \neq i$; π' looks like this:

$$\begin{pmatrix} 1 & \ldots & i-1 & i+1 & \ldots & n \\ j_1 & \ldots & j_{i-1} & j_{i+1} & \ldots & j_n \end{pmatrix}. \qquad (3.2.10)$$

Then the presence of 1 in the ith position in (3.2.9) gives rise to exactly $i-1$ order reversals produced by π and *not* produced by π'. Hence, since $(-1)^{i-1} = (-1)^{i+1}$,

$$sgn(\pi') = (-1)^{i+1} sgn(\pi).$$

If M_{i1} is the matrix arising from A by deleting row i and column 1 then the argument above implies that

$$(-1)^{(i+1)} A_{i1} = det(M_{i1}).$$

Exercise 3.2.5. Show that in the formulae (3.2.5) and (3.2.6) if M_{ij} is the matrix arising from A by deleting row i and column j then

$$A_{ij} = B_{ij} = (-1)^{(i+j)} det(M_{ij}).$$

[*Hint:* For the result concerning A_{ij}, shift rows and columns of A so that in the new matrix A' row 1 is row i of A and column 1 is column j of A and that all other rows and columns in A' are in the same order in which they occur in A. For the result concerning B_{ij} work with A^t.]

The *number* A_{ij} is called the *cofactor* of a_{ij}.

The $(n-1) \times (n-1)$ *matrix* M_{ij} is called the *minor* corresponding to the entry a_{ij}. More generally a $(n-k) \times (n-k)$ minor of A is the $(n-k) \times (n-k)$ matrix remaining after k rows numbered $i_1, \ldots, i_k$ and k columns numbered $j_1, \ldots, j_k$ are deleted from A.

Thus $A_{ij} = (-1)^{(i+j)} det(M_{ij})$ and

$$det(A) = \sum_{i=1}^{n} (-1)^{(i+j)} a_{ij} det(M_{ij}), \ 1 \leq j \leq n; \qquad (3.2.11)$$

$$= \sum_{j=1}^{n} (-1)^{(i+j)} a_{ij} det(M_{ij}), \ 1 \leq i \leq n. \qquad (3.2.12)$$

In (3.2.11) the index of summation is the first index, the *row* index, i.e., the pass is made through the *rows* of column j, and so the expansion of $det(A)$ is

---

    *Chapter 3. DETERMINANTS (**A** direct approach)*

according to the entries of column j; in (3.2.12) the index of summation is the second index, the *column* index, i.e., the pass is made through the entries of row i and so the expansion of $det(A)$ is according to the entries of row i.

The factors $(-1)^{(i+j)}$ in (3.2.11) and (3.2.12) may be presented graphically by associating with the ij entry "+" or "−" according to the following (checkerboard) pattern:

$$\begin{pmatrix} + & - & \cdots & \pm \\ - & + & \cdots & \mp \\ \vdots & \vdots & \ddots & \vdots \\ \cdots & \cdots & \cdots & \cdots \end{pmatrix}.$$

For example, for a 3×3 matrix the pattern is

$$\begin{pmatrix} + & - & + \\ - & + & - \\ + & - & + \end{pmatrix};$$

for a 4×4 matrix the pattern is

$$\begin{pmatrix} + & - & + & - \\ - & + & - & + \\ + & - & + & - \\ - & + & - & + \end{pmatrix};$$

etc.

Exercise 3.2.6. Show that if $i \neq i'$ or $j \neq j'$ then

$$\sum_{j=i}^{n} a_{ij} A_{i'j} = \sum_{i=1}^{n} a_{ij} A_{ij'} = 0.$$

[*Hint:* Let $\widetilde{A}$ be the matrix A with column i' replaced by column i. Then $\widetilde{A} E_{ii'} = \widetilde{A}$. Show $det(\widetilde{A}) = 0$. Show $\widetilde{A}_{ij} = (-1)^{i-i'} \widetilde{A}_{i'j}$. Expand $det(\widetilde{A})$ according to the entries in column i. Examine the situation for 2×2 and 3×3 matrices.]

Exercise 3.2.7. Use **Exercise 3.2.6** to derive *Cramer's rule* given next.

Cramer's Rule:

Let A be the matrix of the system

$$a_{11}x_1 + \cdots + a_{1n}x_n = b_1$$

$$\vdots \qquad \ddots \qquad \vdots \qquad \vdots$$

$$a_{n1}x_1 + \cdots + a_{nn}x_n = b_n$$

Let b be

$$\begin{pmatrix} b_1 \\ \vdots \\ b_n \end{pmatrix}$$

and let $A(j)$ be the matrix A with the jth column replaced by b. If A is nonsingular, i.e., if A is invertible, i.e., if $det(A) \neq 0$, then Cramer's rule reads as follows:

$$x_j = \frac{det(A(j))}{det(A)}, \ 1 \leq j \leq n.$$

[*Hint:* Multiply the ith equation by A_{ij} and then sum over i from 1 to n.]

Exercise 3.2.8. Let A be $(a_{ij})_{i,j=1}^{n,n}$ and let $\mathcal{A}$ be the matrix in which the ij entry is A_{ji} (note the reversal of subscripts!). Show that $A\mathcal{A} = det(A)I$.

[**Remark 3.2.2:** The matrix $\mathcal{A}$ is often called the *classical adjoint* of A. It is to be distinguished from the so-called *adjoint*,

$$A^* \stackrel{\text{def}}{=} \overline{A^t} = \begin{pmatrix} \overline{a}_{11} & \cdots & \overline{a}_{n1} \\ \vdots & \ddots & \vdots \\ \overline{a}_{1n} & \cdots & \overline{a}_{nn} \end{pmatrix},$$

(cf. **Exercise 2.9.24** and **Section 4.4**).]

Example 3.2.3. If

$$A \stackrel{\text{def}}{=} \begin{pmatrix} 1 & 2 & 3 \\ 4 & 5 & 6 \\ 7 & 8 & 9 \end{pmatrix}$$

then the classical adjoint $\mathcal{A}$ of A is

$$\begin{pmatrix} det\left[\begin{pmatrix} 5 & 6 \\ 8 & 9 \end{pmatrix}\right] & -det\left[\begin{pmatrix} 2 & 3 \\ 8 & 9 \end{pmatrix}\right] & det\left[\begin{pmatrix} 2 & 3 \\ 5 & 6 \end{pmatrix}\right] \\ -det\left[\begin{pmatrix} 4 & 6 \\ 7 & 9 \end{pmatrix}\right] & det\left[\begin{pmatrix} 1 & 3 \\ 7 & 9 \end{pmatrix}\right] & -det\left[\begin{pmatrix} 1 & 3 \\ 4 & 6 \end{pmatrix}\right] \\ det\left[\begin{pmatrix} 4 & 5 \\ 7 & 8 \end{pmatrix}\right] & -det\left[\begin{pmatrix} 1 & 2 \\ 7 & 8 \end{pmatrix}\right] & det\left[\begin{pmatrix} 1 & 2 \\ 4 & 5 \end{pmatrix}\right] \end{pmatrix}.$$

In **Part II**, where applications of linear algebra are discussed, some of the rare uses of determinants in applied problems are demonstrated. Even Cramer's rule, despite its elegance, is not a practical device for solving SQUARE systems of linear equations, except perhaps in problems where the "basket-weaving" techniques are available and feasible or where a simple cofactor expansion can be found.

On the other hand, the theoretical aspects of determinants are broad and deep. For example, in **Chapter 4**, in the study of eigenvalues of SQUARE matrices, determinants provide a simple theoretical criterion for eigenvalues. Nevertheless, even there, the utility of the criterion is theoretical and the criterion is applicable really only for 2×2 or 3×3 matrices.

Example 3.2.4. Let A be an $n \times n$ matrix. For the complex variable z

$$det(A - zI) = (-1)^n z^n + c_1 z^{n-1} + \cdots + c_n$$

is a polynomial χ_A of degree n in the variable z. Note that $\chi_A(0) = c_n = det(A)$. (The function χ_A is discussed at greater length in **Section 4.2** where its rôle in the *eigenvalue* problem is explained.)

Let $\mathcal{A}_z$ denote the classical adjoint of $A - zI$. Each *entry* of $\mathcal{A}_z$ is a polynomial of degree not exceeding $n - 1$. Thus there are $n \times n$ *matrices* $M_1, \ldots, M_n$ such that

$$\mathcal{A}_z = z^{n-1} M_1 + z^{n-2} M_2 + \cdots + M_n. \tag{3.2.13}$$

Hence (cf. **Exercise 3.2.8**)

$$\begin{aligned}
(A - zI)\mathcal{A}_z &= (A - zI)(z^{n-1} M_1 + z^{n-2} M_2 + \cdots + M_n) \\
&= det(A - zI)I. \tag{3.2.14}
\end{aligned}$$

The middle member in (3.2.14) is

$$-z^n M_1 + z^{n-1}(AM_1 - M_2) + z^{n-2}(AM_2 - M_3) + \cdots + AM_n.$$

The right member in (3.2.14) is, since $det(A) = c_n$,

$$(-1)^n z^n I + c_1 z^{n-1} I + \cdots + det(A)I.$$

Hence comparing like powers of z in the last two expressions leads to

$$-M_1 = (-1)^n I$$
$$(AM_1 - M_2) = c_1 I$$
$$\vdots \qquad \vdots$$
$$AM_n = det(A)I.$$

Multiply both members of the first equation by A^n, both members of the second equation by A^{n-1}, etc., the last equation by I and then add all the (multiplied) left members and all the (multiplied) right members. On the left the result is O by virtue of "telescoping cancellation." On the right the result is $\chi_A(A)$ and so there emerges the *Cayley-Hamilton* theorem: $\chi_A(A) = O$.

Example 3.2.5. Let A be the matrix

$$\begin{pmatrix} 1 & 2 & 3 \\ 4 & 5 & 6 \\ 7 & 8 & 9 \end{pmatrix}.$$

Then the classical adjoint of $A - \lambda I$ is

$$\begin{pmatrix} \lambda^2 - 14\lambda - 3 & 2\lambda + 6 & 3\lambda - 3 \\ 4\lambda + 6 & \lambda^2 - 10\lambda - 12 & 6\lambda + 6 \\ 7\lambda - 3 & 8\lambda + 6 & \lambda^2 - 6\lambda - 3 \end{pmatrix}$$

$$= \lambda^2 \begin{pmatrix} 1 & 0 & 0 \\ 0 & 1 & 0 \\ 0 & 0 & 1 \end{pmatrix} + \lambda \begin{pmatrix} -14 & 2 & 3 \\ 4 & -10 & 6 \\ 7 & 8 & -6 \end{pmatrix} + \begin{pmatrix} -3 & 6 & -3 \\ 6 & -12 & 6 \\ -3 & 6 & -3 \end{pmatrix}$$

whence

$$M_1 = \begin{pmatrix} 1 & 0 & 0 \\ 0 & 1 & 0 \\ 0 & 0 & 1 \end{pmatrix}, \quad M_2 = \begin{pmatrix} -14 & 2 & 3 \\ 4 & -10 & 6 \\ 7 & 8 & -6 \end{pmatrix}, \quad \text{and } M_3 = \begin{pmatrix} -3 & 6 & -3 \\ 6 & -12 & 6 \\ -3 & 6 & -3 \end{pmatrix}.$$

Furthermore

$$\chi_A(\lambda) = -\lambda^3 + 15\lambda^2 + 18\lambda$$

and on the other hand

$$\chi_A(A) = -A^3 + 15A^2 + 18A$$

$$= \begin{pmatrix} -468 & -576 & -684 \\ -1062 & -1305 & -1548 \\ -1656 & -2034 & -2412 \end{pmatrix}$$

$$+ \begin{pmatrix} 450 & 540 & 630 \\ 990 & 1265 & 1440 \\ 1530 & 1890 & 2250 \end{pmatrix} + \begin{pmatrix} 18 & 36 & 54 \\ 72 & 90 & 108 \\ 126 & 144 & 162 \end{pmatrix} = O,$$

as implied by the Cayley-Hamilton theorem.

Exercise 3.2.9. Let each SQUARE matrix below be called A. Calculate the classical adjoint of $A - \lambda I$, write out the matrices $M_1, \dots$ in (3.2.14), and then verify the Cayley-Hamilton equation.

$$i. \begin{pmatrix} 1 & 2 \\ 3 & 4 \end{pmatrix}; \quad ii. \begin{pmatrix} 1 & 2 & 3 \\ 3 & 1 & 2 \\ 3 & 1 & 2 \end{pmatrix}; \quad iii. \begin{pmatrix} \lambda_1 & & & \\ & \lambda_2 & & \\ & & \ddots & \\ & & & \lambda_n \end{pmatrix}; \quad iv. \begin{pmatrix} 0 & 1 & 2 \\ 0 & 0 & 3 \\ 0 & 0 & 0 \end{pmatrix}.$$

Exercise 3.2.10. Show that if A is a nonsingular SQUARE matrix then

$$A^{-1} = \frac{1}{\det(A)} \mathcal{A}.$$

Exercise 3.2.11. How many 3×3 minors are there in a 4×4 matrix? How many 2×2 minors are there in a 4×4 matrix? If $1 \leq k \leq n$ how many $k \times k$ minors are there in an $n \times n$ matrix ? For the matrix $(a_{ij})_{i,j=1}^{4,4}$ write out at least half of the 3×3 minors and at least one fourth of the 2×2 minors.

Exercise 3.2.12. Expand the determinant of

$$\begin{pmatrix} 1 & 0 & 2 & 1 \\ 3 & 4 & 5 & 0 \\ -1 & 0 & 7 & -8 \\ 0 & 2 & -2 & 5 \end{pmatrix}$$

according to the entries of each row and of each column (eight expansions in all).

Exercise 3.2.13. Assume that in the $n \times n$ matrix $A \overset{\text{def}}{=} (a_{ij})_{i,j=1}^{n,n}$ there is in row i exactly one nonzero entry a_{ij_i} and there is in column j exactly one nonzero entry $a_{i_j j}$, $1 \leq i, j \leq n$. Show that $\det(A) \neq 0$ and find the formula for $\det(A)$.

[*Hint:* The value of $\det(A)$ depends on the sign of the permutation

$$\begin{pmatrix} 1 & 2 & \ldots & n \\ j_1 & j_2 & \ldots & j_n \end{pmatrix}.$$

Try first the cases in which $n = 2$ or 3.]

Exercise 3.2.14. Show that if A is an $n \times n$ matrix then

$$\det(tA) = t^n \det(A).$$

Exercise 3.2.15. Show that if $PP^t = I$ then $|\det(P)| = 1$. Give an example of a matrix P such that $PP^t = I$ and $\det(P) \neq 1$. Show that if Π is a permutation matrix then $\Pi\Pi^t = I$.

Exercise 3.2.16. Assume the A is an $n \times n$ matrix and that $\det(A) = \pm 1$.

---

i. Show:
$$A^{-1} = \pm\mathcal{A} \ (= \pm\text{the classical adjoint of } A).$$

ii. Assume that A is an $n \times n$ matrix that is the product of elementary matrices in which all "parameters" are integers, i.e., each r in an $E_{ij}(r)$ and each s in an $E_i(s)$ is an integer: $r, s \in \mathbf{Z}$. Show that if $det(A) = \pm 1$ then every entry in A^{-1} is also an integer.

iii. Construct a 4×4 matrix $A \stackrel{\text{def}}{=} (a_{ij})_{i,j=1}^{4,4}$ such that each entry a_{ij} is a nonzero integer and such that each entry α_{ij} of $A^{-1} \stackrel{\text{def}}{=} (\alpha_{ij})_{i,j=1}^{4,4}$ is an integer:
$$a_{ij} \in \mathbf{Z} \setminus \{0\}, \ \alpha_{ij} \in \mathbf{Z}.$$

Exercise 3.2.17. Show that a SQUARE matrix A is invertible iff its classical adjoint $\mathcal{A}$ is also invertible.

Exercise 3.2.18. Assume A is a nonsingular $n \times n$ matrix and $det(-A) = det(A)$. What kind of natural number is n?

Exercise 3.2.19. Let $N(n)$ be the number of additions and subtractions performed in evaluating a determinant of a general $n \times n$ matrix according to the entries in a column or in a row. Show that $N(n+1) = (n+1)N(n) + n$. What is $N(10)$?

Exercise 3.2.20. Let A be the matrix
$$\begin{pmatrix} \cos x & \sin x & e^x \\ -\sin x & \cos x & e^{-x} \\ \tan x & \cot x & \sec x \end{pmatrix}$$
and let $W(x)$ be $det(A)$. Calculate
$$\frac{dW}{dx}.$$

Exercise 3.2.21. Let $A \stackrel{\text{def}}{=} (a_{ij})_{i,j=1}^{n,n}$ be an $n \times n$ matrix in which each entry a_{ij} is a differentiable function $a_{ij}(x)$ of the real variable x. Let $W(x)$ be $det(A)$. Show that if $A^{(k)}$ is the matrix identical to A except that the entries in row k of $A^{(k)}$ are the *derivatives* of the entries in row k of A then
$$\frac{dW}{dx} = \sum_{k=1}^{n} det(A^{(k)}).$$

[*Hint:* Use mathematical induction.]

---

 *Chapter 3. DETERMINANTS (**A direct approach**)*

Exercise 3.2.22. The cofactor expansion of the determinant of a SQUARE matrix $A \stackrel{\text{def}}{=} (a_{ij})_{i,j=1}^{n,n}$ can be derived by use of D1-D3. Justify each of the statements below. Together they constitute an alternative proof of the cofactor formula.

i. If $a_{11} \neq 0$ let A^1 be the result of applying the GEM to A up to the point where all entries below the first in column 1 are 0's. Let M be the $n-1 \times n-1$ matrix remaining after the first row and first column of A^1 are deleted. Then $det(A) = a_{11} det(M)$.

ii. The ith row of M comes about by adding to the ith row of M_{11} the row vector
$$-\frac{a_{i+1,1}}{a_{11}}(a_{12}, a_{13}, \ldots, a_{1n}).$$

iii.
$$det(M) = A_{11} - \frac{a_{21}}{a_{11}}A_{21} + \cdots + (-1)^{n-1}\frac{a_{n1}}{a_{11}}A_{n1}. \quad [\text{D1-D3}]$$

iv.
$$det(A) = a_{11}A_{11} - a_{21}A_{21} + \cdots + (-1)^{n+1}a_{n1}A_{n1}. \quad [(-1)^{n-1} = (-1)^{n+1}]$$

v. If $a_{11} = 0$ and $det(A) \neq 0$ some $a_{i1} \neq 0$.

vi. If $a_{11} = 0$ and $det(A) \neq 0$ a row exchange justifies statements like those in *ii - iv*.

vii. If all entries in the first column of A are 0's the cofactor expansion of A according to the elements of the first column is valid.

viii. The cofactor expansion according to the elements of *any* column is valid.

[*Hint:* Use COLUMN EXCHANGEs.]

ix. The cofactor expansion according to the elements of *any* row is valid.

[*Hint:* Use transposes.]

Exercise 3.2.23. Let $A \stackrel{\text{def}}{=} (a_{ij})_{i,j=1}^{n,n}$ be a SQUARE matrix and let $A^{(k)}$ be the $k \times k$ submatrix $(a_{ij})_{i,j=1}^{k,k}$. Show that in the application of the GEM no ROW EXCHANGEs are performed iff $det(A^{(k)}) \neq 0$, $1 \leq k \leq n$.

Exercise 3.2.24. Assume $E^{(1)} \ldots, E^{(K)}$ and $F^{(1)}, \ldots, F^{(L)}$ are elementary permutation matrices, none is I, and that
$$\prod_{k=1}^{K} E^{(k)} = \prod_{l=1}^{L} F^{(l)}.$$

Use the product rule for determinants to show that $K - L$ is an even integer (cf. **Exercise 2.7.14**).

Direct consequences of the properties of determinants are the next results. These results are the *"working rules"* or the *"clean-up"* operations used for manipulating SQUARE matrices to simplify the evaluation of their determinants.

The working rules say essentially that the basic operations ROW COMBINATION, ROW EXCHANGE, and ROW SCALING, used to varying degrees in the *PROCESS*, the GEM, and the GJM, may be used to help calculate determinants. The counterparts COLUMN COMBINATION, COLUMN EXCHANGE, and COLUMN SCALING, are equally applicable. The justification for the rules lies in:

P. the Product rule for determinants;

M. the fact that the row resp. column operations are achieved via pre- resp. post-Multiplication by elementary matrices;

T. the Transpose equation: $det(A^t) = det(A)$.

THE WORKING RULES FOR CALCULATING DETERMINANTS

i. If two rows or two columns of a SQUARE matrix A are interchanged to yield the matrix A' then $det(A') = -det(A)$. [P]

ii. If two rows or two columns of a SQUARE matrix A are the same then $det(A) = 0$. [*i*, T]

iii. If any column or any row of A consists entirely of 0's then $det(A) = 0$. [M, T]

iv. If A is an upper triangular or lower triangular matrix then $det(A)$ is the product of the diagonal entries in A: if $A \stackrel{\text{def}}{=} (a_{ij})_{i,j=1}^{n,n}$ is upper or lower triangular then

$$det(A) = \prod_{i=1}^{n} a_{ii}. \quad [\text{M, T}]$$

v. If $\mathbf{c}_1, \ldots, \mathbf{c}_n$ are the n column vectors that are the columns of $A \stackrel{\text{def}}{=} (a_{ij})_{i,j=1}^{n,n}$ and if all columns save the ith are held fixed and the ith varies, then $det(A)$ is a linear function of the ith column. [M, T]

vi. If some row of A is modified by the addition of some numerical multiple of a *different* row the resulting matrix A' is such that $det(A') = det(A)$ (the determinant remains unchanged). [M]

vii. If some column of A is modified by the addition of some numerical multiple of a *different* column the resulting matrix A' is such that $det(A') = detA$ (the determinant remains unchanged). [T]

In particular, if A is a triangular matrix then $det(A) \neq 0$ iff *each* diagonal entry is not 0; $det(A) = 0$ iff some diagonal entry is 0. Hence a triangular matrix A is nonsingular iff each diagonal entry is nonzero; A is singular iff *some* diagonal entry is 0.

The working rules, properly and ingeniously applied, cut down the tedious labor required in the evaluation of a determinant by direct application of the classical formula (3.2.4). In that formula for an $n \times n$ matrix there are $n!$ terms, each consisting of the product of n factors prefixed by a "+" or "−." If $n \geq 100$ the computational difficulties in the direct application of the classical formula are physically insuperable. However, even when the working rules are applied to shorten the job, they are genuinely useful only for determinants of modest size $(n \leq 5)$.

Exercise 3.2.25. Show that

$$f(t_1,t_2,t_3) \overset{\text{def}}{=} det \left[\begin{pmatrix} 1 & 1 & 1 \\ t_1 & t_2 & t_3 \\ t_1^2 & t_2^2 & t_3^2 \end{pmatrix} \right] = - \prod_{i<j}(t_i - t_j).$$

[*Hint:* Show that $f(x,t_1,t_2)$ is a quadratic polynomial function of x and that $f(x,t_1,t_2) = c(x-t_1)(x-t_2)$. Then evaluate c, the coefficient of x^2.]

Generalize the result to the following formula:

$$f(t_1,t_2\ldots,t_n) \overset{\text{def}}{=} det \left[\begin{pmatrix} 1 & 1 & \cdots & 1 \\ t_1 & t_2 & \cdots & t_n \\ \vdots & \vdots & \ddots & \vdots \\ t_1^{n-1} & t_2^{n-1} & \cdots & t_n^{n-1} \end{pmatrix} \right] = (-1)^n \prod_{1 \leq i < j \leq n}(t_i - t_j).$$

[*Hint:* Show that $f(x,t_2,\ldots,t_n)$ is a polynomial of degree $n-1$ and factor it in the form $c(x-t_2)\cdots(x-t_n)$ (cf. **Exercise 4.8.2**).]

3.3. Complements

I *The product rule revisited*

The derivation of the *product rule* $det(AB) = det(A)det(B)$ from the classical formula (3.2.4) can be carried out in several ways. A rather ingenious proof to be found in [**Bô**] is given below. Throughout only the classical formula (3.2.4) and the "working rules" are used in the derivation.

Let A and B be $n \times n$ matrices and let C be the $2n \times 2n$ *block matrix*

$$\begin{pmatrix} A & O \\ -I & B \end{pmatrix}.$$

In other words the matrix C consists of A in the "northwest" corner, O in the "northeast" corner, $-I$ in the "southwest" corner and B in the "southeast" corner.

The reader can refresh his understanding of block matrices and block multiplication of matrices by consulting **Example 2.8.8**.

Example 3.3.1. If $A = (a_{ij})_{i,j=1}^{n,n}$ and $B = (b_{ij})_{i,j=1}^{n,n}$ then

$$C \overset{\text{def}}{=} \begin{pmatrix} a_{11} & \cdots & a_{1n} & 0 & \cdots & 0 \\ \vdots & \ddots & \vdots & \vdots & \ddots & \vdots \\ a_{n1} & \cdots & a_{nn} & 0 & \cdots & 0 \\ -1 & \cdots & 0 & b_{11} & \cdots & b_{1n} \\ \vdots & \ddots & \vdots & \vdots & \ddots & \vdots \\ 0 & \cdots & -1 & b_{n1} & \cdots & b_{nn} \end{pmatrix}. \tag{3.3.1}$$

Exercise 3.3.1. Calculate $det(C)$ if

$$A \overset{\text{def}}{=} \begin{pmatrix} 5 & 2 \\ 3 & 4 \end{pmatrix} \text{ and } B \overset{\text{def}}{=} \begin{pmatrix} a & b \\ c & d \end{pmatrix}$$

and show, without reference to the product rule for determinants, that

$$det(C) = det(A)det(B) = det(AB).$$

The result above holds for general C because when the classical formula (3.2.4) is invoked any product involving an entry from $-I$ must contain as well an entry from O and hence must be 0. Each other product comes from a product involved in forming $det(A)$ and a product involved in forming $det(B)$. Furthermore because of the relative positions of A and B the signs of the two products yield the proper sign for the overall product.

Hence in general

$$det(C) = det(A)det(B). \tag{3.3.2}$$

Exercise 3.3.2. For C as in **Exercise 3.3.1** show that

$$E_{42}(4)E_{32}(3)E_{41}(2)E_{31}(5)C = \begin{pmatrix} O & AB \\ -I & B \end{pmatrix}.$$

As in **Exercise 3.3.2** a sequence of ROW COMBINATIONS can reduce the matrix C in (3.3.1) to the block matrix

$$\widetilde{C} \overset{\text{def}}{=} \begin{pmatrix} O & AB \\ -I & B \end{pmatrix}.$$

---

     *Chapter 3. DETERMINANTS (**A direct approach***)

The essence of the idea is to ELIMINATE by ROW COMBINATIONs only the entries in A in the "northwest" corner of C.

The reduction of C to $\widetilde{C}$ can be accomplished by block multiplication as follows:

$$\widetilde{C} = \begin{pmatrix} I & A \\ O & I \end{pmatrix} \begin{pmatrix} A & O \\ -I & B \end{pmatrix} \overset{\text{def}}{=} \mathcal{A}C$$

The matrix

$$\mathcal{A} \overset{\text{def}}{=} \begin{pmatrix} I & A \\ O & I \end{pmatrix}$$

may be viewed as a block matrix form of a ROW COMBINATION matrix. By abuse of notation, in which Roman numerals are used for "row" indices of the block matrix,

$$\mathcal{A} = E_{II,I} \cdot (A).$$

By a sequence of ROW EXCHANGEs $\widetilde{C}$ can be transformed to

$$\widetilde{C}' \overset{\text{def}}{=} \begin{pmatrix} -I & B \\ O & AB \end{pmatrix}$$

These ROW EXCHANGEs can and should be performed so that the order of the rows within AB is preserved. Thus

$$\widetilde{C}' = E_{2n,n} E_{2n-1,n-1} \cdots E_{n+2,2} E_{n+1,1} \widetilde{C}$$

and, so since n ROW EXCHANGES are performed,

$$det(\widetilde{C}') = (-1)^n det(\widetilde{C}). \tag{3.3.3}$$

On the other hand the "working rules" imply that

$$det(\widetilde{C}') = (-1)^n det(AB). \tag{3.3.4}$$

From (3.3.2) - (3.3.4) it follows that $det(AB) = det(A)det(B)$.

Exercise 3.3.3. The shift from

$$\begin{pmatrix} O & AB \\ -I & B \end{pmatrix} \text{ to } \begin{pmatrix} -I & B \\ O & AB \end{pmatrix}$$

can also be accomplished by a sequence of ROW EXCHANGEs that switch row $n+1$ with row n, then new row n with row $n-1, \ldots$, and then new row 2 with row 1. These ROW EXCHANGEs can be followed by ROW EXCHANGEs between new row $n + 2$ and new row $n + 1$, then with new row $n, \ldots$, and then with new row 2, etc. Done this way, the procedure involves n ROW EXCHANGEs

---

for each of n rows originally numbered $n + 1, \ldots 2n$. The total number of row exchanges is thus n^2. Show

$$(-1)^{n^2} = (-1)^n$$

and thus again that the conclusion $det(AB) = det(A)det(B)$ is validly reached.

II *Laplace's expansion*

A generalization of the argument in **I** leads to the *Laplace expansion of a determinant*, to be derived next.

Let A be an $n \times n$ matrix and assume $1 < k < n$. Let the $k \times k$ matrix in the "northwest" corner of A be denoted

$$M_{12\ldots k}^{12\ldots k} \tag{3.3.5}$$

and let the $(n - k) \times (n - k)$ matrix in the "southeast" corner of A be denoted

$$M_{k+1,\ldots,n}^{k+1,\ldots,n}. \tag{3.3.6}$$

The matrices in (3.3.5) and (3.3.6) are examples of *complementary minors*.

More generally if

$$1 \le j_1 < j_2 < \cdots < j_k \le n, \tag{3.3.7}$$

if

$$1 \le j_{k+1} < j_{k+2} < \cdots < j_n \le n, \tag{3.3.8}$$

and if the numbers

$$j_1, \ldots, j_k, j_{k+1}, \ldots, j_n$$

constitute the set

$$\{1, 2, \ldots, n\},$$

in other words if the sets in (3.3.7) and (3.3.8) have no elements in common and together constitute $\{1, 2, \ldots, n\}$, then the sets described in (3.3.7) and (3.3.8) are called *complementary*. Correspondingly there are matrices

$$M_{12\ldots k}^{j_1 j_2 \ldots j_k} \tag{3.3.9}$$

and

$$M_{k+1,,\ldots,n}^{j_{k+1},\ldots,j_n} \tag{3.3.10}$$

and these are again *complementary minors*.

According to the classical rules the expansion of

$$det(M_{12\ldots k}^{12\ldots k})$$

---

is a sum of $k!$ products. Similarly the expansion of

$$det(M_{k+1,\ldots,n}^{k+1,\ldots,n})$$

is a sum of $(n-k)!$ products. Hence the *product*

$$det(M_{12\ldots k}^{12\ldots k})det(M_{k+1,\ldots,n}^{k+1,\ldots,n})$$

is a sum of $k!(n-k)!$ products each of which appears somewhere in the classical formula for $det(A)$. Furthermore, because of the relative positions of

$$M_{12\ldots k}^{12\ldots k} \text{ and } M_{k+1,\ldots,n}^{k+1,\ldots,n}$$

in A, it follows that even the sign of each term in

$$det(M_{12\ldots k}^{12\ldots k})det(M_{k+1,\ldots,n}^{k+1,\ldots,n})$$

is exactly the sign it gets in the classical formula for $det(A)$.

Since there are

$$\binom{n}{k} \equiv \frac{n!}{k!(n-k)!}$$

pairs of complementary minors like those in (3.3.9) and (3.3.10) it follows that if one sums all products of the form

$$\pm det(M_{12\ldots k}^{j_1 j_2 \ldots j_k})det(M_{k+1,,\ldots,n}^{j_{k+1},\ldots,j_n}) \tag{3.3.11}$$

the result will consist of

$$\frac{n!}{k!(n-k)!} \cdot k!(n-k)! = n!$$

terms and so at least the *number of terms* in the result will be correct. Furthermore the very definition of what is going on in (3.3.11) implies that the $n!$ products found will all be different and hence all that remains is to show what "$\pm$" in (3.3.11) should be.

Let the columns of A be rearranged so that column j_1 is shifted to column 1, column j_2 is shifted to column 2, ..., column j_k is shifted to column k, *while the order among the other columns is preserved.* Then the product in (3.3.11) is not affected and the sign affixed to it because of the new relative positions ("northwest - southeast") is now "+."

To achieve this new arrangement column j_1 is shifted $j_1 - 1$ times, column j_2 is shifted $j_2 - 2$ times, ..., column j_k is shifted $j_k - k$ times. In all, $j_1 + \cdots j_k - (1 + 2 + \cdots k)$ column shifts are performed and correspond to affixing

$$(-1)^{j_1 + \cdots + j_k - (1 + \cdots + k)} = (-1)^{j_1 + \cdots + j_k + 1 + \cdots + k}$$

before the product calculated for the complementary minors in their shifted positions. Hence "$\pm$" is

$$(-1)^{j_1+\cdots+j_k+1+\cdots+k}$$

and Laplace's expansion may be written

$$det(A) = \sum_{j_1<\cdots<j_k} (-1)^{j_1+\cdots+j_k+1+\cdots+k} \left(det(M_{12\ldots k}^{j_1 j_2 \cdots j_k}) det(M_{k+1,\ldots,n}^{j_{k+1},\ldots,j_n}) \right). \quad (3.3.12)$$

One further generalization is possible. Assume $i_1 < i_2 < \cdots < i_k$ and $i_{k+1} < i_{k+2} < \cdots < i_n$ are complementary sets of *row* indices and that

$$M_{i_1 i_2 \ldots i_k}^{j_1 j_2 \cdots j_k}$$

and

$$M_{i_{k+1} i_{k+2} \ldots i_n}^{j_{k+1} j_{k+2} \cdots j_n}$$

are corresponding complementary minors. The argument leading to to formula (3.3.12) can be repeated with the extra twist needed to bring the rows numbered $i_1, i_2, \ldots i_k$ into rows $1, 2, \ldots, k$ while the order of the other rows is preserved. Since

$$(-1)^{i_1+\cdots+i_k+1+\cdots+k+j_1+\cdots+j_k+1+\cdots+k}$$
$$= (-1)^{i_1+\cdots+i_k+j_1+\cdots+j_k+2(1+\cdots+k)}$$
$$= (-1)^{i_1+\cdots+i_k+j_1+\cdots+j_k}$$

the grand finale is the formula:

$$det(A) = \sum_{\left(\begin{array}{c} i_1<\cdots<i_k \\ j_1<\cdots<j_k \end{array}\right)} (-1)^{i_1+\cdots+i_k+j_1+\cdots+j_k} det(M_{i_1\ldots i_k}^{j_1\ldots j_k}) det(M_{i_{k+1}\ldots i_n}^{j_{k+1}\ldots j_n}).$$

$$(3.3.13)$$

The formula in (3.3.13) is the general Laplace expansion of $det(A)$ in terms of the determinants of the pairs of an appropriate set of complementary minors.

Exercise 3.3.4. Calculate the determinant of the matrix displayed below:

$$\begin{pmatrix} 1 & -1 & 2 & -2 & 3 \\ 2 & 3 & 4 & 5 & 6 \\ 1 & 3 & 5 & 7 & 9 \\ 1 & 2 & 3 & 4 & 5 \\ 1 & -1 & 1 & -1 & 1 \end{pmatrix}.$$

Use the Laplace expansion based on rows 2 and 4.

Exercise 3.3.5. Show how Laplace's expansion can be used to justify directly the argument in **I**.

III *Determinants and area* If, in **Figure 3.3.1** $\theta \overset{\text{def}}{=} \angle(AOC)$ then

$$area(OABC) = \sqrt{x_1^2 + y_1^2}\sqrt{x_2^2 + y_2^2}\sin\theta \overset{\text{def}}{=} |OA| \cdot |OC|\sin\theta$$

$$\sin\theta = \frac{x_1}{|OA|}\frac{y_2}{|OC|} - \frac{y_1}{|OA|}\frac{x_2}{|OC|}$$

whence $area(OABC) = x_1 y_2 - y_1 x_2 = det(M)$.

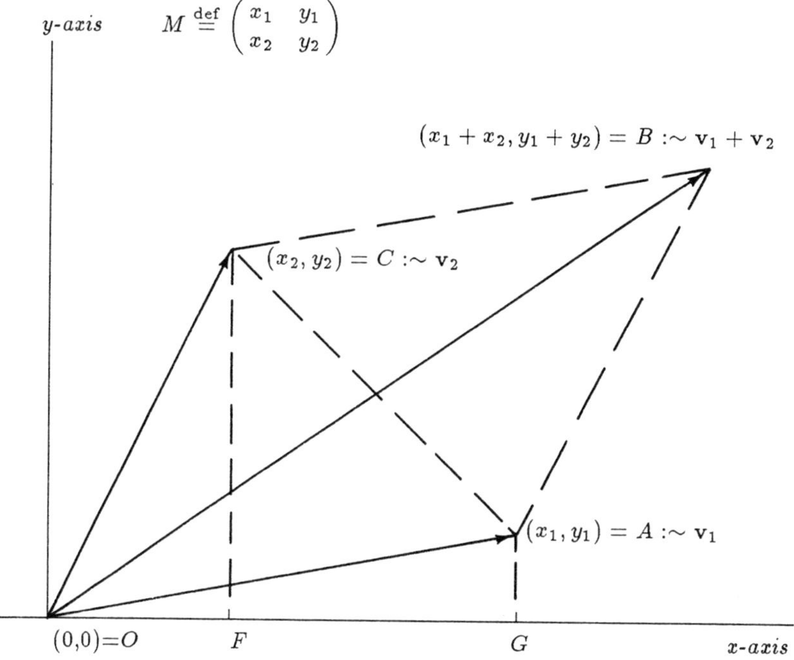

Figure 3.3.1.
$$area(OABC) = 2\,area(OAC)$$
$$= 2\,area(OFC) + 2\,area(FGAC) - 2\,area(OGA)$$
$$= x_2 y_2 + (x_1 - x_2)(y_1 + y_2) - x_1 y_1$$
$$= x_1 y_2 - x_2 y_1 = det(M)$$

The ingredients of an alternative area computation are laid out and related to a determinant in the caption to **Figure 3.3.1**.

There are generalizations of the last formula for "volume" in n-dimensional space, i.e., in $\mathbf{R}^n$ consisting of all $n \times 1$ column vectors $(x_1, \ldots, x_n)^t$. Thus if $\mathbf{v}_k = (x_{k1}, \ldots, x_{kn})$, $1 \leq k \leq n$ then those vectors determine an n-dimensional *parallelotope* P in the same way that $\mathbf{v}_1$ and $\mathbf{v}_2$ in 2-dimensional space $\mathbf{R}_2$ determine the parallelogram $OABC$.

More precisely define Q by the equation

$$Q \overset{\text{def}}{=} \left\{ \begin{pmatrix} x_1 \\ \vdots \\ x_n \end{pmatrix} : 0 \leq x_k \leq 1, \ 1 \leq k \leq n \right\}.$$

Thus Q is a *"unit cube"* in $\mathbf{R}^n$. Let M be the $n \times n$ matrix $(m_{ij})_{i,j=1}^{n,n}$. Then the parallelotope P is

$$\left\{ M \begin{pmatrix} x_1 \\ \vdots \\ x_n \end{pmatrix} : \begin{pmatrix} x_1 \\ \vdots \\ x_n \end{pmatrix} \in Q \right\}.$$

By abuse of notation P may be described by the equation $P = MQ$. The natural extension of the notion of the area of the 2-dimensional "unit cube," i.e., the "unit square," may be made analogously simply by defining the "volume" of Q to be 1 and the volume of a "cube" Q_t consisting of all vectors $(x_1, \ldots, x_n)^t$ such that $0 \leq x_k \leq t_k$, $1 \leq k \leq n$ to be

$$\prod_{k=1}^{n} t_k \overset{\text{def}}{=} t_1 t_2 \cdots t_n.$$

If $\mathbf{v} \overset{\text{def}}{=} (v_1, \ldots, v_n)^t$ is an arbitrary fixed vector then $\mathbf{v} + Q_t \overset{\text{def}}{=} Q_{t,\mathbf{v}}$ denotes the set of all points

$$(v_1 + x_1, \ldots, v_n + x_n)^t, \ 0 \leq x_k \leq t_k, 1 \leq k \leq n.$$

By definition the volume of $Q_{t,\mathbf{v}}$ is the same as the volume of Q_t. The set $Q_{t,\mathbf{v}}$ may viewed as Q_t moved, shifted or *translated* by the vector $\mathbf{v}$. In other words, volume is defined to be *translation invariant*.

The volume of a figure F that can be constructed of finite or countable union of pairwise disjoint "cubes" like $Q_{t,\mathbf{x}}$ is defined to be the sum of the constituent "cubes:"

$$F \overset{\text{def}}{=} \bigcup_{m=1}^{\infty} Q_{\mathbf{x}_m, t_m}$$

$$volume(F) \overset{\text{def}}{=} \sum_{1=m}^{\infty} volume(Q_{t_m})$$

$$= \sum_{1=m}^{\infty} t_{m1} \cdots t_{mn}.$$

  *Chapter 3. DETERMINANTS (**A direct approach**)*

In these circumstances it can be shown that

$$volume(P) \; [= volume(MQ)] \; = det(M).$$ \hfill (3.3.14)

Exercise 3.3.6. Using the standard definition of the volume of a cube in 3-dimensional space $\mathbf{R}^3$ show that formula (3.3.14) is valid.

Exercise 3.3.7. If in **Figure 3.3.1** the pairs (x_1, y_1) and (x_2, y_2) are proportional then OA and OC are collinear and the area of $OABC$ is 0. The parallelogram is *degenerate*. Is the formula $area(OABC) = det(M)$ still valid? Why?

Exercise 3.3.8. Let the rows of $M \overset{\text{def}}{=} (m_{ij})_{i,j=1}^{3,3}$ with *real* entries be regarded as coordinate triples representing points P_1, P_2, and P_3 in $\mathbf{R}^3$. Show:

  i. If there is exactly one row of 0's in the echelon form M_E of M then P_1, P_2, and P_3 lie in a plane passing through the origin $(0, 0, 0) \overset{\text{def}}{=} \mathbf{O}$.
  ii. If there are exactly two rows of 0's in M_E then P_1, P_2, and P_3 lie on a line passing through $\mathbf{O}$.

Interpret the results in i and ii in terms of the volume of the parallelopiped determined by $\mathbf{O}$, P_1, P_2, and P_3.

Exercise 3.3.9. Let P_1 have coordinates $(3, 20)$ and let P_2 have coordinates $(5, 18)$. The formula for the area of the parallelogram determined by $\mathbf{O}$, P_1, and P_2 yields a negative number. Explain this phenomenon in terms of the relative *orientation* of the two vectors

$$\mathbf{x}_1 \overset{\text{def}}{=} (3, 20)^t \text{ and } \mathbf{x}_2 \overset{\text{def}}{=} (5, 18)^t.$$

What is the area of the parallelogram determined by $\mathbf{O}$, P_2, and P_1 *in that order*?

Exercise 3.3.10. Let A be an $n \times n$ matrix. Assume that for some k in $\{1, 2, \ldots, n\}$ there is a $k \times (n - k + 1)$ submatrix $O_{k \times n - k + 1}$ in which all entries are 0's. (For example, in the 4×4 matrix

$$A \overset{\text{def}}{=} \begin{pmatrix} 0 & 3 & 0 & 4 \\ 0 & 1 & 0 & 5 \\ 1 & 6 & 1 & 6 \\ 0 & 3 & 0 & 7 \end{pmatrix}$$

there is a 3×2, i.e., $3 \times (4 - 3 + 1)$ submatrix

$$O_{3 \times 4 - 3 + 1} \overset{\text{def}}{=} \begin{pmatrix} 0 & 0 \\ 0 & 0 \\ 0 & 0 \end{pmatrix}$$

consisting of the zero entries in positions (1,1), (1,3), (2,1), (2,3), (4,1), and (4,3).)

i. Interpret the formula $k \times (n - k + 1)$ for the special cases in which $k = 1$, $k = 4$, and $k = n$. Show that for the 4×4 matrix A, as above, each term is 0 in the right member of the formula

$$
det \left[\begin{pmatrix} a_{11} & \cdots & a_{1n} \\ \vdots & \ddots & \vdots \\ a_{n1} & \cdots & a_{nn} \end{pmatrix} \right]
$$
$$
= \sum_{\pi \in S_{1,2,\ldots,n}} sgn(\pi) a_{1\pi(1)} a_{2\pi(2)} \cdots a_{n\pi(n)} \qquad (3.3.15)
$$

(cf. (3.2.4)).

ii. Show that for the general $n \times n$ matrix A containing a submatrix $O_{k \times n - k + 1}$ each term is 0 in the right member of (3.3.15).

ii. State the converse of the assertion in *ii*. Prove the converse for 2×2 and 3×3 matrices. (The original proof of the converse for an arbitrary $n \times n$ matrix is due to Frobenius. A proof is given in **Section 5.8** in connection with the applications of linear programming and the *Marriage/Assignment Theorem*.)

[*Hint:* For *ii* show first that permutations of columns and rows of A do not affect the terms in the right member of (3.3.15). Thus it may be assumed that there is a $k \times n - k + 1$ block of 0's in the northwest corner of A. Consider in (3.3.15) all terms in which the last $n - k$ factors are in the $n - k \times n - k$ block in the southwest corner of A.]

Exercise 3.3.11. Let a triangle T in $\mathbf{R}^2$ have vertices at (x_1, y_1), (x_2, y_2), and (x_3, y_3). Show that the area of T is

$$
\frac{1}{2} det \left[\begin{pmatrix} x_1 & y_1 & 1 \\ x_2 & y_2 & 1 \\ x_3 & y_3 & 1 \end{pmatrix} \right].
$$

Exercise 3.3.12. Let T be a triangle with side lengths a, b, and c and opposing angles of size α, β, and γ.

i. Show that

$$
a = b \cos \gamma + c \cos \beta
$$
$$
b = c \cos \alpha + a \cos \gamma
$$
$$
c = a \cos \beta + b \cos \alpha. \qquad (3.3.16)
$$

ii. Regard (3.3.16) as a system of three equations in the three unknowns $\cos\alpha$, $\cos\beta$, and $\cos\gamma$. Use Cramer's rule to solve for the unknowns in terms of a, b, and c. The results are three versions of the *law of cosines* (cf. **Section 4.4**).

Exercise 3.3.13. Assume $A \overset{\text{def}}{=} (a_{ij})_{i,j=1}^{n,n}$ is a SQUARE matrix with minors M_{ij}. Show that $AA^t = I$ iff $det(A) \neq 0$ and

$$a_{ij} = (-1)^{i+j} \frac{det(M_{ji})}{det(A)}.$$

Exercise 3.3.14. Let $A \overset{\text{def}}{=} (a_{ij})_{i,j=1}^{n,n}$ and $B \overset{\text{def}}{=} (b_{ij})_{i,j=1}^{n,n}$ be two SQUARE matrices. The *Hadamard product* of A and B is the matrix $C \overset{\text{def}}{=} (c_{ij})_{i,j=1}^{n,n}$ such that $c_{ij} = a_{ij}b_{ij}$: $C \overset{\text{def}}{=} A \circ B$. Assume A is invertible. Show that the sum of the entries in any column or in any row of $A \circ (A^{-1})^t$ is 1.

Here are some highlights from **Chapters 1 - 3**. They relate matrix theory, determinant theory, and the analysis of systems of linear equations.

Let M be an $m \times n$ matrix $(m_{ij})_{i,j=1}^{m,n}$ and let its rank be r.

   i. $r \leq minimum(m, n)$.

   ii. There is in M an $r \times r$ submatrix M_1 such that $det(M_1) \neq 0$.

   iii. For every SQUARE submatrix M_2 of size greater than $r \times r$, $det(M_2) = 0$.

   iv. In M there are r linearly independent rows and r linearly independent columns.

   v. Any $r + 1$ rows are linearly dependent; any $r + 1$ columns are linearly dependent.

   vi. Every row is a linear combination of any r linearly independent rows.

   vii. Every column is a linear combination of any r linearly independent columns.

   viii. The system $M\mathbf{x} = \mathbf{b}$ has a solution iff $rank(M|\mathbf{b}) = rank(M) = r$. In particular $M\mathbf{x} = \mathbf{O}$ has at least ONE SOLUTION and if M is WIDE $M\mathbf{x} = \mathbf{b}$ has either NO SOLUTION or INFINITELY MANY SOLUTIONS.

   ix. If M is SQUARE then $M\mathbf{x} = \mathbf{b}$ has one and only one solution (UNISOL) for every $\mathbf{b}$ iff $det(M) \neq 0$.

Exercise 3.3.15. If $\mathbf{v}_i \overset{\text{def}}{=} (v_{i1}, \ldots, v_{im})$, $1 \leq i \leq n$, $A \overset{\text{def}}{=} (v_{ij})_{i,j=1}^{n,m}$, and $A^* \overset{\text{def}}{=} ((\overline{v}_{ij})_{i,j=1}^{n,m})^t$ then the ij entry of the $n \times n$ matrix $\mathcal{G} \overset{\text{def}}{=} AA^*$, the *Gramian*

---

of the set $\{\mathbf{v}_1, \ldots, \mathbf{v}_n\}$, is

$$g_{ij} \overset{\text{def}}{=} \sum_{k=1}^{m} v_{ik}\overline{v}_{jk}, \ \ 1 \le i, j \le n,$$

and is denoted $(\mathbf{v}_i, \mathbf{v}_j)$ (cf. **Section 4.4**).

   i. Show that $rank(\mathcal{G}) \le minimum\{m, n\}$. (Hence if $n > m$ then $det(\mathcal{G}) = 0$.)

   ii. Show that if $n \le m$ then

$$det(\mathcal{G}) = \sum_{1 \le k_1 < \cdots < k_n \le m} \left| det \left[\begin{pmatrix} v_{1k_1} & \cdots & v_{1k_n} \\ \vdots & \ddots & \vdots \\ v_{nk_1} & \cdots & v_{nk_n} \end{pmatrix} \right] \right|^2 . \tag{3.3.17}$$

[*Hint:* Use the multilinearity of *det* and the break-out technique.]

   iii. Show that $det(\mathcal{G}) = 0$ iff the vectors $\mathbf{v}_1, \ldots, \mathbf{v}_n$ are linearly dependent.

Exercise 3.3.16. Assume that A as a SQUARE block matrix $(A_{ij})_{i,j=1}^{n,n}$ has SQUARE "diagonal" blocks A_{ii} and "subdiagonal" blocks A_{ij}, $i > j$ that are O's. Show $det(A) = \prod_{i=1}^{n} det(A_{ii})$.

CHAPTER 4

USEFUL FORMS FOR MATRICES

4.1. Introduction

Matrix addition is computationally simple. Matrix multiplication, in particular the formation of powers of a matrix, is computationally more difficult.

Let A and B be $n \times n$ matrices.

Finding $A + B$ requires at most the performance of n^2 *numerical* addition operations.

Finding AB or A^2 according to the laws of matrix multiplication involves the formation of n^2 entries, each the sum of n numbers, each number the product of two numbers. All told there are performed n^3 multiplication operations and $n^2(n-1)$ numerical addition operations.

In **Example 1.1.2** there arises a power series

$$I + tC + \frac{t^2 C^2}{2!} + \ldots$$

in which the terms contain powers of a matrix C. More generally if A is an $n \times n$ matrix and if each of the entries in

$$\sum_{k=0}^{\infty} a_k A^k \tag{4.1.1}$$

is a convergent series the calculation of (4.1.1) involves finding a number K such that all the entries in the *remainder*

$$\sum_{k=K+1}^{\infty} a_k A^k$$

are negligible for practical purposes. The calculation of

$$\sum_{k=0}^{K} a_k A^k, \tag{4.1.2}$$

involves about $(K+1)n^3$ multiplication operations.

On the other hand, if there is a matrix B that is *sparse*, i.e., such that most of its entries are 0's, and if there is an invertible matrix P such that $B = P^{-1}AP$ then the calculation of (4.1.2) is simplified because the presence of many 0 entries in B makes the calculation of B^k easy and the calculation of (4.1.2) can be achieved as follows:

$$A = PBP^{-1}$$
$$A^2 = PBP^{-1} \cdot PBP^{-1} = PB^2P^{-1}$$
$$\cdots$$
$$A^k = PB^kP^{-1}$$
$$\cdots$$
$$\sum_{k=0}^{K} a_k A^k = P\left(\sum_{k=0}^{K} a_k B^k\right)P^{-1}.$$

Thus there is some merit in finding an invertible matrix P such that

$$P^{-1}AP \stackrel{\text{def}}{=} B$$

is sparse.

For example, if B is triangular about half its entries are 0's and the number of multiplication operations used to find B^2 is about

$$\frac{n(n+1)(n+2)}{6}. \tag{4.1.3}$$

For large n (4.1.3) is about one-sixth of n^3.

If B is a diagonal matrix the number of multiplication operations used to find B^2 is n. If $n = 10$ then the number of multiplication operations used to find B^2 is 10, and for A^2 the corresponding number might be 1,000.

In fact B can always be found so that at worst it has n nonzero diagonal entries, at worst $n-1$ 1's lying just above the diagonal, and 0's elsewhere. Such a B looks like this:

$$\begin{pmatrix} \lambda_1 & * & & & & \\ & \lambda_2 & * & & & \\ & & \ddots & * & & \\ & & & \lambda_{n-1} & * & \\ & & & & \lambda_n & \end{pmatrix}.$$

Each * above represents either a 1 or a 0. All entries not printed are 0's. In other words, if $B \stackrel{\text{def}}{=} (b_{ij})_{i,j=1}^{n,n}$ then

$$b_{ij} = \begin{cases} \lambda_i & \text{if } i = j; \\ * = 1 \text{ or } 0 & \text{if } j = i+1; \\ 0 & \text{otherwise.} \end{cases}$$

A matrix B like that in (4.1.4) is said to be in *Jordan normal form*.

If all the λ_i are different from each other then all the *'s are 0's and B is a diagonal matrix.

More generally there is a special kind of $k \times k$ matrix

$$J_k(\lambda) \stackrel{\text{def}}{=} \begin{pmatrix} \lambda & 1 & & & \\ & \lambda & 1 & & \\ & & \ddots & 1 & \\ & & & \lambda & 1 \\ & & & & \lambda \end{pmatrix}$$

called a *Jordan block* and, at worst, B can be found as a block matrix

$$\begin{pmatrix} J_{k_1}(\lambda_1) & & & \\ & J_{k_2}(\lambda_2) & & \\ & & \ddots & \\ & & & J_{k_m}(\lambda_m) \end{pmatrix} \tag{4.1.4}$$

in which the λ_i are not necessarily different from each other. Here is how such a B might look:

$$\begin{pmatrix} 5 & 1 & & & & & & & & \\ & 5 & 1 & & & & & & & \\ & & 5 & 1 & & & & & & \\ & & & 5 & & & & & & \\ & & & & 5 & 1 & & & & \\ & & & & & 5 & 1 & & & \\ & & & & & & 5 & & & \\ & & & & & & & 2 & 1 & \\ & & & & & & & & 2 & \\ & & & & & & & & & 6 \end{pmatrix} . \tag{4.1.5}$$

In (4.1.5)

$$n = 10$$
$$k_1 = 4, \ \lambda_1 = 5$$
$$k_2 = 3, \ \lambda_2 = 5$$
$$k_3 = 2, \ \lambda_3 = 2$$
$$k_4 = 1, \ \lambda_4 = 6$$
$$k_1 + k_2 + k_3 + k_4 = 10 = n.$$

The Jordan blocks in (4.1.5) are

$$\begin{pmatrix} 5 & 1 & & \\ & 5 & 1 & \\ & & 5 & 1 \\ & & & 5 \end{pmatrix}, \ \begin{pmatrix} 5 & 1 & \\ & 5 & 1 \\ & & 5 \end{pmatrix}, \ \begin{pmatrix} 2 & 1 \\ & 2 \end{pmatrix}, \ \text{and} \ (6).$$

Exercise 4.1.1. Let B be the matrix in (4.1.5). Calculate B^2, B^3, and B^{10}. Let K_2, K_3, and K_{10} be the numbers of multiplication operations used in these calculations. Calculate K_2, K_3, and K_{10} and compare them with 1,000, 2,000, and 4,000, which are roughly the counts of the multiplication operations that can occur in raising a 10×10 matrix to the powers 2, 3, and 10. Why is 4,000 rather than 9,000 an upper bound corresponding to the power 10?

An advantage to the Jordan normal form in the calculation of power series is revealed next.

Let f be a function such that for all z and any a there is a valid power series (Taylor series) representation

$$f(z) = \sum_{n=0}^{\infty} \frac{f^{(n)}(a)}{n!}(z - a)^n.$$

Then the existence of a Jordan normal form, e.g., B as in (4.1.5), yields a formula for $f(B)$ (cf. **Section 4.7**), namely

$$\begin{pmatrix} f(5) & f'(5) & \frac{f''(5)}{2!} & \frac{f'''(5)}{3!} & & & & & & \\ & f(5) & f'(5) & \frac{f''(5)}{2!} & & & & & & \\ & & f(5) & f'(5) & & & & & & \\ & & & f(5) & & & & & & \\ & & & & f(5) & f'(5) & \frac{f''(5)}{2!} & & & \\ & & & & & f(5) & f'(5) & & & \\ & & & & & & f(5) & & & \\ & & & & & & & f(2) & f'(2) & \\ & & & & & & & & f(2) & \\ & & & & & & & & & f(6) \end{pmatrix}.$$

If B is the Jordan normal form of a matrix A and if $B = P^{-1}AP$ then $f(A)$ is naturally defined to be $Pf(A)P^{-1}$.

Exercise 4.1.2. Let B be the matrix

$$\begin{pmatrix} 2 & 1 & \\ & 2 & \\ & & 3 \end{pmatrix}.$$

Calculate $\cos B$, e^B, and $\log B$.

If P is a matrix such that $P^{-1}AP \stackrel{\text{def}}{=} B$ is in Jordan normal form, let $\mathbf{p}_1, \ldots, \mathbf{p}_n$ be the column vectors that are the columns of P. Thus if

$$P = \begin{pmatrix} p_{11} & \cdots & p_{1n} \\ \vdots & \ddots & \vdots \\ p_{n1} & \cdots & p_{nn} \end{pmatrix}$$

Chapter 4. Useful forms for matrices

then

$$\mathbf{p}_j = \begin{pmatrix} p_{1j} \\ \vdots \\ p_{nj} \end{pmatrix}, \ 1 \le j \le n.$$

Since $AP = PB$, comparison of the first columns of AP and PB shows that

$$A\mathbf{p}_1 = \lambda_1 \mathbf{p}_1. \tag{4.1.6}$$

Since P is invertible, $\mathbf{p}_1 \ne \mathbf{O}$. The number λ_1 is called an *eigenvalue* of A and the *nonzero* vector $\mathbf{p}_1$ is called an *eigenvector* corresponding to the eigenvalue λ_1. Similarly, each diagonal entry (4.1.5) is an eigenvalue of A and some of, *but not necessarily all*, the columns of P, are eigenvectors that correspond to λ_i.

Furthermore, if λ is an eigenvalue of A and if $\mathbf{x}$ is a corresponding (nonzero!) eigenvector then

$$A\mathbf{x} = \lambda \mathbf{x}$$
$$(A - \lambda I)\mathbf{x} = \mathbf{O}$$
$$P^{-1}(A - \lambda I)\mathbf{x} = P^{-1}(A - \lambda I)PP^{-1}\mathbf{x} = P^{-1}\mathbf{O} = \mathbf{O}$$
$$(B - \lambda I)P^{-1}\mathbf{x} = \mathbf{O}. \tag{4.1.7}$$

Thus if $\mathbf{y} \overset{\text{def}}{=} P^{-1}\mathbf{x}$ then $\mathbf{y} \ne \mathbf{O}$ (because P^{-1} is invertible and $\mathbf{x} \ne \mathbf{O}$) and λ is an eigenvalue for B as well. The corresponding eigenvector $\mathbf{y}$ (for B) is not necessarily the same as the eigenvector $\mathbf{x}$ (for A). Because $(B - \lambda I)\mathbf{y} = \mathbf{O}$ and $\mathbf{y} \ne \mathbf{O}$ it follows that $B - \lambda I$ is singular. Since $B - \lambda I$ is triangular one of its diagonal entries must be 0. Hence λ is some eigenvalue λ_i.

A number λ is an eigenvalue of A iff λ is one of the diagonal entries of the Jordan normal form B of A.

It appears that if, in fact, there is a Jordan normal form for A then one way to find it might well involve finding the eigenvalues of A.

Note that $B - \lambda I$ is singular iff $A - \lambda I$ is singular iff $det(A - \lambda I) = 0$ Thus λ is an eigenvalue of A iff

$$det(A - \lambda I) = 0. \tag{4.1.8}$$

The last statement is of theoretical interest. If A is a large matrix, e.g., a 10×10 matrix, the use of (4.1.8) to find the eigenvalues of A is most impractical.

In the following **Sections** questions about eigenvalues, whether they exist (they do), how to find them, etc. are addressed in some detail.

Exercise 4.1.3. Find the eigenvalues and for each eigenvalue a correspond-

---

ing eigenvector for each of the following matrices:

$$i. \quad \begin{pmatrix} 1 & 2 \\ 3 & 4 \end{pmatrix}$$

$$ii. \quad \begin{pmatrix} a & b \\ c & d \end{pmatrix}$$

$$iii. \quad \begin{pmatrix} \lambda_1 & & & \\ & \lambda_2 & & \\ & & \ddots & \\ & & & \lambda_n \end{pmatrix}$$

$$iv. \quad \begin{pmatrix} a_{11} & a_{12} & \cdots & a_{1n} \\ & a_{22} & \cdots & a_{2n} \\ & & \ddots & \vdots \\ & & & a_{nn} \end{pmatrix}$$

$$v. \quad \begin{pmatrix} 1 & -3 \\ 1 & 6 \end{pmatrix}.$$

Assume A is such that for some invertible matrix $P \overset{\text{def}}{=} (p_{ij})_{i,j=1}^{n,n}$

$$P^{-1}AP = \begin{pmatrix} \lambda_1 & & & \\ & \lambda_2 & & \\ & & \ddots & \\ & & & \lambda_n \end{pmatrix} \overset{\text{def}}{=} D.$$

If

$$P^{-1} = \begin{pmatrix} q_{11} & \cdots & q_{1n} \\ \vdots & \ddots & \vdots \\ q_{n1} & \cdots & q_{nn} \end{pmatrix}$$

then a reasonable formula for $f(A)$ is

$$f(A) = Pf(D)P^{-1},$$

i.e.,

$$f\left[\begin{pmatrix} a_{11} & \cdots & a_{1n} \\ \vdots & \ddots & \vdots \\ a_{n1} & \cdots & a_{nn} \end{pmatrix}\right] = \begin{pmatrix} p_{11} & \cdots & p_{1n} \\ \vdots & \ddots & \vdots \\ p_{n1} & \cdots & p_{nn} \end{pmatrix} \begin{pmatrix} f(\lambda_1) & & & \\ & f(\lambda_2) & & \\ & & \ddots & \\ & & & f(\lambda_n) \end{pmatrix} \begin{pmatrix} q_{11} & \cdots & q_{1n} \\ \vdots & \ddots & \vdots \\ q_{n1} & \cdots & q_{nn} \end{pmatrix}.$$

     Chapter 4. Useful forms for matrices

Once P and P^{-1} are found they serve for all functions f that are well defined at the numbers $\lambda_1, \ldots, \lambda_n$. For example, if no $\lambda_i = 0$, then the function f such that if $z \neq 0$

$$f(z) = \frac{1}{z}$$

can be used to calculate A^{-1} according to the formula

$$A^{-1} = f(A) = Pf(D)P^{-1}$$

$$= \begin{pmatrix} p_{11} & \cdots & p_{1n} \\ \vdots & \ddots & \vdots \\ p_{n1} & \cdots & p_{nn} \end{pmatrix} \times \begin{pmatrix} \frac{1}{\lambda_1} & & & \\ & \frac{1}{\lambda_2} & & \\ & & \ddots & \\ & & & \frac{1}{\lambda_n} \end{pmatrix} \times \begin{pmatrix} q_{11} & \cdots & q_{1n} \\ \vdots & \ddots & \vdots \\ q_{n1} & \cdots & q_{nn} \end{pmatrix}.$$

Indeed,

$$APf(D)P^{-1} = PDP^{-1}Pf(D)P^{-1} = PDf(D)P^{-1}$$

$$= \begin{pmatrix} p_{11} & \cdots & p_{1n} \\ \vdots & \ddots & \vdots \\ p_{n1} & \cdots & p_{nn} \end{pmatrix} \times \begin{pmatrix} \lambda_1 & & & \\ & \lambda_2 & & \\ & & \ddots & \\ & & & \lambda_n \end{pmatrix}$$

$$\times \begin{pmatrix} \frac{1}{\lambda_1} & & & \\ & \frac{1}{\lambda_2} & & \\ & & \ddots & \\ & & & \frac{1}{\lambda_n} \end{pmatrix} \times \begin{pmatrix} q_{11} & \cdots & q_{1n} \\ \vdots & \ddots & \vdots \\ q_{n1} & \cdots & q_{nn} \end{pmatrix}$$

$$= PIP^{-1} = I.$$

By the same token

$$\sin(A) = P \sin(D) P^{-1}$$

$$= \begin{pmatrix} p_{11} & \cdots & p_{1n} \\ \vdots & \ddots & \vdots \\ p_{n1} & \cdots & p_{nn} \end{pmatrix} \times \begin{pmatrix} \sin(\lambda_1) & & & \\ & \sin(\lambda_2) & & \\ & & \ddots & \\ & & & \sin(\lambda_n) \end{pmatrix}$$

$$\times \begin{pmatrix} q_{11} & \cdots & q_{1n} \\ \vdots & \ddots & \vdots \\ q_{n1} & \cdots & q_{nn} \end{pmatrix}$$

and if each λ_i is real then

$$(I + A^2)^{-1} = \begin{pmatrix} p_{11} & \cdots & p_{1n} \\ \vdots & \ddots & \vdots \\ p_{n1} & \cdots & p_{nn} \end{pmatrix} \times \begin{pmatrix} \frac{1}{1+\lambda_1^2} & & & \\ & \frac{1}{1+\lambda_2^2} & & \\ & & \ddots & \\ & & & \frac{1}{1+\lambda_n^2} \end{pmatrix}$$

$$\times \begin{pmatrix} q_{11} & \cdots & q_{1n} \\ \vdots & \ddots & \vdots \\ q_{n1} & \cdots & q_{nn} \end{pmatrix}.$$

[**Remark 4.1.1:** In **Section 4.7** it is shown that for any reasonable function f the value of $f(A)$ can be both defined and calculated by means of a *polynomial p* that *"agrees"* with f in a manner described there: $f(A) = p(A)$. This aspect of linear algebra is known as the *functional calculus* for matrices.]

The matrix A and the matrix $P^{-1}AP$ are called *similar*. To say that two $n \times n$ matrices A and B are similar is to say that for some invertible $n \times n$ matrix P: $B = P^{-1}AP$. If $Q \overset{\text{def}}{=} P^{-1}$ then it is also true that

$$A = PBP^{-1}, \text{ i.e., } A = Q^{-1}BQ.$$

If A and B are similar one writes: $A \sim B$. Hence if P is an invertible $n \times n$ matrix one writes: $A \sim P^{-1}AP$.

Exercise 4.1.4. Show that "$\sim$" is an *equivalence relation*, i.e.,

i. $A \sim A$;
ii. if $A \sim B$ then $B \sim A$;
iii. if $A \sim B$ and $B \sim C$ then $A \sim C$.

Exercise 4.1.5. Let A and P be defined as follows:

$$A \overset{\text{def}}{=} \begin{pmatrix} 3 & 4 \\ 1 & 6 \end{pmatrix}, \; P \overset{\text{def}}{=} \begin{pmatrix} 1 & -4 \\ 1 & 1 \end{pmatrix}.$$

i. Find P^{-1} and then calculate $P^{-1}AP \overset{\text{def}}{=} D$.
ii. Use D to calculate

$$A^{-1}, \; A^{-3}, \; A^2 - 9A + 14I, \; (I + A^2)^{-1}, \; A^A, \text{ and } \log_{10}(A).$$

iii. Compare $det(A)$ with $det(P^{-1}AP)$.

Exercise 4.1.6. Show that generally if A is an $n \times n$ matrix and P is an invertible $n \times n$ matrix then

$$det(A) = det(P^{-1}AP).$$

[**Remark 4.1.2:** The result in **Exercise 4.1.6** may be paraphrased by the statement: the determinant function is a *similarity invariant*.]

Exercise 4.1.7. Let M be the matrix

$$\begin{pmatrix} \frac{1}{5}x_1 + \frac{4}{5}x_2 & \frac{4}{5}x_1 - \frac{4}{5}x_2 \\ \frac{1}{5}x_1 - \frac{1}{5}x_2 & \frac{4}{5}x_1 + \frac{1}{5}x_2 \end{pmatrix}$$

and let P be the matrix in **Exercise 4.1.5.**

i. Show that

$$P^{-1}MP \stackrel{\text{def}}{=} D = \begin{pmatrix} x_1 & \\ & x_2 \end{pmatrix}.$$

ii. Solve the matrix equation:

$$3M^2 - 6M + 2I = O.$$

How many solutions are there?

Example 4.1.1. Let A be the matrix

$$\begin{pmatrix} 0 & 1 \\ 0 & 0 \end{pmatrix}.$$

In **Section 2.8** it is shown that there is *no* invertible 2×2 matrix P such that $P^{-1}AP$ is a diagonal matrix. The matrix A is such that $A^2 = O$. In fact if N is a nonzero *nilpotent* matrix, i.e., if $N \neq O$ and if there is a natural number k such that $N^k = O$, then there is no diagonal form for N: N is not *diagonable*.

Indeed, if

$$P^{-1}NP = \begin{pmatrix} \lambda_1 & & & \\ & \lambda_2 & & \\ & & \ddots & \\ & & & \lambda_n \end{pmatrix} \stackrel{\text{def}}{=} D$$

then

$$(P^{-1}NP)^k = P^{-1}N^kP = O = D^k = \begin{pmatrix} \lambda_1^k & & & \\ & \lambda_2^k & & \\ & & \ddots & \\ & & & \lambda_n^k \end{pmatrix}.$$

Hence each $\lambda_i = 0$, $D = O$, $N = PDP^{-1} = O$, a contradiction.

It follows that not every SQUARE matrix A can be *diagonalized*, i.e., it is not necessarily true that there is an invertible matrix P such that $P^{-1}AP$ is a diagonal matrix. On the other hand, just as it is true that arbitrarily "near" any SQUARE matrix there is an invertible matrix, so also is it true that arbitrarily near any SQUARE matrix there is a *diagonable* matrix. Thus the search for methods that yield a *diagonalizing* matrix P is far from vain or foolhardy. Just as it is true that the "chances are good" that a "randomly chosen" SQUARE matrix A is invertible, so also is it true that the "chances are good" that a "randomly chosen" SQUARE matrix A is diagonable.

[**Remark 4.1.3:** A SQUARE matrix that is not diagonable is often called *defective*. If a matrix is defective it can be brought to Jordan normal form, as the discussion in **Section 4.6** shows.]

Let it be stressed that if A is an $n \times n$ matrix, finding an invertible matrix P such that $P^{-1}AP$ has a simple form is equivalent to finding n column vectors $\mathbf{p}_1, \ldots, \mathbf{p}_n$ such that:

i. they constitute a basis for $\mathbf{C}^n$;

ii. they are the columns of P;

iii. $A\mathbf{p}_i$, $1 \leq i \leq n$ are expressible as simply as possible as linear combinations of $\mathbf{p}_1, \ldots, \mathbf{p}_n$.

To see what these linear combinations are note that since P is invertible:

$$\widetilde{A} = P^{-1}AP \Leftrightarrow P\widetilde{A} = AP. \tag{4.1.9}$$

If A is diagonable then $\widetilde{A}$ is a diagonal matrix:

$$\widetilde{A} = \begin{pmatrix} \lambda_1 & & & \\ & \lambda_2 & & \\ & & \ddots & \\ & & & \lambda_n \end{pmatrix}$$

and so (4.1.6) implies each of the simple linear combinations is a multiple of just one of the $\mathbf{p}_i$:

$$A\mathbf{p}_i = \lambda_i \mathbf{p}_i, \ 1 \leq i \leq n.$$

If A is not diagonable then $\widetilde{A} \overset{\text{def}}{=} P^{-1}AP$ has the form exemplified by (4.1.5). More generally if $\widetilde{A}$ is the matrix

$$\begin{pmatrix} \lambda_1 & 1 & & & & & & & & \\ & \lambda_1 & 1 & & & & & & & \\ & & \lambda_1 & 1 & & & & & & \\ & & & \lambda_1 & & & & & & \\ & & & & \lambda_2 & 1 & & & & \\ & & & & & \lambda_2 & 1 & & & \\ & & & & & & \lambda_2 & & & \\ & & & & & & & \lambda_3 & 1 & \\ & & & & & & & & \lambda_3 & \\ & & & & & & & & & \lambda_4 \end{pmatrix} \tag{4.1.10}$$

---

then the equation $AP = P\widetilde{A}$ implies:

$$
\begin{aligned}
A\mathbf{p}_1 &= \lambda_1\mathbf{p}_1 \\
A\mathbf{p}_2 &= \mathbf{p}_1 + \lambda_1\mathbf{p}_2 \\
A\mathbf{p}_3 &= \mathbf{p}_2 + \lambda_1\mathbf{p}_3 \\
A\mathbf{p}_4 &= \mathbf{p}_3 + \lambda_1\mathbf{p}_4 \\
A\mathbf{p}_5 &= \lambda_2\mathbf{p}_5 \\
A\mathbf{p}_6 &= \mathbf{p}_5 + \lambda_2\mathbf{p}_6 \\
A\mathbf{p}_7 &= \mathbf{p}_6 + \lambda_2\mathbf{p}_7 \\
A\mathbf{p}_8 &= \lambda_3\mathbf{p}_8 \\
A\mathbf{p}_9 &= \mathbf{p}_8 + \lambda_3\mathbf{p}_9 \\
A\mathbf{p}_{10} &= \lambda_4\mathbf{p}_{10}.
\end{aligned}
\tag{4.1.11}
$$

The pattern in (4.1.11) suggests that if A is not diagonable each of the simple linear combinations consists of a multiple of one of the $\mathbf{p}_i$ *or* is the sum of a multiple of one of the $\mathbf{p}_i$ and 1 times $\mathbf{p}_{i-1}$:

$$
\begin{aligned}
&\text{EITHER for some } \lambda \\
&\qquad A\mathbf{p}_i = \lambda\mathbf{p}_i \\
&\text{OR for some } \lambda \\
&\qquad A\mathbf{p}_i = \mathbf{p}_{i-1} + \lambda\mathbf{p}_i.
\end{aligned}
\tag{4.1.12}
$$

In the language of **Section 2.10**, if $\mathcal{P}$ is the vectorial column vector

$$
\begin{pmatrix} \mathbf{p}_1 \\ \vdots \\ \mathbf{p}_n \end{pmatrix}
$$

then the relations in (4.1.12) may be reinterpreted as follows:

$$
\begin{pmatrix} A\mathbf{p}_1 \\ \vdots \\ A\mathbf{p}_n \end{pmatrix} = \widetilde{A}^t\mathcal{P}.
\tag{4.1.13}
$$

If T is the linear transformation corresponding to the action of A on the E-components of a vector then (4.1.13) may be written even more succinctly as:

$$
T(\mathcal{P}) = \widetilde{A}^t\mathcal{P}.
\tag{4.1.14}
$$

Notice the appearance of the transpose in (4.1.13) and (4.1.14). The matrix $\widetilde{A}^t$ is the subdiagonal version of the Jordan normal form of A.

Exercise 4.1.8. Show that if the SQUARE matrix A is invertible then A is diagonable iff A^{-1} is diagonable.

Exercise 4.1.9. Let A be

$$\begin{pmatrix} -8 & 5 \\ 10 & 7 \end{pmatrix}.$$

Find a diagonal form for A and thereby simplify the calculation of A^6.

Exercise 4.1.10. Assume $-\infty < \theta < \infty$. Find the values of λ for which

$$\begin{pmatrix} \cos\theta - \lambda & \sin\theta \\ -\sin\theta & \cos\theta - \lambda \end{pmatrix}$$

is singular.

Let $A \overset{\text{def}}{=} (a_{ij})_{i,j=1}^{n,n}$ be an $n \times n$ SQUARE matrix. The *trace* of A, denoted $tr(A)$, is the *number*

$$a_{11} + \cdots + a_{nn} = \sum_{i=1}^{n} a_{ii},$$

i.e., the sum of the diagonal entries of A.

Example 4.1.2.

$$tr\left[\begin{pmatrix} 1 & 2 \\ 3 & 4 \end{pmatrix}\right] = 1 + 4 = 5$$

$$tr\left[\begin{pmatrix} d_1 & & & \\ & d_2 & & \\ & & \ddots & \\ & & & d_n \end{pmatrix}\right] = d_1 + \cdots + d_n = \sum_{i=1}^{n} d_i$$

$$tr\left[\begin{pmatrix} 1 & 2 \\ 3 & 4 \end{pmatrix}\begin{pmatrix} 5 & 6 \\ 7 & 8 \end{pmatrix}\right] = (5 + 14) + (18 + 32) = 69$$

$$tr\left[\begin{pmatrix} 5 & 6 \\ 7 & 8 \end{pmatrix}\begin{pmatrix} 1 & 2 \\ 3 & 4 \end{pmatrix}\right] = (5 + 18) + (14 + 32) = 69$$

$$\neq tr\left[\begin{pmatrix} 5 & 6 \\ 7 & 8 \end{pmatrix}\right] tr\left[\begin{pmatrix} 1 & 2 \\ 3 & 4 \end{pmatrix}\right] = 65$$

Exercise 4.1.11. Show:

i. if B is an arbitrary $n \times n$ matrix then $tr(AB) = tr(BA)$;

ii. the trace of A is a similarity invariant, i.e., if P is an invertible $n \times n$ matrix then $tr(P^{-1}AP) = tr(A)$.

     Chapter 4. Useful forms for matrices

Exercise 4.1.12. Show that if

$$A \stackrel{\text{def}}{=} \begin{pmatrix} a & b \\ c & d \end{pmatrix}$$

then

$$\chi_A(\lambda) = \lambda^2 - tr(A)\lambda + det(A).$$

Exercise 4.1.13. Show that if

$$A \stackrel{\text{def}}{=} \begin{pmatrix} a_{11} & \cdots & a_{1n} \\ \vdots & \ddots & \vdots \\ a_{n1} & \cdots & a_{nn} \end{pmatrix}$$

then

$$\chi_A(\lambda) = (-1)^n \lambda^n - tr(A)\lambda^{n-1} + \cdots + det(A).$$

Exercise 4.1.14. Let A be the matrix

$$\begin{pmatrix} 1 - \sin\theta\cos\theta & \cos^2\theta \\ -\sin^2\theta & 1 + \sin\theta\cos\theta \end{pmatrix}.$$

i. Show that A is not diagonable.
ii. Find the value(s) of λ for which $A - \lambda I$ is singular.
iii. Find a nonzero solution $\mathbf{p}_1$ of the system: $A\mathbf{x} = \lambda\mathbf{x}$.
iv. Find a vector $\mathbf{p}_2$ for which

$$A\mathbf{p}_2 = \mathbf{p}_1 + \lambda\mathbf{p}_2.$$

v. Find an invertible 2×2 matrix P such that $\widetilde{A} \stackrel{\text{def}}{=} P^{-1}AP$ is in Jordan normal form.

Exercise 4.1.15. Assume that in the matrix $\widetilde{A}$ in (4.1.10) each $\lambda_i = 0$. Show:

i. $\widetilde{A}^3 \neq O = \widetilde{A}^4$ ($\widetilde{A}$ is nilpotent of order 4);
ii. the vectors

$$\mathbf{p}_4, A\mathbf{p}_4, A^2\mathbf{p}_4, A^3\mathbf{p}_4;$$

$$\mathbf{p}_7, A\mathbf{p}_7, A^2\mathbf{p}_7,;$$

$$\mathbf{p}_9, A\mathbf{p}_9$$

$$\mathbf{p}_{10}$$

---

constitute a basis for $\mathbf{C}^{10}$;

iii. the vectors

$$\mathbf{p}_1, \mathbf{p}_5, \mathbf{p}_8, \mathbf{p}_{10}$$

constitute a basis for $ker(A)$;

iv. the vectors

$$\mathbf{p}_1, \mathbf{p}_5, \mathbf{p}_8, \mathbf{p}_{10}, \mathbf{p}_2, \mathbf{p}_6, \mathbf{p}_9$$

constitute a basis for $ker(A^2)$;

v. the vectors

$$\mathbf{p}_1, \mathbf{p}_5, \mathbf{p}_8, \mathbf{p}_{10}, \mathbf{p}_2, \mathbf{p}_6, \mathbf{p}_9, \mathbf{p}_3, \mathbf{p}_7$$

constitute a basis for $ker(A^3)$;

vi. the vectors

$$\mathbf{p}_1, \mathbf{p}_5, \mathbf{p}_8, \mathbf{p}_{10}, \mathbf{p}_2, \mathbf{p}_6, \mathbf{p}_9, \mathbf{p}_3, \mathbf{p}_7, \mathbf{p}_4$$

constitute a basis for $ker(A^4) = \mathbf{C}^{10}$.

Exercise 4.1.16. Let A be a 3×3 matrix such that for the invertible matrix

$$P \overset{\text{def}}{=} \begin{pmatrix} 1 & 2 & 3 \\ 3 & 1 & 2 \\ 2 & 3 & 1 \end{pmatrix}$$

$P^{-1}AP$ is the matrix

$$\begin{pmatrix} 1 & & \\ & i\pi & \\ & & -2 \end{pmatrix}.$$

Calculate A^3 and e^A.

There is a recurrent theme in the development of linear algebra and that theme is the *factorization* or *decomposition* of a matrix A into a product of simpler matrices. Thus in the study of an arbitrary $m \times n$ matrix A there emerges the factorization:

$$A = \Pi^{-1}LDU. \tag{4.1.15}$$

The diagonalization of a SQUARE matrix A may be viewed as a factorization:

$$A = PDP^{-1}. \tag{4.1.16}$$

The Jordan normal form of a matrix may be viewed as a factorization:

$$A = P\widetilde{A}P^{-1}. \tag{4.1.17}$$

In each case once the factorization is achieved it may be applied in all situations in which A occurs.

---

i. The factorization (4.1.15) is used to attack every system $A\mathbf{x} = \mathbf{b}$, even if A is not SQUARE.

ii. The factorizations (4.1.16) and (4.1.17) are the bases for every application of the functional calculus.

In **Part II**, where applications of linear algebra are studied, there are other decompositions and factorizations that are designed to cope with the numerical (computational) problems that arise when linear algebra is used in the "real world."

Exercise 4.1.17. Assume $\lambda \neq 0$. Show that the Jordan block $J_k(\lambda)$ is invertible and find its inverse.

Exercise 4.1.18. For m in $\mathbf{N}$ calculate $J_k^m(\lambda)$. For p a polynomial calculate $p(J_k(\lambda))$.

Exercise 4.1.19. Assume that f is a function having derivatives of all orders. On the basis of the results in **Exercise 4.1.17** and **4.1.18** suggest a reasonable formula for $f(J_k(\lambda))$. In particular, use the power series formulae for the functions sin and exp to suggest a formula for $\sin(J_k(\lambda))$ and $\exp(J_k(\lambda))$.

[*Hint:* Examine the cases for which $k = 2$ and $k = 3$.]

Exercise 4.1.20. Let A be a SQUARE matrix and let $\mathcal{A}_\lambda$ be the classical adjoint of $A - \lambda I$. Show that the derivative of $\chi_A(\lambda)$ is the negative of the trace of $\mathcal{A}_\lambda$:

$$\frac{d\chi_A(\lambda)}{d\lambda} = -tr(\mathcal{A}_\lambda).$$

[*Hint:* See **Exercise 3.2.21.**]

4.2. Eigenvalues and eigenvectors

The discussion in **Section 4.1** shows that the finding of a Jordan normal form and in particular a diagonal form, if it exists, for a SQUARE matrix A entails at least finding its eigenvalues and corresponding eigenvectors. These terms are defined formally below and, following the DEFINITION, there is a list of some of its important consequences. These are essential ingredients of much of the discussion in the remaining portions of this book.

DEFINITION 4.2.1. LET A BE A SQUARE MATRIX. A NUMBER λ FOR WHICH $A - \lambda I$ IS SINGULAR IS CALLED AN *eigenvalue*. A *nonzero* VECTOR $\mathbf{x}$ SUCH THAT $(A - \lambda I)\mathbf{x} = 0$ IS CALLED AN *eigenvector* (CORRESPONDING TO THE *eigenvalue* λ).

i. A vector $\mathbf{x}$ is an eigenvector for A iff $\mathbf{x} \neq \mathbf{O}^c$ and, for some eigenvalue λ (which may be 0), $(A - \lambda I)\mathbf{x} = \mathbf{O}^c$, i.e., $A\mathbf{x} = \lambda\mathbf{x}$;

---

ii. The number λ is an eigenvalue of A iff $A - \lambda I$ is singular (cf. **Exercise 2.8.2***vii*).

iii. If $\mathbf{x}$ and $\mathbf{y}$ are eigenvectors corresponding to the eigenvalue λ and if $p, q \in \mathbf{C}$ then $p\mathbf{x} + q\mathbf{y}$ is also an eigenvector (corresponding to λ).

iv. The set of eigenvalues of an upper or lower triangular matrix (in particular, of a diagonal matrix) is the set of different entries on the diagonal of A (cf. *ii* and **Exercise 2.6.17**).

[The German adjective *eigen* and its inflected forms, as in mein *eigenes* Kind — my *own* child — is used to modify a noun representing what is peculiarly, specially, *singularly* the property or nature of the person or thing that owns what the noun represents. Thus, with a slight twist in translation, eigenvalues and eigenvectors of a matrix A are *owned* by or are *singularly* the property of the matrix A with which they are associated or to which they *belong*. The eigenvalues of A are the numbers λ that force $A - \lambda I$ to be *singular*. The (nonzero) eigenvectors $\mathbf{x}$ associated with or corresponding to eigenvalues are also singularly the property of the matrix A.]

Example 4.2.1. Let A be the 2×2 matrix

$$\begin{pmatrix} 5 & 4 \\ 10 & 2 \end{pmatrix}.$$

Then

$$A - \lambda I = \begin{pmatrix} 5 - \lambda & 4 \\ 10 & 2 - \lambda \end{pmatrix}$$

and $A - \lambda I$ is singular iff $det(A - \lambda I) = 0$. Since

$$det(A - \lambda I) = (5 - \lambda)(2 - \lambda) - 10 \times 4 = \lambda^2 - 7\lambda - 30 = (\lambda - 10)(\lambda + 3)$$

it follows that $A - \lambda I$ is singular iff $\lambda = 10$ or $\lambda = -3$. In other words the eigenvalues of A are 10 and -3.

Corresponding to the eigenvalue 10 one finds *an* eigenvector by finding an $\mathbf{x}$ such that $\mathbf{x} \neq \mathbf{O}^c$ and

$$(A - 10I)\mathbf{x} = \mathbf{O}^c.$$

If

$$\mathbf{x} = \begin{pmatrix} x_1 \\ x_2 \end{pmatrix},$$

then at least one of the components x_1, x_2 must not be 0 and they should satisfy

$$-5x_1 + 4x_2 = 0$$
$$10x_1 - 8x_2 = 0. \tag{4.2.1}$$

The left and right members of the second equation in (4.2.1) are multiples (by the factor -2) of the left and right members of the first equation in (4.2.1). Hence a solution to either equation is a solution of the other and hence of the whole system. For example, if

$$\mathbf{x} = \begin{pmatrix} x_1 \\ x_2 \end{pmatrix} = \begin{pmatrix} 4 \\ 5 \end{pmatrix}$$

then $\mathbf{x}$ is a nonzero solution of (4.2.1). Indeed, any nonzero multiple,

$$t\mathbf{x} = \begin{pmatrix} 4t \\ 5t \end{pmatrix}, \quad t \neq 0,$$

is a nonzero solution of (4.2.1), that is to say is an eigenvector corresponding to the eigenvalue 10.

Similarly, corresponding to the eigenvalue -3 is the system

$$8x_1 + 4x_2 = 0$$
$$10x_1 + 5x_2 = 0$$

in which the left and right members of the second equation are multiples (by the factor $\frac{5}{4}$) of the left and right members of the first equation. Thus any vector

$$\begin{pmatrix} x_1 \\ x_2 \end{pmatrix} = \begin{pmatrix} s \\ 2s \end{pmatrix}, \quad s \neq 0$$

is an eigenvector corresponding to the eigenvalue -3.

It is occasionally helpful to display the eigenvalues and eigenvectors in tabular form as follows:

$$
\begin{array}{cc}
\text{EIGENVALUES} & \text{EIGENVECTORS} \\[2ex]
10 & \begin{pmatrix} 4t \\ 3t \end{pmatrix}, \ t \neq 0 \\[3ex]
-3 & \begin{pmatrix} s \\ 3s \end{pmatrix} \ s \neq 0.
\end{array}
$$

$$(4.2.2)$$

Exercise 4.2.1. For any n what is the only eigenvalue of the $n \times n$ identity matrix I? For I make a table like the one in (4.2.2). Make a similar table for each of the three kinds of $n \times n$ elementary matrices.

Exercise 4.2.2. For each matrix below make a table like the one in (4.2.2).

$$i. \quad \begin{pmatrix} -3 & 22 \\ 2 & 4 \end{pmatrix} ;$$

$$ii. \quad \begin{pmatrix} 0 & 5 \\ 0 & 0 \end{pmatrix} ;$$

$$iii. \quad \begin{pmatrix} 0 & 1 \\ -1 & 0 \end{pmatrix} ;$$

$$iv. \quad \begin{pmatrix} 1 & 2 & 3 \\ 4 & 5 & 6 \\ 7 & 8 & 9 \end{pmatrix} ;$$

$$v. \quad \begin{pmatrix} 0 & 0 & 1 \\ 1 & 0 & 0 \\ 0 & 1 & 0 \end{pmatrix} \quad \text{(a permutation matrix).}$$

A number λ is an eigenvalue of the $n \times n$ matrix A iff $A - \lambda I$ is singular iff $det(A - \lambda I) = 0$. The classical formula for

$$\chi_A(\lambda) \overset{\text{def}}{=} det(A - \lambda I)$$

shows that $\chi_A(\lambda)$ is a polynomial of degree n (in λ) (cf. **Section 3.2**). It is called the *characteristic* polynomial of A. Indeed

$$\chi_A(\lambda) = (-1)^n(\lambda^n - (a_{11} + \ldots + a_{nn})\lambda^{n-1} + \cdots + (-1)^n det(A))$$
$$= (-1)^n(\lambda^n - tr(A)\lambda^{n-1} + \cdots + (-1)^n det(A)).$$

(The constant term is $(-1)^n(-1)^n det(A) = det(A)$ since $\chi_A(0) = det(A)$.)

A number λ is an eigenvalue of A iff $\chi_A(\lambda) = 0$.

It is now possible to demonstrate that for any SQUARE matrix there is at least one eigenvalue. The proof depends ultimately upon the FUNDAMENTAL THEOREM OF ALGEBRA, namely:

IF p IS A POLYNOMIAL FUNCTION, i.e., IF

$$p(z) = t_m z^m + t_{m-1} z^{m-1} + \cdots + t_1 z + t_0, \ m \geq 1, \ t_m \neq 0,$$

THEN THERE IS AT LEAST ONE NUMBER z_0 SUCH THAT $p(z_0) = 0$.

Related to the Fundamental Theorem of Algebra is the Factor Theorem, viz.:

     Chapter 4. Useful forms for matrices

z_0 is a zero of p iff there is a polynomial q such that

$$p(z) = (z - z_0)q(z). \qquad (4.2.3)$$

[*Proof.* If (4.2.3) holds then $p(z_0) = 0q(z_0) = 0$. Conversely if $p(z_0) = 0$ then "long division" shows there is a polynomial q and a *number* r such that

$$p(z) = (z - z_0)q(z) + r.$$

But then since $p(z_0) = 0$ it follows that $r = 0$.]

Finally, repeated application of the Factor Theorem yields, by mathematical induction, the Unique Factorization Theorem, viz.:

There are complex numbers

$$z_1, \ldots, z_k$$

and natural numbers

$$h_1, \ldots, h_k$$

such that p can be factored in the form:

$$p(z) = t_m(z - z_1)^{h_1} \cdots (z - z_k)^{h_k}.$$

If, also,

$$p(z) = \tau_\mu(z - w_1)^{j_1} \cdots (z - w_l)^{j_l}$$

then $t_m = \tau_\mu$, $l = k$, the numbers $j_1, \ldots, j_l$ are the numbers $h_1, \ldots, h_k$ written in some order, and the numbers $w_1, \ldots, w_l$ are the numbers $z_1, \ldots, z_k$ written in some order. Furthermore $h_1 + \cdots h_k = m \stackrel{\text{def}}{=}$ the degree of p. (Frequently the degree of p is denoted $deg(p)$.)

A relatively elementary proof of the Fundamental Theorem of Algebra is offered at the end of this **Chapter**.

THEOREM 4.2.1. IF A IS AN $n \times n$ SQUARE MATRIX THEN THERE IS AT LEAST ONE NUMBER λ SUCH THAT $A - \lambda I$ IS SINGULAR, i.e., THERE IS AT LEAST ONE EIGENVALUE FOR A.

[**Remark 4.2.1:** The Fundamental Theorem of Algebra implies that the characteristic polynomial χ_A has at least one zero, i.e., there is at least one number λ_0 such that $\chi_A(\lambda_0) = 0$. Hence λ_0 is an eigenvalue.

Applied to the polynomial χ_A the Unique Factorization Theorem says that $\chi_A(z)$ is uniquely representable in the factored form:

$$\chi_A(z) \stackrel{\text{def}}{=} c(z - \lambda_1)^{s_1} \cdots (z - \lambda_k)^{s_k}, \quad c \in \mathbf{C}.$$

Thus a number λ is an eigenvalue of A iff λ is some λ_i — a zero of the polynomial function χ_A.

However, the proof above uses determinant theory. Since determinants are awkward tools for dealing with large practical problems in linear algebra, a proof independent of determinant theory is desirable. The proof below goes back to first principles and makes no appeal to determinant theory.

The proof is motivated as follows. The matrix $A - \lambda I$ is a polynomial function of the matrix A:

$$A - \lambda I = A^1 - \lambda A^0.$$

The equation

$$(A - \lambda I)\mathbf{x} = \mathbf{O}^c$$

is a special case of the polynomial equation

$$(t_k A^k + t_{k-1} A^{k-1} + \cdots + t_0 A^0)\mathbf{x} = \mathbf{O}^c.$$

or

$$t_k A^k \mathbf{x} + \cdots + t_0 I \mathbf{x} = \mathbf{O}^c.$$

Therefore it makes sense to seek a polynomial p such that

$$p(A)\mathbf{x} = \mathbf{O}^c,$$

to factor p, and to seek a linear factor $(A - \lambda_i I)$ of $p(A)$ and a nonzero vector $\mathbf{y}$ such that $(A - \lambda_i I)\mathbf{y} = \mathbf{O}^c$. The argument that follows successfully pursues this line.]

PROOF. Let $\mathbf{x}$ be any nonzero column vector:

$$\mathbf{x} = \begin{pmatrix} x_1 \\ \vdots \\ x_n \end{pmatrix}.$$

The $n+1$ column vectors $I\mathbf{x}, A\mathbf{x}, A^2\mathbf{x}, \ldots, A^n\mathbf{x}$ can be used to form the columns of an $n \times n+1$ matrix

$$\mathcal{A} \overset{\mathrm{def}}{=} \begin{pmatrix} a_{11} & \cdots & a_{1n+1} \\ \vdots & \ddots & \vdots \\ a_{n1} & \cdots & a_{nn+1} \end{pmatrix} \quad \text{in which} \quad \begin{pmatrix} a_{1j} \\ \vdots \\ a_{nj} \end{pmatrix} = A^{j-1} \begin{pmatrix} x_1 \\ \vdots \\ x_n \end{pmatrix}.$$

Hence, since $n + 1$ $n \times 1$ vectors are necessarily linearly dependent, there is a nonzero $(n + 1) \times 1$ column vector

$$\begin{pmatrix} t_0 \\ t_1 \\ \vdots \\ t_n \end{pmatrix}$$

     Chapter 4. Useful forms for matrices

such that

$$t_0 I \mathbf{x} + t_1 A \mathbf{x} + t_2 A^2 \mathbf{x} + \cdots + t_n A^n \mathbf{x} = \mathbf{O}^c. \qquad (4.2.4)$$

The equation (4.2.4) may be rewritten

$$(t_n A^n + t_{n-1} A^{n-1} + \cdots + t_2 A^2 + t_1 A + t_0 I)\mathbf{x} = \mathbf{O}^c.$$

The polynomial p such that

$$p(z) = t_n z^n + \cdots + t_1 z + t_0$$

may, owing to the Fundamental Theorem of Algebra and its corollary, the Factor Theorem, be factored:

$$p(z) = t_n (z - \lambda_1)^{r_1} (z - \lambda_2)^{r_2} \cdots (z - \lambda_k)^{r_k}$$

and $r_1 + \cdots + r_k = n$.

Correspondingly,

$$\begin{aligned}
(t_n A^n + &\cdots + t_0 I) \\
&= t_n (A - \lambda_1 I)^{r_1} (A - \lambda_2 I)^{r_2} \cdots (A - \lambda_k I)^{r_k}. \qquad (4.2.5)
\end{aligned}$$

There are n not necessarily distinct factors in the right member of (4.2.5). Their product applied to $\mathbf{x}$ produces $\mathbf{O}^c$. There are at most $2^n - 1$ different polynomials that can be formed with n or fewer factors drawn from the factors of p. Among these is some polynomial q involving the smallest number ν of such factors and such that $B \overset{\text{def}}{=} q(A)$ applied to $\mathbf{x}$ also produces $\mathbf{O}^c$:

$$B\mathbf{x} = \mathbf{O}^c.$$

Since $\mathbf{x} \neq \mathbf{O}^c$ it follows that $n \geq \nu \geq 1$.

If $\nu = 1$ and the single factor of B is $A - \lambda_i I$ then $B = A - \lambda_i I$, whence

$$(A - \lambda_i I)\mathbf{x} = \mathbf{O}^c,$$

λ_i is an eigenvalue of A, and $\mathbf{x}$ is a corresponding eigenvector.

If $\nu > 1$ let $A - \lambda_i I$ be *any* one of the factors in B and let B_i be the product of the *other* factors in B. Owing to the minimality of ν it follows that

$$\mathbf{y}_i \overset{\text{def}}{=} B_i \mathbf{x} \neq \mathbf{O}^c$$

and yet

$$(A - \lambda_i I)\mathbf{y}_i = (A - \lambda_i I)B_i \mathbf{x} = B\mathbf{x} = \mathbf{O}^c.$$

Hence λ_i is an eigenvalue of A and $\mathbf{y}_i$ is an eigenvector corresponding to λ_i.
$$\Omega$$

[**Note 4.2.1:**   *Each* factor $A - \lambda_i I$ of B gives rise to a B_i and an eigenvector $\mathbf{y}_i$. It is important to note that, as the following illustration shows, *not every* λ_j in (4.2.5) is necessarily an eigenvalue of A.

For example, if $A = I$ then for *any* vector $\mathbf{x}$

$$(I^2 - I)\mathbf{x} = O\mathbf{x} = \mathbf{O}^c.$$

The polynomial p for which

$$p(z) = z^2 - 1 = (z - 1)(z + 1)$$

and the polynomial q for which

$$q(z) = z^2 - z = z(z - 1)$$

are such that for any vector $\mathbf{x}$

$$p(I)\mathbf{x} = q(I)\mathbf{x} = I^2 - I\mathbf{x} = \mathbf{O}^c.$$

Nevertheless only the zero 1 of p and of q is an eigenvalue of I and *neither* the zero -1 of p *nor* the zero 0 of q is an eigenvalue of I (cf. **Section 4.3**).]

[**Note 4.2.2:**   If A is a SQUARE matrix, if $\mathbf{x}$ is a nonzero vector, and if

$$p(A)\mathbf{x} = \mathbf{O}^c$$

there are infinitely many different polynomials q such that

$$q(A)\mathbf{x} = \mathbf{O}^c.$$

Indeed, if r is any polynomial and $q = rp$ then

$$q(A)\mathbf{x} = \mathbf{O}^c.$$

In **Section 4.3** there is singled out a particular uniquely defined polynomial $m_{A,\mathbf{x}}$ such that

$$m_{A,\mathbf{x}}(A)\mathbf{x} = \mathbf{O}^c.$$

Each of the zeros of $m_{A,\mathbf{x}}$ is an eigenvalue of A although there can be eigenvalues of A that are not zeros of $m_{A,\mathbf{x}}.$]

Example 4.2.2. Let A be the matrix

$$\begin{pmatrix} 1 & 2 \\ 3 & 4 \end{pmatrix}$$

---

and let $\mathbf{x}$ be

$$\begin{pmatrix} 1 \\ 0 \end{pmatrix} \stackrel{\text{def}}{=} \mathbf{e}_1.$$

Then

$$A\mathbf{x} = \begin{pmatrix} 1 \\ 3 \end{pmatrix}, \quad A^2\mathbf{x} = \begin{pmatrix} 7 & 10 \\ 15 & 22 \end{pmatrix}\begin{pmatrix} 1 \\ 0 \end{pmatrix} = \begin{pmatrix} 7 \\ 15 \end{pmatrix}.$$

Hence

$$(-2I - 5A + A^2)\mathbf{x} = \mathbf{O}^c.$$

The quadratic formula shows that

$$(A - \frac{1}{2}(5 + \sqrt{33})I)(A - \frac{1}{2}(5 - \sqrt{33})I)\mathbf{x} = \mathbf{O}^c. \tag{4.2.6}$$

Direct calculation shows that

$$\mathbf{y} \stackrel{\text{def}}{=} (A - \frac{1}{2}(5 + \sqrt{33}I)\mathbf{x} = \begin{pmatrix} \frac{1}{2}(-3 - \sqrt{33}) \\ 3 \end{pmatrix} \neq \mathbf{O}^c.$$

On the other hand

$$(A - \frac{1}{2}(5 - \sqrt{33})I)\mathbf{y} = \mathbf{O}^c$$

and thus $\frac{1}{2}(5 - \sqrt{33})$ is an eigenvalue and $\mathbf{y}$ is an eigenvector.

Exercise 4.2.3. Let $\mathbf{w}$ be

$$\begin{pmatrix} 0 \\ 1 \end{pmatrix} \stackrel{\text{def}}{=} \mathbf{e}_2.$$

Calculate

$$(A - \frac{1}{2}(5 - \sqrt{33})I)\mathbf{w} \stackrel{\text{def}}{=} \mathbf{z}$$

and show that $\mathbf{z}$ is an eigenvector corresponding to the eigenvalue $\frac{1}{2}(5 + \sqrt{33})$.

Exercise 4.2.4. For A as in **Example 4.2.2** find the eigenvalues of A in an alternative way by solving the equation $det(A - \lambda I) = 0$.

Exercise 4.2.5. Show that the vectors $\mathbf{y}$ and $\mathbf{z}$ above are linearly independent.

Exercise 4.2.6. Let P be the matrix in which the first column is $\mathbf{y}$ above and the second column is $\mathbf{z}$ above. Show P is invertible and that

$$P^{-1}AP = \begin{pmatrix} \frac{1}{2}(5 - \sqrt{33}) & \\ & \frac{1}{2}(5 + \sqrt{33}) \end{pmatrix}.$$

Exercise 4.2.7. Use the method illustrated in **Example 4.2.2** to find the eigenvalues, and for each eigenvalue an eigenvector for the matrix

$$A \stackrel{\text{def}}{=} \begin{pmatrix} 7 & 5 \\ -2 & 1 \end{pmatrix}.$$

Then find an invertible 2×2 matrix P such that $P^{-1}AP$ is a diagonal matrix.

[**Remark 4.2.2:** If A is an $n \times n$ SQUARE matrix and if P is an invertible $n \times n$ SQUARE matrix let B be $P^{-1}AP$. Then

$$\begin{aligned}
\chi_B &= det(P^{-1}AP - \lambda I) \\
&= det(P^{-1}AP - P^{-1}\lambda IP) \\
&= det(P^{-1}(A - \lambda I)P).
\end{aligned}$$

The right member of the last equation is, according to the product rule for determinants,

$$\begin{aligned}
det(P^{-1})det(A - \lambda I)det(P) &= det(P^{-1})det(P)det(A - \lambda I) \\
&= det(P^{-1}P)\chi_A \\
&= \chi_A.
\end{aligned}$$

Hence $\chi_B = \chi_A$, i.e., since A and B are similar and P is an arbitrary invertible $n \times n$ matrix, the characteristic polynomial is a *similarity invariant*. In particular the zeros of χ_A resp. of χ_B are the same, and since these zeros are the eigenvalues of A resp. of B it follows that the set of eigenvalues of A is a similarity invariant. (The argument shows also that the function *det* is a similarity invariant: $det(P^{-1}AP) = det(A)$.)

If $P^{-1}AP \stackrel{\text{def}}{=} D$ is a *diagonal* matrix, i.e., if

$$D = \begin{pmatrix} \lambda_1 & & & \\ & \lambda_2 & & \\ & & \ddots & \\ & & & \lambda_n \end{pmatrix}$$

then although its eigen*values* are the same as the eigenvalues of A, its eigen*vectors* are usually different from the eigenvectors of A. In fact the standard basis vectors $e_1, \ldots, e_n$ are the eigenvectors of any $n \times n$ diagonal matrix D. They constitute the standard basis E. Thus to diagonalize a matrix A is essentially to find a matrix D for which:

i. D is *similar* to A, i.e., for some invertible matrix P, $D = P^{-1}AP$;

     Chapter 4. Useful forms for matrices

ii. the standard basis vectors $\mathbf{e}_1,\ldots,\mathbf{e}_n$ are eigenvectors for D: E is an *eigenbasis* for D.

A matrix D for which *ii* holds is automatically a diagonal matrix.]

The following observations may serve as a guide to what may be expected in the next developments.

i. Thus far, for each 2×2 matrix for which there are two distinct eigenvalues, there are two linearly independent eigenvectors. The matrix formed of columns that are linearly independent eigenvectors diagonalizes the given matrix.

ii. Neither of the 2×2 matrices

$$\begin{pmatrix} 0 & 1 \\ 0 & 0 \end{pmatrix} \text{ and } \begin{pmatrix} 1 & 1 \\ 0 & 1 \end{pmatrix}$$

is diagonable and each has only *one* eigenvalue.

iii. Finally, the diagonal, hence diagonable, matrix I has only one eigenvalue.

Some clarification is in the next result.

THEOREM 4.2.2. LET A BE A SQUARE $n \times n$ MATRIX FOR WHICH THERE ARE k PAIRWISE DIFFERENT EIGENVALUES $\lambda_1,\ldots,\lambda_k$. LET $\mathbf{x}_1,\ldots,\mathbf{x}_k$ BE CORRESPONDING EIGENVECTORS. THEN THOSE EIGENVECTORS ARE LINEARLY INDEPENDENT.

PROOF. Assume that the eigenvectors are linearly dependent. Then there are numbers $t_1,\ldots,t_k$, *at least one of which is not 0*, and such that

$$t_1\mathbf{x}_1 + \cdots + t_k\mathbf{x}_k = \sum_{j=1}^{k} t_j\mathbf{x}_j = \mathbf{O}^c.$$

Among all such (nontrivial) linear combinations there is one involving a minimal number, say l, of the vectors. It may be assumed that these are $\mathbf{x}_1,\ldots,\mathbf{x}_l$. Thus there are numbers $s_1,\ldots,s_l$ such that

$$s_1\mathbf{x}_1 + \cdots + s_l\mathbf{x}_l = \mathbf{O}^c \tag{4.2.7}$$

and *none* of the s_i is 0. Both members of (4.2.7) may be multiplied by the eigenvalue λ_1 or by the matrix A. The respective results are

$$s_1\lambda_1\mathbf{x}_1 + s_2\lambda_1\mathbf{x}_2 + \cdots + s_l\lambda_1\mathbf{x}_l = \mathbf{O}^c \tag{4.2.8}$$

$$A(s_1\mathbf{x}_1 + s_2\mathbf{x}_2 + \cdots + s_l\mathbf{x}_l) = \mathbf{O}^c$$

$$s_1 A\mathbf{x}_1 + s_2 A\mathbf{x}_2 \cdots + s_l A\mathbf{x}_l = \mathbf{O}^c$$

$$s_1\lambda_1\mathbf{x}_1 + s_2\lambda_2\mathbf{x}_2 + \cdots + s_l\lambda_l\mathbf{x}_l = \mathbf{O}^c. \tag{4.2.9}$$

From (4.2.8) and (4.2.9) it follows by subtraction of corresponding members that

$$s_2(\lambda_1 - \lambda_2)\mathbf{x}_2 + \cdots + s_l(\lambda_1 - \lambda_l)\mathbf{x}_l = \mathbf{O}^c. \tag{4.2.10}$$

In (4.2.10) there are fewer than l terms. Because l is minimal each coefficient in (4.2.10) must be 0. Since no s_i is 0 it follows that

$$\lambda_1 - \lambda_2 = \lambda_1 - \lambda_3 = \cdots = \lambda_1 - \lambda_l = 0,$$

in contradiction of the assumption: $\lambda_i \neq \lambda_j$ if $i \neq j$.

$$\Omega$$

[**Remark 4.2.3:**   In particular, no vector can serve as an eigenvector for two different eigenvalues.]

Example 4.2.3. If A is an $n \times n$ matrix and if P is an invertible matrix such that $P^{-1}AP \overset{\text{def}}{=} D$ is a diagonal matrix then each column of P is an eigenvector of A. Indeed, $AP = PD$. If

$$D \overset{\text{def}}{=} \begin{pmatrix} \lambda_1 & & & \\ & \lambda_2 & & \\ & & \ddots & \\ & & & \lambda_n \end{pmatrix}$$

then the columns of D are

$$\lambda_1\mathbf{e}_1, \ldots, \lambda_n\mathbf{e}_n.$$

If $\mathbf{p}_1, \ldots, \mathbf{p}_n$ are the column vectors making up the columns of P then the jth column of AP is $A\mathbf{p}_j$ and the jth column of PD is $P(\lambda_j\mathbf{e}_j) = \lambda_j\mathbf{p}_j$. Thus

$$A\mathbf{p}_j = \lambda_j\mathbf{p}_j, \quad ,1 \leq j \leq n.$$

In other words $\mathbf{p}_j$ is an eigenvector corresponding to the eigenvalue λ_j. Hence:

An $n \times n$ invertible matrix P "diagonalizes" an $n \times n$ matrix A, i.e.,

$$P^{-1}AP = D \overset{\text{def}}{=} \begin{pmatrix} \lambda_1 & & & \\ & \lambda_2 & & \\ & & \ddots & \\ & & & \lambda_n \end{pmatrix},$$

iff $AP = PD$ and the jth column of P is for A an eigenvector corresponding to the eigenvalue λ_j: $A\mathbf{p}_j = \lambda_j\mathbf{p}_j$.

Exercise 4.2.8. Prove that if a SQUARE $n \times n$ matrix A has n different eigenvalues then A is diagonable, i.e., that there is an invertible $n \times n$ matrix P such that $P^{-1}AP$ is a diagonal matrix.

[*Hint:* Such a P exists iff there is an *eigenbasis*, i.e., basis consisting of eigenvectors of A.]

Exercise 4.2.9. Prove that if A is an $n \times n$ matrix then there are at most n different eigenvalues for A.

[*Hint:* No linearly independent set of $n \times 1$ vectors contains more than n members (elements, vectors).]

Exercise 4.2.10. Show that if A is a SQUARE matrix then λ is an eigenvalue of A iff λ is an eigenvalue of A^t.

The next result is relatively simple and provides an entering wedge for finding the most general useful form for any matrix. The conclusion is that every $n \times n$ SQUARE matrix A is similar to an upper triangular matrix (and also to a lower triangular matrix). The triangular matrix has at most $\frac{n(n+1)}{2}$ rather than as many as n^2 nonzero entries. For large n the "discount" can be almost as large as 50%.

THEOREM **4.2.3.** LET A BE THE $n \times n$ MATRIX $(a_{ij})_{i,j=1}^{n,n}$.

i. THERE IS AN INVERTIBLE $n \times n$ MATRIX P SUCH THAT $P^{-1}AP$ IS AN UPPER TRIANGULAR MATRIX A_U.

ii. EACH DIAGONAL ENTRY OF A_U IS AN EIGENVALUE OF A AND EVERY EIGENVALUE OF A IS FOUND SOMEWHERE ON THE DIAGONAL OF A_U.

PROOF.

ad*i*: The statement is certainly true if $n = 1$ since every 1×1 matrix is in diagonal form, without further adjustment. If the statement is not generally true, there is some least natural number n for which there is an $n \times n$ matrix A that is not similar to an upper triangular matrix and if $k < n$ any $k \times k$ matrix *is* similar to an upper triangular matrix.

Let λ_1 be an eigenvalue of A and let $\mathbf{x}_1$ be a corresponding eigenvector. Then fill $\{\mathbf{x}_1\}$ out to a basis $X \stackrel{\text{def}}{=} \{\mathbf{x}_1, \mathbf{y}_2, \ldots, \mathbf{y}_n\}$ for $\mathbf{C}^n$. Let Q be the matrix for which the first column is the column vector $\mathbf{x}_1$ and, if $j > 1$, the jth column is $\mathbf{y}_j$. Then Q is invertible since the columns in Q are linearly independent. Furthermore the first column of AQ is $\lambda_1 \mathbf{x}_1$ and thus the first column of $Q^{-1}AQ$ is $\lambda_1 \mathbf{e}_1$, i.e.,

$$\begin{pmatrix} \lambda_1 \\ 0 \\ \vdots \\ 0 \end{pmatrix}.$$

Let B be the $n-1 \times n-1$ matrix made up of the entries in rows and columns

---

numbered $2, 3, \ldots, n$ of $Q^{-1}AQ$. By hypothesis there is an invertible $n-1 \times n-1$ matrix R such that $R^{-1}BR$ is upper triangular. If P is the block matrix

$$
\begin{array}{cc}
 & \begin{array}{cc} 1 & n-1 \end{array} \\
\begin{array}{c} 1 \\ n-1 \end{array} & \begin{pmatrix} I & O \\ O & R \end{pmatrix}
\end{array}
$$

then P is invertible,

$$
P^{-1} = \begin{array}{cc}
 & \begin{array}{cc} 1 & n-1 \end{array} \\
\begin{array}{c} 1 \\ n-1 \end{array} & \begin{pmatrix} I & O \\ O & R^{-1} \end{pmatrix}
\end{array}
$$

and $A_U \stackrel{\text{def}}{=} P^{-1}Q^{-1}AQP$ is upper triangular.

At this point the nontriangulability of A is contradicted and i is proved.

adii: Since A_U is upper triangular $A_U - \lambda I$ is also upper triangular. According to **Exercise 2.6.17** $A_U - \lambda I$ is singular iff some diagonal entry of $A_U - \lambda I$ is 0. It follows that every diagonal entry of A_U is an eigenvalue of A_U and that every eigenvalue of A_U is some diagonal entry of A_U.

Furthermore since $P^{-1}(A - \lambda I)P = A_U - \lambda I$ it follows that $A - \lambda I$ is singular iff $A_U - \lambda I$ is singular. Hence a number λ is an eigenvalue of A_U iff λ is an eigenvalue of A. Thus every diagonal entry of A_U is an eigenvalue of A and every eigenvalue of A is a diagonal entry of A_U.

In other words the diagonal entries of A_U constitute precisely the set of eigenvalues of A.

$$\Omega$$

The diagonal entries of A_U need not be distinct, e.g., if $A = I$ then $A = A_U$ and each diagonal entry of A_U is 1.

There is an important difference between the THEOREM **4.2.3** just proved and the earlier results about the echelon form, the Jordan elimination form, etc. The matrices A_E and A_J corresponding to a SQUARE matrix A are *similar* to A only in special circumstances, e.g., if A is upper triangular or diagonal then $A = A_E$, if $A = I$ then $A = A_J$.

For example, if

$$
A = \begin{pmatrix} 1 & 1 \\ 1 & 2 \end{pmatrix}
$$

then

$$
A_E = \begin{pmatrix} 1 & 1 \\ 0 & 1 \end{pmatrix}.
$$

But if there is a 2×2 matrix

$$
P \stackrel{\text{def}}{=} \begin{pmatrix} a & b \\ c & d \end{pmatrix}
$$

such that $A_E = P^{-1}AP$, i.e., such that

$$
PA_E = AP \tag{4.2.11}
$$

then entry-by-entry comparison of the left and right members of (4.2.11) shows
that the following equations must hold:

$$a + c = a$$
$$b + d = a + b$$
$$a + 2c = c$$
$$b + 2d = c + d.$$

The first equation implies that $c = 0$. The third equation then implies that
$a = 0$, whence from the second equation it follows that $d = 0$, and then the last
equation implies that $b = 0$. In other words, $P = O$, and P is not invertible, a
contradiction.

Nevertheless, if P is an invertible matrix such that $P^{-1}AP$ *is* an upper
triangular matrix A_U, then the system

$$A\mathbf{x} = \mathbf{b} \tag{4.2.12}$$

can be analyzed quickly. Indeed, since $A = PA_U P^{-1}$, the system (4.2.12) may
be written

$$PA_U P^{-1}\mathbf{x} = \mathbf{b}$$

or

$$A_U P^{-1}\mathbf{x} = P^{-1}\mathbf{b}.$$

Since P and P^{-1} are known, let $P^{-1}\mathbf{b}$ be $\mathbf{c}$, a new *known*, and let $P^{-1}\mathbf{x}$ be $\mathbf{y}$,
a new *unknown*. It follows that

$$A_U\mathbf{y} = \mathbf{c}. \tag{4.2.13}$$

Since A_U is upper triangular it is in echelon form and thus (4.2.13) can be
analyzed with ease. If (4.2.13) has NO solution then (4.2.12) has NO solution. If
(4.2.13) has a solution $\mathbf{y}$, then $P\mathbf{y}$ is a solution of (4.2.12). In sum, the systems
(4.2.12) and (4.2.13) are logically equivalent.

Unless an upper triangular form *similar* to A can be found easily, the *PRO-
CESS*, the GEM, or the GJM provide techniques quicker than "upper trian-
gularization by similarity" for analyzing (4.2.12). Furthermore, "upper trian-
gularization by similarity" is available only for SQUARE systems whereas the
PROCESS, the GEM and the GJM work for all systems, HIGH, WIDE, and
SQUARE.

Example 4.2.4. Let A be the matrix

$$\begin{pmatrix} 1 & 1 \\ -1 & 3 \end{pmatrix}.$$

Then

$$\chi_A(\lambda) = (\lambda - 2)^2$$

---

and so its only eigenvalue is $\lambda_1 \stackrel{\text{def}}{=} 2$. Any corresponding eigenvector is a multiple of

$$\mathbf{w} \stackrel{\text{def}}{=} \begin{pmatrix} 1 \\ 1 \end{pmatrix}.$$

Thus

$$W_1 = ker(A - 2I) = \{\, \mathbf{x} \; : \; \mathbf{x} = c\mathbf{w}, c \in \mathbf{C} \,\}.$$

For the basis X use the set

$$\left\{ \mathbf{y}_1 \stackrel{\text{def}}{=} \mathbf{w}, \mathbf{z}_1 \stackrel{\text{def}}{=} \begin{pmatrix} 2 \\ 0 \end{pmatrix} \right\}.$$

If

$$Q \stackrel{\text{def}}{=} \begin{pmatrix} 1 & 2 \\ 1 & 0 \end{pmatrix} \quad \text{then } Q^{-1} = \begin{pmatrix} 0 & 1 \\ \frac{1}{2} & -\frac{1}{2} \end{pmatrix}$$

and

$$A_U \stackrel{\text{def}}{=} Q^{-1}AQ = \begin{pmatrix} 2 & -2 \\ 0 & 2 \end{pmatrix},$$

an upper triangular matrix. On the diagonal of A_U each entry is an eigenvalue (2) of A. Since 2 is the only eigenvalue of A it follows that A_U has the required form.

Exercise 4.2.11. In the matrix $A_U \stackrel{\text{def}}{=} (\xi_{ij})_{i,j=1}^{n,n}$ each entry $\xi_{ij} = 0$ if $i > j$. Thus A_U is the sum of two matrices:

$$\begin{pmatrix} \xi_{11} & \cdots & \xi_{1n} \\ \vdots & \ddots & \vdots \\ \xi_{n1} & \cdots & \xi_{nn} \end{pmatrix} = \begin{pmatrix} \xi_1 & & \\ & \xi_2 & \\ & & \ddots \\ & & & \xi_n \end{pmatrix} + \begin{pmatrix} 0 & \xi_{12} & \cdots & \xi_{1n} \\ & 0 & \cdots & \xi_{2n} \\ \vdots & & \ddots & \vdots \\ & & & 0 \end{pmatrix}$$

$$\stackrel{\text{def}}{=} D + N.$$

Show that $N^n = O$.

A SQUARE matrix N such that $N^k = O$ for some natural number k is called *nilpotent*. If, to boot, whenever $l < k$, $N^l \neq O$ then N is called *nilpotent of order* k. Thus N in **Exercise 4.2.11** is a *nilpotent* matrix of order not exceeding n.

Nilpotent matrices constitute a very important class of SQUARE matrices. They are treated at length in **Section 4.5.**

[The prefix and word *nil* stems from the Indo-European root
— *ne* — not. The origin of *potent* is the Indo-European root —
poti — powerful, lord. Some cognates are *potential*, *power*, *posse*,
and *possible*.]

Example 4.2.5. Suppose that A_U in THEOREM **4.2.3** looks like this:

$$\begin{pmatrix} 1 & & & & & & & & \\ & 1 & & & & & & & \\ & & 1 & & & & & & \\ & & & 2 & & & & & \\ & & & & 2 & & & & \\ & & & & & 1 & 6 & & \\ & & & & & & 1 & & \\ & & & & & & & 2 & 7 \\ & & & & & & & & 2 \end{pmatrix}.$$

Then shifting the (1,1) entry 1 five times downward and five times to the right leads to

$$\begin{pmatrix} 1 & & & & & & & & \\ & 1 & & & & & & & \\ & & 2 & & & & & & \\ & & & 2 & & & & & \\ & & & & 1 & 6 & & & \\ & & & & & 1 & & & \\ & & & & & & 1 & & \\ & & & & & & & 2 & 7 \\ & & & & & & & & 2 \end{pmatrix}.$$

The result is achieved by multiplying A_U on the left by $E_{45}E_{34}$ and on the right by $E_{34}E_{45}$, the *inverse* of $E_{45}E_{34}$. Further shifts can be achieved similarly. The final result is a block matrix consisting of the following blocks:

one block: $\begin{pmatrix} 1 & 6 \\ 0 & 1 \end{pmatrix}$; three blocks: (1); one block: $\begin{pmatrix} 2 & 7 \\ 0 & 2 \end{pmatrix}$; two blocks:(2);

and it looks like this

$$\begin{pmatrix} 1 & 6 & & & & & & & \\ & 1 & & & & & & & \\ & & 1 & & & & & & \\ & & & 1 & & & & & \\ & & & & 1 & & & & \\ & & & & & 2 & 7 & & \\ & & & & & & 2 & & \\ & & & & & & & 2 & \\ & & & & & & & & 2 \end{pmatrix}.$$

The diagonal entries in each block are all the same. In fact the pattern just observed is nearly the model for the most general *Jordan form* to be treated in Section **4.6**.

Exercise 4.2.12. In **Example 4.2.4** let z_1 be

$$\begin{pmatrix} 0 \\ 2 \end{pmatrix}.$$

Calculate the corresponding Q and Q^{-1} and $Q^{-1}AQ$. (The result is again upper triangular but different from the result in **Example 4.2.4**. In both cases, because $n = 2$ there is no need to find an R and ultimately a (different) P, i.e., $P = Q$.)

Exercise 4.2.13. Let A be an $n \times n$ matrix. Let A_U be an upper triangular matrix similar to A^t. Thus for some invertible $n \times n$ matrix P, $P^{-1}A^tP = A_U$.

Show how to find an invertible $n \times n$ matrix R such that $R^{-1}AR \stackrel{\text{def}}{=} L$ is *lower triangular*.

[*Hint:* Consider A^t.]

Exercise 4.2.14. Let A be

$$\begin{pmatrix} -3 & 5 & 9 \\ -2 & 4 & 6 \\ -5 & 5 & 7 \end{pmatrix}.$$

Find an upper triangular form and a lower triangular form for A.

Example 4.2.6. An "application" of the triangulability theorem (THEOREM **4.2.3**) is a second proof of the Cayley-Hamilton theorem (cf. **Section 3.2**).

Let A be an $n \times n$ matrix and let A_U be one of its triangular forms. Thus, $\lambda_1, \ldots, \lambda_n$ being the (not necessarily different) eigenvalues of A, for some invertible $n \times n$ matrix P

$$P^{-1}AP = A_U = \begin{pmatrix} \lambda_1 & t_{12} & t_{13} & \cdots & t_{1n} \\ & \lambda_2 & t_{23} & \cdots & t_{2n} \\ & & \ddots & & \vdots \\ & & & \ddots & \vdots \\ & & & & \lambda_n \end{pmatrix}.$$

A direct calculation shows that all entries of the first column of the upper triangular matrix $A_U - \lambda_1 I$ are 0's and all entries in the first two columns of the upper triangular matrix $(A_U - \lambda_1 I)(A_U - \lambda_2 I)$ are 0's. Mathematical induction then shows that if $1 \le k \le n$, then all the entries in the first k columns of the upper triangular matrix

$$(A_U - \lambda_1 I) \cdots (A_U - \lambda_k I) \stackrel{\text{def}}{=} \prod_{i=1}^{k}(A_U - \lambda_i I)$$

---

are 0's. Hence

$$\prod_{i=1}^{n}(A_U - \lambda_i I) = O, \text{ i.e. } \chi_{A_U}(A_U) = O.$$

Since $\chi_A = \chi_{A_U}$ it follows that

$$O = \chi_{A_U}(A_U) = \chi_A(A_U) = \chi_A(P^{-1}AP) = P^{-1}\chi_A(A)P$$

whence $\chi_A(A) = O$, which is the burden of the *Cayley-Hamilton* theorem.

Exercise 4.2.15. For the matrix A in **Exercise 4.2.14** verify that if λ_1, λ_2, and λ_3 are the (not necessarily different) eigenvalues of A and if A_U is the upper triangular form found for A in **Exercise 4.2.14** then the first column of $A_U - \lambda_1 I$ consists of 0's, the first two columns of $(A_U - \lambda_1 I)(A_U - \lambda_2 I)$ consists of 0's and that $(A_U - \lambda_1 I)(A_U - \lambda_2 I)(A_U - \lambda_3 I) = O$.

Exercise 4.2.16. Let E be a SQUARE matrix such that $E^2 = E \neq I$. Show that E is singular and that if $E - \lambda I$ is singular then $\lambda = 0$ or $\lambda = 1$.

Exercise 4.2.17. Let p be a polynomial such that

$$p(\lambda) = (-1)^n(\lambda^n - c_{n-1}\lambda^{n-1} - \cdots - c_1\lambda - c_0).$$

Show that the matrix

$$C \stackrel{\text{def}}{=} \begin{pmatrix} c_{n-1} & c_{n-2} & \cdots & c_1 & c_0 \\ 1 & 0 & \cdots & 0 & 0 \\ 0 & 1 & \cdots & 0 & 0 \\ \vdots & \vdots & \ddots & \vdots & \vdots \\ 0 & 0 & \cdots & 1 & 0 \end{pmatrix}$$

is such that

$$\chi_C(\lambda) = p(\lambda).$$

The matrix C is called the *companion matrix* of p.

[*Hint:* Explore the case for $n = 3$ and then use mathematical induction.]

Exercise 4.2.18. Let A and B be two $n \times n$ matrices.

i. Show that if t is fixed and $|\lambda|$ is sufficiently large then

$$\det[tI - (A - \lambda I)B] = \det[tI - B(A - \lambda I)]. \tag{4.2.14}$$

[*Hint:* For large $|\lambda|$, the matrix $A - \lambda I = \lambda(\frac{A}{\lambda} - I)$ is invertible.]

ii. Show that (4.2.14) is valid for all t and all λ.

[*Hint:* Two polynomials that are equal for infinitely many values of the variable are equal for all values of the variable.]

iii. Use *i* and *ii* to show that always $\chi_{AB} = \chi_{BA}$.

Exercise 4.2.19. Find a matrix A for which the following is the table of corresponding eigenvalues and eigenvectors:

$$
\begin{array}{cc}
\text{EIGENVALUES} & \text{EIGENVECTORS} \\[1em]
1 & t \begin{pmatrix} -4 \\ 3 \end{pmatrix} \ (t \neq 0), \\[1.5em]
3 & s \begin{pmatrix} 2 \\ -1 \end{pmatrix} \ (s \neq 0).
\end{array}
$$

$$(4.2.15)$$

Exercise 4.2.20. Assume A and B are $n \times n$ matrices such that there is a basis $X \overset{\text{def}}{=} \{x_1, \ldots, x_n\}$ consisting of eigenvectors for *both* A and B. Show $AB = BA$.

Exercise 4.2.21. Show that if A is an $n \times n$ matrix having n different eigenvalues, $\lambda_1, \ldots, \lambda_n$, then

$$\lambda_j = tr(A) - \sum_{i \neq j} \lambda_i, \ 1 \leq j \leq n$$

(cf. **Exercise 4.6.23**).

Exercise 4.2.22. Let A be a 3×3 matrix and let P be the invertible matrix

$$\begin{pmatrix} 1 & 2 & 3 \\ -3 & 1 & -2 \\ 2 & -3 & 1 \end{pmatrix}.$$

Assume

$$P^{-1}AP = \begin{pmatrix} i & 1 & \\ & i & \\ & & 2 \end{pmatrix}.$$

Calculate A^2 and A^3.

Exercise 4.2.23. Assume A is the matrix

$$\begin{pmatrix} a & b \\ c & d \end{pmatrix}$$

---

and that $A^2 = O$. Describe all the possibilities for A. Prove that if $A^2 \neq O$ then A is invertible.

[*Hint:* Use THEOREM **4.2.3**.]

If A is an $n \times n$ matrix and if λ is an eigenvalue for A then

$$W_\lambda \overset{\text{def}}{=} \{\, \mathbf{x} \;:\; A\mathbf{x} = \lambda\mathbf{x} \,\} = ker(A - \lambda I)$$

is a special subspace called the *eigenspace* corresponding to the eigenvalue λ.

Exercise 4.2.24. Assume $\mathbf{x} \in W_\lambda$ and that p is a polynomial. Show:

i. $A\mathbf{x} \in W_\lambda$, i.e., "$AW_\lambda \subset W_\lambda$;"

ii.
$$p(A)\mathbf{x} \in W_\lambda. \tag{4.2.16}$$

A subspace W such that $AW \subset W$ is called an *invariant* subspace. Thus $ker(A)$, $im(A)$, and (in i above) W_λ are invariant subspaces.

Exercise 4.2.25. Show that if W is an invariant subspace containing one nonzero vector then W contains an eigenvector for A.

[*Hint:* See the proof of THEOREM **4.2.1**.]

Exercise 4.2.26. Show that a SQUARE matrix A is singular iff 0 is one of its eigenvalues.

If A is a SQUARE matrix and $\lambda \in \mathbf{C}$ then $ker[(A - \lambda I)^k]$ is called a *generalized eigenspace* of A. According to **Exercise 2.9.34** there is a natural number $k_{A-\lambda I} \overset{\text{def}}{=} k_\lambda$ such that for all n in $\mathbf{N}$

$$ker[(A - \lambda I)^{k_\lambda - 1}] \subsetneq ker[(A - \lambda i)^{k_\lambda}] = ker[(A - \lambda I)^{k_\lambda + n}].$$

Exercise 4.2.27. Let each of the matrices below be A. For each number λ find the sequence of eigenspaces of A:

$$i. \begin{pmatrix} 1 & 2 \\ -2 & 4 \end{pmatrix} \quad ii. \begin{pmatrix} 2 & 1 \\ 0 & 2 \end{pmatrix} \quad iii. \begin{pmatrix} \lambda_1 & & & \\ & \lambda_2 & & \\ & & \ddots & \\ & & & \lambda_n \end{pmatrix}.$$

Exercise 4.2.28. Let A be an $n \times n$ matrix such that every nonzero vector in $\mathbf{C}^n$ is an eigenvector for A. Show that the matrix A is a multiple of I: $A = tI$.

4.3. Minimal polynomials

Let A be an $n \times n$ SQUARE matrix. If $\mathbf{x}$ is a nonzero vector in $\mathbf{C}^n$, then (cf. **Section 4.2**) for some nonzero polynomial p, it is true that $p(A)\mathbf{x} = \mathbf{O}^c$. But if r is *any* polynomial then also

$$r(A)p(A)\mathbf{x} = \mathbf{O}^c.$$

However, r might be the product of $n + 1$ different factors:

$$r(z) = (z - \sigma_1) \cdots (z - \sigma_{n+1})$$

whence if

$$p(z) = (z - \lambda_1)^{r_1} \cdots (z - \lambda_k)^{r_k}$$

then the polynomial rp may be factored:

$$r(z)p(z) = (z - \sigma_1) \cdots (z - \sigma_{n+1})(z - \lambda_1)^{r_1} \cdots (z - \lambda_k)^{r_k}$$

and yet at most n of the different numbers σ_j can be eigenvalues of A. In other words if $p(A)\mathbf{x} = \mathbf{O}$ and $\mathbf{x} \neq \mathbf{O}$ it is not necessarily true that each zero of p is an eigenvalue of A. In what follows there is singled out a special polynomial m_A such that $m_A(\lambda) = 0$ iff λ is an eigenvalue of A. Furthermore, it is feasible to calculate m_A without prior knowledge of the eigenvalues of A.

Any nonzero polynomial p such that

$$p(A)\mathbf{x} \overset{\text{def}}{=} (a_0 + a_1 A + \cdots + a_{m-1} A^{m-1} + a_m A^m)\mathbf{x} = \mathbf{O}^c$$

is called an *annihilating* or *nullifying* polynomial of degree m, for A and $\mathbf{x}$. The coefficient a_m of z^m, the highest power of z in p is called the *leading coefficient* of the annihilating polynomial. By definition, the leading coefficient is not 0: $a_m \neq 0$.

There are infinitely many nonzero annihilating polynomials for A and $\mathbf{x}$ and among them are some of least degree. Among those of least degree there is at least one for which the *leading coefficient* a_m is 1. (If p is an annihilating polynomial of least degree and if its leading coefficient is a_m then

$$\frac{1}{a_m} p$$

is also of least degree and its leading coefficient is

$$\frac{a_m}{a_m} = 1.)$$

In fact, for given A and $\mathbf{x}$, there is precisely one nonzero annihilating polynomial of least degree and for which the leading coefficient is 1. Indeed, if p

---

and q are two such polynomials let r be $p - q$. The degree of r is less than the (common) degree of p and q (because each is of (the same) least degree and has 1 as leading coefficient). Furthermore, for any vector $\mathbf{x}$,

$$r(A)\mathbf{x} = p(A)\mathbf{x} - q(A)\mathbf{x} = \mathbf{O}^c - \mathbf{O}^c = \mathbf{O}^c.$$

Hence r is either 0 or another annihilating polynomial. Since p and q are of least degree and the degree of r is less than the degree of p, it follows that $r = 0$, i.e., $p = q$. For given A and $\mathbf{x}$, the *unique* nonzero annihilating polynomial of least degree and with leading coefficient 1 is called the their *minimal polynomial* and is denoted $m_{A,\mathbf{x}}$ to indicate its association with both A and $\mathbf{x}$.

The *degree* of $m_{A,\mathbf{x}}$ is denoted $\mu_{A,\mathbf{x}}$ and, since the $n + 1$ vectors

$$\mathbf{x}, A\mathbf{x}, \ldots, A^n\mathbf{x}$$

are linearly dependent, it follows that $1 \leq \mu_{A,\mathbf{x}} \leq n$.

The minimal polynomial for A and $\mathbf{x}$ has some remarkable and useful properties. The factored form of $m_{A,\mathbf{x}}$ is most helpful for the study of $m_{A,\mathbf{x}}$. Thus if the different zeros of $m_{A,\mathbf{x}}$ are $\lambda_1, \ldots, \lambda_k$, then

$$m_{A,\mathbf{x}}(z) = (z - \lambda_1)^{r_1} \cdots (z - \lambda_k)^{r_k} \tag{4.3.1}$$

and $r_1 + \cdots + r_k = \mu_{A,\mathbf{x}}$.

The following observations are noteworthy:

i. If any factor $z - \lambda_i$ is dropped from the right member of (4.3.1) the resulting polynomial, say B_i, is of degree $\mu_{A,\mathbf{x}} - 1$. If $\mu_{A,\mathbf{x}} = 1$ then $B_i = 1$, λ_i is an eigenvalue of A and $\mathbf{x}$ is a corresponding eigenvector. If $\mu_{A,\mathbf{x}} > 1$ then B_i is a polynomial of positive degree $\mu_{A,\mathbf{x}} - 1$. Hence

$$\mathbf{y}_i \stackrel{\text{def}}{=} B_i\mathbf{x} \neq \mathbf{O}^c \text{ and } (A - \lambda_i I)\mathbf{y}_i = m_{A,\mathbf{x}}(A)\mathbf{x} = \mathbf{O}^c,$$

i.e., again λ_i is an eigenvalue but this time $\mathbf{y}_i$ is an eigenvector corresponding to λ_i. Thus:

Each zero λ_i of the minimal polynomial $m_{A,\mathbf{x}}$ is an eigenvalue of A.

ii. There are at most n linearly independent vectors in any set of $n \times 1$ vectors. If there are k different eigenvalues of A then any k eigenvectors corresponding to these k different eigenvalues are linearly independent. Hence there can be not more than n different eigenvalues of A (cf. **Exercise 4.2.9**).

iii. Any polynomial p of degree not more than n, with leading coefficient 1, and having as its only zeros some of or all the numbers $\lambda_1, \ldots, \lambda_k$ is of the form

$$p(z) = (z - \lambda_1)^{s_1}(z - \lambda_2)^{s_2} \cdots (z - \lambda_k)^{s_k},$$

---

and $0 \leq r_i \leq \sum_{i=1}^{k} s_i \leq n$.

Thus there there are only *finitely many* polynomials of degree not exceeding n, with leading coefficient 1, and having as their zeros some of or all the finitely many eigenvalues of A. Hence:

> *The set of all $m_{A,\mathbf{x}}$ formed for* all $n \times 1$ *vectors* $\mathbf{x}$ *and a fixed matrix A is* finite.

iv. If p is a polynomial such that $p(A)\mathbf{x} = \mathbf{O}$ then there is a polynomial q such that $p = q m_{A,\mathbf{x}}$. This is true because the degree of p is at least that of $m_{A,\mathbf{x}}$ and so if p is divided by $m_{A,\mathbf{x}}$ there is a quotient polynomial q and a remainder polynomial r:

$$p = q m_{A,\mathbf{x}} + r.$$

Then the degree of r is *less* than the degree of $m_{A,\mathbf{x}}$. On the other hand

$$r(A)\mathbf{x} = p(A)\mathbf{x} - q(A)m_{A,\mathbf{x}}(A)\mathbf{x} = \mathbf{O}.$$

Since the degree of r is less than the degree of $m_{A,\mathbf{x}}$ the minimality of the degree of $m_{A,\mathbf{x}}$ is contradicted unless $r = 0$, i.e., unless $p = q m_{A,\mathbf{x}}$.

Let $\mathbf{x}_1, \ldots, \mathbf{x}_L$ be nonzero vectors such that

$$\{m_{A,\mathbf{x}_1}, \ldots, m_{A,\mathbf{x}_L}\}$$

is the (*finite*) set of all *different* minimal polynomials found (for A) by considering the minimal polynomials $m_{A,\mathbf{x}}$ for all vectors $\mathbf{x}$. For each i let W_i be the set of all vectors $\mathbf{w}$ such that $m_{A,\mathbf{x}_i}(A)\mathbf{w} = \mathbf{O}^c$. In other words, W_i is the set of all vectors annihilated by $m_{A,\mathbf{x}_i}(A)$:

$$W_i = ker(m_{A,\mathbf{x}_i}(A)).$$

Then each vector $\mathbf{x}$ is annihilated by $m_{A,\mathbf{x}}(A)$, which must be some $m_{A,\mathbf{x}_i}(A)$, and so each vector $\mathbf{x}$ is in some W_i. The last conclusion may be expressed by the equation:

$$\mathbf{C}^n = \bigcup_{i=1}^{L} W_i.$$

Since each W_i is a kernel, each W_i is a subspace of $\mathbf{C}^n$.

The next result shows that some one of the W_i is in fact *all* $\mathbf{C}^n$.

LEMMA **4.3.1.** LET $W_1, \ldots, W_L$ BE FINITELY MANY ARBITRARY SUB-SPACES OF $\mathbf{C}^n$. IF

$$\mathbf{C}^n = \bigcup_{i=1}^{L} W_i$$

THEN THERE IS AN i_0 SUCH THAT $\mathbf{C}^n = W_{i_0}$.

PROOF. Indeed, first let

$$\widetilde{W_1}, \ldots, \widetilde{W_l}, \ (l \leq L). \tag{4.3.2}$$

be a minimal collection of the W_i such that

$$\bigcup_{j=1}^{l} \widetilde{W_j} = \mathbf{C}^n. \tag{4.3.3}$$

In other words, if one of the subspaces listed in (4.3.3) is deleted from the list, the union of the others in the list is *not* $\mathbf{C}^n$. In particular no two of the $\widetilde{W_j}$ are the same.

If $l = 1$ the argument is finished: $\widetilde{W_1} = \mathbf{C}^n$. If $l > 1$ then in each $\widetilde{W_i}$ there is a $\mathbf{w}_i$ not in any other $\widetilde{W_j}$. (Otherwise, for some i,

$$\widetilde{W_i} \subset \bigcup_{j \neq i} \widetilde{W_j}$$

$$\bigcup_{j=1}^{l} \widetilde{W_j} = \bigcup_{j \neq i} \widetilde{W_j}$$

i.e., the collection (4.3.3) is not minimal, a contradiction.)

For any real number t other than 0 or 1 the vector

$$\mathbf{w}(t) \stackrel{\text{def}}{=} t\mathbf{w}_1 + (1 - t)\mathbf{w}_2$$

is in neither $\widetilde{W_1}$ nor $\widetilde{W_2}$: if $\mathbf{w}(t) \in \widetilde{W_1}$ then

$$t\mathbf{w}_1 \in \widetilde{W_1},$$

$$(1 - t)\mathbf{w}_2 = \mathbf{w}(t) - t\mathbf{w}_1 \in \widetilde{W_1},$$

and since $1 - t \neq 0$, $\mathbf{w}_2 \in \widetilde{W_1}$, a contradiction; similarly, if $\mathbf{w}(t) \in \widetilde{W_2}$, then,

$$(1 - t)\mathbf{w}_2 \in \widetilde{W_2},$$

$$t\mathbf{w}_1 = \mathbf{w}(t) - (1 - t)\mathbf{w}_2 \in \widetilde{W_2},$$

and since $t \neq 0$, $\mathbf{w}_1 \in \widetilde{W_2}$, another contradiction.

Since there are infinitely many t that are neither 0 nor 1, the "pigeonhole principle" implies that there are two such, say t_1 and t_2, such that *both* $\mathbf{w}(t_1)$ and $\mathbf{w}(t_2)$ are in some $\widetilde{W_i}$ and consequently Gaussian elimination shows

$$\mathbf{w}_1 = \frac{(1 - t_2)\mathbf{w}(t_1) - (1 - t_1)\mathbf{w}(t_2)}{t_1 - t_2} \in \widetilde{W_i}$$

$$\mathbf{w}_2 = \frac{t_2\mathbf{w}(t_1) - t_1\mathbf{w}(t_2)}{t_2 - t_1} \in \widetilde{W_i}.$$

Since $\mathbf{w}_1$ belongs only to $\widetilde{W_1}$ and since $\mathbf{w}_2$ belongs only to $\widetilde{W_2}$ then $\widetilde{W_i} = \widetilde{W_1} = \widetilde{W_2}$, a final contradiction since no two of the $\widetilde{W_j}$ are the same. Hence $l = 1$ and $\widetilde{W_1} = \mathbf{C}^n$.

Ω

[**Remark 4.3.1:** In **Part III** there is an abstract approach to linear algebra. In that context generalizations of $\mathbf{C}^n$ and $\mathbf{C}_n$ are studied. Such generalizations, like $\mathbf{C}^n$ and $\mathbf{C}_n$ themselves, are called *vector spaces* and their subsets that are also vector spaces are called *subspaces*. A careful reading of the argument just presented shows that:

If a vector space V is the union of finitely many subspaces $\widetilde{W_j}$, i.e.,

$$V = \bigcup_{j=1}^{J} \widetilde{W_j},$$

then one of the $\widetilde{W_j}$ is actually all V.]

If $W_{i_0} = \mathbf{C}^n$ let m_A be $m_{A,\mathbf{x}_{i_0}}$ and let μ_A be $\mu_{A,\mathbf{x}_{i_0}}$. Then it follows that for every vector $\mathbf{x}$ in $\mathbf{C}^n$, $\mathbf{x}$ is in W_{i_0}, i.e.,

$$m_A(A)\mathbf{x} = \mathbf{O}^c.$$

The polynomial m_A has the form

$$m_A(z) \stackrel{\text{def}}{=} z^{\mu_A} + a_{\mu_A-1} z^{\mu_A-1} + \cdots + a_1 z + a_0$$

and for every vector $\mathbf{x}$

$$m_A(A)\mathbf{x} \stackrel{\text{def}}{=} (A^{\mu_A} + a_{\mu_A-1} A^{\mu_A-1} + \cdots + a_1 A + a_0 I)\mathbf{x} = \mathbf{O}^c,$$

i.e., $m_A(A)$ annihilates every vector $\mathbf{x}$ and the leading coefficient of m_A is 1. Since m_A is one of the $m_{A,\mathbf{x}_i}$ there is no polynomial p that is of smaller degree and that annihilates all $\mathbf{x}$. (Otherwise $m_{A,\mathbf{x}_i}$ is not of minimal degree for $\mathbf{x}_i$.)

There is only one such "all-annihilating" polynomial with leading coefficient 1 and of minimal degree. (If there is another, say p, then $p - m_A$ is also "all-annihilating." Since both p and m_A are of minimal degree, they are both of degree μ_A. Since both have leading coefficient 1, $p - m_A$ is of degree lower than μ_A and so, if $p - m_A \neq 0$ there emerges the contradiction: μ_A is not the minimal degree for "all-annihilating" polynomials.)

Since each $\mu_{A,\mathbf{x}_i} \leq n$ it follows that $\mu_A \leq n$.

The (unique) "all-annihilating" nonconstant polynomial for A, m_A, of minimal degree μ_A and with leading coefficient 1, is called the *minimal polynomial* of A.

Example 4.3.1. If m_I is the minimal polynomial for $n \times n$ identity then

$$m_I(z) = z - 1$$

---

since m_I is nonconstant, is of minimal degree (1), and, for any vector $\mathbf{x}$,

$$(I - 1I)\mathbf{x} = \mathbf{x} - \mathbf{x} = \mathbf{O}^c.$$

If m_A is the minimal polynomial for the matrix

$$A \stackrel{\text{def}}{=} \begin{pmatrix} 0 & 1 \\ 0 & 0 \end{pmatrix}$$

then $m_A(z) = z^2$.

Indeed, since $A^2 = O$ it follows that $\mu_A \leq 2$. For any number λ if $y \neq 0$ then

$$(A - \lambda I)\mathbf{x} \stackrel{\text{def}}{=} \begin{pmatrix} -\lambda & 1 \\ 0 & -\lambda \end{pmatrix} \begin{pmatrix} x \\ y \end{pmatrix} = \begin{pmatrix} -\lambda x + y \\ -\lambda y \end{pmatrix} \neq \mathbf{O}^c.$$

Hence $\mu_A > 1$, whence $\mu_A = 2$.

If m_B is the minimal polynomial for the matrix

$$B \stackrel{\text{def}}{=} \begin{pmatrix} 3 & 12 \\ 8 & 7 \end{pmatrix}$$

then

$$m_B(z) = z^2 - 10z - 75 = (z - 15)(z + 5). \tag{4.3.4}$$

In this instance the proof that m_B is as described in (4.3.4) is carried out according to the following outline:

   i. For no value of λ is $B - \lambda I$ "all-annihilating;"
  ii. Hence m_B must be a polynomial of the second degree;
iii. The eigenvalues of B are -5 resp. 15 and corresponding eigenvectors are

$$\mathbf{x}_1 \stackrel{\text{def}}{=} \begin{pmatrix} 3 \\ -2 \end{pmatrix} \text{ and } \mathbf{x}_2 \stackrel{\text{def}}{=} \begin{pmatrix} 1 \\ 1 \end{pmatrix}$$

which are linearly independent. If

$$\mathbf{v} \stackrel{\text{def}}{=} \begin{pmatrix} x \\ y \end{pmatrix}$$

is an arbitrary 2×1 vector then

$$\mathbf{v} = \frac{(x - y)}{5}\mathbf{x}_1 + \frac{(2x + 3y)}{5}\mathbf{x}_2.$$

Thus (see (4.3.4))

$$(A - 15I)(A + 5I)\mathbf{v}$$

$$= \frac{(x - y)}{5}(A - 15I)(A + 5I)\mathbf{x}_1$$

$$+ \frac{(2x + 3y)}{5}(A - 15I)(A + 5I)\mathbf{x}_2$$

$$= \frac{(x - y)}{5}(A - 15I)\mathbf{O}^c + \frac{(2x + 3y)}{5}(A + 5I)(A - 15I)\mathbf{x}_2$$

$$= \frac{(x - y)}{5}(A - 15I)\mathbf{O}^c + \frac{(2x + 3y)}{5}(A + 5I)\mathbf{O}^c$$

$$= \mathbf{O}^c$$

and so m_B is indeed the minimal polynomial for B.

In the examples above the determination of each minimal polynomial is carried out in a manner tailored to the matrix under consideration. It appears that a routine or procedure for finding the minimal polynomial of an arbitrary SQUARE matrix A can be useful. In the following lines just such a device is described.

AN ALGORITHM FOR THE MINIMAL POLYNOMIAL

i. Write each of the matrices I and A as a $1 \times n^2$ (row) vector in which the first n components are the entries in the first row of the matrix, the next n components are the entries in the second row of the matrix, ..., the last n components are the entries in the last row of the matrix. Thus if $\mathbf{I}$ and $\mathbf{A}$ are these $1 \times n^2$ vectors then,

$$
\begin{array}{lllll}
\mathbf{I} & = 1,0,\ldots,0, & 0,1,0,\ldots,0, & \ldots, & 0,0,\ldots 1 & \leftarrow (\sim I) \\
\mathbf{A} & = a_{11},\ldots,a_{1n}, & a_{21},\ldots,a_{2n}, & \ldots, & a_{n1},\ldots,a_{nn} & \leftarrow (\sim A),
\end{array}
$$

etc.

ii. Let $\mathcal{B}_2$ denote the $2 \times n^2$ matrix in which row 1 is $\mathbf{I}$ and row 2 is $\mathbf{A}$.

iii. Apply the GEM to the $2 \times n^2$ matrix $\mathcal{B}_2$. The result is $\mathcal{B}_{2E}$ and there is a 2×2 GEM matrix $\mathcal{E}_2$ such that

$$\mathcal{E}_2 \mathcal{B}_2 = \mathcal{B}_{2E}.$$

If the bottom row of $\mathcal{B}_{2E}$ consists of 0's, then for some number t_1

$$A + t_1 I = O$$

and $m_A(A) = A + t_1 I$.

iv. If the bottom row of $\mathcal{B}_{2E}$ does not consist of zeros, enlarge $\mathcal{B}_{2E}$ to a $3 \times n^2$ matrix $\mathcal{B}_3$ by appending the row $\mathbf{A}^2$. Apply the GEM to $\mathcal{B}_3$ so that for 3×3 GEM matrix $\mathcal{E}_3$

$$\mathcal{E}_3 \mathcal{B}_3 = \mathcal{B}_{3E}.$$

If the bottom row of $\mathcal{B}_{3E}$ consists of 0's then there are numbers s_1, s_2 such that

$$A^2 + s_1 A + s_2 I = O.$$

and $m_A(A) = A^2 + s_1 A + s_2 I$.

v. Continuing in this way find m_A in not more than n performances of the procedure described.

Following the next **Example** there is a discussion of a similar algorithm for finding $m_{A,\mathbf{x}}$ for a given $n \times n$ matrix A and a given $n \times 1$ vector $\mathbf{x}$.

Example 4.3.2. Let A be the matrix

$$\begin{pmatrix} 5 & -1 & 0 \\ 1 & 5 & 2 \\ -1 & 1 & 4 \end{pmatrix}.$$

Then

$$\mathcal{B}_2 = \begin{pmatrix} 1 & 0 & 0 & 0 & 1 & 0 & 0 & 0 & 1 & \leftarrow (\sim I) \\ 5 & -1 & 0 & 1 & 5 & 2 & -1 & 1 & 4 & \leftarrow (\sim A) \end{pmatrix}$$

$$E_{12}(t_1) = E_{12}(-5) = \begin{pmatrix} 1 & 0 \\ -5 & 0 \end{pmatrix}$$

$$\mathcal{B}_{2E} = E_{12}(-5)\mathcal{B}_2 = \begin{pmatrix} 1 & 0 & 0 & 0 & 1 & 0 & 0 & 0 & 1 & \leftarrow (\sim I) \\ 0 & -1 & 0 & 1 & 0 & 2 & -1 & 1 & -1 & \leftarrow (\sim A - 5I) \end{pmatrix}.$$

$$A^2 = \begin{pmatrix} 24 & -10 & -2 \\ 8 & 26 & 18 \\ -8 & 10 & 18 \end{pmatrix}$$

$$\mathcal{B}_3 = \begin{pmatrix} 1 & 0 & 0 & 0 & 1 & 0 & 0 & 0 & 1 & \leftarrow (I) \\ 0 & -1 & 0 & 1 & 0 & 2 & -1 & 1 & -1 & \leftarrow (A - 5I) \\ 24 & -10 & -2 & 8 & 26 & 18 & -8 & 10 & 18 & \leftarrow (A^2) \end{pmatrix}$$

$$E_{13}(-24) = \begin{pmatrix} 1 & 0 & 0 \\ 0 & 1 & 0 \\ -24 & 0 & 1 \end{pmatrix} \quad E_{23}(-10) = \begin{pmatrix} 1 & 0 & 0 \\ 0 & 1 & 0 \\ 0 & -10 & 1 \end{pmatrix}$$

$$\mathcal{B}_{3E} = E_{23}(-10)\,E_{13}(-24)\mathcal{B}_3$$

$$= \begin{pmatrix} 1 & 0 & 0 & 0 & 1 & 0 & 0 & 0 & 1 & & \leftarrow (I) \\ 0 & -1 & 0 & 1 & 0 & 2 & -1 & 1 & -1 & & \leftarrow (A - 5I) \\ 0 & 0 & -2 & -2 & 2 & -2 & 2 & 0 & 4 & \leftarrow & \begin{pmatrix} A^2 - 24I \\ -10(A - 5I) \end{pmatrix} \end{pmatrix}$$

$$A^3 = \begin{pmatrix} 112 & -76 & -28 \\ 48 & 140 & 124 \\ -48 & 76 & 92 \end{pmatrix}$$

$$\mathcal{B}_4 = \begin{pmatrix} 1 & 0 & 0 & 0 & 1 & 0 & 0 & 0 & 1 \\ 0 & -1 & 0 & 1 & 0 & 1 & -1 & 1 & -1 \\ 0 & 0 & -2 & -2 & 2 & -2 & 2 & 0 & 4 \\ 112 & -76 & -28 & 48 & 140 & 124 & -48 & 76 & 92 \end{pmatrix} \begin{matrix} \leftarrow (I) \\ \leftarrow (A-5I) \\ \leftarrow \left(\begin{smallmatrix} A^2 - 10A \\ +26I \end{smallmatrix} \right) \\ \leftarrow (A^3) \end{matrix}$$

$$E_{14}(-112) = \begin{pmatrix} 1 & 0 & 0 & 0 \\ 0 & 1 & 0 & 0 \\ 0 & 0 & 1 & 0 \\ -112 & 0 & 0 & 1 \end{pmatrix}$$

$$E_{24}(-76) = \begin{pmatrix} 1 & 0 & 0 & 0 \\ 0 & 1 & 0 & 0 \\ 0 & 0 & 1 & 0 \\ 0 & -76 & 0 & 1 \end{pmatrix}$$

$$E_{34}(-14) = \begin{pmatrix} 1 & 0 & 0 & 0 \\ 0 & 1 & 0 & 0 \\ 0 & 0 & 1 & 0 \\ 0 & 0 & -14 & 1 \end{pmatrix}$$

$$\mathcal{B}_{4E} = E_{34}(-14)E_{24}(-76)E_{14}(-112)\mathcal{B}_4$$

$$= \begin{pmatrix} 1 & 0 & 0 & 0 & 1 & 0 & 0 & 0 & 1 \\ 0 & -1 & 0 & 1 & 0 & 1 & -1 & 1 & -1 \\ 0 & 0 & -2 & -2 & 2 & -2 & 2 & 0 & 4 \\ 0 & 0 & 0 & 0 & 0 & 0 & 0 & 0 & 0 \end{pmatrix} \begin{matrix} \leftarrow (I) \\ \leftarrow (A-5I) \\ \leftarrow \left(\begin{smallmatrix} A^2 - 10A \\ +26I \end{smallmatrix} \right) \\ \leftarrow \left(\begin{smallmatrix} A^3 \\ -112I \\ -76(A-5I) \\ -14\left(\begin{smallmatrix} A^2 - 10A \\ +26I \end{smallmatrix} \right) \end{smallmatrix} \right) \end{matrix}$$

Consequently

$$m_A(A) = A^3 - 112I - 76(A-5I) - 14(A^2 - 10A + 26I))$$
$$= A^3 - 14A^2 + 64A - 96I.$$

The coefficients -112, -76, -14, -24, -10, and -5 above come respectively from the crucial entries in the corresponding elementary matrices

$$E_{14}(\mathbf{-112}), E_{24}(\mathbf{-76}), E_{34}(\mathbf{-14}), E_{13}(\mathbf{-24}), E_{23}(\mathbf{-10}), \text{ and } E_{12}(\mathbf{-5}).$$

     Chapter 4. Useful forms for matrices

[**Note 4.3.1:** The calculations stop as soon as a row of 0's is found in some $\mathcal{B}_{kE}$. Thus the matrices $A^2, A^3, \ldots$ are calculated only as they are needed.

At most the algorithm requires the calculation of A^n since the degree of m_A is never more than n. On the other hand if the degree μ_A of m_A is less than n then, for the first time, there appears a row of 0's in $\mathcal{B}_{\mu_A E}$. Hence only $A, A^2, \ldots, A^{\mu_A}$ need be calculated to carry out the steps of the algorithm to find m_A.

Just how many of $A, A^2, \ldots$ are to be calculated is not known when the algorithm is begun. Nevertheless, only the minimum number required of the $A, A^2, \ldots$ ever enter in the algorithm.

Since $dim(\mathbf{C}^{n^2}) = n^2$ the algorithm described above provides a *proof* that the minimal polynomial m_A exists. However, the algorithm does *not* show that the degree μ_A of m_A is not more than n, only that $\mu_A \leq n^2$.]

To calculate $m_{A,\mathbf{x}}$ use a similar algorithm.

i. Write $\mathbf{x}, A\mathbf{x}$ as $1 \times n$ row vectors. Form the $2 \times n$ matrix $\mathcal{B}_2(\mathbf{x})$ in which row 1 is $\mathbf{x}$ and row 2 is $A\mathbf{x}$.

ii. Apply the GEM to $\mathcal{B}_2(\mathbf{x})$ and produce $\mathcal{B}_{2E}(\mathbf{x})$. If the bottom row of $\mathcal{B}_{2E}(\mathbf{x})$ consists exclusively of 0's then for some number t_1

$$(A + t_1 I)\mathbf{x} = \mathbf{O}$$

and

$$m_{A,\mathbf{x}}(A) = A + t_1 I.$$

iii. If the bottom row of $\mathcal{E}_{2E}(\mathbf{x})$ has a nonzero entry, form the $3 \times n$ matrix $\mathcal{B}_3(\mathbf{x})$ by appending to $\mathcal{B}_{2E}(\mathbf{x})$ the $1 \times n$ vector $A^2\mathbf{x}$ written as a row vector. Apply the GEM to $\mathcal{B}_3(\mathbf{x})$ to produce $\mathcal{B}_{3E}(\mathbf{x})$. If the bottom row of $\mathcal{B}_{3E}(\mathbf{x})$ consists exclusively of 0's there are numbers s_1 and s_2 such that

$$m_{A,\mathbf{x}} = A^2 + s_1 A + s_2 I.$$

Continuing in this way, in not more than n steps, find $m_{A,\mathbf{x}}$.

Thus in about n^4 calculations there can be found a polynomial for which the zeros are eigenvalues of A. Using determinants to find χ_A and its zeros can entail about $n!$ calculations. If $n = 10$ then

$$\frac{n^4}{n!} = \frac{10^4}{10!} \approx 0.002756.$$

The advantage of alternatives to the use of determinants for finding eigenvalues is clear. The calculation of the minimal polynomial and the subsequent finding of its zeros (the eigenvalues) is an option, providing as well some insight into the diagonability of the matrix. In the application of linear algebra eigenvalues are calculated by approximative methods some of which are described in **Section 5.3.**

Exercise 4.3.1.

i. Use mathematical induction to describe precisely what "continuing in this way" means in the description above of the algorithm for finding the minimal polynomial of a SQUARE matrix.

ii. Describe in detail the algorithm for calculating $m_{A,\mathbf{x}}$.

iii. Show that the degree μ_A of m_A is the maximum among the degrees of the $m_{A,\mathbf{x}_i}$, $1 \le i \le l$, used to derive the existence of m_A.

iv. Show that the degree μ_A of the minimal polynomial m_A found by means of the algorithm is not more than n.

v. For the matrix A in **Example 4.3.2** and the vector

$$\mathbf{x} \stackrel{\text{def}}{=} \begin{pmatrix} 1 \\ 2 \\ 3 \end{pmatrix}$$

use the appropriate algorithm to find $m_{A,\mathbf{x}}$.

The next set of **Exercises** deals with the explicit calculation of the minimal polynomial for each kind of $n \times n$ elementary matrix. The results, together with THEOREM **4.3.2** below, show that E_{ij} is diagonable. At the same time they show that, in the important instances where $E_{ij}(r)$ is used, it is *not* diagonable. As a diagonal matrix $E_i(s)$ is automatically diagonable.

Exercise 4.3.2. Calculate

$$m_{E_i(6)}.$$

Exercise 4.3.3. Show that if $n > 1$

$$m_{E_{ij}}(z) = \begin{cases} z^2 - 1 & \text{if } i \neq j \\ z - 1 & \text{if } i = j. \end{cases}$$

Exercise 4.3.4. Show that if $n > 1$

$$m_{E_{ij}(r)}(z) = \begin{cases} (z-1)^2 & \text{if } i \neq j \text{ and } r \neq 0 \\ (z-1)(z-(r+1)) & \text{if } i = j \text{ and } r \neq 0 \\ z-1 & \text{if } r = 0. \end{cases}$$

[*Hint:* Write $E_{ij}(r)$ as $I + rU_{ij}$ (cf. **Exercise 1.1.15**).]

Exercise 4.3.5. Show that if $n > 1$

$$m_{E_i(r)}(z) = \begin{cases} (z-1)(z-r) & \text{if } r \neq 1 \\ z-1 & \text{if } r = 1. \end{cases}$$

[*Hint:* Write $E_i(r)$ as $I + (r-1)U_{ii}$ (cf. **Exercise 1.1.15**).]

Exercise 4.3.6. Show that if p is a polynomial such that $p(A) = O$, i.e., $p(A)$ is "all annihilating," then for some polynomial q: $p = qm_A$.

Let A be an $n \times n$ matrix and let m_A be its minimal polynomial (of degree μ_A). If $m > \mu_A$ there is a polynomial r_m of degree less than μ_A and a polynomial Q_m such that

$$z^m = m_A(z)Q_m(z) + r_m(z).$$

Hence

$$A^m = m_A(A)Q_m(A) + r_m(A) = OQ_m(A) + r_m(A) = r_m(A).$$

It follows that for *every* polynomial p there is a polynomial p_m *of degree less than* μ_A and such that

$$p(A) = p_m(A).$$

Thus calculation of polynomial functions of A is never more difficult than the calculation of $A^2, \ldots, A^{\mu_A - 1}$. A more detailed study shows that for "well-behaved" functions f the calculation of $f(A)$ is no more complicated than the calculation of $A^2, \ldots, A^{\mu_A - 1}$ (cf. **Section 4.7**).

Exercise 4.3.7. Let N be an $n \times n$ matrix such that $N^k = O \neq N^{k-1}$. Show that $m_N(z) = z^k$ and that the only eigenvalue of N is 0.

[*Hint:* If $m_N(z) = z^{\mu_N} + \cdots$ then $\mu_N \leq k$. Derive a contradiction if $\mu_N < k$.]

THEOREM **4.3.1.** IF A IS AN $n \times n$ MATRIX THEN λ IS AN EIGENVALUE OF A IFF $m_A(\lambda) = 0$.

PROOF. If $m_A(\lambda) = 0$ then $z - \lambda$ is a factor of $m_A(z)$. Hence for some polynomial q

$$m_A(A) = (A - \lambda I)q(A).$$

Since the degree of q is less than the degree of m_A, for some vector $\mathbf{x}$

$$\mathbf{y} \stackrel{\mathrm{def}}{=} q(A)\mathbf{x} \neq \mathbf{O}^c.$$

Since m_A is "all-annihilating,"

$$m_A(A)\mathbf{x} = (A - \lambda I)q(A)\mathbf{x} = (A - \lambda I)\mathbf{y} = \mathbf{O}^c,$$

i.e., λ is an eigenvalue and the nonzero vector $\mathbf{y}$ is an eigenvector corresponding to the eigenvalue λ.

Conversely, if λ is an eigenvalue of A and if $\mathbf{x}$ is a corresponding (nonzero) eigenvector then $A\mathbf{x} = \lambda\mathbf{x}$ and more generally $A^k\mathbf{x} = \lambda^k\mathbf{x}$ whence, since m_A is "all-annihilating,"

$$m_A(A)\mathbf{x} = m_A(\lambda)\mathbf{x} = \mathbf{O}^c.$$

Since $\mathbf{x} \neq \mathbf{O}^c$ it follows that $m_A(\lambda) = 0$.

$$\Omega$$

Exercise 4.3.8. Show that if A is an $n \times n$ matrix having n different eigenvalues then $m_A = (-1)^n \chi_A$. (This is one of a number of situations in which the validity of the Cayley-Hamilton theorem (cf. **Section 4.2**) is rather easily demonstrated.)

[**Remark 4.3.2:** In **Section 4.2** it is shown that the characteristic polynomial χ_A and the set of eigenvalues of A are similarity invariants. The minimal polynomial m_A is also a similarity invariant.

Indeed, if P is invertible and $B \stackrel{\mathrm{def}}{=} P^{-1}AP$ then

$$m_A(B) = P^{-1}m_A(A)P = O$$
$$m_B(A) = Pm_B(B)P^{-1} = O.$$

Hence $m_B(A) = O = m_A(B)$ and according to **Exercise 4.3.6** each of the *polynomials* m_A and m_B is a factor of the other. Hence their degrees are the same and so one is a constant multiple of the other. Since their leading coefficients are 1's it follows that $m_A = m_B$.

In particular, since $\chi_A(A) = O$ (cf. **Section 4.2**) it follows that m_A is a factor of χ_A. Since the only zeros of m_A are eigenvalues and the only zeros of χ_A are eigenvalues it follows that if

$$m_A(z) = (z - \lambda_1)^{r_1} \cdots (z - \lambda_k)^{r_k} \tag{4.3.5}$$

then for some constant c and natural numbers $s_1, \ldots, s_k$

$$\chi_A(z) = c(z - \lambda_1)^{s_1} \cdots (z - \lambda_k)^{s_k}. \tag{4.3.6}$$

     Chapter 4. Useful forms for matrices

Furthermore, because m_A is a factor of χ_A,

$$r_1 \leq s_1, \ldots, r_k \leq s_k, \qquad\qquad (4.3.7)$$

and since

$$\chi_A(z) = (-1)^n z^n + \cdots$$

it follows that $c = (-1)^n$.]

Exercise 4.3.9. Show, without reference to the general Cayley-Hamilton theorem, that if the SQUARE matrix A is diagonable then

$$\chi_A(A) = O.$$

The theoretical value of the minimal polynomial m_A is highlighted in THEOREM **4.3.2** below. It deals with the notion of *simple zeros* of a polynomial.
If

$$p(z) = t_m z^m + t_{m-1} z^{m-1} + \cdots + t_1 z + t_0, \ t_m \neq 0,$$

then p can be factored uniquely:

$$p(z) = t_m (z - \lambda_1)^{r_1} (z - \lambda_2)^{r_2} \cdots (z - \lambda_k)^{r_k}.$$

The numbers $\lambda_1, \ldots \lambda_k$ are the zeros of p, the numbers $r_1, \ldots, r_k$ are their *multiplicities*, and $\sum_{i=1}^{k} r_i = m$. If $r_i = 1$, $1 \leq i \leq k$, the zeros are called *simple*.

THEOREM **4.3.2.** AN $n \times n$ MATRIX A IS DIAGONABLE IFF ALL THE ZEROS OF ITS MINIMAL POLYNOMIAL m_A ARE SIMPLE.

PROOF. If A is diagonable then for some invertible matrix P

$$P^{-1}AP \overset{\text{def}}{=} D \overset{\text{def}}{=} \begin{pmatrix} d_1 & & & \\ & d_2 & & \\ & & \ddots & \\ & & & d_n \end{pmatrix}$$

is a diagonal matrix. Assume that among the diagonal entries $d_1, \ldots, d_n$ there are precisely k different ones. In other words there are k numbers $\lambda_1, \ldots, \lambda_k$, no two the same, and such that each d_i is some λ_j.

Hence it may be assumed that there are identity matrices I_i of various sizes such that D is the block matrix

$$D = \begin{pmatrix} \lambda_1 I_1 & \cdots & O \\ \vdots & \ddots & \vdots \\ O & \cdots & \lambda_k I_k \end{pmatrix}.$$

Then $(D - \lambda_1 I)\cdots(D - \lambda_k I) = O$ whence, since $(D - \lambda_i I) = P^{-1}(A - \lambda_i I)P$, it follows that $P^{-1}(A - \lambda_1 I)\cdots(A - \lambda_k I)P = O$, and finally

$$p(A) \stackrel{\text{def}}{=} (A - \lambda_1 I)\cdots(A - \lambda_k I) = O.$$

Since p is "all-annihilating" and since all the different eigenvalues of A are the zeros of p it follows that $p = m_A$ and that the zeros of m_A are simple.

For ease in understanding, the proof of the converse is motivated by the following considerations.

If A is indeed diagonable and for some invertible matrix P

$$P^{-1}AP = D \stackrel{\text{def}}{=} \begin{pmatrix} \lambda_1 I_1 & \cdots & O \\ \vdots & \ddots & \vdots \\ O & \cdots & \lambda_k I_k \end{pmatrix}$$

and if

$$\pi_i(\lambda) \stackrel{\text{def}}{=} \prod_{j \neq i}(\lambda - \lambda_j)$$

then

$$\frac{1}{\pi_i(\lambda_i)}\pi_i(D) = \begin{pmatrix} 0I_1 & & & & \\ & \ddots & & & \\ & & I_i & & \\ & & & \ddots & \\ & & & & 0I_k \end{pmatrix} \stackrel{\text{def}}{=} E_i.$$

Furthermore direct calculation shows that

$$E_1 + \cdots + E_k = I$$
$$E_i E_{i'} = \begin{cases} E_i & \text{if } i = i', \\ O & \text{if } i \neq i'. \end{cases} \tag{4.3.8}$$

Hence if

$$F_i \stackrel{\text{def}}{=} PE_i P^{-1} = \frac{1}{\pi_i(\lambda_i)}\pi_i(A)$$

then

$$F_1 + \cdots + F_k = I \tag{4.3.9}$$

$$F_i F_{i'} = \begin{cases} F_i & \text{if } i = i', \\ O & \text{if } i \neq i'. \end{cases} \tag{4.3.10}$$

Equations (4.3.9) and (4.3.10) are quite useful as the next developments show.

Let $im(F_i)$ be denoted K_i.

   Chapter 4. Useful forms for matrices

i. If $\mathbf{x} \in K_i$ then $F_i\mathbf{x} = \mathbf{x}$. [*Proof.* There is a $\mathbf{y}$ such that $\mathbf{x} = F_i\mathbf{y}$; $F_i\mathbf{x} = F_i^2\mathbf{y} = F_i\mathbf{y} = \mathbf{x}$.]

ii. If $\mathbf{x}_i \in K_i$ and

$$\sum_{i=1}^{k} \mathbf{x}_i = \mathbf{O}$$

then each $\mathbf{x}_i = \mathbf{O}$. [*Proof.*

$$\sum_{i=1}^{k} \mathbf{x}_i = \sum_{i=1}^{k} F_i\mathbf{x}_i = \mathbf{O}$$

$$\sum_{i=1}^{k} F_j F_i \mathbf{x}_i = F_j\mathbf{x}_j = \mathbf{x}_j = \mathbf{O}.]$$

iii. $K_i \subset \ker(A - \lambda_i I)$. [*Proof.* If $\mathbf{x} \in K_i$ then

$$(A - \lambda_i I)\mathbf{x} = (A - \lambda_i I)F_i\mathbf{x} = m_A(A)\mathbf{x} = \mathbf{O}.]$$

iv. $K_i \supset \ker(A - \lambda_i I)$. [*Proof.* If $(A - \lambda_i I)\mathbf{x} = \mathbf{O}$ then

$$\mathbf{O} = \sum_{j=1}^{k}(A - \lambda_i I)F_j\mathbf{x} = \sum_{j \neq i} F_j(A - \lambda_i I)\mathbf{x}.$$

Since $F_j(A - \lambda_i I)\mathbf{x} \in K_j$ and $F_j\mathbf{x} \in K_j$ it follows that if if $j \neq i$ and $F_j\mathbf{x} \neq \mathbf{O}$ then $F_j\mathbf{x}$ is an eigenvector for two different eigenvalues, λ_i and λ_j, an impossibility. Hence if $j \neq i$ then $F_j\mathbf{x} = \mathbf{O}$ and so $\mathbf{x} \in K_i$.]

v. $K_i = \ker(A - \lambda_i I)$ [*Proof.* See *iii* and *iv.*]

vi. If $i \neq j$ then $K_i \cap K_j = \{\mathbf{O}\}$. [*Proof.* If $\mathbf{x} \in K_i \cap K_j$ then $\mathbf{x} = F_i\mathbf{x} = F_j\mathbf{x} = F_j F_i\mathbf{x} = O\mathbf{x} = \mathbf{O}.]$

Now let $X_j \stackrel{\text{def}}{=} \{\mathbf{x}_{j1} \ldots, \mathbf{x}_{jd_j}\}$ be a basis for K_j. It is shown next that

$$X \stackrel{\text{def}}{=} \bigcup_{j=1}^{k} X_j$$

is a basis for $\mathbf{C}^n$.

Indeed, if $\mathbf{x} \in \mathbf{C}^n$ then

$$\mathbf{x} = \sum_{i=1}^{k} F_i\mathbf{x}$$

$$= \sum_{i=1}^{k}(\sum_{j=1}^{d_i} a_{ij}\mathbf{x}_{ij})$$

---

and so $\mathbf{C}^n \subset span(X) \subset \mathbf{C}^n$, i.e., $\mathbf{C}^n = span(X)$.

If

$$\sum_{i=1}^{k}(\sum_{j=1}^{d_i} b_{ij}\mathbf{x}_{ij}) = \mathbf{O}$$

then

$$\sum_{i=1}^{k} F_p(\sum_{j=1}^{d_i} b_{ij}\mathbf{x}_{ij}) = \sum_{j=1}^{d_p} b_{ip}\mathbf{x}_{ij} = \mathbf{O}$$

and so for each p and each i, $b_{ip} = 0$: X is linearly independent, X is basis for $\mathbf{C}^n$.

Since each vector in X is a nonzero vector in some K_i it follows that each vector in X is an eigenvector of A: X is an *eigenbasis* for $\mathbf{C}^n$. In particular, $d_1 + \cdots + d_k = n$ and

$$d_i = dim(K_i) = dim[ker(A - \lambda_i I)].$$

If P is the matrix in which

columns $1, \ldots, d_1$ are the vectors $\mathbf{x}_{11}, \ldots, \mathbf{x}_{1d_1}$,

$$\vdots$$

columns $d_1 + \cdots + d_{k-1} + 1, \ldots, n$ are the vectors $\mathbf{x}_{k1}, \ldots, \mathbf{x}_{kd_k}$,

then $P^{-1}AP \stackrel{\text{def}}{=} D$ is a block diagonal matrix in which block j is the $d_j \times d_j$ diagonal matrix $\lambda_j I$.

$$\Omega$$

Since $\chi_D = \chi_A$ it follows that if

$$\chi_A(z) \stackrel{\text{def}}{=} \prod_{i=1}^{k}(z - \lambda_i I)^{s_i}$$

then the diagonability of A implies $s_i = d_i$.

[**Remark 4.3.3:** The number d_i as the dimension of the subspace $ker(A - \lambda_i I)$ is regarded as a *geometric multiplicity* associated with the matrix A. The number s_i as the multiplicity of the zero λ_i of the purely algebraic object χ_A is regarded as an *algebraic multiplicity* associated with the matrix A.]

The conclusion following the proof of THEOREM **4.3.2** can be strengthened as follows.

THEOREM **4.3.3**. THE SQUARE MATRIX A IS DIAGONABLE IFF EACH GEOMETRIC MULTIPLICITY OF AN EIGENVALUE IS THE SAME AS THE ALGEBRAIC MULTIPLICITY OF THE SAME EIGENVALUE.

   Chapter 4. Useful forms for matrices

PROOF. If A is diagonable then, as noted above, the geometric and algebraic multiplicities of each eigenvalue are equal.

Conversely if $d_i = s_i$, $1 \le i \le k$, then the subspaces $ker(A - \lambda_i I)$ have bases X_i. Since $s_1 + \cdots + s_k = n$ it follows that

$$X \overset{\text{def}}{=} \bigcup_{i=1}^{k} X_i$$

consists of n different vectors. The linear independence of eigenvectors corresponding to different eigenvalues then shows that X consists of n linearly independent vectors and so X is an eigenbasis for A: A is diagonable.

$$\Omega$$

Exercise 4.3.10. Let the geometric multiplicity of the eigenvalue λ_i of the matrix A be d_i. Modify the proof of THEOREM **4.2.3** to prove there is an invertible matrix P such that $P^{-1}AP$ has the form of a block matrix in which one block is the $d_i \times d_i$ matrix $\lambda_i I$, one block U is upper triangular, one block is O and one block B is rectangular:

$$P^{-1}AP = \begin{pmatrix} \lambda_i I & B \\ O & U \end{pmatrix}.$$

Use the conclusion to prove that the geometric multiplicity of λ_i does not exceed the algebraic multiplicity: $d_i \le s_i$.

[**Note 4.3.2:** Which of the two criteria THEOREM **4.3.2** and THEOREM **4.3.3** is easier to apply?]

The use of the first criterion involves about n^4 arithmetical operations to find m_A and then the work required to find whether m_A has multiple zeros: m_A has multiple zeros iff the derivative m'_A of m_A and m_A have a nonconstant common factor.

There is no need to find the eigenvalues themselves. The *Euclidean algorithm:*

$$m_A = q_1 m'_A + r_1, \ deg(r_1) < deg(m'_A)$$
$$m'_A = q_2 r_1 + r_2, \ deg(r_2) < deg(r_1)$$
$$r_1 = q_3 r_2 + r_3, \ deg(r_3) < deg(r_2)$$
$$\vdots$$
$$r_{d-2} = q_d r_{d-1} + r_d, \ deg(r_d) < deg(r_{d-1})$$
$$r_{d-1} = q_{d+1} r_d + 0$$
$$r_{d-2} = (q_d q_{d+1} + 1) r_d$$
$$\vdots$$

(for some polynomial function R of d variables)

$$m'_A = R(q_2, \ldots, q_{d+1}) r_d$$

(and for some polynomial function S of $d + 1$ variables)

$$m_A = S(q_1, \ldots, q_{d+1}) r_d$$

yields the *greatest common divisor* (GCD) r_d of m_A and m'_A. The polynomial m_A has only simple zeros iff r_d is a constant.

The use of the second criterion involves the calculation of χ_A, a daunting task in itself, and then its factorization, even more work.

Assume

$$p(x) = \sum_{i=0}^{n} a_i x^i, \ a_n \neq 0$$
$$q(x) = \sum_{j=0}^{m} b_j x^j, \ b_m \neq 0.$$

The polynomials p and q have a nonconstant common divisor iff there are polynomials r and s not both 0 and such that

$$r(x) = \sum_{\alpha=0}^{m-1} c_\alpha x^\alpha$$
$$s(x) = \sum_{\beta=0}^{n-1} d_\beta x^\beta$$
$$rp = sq.$$

The equation $rp = sq$ is equivalent to an homogeneous system of linear equations that comes about from identifying the coefficients of like powers of x in the two members of $rp = sq$. This system consists of $m + n$ equations in the $m + n$ unknowns $c_0, \ldots, c_{m-1}, d_0, \ldots, d_{n-1}$.

Let $\mathcal{A}$ be the $n \times m + n$ matrix

$$\begin{pmatrix} a_n & a_{n-1} & \cdots & a_0 & & & \\ & a_n & a_{n-1} & \cdots & a_0 & & \\ \cdots & \cdots & \cdots & \cdots & \cdots & \cdots & \cdots \\ & & & a_n & a_{n-1} & \cdots & a_0 \end{pmatrix}$$

and let $\mathcal{B}$ be the $m \times m + n$ matrix

$$\begin{pmatrix} b_m & b_{m-1} & \cdots & b_0 & & & \\ & b_m & b_{m-1} & \cdots & b_0 & & \\ \cdots & \cdots & \cdots & \cdots & \cdots & \cdots & \cdots \\ & & & b_m & b_{m-1} & \cdots & b_0 \end{pmatrix}.$$

The matrix $\mathcal{R}$ of the system may be written in the block matrix form:

$$\begin{pmatrix} \mathcal{A} \\ \mathcal{B} \end{pmatrix}$$

The singularity of $\mathcal{R}$, the *resultant* of p and q, is equivalent to the existence of a nonzero solution for the system and hence is equivalent to the existence of a nonconstant divisor of p and q.

Applied to the polynomials m_A and m'_A the criterion just given may be given as the singularity of the $2\mu_A - 1 \times 2\mu_A - 1$ matrix

$$\begin{pmatrix} 1 & & a_{\mu_A-1} & \cdots & a_0 & & & \\ & 1 & & a_{\mu_A-1} & \cdots & a_0 & & \\ \cdots & & \cdots & & \cdots & & & \\ & & & & 1 & a_{\mu_A-1} & \cdots & a_0 \\ (\mu_A-1)a_{\mu_A-1} & & & \cdots & a_1 & & & \\ & (\mu_A-1)a_{\mu_A-1} & & & \cdots & a_1 & & \\ \cdots & & \cdots & & \cdots & & \cdots & \cdots \\ & & & (\mu_A-1)a_{\mu_A-1} & & & \cdots & a_1 \end{pmatrix},$$

the *resultant* associated with A.

Exercise 4.3.11.

i. Show that if $r \neq 0$ and $i \neq j$ then $E_{ij}(r)$ is *not* diagonable. What are its eigenvalues?

ii. Show that the matrix

$$\Pi \overset{\mathrm{def}}{=} \begin{pmatrix} 0 & 1 & 0 \\ 0 & 0 & 1 \\ 1 & 0 & 0 \end{pmatrix},$$

a permutation matrix, is diagonable.

[*Hint:* Show $\Pi^3 = I$. Then note that the zeros of $z^3 - 1$ are simple.]

iii. Let Π be *any* permutation matrix (the product of elementary permutation matrices E_{ij}).

    a. Show that 1 is always an eigenvalue of Π.

    b. Describe the eigenvalues of Π by writing each in the form

$$re^{i\theta}, \ 0 \le r, \ 0 \le \theta < 2\pi.$$

[*Hint:* Show that for some natural number k, $\Pi^k = I$.]

    c. Show Π is diagonable.

    d. For each eigenvalue of Π find a maximal linearly independent set of eigenvectors.

    e. Find a matrix P such that $P^*P = I$ and such that $P^*\Pi P$ is a diagonal matrix (cf. **Section 4.4**).

[**Remark 4.3.4:**   If $p_i \overset{\text{def}}{=} \frac{\pi_i}{\pi_i(\lambda_i)}$ then

$$p_i(\lambda_{i'}) = \delta_{ii'} \overset{\text{def}}{=} \begin{cases} 1 & \text{if } i = i' \\ 0 & \text{if } i \ne i'. \end{cases}$$

Hence if

$$q(z) \overset{\text{def}}{=} p_1(z) + \cdots + p_k(z) - 1$$

then $deg(q) \le \mu_A - 1 = k - 1$ and yet

$$q(\lambda_i) = 0, \ 1 \le i \le k.$$

Hence as a polynomial of degree not more than $k - 1$ and having at least k zeros $q \equiv 0$ (cf. (4.3.9)):

$$p_1(A) + \cdots + p_k(A) = F_1 + \cdots + F_k = I.$$

Furthermore if $i \ne i'$ then m_A is a factor of $p_i p_{i'}$ whence

$$p_i(A)p_{i'}(A) = F_i F_{i'} = \begin{cases} F_i & \text{if } i = i' \\ O & \text{if } i \ne i' \end{cases}$$

(cf. (4.3.10)).

The equation $p_1(z) + \cdots + p_k(z) \equiv 1$ is related to the decomposition of $\frac{1}{m_A(z)}$ into partial fractions, a device used in the study of integral calculus, for finding

$$\int \frac{1}{m_A(z)} dz.$$

If

$$a_i \stackrel{\text{def}}{=} \frac{1}{\pi_i(\lambda_i)}$$

then

$$\sum_{i=1}^{k} \frac{a_i}{(z - \lambda_i)} = \frac{\sum_{i=1}^{k} a_i \pi_i(z)}{m_A(z)} = \frac{1}{m_A(z)},$$

the partial fraction decomposition of $\frac{1}{m_A(z)}$. Hence, comparison of numerators leads to

$$\sum_{i=1}^{k} a_i \pi_i(z) \equiv 1$$

and if $p_i(z) \stackrel{\text{def}}{=} a_i \pi_i(z)$ then

$$\sum_{i=1}^{k} \pi_i(z) \equiv 1.]$$

A SQUARE matrix M such that $M^2 = M$ is called an *idempotent* matrix. Thus each F_i is an idempotent matrix.

> [The origin of *idem* is the Latin — *idem* — same. The Indo-European root is — *i* — a pronominal stem that appears in *i*lk, *identity*, *iterate*, etc.]

Exercise 4.3.12. Show that if A is a SQUARE matrix and p and q are polynomials then $p(A)q(A) = q(A)p(A)$.

Exercise 4.3.13. Show that if M is idempotent, $M \neq I$, and $M \neq O$, then:

i.

$$m_M(z) = z^2 - z;$$

ii. M is diagonable;

iii. M is not invertible.

Exercise 4.3.14. Let $F_1, \ldots, F_k$ be as in (4.3.10) and let $t_i, 1 \leq i \leq k$, be numbers. Let C be the matrix

$$\sum_{i=1}^{k} t_i F_i.$$

Show that if g is a polynomial then

$$g(C) = \sum_{i=1}^{k} g(t_i) F_i$$

$$g(A) = \sum_{i=1}^{k} g(\lambda_i) F_i. \qquad (4.3.11)$$

[**Remark 4.3.5:** Equation (4.3.11) is the start of a *functional calculus* for matrices. It will be developed further in **Sections 4.4 – 4.7**.]

Exercise 4.3.15. As in **Exercise 1.1.15** let U_{ij} represent the $n \times n$ matrix unit in which the ij entry is 1 and all other entries are 0's. Show that:

i. if no two summands are equal then the sum

$$\sum_{k=1}^{m} U_{i_k i_k}$$

is an idempotent matrix;

ii. if M is an idempotent matrix and P is an invertible matrix then $P^{-1}MP$ is also an idempotent matrix;

iii. every diagonable SQUARE matrix is a linear combination of idempotent matrices.

Exercise 4.3.16. Assume M is a SQUARE matrix and that:

EITHER

i. $1 < k \in \mathbf{N}$ and $M^k = M$;
OR

ii. $k \in \mathbf{N}$, D is a diagonable matrix, and $M^k = D$.

Show M is diagonable.

[*Hint:* For *ii* show that there is a polynomial q such that the minimal polynomials m_D and m_M satisfy: $m_D(z^k) = q m_M(z)$.]

Exercise 4.3.17. Assume A and B are $n \times n$ matrices such that $AB = BA$ and such that both A and B are diagonable.

i. Let F_i resp. G_j be polynomial functions of the matrix A resp. B and assume

---

that

$$\sum_{i=1}^{k} F_i = I$$

$$F_i F_{i'} = \begin{cases} F_i & \text{if } i = i'; \\ O & \text{if } i \neq i'; \end{cases}$$

$$\sum_{j=1}^{m} G_j = I$$

$$G_j G_{j'} = \begin{cases} G_j & \text{if } j = j'; \\ O & \text{if } j \neq j'. \end{cases}$$

Show that for all i and j: $F_i G_j = G_j F_i$ and that

$$\sum_{i,j} F_i G_j = I$$

$$F_i G_j F_{i'} G_{j'} = \begin{cases} F_i G_j & \text{if } i = i' \text{ and } j = j'; \\ O & \text{otherwise.} \end{cases}$$

ii. Let W_{ij} be $im(F_i G_j)$. Show that if $\mathbf{x} \in \mathbf{C}^n$ then there is in each W_{ij} a unique $\mathbf{x}_{ij}$ such that

$$\mathbf{x} = \sum_{i,j} \mathbf{x}_{ij}.$$

iii. Let m_A and m_B have the unique factorizations

$$m_A(z) = \prod_{i=1}^{k} (z - \lambda_i)^{s_i}$$

$$m_B(z) = \prod_{j=1}^{m} (z - \mu_j)^{t_j}.$$

Show $s_i = t_j = 1$, $1 \leq i \leq k$, $1 \leq j \leq m$. Assume

$$\pi_i(z) = \prod_{i' \neq i} (z - \lambda_{i'})$$

$$\phi_j(z) = \prod_{j' \neq j} (z - \mu_{j'})$$

$$\sum_{i=1}^{k} a_i \pi_i = 1$$

$$\sum_{j=i}^{m} b_j \phi_j = 1$$

$$p_i = a_i \pi_i$$

$$q_j = b_j \phi_j$$

$$F_i = p_i(A) \tag{4.3.12}$$

$$G_j = q_j(B). \tag{4.3.13}$$

Show that the F_i and G_j in (4.3.12) and (4.3.13) behave according to the pattern in i.

$iv.$ Show that every nonzero vector in $im(F_iG_j)$ is an eigenvector of *both* A and B.

$v.$ Show that if X_{ij} is a basis for $im(F_iG_j)$ then

$$X \overset{\text{def}}{=} \bigcup_{i,j} X_{ij}$$

is an eigenbasis (a basis consisting of eigenvectors) for both A and B.

$vi.$ Show that there is an invertible matrix P such that whenever H is a polynomial function of two variables:

$$H(x,y) \overset{\text{def}}{=} \sum_{i,j} c_{ij} x^i y^j$$

then the *matrix* $P^{-1}H(A,B)P$ is a diagonal matrix. (In particular, $P^{-1}AP$, $P^{-1}BP$, and $P^{-1}(A-B)P$ are diagonal matrices.)

$vii.$ Repeat the analysis above for a finite set $\{A_1, \ldots, A_r\}$ of commuting diagonable $n \times n$ matrices and show thereby that they are simultaneously diagonable.

[**Note 4.3.3:** In iv and v above $im(F_iG_j)$ might well be $\{O\}$ whence, e.g., there is no basis in $im(F_iG_j)$. Similarly comments apply to $vii.$]

Exercise 4.3.18. Assume that m_A has at least two different zeros:

$$m_A(z) = \prod_{j=1}^{k}(z - r_j)^{s_j}, \ k > 1,$$

$$i \neq j \Rightarrow r_i \neq r_j.$$

For each i define π_i to be the polynomial such that

$$\pi_i(z) = \prod_{j \neq i}(z - r_j)^{s_j}.$$

Here are steps leading to the derivation of the existence of *polynomials* q_i such that for all z

$$\sum_{i=1}^{k} q_i(z)\pi_i(z) = 1$$

and

$$\sum_{i=1}^{k} q_i(A)\pi_i(A) = I.$$

  Chapter 4. Useful forms for matrices

Let $\mathcal{I}$ be the set of all polynomials p that are linear combinations *with polynomial coefficients* q_i of the polynomials π_i:

$$\mathcal{I} \stackrel{\text{def}}{=} \left\{ p \; : \; p(z) = \sum_{i=1}^{k} q_i(z)\pi_i(z) \right\}.$$

In $\mathcal{I}$ there are polynomials of least degree and among those of least degree there are those with leading coefficient 1.

- *i.* Show that if p_1 and p_2 are in $\mathcal{I}$ and if q_1 and q_2 are any polynomials then $q_1 p_1 + q_2 p_2$ is also in $\mathcal{I}$.
- *ii.* Show that there is in $\mathcal{I}$ precisely one polynomial δ of least degree and with leading coefficient 1.
- *iii.* Let δ be the unique polynomial described in *ii* and let p be any polynomial in $\mathcal{I}$. Show, by considering the "long division" equation

$$p = \delta q + r, \; deg(r) < deg(\delta)$$

  that $r \in \mathcal{I}$, hence that $r = 0$, and thus that δ is a factor of p. In particular, for each i there is a polynomial q_i such that

$$\pi_i = \delta q_i.$$

- *iv.* Use *iii* to show that if $\delta(\lambda) = 0$ then

$$\pi_i(\lambda) = 0, \; 1 \leq i \leq k.$$

  Show that, on the other hand, the π_i have no zero in common, hence that δ is a constant, i.e., $\delta = 1$.
- *v.* Use *iv* to show that there are polynomials q_i such that for all z

$$1 = \sum_{i=1}^{k} q_i(z)\pi_i(z)$$

$$I = \sum_{i=1}^{k} q_i(A)\pi_i(A).$$

  In analogy with the pattern in **THEOREM 4.3.2** let p_i be $q_i \pi_i$ so that

$$I = \sum_{i=1}^{k} p_i(A).$$

- *vi.* Now assume that $s_i = 1$, $1 \leq i \leq k$. If q_i is not a constant then there is a constant a_i such that for some polynomial Q_i

$$q_i(z) = Q_i(z)(z - r_i) + a_i.$$

Then

$$q_i(z)\pi_i(z) = Q_i(z)m_A(z) + a_i\pi_i(z)$$
$$q_i(A)\pi_i(A) = a_i\pi_i(A)$$

$$I = \sum_{i=1}^{k} q_i(A)\pi_i(A)$$
$$= \sum_{i=1}^{k} a_i\pi_i(A)$$

and so if each s_i is 1 I is a linear combination (with *constant coefficients*) of the polynomials π_i.

Exercise 4.3.19. Refer to **Exercise 4.3.18** for notation. In the general case, where no restriction is placed on the s_i, let p_i be the polynomial $q_i\pi_i$ and let F_i be the *matrix* $p_i(A)$.

i. Show that

$$I = \sum_{i=1}^{k} F_i$$
$$F_i F_{i'} = \begin{cases} F_i, & \text{if } i = i'; \\ O & \text{if } i \neq i'. \end{cases}$$

ii. Let W_i be $im(F_i)$. Show that $AW_i \subset W_i$, i.e., that each W_i is an invariant subspace, and that if $\mathbf{x} \in W_i$ then $(A - r_iI)^{s_i}\mathbf{x} = \mathbf{O}$. Show also that there is in W_i a vector $\mathbf{y}$ such that $(A - r_iI)^{s_i-1}\mathbf{y} \neq \mathbf{O}$. ("Restricted to W_i the matrix $A - r_iI$ is *nilpotent of order s_i*." For further discussion of "nilpotence" see **Section 4.5**.)

iii. Show that if $i \neq i'$ then $W_i \cap W_{i'} = \{\mathbf{O}\}$.

iv. Let X_i be a basis for W_i. Use i to show that

$$X \overset{\text{def}}{=} \bigcup_{i=1}^{k} X_i$$

is a basis for $\mathbf{C}^n$. Show that

$$Z_i \overset{\text{def}}{=} \bigcup_{j \neq i} X_j$$

is a basis for $ker(F_i)$.

v. Let Y_{ij} be a basis for $ker[(A - r_iI)^j F_i] \cap W_i$. Show that $Z_i \cup Y_{ij}$ is a basis for $ker[(A - r_iI)^j F_i]$.

Exercise 4.3.20. Show that the q_i in **Exercise 4.3.18** can always be chosen so that

$$deg(q_i) < s_i.$$

Example 4.3.3. Assume

$$m_A(z) = (z-1)(z-2).$$

Then for any a and all z

$$(az + 1 - 2a)(z-1) + (-az + a - 1)(z-2) = 1.$$

and if $a = 0$ then

$$1(z-1) + (-1)(z-2) = 1.$$

If, for all z, and constants b and c

$$b(z-1) + c(z-2) = 1 \qquad\qquad (4.3.14)$$

then setting z to be 1 shows $c = -1$ and setting z to be 2 shows $b = 1$. In other words, even though there are many sets of *polynomials* $\{q_1, q_2\}$ such that for all z

$$q_1(z)(z-1) + q_2(z)(z-2) = 1$$

there is only one set of *constants* for which (4.3.14) is valid for all z.

Exercise 4.3.21. Show that if, in the notation used for m_A, each $s_i = 1$ then there are *unique* constants a_i such that

$$\sum_{i=1}^{k} a_i \pi_i(z) \equiv 1.$$

Exercise 4.3.22. Show how to derive the result in **Exercise 4.2.8** from Theorem **4.3.2**.

Exercise 4.3.23. Assume that A is a 3×3 matrix for which the Jordan normal form is

$$\mathcal{A} \stackrel{\text{def}}{=} \begin{pmatrix} i & 1 & \\ & i & \\ & & 1 \end{pmatrix},$$

i.e., for some invertible matrix P: $P^{-1}AP = \mathcal{A}$. Calculate m_A.

Exercise 4.3.24. Use the Cayley-Hamilton Theorem to prove that if A is an $n \times n$ matrix then the degree μ_A of m_A does not exceed n.

Exercise 4.3.25. Assume B and C are *commuting* SQUARE matrices, i.e., $BC = CB$. Show that if $k \in \mathbf{N}$ then $(BC)^k = B^k C^k$. Show in particular that if A is a SQUARE matrix and if p and q are polynomials then $(p(A)q(A))^k = (p(A))^k(q(A))^k$.

Exercise 4.3.26. Associated with each matrix below there is a SQUARE matrix A and an invertible matrix P such that the given matrix is $P^{-1}AP$. In each case determine m_A, χ_A, the set of eigenvalues of A, $tr(A)$, $det(A)$, and whether A is diagonable.

$$i. \quad \begin{pmatrix} 1 & & \\ & 1 & \\ & & 2 \end{pmatrix}; \quad ii. \quad \begin{pmatrix} 1 & 1 & \\ & 1 & \\ & & 2 \end{pmatrix}; \quad iii. \quad \begin{pmatrix} 3 & 1 & & \\ & 3 & & \\ & & i & 1 \\ & & & i \end{pmatrix}.$$

Exercise 4.3.27. Let P be

$$\begin{pmatrix} 1 & 2 \\ 3 & 4 \end{pmatrix}.$$

For each of the matrices below there is a SQUARE matrix A such that the given matrix is $P^{-1}AP$. Determine A^{10}, m_A, χ_A, the set of eigenvalues of A, $tr(A)$, $det(A)$, and whether A is diagonable.

$$i. \quad \begin{pmatrix} 2 & \\ & i \end{pmatrix}; \quad ii. \quad \begin{pmatrix} 1 & 1 \\ & 1 \end{pmatrix}; \quad iii. \quad \begin{pmatrix} 2 & \\ & 2 \end{pmatrix}.$$

The algorithm for finding $m_{A,\mathbf{x}}$ can be used as follows to find m_A.

$i.$ For any nonzero vector $\mathbf{x}_1$ find $m_{A,\mathbf{x}_1}$.

$ii.$ If
$$span(\{\mathbf{x}_1, A\mathbf{x}_1, A^2\mathbf{x}_1, \ldots\}) = \mathbf{C}^n$$

then $m_A = m_{A,\mathbf{x}_1}$. Otherwise find a vector $\mathbf{x}_2$ *not* in

$$span(\{\mathbf{x}_1, A\mathbf{x}_1, A^2\mathbf{x}_1, \ldots\})$$

Find $m_{A,\mathbf{x}_2}$.

$iii.$ If the
$$span(\{\mathbf{x}_1, A\mathbf{x}_1, \ldots, \mathbf{x}_2, A\mathbf{x}_2, \ldots\}) = \mathbf{C}^n$$

then m_A is the *least common multiple* (LCM) of $m_{A,\mathbf{x}_1}$ and $m_{A,\mathbf{x}_2}$. Otherwise find a vector $\mathbf{x}_3$ *not* in

$$span(\{\mathbf{x}_1, A\mathbf{x}_1, \ldots, \mathbf{x}_2, A\mathbf{x}_2, \ldots\}).$$

iv. Continue in this way until for the first time

$$span(\{\, A^i \mathbf{x}_j \ : \ 0 \le i \le i_j, 1 \le j \le k \,\}) = \mathbf{C}^n.$$

Then

$$m_A = LCM(\{m_{A,\mathbf{x}_1}, \ldots, m_{A,\mathbf{x}_k}\}).$$

Exercise 4.3.28. Fill in the details of the procedure just outlined. In particular:

i. using mathematical induction, describe what "continue in this way" means;

ii. show how to use the GEM to find $\mathbf{x}_2$, $\mathbf{x}_3$, etc.;

iii. show why the LCM of the $m_{A,\mathbf{x}_i}$ is m_A.

Exercise 4.3.29. Prove that if m and n are in $\mathbf{N}$ and if $m \le n$ then there are infinitely many $m \times n$ matrices A such that $rank(A) = m$ and such that the rank of *every* $m \times m$ submatrix of A is also m.

[*Hint:* Use LEMMA **4.3.1.**]

Exercise 4.3.30. Assume A is a nonsingular matrix. Use the existence of m_A to prove that there is a polynomial p such that $p(A) = A^{-1}$.

[*Hint:* Show that if

$$m_A(t) = t^k + \cdots + a_1 t + a_0$$

then $a_0 \neq 0$. Then consider $A^{-1}m_A(A)$.]

4.4. Special kinds of diagonable matrices

Some $n \times n$ matrices are "visibly" diagonable. For example every diagonal matrix is clearly diagonable. But there are less trivial instances of matrices whose very forms reveal their diagonability.

If

$$A = \begin{pmatrix} a_{11} & \cdots & a_{1n} \\ \vdots & \ddots & \vdots \\ a_{m1} & \cdots & a_{mn} \end{pmatrix}$$

then A^* is the matrix

$$\begin{pmatrix} \overline{a}_{11} & \cdots & \overline{a}_{n1} \\ \vdots & \ddots & \vdots \\ \overline{a}_{1n} & \cdots & \overline{a}_{nn} \end{pmatrix}$$

in which the ij entry is $\overline{a}_{ji}$. Since $(A^t)^t = A$ and $\overline{\overline{z}} = z$ it follows that $(A^*)^* = A$. Furthermore if M and N are $n \times n$ matrices then $(MN)^* = N^*M^*$.

[**Remark 4.4.1:** If A is SQUARE simple inspection of A reveals whether $A = A^*$:

$$A = A^* \Leftrightarrow a_{ij} = \overline{a}_{ji}.$$

A slightly more tedious inspection reveals whether $AA^* = A^*A$:

$$AA^* = A^*A \Leftrightarrow \sum_{j=1}^{n} a_{ij}\overline{a_{kj}} = \sum_{p=1}^{n} \overline{a_{pi}}a_{pk},$$

a criterion that is rarely applied (it is a thinly disguised and relatively useless tautology).]

If $A = A^*$ then $AA^* = AA = A^*A$ whence the following development showing that A is diagonable if $AA^* = A^*A$ also implies that A is diagonable if $A = A^*$.

To proceed it will be helpful to begin with the elements of theory of the *inner product* in $\mathbf{C}^n$, the set of all $n \times 1$ column vectors. A motivation for DEFINITION **4.4.1** below emerges from the following geometric considerations.

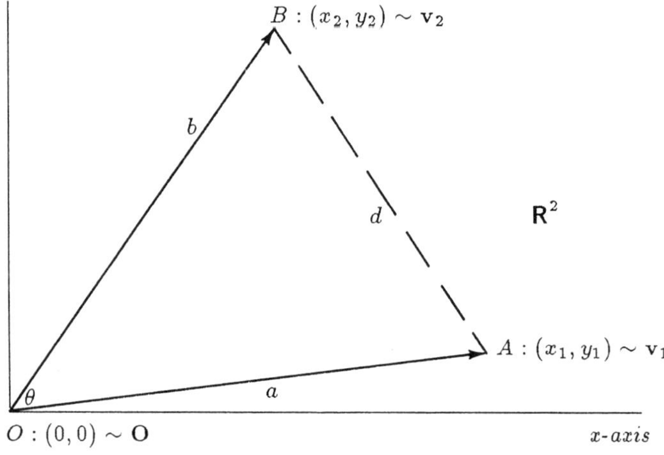

Figure 4.4.1. The geometry of the inner product

In **Figure 4.4.1** there is a representation of $\mathbf{R}^2$ as the (real) Euclidean plane. A similar representation of $\mathbf{C}^n$ is impossible if $n > 1$. Since $\mathbf{C}$ does not provide sufficient insight for the study of inner products the choice of $\mathbf{R}^2$ appears to be the most feasible.

The distance between the points representing $\mathbf{v}_1$ and $\mathbf{v}_2$ is

$$d \stackrel{\text{def}}{=} [|x_1 - x_2|^2 + |y_1 - y_2|^2]^{\frac{1}{2}}. \tag{4.4.1}$$

According to the law of cosines

$$d^2 = a^2 + b^2 - 2ab\cos\theta$$
$$d^2 = (x_1^2 + y_1^2) + (x_2^2 + y_2^2) - 2(x_1^2 + y_1^2)^{\frac{1}{2}}(x_2^2 + y_2^2)^{\frac{1}{2}}\cos\theta. \qquad (4.4.2)$$

In light of (4.4.1) and upon simplification (4.4.2) reduces to

$$x_1 y_1 + x_2 y_2 = (x_1^2 + y_1^2)^{\frac{1}{2}}(y_1^2 + y_2^2)^{\frac{1}{2}}\cos\theta. \qquad (4.4.3)$$

Since $x_1, \ldots, y_2$ are real the left member of (4.4.3) is also

$$x_1\overline{y_1} + x_2\overline{y_2}, \qquad (*)$$

which is the formula to be used when vectors with complex components are under study. The left member above is called the *inner product* of $\mathbf{x}$ and $\mathbf{y}$. Their inner product is denoted $(\mathbf{x},\mathbf{y})$ and $(*)$ serves as the motivating pattern for the general case taken up next.

$\mathbf{C}^n$ and $\mathbf{C}_n$ as EUCLIDEAN VECTOR SPACES

DEFINITION 4.4.1. IF $\mathbf{x}$ AND $\mathbf{y}$ ARE $n \times 1$ COLUMN VECTORS THEIR *inner product*, DENOTED $(\mathbf{x},\mathbf{y})$, IS THE *number* $\mathbf{y}^*\mathbf{x}$. IF $\mathbf{u}$ AND $\mathbf{v}$ ARE $1 \times n$ ROW VECTORS THEIR INNER PRODUCT, DENOTED $(\mathbf{u},\mathbf{v})$, IS THE NUMBER $\mathbf{uv}^*$.

Example 4.4.1. If

$$\mathbf{x} \stackrel{\text{def}}{=} \begin{pmatrix} 1 \\ 2 \\ 3 \end{pmatrix}, \quad \mathbf{y} \stackrel{\text{def}}{=} \begin{pmatrix} 1 - 2i \\ 2 + 4i \\ 6 \end{pmatrix}$$

$$\mathbf{u} \stackrel{\text{def}}{=} (-1, 3i, 7, 2 + 5i), \quad \mathbf{v} \stackrel{\text{def}}{=} (i, -i, 1 + 2i, -5)$$

then

$$
\begin{aligned}
(\mathbf{x},\mathbf{y}) &= 1 \times (1 + 2i) + 2 \times (2 - 4i) + 3 \times 6 = 23 - 6i \\
(\mathbf{y},\mathbf{x}) &= (1 - 2i) \times 1 + (2 + 4i) \times 2 + 6 \times 3 = 23 + 6i \\
&= \overline{(\mathbf{x},\mathbf{y})} \\
(\mathbf{u},\mathbf{v}) &= -1 \times (-i) + 3i \times (i) + 7 \times (1 - 2i) + (2 + 5i) \times (-5) = -6 - 38i. \\
(\mathbf{v},\mathbf{u}) &= i \times (-1) + (-i) \times (-3i) + (1 + 2i) \times 7 + (-5) \times (2 - 5i) = -6 + 38i \\
&= \overline{(\mathbf{u},\mathbf{v})}.
\end{aligned}
$$

Exercise 4.4.1. Show that if

$$\mathbf{x} = \begin{pmatrix} x_1 \\ \vdots \\ x_n \end{pmatrix} \text{ and } \mathbf{y} = \begin{pmatrix} y_1 \\ \vdots \\ y_n \end{pmatrix}$$

then

$$(\mathbf{x}, \mathbf{y}) = \sum_{i=1}^{n} x_i \overline{y_i}.$$

Find a similar formula for the inner product of two row vectors.

If the vectors $\mathbf{x}$ and $\mathbf{y}$ in $\mathbf{R}^2$ are nonzero they are perpendicular iff $\cos\theta = 0$. Hence they are perpendicular iff $(\mathbf{x},\mathbf{y})=0$. An alternative notation expressing the perpendicularity of $\mathbf{x}$ and $\mathbf{y}$ is $\mathbf{x} \perp \mathbf{y}$.

Two lines are perpendicular to each other iff the angle they form is a right angle. The Greek-rooted English word for *right-angled* is *orthogonal*. Custom decrees that *orthogonal* rather than *perpendicular* be used to describe two vectors $\mathbf{x}$ and $\mathbf{y}$ such that their inner product $(\mathbf{x}, \mathbf{y}) = 0$. The situation in $\mathbf{R}^2$ motivates the following DEFINITION.

DEFINITION **4.4.2.** THE VECTORS $\mathbf{x}$ AND $\mathbf{y}$ IN $\mathbf{C}^n$ ARE *orthogonal*, i.e., $\mathbf{x} \perp \mathbf{y}$, IFF $(\mathbf{x}, \mathbf{y}) = 0$. IF S IS A SET IN $\mathbf{C}^n$ THEN $S^\perp$ IS THE SET OF ALL VECTORS ORTHOGONAL TO ALL VECTORS IN S:

$$S^\perp \overset{\text{def}}{=} \{\mathbf{x} \; : \; (\mathbf{x}, \mathbf{y}) = 0 \text{ FOR ALL } \mathbf{y} \text{ IN } S\}.$$

Example 4.4.2.

i. The zero vector $\mathbf{O}$ is the only vector orthogonal to every vector:

$$(\mathbf{C}^n)^\perp = \{\mathbf{O}\}$$

i.e.,

$$[(\mathbf{x}, \mathbf{y}) = 0 \text{ for all } \mathbf{y} \text{ in } \mathbf{C}^n] \Rightarrow \mathbf{x} = \mathbf{O}.$$

ii. Every vector is orthogonal to $\mathbf{O}$: $\mathbf{O}^\perp = \mathbf{C}^n$.

iii. The only vector orthogonal to itself is the zero vector $\mathbf{O}$:

$$\mathbf{x} \perp \mathbf{x} \text{ iff } \mathbf{x} = \mathbf{O}.$$

iv. In $\mathbf{C}^n$ the vectors $\mathbf{e}_i$ and $\mathbf{e}_j$ are orthogonal iff $i \neq j$: $\mathbf{e}_i \perp \mathbf{e}_j$ iff $i \neq j$.

v. In $\mathbf{R}^2$ the vectors

$$\begin{pmatrix} a \\ b \end{pmatrix} \text{ and } \begin{pmatrix} -b \\ a \end{pmatrix}$$

are orthogonal.

vi. In $\mathbf{C}^2$ the vectors

$$\begin{pmatrix} a \\ \overline{b} \end{pmatrix} \text{ and } \begin{pmatrix} -b \\ \overline{a} \end{pmatrix}$$

are orthogonal.

vii. If $\sigma \overset{\text{def}}{=} \{i_1, \ldots, i_k\}$ and $\tau \overset{\text{def}}{=} \{j_1, \ldots, j_l\}$ are two sets of natural numbers such that

$$\sigma \cup \tau \subset \{1, \ldots, n\}$$
$$\sigma \cap \tau = \emptyset$$

and if

$$(a_1, \ldots, a_n) \in \mathbf{C}$$
$$\mathbf{x} \overset{\text{def}}{=} \sum_{s \in \sigma} a_s \mathbf{e}_s$$
$$\mathbf{y} \overset{\text{def}}{=} \sum_{t \in \tau} a_t \mathbf{e}_t$$

then $\mathbf{x} \perp \mathbf{y}$. For example if

$$\sigma \overset{\text{def}}{=} \{ s \ : \ s \text{ odd}, 1 \leq s \leq n \}, \ \tau \overset{\text{def}}{=} \{ t \ : \ t \text{ even}, 2 \leq t \leq n \}$$
$$(a_1, \ldots, a_n) \in \mathbf{C}$$
$$\mathbf{x} \overset{\text{def}}{=} \sum_{s \in \sigma} a_s \mathbf{e}_s, \ \mathbf{y} \overset{\text{def}}{=} \sum_{t \in \tau} a_t \mathbf{e}_t$$

then $\mathbf{x} \perp \mathbf{y}$.

viii. If S is any subset of $\mathbf{C}^n$ then $S^\perp$ is a sub*space* of $\mathbf{C}^n$.

[*Orthogonality* is derived from the Greek roots *orthos*, — correct, straight, upright, right — and *gonia* — angle. These roots are derived from the Indo-European roots *werdh* — to grow — and *genu* — knee or angle. *Ortho-* appears in *orthodoxy* — right teaching, — *orthodontia* — straightening teeth, etc., and *gon-* is seen in tri*gon*ometry — the measurement of three angles, etc. *Genu* is seen in *genuflection* — making an angle with or bending the knee.]

[*Perpendicularity* stems from Latin *pendere* — to hang (hence to make a right angle with the floor or the surface of the earth) — and the Indo-European root *spen* — to spin, draw or stretch — as in *spin*, su*spen*se, su*spen*d.]

Exercise 4.4.2. Derive the analog of (4.4.2) for two vectors in $\mathbf{R}^3$.

Exercise 4.4.3. Show that the following relationships hold for column vectors in $\mathbf{C}^n$:

E1.
$$(\mathbf{x}, \mathbf{y}) = \overline{(\mathbf{y}, \mathbf{x})};$$

E2.
$$(\mathbf{x} + \mathbf{z}, \mathbf{y}) = (\mathbf{x}, \mathbf{y}) + (\mathbf{z}, \mathbf{y});$$

E3. if t is a number then
$$(t\mathbf{x}, \mathbf{y}) = t(\mathbf{x}, \mathbf{y});$$

E4.
$$(\mathbf{x}, \mathbf{x}) \stackrel{\text{def}}{=} \|\mathbf{x}\|^2 \geq 0$$

and equality holds iff $\mathbf{x} = \mathbf{O}$.

(The quantity $\|\mathbf{x}\| \stackrel{\text{def}}{=} \sqrt{(\mathbf{x}, \mathbf{x})}$ is called the *norm* of $\mathbf{x}$. If

$$\mathbf{x} \stackrel{\text{def}}{=} \begin{pmatrix} x_1 \\ \vdots \\ x_n \end{pmatrix} \text{ and } \mathbf{y} \stackrel{\text{def}}{=} \begin{pmatrix} y_1 \\ \vdots \\ y_n \end{pmatrix}$$

then
$$\|\mathbf{x} - \mathbf{y}\| = \sqrt{|x_1 - y_1|^2 + \cdots + |x_n - y_n|^2},$$

which is the (Euclidean) distance between the points $(x_1, \ldots, x_n)$ and $(y_1, \ldots, y_n)$ in $\mathbf{C}_n$. Hence $\|\mathbf{x} - \mathbf{y}\|$ is called the *distance* between $\mathbf{x}$ and $\mathbf{y}$.)

[**Note 4.4.1:** The labels E1-E4 are used to call to call attention to the fact that *E*uclidean geometry lies at the heart of the notion of *inner product*. Alternative names for *inner product* are *scalar product*, *dot product*, and *Hermitian positive definite conjugate bilinear form*. In this book the terms *inner product* and *scalar product* are favored.]

The relations E1-E4 are central in a general development of the theory of *E*uclidean vector spaces. Together with some topological conditions ·E1-E4 are the formalizations due to von Neumann (cf. [**Be, v.N1, v.N2**]) for the theory of *Hilbert* spaces, a theory that is central in modern studies of quantum mechanics. There is some further discussion of these matters in **Part III**.

Other properties of the inner product (,) are:

E5. for any $n \times n$ matrix A
$$(A\mathbf{x}, \mathbf{y}) = (\mathbf{x}, A^*\mathbf{y});$$

E6.
$$(\mathbf{x}, \mathbf{y} + \mathbf{z}) = (\mathbf{x}, \mathbf{y}) + (\mathbf{x}, \mathbf{z})$$

    Chapter 4. Useful forms for matrices

(deducible from E1 and E4);

E7.
$$(\mathbf{x}, t\mathbf{y}) = \bar{t}(\mathbf{x}, \mathbf{y})$$

(deducible from E1 and E3);

E8.
$$\tfrac{1}{4}[\|\mathbf{x} + \mathbf{y}\|^2 - \|\mathbf{x} - \mathbf{y}\|^2$$
$$+ i\|\mathbf{x} + i\mathbf{y}\|^2 - i\|\mathbf{x} - i\mathbf{y}\|^2] = (\mathbf{x}, \mathbf{y})$$

(deducible from E1-E4).

[**Remark 4.4.2:** There is a striking parallel between the penetrating quality of E1-E4 and the penetrating quality of D1-D3 (cf. **Section 3.1**). Each group, by characterizing the elements in its field, provides a controlling influence over the developments in its field, Euclidean vector spaces for E1-E4 (as the development following shows) and multilinear functions on vector spaces for D1-D3.]

THEOREM **4.4.1.** THE $n \times n$ MATRIX $A \overset{\text{def}}{=} (a_{ij})_{i,j=1}^{n,n}$ IS SUCH THAT $A = A^*$ IFF FOR ALL $n \times 1$ COLUMN VECTORS $\mathbf{x}$ AND $\mathbf{y}$

$$(A\mathbf{x}, \mathbf{y}) = (\mathbf{x}, A\mathbf{y}). \tag{4.4.4}$$

THE MATRIX A IS SUCH THAT $AA^* = A^*A$ IFF FOR EACH $n \times 1$ COLUMN VECTOR $\mathbf{x}$

$$(A\mathbf{x}, A\mathbf{x}) = (A^*\mathbf{x}, A^*\mathbf{x}), \tag{4.4.5}$$

i.e., $\|A\mathbf{x}\| = \|A^*\mathbf{x}\|$.

PROOF. Assume

$$\mathbf{x} = \begin{pmatrix} x_1 \\ \vdots \\ x_n \end{pmatrix} \quad \text{and} \quad \mathbf{y} = \begin{pmatrix} y_1 \\ \vdots \\ y_n \end{pmatrix}.$$

If $A = A^*$ then

$$(A\mathbf{x}, \mathbf{y}) = \sum_{i=1}^{n}\Big(\sum_{j=1}^{n} a_{ij} x_j\Big)\overline{y_i} = \sum_{j=1}^{n}\Big(\sum_{i=1}^{n} a_{ij}\overline{y_i}\Big)x_j$$
$$= \sum_{j=1}^{n}\Big(\sum_{i=1}^{n} \overline{a_{ji}y_i}\Big)x_j = \sum_{j=1}^{n} x_j\overline{\Big(\sum_{i=1}^{n} a_{ji}y_i\Big)} = (\mathbf{x}, A\mathbf{y}).$$

Conversely assume (4.4.4) is true for all $\mathbf{x}$ and for all $\mathbf{y}$. Then for any pair ij, say $i < j$,

$$
\begin{aligned}
a_{ij} &= (A\mathbf{e}_i, \mathbf{e}_j) \\
&= (\mathbf{e}_i, A\mathbf{e}_j) \\
&= \overline{(A\mathbf{e}_j, \mathbf{e}_i)} \\
&= \overline{a_{ji}}.
\end{aligned}
$$

If $AA^* = A^*A$ then

$$
\begin{aligned}
(A\mathbf{x}, A\mathbf{x}) &= (\mathbf{x}, A^*A\mathbf{x}) = (\mathbf{x}, AA^*\mathbf{x}) \\
(A^*\mathbf{x}, A^*\mathbf{x}) &= (\mathbf{x}, (A^*)^*A^*\mathbf{x}) = (\mathbf{x}, AA^*\mathbf{x}) \\
(A\mathbf{x}, A\mathbf{x}) &= (A^*\mathbf{x}, A^*\mathbf{x}).
\end{aligned}
$$

Conversely if for all $\mathbf{x}$, $(A\mathbf{x}, A\mathbf{x}) = (A^*\mathbf{x}, A^*\mathbf{x})$ the following *polarization* trick shows that $AA^* = A^*A$. Indeed

$$
(A\mathbf{x}, A\mathbf{x}) = (A^*\mathbf{x}, A^*\mathbf{x}) \Leftrightarrow (A^*A\mathbf{x}, \mathbf{x}) = (AA^*\mathbf{x}, \mathbf{x}).
$$

Hence

$$
\begin{aligned}
(A^*A(\mathbf{x} + i\mathbf{y}), \mathbf{x} + i\mathbf{y}) &= (AA^*(\mathbf{x} + i\mathbf{y}), \mathbf{x} + i\mathbf{y}) \\
(A^*A\mathbf{x}, \mathbf{x}) + (A^*A\mathbf{y}, \mathbf{y}) + i[(A^*A\mathbf{y}, \mathbf{x}) - (A^*A\mathbf{x}, \mathbf{y})] & \\
= (AA^*\mathbf{x}, \mathbf{x}) + (AA^*\mathbf{y}, \mathbf{y}) &+ i[(AA^*\mathbf{y}, \mathbf{x}) - (AA^*\mathbf{x}, \mathbf{y})] \\
i[(A^*A\mathbf{y}, \mathbf{x}) - (A^*A\mathbf{x}, \mathbf{y})] &= i[(AA^*\mathbf{y}, \mathbf{x}) - (AA^*\mathbf{x}, \mathbf{y})]. \qquad (4.4.6)
\end{aligned}
$$

Denote $A^*A - AA^*$ by B. Then $B^* = (A^*A - AA^*)^* = A^*A - AA^* = B$. Hence from (4.4.6) it follows that

$$
\begin{aligned}
(B\mathbf{y}, \mathbf{x}) &= (B\mathbf{x}, \mathbf{y}) \\
&= (\mathbf{x}, B^*\mathbf{y}) = (\mathbf{x}, B\mathbf{y}) \\
&= \overline{(B\mathbf{y}, \mathbf{x})}.
\end{aligned}
$$

Hence for all $\mathbf{x}$ and $\mathbf{y}$ $(B\mathbf{y}, \mathbf{x})$ is real. If $\mathbf{y}$ is replaced by $i\mathbf{y}$ then $(B\mathbf{y}, \mathbf{x})$ is replaced by $i(B\mathbf{y}, \mathbf{x})$ which should also be real. This is possible iff $(B\mathbf{y}, \mathbf{x}) = 0$. For any $\mathbf{y}$ if $\mathbf{x} = B\mathbf{y}$ then

$$
\|B\mathbf{y}\|^2 = 0,
$$

i.e., $B = A^*A - AA^* = O$, as promised.

$$\Omega$$

DEFINITION **4.4.3.** AN $n \times n$ MATRIX A SUCH THAT $A = A^*$ IS CALLED *self-adjoint*. IF $AA^* = A^*A$ THEN A IS CALLED *normal*.

    Chapter 4. Useful forms for matrices

[**Note 4.4.2:**   If A is self-adjoint then A is normal: $A^*A = A^2 = AA^*$. The converse is false, e.g., if

$$A = \begin{pmatrix} 0 & 1 \\ i & 0 \end{pmatrix}$$

Then $A \neq A^*$, i.e., A is *not* self-adjoint, although $AA^* = I$ and so $A^* = A^{-1}$ whence $A^*A = I = AA^*$, i.e., A is normal.]

[**Note 4.4.3:**   If A is any SQUARE matrix then much as a complex number z may be written as

$$\frac{(z + \overline{z})}{2} + i\frac{(z - \overline{z})}{2i} \overset{\text{def}}{=} x + iy$$

so the matrix A may be written as

$$\frac{(A + A^*)}{2} + i\frac{(A - A^*)}{2i} \overset{\text{def}}{=} M + iN.$$

Just as the numbers x and y are real the matrices M and N are self-adjoint. Although the (real) numbers x and y always commute, i.e., $xy = yx$ the matrices M and N do not necessarily commute. In fact M and N commute iff the matrix A is normal: A is normal iff $MN = NM$.]

The next THEOREM and its generalizations are at the heart of great blocks of modern mathematics and mathematical physics, in particular quantum mechanics. In the circumstances, it is surprisingly simple to prove the finite-dimensional version below.

THEOREM 4.4.2. (THE SPECTRAL THEOREM) IF A IS A NORMAL $n \times n$ MATRIX THERE IS AN INVERTIBLE MATRIX P SUCH THAT $(i.)$ $P^{-1}AP \overset{\text{def}}{=} D$ IS DIAGONAL AND $(ii.)$ $P^* = P^{-1}$.

CONVERSELY, IF A IS AN $n \times n$ MATRIX, IF P IS AN INVERTIBLE MATRIX SUCH THAT $P^* = P^{-1}$, AND IF $P^*AP \overset{\text{def}}{=} D$ IS DIAGONAL THEN A IS NORMAL.

Of the two assertions in the THEOREM, the converse is the simpler one to prove:

$$
\begin{aligned}
AA^* &= PDP^*(PDP^*)^* & A^*A &= (PDP^*)^*PDP^* \\
&= PDP^*PD^*P^* & &= PD^*P^*PDP^* \\
&= PDD^*P^*; & &= PD^*DP^*.
\end{aligned}
$$

Since multiplication of compatible SQUARE diagonal matrices is commutative it follows that $DD^* = D^*D$ and so $AA^* = A^*A$, as required.

The first (direct) statement ($i.$) is very important. Hence for it there follow two entirely different proofs. The first uses the characterization THEOREM **4.3.2** of diagonability in terms of the simplicity of the zeros of the minimal polynomial m_A. The second develops further the theory of inner products and uses the additional results to derive not only the diagonability of A but also the equation $P^* = P^{-1}$.

PROOF 1 of i (the diagonability of A).

The first step is to show that if p is a polynomial and $\mathbf{x} \in V$ then

$$p(A)\mathbf{x} = \mathbf{O} \Leftrightarrow (p(A))^*\mathbf{x} = \mathbf{O}. \tag{4.4.7}$$

Indeed, because A is normal, $p(A)p(A)^* = p(A)^*p(A)$ and so $p(A)\mathbf{x} = \mathbf{O}$ iff

$$\begin{aligned}
0 &= (p(A)\mathbf{x}, p(A)\mathbf{x}) \\
&= (p(A)^*p(A)\mathbf{x}, \mathbf{x}) = (p(A)p(A)^*\mathbf{x}, \mathbf{x}) \\
&= (p(A)^*\mathbf{x}, p(A)^*\mathbf{x})
\end{aligned}$$

iff $p(A)^*\mathbf{x} = \mathbf{O}$.

Now assume that $(\lambda - \lambda_i)^2$ is a factor of $m_A(\lambda)$. Then

$$\begin{aligned}
m_A(\lambda) &\stackrel{\text{def}}{=} (\lambda - \lambda_i)m_1(\lambda) \\
&\stackrel{\text{def}}{=} (\lambda - \lambda_i)^2 m_2(\lambda) \\
m_1(\lambda) &= (\lambda - \lambda_i)m_2(\lambda).
\end{aligned}$$

For all vectors $\mathbf{x}$, $\mathbf{y}$, $(m_A(A)\mathbf{x}, \mathbf{y}) = 0$. Because

$$\text{degree of } m_1 = \mu_A - 1 < \text{degree of } m_A,$$

for some $\mathbf{z}$, $\mathbf{w} \stackrel{\text{def}}{=} m_1(A)\mathbf{z} \neq \mathbf{O}$. Since $(A - \lambda_i I)\mathbf{w} = m_A(A)\mathbf{z} = \mathbf{O}$, it follows from (4.4.7) that $(A - \lambda_i I)^*\mathbf{w} = (A - \lambda_i I)^* m_1(A)\mathbf{z} = \mathbf{O}$ and so

$$\begin{aligned}
0 &= ((A - \lambda_i I)^* m_1(A)\mathbf{z}, m_2(A)\mathbf{z}) \\
&= (m_1(A)\mathbf{z}, (A - \lambda_i I)m_2(A)\mathbf{x}) \\
&= \| m_1(A)\mathbf{z} \|^2.
\end{aligned}$$

There emerges the contradiction:

$$\mathbf{O} \neq \mathbf{w} = m_1(A)\mathbf{z} = \mathbf{O}$$

and the result follows.

$$\Omega$$

PROOF 2 of i (the diagonability of A).

[**Remark 4.4.3:** In the argument below some additional properties of the inner product are derived and these properties together with some

---

    Chapter 4. Useful forms for matrices

discussion of maximally linearly independent sets and mathematical induction permit the proof to be completed. The virtue of this longer proof is that it serves to introduce a number of ideas and techniques that prove useful in other parts of linear algebra.]

The set

$$S \stackrel{\text{def}}{=} \{e_1, \ldots, e_n\}$$

is a basis for $\mathbf{C}^n$ and

$$(e_i, e_j) = \begin{cases} 1 & \text{if } i = j; \\ 0 & \text{if } i \neq j. \end{cases}$$

If

$$X \stackrel{\text{def}}{=} \{\mathbf{x}_1, \ldots, \mathbf{x}_m\}$$

is any set of vectors in $\mathbf{C}^n$ and if

$$(\mathbf{x}_i, \mathbf{x}_j) = \begin{cases} 1 & \text{if } i = j, \\ 0 & \text{if } i \neq j, \end{cases} \tag{4.4.8}$$

then X is called an *orthonormal* set of vectors. Thus S above is an *orthonormal basis* for $\mathbf{C}^n$.

Example 4.4.3. Each of the following is an orthonormal set of vectors:

i. $\{\frac{1}{\sqrt{2}}(e_1 + e_2), \frac{1}{\sqrt{2}}(e_1 - e_2)\}$ (in $\mathbf{C}^n$);

ii. $\{(\cos\theta, \sin\theta), (-\sin\theta, \cos\theta)\}$ (in $\mathbf{C}_2$);

iii. $\{(\frac{i}{\sqrt{2}}, \frac{1}{\sqrt{2}}), (\frac{1}{\sqrt{2}}, \frac{i}{\sqrt{2}})\}$ (in $\mathbf{C}_2$).

LEMMA 4.4.1. EVERY ORTHONORMAL SET $X \stackrel{\text{def}}{=} \{\mathbf{x}_1, \ldots, \mathbf{x}_m\}$ IS LINEARLY INDEPENDENT.

[**Remark 4.4.4:** Hence every orthonormal set in $\mathbf{C}^n$ contains not more than n vectors.]

PROOF. If

$$\sum_{i=1}^{m} a_i \mathbf{x}_i = \mathbf{O}$$

then, by virtue of E1-E4 and the orthonormality of X,

$$a_j = \sum_{i=1}^{m} a_i(\mathbf{x}_j, \mathbf{x}_i) = (\mathbf{x}_j, \sum_{i=1}^{m} \overline{a}_i \mathbf{x}_i) = \overline{(\mathbf{O}, \mathbf{x}_j)} = 0.$$

$$\Omega$$

LEMMA **4.4.2.** IF W IS A SUBSPACE OF $\mathbf{C}^n$ AND $W \neq \{\mathbf{O}\}$ THEN THERE IS IN W AN ORTHONORMAL BASIS FOR W.

PROOF. Since $W \neq \{\mathbf{O}\}$ it follows that there is in W some nonzero vector $\mathbf{v}$. Hence $\|\mathbf{v}\| \neq 0$,

$$\mathbf{u} \overset{\text{def}}{=} \frac{\mathbf{v}}{\|\mathbf{v}\|}$$

is well-defined, and $\|\mathbf{u}\| = 1$. The singleton set $\{\mathbf{u}\}$ is *an* orthonormal subset of W.

If Y is an orthonormal set in W and if Y consists of k vectors then $k \leq n$. Assume that m is the largest number of vectors to be found in any orthonormal set in W and let $X \overset{\text{def}}{=} \{\mathbf{x}_1, \ldots, \mathbf{x}_m\}$ be an orthonormal set in W. If X is not a basis for W there is in W some vector $\mathbf{z}$ such that $\mathbf{z}$ can*not* be written as a linear combination of the vectors in X, i.e., $\mathbf{z} \notin span(X)$.

Hence the vector

$$\mathbf{w} \overset{\text{def}}{=} \mathbf{z} - \sum_{i=1}^{m} (\mathbf{z}, \mathbf{x}_i)\mathbf{x}_i$$

is not $\mathbf{O}$. As a linear combination of vectors in W the vector $\mathbf{w}$ is in W and, owing to the orthonormality of X,

$$(\mathbf{w}, \mathbf{x}_j) = (\mathbf{z}, \mathbf{x}_j) - \sum_{i=1}^{m} (\mathbf{z}, \mathbf{x}_i)(\mathbf{x}_i, \mathbf{x}_j)$$
$$= (\mathbf{z}, \mathbf{x}_j) - (\mathbf{z}, \mathbf{x}_j) = 0, \ 1 \leq j \leq m.$$

Hence

$$\widetilde{X} \overset{\text{def}}{=} \left\{\frac{\mathbf{w}}{\|\mathbf{w}\|}, \mathbf{x}_1, \ldots, \mathbf{x}_m\right\}$$

is an orthonormal set in W, and yet contains $m + 1$ vectors, in contradiction of the maximality of m.

$$\Omega$$

LEMMA **4.4.3.** IF A IS NORMAL, i.e., IF $AA^* = A^*A$, IF λ IS AN EIGENVALUE OF A, AND IF $\mathbf{x}$ IS A CORRESPONDING EIGENVECTOR THEN $\overline{\lambda}$ IS AN EIGEN*value* FOR A^* AND THE SAME $\mathbf{x}$ IS AN EIGEN*vector* FOR A^*: $A^*\mathbf{x} = \overline{\lambda}\mathbf{x}$.

    Chapter 4. Useful forms for matrices

PROOF. Since $(A - \lambda I)\mathbf{x} = \mathbf{O}$,

$$
\begin{aligned}
\|(A^* - \overline{\lambda} I)\mathbf{x}\|^2 &= ((A^* - \overline{\lambda} I)\mathbf{x}, (A^* - \overline{\lambda} I)\mathbf{x}) \\
&= (\mathbf{x}, (A - \lambda I)(A^* - \overline{\lambda} I)\mathbf{x}) \\
&= (\mathbf{x}, (A^* - \overline{\lambda} I)(A - \lambda I)\mathbf{x}) = ((A - \lambda I)\mathbf{x}, (A - \lambda I)\mathbf{x}) \\
&= \|(A - \lambda I)\mathbf{x}\|^2 = \|\mathbf{O}\|^2 = 0
\end{aligned}
$$

whence

$$
(A^* - \overline{\lambda} I)\mathbf{x} = \mathbf{O}.
$$

$$\Omega$$

[**Remark 4.4.5:** LEMMA **4.4.3** is also a direct consequence of the first result in PROOF 1 of THEOREM **4.4.2**.]

The method in PROOF 2 is that of mathematical induction. If A is an $n \times n$ matrix and $n = 1$ then A is not only diagonable but is actually a diagonal matrix. If the THEOREM is false then for some n greater than 1 there is a normal $n \times n$ matrix A that is not diagonable, but if $m < n$ then every $m \times m$ normal matrix *is* diagonable.

There is always at least one eigenvalue, say λ_1, for A. Let k be the number of pairwise different eigenvalues

$$
\{\lambda_1, \ldots, \lambda_k\}
$$

of A and let W be the span of the set of all eigenvectors corresponding to all eigenvalues of A.

If $W = \mathbf{C}^n$ then there is in W a basis

$$
Y \overset{\text{def}}{=} \{\mathbf{y}_1, \ldots, \mathbf{y}_n\}
$$

consisting of eigenvectors for A.

If Q is the $n \times n$ matrix in which the columns are the column vectors in Y then AQ is a matrix in which the jth column is some (eigenvalue) multiple of $\mathbf{y}_j$ and $Q^{-1}AQ$ is a diagonal matrix D in which the jth column is some multiple of $\mathbf{e}_j$: A is diagonable, a contradiction.

Thus $W \neq \mathbf{C}^n$ whence $k < n$ and there is some nonzero vector $\mathbf{z}$ not in W, and there is in W an orthonormal basis $Y \overset{\text{def}}{=} \{\mathbf{y}_1, \ldots, \mathbf{y}_m\}$. As in the proof of LEMMA **4.4.2**

$$
\mathbf{u} \overset{\text{def}}{=} \mathbf{z} - \sum_{i=1}^{m}(\mathbf{z}, \mathbf{y}_i)\mathbf{y}_i \notin W
$$

and if $\mathbf{w} \in W$ then $(\mathbf{u}, \mathbf{w}) = 0$. Since $\mathbf{O} \neq \mathbf{u} \in W^\perp$ it follows that $W^\perp \neq \{\mathbf{O}\}$.

If $\mathbf{v} \in W^\perp$ and $\mathbf{w} \in W$ then for some $\lambda : A\mathbf{w} = \lambda\mathbf{w}$, $A^*\mathbf{w} = \overline{\lambda}\mathbf{w}$, and

$$
(A\mathbf{v}, \mathbf{w}) = (\mathbf{v}, A^*\mathbf{w}) = (\mathbf{v}, \overline{\lambda}\mathbf{w}) = \lambda(\mathbf{v}, \mathbf{w}) = \lambda \cdot 0 = 0.
$$

---

i.e., $A\mathbf{v} \in W^\perp$. Thus $W^\perp$ is an invariant subspace: $AW^\perp \subset W^\perp$.

The basic argument showing that a matrix has an eigenvalue and a corresponding eigenvector and the fact that $AW^\perp \subset W^\perp$ implies for some eigenvalue μ and some nonzero vector $\mathbf{x}$ in $W^\perp$ that $A\mathbf{x} = \mu\mathbf{x}$. As an eigenvalue of A μ must be some λ_j. As a corresponding eigenvector $\mathbf{x}$ must be in W. By assumption $\mathbf{x} \in W^\perp$. Hence $\mathbf{x}$ is *orthogonal to itself*: $(\mathbf{x}, \mathbf{x}) = \|\mathbf{x}\|^2 = 0$. Thus $\mathbf{x} = \mathbf{O}$, a contradiction of the fact that $\mathbf{x}$ as an eigenvector is necessarily nonzero.

The assumption that A is not diagonable leads, in every case, to a contradiction and so $W = \mathbf{C}^n$. The basis Y for W is a basis for $\mathbf{C}^n$, Y is an eigenbasis for $\mathbf{C}^n$ and the corresponding matrix Q does indeed diagonalize A: $Q^{-1}AQ \stackrel{\text{def}}{=} D$ is a diagonal matrix.

The proof of *ii* is the burden of the next **Exercise**.

$$\Omega$$

Exercise 4.4.4. Assume A is a normal $n \times n$ matrix and that $\{\lambda_1, \ldots, \lambda_k\}$ is the set of pairwise different eigenvalues of A. Let W_j be the span of the eigenvectors corresponding to the eigenvalue λ_j and let X_j be an orthonormal basis for W_j. Show:

   i. if $\mathbf{x}_j \in W_j$, $1 \le j \le k$, then $A\mathbf{x}_j = \lambda_j\mathbf{x}_j$ and if $j \ne j'$ then $(\mathbf{x}_j, \mathbf{x}_{j'}) = 0$;

   [*Hint:* Calculate $(A\mathbf{x}_i, \mathbf{x}_j)$ in two different ways.]

  ii. if $j \ne j'$ then $X_j \cap X_{j'} = \emptyset$ and

$$X \stackrel{\text{def}}{=} \bigcup_{j=1}^{k} X_j$$

is an orthonormal basis consisting of eigenvectors for A;

 iii. there is a *unitary* matrix P (a matrix P such that $P^*P = I$) for which $P^{-1}AP \stackrel{\text{def}}{=} D$ is a diagonal matrix;

 iv. an $n \times n$ matrix P is unitary iff $(P\mathbf{x}, P\mathbf{y}) = (\mathbf{x}, \mathbf{y})$ for all vectors $\mathbf{x}$ and $\mathbf{y}$ in $\mathbf{C}^n$ iff $\|P\mathbf{x}\| = \|\mathbf{x}\|$ for all vectors $\mathbf{x}$ in $\mathbf{C}^n$;

  v. a unitary matrix is normal;

 vi. if P is a unitary matrix then $|det(P)| = 1$;

   [*Hint:* Use the product rule for determinants.]

 vii. if A is self-adjoint then for any vector $\mathbf{x}$ the number $(A\mathbf{x}, \mathbf{x})$ is real (it is in $\mathbf{R}$);

viii. the eigenvalues of a self-adjoint matrix are real;

   [*Hint:* For an eigenvalue λ and corresponding eigenvector $\mathbf{x}$ calculate $(A\mathbf{x}, \mathbf{x})$ in two ways.]

[**Remark 4.4.6:** Owing to *iv* above, a unitary matrix preserves the length of a vector. Hence unitary matrices are visualized by some as "rotations" in $\mathbf{C}^n$.]

SUMMARY OF DIAGONABILITY

A matrix A is *unitarily diagonable* iff there is a unitary matrix U such that $U^{-1}AU$ is a diagonal matrix.

A MATRIX IS DIAGONABLE IFF THE ZEROS OF ITS MINIMAL POLYNOMIAL ARE SIMPLE
EVERY $n \times n$ MATRIX HAVING n PAIRWISE DIFFERENT EIGENVALUES IS DIAGONABLE
EVERY NORMAL MATRIX A IS UNITARILY DIAGONABLE
EVERY SELF-ADJOINT MATRIX A IS UNITARILY DIAGONABLE.
EVERY REAL MATRIX A SUCH THAT $A = A^t$ IS UNITARILY DIAGONABLE.

[**Note 4.4.4:** The matrix

$$A \stackrel{\mathrm{def}}{=} \begin{pmatrix} 1 & 2 \\ 3 & 4 \end{pmatrix}$$

is such that it has two different eigenvalues. Hence A is diagonable. Since

$$AA^* = \begin{pmatrix} 5 & 11 \\ 11 & 25 \end{pmatrix} \neq \begin{pmatrix} 10 & 14 \\ 14 & 18 \end{pmatrix} = A^*A,$$

A is not normal and (hence) not self-adjoint; A is not unitarily diagonable.]

LEMMA 4.4.4. IF $O \neq \mathbf{u} \in \mathbf{C}^n$ AND IF

$$W \stackrel{\mathrm{def}}{=} \{\, \mathbf{w} \ : \ (\mathbf{w}, \mathbf{u}) = 0 \,\} \ \left(\stackrel{\mathrm{def}}{=} \{\mathbf{u}\}^{\perp} \right)$$

THERE IS IN W A UNIQUE VECTOR $\mathbf{y}$ AND THERE IS A UNIQUE NUMBER a SUCH THAT $\mathbf{x} = a\mathbf{u} + \mathbf{y}$.

PROOF. It may be and is assumed that $\|\mathbf{u}\| = 1$. The number a is $(\mathbf{x}, \mathbf{u})$ and the vector $\mathbf{y}$ is $\mathbf{x} - (\mathbf{x}, \mathbf{u})\mathbf{u}$.
Indeed,

$$\begin{aligned}
(\mathbf{u}, \mathbf{x} - (\mathbf{x}, \mathbf{u})\mathbf{u}) &= (\mathbf{u}, \mathbf{x}) - \overline{(\mathbf{x}, \mathbf{u})}(\mathbf{u}, \mathbf{u}) \\
&= (\mathbf{u}, \mathbf{x}) - (\mathbf{u}, \mathbf{x}) \cdot 1 = 0.
\end{aligned}$$

Furthermore if $\mathbf{y}' \in W$, if b is a number, and if $\mathbf{x} = b\mathbf{u} + \mathbf{y}'$ then

$$a = (\mathbf{x}, \mathbf{u}) = b(\mathbf{u}, \mathbf{u}) + (\mathbf{y}', \mathbf{u}) = b + 0 = b$$

$$\mathbf{y} = \mathbf{x} - a\mathbf{u}$$

$$\mathbf{y}' = \mathbf{x} - b\mathbf{u} = \mathbf{x} - a\mathbf{u} = \mathbf{y}.$$

$$\Omega$$

Exercise 4.4.5. Show that any 2×2 matrix of the form

$$\begin{pmatrix} a & b \\ b & a \end{pmatrix}$$

is normal and that it is self-adjoint iff a and b are real.

Exercise 4.4.6. Show that if $\theta \in \mathbf{R}$ then the "rotation matrix"

$$\begin{pmatrix} \cos\theta & \sin\theta \\ -\sin\theta & \cos\theta \end{pmatrix}$$

is a unitary 2×2 matrix.

Exercise 4.4.7. Assume c and s are complex numbers such that

$$|c|^2 + |s|^2 = 1 \text{ and } c\bar{s} + \bar{c}s = 0.$$

According to **Exercise 4.4.5** if p and q are complex numbers then any 2×2 matrix

$$A \stackrel{\text{def}}{=} \begin{pmatrix} p|c|^2 + q|s|^2 & p\bar{c}s + qc\bar{s} \\ p\bar{c}s + qc\bar{s} & p|c|^2 + q|s|^2 \end{pmatrix}$$

is normal. Show that if

$$P \stackrel{\text{def}}{=} \begin{pmatrix} c & s \\ s & c \end{pmatrix}$$

then P is invertible, $P^* = P^{-1}$ (P is unitary), and

$$P^* A P = \begin{pmatrix} p & 0 \\ 0 & q \end{pmatrix}.$$

Exercise 4.4.8. Show that if $p, q, x,$ and y are real numbers, i.e.,

$$p, q, x, y \in \mathbf{R},$$

then

$$A \stackrel{\text{def}}{=} \begin{pmatrix} x + iy & p + iq \\ q + ip & x + iy \end{pmatrix}$$

is normal.

Exercise 4.4.9. Show that each of the following matrices is self-adjoint and then diagonalize it:

$$i. \ \begin{pmatrix} 1 & 2 + 3i \\ 2 - 3i & 6 \end{pmatrix}; \ ii. \ \begin{pmatrix} 6 & -4 - i \\ -4 + i & -2 \end{pmatrix}.$$

   Chapter 4. Useful forms for matrices

Exercise 4.4.10. Show that each of the following matrices is normal and then diagonalize it:

$$i. \quad \begin{pmatrix} 1+2i & 3+4i \\ 4+3i & 1+2i \end{pmatrix} ; \quad ii. \quad \begin{pmatrix} -2-5i & 2+3i \\ 3+2i & -2-5i \end{pmatrix} .$$

Exercise 4.4.11. Let $\{x_1, \ldots, x_n\}$ be a set of n orthonormal $n \times 1$ vectors and let U be the $n \times n$ matrix in which the jth column is x_j. Show that just as the columns of U constitute a set of n orthonormal $n \times 1$ vectors, the *rows* of U constitute an orthonormal set of $1 \times n$ vectors.

Example 4.4.4. From the preceding developments emerges a method for constructing the most general normal $n \times n$ matrix A:

i. choose any diagonal matrix

$$D \stackrel{\text{def}}{=} \begin{pmatrix} d_1 & & & \\ & d_2 & & \\ & & \ddots & \\ & & & d_n \end{pmatrix} ;$$

ii. choose any set $\{x_1, \ldots, x_n\}$ of n orthonormal $n \times 1$ vectors and form the $n \times n$ matrix U in which the jth column is x_j;

iii. form the matrix $A \stackrel{\text{def}}{=} UDU^*$.

Then

$$U^*U = I$$
$$AA^* = UDU^*(UDU^*)^* = UDU^*UD^*U^* = UDD^*U^*$$
$$A^*A = (UDU^*)^*UDU^* = UD^*U^*UDU^* = UDD^*U^*.$$

Since D and D^* are diagonal matrices,

$$DD^* = D^*D$$

whence $AA^* = A^*A$.

In view of **Lemma 4.4.2** there is some value in devising a constructive method for producing a set of n orthonormal vectors in $\mathbf{C}^n$. Such a constructive method can be motivated by the next considerations.

In **Figure 4.4.2** x_1 and x_2 are given nonzero and noncollinear vectors. The geometry suggests how the vectors u_1 and u_2 can be constructed so that $\{u_1, u_2\}$ is an orthonormal set.

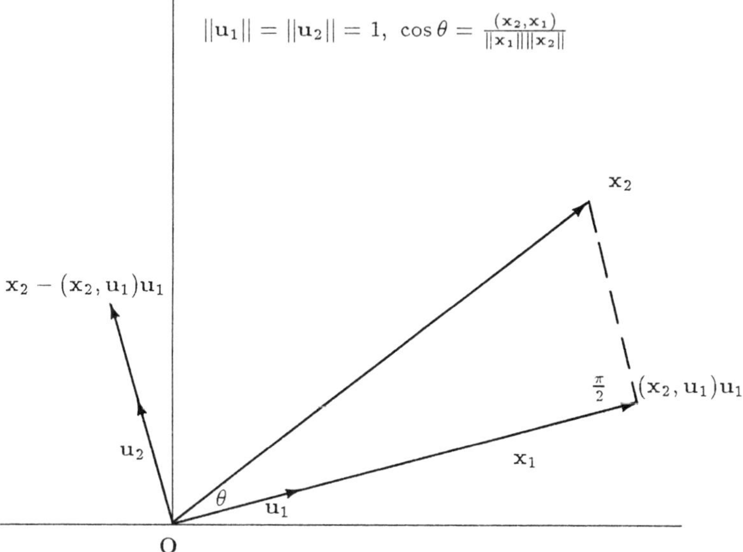

Figure 4.4.2.   The Gram-Schmidt process

Since the vector $x_1 \neq O$ the vector

$$u_1 \stackrel{\text{def}}{=} \frac{x_1}{\|x_1\|}$$

is defined and its norm is 1. The *length* of the orthogonal (perpendicular) projection of x_2 onto the line through x_1 is

$$l \stackrel{\text{def}}{=} \begin{cases} \|x_2\|\cos\theta & \text{if } \cos\theta \geq 0 \\ -\|x_2\|\cos\theta & \text{if } \cos\theta < 0 \end{cases} = \pm\frac{(x_2, x_1)}{\|x_1\|}.$$

Since $\|u_1\| = 1$ the vector from the origin (O) to the end of the projection is $u_1 = \pm(x_2, u_1)u_1$.

In any event, the vector $x_2 - (x_2, u_1)u_1$ as indicated in **Figure 4.4.2**, is the projection of x_2 onto the line containing x_1.

If $\mathbf{w}_2 \stackrel{\text{def}}{=} \mathbf{x}_2 - (\mathbf{x}_2, \mathbf{u}_1)\mathbf{u}_1$ then $\mathbf{w}_2$ is in fact a linear combination of $\mathbf{x}_1$ and $\mathbf{x}_2$ and in that linear combination the coefficient of $\mathbf{x}_2$ is 1. Since the vectors $\mathbf{x}_1$ and $\mathbf{x}_2$ are not collinear they are linearly independent and so $\mathbf{w}_2 \neq \mathbf{O}$. Hence

$$\mathbf{u}_2 \stackrel{\text{def}}{=} \frac{\mathbf{w}_2}{\|\mathbf{w}_2\|}$$

is defined and $\|\mathbf{u}_2\| = 1$. Finally,

$$
\begin{aligned}
(\mathbf{u}_2, \mathbf{u}_1) &= \frac{(\mathbf{w}_1, \mathbf{u}_1)}{\|\mathbf{w}_2\|} \\
&= \frac{(\mathbf{x}_2 - (\mathbf{x}_2, \mathbf{u}_1)\mathbf{u}_1, \mathbf{u}_1)}{\|\mathbf{w}_2\|} \\
&= \frac{(\mathbf{x}_2, \mathbf{u}_1) - (\mathbf{x}_2, \mathbf{u}_1)\|\mathbf{u}_1\|^2}{\|\mathbf{w}_2\|} \\
&= \frac{(\mathbf{x}_2, \mathbf{u}_1) - (\mathbf{x}_2, \mathbf{u}_1) \cdot 1}{\|\mathbf{w}_2\|} \\
&= 0.
\end{aligned}
$$

As the **Figure 4.4.2** suggests and as the calculations just concluded show, $\mathbf{u}_1$ and $\mathbf{u}_2$ are orthogonal vectors, each of norm 1.

The generalization of the procedure described geometrically above is the *Gram-Schmidt process* to be described next.

Let $X \stackrel{\text{def}}{=} \{\mathbf{x}_1, \ldots, \mathbf{x}_m\}$ be a linearly independent set of $n \times 1$ vectors. Hence $m \leq n$. If $m < n$ then X may be filled out to a maximally linearly independent set of vectors. Hence it may be assumed that $m = n$, i.e., that X is a basis for $\mathbf{C}^n$. By mathematical induction define:

$$
\begin{aligned}
\mathbf{u}_1 &= \frac{\mathbf{x}_1}{\|\mathbf{x}_1\|}; \\
\mathbf{w}_{k+1} &= \mathbf{x}_{k+1} - \sum_{i=1}^{k} (\mathbf{x}_{k+1}, \mathbf{u}_i)\mathbf{u}_i; \\
\mathbf{u}_{k+1} &= \frac{\mathbf{w}_{k+1}}{\|\mathbf{w}_{k+1}\|}; \\
k &= 1, \ldots, n-1.
\end{aligned}
\qquad (4.4.9)
$$

Then:

i. Since X is linearly independent $\|\mathbf{x}_1\| > 0$ and so $\mathbf{u}_1$ is well-defined. Furthermore $\|\mathbf{u}_1\| = 1$.

ii. Mathematical induction shows that:

    a. each $\mathbf{w}_{k+1}$ is a linear combination of $\mathbf{x}_1, \ldots, \mathbf{x}_{k+1}$ and that in this linear combination the coefficient of $\mathbf{x}_{k+1}$ is 1, hence not 0;

---

b. each $\mathbf{w}_{k+1} \neq \mathbf{O}$ and so the denominator in the definition of $\mathbf{u}_{k+1}$ is not 0, i.e., $\mathbf{u}_{k+1}$ is well-defined and $\| \mathbf{u}_{k+1} \| = 1$.

iii.

$$(\mathbf{w}_2, \mathbf{u}_1) = (\mathbf{x}_2 - (\mathbf{x}_2, \mathbf{u}_1)\mathbf{u}_1, \mathbf{u}_1)$$
$$= (\mathbf{x}_2, \mathbf{u}_1) - (\mathbf{x}_2, \mathbf{u}_1)(\mathbf{u}_1, \mathbf{u}_1)$$
$$= (\mathbf{x}_2, \mathbf{u}_1) - (\mathbf{x}_2, \mathbf{u}_1)1 = 0. \tag{4.4.10}$$

More generally, if it is assumed that

$$(\mathbf{u}_i, \mathbf{u}_j) = \begin{cases} 1, & \text{if } i = j \leq k \\ 0, & \text{if } i \neq j, \ i, j \leq k \end{cases}$$

then

$$(\mathbf{w}_{k+1}, \mathbf{u}_j) = (\mathbf{x}_{k+1} - \sum_{i=1}^{k}(\mathbf{x}_{k+1}, \mathbf{u}_i)\mathbf{u}_i, \mathbf{u}_j)$$
$$= (\mathbf{x}_{k+1}, \mathbf{u}_j) - \sum_{i=1}^{k}(\mathbf{x}_{k+1}, \mathbf{u}_i)(\mathbf{u}_i, \mathbf{u}_j)$$
$$= (\mathbf{x}_{k+1}, \mathbf{u}_j) - (\mathbf{x}_{k+1}, \mathbf{u}_j)(\mathbf{u}_j, \mathbf{u}_j)$$
$$= (\mathbf{x}_{k+1}, \mathbf{u}_j) - (\mathbf{x}_{k+1}, \mathbf{u}_j)1 = 0.$$

Hence $(\mathbf{u}_{k+1}, \mathbf{u}_j) = 0$ if $j \leq k$ and $U \stackrel{\text{def}}{=} \{\mathbf{u}_1, \ldots, \mathbf{u}_n\}$ is a set of n orthonormal vectors.

Exercise 4.4.12. Let X be

$$\{(1, 2, 3), \ (2 + 1, 3 - 2i, 4i), \ \text{and} \ (3, 2, 1)\} \ \text{in} \ \mathbf{C}_3.$$

Apply the Gram-Schmidt process to X and produce thereby an orthonormal basis U for $\mathbf{C}_3$.

Exercise 4.4.13. Show that in the notation used in the Gram-Schmidt process if $1 \leq i \leq n$ then $\mathbf{x}_i$ is a linear combination of $\mathbf{u}_1, \ldots, \mathbf{u}_i$.

Exercise 4.4.14. Show that if S is any set of vectors in $\mathbf{C}^n$ then S is linearly dependent iff there is in S a finite set of vectors, $\mathbf{x}_1, \ldots, \mathbf{x}_m$ such that in the application of the Gram-Schmidt process to these vectors,

$$\mathbf{x}_m - \sum_{i=1}^{m-1}(\mathbf{x}_m, \mathbf{u}_i)\mathbf{u}_i = \mathbf{O}. \tag{4.4.11}$$

Assume that X is finite: $X = \{\mathbf{x}_1, \ldots, \mathbf{x}_n\}$, but that X is not necessarily linearly independent. If for some m strictly less than n (4.4.11) is valid then S is

---

linearly dependent and (4.4.11) may be viewed as marking a *degeneracy*. Each time a degeneracy occurs the "degenerate" vector $(\mathbf{x}_m)$ is removed from X and the Gram-Schmidt process continues. After at most n steps of the procedure there emerges a set $\tilde{U}$ consisting of (possibly fewer than n) orthonormal vectors.

The Gram-Schmidt process plays an important rôle in many parts of linear algebra. For example, the Gram-Schmidt process applied to a basis X for a subspace W yields an orthonormal basis Y for W. Thus it follows for a second time that every subspace W of a finite-dimensional Euclidean vector space contains an orthonormal basis.

In **Figure 4.4.2** the vector

$$(\mathbf{x}_2, \mathbf{u}_1)\mathbf{u}_1$$

is, among all multiples of $\mathbf{x}_1$, i.e., among all vectors lying on the line containing $\mathbf{x}_1$, the vector nearest to $\mathbf{x}_2$, i.e., for all numbers a

$$\|\mathbf{x}_2 - (\mathbf{x}_2, \mathbf{u}_1)\mathbf{u}_1\| \leq \|\mathbf{x}_2 - a\mathbf{x}_1\|.$$

Equivalently one may say

$$\|\mathbf{x}_2 - (\mathbf{x}_2, \mathbf{u}_1)\mathbf{u}_1\| = minimum\{\|\mathbf{x}_2 - a\mathbf{x}_1\| \mid a \in \mathbf{C}\}.$$

The essential generalization of the last statement is easily phrased in terms of the notion of the span of a set of vectors.

Since
$$\{\,a\mathbf{x}_1 \;:\; a \in \mathbf{C}\,\} = span(\mathbf{x}_1) = span(\mathbf{u}_1)$$

it follows that

$$\|\mathbf{x}_2 - (\mathbf{x}_2, \mathbf{u}_1)\mathbf{u}_1\| = minimum\{\|\mathbf{x}_2 - \mathbf{w}\| \mid \mathbf{w} \in span(\mathbf{u}_1)\}.$$

The generalization mentioned is the following.

THEOREM **4.4.3.** LET

$$U \overset{\text{def}}{=} \{\mathbf{u}_1, \ldots, \mathbf{u}_m\}$$

BE AN ORTHONORMAL SET OF VECTORS IN $\mathbf{C}^n$ AND LET $\mathbf{x}$ BE AN $n \times 1$ VECTOR. IF W IS THE SPAN OF U: $W = span(U) = span(\{\mathbf{u}_1, \ldots, \mathbf{u}_m\})$, THEN THERE IS IN W A UNIQUE VECTOR $\mathbf{v}$ SUCH THAT

$$\|\,\mathbf{x} - \mathbf{v}\,\| = minimum\,\{\,\|\,\mathbf{x} - \mathbf{w}\,\| \;:\; \mathbf{w} \in W\,\}.$$

Proof. Consider the set

$$H \stackrel{\text{def}}{=} \left\{ \mathbf{h} \stackrel{\text{def}}{=} \mathbf{x} - \sum_{i=1}^{m} a_i \mathbf{u}_i \; : \; \mathbf{a} = (a_1, \ldots, a_m) \in \mathbf{C}_m \right\}.$$

If $\mathbf{h} \in H$ and if $b_i = (\mathbf{x}, \mathbf{u}_i)$ then

$$\| \mathbf{h} \|^2 = \|\mathbf{x}\|^2 - \sum_{i=1}^{m} \overline{a}_i (\mathbf{x}, \mathbf{u}_i)$$

$$- \sum_{i=1}^{m} a_i (\mathbf{u}_i, \mathbf{x}) + \sum_{i,j=1}^{m} a_i \overline{a}_j (\mathbf{u}_i, \mathbf{u}_j)$$

$$= \| \mathbf{x} \|^2 + \left(- \sum_{i=1}^{m} \overline{a}_i b_i - \sum_{i=1}^{m} a_i \overline{b}_i + \sum_{i=1}^{m} |a_i|^2 \right). \tag{4.4.12}$$

The quantity in parentheses in the right member of (4.4.12) may be rewritten as follows:

$$\left(\sum_{i=1}^{m} |a_i - b_i|^2 - \sum_{i-1}^{m} |b_i|^2 \right)$$

and so

$$\| \mathbf{h} \|^2 = \| \mathbf{x} \|^2 - \sum_{i=1}^{m} |b_i|^2 + \sum_{i=1}^{m} |a_i - b_i|^2. \tag{4.4.13}$$

Since $\mathbf{x}$ is given, it determines the b_i and so the only free variables to which values may be assigned at will are the a_i. Thus $\| \mathbf{h} \|^2$ is least iff $a_i = b_i$, i.e., iff

$$a_i = b_i = (\mathbf{x}, \mathbf{u}_i), \; 1 \leq i \leq m.$$

Thus the vector $\mathbf{v}$ in W and nearest to $\mathbf{x}$ is

$$\sum_{i=1}^{m} (\mathbf{x}, \mathbf{u}_i) \mathbf{u}_i. \tag{4.4.14}$$

$$\Omega$$

The formula in (4.4.14) further illuminates the geometry in **Figure 4.4.2** and the expression

$$\sum_{i=1}^{k} (\mathbf{x}_{k+1}, \mathbf{u}_i) \mathbf{u}_i$$

in (4.4.9).

Theorem **4.4.3** may be applied as follows:

Example 4.4.5. Let S be any set of vectors in $\mathbf{C}^n$. Then S contains a maximally linearly independent set $X \stackrel{\text{def}}{=} \{\mathbf{x}_1, \ldots, \mathbf{x}_m\}$ and $m \leq n$. Furthermore

$$W \stackrel{\text{def}}{=} span(S) = span(X)$$

and if $U \overset{\text{def}}{=} \{u_1, \ldots, u_m\}$ is the orthonormal set arising from the Gram-Schmidt process applied to X then $W = span(U)$. If $\mathbf{x}$ is *any* vector, in W or not in W,

$$\left\| \mathbf{x} - \sum_{i=1}^{m} (\mathbf{x}, u_i) u_i \right\| = minimum \{ \|\mathbf{x} - \mathbf{w}\| \ : \ \mathbf{w} \in W \}.$$

The minimum of all such distances between $\mathbf{x}$ and vectors $\mathbf{w}$ in W is called the *distance* of $\mathbf{x}$ from W.

Exercise 4.4.15. Find the distance from the vector

$$(1, 2 - i, 3)$$

to the span of

$$(i, -1, -i) \quad \text{and} \quad (6, 2i, -3)$$

in $\mathbf{C}_3$.

Exercise 4.4.16. Let A be the matrix

$$\begin{pmatrix} 1 & 2 & 3 & 4 \\ -4 & 3 & -2 & 1 \\ 3 & -5 & -1 & -5 \\ 2 & -7 & -4 & -9 \end{pmatrix}.$$

i. Find a maximally linearly independent set X of solutions of the system

$$A\mathbf{x} = \mathbf{O},$$

i.e., find a basis for $ker(A)$.

ii. Apply the Gram-Schmidt process to X and call the result U.

iii. Show that if

$$\mathbf{w} \in W \overset{\text{def}}{=} span(U) \ (= span(X))$$

then $A\mathbf{w} = \mathbf{O}$. (For this reason W, which is $ker(A)$, is called also the *solution space* of the system $A\mathbf{x} = \mathbf{O}$.)

iv. Find the distance of

$$\mathbf{b} \overset{\text{def}}{=} \begin{pmatrix} 10 \\ 17 \\ 3 \\ -11 \end{pmatrix}$$

from W.

v. Fill out the set $S \overset{\text{def}}{=} \{(1, 2, 3, 4), (-4, 3, -2, 1)\}$ consisting of the first two rows of A to a basis Y for $\mathbf{C}_4$. Apply the Gram-Schmidt process to Y and call the result V: $V = \{v_1, \ldots, v_4\}$. Show that $W = span(\{v_3^*, v_4^*\})$.

The results in **Exercise 4.4.16** may be summarized in the following general terms.

Let $A \overset{\text{def}}{=} (a_{ij})_{i,j=1}^{m,n}$ be an $m \times n$ matrix and let W be the solution space of the system $A\mathbf{x} = \mathbf{O}$. Let V be the row space of A: $V = span(\{\mathbf{a}_1, \ldots, \mathbf{a}_m\})$. Here are some useful connections between W and V.

   i. The solution space W consists of all $n \times 1$ vectors

$$\mathbf{w} \overset{\text{def}}{=} \begin{pmatrix} w_1 \\ \vdots \\ w_n \end{pmatrix}$$

such that $\mathbf{w}^*$ is orthogonal to all vectors V. In other words, if $\mathbf{v} \in V$ then $\mathbf{vw} = (\mathbf{v}, \mathbf{w}^*) = 0$. Hence, with abuse of notation, $W^* = V^{\perp}$.

   ii. Let $U \overset{\text{def}}{=} \{\mathbf{u}_1, \ldots, \mathbf{u}_k\}$, $1 \le k \le n$, be *any* maximally orthonormal (hence maximally linearly independent) set in W. The distance of a vector $\mathbf{y}$ in $\mathbf{C}^n$ from W is

$$\left\| \mathbf{y} - \sum_{i=1}^{k} (\mathbf{y}, \mathbf{u}_i)\mathbf{u}_i \right\|.$$

Let $U \overset{\text{def}}{=} \{\mathbf{u}_1, \ldots, \mathbf{u}_m\}$ be an orthonormal set in $\mathbf{C}^n$. Since U is linearly independent, $m \le n$. If $\mathbf{x} \in \mathbf{C}^n$ then

$$\left\| \mathbf{x} - \sum_{i=1}^{m} (\mathbf{x}, \mathbf{u}_i)\mathbf{u}_i \right\|^2 = \|\mathbf{x}\|^2 - \sum_{i=1}^{m} |(\mathbf{x}, \mathbf{u}_i)|^2 \ge 0.$$

Hence there emerges:

$$\sum_{i=1}^{m} |(\mathbf{x}, \mathbf{u}_i)|^2 \le \|\mathbf{x}\|^2 \ \textit{Bessel's inequality}.$$

Furthermore equality holds in Bessel's inequality iff

$$\mathbf{x} = \sum_{i=1}^{m} (\mathbf{x}, \mathbf{u}_i)\mathbf{u}_i.$$

Hence Bessel's inequality is an equality for *all* $\mathbf{x}$ in $\mathbf{C}^n$ iff $m = n$ iff U is not merely an orthonormal set in $\mathbf{C}^n$ but actually an orthonormal *basis* in $\mathbf{C}^n$. (The last sentence is sometimes called the *Parseval theorem*.)

The considerations above lead to the following *alternative* for an $m \times n$ matrix $A \overset{\text{def}}{=} (a_{ij})_{i,j=1}^{m,n}$.

THEOREM 4.4.4. IF $A \overset{\text{def}}{=} (a_{ij})_{i,j=1}^{m,n}$ AND IF $\mathbf{b} \in \mathbf{C}_n$ THEN:

EITHER

THERE IS IN $\mathbf{C}_m$ AN $\mathbf{x}$ SUCH THAT

$$\mathbf{x}A = \mathbf{b};$$

OR

THERE IS IN $\mathbf{C}^n$ A $\mathbf{y}$ SUCH THAT

$$A\mathbf{y} = \mathbf{O}, \quad \text{AND} \quad \mathbf{b}\mathbf{y} = 1.$$

PROOF. Let $\mathbf{a}_1, \ldots, \mathbf{a}_m$ be the $1 \times n$ vectors that are the rows of A. To say the $\mathbf{x}A = \mathbf{b}$ is to say that $\mathbf{b}$ is in row space of A, i.e., that

$$\mathbf{b} \in span(\{\mathbf{a}_1, \ldots, \mathbf{a}_m\}) \overset{\text{def}}{=} W \subset \mathbf{C}_n.$$

Hence if no such $\mathbf{x}$ exists then $\mathbf{b} \notin W$ and if $U \overset{\text{def}}{=} \{\mathbf{u}_1, \ldots, \mathbf{u}_k\}$ is an orthonormal basis for W then $k \leq m$ and

$$\mathbf{b} - \sum_{i=1}^{k} (\mathbf{b}, \mathbf{u}_i)\mathbf{u}_i \overset{\text{def}}{=} \mathbf{z} \neq \mathbf{O}.$$

Then

$$(\mathbf{z}, \mathbf{u}_j) = (\mathbf{b}, \mathbf{u}_j) - \sum_{i=1}^{k} (\mathbf{b}, \mathbf{u}_i)(\mathbf{u}_i, \mathbf{u}_j)$$

$$= (\mathbf{b}, \mathbf{u}_j) - (\mathbf{b}, \mathbf{u}_j) = 0, \text{ i.e., } \mathbf{z} \in W^{\perp}$$

$$(\mathbf{z}, \mathbf{b}) \overset{\text{def}}{=} a = \|\mathbf{b}\|^2 - \sum_{i=1}^{k} |(\mathbf{b}, \mathbf{u}_i)|^2 > 0$$

according to Bessel's inequality. Hence if

$$\mathbf{y} \overset{\text{def}}{=} \frac{1}{a}\mathbf{z}$$

then, since $\mathbf{z} \in W^{\perp}$ it follows that

$$Ay = \mathbf{O}$$
$$by = 1.$$

$$\Omega$$

Exercise 4.4.17. In the notation used in describing the Gram-Schmidt process write X as a vectorial column vector $\mathcal{X}$ in which the components are the *vectors* that constitute X. Similarly write U as a vectorial column vector $\mathcal{U}$ in which the components are the *vectors* that constitute U. Thus there is a matrix $\Gamma \overset{\text{def}}{=} (\gamma_{ij})_{i,j=1}^{n,n}$ such that

$$\mathcal{U} \overset{\text{def}}{=} \begin{pmatrix} \mathbf{u}_1 \\ \vdots \\ \mathbf{u}_n \end{pmatrix} = \begin{pmatrix} \gamma_{11} & \cdots & \gamma_{1n} \\ \vdots & \ddots & \vdots \\ \gamma_{n1} & \cdots & \gamma_{nn} \end{pmatrix} \begin{pmatrix} \mathbf{x}_1 \\ \vdots \\ \mathbf{x}_n \end{pmatrix} \overset{\text{def}}{=} \Gamma \mathcal{X}.$$

i. Show that Γ is lower triangular and invertible. (Hence

$$\mathbf{u}_i \in span(\{\mathbf{x}_1,\ldots,\mathbf{x}_i\}), \ 1 \leq i \leq n;$$

and conversely

$$\mathbf{x}_i \in span(\{\mathbf{u}_1,\ldots,\mathbf{u}_i\}), \ 1 \leq i \leq n.)$$

ii. Show that the "volume" of the parallelotope corresponding to the vectors in X is $[det(\Gamma)]^{-1}$ (cf. **Section 3.3 III**).

iii. Let A be an $m \times n$ matrix in which all the rows are linearly independent ($m \leq n$, A is WIDE or SQUARE). Show that there is an invertible lower triangular $m \times m$ matrix R and an $m \times n$ matrix Q in which the rows are orthonormal and for which $A = RQ$. (If A is SQUARE then Q is unitary.)

iv. Let A be an $m \times n$ matrix in which the columns are linearly independent ($m \geq n$, A is HIGH or SQUARE). Show there is an invertible upper triangular $n \times n$ matrix R and an $m \times n$ matrix Q in which the rows are orthonormal and for which $A = QR$.

v. Show that if A is a SQUARE matrix (possibly singular) there is an upper triangular matrix R and a unitary matrix Q such that $A = QR$.

[*Hint:* Assume the *unknown* matrix $R = (r_{ij})_{i,j=1}^{n,n}$. Let the columns of A be the *known* vectors $\mathbf{a}_1,\ldots,\mathbf{a}_n$ and let the columns of Q be the *unknown* vectors $\mathbf{q}_1,\ldots,\mathbf{q}_n$. Show that there is a number r_{11} and a vector $\mathbf{q}_1$ of length 1 and such that $\mathbf{a}_1 = r_{11}\mathbf{q}_1$. Then show there is a pair r_{12}, r_{22} of numbers and a vector $\mathbf{q}_2$ of length 1 so that $\mathbf{a}_2 = r_{12}\mathbf{q}_1 + r_{22}\mathbf{q}_2$ *and* $(\mathbf{q}_1, \mathbf{q}_2) = 0$. Continue by mathematical induction and thereby define R as an upper triangular matrix and Q as a matrix in which the rows are orthonormal. (The last step involves finding numbers $r_{1n},\ldots,r_{nn}$ and a vector $\mathbf{q}_n$ of length 1 so that

$$\mathbf{a}_n = \sum_{i=1}^{n} r_{in}\mathbf{q}_i$$
$$(\mathbf{q}_n, \mathbf{q}_i) = 0, \ 1 \leq i \leq n - 1.)]$$

[**Remark 4.4.7:** The representation $A = QR$ in *iv* above is particularly useful when A is SQUARE, since by a simple trick it leads in some instances to an iteration algorithm for approximating the eigenvalues of A (cf. **Section 5.3**).]

Exercise 4.4.18. Assume $m < n$ and that

$$S \stackrel{\text{def}}{=} \{\mathbf{x}_1, \ldots, \mathbf{x}_m\}$$

is an orthonormal set of $n \times 1$ vectors. Let X be

$$\{\mathbf{x}_1, \ldots, \mathbf{x}_m, \mathbf{e}_1, \ldots, \mathbf{e}_n\}.$$

Show that the result U of applying the Gram-Schmidt process to X fills S out to a maximally linearly independent set in $\mathbf{C}^n$.

Example 4.4.6. Let A be the matrix

$$\begin{pmatrix} 3 & 8 \\ 12 & 7 \end{pmatrix}.$$

Write A in columnar form as follows:

$$A \sim \mathbf{A} = \begin{pmatrix} 3 \\ 8 \\ 12 \\ 7 \end{pmatrix}.$$

Similarly I and A^2 may be written in columnar form as follows:

$$I \sim \mathbf{I} = \begin{pmatrix} 1 \\ 0 \\ 0 \\ 1 \end{pmatrix} \quad \text{and} \quad A \sim \mathbf{A}^2 = \begin{pmatrix} 105 \\ 80 \\ 120 \\ 145 \end{pmatrix}.$$

Regarded as vectors in $\mathbf{C}^4$ the columnar forms of I, A, and A^2 may be submitted to the Gram-Schmidt process with these consequences:

$$\mathbf{u}_1 = \begin{pmatrix} \frac{1}{\sqrt{2}} \\ 0 \\ 0 \\ \frac{1}{\sqrt{2}} \end{pmatrix}, \mathbf{u}_2 = \begin{pmatrix} -\frac{1}{3\sqrt{6}} \\ \frac{8}{6\sqrt{6}} \\ \frac{12}{6\sqrt{6}} \\ \frac{1}{3\sqrt{6}} \end{pmatrix}.$$

Since

$$u_1 = \frac{1}{\sqrt{2}}\mathbf{I},$$

$$u_2 = \frac{1}{6\sqrt{6}}(\mathbf{A} - 5\mathbf{I})$$

it follows that

$$\mathbf{A}^2 - (\mathbf{A}^2, u_2)u_2 - (\mathbf{A}^2, u_1)u_1 = \mathbf{A}^2 - (10\mathbf{A} - 50\mathbf{I}) - 125\mathbf{I} = O,$$

i.e., at $\mathbf{A}^2$ a degeneracy occurs. Since for no number a is it true that $A - aI = O$ it follows that $m_A(A) = A^2 - 10A - 75$.

The calculations above can be generalized.

Example 4.4.7. Let $A \overset{\text{def}}{=} (a_{ij})_{i,j=1}^{n,n}$ be an $n \times n$ matrix. Rearrange the entries so that they are lined up in a column:

$$A \sim \mathbf{A} \overset{\text{def}}{=} \begin{pmatrix} a_{11} \\ \vdots \\ a_{1n} \\ a_{21} \\ \vdots \\ a_{2n} \\ \vdots \\ a_{n1} \\ \vdots \\ a_{nn} \end{pmatrix}.$$

Let the entries of $A^2, \ldots, A^k, \ldots$ be arranged similarly. Hence each A^k is to be regarded as an element $\mathbf{A}^k$ of $\mathbf{C}^{n^2}$. Let the Gram-Schmidt process be applied to the set $\{\mathbf{I}, \mathbf{A}, \mathbf{A}^2, \ldots\}$. Denote the orthonormal $n^2 \times 1$ vectors so produced by $u_1, \ldots$. At some point in the procedure there will emerge a degeneracy:

$$\mathbf{A}^\mu - \sum_{i=0}^{\mu-1}(\mathbf{A}^\mu, u_i)u_i = \mathbf{O}. \tag{4.4.15}$$

If each u_i is written as a linear combination of $\mathbf{I}, \mathbf{A}, \ldots \mathbf{A}^i$ there results an equation

$$\mathbf{A}^\mu - \sum_{i=0}^{\mu-1} b_i \mathbf{A}^i = O. \tag{4.4.16}$$

Exercise 4.4.19.

i. Let A and B be $n \times n$ matrices and let $\mathbf{A}$ and $\mathbf{B}$ be their correspondents in $\mathbf{C}^{n^2}$. Show that the scalar product $(\mathbf{A},\mathbf{B})$ in $\mathbf{C}^{n^2}$ is the same as $tr(AB^*)$.

ii. Show that if m_A is the minimal polynomial of A then $\mu = \mu_A$ and

$$m_A(A) = A^\mu - \sum_{i=0}^{\mu-1} b_i A^i$$

corresponding to the left member of (4.4.16) (cf. [**Gep, Le**]). Compare this method with the method given by the algorithm in **Section 4.3**.

Exercise 4.4.20. Use the techniques just explored to find the minimal polynomial for the matrices:

$$B \stackrel{\text{def}}{=} \begin{pmatrix} -2 & 4 \\ 6 & 3 \end{pmatrix} \text{ and } A \stackrel{\text{def}}{=} \begin{pmatrix} 5 & -1 & 0 \\ 1 & 5 & 2 \\ -1 & 1 & 4 \end{pmatrix}$$

Compare the result for A with the result found for it in **Section 4.3**.

Exercise 4.4.21. Review the proof of THEOREM 4.2.3 on the upper triangulability of a SQUARE matrix A. Show that the matrix P such that $P^{-1}AP$ is upper triangular may be chosen so that $P^{-1} = P^*$, i.e., so that P is unitary. This result is one of many contributions of I. Schur.

[*Hint:* Use the fact that the first vector $\mathbf{x}_1$ chosen is an eigenvector of A and hence may be chosen so that its length is 1. Then show that the other vectors $\mathbf{y}_2, \ldots, \mathbf{y}_n$ may be chosen so that X is an orthonormal basis. Hence Q is unitary. Use mathematical induction.]

Exercise 4.4.22. Let A be the matrix

$$\begin{pmatrix} -3 & 5 & 9 \\ -2 & 4 & 6 \\ -5 & 5 & 7 \end{pmatrix}.$$

Find a unitary matrix P such that P^*AP is upper triangular (cf. **Exercise 4.2.14**).

Exercise 4.4.23.

i. Show that a normal upper triangular matrix A is a diagonal matrix.

---

ii. Show that if P is unitary and P^*AP is upper triangular (cf. **Exercise 4.4.21**) then A is normal iff P^*AP is diagonal.

Exercise 4.4.24. Combine Exercises **4.4.21** and **4.4.23** to give a third proof of the SPECTRAL THEOREM (THEOREM **4.4.2**), i.e., a proof that if A is a normal matrix there is a unitary matrix P such that $P^{-1}AP$ $(=P^*AP)$ is a diagonal matrix.

In the following paragraphs there are brought together many of the ideas and techniques encountered in the work to this point.

THE STRUCTURE OF EUCLIDEAN SPACES

Let A be an $m \times n$ matrix. Then $ker(A) \subset \mathbf{C}^n$ and $im(A) \subset \mathbf{C}^m$. On the other hand, $ker(A^*) \subset \mathbf{C}^m$ and $im(A^*) \subset \mathbf{C}^n$. Thus $im(A)$ and $ker(A^*)$, as subspaces of $\mathbf{C}^m$, ought to be related; similarly $im(A^*)$ and $ker(A)$, as subspaces of $\mathbf{C}^n$ ought to be related. They, $rank(A)$, and $rank(A^*)$ are related in the following easily remembered ways:

E9. $rank(A) = rank(A^*)$
E10. $rank(A^*) = dim(im(A^*))$
E11. $dim(ker(A)) = n - dim(im(A)) = n - rank(A)$

Furthermore, the items listed next are regularly encountered phenomena in Euclidean spaces.

E12. If S is a subset of $\mathbf{C}^n$ then $S \cap S^\perp = \{\mathbf{O}\}$. (If $\mathbf{y} \in S \cap S^\perp$ then $(\mathbf{y}, \mathbf{y}) = 0$, whence, according to E4, $\mathbf{y} = \mathbf{O}$.)
E13. If V is a subspace of $\mathbf{C}^n$ then for every vector $\mathbf{x}$ in $\mathbf{C}^n$ there are unique vectors $\mathbf{v}$ in V and $\mathbf{w}$ in $V^\perp$ such that

$$\mathbf{x} = \mathbf{v} + \mathbf{w}.$$

Thus there is for $\mathbf{C}^n$ an *orthogonal decomposition* into the *direct sum* of the subspaces V and $V^\perp$: $\mathbf{C}^n = V \oplus V^\perp$. Furthermore $\mathbf{v}$ is characterized by the equation

$$\|\mathbf{x} - \mathbf{v}\| = minimum\{\|\mathbf{x} - \mathbf{z}\| : \mathbf{z} \in V\}$$

as in THEOREM **4.4.3**. [*Proof.* Let $X \stackrel{\text{def}}{=} \{\mathbf{u}_1, \ldots, \mathbf{u}_k\}$ be an orthonormal basis for V. Then $\mathbf{v}$ and $\mathbf{w}$ may be defined according to

$$\mathbf{v} \stackrel{\text{def}}{=} \sum_{i=1}^{k} (\mathbf{x}, \mathbf{u}_i)\mathbf{u}_i \in V$$

$$\mathbf{w} \stackrel{\text{def}}{=} \mathbf{x} - \sum_{i=1}^{k} (\mathbf{x}, \mathbf{u}_i)\mathbf{u}_i$$

whereupon $\mathbf{x} = \mathbf{v} + \mathbf{w}$. Furthermore, if $\mathbf{z} \in V$ then

$$(\mathbf{z}, \mathbf{w}) = \left(\sum_{j=1}^{k}(\mathbf{z}, \mathbf{u}_j)\mathbf{u}_j, \mathbf{x} - \sum_{i=1}^{k}(\mathbf{x}, \mathbf{u}_i)\mathbf{u}_i\right)$$

$$= \sum_{j=1}^{k}(\mathbf{z}, \mathbf{u}_j)(\mathbf{u}_j, \mathbf{x}) - \sum_{i,j}[(\mathbf{z}, \mathbf{u}_j)\overline{(\mathbf{x}, \mathbf{u}_i)}(\mathbf{u}_j, \mathbf{u}_i)$$

$$= \sum_{j=1}^{k}[(\mathbf{z}, \mathbf{u}_j)(\mathbf{u}_j, \mathbf{x}) - (\mathbf{z}, \mathbf{u}_j)(\mathbf{u}_j, \mathbf{x})] = 0.$$

Hence $\mathbf{w} \in V^{\perp}$. If $\mathbf{v}' \in V$, $\mathbf{w}' \in V^{\perp}$, and $\mathbf{x} = \mathbf{v}' + \mathbf{w}'$ then

$$\mathbf{v} - \mathbf{v}' = \mathbf{w}' - \mathbf{w} \in V \cap V^{\perp} = \{\mathbf{O}\}.]$$

If P is defined by the equation $P\mathbf{x} = \mathbf{v}$ and p and q are numbers then by direct calculation it follows that:

$$P(p\mathbf{x} + q\mathbf{y}) = pP\mathbf{x} + qP\mathbf{y} \ (P \text{ is linear});$$

$$P^2 = P \ (P \text{ is an } idempotent);$$

$$(I - P)^2 = I - P \ (I - P \text{ is an idempotent});$$

$$\|\mathbf{x}\|^2 = \|\mathbf{v}\|^2 + \|\mathbf{w}\|^2;$$

$$\|\mathbf{v}\|, \|\mathbf{w}\| \le \|\mathbf{x}\|.$$

E14. If V is a subspace of $\mathbf{C}^n$ then

$$V = (V^{\perp})^{\perp}.$$

[*Proof.* If $\mathbf{x} \in V$ and $\mathbf{y} \in V^{\perp}$ then $(\mathbf{x}, \mathbf{y}) = 0$ whence

$$V \subset (V^{\perp})^{\perp}.$$

From E13 it follows that $dim(V) + dim(V^{\perp}) = n$ and that also

$$dim(V^{\perp}) + dim((V^{\perp})^{\perp}) = n.$$

Hence $dim((V^{\perp})^{\perp}) = dim(V)$. Since $V \subset (V^{\perp})^{\perp}$ it follows that $V = (V^{\perp})^{\perp}$.]

E15. $dim((im(A^*))^{\perp}) = n - dim(im(A^*)) = n - rank(A^*)$

Among the consequences of E9 -E15 are:

$$ker(A) = (im(A^*))^{\perp}; \tag{4.4.17}$$

$$ker(A^*) = (im(A))^{\perp}; \tag{4.4.18}$$

$$im(A) = (ker(A^*))^{\perp}; \tag{4.4.19}$$

$$im(A^*) = (ker(A))^{\perp}. \tag{4.4.20}$$

[**Remark 4.4.8:**   The relations in (4.4.17) - (4.4.20) should be compared with the discussion of the four fundamental subspaces in **Exercise 2.9.24.**]

Here are the proofs.

(4.4.17): If $\mathbf{x} \in ker(A)$ then for all vectors $\mathbf{y}$

$$(A\mathbf{x}, \mathbf{y}) = (\mathbf{O}, \mathbf{y}) = 0$$
$$= (\mathbf{x}, A^*\mathbf{y})$$

whence $ker(A) \subset (im(A^*))^\perp$. From E9, E11, and E15 it follows that

$$dim[ker(A)] = dim[(im(A^*))^\perp]$$

and so (4.4.17) is true.

(4.4.18): Replacing A by A^* in (4.4.17) yields (4.4.18).

(4.4.19): See (4.4.18) and E14.

(4.4.20): See (4.4.17) and E14.

The geometric relations in $\mathbf{R}^2$ suggest that if $\mathbf{x}$ and $\mathbf{y}$ are vectors in $\mathbf{R}^2$ then $(\mathbf{x}, \mathbf{y}) = ||\mathbf{x}|| \cdot ||\mathbf{y}|| \cos\theta$. Since $|\cos\theta| \leq 1$ it follows that

$$|(\mathbf{x}, \mathbf{y})| \leq ||\mathbf{x}|| \cdot ||\mathbf{y}||, \text{ the Schwarz inequality.} \tag{4.4.21}$$

The next **Exercise** leads to the conclusion that the Schwarz inequality (4.4.21) is valid not only for vectors in $\mathbf{R}^2$ but in general. Furthermore the deduction is based entirely on the properties E1-E4.

Exercise 4.4.25. Let $\mathbf{x}$ and $\mathbf{y}$ be vectors in $\mathbf{C}^n$. Show:

i. E3 implies that for any vectors $\mathbf{x}$ and $\mathbf{y}$: $(\mathbf{x}, \mathbf{O}) = (\mathbf{O}, \mathbf{y}) = 0$;

ii. E3 and E4 imply (4.4.21) is valid if $\mathbf{x}=\mathbf{O}$ or if $\mathbf{y}=\mathbf{O}$;

iii. E1-E4 imply that for any number a

$$(\mathbf{x} + a\mathbf{y}, \mathbf{x} + a\mathbf{y}) = ||\mathbf{x}||^2 + \overline{a}(\mathbf{x}, \mathbf{y}) + a(\mathbf{y}, \mathbf{x}) + ||\mathbf{y}||^2$$
$$\geq 0;$$

iv. if, in *iii*, $\mathbf{y} \neq \mathbf{O}$, the number

$$a \overset{\text{def}}{=} -\frac{(\mathbf{x}, \mathbf{y})}{||\mathbf{y}||^2}$$

is well defined;

v. if a as in *iv* is substituted in the inequality in *iii* and the result is simplified the conclusion is (4.4.21);

vi. equality holds in the Schwarz inequality iff $\mathbf{x}$ and $\mathbf{y}$ are linearly dependent.

Exercise 4.4.26. Use the conclusion in **Exercise 4.4.25** to prove

$$\|\mathbf{x} + \mathbf{y}\|^2 \leq (\|\mathbf{x}\| + \|\mathbf{y}\|)^2,$$

i.e., that

$$\|\mathbf{x} + \mathbf{y}\| \leq \|\mathbf{x}\| + \|\mathbf{y}\|, \text{ the Minkowski inequality.} \qquad (4.4.22)$$

The Minkowski inequality (4.4.22) is known also as the *triangle inequality*. The Schwarz and Minkowski inequalities play important rôles in the computational aspects of linear algebra and in the infinite-dimensional versions of Euclidean vector spaces, i.e., in *Hilbert spaces*.

Exercise 4.4.27. Use **Figure 4.4.1** to interpret the triangle inequality (4.4.22) as follows:

The sum of the lengths of two sides of a triangle is at least as great as the length of the third side.

Exercise 4.4.28. Let **f** be a linear functional defined on $\mathbf{C}^n$. Assume

$$\mathbf{f}(\mathbf{e}_i) = \overline{b}_i,$$

and define **y** by the equation

$$\mathbf{y} \stackrel{\text{def}}{=} \sum_{i=1}^{n} b_i \mathbf{e}_i.$$

(Note that **y** depends on **f**: $\mathbf{y} = T(\mathbf{f})$.) Show that for any vector **x** in $\mathbf{C}^n$

$$\mathbf{f}(\mathbf{x}) = (\mathbf{x}, \mathbf{y}).$$

Show also that T is a linear transformation: for numbers a, b and linear functionals **f**, **g**, $T(a\mathbf{f} + b\mathbf{g}) = aT(\mathbf{f}) + bT(\mathbf{g})$.

[**Remark 4.4.9:** The last result is one of several *Riesz representation theorems*.]

Example 4.4.8. Let

$$X \stackrel{\text{def}}{=} \{\mathbf{x}_1, \ldots, \mathbf{x}_n\}$$

be an orthonormal basis for $\mathbf{C}^n$. If $\mathbf{x} \in \mathbf{C}^n$ then there are numbers a_i such that

$$\mathbf{x} = \sum_{i=1}^{n} a_i \mathbf{x}_i.$$

Owing to the orthonormality of the set X,

$$(\mathbf{x}, \mathbf{x}_j) = \left(\sum_{i=1}^{n} a_i \mathbf{x}_i, \mathbf{x}_j\right) = \sum_{i=1}^{n} \overline{a}_i (\mathbf{x}_i, \mathbf{x}_j)$$
$$= \sum_{i=1}^{n} a_i \delta_{ij}$$
$$= a_j.$$

If

$$\mathbf{y} = \sum_{i=1}^{n} b_i \mathbf{x}_i$$

then $b_i = (\mathbf{y}, \mathbf{x}_i)$. Furthermore, direct calculation shows that

$$(\mathbf{x}, \mathbf{y}) = \sum_{i=1}^{n} a_i \overline{b}_i$$
$$\|\mathbf{x}\|^2 = \sum_{i=1}^{n} |a_i|^2$$
$$\|\mathbf{y}\|^2 = \sum_{i=1}^{n} |b_i|^2.$$

Let $\{\mathbf{f}_j\}$, $1 \leq j \leq n$, be the *coefficient functionals* such that

$$\mathbf{f}_j(\mathbf{x}) = a_i = (\mathbf{x}, \mathbf{x}_j). \tag{4.4.23}$$

Then, according to the Riesz representation theorem above, there are vectors $\mathbf{y}_i$ such that

$$\mathbf{f}_j(\mathbf{x}) = (\mathbf{x}, \mathbf{y}_j). \tag{4.4.24}$$

Comparison of (4.4.23) and (4.4.24) shows that the "Riesz representing" vector $\mathbf{y}_j$ for $\mathbf{f}_j$ is $\mathbf{x}_j$.

Exercise 4.4.29. Let S be the set

$$\{(1, 2, 3), (4, 5, 6), (7, 8, 6)\} \stackrel{\text{def}}{=} \{\mathbf{z}_1, \mathbf{z}_2, \mathbf{z}_3\}.$$

i. Show S is linearly independent and a basis for $\mathbf{C}_3$.

ii. Find the coefficient functionals for S. (For an arbitrary vector (a_1, a_2, a_3) in $\mathbf{C}_3$

$$(a_1, a_2, a_3) = \alpha_1 \mathbf{z}_1 + \alpha_2 \mathbf{z}_2 + \alpha_3 \mathbf{z}_3.$$

Give formulae for $\alpha_1, \alpha_2, \alpha_3$ in terms of a_1, a_2, a_3.)

    Chapter 4. Useful forms for matrices

iii. Apply the Gram-Schmidt process to S to produce an orthonormal basis Σ for $\mathbf{C}_3$.

iv. Represent $\mathbf{x} \overset{\text{def}}{=} (-1, 2, -3)$ in terms of the orthonormal basis Σ.

v. Repeat *iii* for Σ.

Let

$$\mathbf{a} \overset{\text{def}}{=} \begin{pmatrix} a_1 \\ a_2 \\ a_3 \end{pmatrix}$$

be a fixed vector in $\mathbf{C}^3$. For any vector

$$\mathbf{x} \overset{\text{def}}{=} \begin{pmatrix} x_1 \\ x_2 \\ x_3 \end{pmatrix}$$

in $\mathbf{C}^3$ the vector

$$\mathbf{x} \times \mathbf{a} \overset{\text{def}}{=} \begin{pmatrix} x_2 a_3 - x_3 a_2 \\ x_3 a_1 - x_1 a_3 \\ x_1 a_2 - x_2 a_1 \end{pmatrix}$$

is called the *cross-product* of $\mathbf{x}$ and $\mathbf{a}$ (in that order).

Exercise 4.4.30. Calculate

$$\begin{pmatrix} 1 \\ 2 \\ 3 \end{pmatrix} \times \begin{pmatrix} 4 \\ 5 \\ 6 \end{pmatrix}.$$

Exercise 4.4.31. Show that:

i. $\mathbf{x} \times \mathbf{y} = -(\mathbf{y} \times \mathbf{x}) = (-\mathbf{y}) \times \mathbf{x}$;

ii. if the components of $\mathbf{x}$ and $\mathbf{y}$ are real then $\mathbf{x} \times \mathbf{y}$ is orthogonal to both $\mathbf{x}$ and $\mathbf{y}$;

iii. $\mathbf{x}$ and $\mathbf{y}$ are linearly dependent iff if $\mathbf{x} \times \mathbf{y} = \mathbf{O}$.

Exercise 4.4.32. Let $\mathbf{a} \overset{\text{def}}{=} (a_1, a_2, a_3)^t$ be a fixed *nonzero* vector in $\mathbf{C}^3$ and let T be the transformation that carries each $\mathbf{x}$ into $\mathbf{x} \times \mathbf{a}$:

$$T(\mathbf{x}) \overset{\text{def}}{=} \mathbf{x} \times \mathbf{a}.$$

i. Show that T is a linear transformation.

ii. Find the matrix A that acts on the E-components of a vector $\mathbf{x}$ and yields the E-components of $\mathbf{x} \times \mathbf{a}$.

iii. Describe $\chi_A(\lambda)$.

iv. Assume
$$a_1^2 + a_2^2 + a_3^2 \neq 0.$$
Find the eigenvalues of A, for each eigenvalue a corresponding eigenvector, and a matrix P such that $P^{-1}AP$ is a diagonal matrix D.

v. Show that 0 is always an eigenvalue for A and that $\mathbf{a}$ is always a corresponding eigenvector for A.

vi. Show that if $a_1^2 + a_2^2 + a_3^2 = 0$ then A is not diagonable.

Exercise 4.4.33. Let x, y, and r be real numbers $(x, y, r \in \mathbf{R})$ and assume $(x - r)^2 + y^2 = r^2$. Show that
$$\begin{pmatrix} x \\ y \end{pmatrix} \quad \text{and} \quad \begin{pmatrix} x - 2r \\ y \end{pmatrix}$$
are orthogonal. Interpret this result as showing algebraically that if P and Q are diametrically opposite points on a circle C and if R is a third point on C and different from both P and Q then the line segments $\overline{PR}$ and $\overline{QR}$ are perpendicular. ("An angle inscribed in a semicircle is a right angle.")

Let $X \stackrel{\text{def}}{=} \{\mathbf{x}_1, \ldots, \mathbf{x}_m\}$ be a set of vectors in $\mathbf{C}^n$. No relation is assumed to exist between m and n: $m < n$, $m = n$, or $m > n$. Form the $m \times m$ matrix $\mathcal{G} \stackrel{\text{def}}{=} (g_{ij})_{i,j=1}^{m,m}$ in which $g_{ij} = (\mathbf{x}_i, \mathbf{x}_j)$. The matrix $\mathcal{G}$ is called the *Gramian* of the set X.

Exercise 4.4.34. Show that for each $m \times 1$ vector $\mathbf{y}$,
$$(\mathcal{G}\mathbf{y}, \mathbf{y}) \geq 0 \tag{4.4.25}$$
and that (4.4.25) implies that $\mathcal{G}$ is self-adjoint.

[*Hint:* Let G be the $m \times n$ matrix in which the ith row is the row vector $\mathbf{x}_i^t$. Show $\mathcal{G} = GG^\times$.]

Let $P \stackrel{\text{def}}{=} (p_{ij})_{i,j=1}^{m,m}$ be an $m \times m$ unitary matrix such that $P^{-1}\mathcal{G}P \stackrel{\text{def}}{=} D$ is a diagonal matrix:
$$D \stackrel{\text{def}}{=} \begin{pmatrix} \lambda_1 & & & \\ & \lambda_2 & & \\ & & \ddots & \\ & & & \lambda_m \end{pmatrix}.$$
Since $\mathcal{G}$ is self-adjoint its eigenvalues λ_j are real and hence they may be indexed so that $\lambda_1 \geq \lambda_2 \geq \cdots \geq \lambda_m$. In accordance with the notation introduced in **Section 2.10** let $\mathcal{X}$ denote the vectorial column vector
$$\begin{pmatrix} \mathbf{x}_1 \\ \vdots \\ \mathbf{x}_m \end{pmatrix}.$$

Exercise 4.4.35. Refer to **Exercise 4.4.34** and to the paragraph above.

i. Show that if $P^{-1} \stackrel{\text{def}}{=} (q_{ij})_{i,j=1}^{m,m}$ then the vectorial components in the vectorial column vector

$$\mathcal{Y} \stackrel{\text{def}}{=} \begin{pmatrix} \mathbf{y}_1 \\ \vdots \\ \mathbf{y}_m \end{pmatrix} \stackrel{\text{def}}{=} \begin{pmatrix} q_{11} & \cdots & q_{1m} \\ \vdots & \ddots & \vdots \\ q_{m1} & \cdots & q_{mm} \end{pmatrix} \begin{pmatrix} \mathbf{x}_1 \\ \vdots \\ \mathbf{x}_m \end{pmatrix} = P^{-1}\mathcal{X}$$

are pairwise orthogonal and in fact

$$(\mathbf{y}_i, \mathbf{y}_j) = \begin{cases} \lambda_i & \text{if } i = j; \\ 0 & \text{if } i \neq j. \end{cases}$$

ii. Show that each $\lambda_i \geq 0$ and that if $m > n$ then $\lambda_{n+1} = \cdots = \lambda_m = 0$.

iii. Assume $\lambda_k \neq 0 = \lambda_{k+1}$. Show that

$$S \stackrel{\text{def}}{=} \{\frac{\mathbf{y}_1}{\sqrt{\lambda_1}}, \ldots, \frac{\mathbf{y}_k}{\sqrt{\lambda_k}}\}$$

is an orthonormal set.

Exercise 4.4.36. Show that the following matrices are normal and invertible.

$$i. \quad \begin{pmatrix} 2 + 21 & 2 - i & -1 - i \\ 2 - i & 2 + 2i & -1 - i \\ -1 - i & -1 - i & 5 + 2i \end{pmatrix};$$

$$ii. \quad \begin{pmatrix} \frac{1}{\sqrt{3}} & \frac{1}{\sqrt{3}} & \frac{1}{\sqrt{3}} \\ -\frac{1}{\sqrt{2}} & \frac{1}{\sqrt{2}} & 0 \\ -\frac{1}{\sqrt{6}} & -\frac{1}{\sqrt{6}} & \frac{2}{\sqrt{6}} \end{pmatrix};$$

$$iii. \quad \begin{pmatrix} \frac{1}{\sqrt{3}} & -\frac{1}{\sqrt{3}} & -\frac{1}{\sqrt{3}} \\ -\frac{1}{\sqrt{6}} & \frac{1}{\sqrt{6}} & -\frac{2}{\sqrt{6}} \\ \frac{1}{\sqrt{2}} & \frac{1}{\sqrt{2}} & 0 \end{pmatrix}.$$

Find a diagonal form for each matrix.

Exercise 4.4.37. Let $A \stackrel{\text{def}}{=} (a_{ij})_{i,j=1}^{m,n}$ be an $m \times n$ matrix. Then the system $Ax = b$ has a solution iff $b \in im(A)$. Show that:

i. if $b \notin im(A)$ there is in $im(A)$ a unique vector $\mathbf{y}$ such that

$$\|\mathbf{b} - \mathbf{y}\|^2 = minimum_{\mathbf{z} \in im(A)}\|\mathbf{b} - \mathbf{z}\|^2;$$

ii. there is a vector $\mathbf{x}$ such that $A\mathbf{x} = \mathbf{y}$;

iii. if $m < n$, i.e., if A is WIDE, then the $\mathbf{x}$ in *ii* is not unique;

iv. if $m \geq n$, i.e., if A is SQUARE or HIGH, then the $\mathbf{x}$ in *ii* is unique iff $rank(A) = n$.

[**Remark 4.4.10:** Any vector $\mathbf{x}$ as in *ii* above is called a *least squares* solution to the system $A\mathbf{y} = \mathbf{b}$ because $\mathbf{x}$ minimizes

$$\|\mathbf{b} - A\mathbf{y}\|^2. \tag{4.4.26}$$

Such an $\mathbf{x}$ is an actual solution iff the minimum of (4.4.26) is 0.]

Exercise 4.4.38. Assume A is an $m \times n$ matrix and that the columns of A are linearly independent. Show:

   i. A is HIGH or SQUARE $(m \geq n)$ and $rank(A) = n$;

   ii. $A^* A$ is invertible;

iii. if $\mathbf{y}$ is a least squares solution of $A\mathbf{x} = \mathbf{b}$ then $\mathbf{y} = (A^* A)^{-1} A^* \mathbf{b}$.

Let A be an $m \times n$ matrix and let $\mathbf{y}$ be a least squares solution of the equation $A\mathbf{x} = \mathbf{b}$. According to the orthogonal decomposition result (E13) there are unique vectors $\mathbf{v}$ and $\mathbf{w}$ such that

$$\mathbf{v} \in ker(A), \ \mathbf{w} \in (ker(A))^t) \ (= im(A^*)), \text{ and } \mathbf{y} = \mathbf{v} + \mathbf{w}.$$

Hence $A\mathbf{y} = A\mathbf{w}$, i.e., $\mathbf{w}$ is also a least squares solution of $A\mathbf{x} = \mathbf{b}$. The vector $A\mathbf{y}$ is the unique vector that is in $im(A)$ and closest to $\mathbf{b}$. The vector $\mathbf{w}$ is the unique vector that is in $im(A^*)$ and also a least squares solution of $A\mathbf{x} = \mathbf{b}$. Indeed, if $\widetilde{\mathbf{w}} \in im(A^*)$ and if $\widetilde{\mathbf{w}}$ is also a least squares solution of $A\mathbf{x} = \mathbf{b}$ then

$$A\mathbf{w} = A\mathbf{y} = A\widetilde{\mathbf{w}}$$
$$A(\mathbf{w} - \widetilde{\mathbf{w}}) = \mathbf{O}$$
$$\mathbf{w} - \widetilde{\mathbf{w}} \in ker(A) \cap im(A^*)$$
$$= ker(A) \cap ker(A)^\perp = \{\mathbf{O}\}$$
$$\mathbf{w} = \widetilde{\mathbf{w}}.$$

The unique vector $\mathbf{w}$ that is in $im(A^*)$ and is a least squares solution of $A\mathbf{x} = \mathbf{b}$ is denoted $T^+(\mathbf{b})$ and is regarded as the *optimal* least squares solution of $A\mathbf{y} = \mathbf{b}$.

Exercise 4.4.39. Show that $T^+ : \mathbf{C}^m \ni \mathbf{b} \mapsto T^+(\mathbf{b}) \in \mathbf{C}^n$ is a linear transformation, i.e., show that if p and q are numbers and if $\mathbf{a}$ and $\mathbf{b}$ are vectors in $\mathbf{C}^m$ then

$$\mathbf{C}^n \ni T^+(p\mathbf{a} + q\mathbf{b}) = pT^+(\mathbf{a}) + qT^+(\mathbf{b}).$$

Show also that $T^+(\mathbf{b})$ is the *shortest* least squares solution of $A\mathbf{x} = \mathbf{b}$. (That is why $T^+(\mathbf{b})$ is regarded as optimal.)

 Chapter 4. Useful forms for matrices

In the defining equation below the $\mathbf{e}_i$ in the left member are in $\mathbf{C}^m$ whereas the $\mathbf{e}_j$ in the right member are in $\mathbf{C}^n$:

$$T^+(\mathbf{e}_i) \stackrel{\text{def}}{=} \sum_{j=1}^{n} t_{ij}\mathbf{e}_j, 1 \leq i \leq m.$$

Let $((t_{ij})_{i,j=1}^{n,m})^t$ be denoted A^+. It is called the *Moore-Penrose* inverse or the *pseudo-inverse* of A

Exercise 4.4.40. Show that:

i. A^*A is invertible iff the columns of A are linearly independent;
ii. if A^*A is invertible then $A^+ = (A^*A)^{-1}A^*$;
iii. if A is invertible (hence A is SQUARE) then $A^+ = A^{-1}$.

DEFINITION **4.4.4.** AN $m \times n$ MATRIX A IS OF *full rank* IFF $rank(A) = minimum(m, n) \stackrel{\text{def}}{=} \mu$.

Exercise 4.4.41. Assume that $A \stackrel{\text{def}}{=} (a_{ij})_{i,j=1}^{m,n}$ is of full rank, i.e., that $rank(A) = minimum(m, n) \stackrel{\text{def}}{=} \mu$. Show that:

i. AA^* is invertible iff $\mu = m$;
ii. A^*A is invertible iff $\mu = n$.

Exercise 4.4.42. For each of the matrices below find its pseudo-inverse.

$$i. \ \begin{pmatrix} 1 & 2 \\ 3 & 4 \\ 5 & 6 \end{pmatrix}, \ ii. \ \begin{pmatrix} 1 & 2 & 4 \\ 3 & 4 & 10 \\ 5 & 6 & 16 \end{pmatrix}.$$

[*Hint:* Let A be one of the matrices above. If the columns are *not* linearly independent let $\mathbf{e}_i$ be the standard basis elements and find shortest least squares solutions of $A\mathbf{y} = \mathbf{e}_i$.]

Exercise 4.4.43.

i. Show that if A is an $m \times n$ matrix then $AA^+A = A$ (cf. THEOREM **2.10.1**). The last equation explains the choice of the name *pseudo-inverse*: the matrix AA^+ multiplying A acts like I multiplying A and so AA^+ is, from the standpoint of A, the same as I : AA^+ "$=$" I
ii. Show that $P \stackrel{\text{def}}{=} AA^+$ is idempotent ($P^2 = P$) and self-adjoint ($P = P^*$).

---

iii. Show that $im(P) = im(A)$ $(= (ker(A^*))^\perp)$. (Hence $(I - A^*(A^*)^+)$ is the self-adjoint idempotent projection into $ker(A)$.)

Exercise 4.4.44. The items below lead to an interesting result of Hadamard. Prove the assertions in each item.

i. Let $\{u_i\}_{i=1}^n$ be an orthonormal basis for $\mathbf{C}^n$ and let $\mathbf{x}$ be a vector in $\mathbf{C}^n$. Then if $1 \le k \le n$

$$\|\sum_{i=1}^{k}(\mathbf{x}, \mathbf{u}_i)\mathbf{u}_i\|^2 \le \|\mathbf{x}\|^2 \text{ (Bessel's inequality)}$$

$$\|\mathbf{x} - \sum_{i=1}^{k}(\mathbf{x}, \mathbf{u}_i)\mathbf{u}_i\|^2 \le \|\mathbf{x}\|^2.$$

("An orthogonal projection of a vector is not bigger than the vector.")

ii. Let $\{\mathbf{x}_i\}_{i=1}^n$ be an arbitrary basis for $\mathbf{C}^n$ and let $\{\mathbf{u}_i\}_{i=1}^n$ be the orthonormal basis arising from the application of the Gram-Schmidt process to $\{\mathbf{x}_i\}_{i=1}^n$. In accordance with the contents of **Exercise 4.4.17** there is an equation

$$\mathcal{U} \overset{\text{def}}{=} \begin{pmatrix} \mathbf{u}_1 \\ \vdots \\ \mathbf{u}_n \end{pmatrix} = \begin{pmatrix} \gamma_{11} & \cdots & \gamma_{1n} \\ \vdots & \ddots & \vdots \\ \gamma_{n1} & \cdots & \gamma_{nn} \end{pmatrix} \begin{pmatrix} \mathbf{x}_1 \\ \vdots \\ \mathbf{x}_n \end{pmatrix} \overset{\text{def}}{=} \Gamma \mathcal{X}.$$

The matrix Γ is lower triangular and invertible and

$$|\gamma_{ii}| = \frac{1}{\|\mathbf{x}_i - \sum_{j=1}^{i-1}(\mathbf{x}_i, \mathbf{u}_j)\mathbf{u}_j\|} \ge \frac{1}{\|\mathbf{x}_i\|}$$

$$det(\Gamma) = \prod_{i=1}^{n}|\gamma_{ii}| \ge \prod_{i=1}^{n}\frac{1}{\|\mathbf{x}_i\|}. \tag{4.4.27}$$

iii. If $A \overset{\text{def}}{=} (a_{ij})_{i,j=1}^{n,n}$ and $det(A) \ne 0$ then the rows $\mathbf{a}_i$ of A are linearly independent. If the Gram-Schmidt process is applied to the $\mathbf{a}_i$ it produces an orthonormal basis $\{\mathbf{u}_i\}_{i=1}^n$. If U is the unitary matrix in which row i is $\mathbf{u}_i$ and if Γ is the "Gram-Schmidtifying" matrix for the rows of A then $U = \Gamma A$. Hence

$$|det(A)| \le \prod_{i=1}^{n}\|\mathbf{a}_i\|. \tag{4.4.28}$$

The inequality (4.4.28) is known is *Hadamard's inequality*. If the volume interpretation is given for a determinant then (4.4.27) has the appealing interpretation that says the volume of a parallelotope is not greater then

---

     Chapter 4. Useful forms for matrices

the product of the lengths of its sides. Furthermore, for given side lengths greatest volume is achieved if the sides are pairwise orthogonal. The pictures in $\mathbf{C}^2$ and $\mathbf{C}^3$ give the story away.

Exercise 4.4.45. Let $X \stackrel{\text{def}}{=} \{\mathbf{x}_i\}_{i=1}^m$ be a set of vectors in $\mathbf{C}^n$. Show that for their Gramian $\mathcal{G}$ there obtains the inequality

$$0 \leq det(\mathcal{G}) \leq \prod_{i=1}^m \|\mathbf{x}_i\|^2 \tag{4.4.29}$$

Interpret the first inequality in (4.4.29) as a generalization of the Schwarz inequality (4.4.21). Show that the first inequality is an equality iff X is linearly dependent. Use the second inequality in (4.4.29) to give a second proof of Hadamard's inequality (4.4.28). Show that the second inequality is an equality iff the vectors $\mathbf{x}_i$ are pairwise orthogonal.

[*Hint:* Use **Exercise 4.4.34** and the similarity invariance of *det* to prove the first inequality and the assertion about linear dependence of X.

If X is linearly dependent the second inequality is trivial. If X is linearly independent let Γ the "Gram-Schmidtifying" matrix for X. Show $det(\Gamma \mathcal{G} \Gamma^*) = 1$ from which the second inequality and the criterion for equality follow.

To derive Hadamard's inequality from (4.4.29) let A be SQUARE and consider AA^*.]

Exercise 4.4.46. Show that if A is any $m \times n$ matrix, $\mathbf{x} \in \mathbf{C}^m$, and $\Re(\alpha) > 0$ then $(\alpha I + AA^*)\mathbf{x} = \mathbf{O}$ iff $\mathbf{x} = \mathbf{O}$ ($\alpha I + AA^*$ is invertible).

Exercise 4.4.47. In **Exercise 4.4.25** there is the conclusion that the Schwarz inequality is an equality iff the vectors involved are linearly dependent. Show that the Minkowski inequality is an equality iff the vectors involved are linearly dependent.

Exercise 4.4.48. Let A be an invertible $n \times n$ matrix. Let Γ be the matrix such that the rows of ΓA are orthonormal. (The matrix Γ is used to *Gram-Schmidtify* the rows of A.) Show that $A^{-1} = A^* \Gamma^* \Gamma$. Let $\widetilde{\Gamma}$ be the matrix that Gram-Schmidtifies the columns of A. Show that $A^{-1} = \widetilde{\Gamma}^t (\widetilde{\Gamma}^t)^* A^*$.

Exercise 4.4.49. Let A be an invertible $n \times n$ matrix and denote its jth column by $\mathbf{c}_j$. Denote the ith row of A^{-1} by $\mathbf{r}_i$. Show that

$$\mathbf{r}_i \in (span(\{\,\overline{\mathbf{c}}_j \; : \; j \neq i\,\}))^{\perp}.$$

Show further that if the entries of A are real then the angle between $\mathbf{r}_i$ and $\mathbf{c}_i$ is acute. Interpret the results in $\mathbf{R}^2$ and $\mathbf{R}^3$.

Exercise 4.4.50. Let $X \overset{\text{def}}{=} \{\mathbf{x}_1, \ldots, \mathbf{x}_n\}$ be a basis for $\mathbf{C}^n$. If $m \leq n$ let V be $span(\{\mathbf{x}_1, \ldots, \mathbf{x}_m\})$ and let $V \oplus V^{\perp}$ be the corresponding orthogonal decomposition of $\mathbf{C}^n$. Let P be as in E13.

i. Show that if

$$P(\mathbf{e}_i) \overset{\text{def}}{=} \sum_{j=1}^{n} h_{ij} \mathbf{e}_j, \ 1 \leq i \leq n,$$

then
    a. the matrix $H \overset{\text{def}}{=} (h_{ij})_{i,j=1}^{n,n}$ is self-adjoint;
    b. $H^2 = H$;
    c. the only possible eigenvalues of H are 0 and 1 and that if $m = n$ then 1 is the only eigenvalue of H.

ii. Show that $\mathbf{x}$ is an eigenvector for H iff $\mathbf{x} \neq \mathbf{O}$ and $\mathbf{x} \in V \cup V^{\perp}$.

iii. Let $Y \overset{\text{def}}{=} \{\mathbf{y}_1, \ldots, \mathbf{y}_m\}$ resp. $Z \overset{\text{def}}{=} \{\mathbf{z}_1, \ldots, \mathbf{z}_{n-m}\}$ be orthonormal bases for V resp. $V^{\perp}$ and let Q be the matrix in which the first m columns are the column vectors $\mathbf{y}_k$, $1 \leq k \leq m$ and the last $n - m$ columns are the column vectors $\mathbf{z}_l$, $1 \leq l \leq n - m$. Show that:
    a. Q is unitary;
    b.

$$Q^*HQ = \begin{array}{c} \\ m \\ n-m \end{array}\overset{\displaystyle m \quad n-m}{\begin{pmatrix} I & O \\ O & O \end{pmatrix}} \overset{\text{def}}{=} D.$$

iv. Let $U \overset{\text{def}}{=} \{\mathbf{u}_1, \ldots, \mathbf{u}_n\}$ arise by Gram-Schmidtifying X and let R be the matrix in which the jth column is the column vector $\mathbf{u}_j$. Show that R is unitary and that $R^*HR = D$ as in *iii*b.

Exercise 4.4.51. Assume that H is a self-adjoint idempotent $n \times n$ matrix: $H = H^*$ and $H^2 = H$. Let $im(H)$ be V. Describe a method for choosing a basis Y for $im(H)$. Such a basis Y can be filled out to a basis X for $\mathbf{C}^n$. Let Z be the result of Gram-Schmidtifying X and let U be the matrix in which column j is the jth vector of Z. Show that that if $dim(im(H)) = m$ $(m \leq n)$ then U^*HU looks like D in *iii*b of **Exercise 4.4.50**.

Let A be a self-adjoint $n \times n$ matrix and let its (necessarily real) eigenvalues λ_i be arranged so that $\lambda_1 \geq \lambda_2 \geq \cdots \geq \lambda_n$. Hence there is unitary matrix U such that

$$U^*AU \overset{\text{def}}{=} D \overset{\text{def}}{=} \begin{pmatrix} \lambda_1 & & & \\ & \lambda_2 & & \\ & & \ddots & \\ & & & \lambda_n \end{pmatrix}.$$

Furthermore there is an orthonormal eigenbasis $Z \stackrel{\text{def}}{=} \{z_1, \dots, z_n\}$ such that $Az_k = \lambda_k z_k$. Let S_0 be $\{O\}$ and let S_k be the span of the first k basis vectors:

$$S_k \stackrel{\text{def}}{=} span\{z_1, \dots, z_k\}, \ 1 \le k \le n.$$

Exercise 4.4.52. Show:

i. $\lambda_1 = maximum_{\|\mathbf{x}\|=1}(A\mathbf{x}, \mathbf{x})$;

[*Hint:* Since U is unitary for each $\mathbf{x}$ there is a unique $\mathbf{y}$ such that $U\mathbf{y} = \mathbf{x}$. Furthermore $\|\mathbf{y}\| = \|\mathbf{x}\|$ and

$$(A\mathbf{x}, \mathbf{x}) = (AU\mathbf{y}, U\mathbf{y}) = (D\mathbf{y}, \mathbf{y}).]$$

ii. $l_k = maximum\left\{ (A\mathbf{x}, \mathbf{x}) \ : \ \|\mathbf{x}\| = 1 \text{ and } \mathbf{x} \in S_{k-1}^{\perp} \right\}, \ 1 \le k \le n$;

[*Hint:* The vector $\mathbf{x}$ is in $S_{k-1}^{\perp}$ iff $\mathbf{x}$ is in the span of $\{z_k, \dots, z_n\}$, an $(n - k + 1)$-dimensional subspace of $\mathbf{C}^n$.]

iii.
$$S_{k-1}^{\perp} = \left\{ \mathbf{x} \ : \ (\mathbf{x}, z_1) = \cdots = (\mathbf{x}, z_{k-1}) = 0 \right\}.$$

Retain the notation of **Exercise 4.4.50.** Let V_{k-1} be a set $\{\mathbf{v}_1, \dots, \mathbf{v}_{k-1}\}$ of $k-1$ vectors and consider the maximum of $(A\mathbf{x}, \mathbf{x})$ when $\mathbf{x}$ is restricted to be of norm 1 and to lie in the subspace $W \stackrel{\text{def}}{=} [span(V_{k-1})]^{\perp}$:

$$M(V_{k-1}) \stackrel{\text{def}}{=} maximum\left\{ (A\mathbf{x}, \mathbf{x}) \ : \ \|\mathbf{x}\| = 1, \ (\mathbf{x}, \mathbf{v}_1) = \cdots = (\mathbf{x}, \mathbf{v}_{k-1}) = 0 \right\}.$$

Exercise 4.4.53.

i. Show $dim([span(V_{k-1})]^{\perp}) \ge n - k + 1$.
ii. Show that $M(V_{k-1}) \ge \lambda_k$ for any V_{k-1}.

[*Hint:* Since $dim(W) \ge n - k + 1$ it follows that there is in W some $\mathbf{x}$ such that $\|\mathbf{x}\| = 1$, $\mathbf{x} = \sum_{i=1}^{k} a_i z_i + \sum_{j=k+1}^{n} b_j z_j$, and some $a_i \ne 0$. (Otherwise $dim(W) \le n - k$.) Calculate $(A\mathbf{x}, \mathbf{x})$.]

iii. Show that

$$\lambda_k = minimum\left\{ M(V_{k-1}) \ : \ V_{k-1} \text{ is a set of } k - 1 \text{ vectors} \right\}$$
$$= minimum_{V_{k-1}} maximum\left\{ (A\mathbf{x}, \mathbf{x}) \ : \ \|\mathbf{x}\| = 1, \ (\mathbf{x}, \mathbf{v}_1) = \cdots = (\mathbf{x}, \mathbf{v}_{k-1}) = 0 \right\}.$$

[**Remark 4.4.11:** The final result in **Exercise 4.4.53** is the Weyl minmax principle for Hermitian matrices.]

The following **Exercise** offers an opportunity to bring to bear some of the high points of this and earlier **Sections**.

Exercise 4.4.54. Fill in the details of the sketch below of yet another proof of Hadamard's inequality: $|det[(a_{ij})_{i,j=1}^{n,n}]| \leq \prod_{i=1}^{n} \|\mathbf{a}_i\|$ (cf. (4.4.28)). Assume that $A \stackrel{\text{def}}{=} (a_{ij})_{i,j=1}^{n,n}$, that A_{ij} is the cofactor of a_{ij}, that $\mathbf{a}_i = (a_{11}, \ldots, a_{in})$, that $\mathbf{A}_i = (\overline{A}_{i1}, \ldots, \overline{A}_{in})$, and that $det(A) \neq 0$.

i. $|det(A)| \leq \|\mathbf{a}_i\| \cdot \|\mathbf{A}_i\|$.

ii. For each i there is a vector $\mathbf{b}_i \stackrel{\text{def}}{=} (b_{i1}, \ldots, b_{in})$ such that $\|\mathbf{b}_i\| = \|\mathbf{a}_i\|$ and, for some number t_i, $\mathbf{b}_i = t_i \mathbf{A}_i$.

iii. If B_i is the matrix A with the ith row replaced by $\mathbf{b}_i$ then

$$det(B_i) = (\mathbf{b}_i, \mathbf{A}_i)$$
$$|det(B_i)| = \|\mathbf{b}_i\| \cdot \|\mathbf{A}_i\| \ (= \|\mathbf{a}_i\| \cdot \|\mathbf{A}_i\|).$$

iv. In the set S of all $n \times n$ matrices C in which the length of the ith row is $\|\mathbf{a}_i\|$, $1 \leq i \leq n$, there is a C_{max} for which

$$|det(C_{max})| = maximum \{ |det(C)| \ : \ C \in S \}.$$

v. Assume $C_{max} \stackrel{\text{def}}{=} (c_{ij})_{i,j=1}^{n,n}$, and let $\mathbf{c}_i$ and $\mathbf{C}_i$ correspond for C_{max} to $\mathbf{a}_i$ and $\mathbf{A}_i$ for A. Then

$$|det(A)|^2 \leq |det(C_{max})|^2 = \prod_{i=1}^{n} \|\mathbf{c}_i\|^2 = \prod_{i=1}^{n} \|\mathbf{a}_i\|^2.$$

4.5. The Jordan normal form for a nilpotent matrix

In **Section 4.2** it is shown that every SQUARE matrix A is the sum of a diagonal matrix D and a nilpotent matrix N: $A = D + N$. If $D = O$ then A is itself nilpotent. Unfortunately, if a nilpotent matrix is not O it is not diagonable.

Indeed, if N is nilpotent, and nonzero then, for some natural number k greater than 1, $N^k = O$. If P is invertible and $P^{-1}NP = \Delta$, a diagonal matrix, then

$$N = P\Delta P^{-1}, \ O = N^k = P\Delta^k P^{-1}, \ \Delta^k = P^{-1}OP = O$$

whence $\Delta = O$ and so $N = P\Delta P^{-1} = O$, a contradiction. (An alternative proof stems from the equation: $m_N(z) = z^k$. Give the details.)

On the other hand, if N is nilpotent it is similar to a sparse matrix $\mathcal{N}$ consisting of Jordan blocks. More precisely, there is an invertible matrix P such that $P^{-1}NP$ is a block diagonal matrix

$$\begin{pmatrix} J_1 & & & \\ & J_2 & & \\ & & \ddots & \\ & & & J_m \end{pmatrix}$$

in which each J_i is a nilpotent Jordan block. The basis consisting of the columns in P is a *"Jordanizing"* basis for N.

The proof of this result follows the exploration below of a variety of nilpotent matrices. The experience with them provides a guide to the derivation of the general result.

Exercise 4.5.1. Show that if N is an $n \times n$ nilpotent matrix then $N^n = O$, i.e., that the order k of nilpotency does not exceed n.

Exercise 4.5.2. Show that if N is an $n \times n$ nilpotent matrix then:

i. $rank(N) < n$;
ii. $dim[im(N)] < n$;
iii. 0 is the only eigenvalue of N.

The subspace $im(N)$ is a particular case of an *invariant subspace* V, i.e., a subspace such that $NV \subset V$.

Exercise 4.5.3. Let N be a nilpotent matrix and let V be a nonzero invariant subspace: $V \neq \{O\}$, $NV \subset V$. Justify each of the following statements:

i. there is in V an eigenvector $\mathbf{x}$;
ii. there is in V a basis X containing the eigenvector $\mathbf{x}$;
iii. if $NV \neq \{O\}$ then NX contains a basis Y for NV;
iv. $\mathbf{x} \notin Y$;
v. $dim(NV) < dim(V)$;
vi. if $N^p \neq O$ then $im(N^{p+1}) \subsetneq im(N^p)$.

Exercise 4.5.4. Assume that the 2×2 matrix N is such that $N^2 = O \neq N$. Show that $dim[ker(N)] = 1$ and that a set X is a basis for $ker(N)$ iff X is a basis for $im(N)$.

If A is an $m \times n$ matrix and if $E \overset{\text{def}}{=} \{\mathbf{e}_i, \ 1 \leq i \leq n\}$ is the standard basis for $\mathbf{C}^n$ then the columns of A are precisely the vectors in AE. If N is a nilpotent $n \times n$ SQUARE matrix its columns are the vectors in NE. If $rank(N) = r_1$ then $r_1 < n$ and among the columns of N there are r_1 linearly independent columns that constitute a basis X_1 of $im(N) = $ column space of N.

[**Note 4.5.1:** The set X_1 may be one of several different sets, each consisting of r_1 linearly independent columns of N.]

Let the order of nilpotency of N be k: $N^k = O \neq N^{k-1}$. Let $X_{k-1} \overset{\text{def}}{=} \{\mathbf{x}_{k-1,1}, \ldots, \mathbf{x}_{k-1,r_{k-1}}\}$ be a basis for $im(N^{k-1})$. Then, as shown below, there are in $im(N^{k-2})$ vectors $\mathbf{x}_{k-2,j}$ such that

$$\widetilde{X}_{k-2} \overset{\text{def}}{=} \{N\mathbf{x}_{k-2,j}(= \mathbf{x}_{k-1,j}), \ \mathbf{x}_{k-2,j}, \ 1 \leq j \leq r_{k-1}\}$$

---

is linearly independent (cf. **Exercise 2.9.34**). Furthermore, the set $\widetilde{X}_{k-2}$ can be filled out to a basis for $im(N^{k-2})$ by the adjunction, as needed, to $\widetilde{X}_{k-2}$ of appropriate linearly independent vectors in $im(N^{k-2}) \cap ker(N)$. The result X_{k-2} consists of the nonzero vectors in the set

$$\{N\mathbf{x}_{k-2,j}, \mathbf{x}_{k-2,j} \ 1 \le j \le r_{k-2}, \ r_{k-2} \ge r_{k-1}\}.$$

In the display above if $r_{k-2} > r_{k-1}$ then $\mathbf{x}_{k-2,j} \in ker(N)$ whenever $j > r_{k-1}$ and so $N\mathbf{x}_{k-2,j} = \mathbf{O}$.

Repeating the procedure that leads from X_{k-1} to X_{k-2} yields a basis X_{k-3} for $im(N^{k-3})$, ..., and ultimately to a basis X_1 for $im(N)$. In sum

$$X_1 = \text{basis of } im(N) = \{\mathbf{x}_{11}, \ldots, \mathbf{x}_{1r_1}\}$$
$$X_2 = \text{basis of } im(N^2) = \{\mathbf{x}_{21}, \ldots, \mathbf{x}_{2r_2}\}$$
$$\vdots$$
$$X_{k-1} = \text{basis of } im(N^{k-1}) = \{\mathbf{x}_{k-1,1}, \ldots, \mathbf{x}_{k-1,r_{k-1}}\}$$

and, X_0 denoting E, $NX_{j-1} \subsetneq X_j$, $1 \le j \le k-1$.

Example 4.5.1. Here are two 3×3 nilpotent matrices:

$$N_1 \overset{\text{def}}{=} \begin{pmatrix} 0 & 1 & 0 \\ 0 & 0 & 1 \\ 0 & 0 & 0 \end{pmatrix} \text{ and } N_2 \overset{\text{def}}{=} \begin{pmatrix} 0 & 0 & 1 \\ 0 & 0 & 0 \\ 0 & 0 & 0 \end{pmatrix}.$$

Direct calculation shows that N_1 is nilpotent of order 3: $N_1^2 \ne O = N_1^3$, and N_2 is nilpotent of order 2: $N_2 \ne O = N_2^2$.

For N_1 the sequence of bases is

$$\{X_1 \overset{\text{def}}{=} \{\mathbf{e}_1, \mathbf{e}_2\}, \ X_2 \overset{\text{def}}{=} \{\mathbf{e}_1\}.$$

Furthermore $N_1\mathbf{e}_3 = \mathbf{e}_2$, $N_1^2\mathbf{e}_3 = \mathbf{e}_1$ and so if $\mathbf{v}_1 \overset{\text{def}}{=} \mathbf{e}_3$ it follows that

$$Y \overset{\text{def}}{=} \{ N_1^p \mathbf{v}_1 \ : \ N^p \mathbf{v}_1 \ne \mathbf{O}, \ 0 \le p < 3 \}$$

is a basis for $\mathbf{C}^3$.

For N_2 the sequence of bases consists of one basis: $X_1 = \{\mathbf{e}_1\}$. Furthermore $N_2\mathbf{e}_3 = \mathbf{e}_1$ and so if $\mathbf{v}_1 \overset{\text{def}}{=} \mathbf{e}_3$

$$\{ N_2^p \mathbf{v}_1 \ : \ N_2^p \mathbf{v}_1 \ne \mathbf{O}, \ 0 \le p < 2 \} = \{\mathbf{e}_1, \mathbf{e}_3\}$$

is *not* a basis for $\mathbf{C}^3$. On the other hand

$$Z \overset{\text{def}}{=} ker(N_2) \cap \{ N_2^p \mathbf{v}_1 \ : \ N_2^p \mathbf{v}_1 \ne \mathbf{O}, \ 0 \le p < 2 \} = \{N_2\mathbf{v}_1\} = \{\mathbf{e}_1\}$$

may be filled out to a basis for $ker(N_2)$ by adjoining to Z the vector $\mathbf{v}_2 \stackrel{\text{def}}{=} \mathbf{e}_2$. The set

$$X \stackrel{\text{def}}{=} \{\, N_2^p \mathbf{v}_q \;:\; N_2^p \mathbf{v}_q \neq \mathbf{O},\ 0 \leq p < 2,\ 1 \leq q \leq 2 \,\}$$
$$= \{\mathbf{e}_1, \mathbf{e}_3\} \cup \{\mathbf{e}_2\}$$

is a basis for $\mathbf{C}^3$.

Let the vectors in Y be arranged in order so that

$$Y = \{N_1^2 \mathbf{v}_1, N_1 \mathbf{v}_1, \mathbf{v}_1\} = \{\mathbf{y}_1, \mathbf{y}_2, \mathbf{y}_3\}$$

and let P be the 3×3 matrix in which column j is $\mathbf{y}_j$, i.e., $P = I = P^{-1}$. Then

$$\mathcal{N}_1 \stackrel{\text{def}}{=} P^{-1} N_1 P = N_1,$$

a Jordan block.

On the other hand let the vectors in X be arranged in order so that

$$X = \{N\mathbf{v}_1, \mathbf{v}_1, \mathbf{v}_2\} \stackrel{\text{def}}{=} \{\mathbf{x}_1, \mathbf{x}_2, \mathbf{x}_3\}.$$

Let P be the 3×3 matrix in which column j is $\mathbf{x}_j$, i.e.,

$$P = \begin{pmatrix} 1 & 0 & 0 \\ 0 & 0 & 1 \\ 0 & 1 & 0 \end{pmatrix} = P^{-1}.$$

Then

$$\mathcal{N}_2 \stackrel{\text{def}}{=} P^{-1} N_2 P = \begin{pmatrix} 0 & 1 & 0 \\ 0 & 0 & 0 \\ 0 & 0 & 0 \end{pmatrix},$$

a block matrix consisting of nilpotent Jordan blocks

$$J_1 \stackrel{\text{def}}{=} \begin{pmatrix} 0 & 1 \\ 0 & 0 \end{pmatrix} \quad \text{and} \quad J_2 \stackrel{\text{def}}{=} (0)$$

on its diagonal:

$$\mathcal{N}_2 = \begin{pmatrix} J_1 & \\ & J_2 \end{pmatrix}.$$

The matrices $\mathcal{N}_1$ and $\mathcal{N}_2$ are the *Jordan normal forms* of N_1 and N_2. (The matrix N_1 is its own Jordan normal form.)

The experience above leads to the following general result.

THEOREM 4.5.1. LET N BE AN $n \times n$ MATRIX THAT IS NILPOTENT OF ORDER k. THEN THERE IS AN INVERTIBLE $n \times n$ MATRIX P AND NILPOTENT JORDAN BLOCKS $J_1, \ldots, J_{m(N)}$ SUCH THAT

$$\mathcal{N} \stackrel{\text{def}}{=} P^{-1} N P = \begin{pmatrix} J_1 & & & \\ & J_2 & & \\ & & \ddots & \\ & & & J_{m(N)} \end{pmatrix}.$$

PROOF. Let
$$X_{k-1} \overset{\text{def}}{=} \{\mathbf{x}_{k-1,1}, \ldots, \mathbf{x}_{k-1,r_{k-1}}\}$$
be a basis for $im(N^{k-1})$.

Then, N^0 denoting I, the basis X_{k-1} may be described as follows:
$$\begin{aligned}
X_{k-1} &= \{N^0\mathbf{x}_{k-1,1}, \ldots, N^0\mathbf{x}_{k-1,1}\} \\
&= \{N^p\mathbf{x}_{k-1,q} \; : \; N^p\mathbf{x}_{k-1,q} \neq \mathbf{O}, \; 0 \leq p < k, \; 1 \leq q \leq r_{k-1}\}.
\end{aligned}$$

Hence
$$\{N^p\mathbf{x}_{k-1,q} \; : \; N^p\mathbf{x}_{k-1,q} \neq \mathbf{O}, \; 0 \leq p < k, 1 \leq q \leq r_{k-1}\}$$
is a basis for $im(N^{k-1})$.

In $im(N^{k-2})$ there are vectors
$$\mathbf{x}_{k-2,1}, \ldots, \mathbf{x}_{k-2,r_{k-1}}$$
such that $N\mathbf{x}_{k-2,q} = \mathbf{x}_{k-1,q}$.

Define Z_{k-2} by the relation
$$Z_{k-2} \overset{\text{def}}{=} \{N^p\mathbf{x}_{k-2,q} \; : \; N^p\mathbf{x}_{k-2,q} \neq \mathbf{O}, 0 \leq p < k, \; 1 \leq q \leq r_{k-1}\}$$
$$\cap \, ker(N) \cap im(N^{k-2})$$

and adjoin to Z_{k-2} as many vectors
$$\mathbf{x}_{k-2,i}, \; r_{k-1} + 1 \leq i \leq r_{k-1} + \rho_{k-2} \overset{\text{def}}{=} r_{k-2}$$

as are needed to fill out Z_{k-2} to a basis for $ker(N) \cap im(N^{k-2})$. (If, as in the case of N_1 above, none is needed, then $\rho_{k-2} = 0$, there is no i such that $r_{k-1} + 1 \leq i \leq r_{k-1} + \rho_{k-2}$, and $r_{k-2} = r_{k-1}$.)

It is shown next that X_{n-2}, the set of nonzero vectors in
$$\{N\mathbf{x}_{k-2,j}, \mathbf{x}_{k-2,j}, \; 1 \leq j \leq r_{k-1} + \rho_{k-2}\},$$

is a basis for $im(N^{k-2})$. Note that if $r_{k-2} > 0$ and $j > r_{k-1}$ then $N\mathbf{x}_{k-2,j} = \mathbf{O}$. The set X_{k-2} can be described by the equation
$$X_{k-2} = \{N^p\mathbf{x}_{k-2,q} \; : \; N^p\mathbf{x}_{k-2,q} \neq \mathbf{O}, \; 0 \leq p < k, \; 1 \leq q \leq r_{k-2}\}.$$

To show that X_{k-2} is a basis for $im(N^{k-2})$ it suffices to prove:

i. X_{k-2} is linearly independent;
ii. $span(X_{k-2}) = im(N^{k-2})$.

Proof of *i*:
Note that if $1 \leq p$ then
$$N^p\mathbf{x}_{k-2,q} \text{ is } \begin{cases} \text{not } \mathbf{O} \text{ and in } im(N^{k-1}) & \text{if } 1 \leq q \leq r_{k-1} \\ \mathbf{O} & \text{if } r_{k-1} < q \leq r_{k-2}. \end{cases}$$

Hence if

$$\sum_{0 \leq p \leq k, 1 \leq q \leq r_{k-2}} a_{pq} N^p \mathbf{x}_{k-2,q} = \mathbf{O} \tag{4.5.1}$$

multiply both members of (4.5.1) by N. In the left member all terms corresponding to vectors in $ker(N)$ are killed and what remain are terms corresponding to the linearly independent vectors in X_{k-1}. Hence the coefficients of those vectors are 0's. The terms in (4.5.1) that are *not* killed by N are terms corresponding to the pre-images of the vectors in X_{k-1}. Since X_{k-1} is linearly independent, so is the set of pre-images of the vectors in X_{k-1} (cf. **Exercise 2.9.35**). Hence the coefficients of the pre-images are also 0's, i.e., each $a_{pq} = 0$.

$$\omega$$

Proof of *ii*:

Let $\mathbf{x}$ be a vector in $im(N^{k-2})$. Then $N\mathbf{x} \in im(N^{k-1})$ and so, since X_{k-1} is a basis for $im(N^{k-1})$,

$$N\mathbf{x} = \sum_{i=1}^{r_{k-1}} b_i \mathbf{x}_{k-1,i}$$

$$= \sum_{i=1}^{r_{k-1}} b_i N \mathbf{x}_{k-2,i}.$$

It follows that

$$\mathbf{w} \overset{\text{def}}{=} \mathbf{x} - \sum_{i=1}^{r_{k-1}} b_i \mathbf{x}_{k-2,i} \in ker(N) \cap im(N^{k-2})$$

and so $\mathbf{w}$ is a linear combination of the vectors in Z_{k-2} together with the vectors adjoined to it to fill out Z_{k-2} to a basis for $ker(N) \cap im(N^{k-2})$. In short, $\mathbf{x} \in span(X_{k-2})$.

$$\omega$$

Now repeat the argument just given. Find pre-images $\mathbf{x}_{k-3,q}$ in $im(N^{k-3})$ so that

$$N\mathbf{x}_{k-3,q} = \mathbf{x}_{k-2,q}, \ 1 \leq q \leq r_{k-2}.$$

Form

$$Z_{k-3} \overset{\text{def}}{=} \{ N^p \mathbf{x}_{k-3,q} \ : \ N^p \mathbf{x}_{k-3,q} \neq \mathbf{O}, \ 0 \leq p < k, 1 \leq q \leq r_{k-2} \}$$
$$\cap ker(N) \cap im(N^{k-3})$$

and fill out Z_{k-3} with vectors $\mathbf{x}_{k-3,r_{k-2}+i}$ as needed to a basis for

$$ker(N) \cap im(N^{k-3}).$$

Then the previous argument, appropriately modified, shows that

$$X_{k-3} \overset{\text{def}}{=} \{ N^p \mathbf{x}_{k-3,q} \ : \ N^p \mathbf{x}_{k-3,q} \neq \mathbf{O}, \ 0 \leq p < k, \ 1 \leq q \leq r_{k-3} \}$$

---

is a basis for $im(N^{k-3})$.

After finitely many repetitions of the argument there emerges a set of vectors $\mathbf{x}_q$ such that

$$X \stackrel{\text{def}}{=} \{ N^p\mathbf{x}_q \ : \ N^p\mathbf{x}_q \neq \mathbf{O}, \ 0 \leq p < k, \ 1 \leq q \leq r_0 \}$$

is a basis for $\mathbf{C}^n$.

For each q there is a number p_q such that

$$N^{p_q-1}\mathbf{x}_q \neq \mathbf{O} = N^{p_q}\mathbf{x}_q.$$

It may be assumed that $p_1 \geq p_2 \geq \cdots \geq p_{r_0}$. Then arrange the vectors in the basis X according to the following scheme:

$$\mathbf{y}_1 \stackrel{\text{def}}{=} N^{p_1-1}\mathbf{x}_1 \qquad\qquad \mathbf{y}_{p_1+1} \stackrel{\text{def}}{=} N^{p_2-1}\mathbf{x}_2 \ \ldots$$

$$\mathbf{y}_2 \stackrel{\text{def}}{=} N^{p_1-2}\mathbf{x}_1 \qquad\qquad \mathbf{y}_{p_1+2} \stackrel{\text{def}}{=} N^{p_2-2}\mathbf{x}_2 \ \ldots$$

$$\vdots \qquad\qquad\qquad\qquad \vdots$$

$$\mathbf{y}_{p_1-1} \stackrel{\text{def}}{=} N\mathbf{x}_1 \ \ldots \qquad\qquad \mathbf{y}_{p_1+p_2-1} \stackrel{\text{def}}{=} N\mathbf{x}_2$$

$$\mathbf{y}_{p_1} \stackrel{\text{def}}{=} N^0\mathbf{x}_1 = \mathbf{x}_1 \qquad\qquad \mathbf{y}_{p_1+p_2} \stackrel{\text{def}}{=} N^0\mathbf{x}_2 = \mathbf{x}_2 \ \ldots \ .$$

(The last entry is

$$\mathbf{y}_n \stackrel{\text{def}}{=} \mathbf{x}_{r_0}.)$$

A direct calculation taking into account the relations

$$N\mathbf{y}_i = \begin{cases} \mathbf{y}_{i-1} \\ \text{or} \\ \mathbf{O} \end{cases}$$

shows that if P is the $n \times n$ matrix in which column j is $\mathbf{y}_j$ then

$$\text{column } 1 \text{ of } NP \text{ is } \mathbf{O}$$

$$\text{column } 2 \text{ of } NP \text{ is column } 1 \text{ of } P$$

$$\vdots$$

$$\text{column } p_1 \text{ of } NP \text{ is column } p_1 - 1 \text{ of } P$$

$$\text{column } p_1 + 1 \text{ of } NP \text{ is } \mathbf{O}$$

$$\vdots$$

$$\text{column } n \text{ of } NP \text{ is column } n - 1 \text{ of } P.$$

It follows that

> column 1 of $P^{-1}NP$ is $\mathbf{O}$
>
> column 2 of $P^{-1}NP$ is column 1 of I
>
> $\vdots$
>
> column p_1 of $P^{-1}NP$ is column $p_1 - 1$ of I
>
> column $p_1 + 1$ of $P^{-1}NP$ is $\mathbf{O}$
>
> $\vdots$
>
> column n of $P^{-1}NP$ is column $n - 1$ of I.

Thus there are Jordan blocks J_q, $1 \le q \le r_0$ such that

$$P^{-1}NP \overset{\text{def}}{=} \mathcal{N} = \begin{pmatrix} J_1 & & \\ & \ddots & \\ & & J_{r_0} \end{pmatrix}.$$

The size of J_q is $p_q \times p_q$. Owing to the way in which the p_i are arranged, the sizes of the blocks J_q do not increase in the passage down the diagonal of $\mathcal{N}$.

$$\Omega$$

[**Remark 4.5.1:** As the procedure used to derive $\mathcal{N}$ is carried out, i.e., as the various bases X_q are found the numbers p_q crop up automatically, hence the Jordan block sizes $p_q \times p_q$ are visible and the Jordan normal form for N can be calculated *without calculating either P or P^{-1}.*]

Example 4.5.2. Let N be the matrix

$$\begin{pmatrix} 0 & 6 & 0 & 0 \\ 0 & 0 & 0 & 0 \\ 2 & 3 & 0 & 1 \\ 4 & 5 & 0 & 0 \end{pmatrix}.$$

Direct calculation shows

$$N^2 = \begin{pmatrix} 0 & 0 & 0 & 0 \\ 0 & 0 & 0 & 0 \\ 4 & 17 & 0 & 0 \\ 0 & 24 & 0 & 0 \end{pmatrix}$$

$$N^3 = \begin{pmatrix} 0 & 0 & 0 & 0 \\ 0 & 0 & 0 & 0 \\ 0 & 24 & 0 & 0 \\ 0 & 0 & 0 & 0 \end{pmatrix}$$

$$N^4 = O.$$

Hence a basis for $im(N^3)$ is

$$X_3 \stackrel{\text{def}}{=} \{24\mathbf{e}_3\} \stackrel{\text{def}}{=} \{\mathbf{x}_{31}\}.$$

To find $\mathbf{x}_{21}$ observe that the only vectors $\mathbf{v}$ such that $N\mathbf{v} = \mathbf{x}_{31} = 24\mathbf{e}_3$ are of the form

$$\begin{pmatrix} 0 \\ 0 \\ y \\ 24 \end{pmatrix}$$

and that the second column of N^2 may be used for $\mathbf{v}$. Thus let $\mathbf{x}_{21}$ be

$$\begin{pmatrix} 0 \\ 0 \\ 17 \\ 24 \end{pmatrix} = 17\mathbf{e}_3 + 24\mathbf{e}_4$$

and then

$$X_2 \stackrel{\text{def}}{=} \{N\mathbf{x}_{21}, \mathbf{x}_{21}\}$$

$$= \left\{ \begin{pmatrix} 0 \\ 0 \\ 24 \\ 0 \end{pmatrix}, \begin{pmatrix} 0 \\ 0 \\ 17 \\ 24 \end{pmatrix} \right\}$$

is a basis for $im(N^2)$. (No supplements from $ker(N)$ are required to fill out $\{N\mathbf{x}_{21}, \mathbf{x}_{21}\}$ to a basis for $im(N^2)$.)

Similarly the only vectors $\mathbf{u}$ such that $N\mathbf{u} = \mathbf{x}_{21}$ are of the form

$$\begin{pmatrix} 6 \\ 0 \\ y \\ 5 \end{pmatrix}$$

and column 2 of N may be used for $\mathbf{u}$. Thus let $\mathbf{x}_{11}$ be

$$\begin{pmatrix} 6 \\ 0 \\ 3 \\ 5 \end{pmatrix} = 6\mathbf{e}_1 + 3\mathbf{e}_3 + 5\mathbf{e}_4$$

and then

$$X_1 \stackrel{\text{def}}{=} \{N^2\mathbf{x}_{11}, N\mathbf{x}_{11}, \mathbf{x}_{11}\}$$

$$= \left\{ \begin{pmatrix} 0 \\ 0 \\ 24 \\ 0 \end{pmatrix}, \begin{pmatrix} 0 \\ 0 \\ 17 \\ 24 \end{pmatrix}, \begin{pmatrix} 6 \\ 0 \\ 3 \\ 5 \end{pmatrix} \right\}$$

   Chapter 4. Useful forms for matrices

is a basis for $im(N)$. (Again no supplements from $ker(N)$ are needed.)

Finally the only vectors $\mathbf{w}$ such that $N\mathbf{w} = \mathbf{x}_{11}$ are of the form

$$\begin{pmatrix} 0 \\ 1 \\ y \\ 0 \end{pmatrix}$$

and the simplest version of such a $\mathbf{w}$ is $\mathbf{e}_2$. Thus if $\mathbf{x}_1 \stackrel{\text{def}}{=} \mathbf{e}_2$ then

$$X \stackrel{\text{def}}{=} \{N^3\mathbf{x}_1, N^2\mathbf{x}_1, N\mathbf{x}_1, \mathbf{x}_1\}$$

$$= \left\{ \begin{pmatrix} 0 \\ 0 \\ 24 \\ 0 \end{pmatrix}, \begin{pmatrix} 0 \\ 0 \\ 17 \\ 24 \end{pmatrix}, \begin{pmatrix} 6 \\ 0 \\ 3 \\ 5 \end{pmatrix}, \begin{pmatrix} 0 \\ 1 \\ 0 \\ 0 \end{pmatrix} \right\}$$

$$= \{ N^p\mathbf{x}_1 \ : \ N^p\mathbf{x}_q \neq \mathbf{O}, \ 0 \le p < 4, \ 1 \le q \le 1 \}$$

is a basis for $\mathbf{C}^4$.

If

$$
\begin{array}{cccc}
N^3\mathbf{x}_1 & N^2\mathbf{x}_1 & N\mathbf{x}_1 & \mathbf{x}_1
\end{array}
$$
$$P = \begin{pmatrix} 0 & 0 & 6 & 0 \\ 0 & 0 & 0 & 1 \\ 24 & 17 & 3 & 0 \\ 0 & 24 & 5 & 0 \end{pmatrix}$$

then

$$NP = \begin{pmatrix} 0 & 0 & 0 & 6 \\ 0 & 0 & 0 & 0 \\ 0 & 24 & 17 & 3 \\ 0 & 0 & 24 & 5 \end{pmatrix}$$

and the Jordan normal form of N is

$$\mathcal{N} \stackrel{\text{def}}{=} P^{-1}NP = \begin{pmatrix} 0 & 1 & 0 & 0 \\ 0 & 0 & 1 & 0 \\ 0 & 0 & 0 & 1 \\ 0 & 0 & 0 & 0 \end{pmatrix},$$

a single nilpotent Jordan block.

Exercise 4.5.5. Let N be the matrix

$$\begin{pmatrix} 2 & 1 \\ -4 & -2 \end{pmatrix}.$$

$i.$ Show N is nilpotent and find an invertible matrix P such that $P^{-1}NP$ is a nilpotent Jordan block.

---

ii. Let M be the block matrix

$$\begin{pmatrix} N & N \\ N & N \end{pmatrix},$$

a 4×4 matrix. For each of the $4!$ $(= 24)$ 4×4 permutation matrices Π there is the matrix $\Pi^{-1} M \Pi \overset{\text{def}}{=} M_\Pi$. Show that each M_Π is nilpotent.

iii. Choose three essentially different matrices M_Π and for each find an invertible matrix P such that $P^{-1} M_\Pi P$ is in Jordan normal form and find the Jordan normal form.

Exercise 4.5.6. Show that the following procedure and argument yield a Jordan normal form for a matrix N that is nilpotent of order k.

i. Form the sequence

$$ker(N^p), \ 0 \le p \le k$$

of kernels. Let Z_1 be a basis for $ker(N)$. Fill out Z_1 with a set W_2 to form a basis Z_2 for $ker(N^2)$. Fill out Z_2 with a set W_3 to form a basis Z_3 for $ker(N^3)$, ..., until there is produced a basis Z_k for $ker(N^k) = dom(N)$.

ii. Show

$$N W_p \subset \left(kcr(N^{p-1}) \setminus kcr(N^{p-2}) \right)$$

and that $N W_p$ is a linearly independent subset of $ker(N^{p-1})$.

iii. Fill out $Z_{k-2} \cup N W_k$ with a set V_{k-1} to form a basis for $ker(N^{k-1})$. Denote $V_{k-1} \cup N W_k$ by Y_{k-1}.

iv. Show $N Y_{k-1} \subset \left(ker(N^{p-2}) \setminus ker(N^{p-3}) \right)$ and that $N Y_{k-1}$ is a linearly independent subset of $ker(N^{k-2})$.

v. Fill out $Z_{k-3} \cup N Y_{k-1}$ with a set V_{k-2} to form a basis for $ker(N^{k-2})$. Denote $V_{k-2} \cup N Y_{k-1}$ by Y_{k-2}.

vi. Continue to construct sets $N Y_p$, V_{p-1}, Y_{p-1}, etc., until there is produced a set Y_1.

vii. Show that

$$Y \overset{\text{def}}{=} Y_1 \cup Y_2 \cup \cdots \cup Y_{k-1} \cup W_k$$

is a basis for $dom(N)$ and that there is in Y a set $\{\mathbf{x}_1, \ldots, \mathbf{x}_Q\}$ such that

$$Y = \{ N^p \mathbf{x}_q \ : \ N^p \mathbf{x}_q \ne \mathbf{O}, 0 \le p < k, 1 \le q \le Q \}.$$

Example 4.5.3. The matrix N of **Example 4.5.2** may be Jordanized by the procedure described in **Exercise 4.5.6**. The salient points are the following.

i.

$$Z_1 = \{\mathbf{e}_3\}$$
$$W_2 = \{\mathbf{e}_4\}, Z_2 = \{\mathbf{e}_3, \mathbf{e}_4\}$$
$$W_3 = \{\mathbf{e}_1\}, Z_3 = \{\mathbf{e}_3, \mathbf{e}_4, \mathbf{e}_1\}$$
$$W_4 = \{\mathbf{e}_2\}, Z_4 = \{\mathbf{e}_3, \mathbf{e}_4, \mathbf{e}_1, \mathbf{e}_2\};$$

$$NW_4 = \{\mathbf{v} \stackrel{\text{def}}{=} 6\mathbf{e}_1 + 3\mathbf{e}_3 + 5\mathbf{e}_4\}, V_3 = \emptyset, Y_3 = NW_4$$
$$NY_3 = \{17\mathbf{e}_3 + 24\mathbf{e}_4\}, V_2 = \emptyset, Y_2 = NY_3$$
$$NY_2 = \{24\mathbf{e}_3\}, V_1 = \emptyset, Y_1 = NY_2$$
$$Y = Y_1 \cup Y_2 \cup Y_3 \cup W_4$$
$$= \{N^3\mathbf{v}, N^2\mathbf{v}, N\mathbf{v}, \mathbf{v}\}$$
$$P = \begin{pmatrix} 0 & 0 & 6 & 0 \\ 0 & 0 & 0 & 1 \\ 24 & 17 & 3 & 0 \\ 0 & 24 & 5 & 0 \end{pmatrix}$$
$$\mathcal{N} = P^{-1}NP = \begin{pmatrix} 0 & 1 & 0 & 0 \\ 0 & 0 & 1 & 0 \\ 0 & 0 & 0 & 1 \\ 0 & 0 & 0 & 0 \end{pmatrix}.$$

However a Jordanizing basis Y for a nilpotent matrix N is found, there are vectors $\mathbf{y}_1, \ldots, \mathbf{y}_Q$ such that Y can be described as

$$\{\, N^p\mathbf{y}_q \ : \ N^p\mathbf{y}_q \neq \mathbf{O}, \ 0 \leq p < k, \ 1 \leq q \leq Q \,\}.$$

In the notation used earlier, for each q there is a p_q such that

$$N^{p_q-1}\mathbf{y}_q \neq \mathbf{O} = N^{p_q}\mathbf{y}_q$$

and it may be assumed that $p_q \geq p_{q+1}$. Hence the vectors in Y may be enumerated as in **Table 4.5.1** below.

$$
\begin{array}{llllll}
N^{p_1-1}\mathbf{y}_1 & N^{p_1-2}\mathbf{y}_1 & \cdots & N\mathbf{y}_1 & \mathbf{y}_1 & (p_1 \text{ vectors}) \\
N^{p_2-1}\mathbf{y}_2 & N^{p_2-2}\mathbf{y}_2 & \cdots & \mathbf{y}_2 & & (p_2 \text{ vectors}) \\
\vdots & \vdots & & & & \\
N^{p_Q-1}\mathbf{y}_Q & \cdots & & & & (p_Q \text{ vectors})
\end{array}
$$

Table 4.5.1.

The vectors in the first column form a basis for $ker(N)$, the vectors in the first two columns form a basis for $ker(N^2), \ldots$. In particular, $Q = dim[ker(N)]$ and so the number Q depends only on N and not on the vectors $\mathbf{y}_j$ themselves.

Because the rows in **Table 4.5.1** are arranged in order of *nonincreasing* length in the passage from the top of the display to its bottom, the lengths of the columns are nonincreasing in the passage from the left side of the display to the right side, i.e., the length l_j of column j in **Table 4.5.1** is never greater

---

than the length l_{j+1} of column $j + 1$. Hence if $d_j \overset{\text{def}}{=} dim[ker(N^j)]$ then

$$d_s - d_{s-1} = l_s \leq l_{s-1} = d_{s-1} - d_{s-2}$$

$$\frac{1}{2}(d_{s-2} + d_s) \leq d_{s-1}.$$

For any real number x let $[x]$ denote the largest integer not exceeding x, e.g., $[\pi] = 3$, $[-e] = -3$, $[4] = 4$. Let $\mathbf{R}^+$ denote the set of nonnegative real numbers: $\mathbf{R}^+ = \{\, x \; : \; x \in \mathbf{R}, \; x \geq 0 \,\}$. If the function

$$d : \mathbf{R}^+ \ni x \mapsto \begin{cases} d_x & \text{if } x \in \{0\} \cup \mathbf{N}; \\ td_{[x]} + (1-t)d_{[x]+1} & \text{if } x = t[x] + (1-t)([x]+1), \; 0 < t < 1 \end{cases}$$

is plotted as in **Figure 4.5.1** the line segment connecting the immediate neighbors of each vertex of the graph of d lies below that vertex. The function d is an instance of a *concave* function, i.e., a function f such that if $a < b$ and $0 \leq t \leq 1$ then

$$f(ta + (1-t)b) \geq tf(a) + (1-t)f(b).$$

("The graph lies above the chord.")

Exercise 4.5.7. For each matrix N_i below find an invertible matrix P_i such that

$$P_i^{-1}N_iP_i \overset{\text{def}}{=} \mathcal{N}_i, \; 1 \leq i \leq 5,$$

is in Jordan form. Use either the "method of decreasing images" or the "method of increasing kernels" on each matrix and use each method at least once in the course of dealing with $N_1, \ldots, N_5$.

$$N_1 \overset{\text{def}}{=} \begin{pmatrix} 0 & 0 \\ 1 & 0 \end{pmatrix}; \; N_2 = \begin{pmatrix} 0 & 0 & 0 \\ 1 & 0 & 0 \\ 2 & 3 & 0 \end{pmatrix}; \; N_3 = \begin{pmatrix} 1 & -1 \\ 1 & -1 \end{pmatrix};$$

$$N_4 = \begin{pmatrix} 0 & 1 & 1 \\ 0 & 1 & 1 \\ 0 & 1 & 1 \end{pmatrix}; \; N_5 = \begin{pmatrix} -1 & 1 & 0 \\ 1 & 1 & 2 \\ 1 & -1 & 0 \end{pmatrix}.$$

Exercise 4.5.8. For each of the matrices in **Exercise 4.5.5** and **Exercise 4.5.7** calculate the numbers d_s, l_j, p_q, Q. Show that in general

$$\sum_{q=1}^{Q} p_q = n \overset{\text{def}}{=} dim[dom(N)].$$

Show also that if the order of nilpotence of N is k, i.e., if $N^k = O \neq N^{k-1}$, then $k = p_1$.

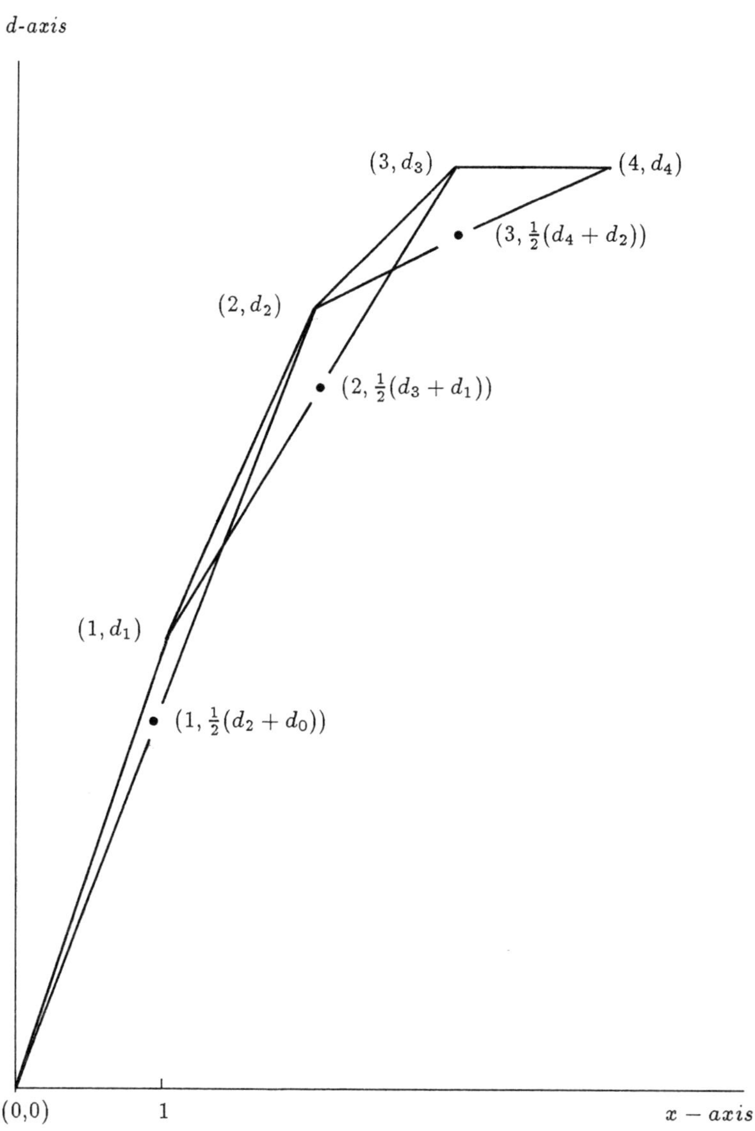

Figure 4.5.1.   The dimension function d

Exercise 4.5.9. Let N be a nilpotent $n \times n$ matrix.

$i.$ Relate the numbers d_s, l_j, p_q, Q, to n, to the order k of nilpotence of N, to
the number $m(N)$ of the nilpotent Jordan blocks J_j in the Jordan normal

form for N, and to the sizes $s_j \times s_j$ of the J_j.

ii. Show that if, in the Jordan normal form $\mathcal{N}$, the nilpotent Jordan blocks on the diagonal of $\mathcal{N}$ are arranged in nonincreasing order of size in the passage down the diagonal then the result is *unique*. In other words, show that if $\mathcal{J}$ is any Jordan normal form for N then there is a permutation matrix Π such that

$$\Pi^{-1}\mathcal{J}\Pi = \mathcal{N}.$$

Exercise 4.5.10. Let N be an $n \times n$ nilpotent matrix such that

$$N^k = O \neq N^{k-1}.$$

Define a dimension function ν according to the equation:

$$\nu : \mathbf{R}^+ \ni x \mapsto \begin{cases} dim[im(N^x)] & \text{if } x \in \{0\} \cup \mathbf{N}; \\ t\nu_{[x]} + (1-t)\nu_{[x]+1} & \text{if } x = t[x] + (1-t)([x]+1),\ 0 < t < 1. \end{cases}$$

Use the relation $dim[im(A)] + dim[ker(A)] = dim[dom(A)]$ to prove that ν is a *convex* function, i.e., that

$$\nu[ta + (1-t)b] \leq t\nu(a) + (1-t)\nu(b),\ \text{if } a < b \text{ and } 0 \leq t \leq 1,$$

i.e., that "the graph lies below the chord." Draw a graph that exemplifies the graph of some ν.

Exercise 4.5.11. Rederive the convexity of the function ν by proving each of the following assertions:

i. In **Table 4.5.1** assume

$$c_k = \text{the number of } p_i \text{ having the value } k$$

$$c_{k-1} = \text{the number of } p_i \text{ having the value } k-1$$

$$\vdots$$

$$c_1 = \text{the number of } p_i \text{ having the value } 1.$$

Then $c_1 \geq 1$ and each $c_j \geq 0$. Furthermore:

a. the first c_1 entries in column 1 of **Table 4.5.1** make up a basis for $im(N^{k-1})$;

b. the first $c_1 + c_2$ entries in column 1 together with the first c_1 entries in column 2 of **Table 4.5.1** make up a basis for $im(N^{k-2})$;

c. etc.

ii.

$$\nu_{k-1} = dim[im(N^{k-1})] = c_1$$

$$\nu_{k-2} = dim[im(N^{k-2})] = 2c_1 + c_2$$

$$\vdots$$

$$\nu_1 = dim[im(N)] = (k-2)c_1 + (k-1)c_2 + \cdots + c_{k-2}.$$

iii.

$$\nu_{j-1} - \nu_{j-2} = c_1 + \cdots + c_{j-1}$$
$$\geq c_1 + \cdots + c_{j-2} = \nu_{j-2} - \nu_{j-3}$$
$$\frac{1}{2}(\nu_{j-3} + \nu_{j-1}) \geq \nu_{j-2}.$$

Exercise 4.5.12. Assume that N is a nilpotent matrix and that

$$N^k = O \neq N^{k-1}.$$

Prove $I + N$ resp. $I - N$ is nonsingular.

[*Hint:* Show, e.g., that

$$(I + N)(I - N + N^2 - \cdots + (-1)^{k-1} N^{k-1}) = I.]$$

4.6. The Jordan normal form of an arbitrary SQUARE matrix

Once the analysis of nilpotent matrices is complete, the next task is to show how to reduce the analysis of an arbitrary matrix to the analysis of a nilpotent matrix. The goal is to prove THEOREM 4.6.1 below. Not unexpectedly the argument uses the existence of matrices F_i such that

$$I = \sum_{i=1}^{k} F_i$$
$$F_i F_{i'} = \begin{cases} F_i & \text{if } i = i' \\ 0 & \text{if } i \neq i'. \end{cases}$$

In connection with such matrices the next result is helpful.

LEMMA 4.6.1. LET F_i BE k $n \times n$ MATRICES SUCH THAT

$$I = \sum_{i=1}^{k} F_i$$
$$F_i F_{i'} = \begin{cases} F_i & \text{if } i = i'; \\ 0 & \text{if } i \neq i'. \end{cases}$$

ASSUME $X_i \overset{\text{def}}{=} \{\mathbf{x}_{i1}, \ldots, \mathbf{x}_{in_i}\}$ IS A BASIS FOR $W_i \overset{\text{def}}{=} im(F_i)$. THEN

$$X \overset{\text{def}}{=} \bigcup_{i=1}^{k} X_i$$

PROOF. The linear independence of X and the validity of the equation $span(X) = \mathbf{C}^n$ are to be proved.

If

$$\sum_{i,j} a_{ij}\mathbf{x}_{ij} = \mathbf{O}$$

then

$$\mathbf{O} = F_m\Big(\sum_{i,j} a_{ij}\mathbf{x}_{ij}\Big) = \sum_{i,j} a_{ij}F_m\mathbf{x}_{ij}$$
$$= \sum_{j=1}^{n_m} a_{mj}\mathbf{x}_{mj}.$$

Since each X_m is a basis $a_{mj} = 0$, $1 \le j \le n_m$, $1 \le m \le k$ and the linear independence of X follows.

If $\mathbf{x} \in \mathbf{C}^n$ then

$$\mathbf{x} = F_1\mathbf{x} + \cdots + F_k\mathbf{x} \stackrel{\text{def}}{=} \sum_{i=1}^{k} \mathbf{x}_i$$

and, since each $\mathbf{x}_i \in im(F_i)$ it follows that each $\mathbf{x}_i \in span(X)$ and thus

$$\mathbf{x} \in span(X).$$

$$\Omega$$

THEOREM **4.6.1.** LET A BE AN $n \times n$ MATRIX AND LET $\{\lambda_1, \ldots, \lambda_k\}$ BE THE SET OF ITS DIFFERENT EIGENVALUES. THERE IS AN INVERTIBLE MATRIX P SUCH THAT $P^{-1}AP$ IS IN JORDAN NORMAL FORM, i.e.,

i. THERE ARE JORDAN BLOCKS

$$J_{iq}(\lambda_i) \stackrel{\text{def}}{=} \begin{pmatrix} \lambda_i & 1 & & & \\ & \lambda_i & 1 & & \\ & & \ddots & & \\ & & & \lambda_i & 1 \\ & & & & \lambda_i \end{pmatrix}, \quad 1 \le q \le Q_i, \ 1 \le i \le k,$$

EACH OF SIZE DENOTED $s_q \times s_q$;

ii. CORRESPONDING TO EACH λ_i THERE ARE EIGENBLOCKS

$$J_i \stackrel{\text{def}}{=} \begin{pmatrix} J_{i1}(\lambda_i) & & & \\ & J_{i2}(\lambda_i) & & \\ & & \ddots & \\ & & & J_{iQ_i}(\lambda_i) \end{pmatrix}, \quad 1 \le i \le k$$

MADE UP OF THE Q_i JORDAN BLOCKS $J_{iq}(\lambda_i)$;

AND

iii.

$$P^{-1}AP = \begin{pmatrix} J_1 & & & \\ & J_2 & & \\ & & \ddots & \\ & & & J_k \end{pmatrix}.$$

[**Note 4.6.1:**   If A is diagonable then $s_q = 1$, $1 \le q \le Q_i$, $1 \le i \le k$.]

PROOF. If A is nilpotent the treatment of **Section 4.5** yields the existence of a Jordanizing basis and a matrix P such that $P^{-1}AP$ is in Jordan normal form. If A is not nilpotent let the factorization of m_A be given according to

$$m_A(z) = (z - \lambda_1)^{s_1} \cdots (z - \lambda_k)^{s_k}$$
$$\stackrel{\mathrm{def}}{=} (z - \lambda_i)^{s_i} \pi_i(z), \ 1 \le i \le k$$
$$\pi_i(z) \stackrel{\mathrm{def}}{=} \prod_{j \neq i} (z - \lambda_j)^{s_j}.$$

As in **Exercises 4.3.17 $-$ 4.3.19** there are polynomials q_i such that if $p_i = q_i \pi_i$ and $p_i(A) \stackrel{\mathrm{def}}{=} F_i$ then

$$I = F_1 + \cdots + F_k$$
$$F_i F_{i'} = \begin{cases} F_i & \text{if } i = i' \\ O & \text{if } i \neq i'. \end{cases} \qquad (4.6.1)$$

Then, since $F_i^2 = F_i$ it follows that $F_i^k = F_i$ for all k in $\mathbb{N}$ and so

$$((A - \lambda_i I)F_i)^{s_i} = (A - \lambda_i I)^{s_i} F_i^{s_i} = (A - \lambda_i I)^{s_i} F_i = m_A(A) = O,$$

i.e., if $N_i \stackrel{\mathrm{def}}{=} (A - \lambda_i I)F_i$ then $N_i^{s_i} = O$, i.e., N_i is nilpotent.

(Indeed the order of nilpotence of N_i is s_i. Otherwise there is in $\mathbb{N}$ an s such that $s < s_i$ and $N_i^s = O$. Since

$$N_i^s = (A - \lambda_i I)^s F_i$$

it follows that $\mu_A \stackrel{\mathrm{def}}{=} deg(m_A) \le s + \sum_{j \neq i} s_j < \sum_{j=1}^k s_j = \mu_A$, a contradiction.)

In **Section 4.5** a Jordan normal form for a nilpotent matrix is derived in two ways:

i. the method of decreasing images;
ii. the method of increasing kernels.

Each of these approaches is applicable in the restricted environment of $W_i \stackrel{\mathrm{def}}{=} im(F_i)$. To simplify the presentation below the subscript i is suppressed.

Thus F is an idempotent, $W \stackrel{\text{def}}{=} im(F)$, and W is an invariant subspace of a nilpotent matrix N: $NW \subset W$, $N^k W = \{\mathbf{O}\} \neq N^{k-1}W$. Each assertion below is demonstrable by an argument almost indistinguishable from the corresponding argument offered in the proof of THEOREM **4.5.1**.

Let X_{k-1} be a basis for $N^{k-1}W$ and let $\widetilde{X}_{k-2}$ in $N^{k-2}W$ be a set of preimage vectors, one for each vector in X_{k-1}. Adjoin to $\widetilde{X}_{k-2}$ a basis for $ker(N) \cap N^{k-2}W$. The set X_{k-2} resulting is a basis for $N^{k-2}W$.

After finitely many repetitions of the procedure just outlined there emerges a basis X for W. Furthermore there are in W finitely many vectors $\mathbf{x}_q$ such that

$$X = \{\, N^p \mathbf{x}_q \ : \ N^p \mathbf{x}_q \neq \mathbf{O}, \ 0 \leq p < \infty, \ 1 \leq q \leq Q \,\}.$$

At this point re-introduce the subscript i. Then for each i there is a basis

$$X_i \stackrel{\text{def}}{=} \{\, N_i^p \mathbf{x}_{iq} \ : \ N_i^p \mathbf{x}_{iq} \neq \mathbf{O}, \ 0 \leq p < \infty, \ 1 \leq q \leq Q_i \,\}$$
$$\stackrel{\text{def}}{=} \{\mathbf{u}_{i1}, \ldots, \mathbf{u}_{in_i}\}$$

for W_i. According to LEMMA **4.6.1**

$$Y \stackrel{\text{def}}{=} \bigcup_{i=1}^{k} X_i$$

is a basis for $\mathbf{C}^n$.

Owing to the relations (4.6.1) The matrix

$$N \stackrel{\text{def}}{=} \sum_{i=1}^{k} N_i$$

is also nilpotent and $N_i = F_i N = N F_i$. Let the vectors in Y be arranged in blocks, each block consisting of the vectors in some X_i, and the vectors $\mathbf{u}_{ij}$ in the ith block arranged so that

$$N_i \mathbf{u}_{ij} = \begin{cases} \mathbf{u}_{i,j-1} \\ \text{or} \\ \mathbf{O}. \end{cases}$$

Then since $\mathbf{u}_{ij} = F_i \mathbf{u}_{ij}$ it follows that $N \mathbf{u}_{ij} = N_i \mathbf{u}_{ij}$ and so Y is Jordanizing basis for N.

Finally,

$$A = N + \sum_{i=1}^{k} \lambda_i F_i \stackrel{\text{def}}{=} N + D.$$

If

$$\mathbf{x} = \sum_{i=1}^{k} F_i \mathbf{x} \stackrel{\text{def}}{=} \sum_{i=1}^{k} \mathbf{x}_i$$

<hr>

then, again, the relations (4.6.1) imply that

$$Dx = \sum_{i=1}^{k} \lambda_i x_i.$$

If P is the matrix in which the columns are the (properly ordered) vectors in Y the circumstances above lead to the conclusion that $P^{-1}AP$ has the Jordan normal form posited in the THEOREM.

$$\Omega$$

Example 4.6.1. For example, the matrix in (4.6.2) below is an eigenblock for the eigenvalue 3. It consists of three Jordan (sub)blocks marked by brackets:

$$
\left(
\begin{array}{ccc}
\left\{\begin{array}{cccc}
3 & 1 & & \\
 & 3 & 1 & \\
 & & 3 & 1 \\
 & & & 3
\end{array}\right\} & & \\
 & \left\{\begin{array}{ccc}
3 & 1 & \\
 & 3 & 1 \\
 & & 3
\end{array}\right\} & \\
 & & \left\{\begin{array}{cc}
3 & 1 \\
 & 3
\end{array}\right\}
\end{array}
\right).
\tag{4.6.2}
$$

If an appropriate order is chosen for the Jordanizing basis, the eigenblocks can be arranged in descending order of size along the diagonal. Within each eigenblock the subblocks can be arranged in descending order of size.

Example 4.6.2. Assume

$$A = \begin{pmatrix} 2 & 0 & 0 & 1 \\ 0 & 3 & 0 & 0 \\ 0 & 1 & 3 & 0 \\ 0 & 0 & 0 & 2 \end{pmatrix}.$$

To find m_A the algorithm in **Section 4.3** is applied. The result is

$$
\begin{aligned}
m_A(z) &= z^4 - 10z^3 + 37z^2 - 60z + 36 \\
&= (z-2)^2(z-3)^2.
\end{aligned}
$$

Consequently

$$\pi_1(z) = (z-3)^2, \quad \pi_2(z) = (z-2)^2$$

and if $q_1(z) = 2z - 3$, $q_2(z) = -2z + 7$ then

$$
\begin{aligned}
&p_1 = q_1 \pi_1, \quad p_2 = q_2 \pi_2 \\
&p_1 + p_2 = (2z-3)(z-3)^2 + (-2z+7)(z-2)^2 \equiv 1.
\end{aligned}
$$

Hence there are two idempotents:

$$F_1 \stackrel{\text{def}}{=} \begin{pmatrix} 1 & & & \\ & 0 & & \\ & & 0 & \\ & & & 1 \end{pmatrix}, \quad F_2 \stackrel{\text{def}}{=} \begin{pmatrix} 0 & & & \\ & 1 & & \\ & & 1 & \\ & & & 0 \end{pmatrix}$$

for which there obtain:

$$F_1 + F_2 = I, \quad F_1^2 = F_1, \quad F_2^2 = F_2, \quad \text{and } F_1 F_2 = O.$$

It follows that

$$\begin{aligned} A = AI &= AF_1 + AF_2 \\ &= (A - 3I)F_1 + (A - 2I)F_2 + 2F_1 + 3F_2 \\ &= \begin{pmatrix} 0 & 0 & 0 & 1 \\ 0 & 0 & 0 & 0 \\ 0 & 1 & 0 & 0 \\ 0 & 0 & 0 & 0 \end{pmatrix} + \begin{pmatrix} 2 & & & \\ & 3 & & \\ & & 3 & \\ & & & 2 \end{pmatrix} \\ &\stackrel{\text{def}}{=} N + D. \end{aligned}$$

The matrix N is such that $N^2 = O$ and the procedure for Jordanizing it leads to the matrix P given by

$$\begin{pmatrix} 1 & 0 & 0 & 0 \\ 0 & 0 & 0 & 1 \\ 0 & 0 & 1 & 0 \\ 0 & 1 & 0 & 0 \end{pmatrix},$$

i.e., $P = E_{24}$. In the circumstances,

$$P^{-1} A P = \begin{pmatrix} 2 & 1 & & \\ & 2 & & \\ & & 3 & 1 \\ & & & 3 \end{pmatrix},$$

and the Jordan normal form for A is achieved.

Exercise 4.6.1. For each of the following matrices find a Jordan normal form.

$$i. \begin{pmatrix} 3 & -1 \\ 1 & 1 \end{pmatrix}; \quad ii. \begin{pmatrix} 3 & -1 \\ -1 & 3 \end{pmatrix}.$$

Exercise 4.6.2. For each of the following matrices find a Jordan normal form.

$$i. \begin{pmatrix} 1 & 0 & -1 \\ -1 & 2 & -1 \\ -1 & 2 & 3 \end{pmatrix}; \quad ii. \begin{pmatrix} 3 & 1 & -1 \\ 0 & 2 & 2 \\ 3 & 3 & 1 \end{pmatrix}; \quad iii. \begin{pmatrix} 4 & 2 & 2 \\ 1 & 3 & -1 \\ 1 & 1 & 5 \end{pmatrix}.$$

  Chapter 4. Useful forms for matrices

Exercise 4.6.3. For each of the following matrices find a Jordan normal form.

$$i. \begin{pmatrix} 65 & -16 & 5 \\ 110 & -16 & 14 \\ 173 & -52 & 41 \end{pmatrix} ; \quad ii. \begin{pmatrix} 15 & 6 & -3 \\ -6 & 30 & -6 \\ 9 & -18 & 27 \end{pmatrix} ;$$

$$iii. \begin{pmatrix} 76 & -32 & 10 \\ 142 & -56 & 22 \\ 190 & -44 & 16 \end{pmatrix} ; \quad iv. \begin{pmatrix} 42 & -6 & 0 \\ 99 & -12 & 3 \\ 174 & 54 & 24 \end{pmatrix} .$$

Exercise 4.6.4. For each of the matrices below find a Jordan normal form.

$$i. \begin{pmatrix} 4 & 0 & -2 \\ -1 & 3 & 1 \\ -1 & -1 & 3 \end{pmatrix} ; \quad ii. \begin{pmatrix} 3 & 1 & -1 \\ 0 & 2 & 0 \\ 1 & 1 & 1 \end{pmatrix} ;$$

$$iii. \begin{pmatrix} 1 & 1 & 1 \\ -1 & 1 & 1 \\ -1 & -1 & 1 \end{pmatrix} ; \quad iv. \begin{pmatrix} 1 & -1 & -1 \\ -1 & 1 & -1 \\ -1 & -1 & 1 \end{pmatrix} .$$

Exercise 4.6.5. For each of the matrices in **Exercises 4.6.1 - 4.6.4** identify the nilpotent and diagonal parts. For each nilpotent part calculate the associated d's, l's, and p's.

Exercise 4.6.6. Describe all kinds of diagonable 2×2 matrices

$$\begin{pmatrix} a & b \\ c & d \end{pmatrix} .$$

Exercise 4.6.7. For each kind of diagonable 2×2 matrix A describe an invertible 2×2 matrix P such that $P^{-1}AP$ is a diagonal matrix.

Exercise 4.6.8. Show that a 2×2 matrix $A \overset{\text{def}}{=} (a_{ij})_{i,j=1}^{2,2}$ is *not* diagonable iff A has only one eigenvalue and A is *not* a constant multiple of the identity I. Prove the result in the following form:

$(a_{ij})_{i,j=1}^{2,2}$ is not diagonable iff $(a_{11} - a_{22})^2 + 4a_{12}a_{21} = 0$ and $|a_{12}| + |a_{21}| > 0$.

Exercise 4.6.9. For each kind of nondiagonable 2×2 matrix A find an invertible matrix P such that $P^{-1}AP$ is in Jordan normal form.

Exercise 4.6.10. Show that if A is an $n \times n$ matrix then $\sum_{i=1}^{k} d_{i\lambda_i} = n$.

Exercise 4.6.11. Let A be an $n \times n$ matrix and let X be a Jordanizing basis for A. Show that if t is an arbitrary nonzero number, then tX is also a Jordanizing basis for A.

Exercise 4.6.12. Assume that for an $n \times n$ matrix A:

$$m_A(\lambda) = \lambda^3(\lambda - 1)^2(\lambda + 2); \ \chi_A(\lambda) = \lambda^4(\lambda - 1)^3(\lambda + 2).$$

Find n and write all possible Jordan normal forms that A might have.

Exercise 4.6.13. Show that N and D of the decomposition $A = N + D$, derived in the proof of THEOREM **4.6.1** are such that $ND = DN$.

The upper triangular form THEOREM **4.2.3** for an arbitrary $n \times n$ matrix A is not necessarily in Jordan normal form. Nevertheless the upper triangular form does permit a decomposition of A. Indeed, let P be such that $P^{-1}AP \overset{\text{def}}{=} A_U$ is the upper triangular form found according to the discussion in THEOREM **4.2.3**. If D is the diagonal matrix in which the entries are the corresponding diagonal entries of A_U and if $N \overset{\text{def}}{=} A_U - D$ then $A = PDP^{-1} + PNP^{-1} \overset{\text{def}}{=} D' + N'$.

Exercise 4.6.14. Show D' is diagonable and N' is nilpotent.

It is not necessarily true that if a SQUARE matrix A is decomposed into the sum of a diagonable matrix D' and a nilpotent matrix N' $(A = D' + N')$ then $D'N' = N'D'$.

Example 4.6.3. If

$$A = \begin{pmatrix} 1 & 2 & 3 \\ 0 & 1 & 4 \\ 0 & 0 & 2 \end{pmatrix} = \begin{pmatrix} 1 & & \\ & 1 & \\ & & 2 \end{pmatrix} + \begin{pmatrix} & 2 & 3 \\ & & 4 \\ & & \end{pmatrix} \overset{\text{def}}{=} D' + N'$$

then

$$D'N' = \begin{pmatrix} & 2 & 3 \\ & & 4 \\ & & \end{pmatrix} \neq \begin{pmatrix} & 2 & 6 \\ & & 8 \\ & & \end{pmatrix} = N'D'.$$

On the other hand, in the proof of THEOREM **4.6.1** there is found a decomposition $A = N + D$ $(= D + N)$ and in that decomposition, since D and N are polynomial functions of A, $DN = ND$. The next result shows in what sense such a decomposition is unique.

THEOREM **4.6.2.** IF A IS AN $n \times n$ SQUARE MATRIX, IF D_1 AND D_2 ARE DIAGONABLE MATRICES, IF N_1 AND N_2 ARE NILPOTENT MATRICES, AND IF

$$A = D_1 + N_1 = D_2 + N_2, \quad D_1 N_1 = N_1 D_1, \quad D_2 N_2 = N_2 D_2,$$

THEN $D_1 = D_2$, $N_1 = N_2$.

PROOF. Let D and N be the matrices found in the proof of THEOREM **4.6.1.**

Then

$$AD_1 = D_1^2 + N_1 D_1 = D_1^2 + D_1 N_1 = D_1 A$$
$$AN_1 = D_1 N_1 + N_1^2 = N_1 D_1 + N_1^2 = N_1 A$$
$$AD = D^2 + ND = D^2 + DN = DA$$
$$AN = DN + N^2 = ND + N^2 = NA.$$

Hence, as polynomial functions of A, the matrices D resp. N commute with D_1 resp. N_1: $D_1 D = DD_1$, $N_1 N = NN_1$. Assume $N^k = O \neq N^{k-1}$ and $N_1^{k_1} = O \neq N_1^{k_1 - 1}$. Then if $K \geq k + k_1$

$$(N_1 - N)^K = N_1^K + \binom{K}{1} N_1^{K-1} N + \cdots + \binom{K}{k} N_1^{K-k} N^k + \cdots \qquad (4.6.3)$$

Every term in the right member of (4.6.3) is O by virtue of the choice of K. Since $D - D_1 = N_1 - N$ it follows that $(D - D_1)^K = O$. As commuting diagonable matrices D and D_1 are simultaneously diagonable (cf. **Exercise 4.3.17**): for some invertible matrix P

$$P^{-1}DP = \begin{pmatrix} \lambda_1 & & & \\ & \lambda_2 & & \\ & & \ddots & \\ & & & \lambda_n \end{pmatrix} \overset{\mathrm{def}}{=} \Delta,$$

$$P^{-1}D_1 P = \begin{pmatrix} \mu_1 & & & \\ & \mu_2 & & \\ & & \ddots & \\ & & & \mu_n \end{pmatrix} \overset{\mathrm{def}}{=} \Delta_1.$$

Hence $(\Delta - \Delta_1)^K = O$ and so $(\lambda_i - \mu_i)^K = 0$, $1 \leq i \leq n$. In other words, $\Delta = \Delta_1$ and so $D = D_1$, whence $N = N_1$. Similarly $D = N_2$ and $N = N_2$.

$$\Omega$$

Exercise 4.6.15. Show that D in the decomposition $A = N + D$ derived in the proof of THEOREM **4.6.1** is diagonable.

Exercise 4.6.16. Let A be an $n \times n$ matrix and let its minimal polynomial m_A be such that

$$m_A(z) = (z - \lambda_1)^{s_1} \cdots (z - \lambda_k)^{s_k}.$$

The decomposition $A = D + N$ used in the derivation of the Jordan normal form for A gives rise to numbers d_s, l_j, p_q, Q associated with N. Relate those numbers, the numbers n, s_i, k, and the sizes and numbers of blocks and subblocks in the Jordan normal form of A (cf. **Exercise 4.5.9**).

Exercise 4.6.17. Let A be an $n \times n$ SQUARE matrix with associated idempotents F_i and eigenvalues λ_i. Show that:

  i. each $W_i \overset{\text{def}}{=} im(F_i)$ is a generalized eigenspace for A;
  ii. each W_i is a *maximal generalized eigenspace* for A in the sense that if W is any generalized eigenspace corresponding to the eigenvalue λ_i then $W \subset W_i$;
  iii. if W is *any* generalized eigenspace then there is an i such that $W \subset W_i$.

If F_i, $1 \leq i \leq k$ are $n \times n$ matrices such that

$$F_i F_{i'} = \begin{cases} F_1 & \text{if } i = i'; \\ O & \text{if } i \neq i'; \end{cases} \tag{4.6.4}$$
$$F_1 + \cdots + F_k = I$$

let W_i denote $im(F_i)$. Such a set of matrices is associated with the derivation of the Jordan normal form of an $n \times n$ matrix A. The matrices themselves emerge directly from consideration of the minimal polynomial m_A of A. The equation (4.6.4) is interpreted as describing the matrices F_i as *orthogonal idempotents*.

Exercise 4.6.18. Assume $\{F_i\}_{i=1}^k$ is a finite set of $n \times n$ orthogonal idempotents and that $W_i = im(F_i)$. Show:

  i. $W_i \cap W_{i'} = \{\mathbf{O}\}$ if $i \neq i'$;
  ii. for each $\mathbf{x}$ in $\mathbf{C}^n$ there is in each W_i a vector $\mathbf{x}_i$ such that $\mathbf{x} = \mathbf{x}_1 + \cdots + \mathbf{x}_k$;
  iii. if $\mathbf{w}_i \in W_i$, $1 \leq i \leq k$ and $\mathbf{w}_1 + \cdots + \mathbf{w}_k = \mathbf{O}$ then each $\mathbf{w}_i = \mathbf{O}$;
  iv. the $\mathbf{x}_i$ in *ii* are unique.

The situation portrayed in $i - iv$ of **Exercise 4.6.18** is described by the sentence:

The vector space $\mathbf{C}^n$ is the direct sum of the subspaces W_i.

In symbols the sentence is: $\mathbf{C}^n = W_1 \oplus \cdots \oplus W_k$ or

$$\mathbf{C}^n = \bigoplus_{i=1}^k W_i.$$

Thus the Jordan normal form result includes the statement that for each $n \times n$ matrix A there are (A-invariant) maximal generalized eigenspaces W_i such that $\mathbf{C}^n = W_1 \oplus \cdots \oplus W_k$.

In **Exercise 4.3.17** there is the statement that commuting diagonable matrices A and B are simultaneously diagonable (cf. also THEOREM **4.6.2**). A related statement is found in the next **Exercise**.

Exercise 4.6.19. Let A and B be commuting $n \times n$ matrices: $AB = BA$. Show that $\mathbf{C}^n$ is the direct sum of (A-invariant and B-invariant) subspaces V_m that are generalized eigenspaces for both A and B.

[*Hint:* Let $\{F_i\}_{i=1}^k$ resp. $\{G_j\}_{j=1}^l$ be the orthogonal idempotents associated with A resp. B. Consider the set $\{F_i G_j\}_{i,j=1}^{k,l}$.]

Example 4.6.4. The matrices A and B displayed below commute: $AB = BA = O$. Despite the discussion preceding **Exercise 4.6.19** and the result in **Exercise 4.6.19** itself there is no invertible matrix P such that *both* $P^{-1}AP$ and $P^{-1}BP$ are in Jordan normal form: A and B are *not* simultaneously Jordanizable. Here are the matrices:

$$A \overset{\text{def}}{=} \begin{pmatrix} 0 & 1 \\ 0 & 0 \end{pmatrix}; \quad B \overset{\text{def}}{=} \begin{pmatrix} 0 & 2 \\ 0 & 0 \end{pmatrix}.$$

Exercise 4.6.20. Derive THEOREM **4.6.1** by means of the increasing kernels method.

Exercise 4.6.21. Let A be the matrix

$$\begin{pmatrix} 2 & 4 \\ -1 & -2 \end{pmatrix}.$$

i. Show that A is not diagonable.
ii. Find an invertible matrix P such that $P^{-1}AP$ is in Jordan normal form.
iii. Show that there is no *unitary* matrix Q such that $Q^{-1}AQ$ is in Jordan normal form.

Exercise 4.6.22. Refer to **Exercises 4.3.3-4.3.5**. Assume $n > 1$, $i \neq j$, and $t \neq 0$. Find a Jordan normal form for $E_{ij}(t)$.

Exercise 4.6.23. Refer to **Exercise 4.2.21**. Derive the conclusion there *without* the hypothesis that the λ_i are pairwise different.

Thus although the upper triangular form can be achieved by means of unitary matrices, the Jordan normal form can*not* always be achieved by means of unitary matrices.

Although polynomials are used to find orthogonal idempotents F_i for a *non*nilpotent SQUARE matrix A the situation for a nilpotent matrix N is somewhat different.

If $Y \stackrel{\text{def}}{=} \{y_i\}_{i=1}^n$ is any basis for $\mathbf{C}^n$ and if $Y_1, \ldots, Y_k$ are subsets of Y such that any two are disjoint and their union is Y then corresponding to each Y_i is an idempotent G_i such that

$$G_1 + \cdots + G_k = I$$
$$G_i G_{i'} = \begin{cases} G_i & \text{if } i = i' \\ O & \text{if } i \neq i'. \end{cases} \tag{4.6.5}$$

Indeed, if $\mathbf{y} \in \mathbf{C}^n$ there are uniquely defined numbers a_j such that

$$\mathbf{y} = \sum_{j=1}^n a_j \mathbf{y}_j.$$

If G_i is defined by the equation

$$G_i \mathbf{y} = \sum_{\mathbf{y}_j \in Y_i} a_j \mathbf{y}_j$$

then a direct calculation shows that the G_i behave according to (4.6.5). In particular if, as in **Section 4.5**, $Y \stackrel{\text{def}}{=} \{\mathbf{y}_j\}_{j=1}^n$ is a Jordanizing basis for a nilpotent matrix N and if

$$Y_1 \stackrel{\text{def}}{=} \{\mathbf{y}_1, \ldots, \mathbf{y}_{p_1}\}$$
$$Y_2 \stackrel{\text{def}}{=} \{\mathbf{y}_{p_1+1}, \ldots, \mathbf{y}_{p_1+p_2}\}$$
$$\vdots$$

then the corresponding $G_1, G_2, \ldots$ are orthogonal idempotents behaving according to (4.6.5). *Nevertheless*, unless there is only one nonzero G_i, there are no polynomials p_i such that $p_i(N) = G_i$.

[*Proof.* If p is a polynomial and $p(N)^2 = p(N)$ then, since N is nilpotent, say of order k,

$$p(N) = a_0 I + \cdots + a_{k-1} N^{k-1} \tag{4.6.6}$$
$$p(N)^2 = a_0 a_0 I + (a_0 a_1 + a_1 a_0)N + (a_0 a_2 + a_1 a_1 + a_2 a_0)N^2 + \cdots$$
$$+ (a_0 a_{k-1} + \cdots + a_{k-1} a_0)N^{k-1}. \tag{4.6.7}$$

Comparison of the coefficients of like powers of N in (4.6.6) and (4.6.7) shows first that $a_0^2 = a_0$ and so $a_0 = 1$ or $a_0 = 0$. In either event further comparison of the other coefficients shows that each is 0. Hence $p(N) = I$ if $a_0 = 1$ and $p(N) = O$ if $a_0 = 0$. If there is more than one nonzero G_i then none is O or I.]

Since N is nilpotent any convergent power series

$$\sum_{n=1}^{\infty} a_n N^n$$

is actually a polynomial $p(N)$ and so there are no convergent power series S_i such that $S_i(N) = G_i$.

4.7. The Jordan normal form and the functional calculus

The principal value of the "useful forms of matrices" is the simplification they provide for the calculation of *functions* of matrices.

Example 4.7.1. Let A be the matrix

$$\begin{pmatrix} 3 & -1 \\ 1 & 1 \end{pmatrix}.$$

Then the system

$$\frac{dx}{dt} = 3x(t) - y(t)$$
$$\frac{dy}{dt} = x(t) + y(t) \tag{4.7.1}$$

of *differential equations* may be written in vector form according to the following procedure.

Let $X(t)$ be the column vector

$$\begin{pmatrix} x(t) \\ y(t) \end{pmatrix}.$$

Then (4.7.1) is really

$$\frac{dX}{dt} = AX(t) \tag{4.7.2}$$

which is reminiscent of and analogous to the ordinary differential equation

$$\frac{dz}{dt} = az(t). \tag{4.7.3}$$

If $z(0)$ is prescribed, the solution to (4.7.3) is

$$z(t) = e^{ta}z(0).$$

By analogy, if $X(0)$ is prescribed, the solution to (4.7.2) should be

$$X(t) = e^{tA}X(0). \tag{4.7.4}$$

The power series formula

$$e^{ta} = 1 + ta + \frac{(ta)^2}{2!} + \cdots + \frac{(ta)^n}{n!} + \cdots$$

$$= \sum_{k=1}^{\infty} \frac{(ta)^k}{k!}$$

suggests that the meaning of e^{tA} be

$$\sum_{k=1}^{\infty} \frac{(tA)^k}{k!}. \tag{4.7.5}$$

Each term of the series in (4.7.5) is a 2×2 matrix and can be calculated. The resulting terms (matrices) can be added together if the additions performed lead to *convergent* series for each of the four entries in each matrix. The computations are absolutely daunting.

On the other hand, the sole eigenvalue of A is 2 and every corresponding eigenvector is a multiple of

$$\begin{pmatrix} 1 \\ 1 \end{pmatrix}.$$

If

$$P \overset{\mathrm{def}}{=} \begin{pmatrix} 1 & 0 \\ 1 & -1 \end{pmatrix}$$

then $P = P^{-1}$ and a Jordan normal form of A is

$$\mathcal{A} \overset{\mathrm{def}}{=} \begin{pmatrix} 2 & 1 \\ 0 & 2 \end{pmatrix} = P^{-1}AP$$

whence $A = P\mathcal{A}P^{-1}$.

Hence $A^k = P\mathcal{A}^k P^{-1}$ and so

$$\sum_{k=0}^{\infty} \frac{(tA)^k}{k!} = P\left(\sum_{k=0}^{\infty} \frac{(t\mathcal{A})^k}{k!}\right)P^{-1}.$$

However,

$$\mathcal{A} = \begin{pmatrix} 2 & 0 \\ 0 & 2 \end{pmatrix} + \begin{pmatrix} 0 & 1 \\ 0 & 0 \end{pmatrix} \overset{\mathrm{def}}{=} 2I + N \overset{\mathrm{def}}{=} D + N.$$

Furthermore $DN = ND = 2N$ and $N \neq O = N^2 = N^{2+m}, 0 \leq m < \infty$. Hence

$$(t\mathcal{A})^k = (tD)^k + k(tD)^{k-1}tN = \begin{pmatrix} (2t)^k & k(2t)^{k-1}t \\ 0 & (2t)^k \end{pmatrix}$$

$$e^{t\mathcal{A}} = \sum_{k=0}^{\infty} \frac{(t\mathcal{A})^k}{k!}$$

$$= \sum_{k=0}^{\infty} \begin{pmatrix} \frac{(2t)^k}{k!} & \frac{k(2t)^k}{2 \cdot k!} \\ 0 & \frac{(2t)^k}{k!} \end{pmatrix}$$

$$= \begin{pmatrix} e^{2t} & te^{2t} \\ 0 & e^{2t} \end{pmatrix}. \tag{4.7.6}$$

Notice that the $(1,2)$ entry te^{2t} in the last matrix is

$$\frac{d(e^{ta})}{da}\bigg|_{a=2}.$$

Finally

$$e^{tA} = Pe^{tA}P^{-1} = e^{2t}\begin{pmatrix} 1+t & -t \\ t & 21-t \end{pmatrix}.$$

A check shows that if

$$X(0) \overset{\text{def}}{=} \mathbf{u} = \begin{pmatrix} v \\ w \end{pmatrix}$$

is any (constant) 2×1 vector then indeed

$$X(t) \overset{\text{def}}{=} e^{tA}\mathbf{u} = e^{2t}\begin{pmatrix} 1+t & -t \\ t & 1-t \end{pmatrix}\begin{pmatrix} v \\ w \end{pmatrix}$$

$$\overset{\text{def}}{=} \begin{pmatrix} x(t) \\ y(t) \end{pmatrix}$$

provides a solution to the system (4.7.1).

Exercise 4.7.1. Let A be the matrix (a Jordan block)

$$\begin{pmatrix} a & 1 & 0 \\ 0 & a & 1 \\ 0 & 0 & a \end{pmatrix}.$$

Calculate:

i. A^2, A^3, and, for any natural number n, A^n;

ii. for any polynomial p such that

$$p(\lambda) = a_0\lambda^n + a_1\lambda^{n-1} + \cdots + a_{n-1}\lambda + a_n$$

the matrix $p(A)$;

iii. the matrix $\sin A$ by using the series representation

$$\sin \lambda = \lambda - \frac{\lambda^3}{3!} + \frac{\lambda^5}{5!} - \cdots ;$$

iv. the matrix $f(A)$ for any function f such that

$$f(\lambda) = \sum_{k=0}^{\infty} \frac{f^{(k)}(a)}{k!}(\lambda - a)^k \quad \text{(the Taylor series in powers of } (\lambda - a))$$

$$\overset{\text{def}}{=} \sum_{k=0}^{\infty} c_k(\lambda - a)^k.$$

Example 4.7.2. For f and A as in **Exercise 4.7.1**iv it should be noted that

$$(A - aI)^k = \begin{cases} \begin{pmatrix} 0 & 0 & 1 \\ 0 & 0 & 0 \\ 0 & 0 & 0 \end{pmatrix}, & \text{if } k = 2; \\[6pt] O, & \text{if } k > 2. \end{cases}$$

whence

$$f(A) = \begin{pmatrix} f(a) & f'(a) & \frac{f''(a)}{2!} \\ 0 & f(a) & f'(a) \\ 0 & 0 & f(a) \end{pmatrix}.$$

Note that if A is an $n \times n$ Jordan block, e.g., if

$$A = \begin{pmatrix} a & 1 \\ & a & 1 \\ & & \ddots & \ddots \\ & & & a & 1 \\ & & & & a \end{pmatrix},$$

then $(A - aI)^k$ is O if $k \geq n$ and that if $k < n$ then $(A - aI)^k$ resembles $A - aI$ except that the supradiagonal string of 1's is shifted k supradiagonals "northeast" of the diagonal. Thus, e.g., if $n = 5$ then

$$N \overset{\text{def}}{=} A - aI = \begin{pmatrix} 0 & 1 \\ & 0 & 1 \\ & & 0 & 1 \\ & & & 0 & 1 \\ & & & & 0 \end{pmatrix}, \quad N^2 = \begin{pmatrix} 0 & 1 \\ & 0 & 1 \\ & & 0 & 1 \\ & & & 0 \end{pmatrix},$$

$$N^3 = \begin{pmatrix} & 0 & 1 \\ & & 0 & 1 \\ & & & 0 \end{pmatrix}, \quad N^4 = \begin{pmatrix} & & 0 & 1 \\ & & & 0 \end{pmatrix}$$

$N^{4+m} \overset{\text{def}}{=} (A - aI)^{4+m} = O$, if $m \geq 1$.

For any n in $\mathbb{N}$ it follows from the Binomial Theorem that if $K \in \mathbb{N}$ then

$$A^K = (aI + N)^K = \begin{cases} a^K I + K a^{K-1} N + \cdots + \binom{K}{n-1} a^{K-n+1} N^{n-1}; & \text{if } K \geq n \\[6pt] a^K I + K a^{K-1} N + \cdots + K a N^{K-1} + N^K; & \text{if } K < n. \end{cases}$$

Since

$$\binom{K}{m} a^{K-m+1} = \frac{1}{m!} \frac{d^m(a^K)}{da^m}$$

<hr>

it follows that for any polynomial p of degree K

$$p(A) = p(aI + N) = p(a)I + p'(a)N + \cdots + p^{(K)}(a)N^K. \qquad (4.7.7)$$

(It is to be understood that in the *finite* series in (4.7.7) if $K > n$ all terms following the nth are O's.)

In general, let A be an $n \times n$ Jordan block

$$\begin{pmatrix} a & 1 & & & \\ & a & 1 & & \\ & & \ddots & \ddots & \\ & & & a & 1 \\ & & & & a \end{pmatrix} \qquad (4.7.8)$$

and let f be representable as a Taylor series in powers of $(\lambda - a)$, i.e.,

$$f(\lambda) = \sum_{k=0}^{\infty} \frac{f^{(k)}(\lambda)}{k!} (\lambda - a)^k.$$

Then since $N^{n+m} = O$ if $m \geq 1$ direct verification shows that

$$f(A) = \begin{pmatrix} f(a) & f'(a) & \frac{f''(a)}{2!} & \cdots & \frac{f^{(n-2)}(a)}{(n-2)!} & \frac{f^{(n-1)}(a)}{(n-1)!} \\ 0 & f(a) & f'(a) & \cdots & \frac{f^{(n-3)}(a)}{(n-3)!} & \frac{f^{(n-2)}(a)}{(n-2)!} \\ 0 & 0 & f(a) & \cdots & \frac{f^{(n-4)}(a)}{(n-4)!} & \frac{f^{(n-3)}(a)}{(n-3)!} \\ \vdots & \vdots & \vdots & \ddots & \vdots & \vdots \\ 0 & 0 & 0 & \cdots & f(a) & f'(a) \\ 0 & 0 & 0 & \cdots & 0 & f(a) \end{pmatrix}.$$

Hence if p is any polynomial such that for the particular number a,

$$p(a) = f(a), p'(a) = f'(a), \ldots p^{(n-1)}(a) = f^{(n-1)}(a)$$

then $p(A) = f(A)$.

Example 4.7.3. Assume $f(x) = \sin x$. If $a \in \mathbf{R}$ to find a polynomial p such that

$$p(a) = f(a) = \sin a \,, p'(a) = f'(a) = \cos a, \ p''(a) = f''(a) = -\sin a,$$

it suffices to assume that p is a polynomial of degree 2:

$$p(x) = \alpha x^2 + \beta x + \gamma.$$

Thus

$$\alpha a^2 + \beta a + \gamma = \sin a$$
$$2\alpha a + \beta = \cos a$$
$$2\alpha a = -\sin a. \tag{4.7.9}$$

The equations (4.7.9) constitute a system of three equations in the three unknowns α, β, and γ. The matrix

$$B(a) \stackrel{\text{def}}{=} \begin{pmatrix} a^2 & a & 1 \\ 2a & 1 & 0 \\ 2a & 0 & 0 \end{pmatrix}$$

is nonsingular iff $a \neq 0$. If $a \neq 0$ the solution is

$$\alpha = -\frac{\sin a}{2a}$$
$$\beta = \cos a + \sin a$$
$$\gamma = \sin a - a \cos a + a \sin a + \frac{a}{2} \sin a.$$

But even when $a = 0$ the system has a solution since the augmented matrix in that event is

$$\left(\begin{array}{ccc|c} 0 & 0 & 1 & 0 \\ 0 & 1 & 0 & 1 \\ 0 & 0 & 0 & 0 \end{array} \right)$$

and has the same rank (2) as the rank of $B(0)$. The value of α is arbitrary (hence, for simplicity, can be chosen to be 0). In any event $\beta = 1$ and $\gamma = 0$. Thus if

$$A = \begin{pmatrix} a & 1 & 0 \\ 0 & a & 1 \\ 0 & 0 & a \end{pmatrix}$$

then

$$\sin A = p(A)$$
$$= \begin{cases} -\frac{\sin a}{2a} A^2 + (\cos a + \sin a)A \\ \qquad + (\sin a - a \cos a - a \sin a + \frac{a}{2} \sin a)I, & \text{if } a \neq 0; \\ A, & \text{if } a = 0. \end{cases}$$

Exercise 4.7.2. Let A be the Jordan block

$$\begin{pmatrix} \pi & 1 & 0 & 0 \\ 0 & \pi & 1 & 0 \\ 0 & 0 & \pi & 1 \\ 0 & 0 & 0 & \pi \end{pmatrix}.$$

Find $\cos A$ as a polynomial in A.

The **Exercises** and **Examples** above show that if the Jordan normal form of a matrix A is a single Jordan block there is for A a *functional calculus*. It makes possible the calculation of trigonometric, exponential, logarithmic, etc., functions of A (so long as these functions are defined at the (single) eigenvalue of A) in terms of *polynomial* functions of A.

In the same vein assume f is merely n times differentiable and that f may be written near a according to the nth order Taylor Formula with Remainder (cf. **Section 4.8**). If p and its first $n-1$ derivatives at a are the same as f and *its* first $n-1$ derivatives at a then again $p(A) = f(A)$. The validity of this conclusion is derived by means of the Jordan normal form. Nevertheless the conclusion is *applicable* without reference to the Jordan normal form.

A summary of experience to this point is the following for an arbitrary $n \times n$ matrix A.

i. If A is diagonable there is an invertible matrix P such that

$$P^{-1}AP = \begin{pmatrix} \lambda_1 & & & \\ & \lambda_2 & & \\ & & \ddots & \\ & & & \lambda_n \end{pmatrix} \stackrel{\text{def}}{=} D$$

is a diagonal matrix. The diagonal entries in D are not necessarily different. For any function f that is defined at each of the numbers $\lambda_1, \ldots, \lambda_n$ one may *define* $f(D)$ by the formula

$$f(D) \stackrel{\text{def}}{=} \begin{pmatrix} f(\lambda_1) & & & & \\ & f(\lambda_2) & & & \\ & & \ddots & & \\ & & & f(\lambda_{n-1}) & \\ & & & & f(\lambda_n) \end{pmatrix}$$

and then *define* $f(A)$ by the formula

$$f(A) \stackrel{\text{def}}{=} Pf(D)P^{-1}.$$

ii. If A is similar to a *single Jordan block* there is an invertible matrix P such that

$$A \stackrel{\text{def}}{=} P^{-1}AP = \begin{pmatrix} a & 1 & & & \\ & a & 1 & & \\ & & \ddots & \ddots & \\ & & & a & 1 \\ & & & & a \end{pmatrix}$$

and a is the sole eigenvalue of A. If f is any function differentiable n times at and near a and if f is representable at and near a by means of the nth

---

order Taylor Formula with a Remainder then $f(\mathcal{A})$ is given by the formula

$$f(\mathcal{A}) \stackrel{\text{def}}{=} \begin{pmatrix} f(a) & f'(a) & \cdots & \frac{f^{(n-2)}(a)}{(n-2)!} & \frac{f^{(n-1)}(a)}{(n-1)!} \\ & f(a) & f'(a) & \cdots & \frac{f^{(n-2)}(a)}{(n-2)!} \\ & & \ddots & \ddots & \vdots \\ & & & f(a) & f'(a) \\ & & & & f(a) \end{pmatrix}.$$

One is led to *define* $f(A)$ as

$$Pf(\mathcal{A})P^{-1}.$$

Furthermore if p is any polynomial such that

$$p(a) = f(a), \ldots, p^{(n-1)}(a) = f^{(n-1)}(a)$$

then

$$p(\mathcal{A}) = f(\mathcal{A})$$

and so

$$p(A) = Pp(\mathcal{A})P^{-1}.$$

However, *and this point cannot be over-emphasized*, because p is a polynomial

$$Pp(\mathcal{A})P^{-1} = p(P\mathcal{A}P^{-1}) = p(A).$$

In other words to calculate $f(A)$ it suffices to find some p such that p and its first $n-1$ derivatives have the same values at a as f and *its* first $n-1$ derivatives have at a. Then $f(A) = p(A)$.

iii. If A is diagonable (as in i) and if f is a function defined at each of the numbers $\lambda_1, \ldots, \lambda_n$ and if p is any polynomial such that $p(\lambda_i) = f(\lambda_i)$, $1 \le i \le n$ then $f(A) = p(A)$. Again the calculation of $f(A)$ can be carried out via the *polynomial p*.

iv. In ii and iii although the Jordan normal form lies behind the conclusions, the Jordan normal form *need not be used* in the calculation of $p(A)$.

Example 4.7.4. If $a \ne 0$ and if p is any polynomial such that

$$p(a) = \log a, \ldots, p^{(n-1)}(a) = (-1)^{n-2}\frac{(n-1)!}{a^{n-1}}$$

then $\log A = p(A)$.

If an $n \times n$ matrix $\mathcal{A}$ is not a single Jordan block but is made up of a number of Jordan blocks, say $\mathcal{J}_1, \ldots \mathcal{J}_K$, let $\mathcal{A}_i$ denote the $n \times n$ matrix written in block form and in which all blocks are O except for the ith diagonal block which is $\mathcal{J}_i$. Then

$$\mathcal{A} = \mathcal{A}_1 + \cdots + \mathcal{A}_K$$
$$\mathcal{A}_i \mathcal{A}_j = O \text{ if } i \neq j$$
$$\mathcal{A}^m = \mathcal{A}_1^m + \cdots + \mathcal{A}_K^m$$

and for any polynomial p,

$$p(\mathcal{A}) = p(\mathcal{A}_1) + \cdots + p(\mathcal{A}_K). \tag{4.7.10}$$

Let the different eigenvalues of A be $\lambda_1, \ldots, \lambda_k$. Then, since several Jordan blocks for A may share an eigenvalue, $k \leq K$. A functional calculus can be developed for A in terms of the functional calculus for each of the Jordan blocks that constitute the Jordan normal form of A. The essential tool in the discussion is the set of different eigenvalues of A and the powers $s_1, \ldots s_k$ found in the factorization of the minimal polynomial m_A of A:

$$m_A(\lambda) = (\lambda - \lambda_1)^{s_1} \cdots (\lambda - \lambda_k)^{s_k}.$$

To extend the functional calculus to A it is helpful to introduce the following definition.

The set $\sigma(A)$ consisting of the different eigenvalues $\lambda_1, \ldots, \lambda_k$ of an $n \times n$ matrix A is called the *spectrum* of A.

Example 4.7.5. The spectrum of a Jordan block is its sole eigenvalue. The spectrum of I is the number 1. The spectrum of a nilpotent matrix is the number 0. The spectrum of

$$D \stackrel{\text{def}}{=} \begin{pmatrix} \lambda_1 & & & \\ & \lambda_2 & & \\ & & \ddots & \\ & & & \lambda_n \end{pmatrix}$$

is the set of the *different* numbers in $\{\lambda_1, \ldots, \lambda_n\}$: $\sigma(D) \subset \{\lambda_1, \ldots, \lambda_n\}$.

Exercise 4.7.3. Find the spectrum of each of the following matrices:

i.

$$\begin{pmatrix} 1 & 2 \\ 3 & 4 \end{pmatrix};$$

ii.

$$\begin{pmatrix} 3 & -1 \\ 1 & 5 \end{pmatrix};$$

iii.

$$\begin{pmatrix} 2 & 0 & 0 \\ 0 & 1 & 9 \\ 0 & -1 & 7 \end{pmatrix}.$$

Exercise 4.7.4. For each matrix, say A, in **Exercise 4.7.3** calculate: $A^2, \ldots, A^k$, $\sin A$, e^A, $\arctan A$.

Let the minimal polynomial of A have the following factored form:

$$m_A(\lambda) \overset{\text{def}}{=} (\lambda - \lambda_1)^{s_1}(\lambda - \lambda_2)^{s_2} \cdots (\lambda - \lambda_k)^{s_k}.$$

Let f be a function such that for each i the function f is s_i times differentiable at and near λ_i and that f may be represented by the Taylor Formula with a Remainder. Suppose also that p is a polynomial such that p and its first $s_i - 1$ derivatives have the same values at λ_i as the values of f and *its* first $s_i - 1$ derivatives at λ_i:

$$p^{(j)}(\lambda_i) = f^{(j)}(\lambda_i), \ 1 \leq j \leq s_i - 1, 1 \leq i \leq k. \tag{4.7.11}$$

Although each $\mathcal{A}_j$ in (4.7.10) is not a Jordan block it *behaves* like the Jordan block $\mathcal{J}_j$, its only significant block. The matrix $f(\mathcal{J}_j)$ is well-defined. One *defines* $f(\mathcal{A}_j)$ to be the counterpart of $\mathcal{A}_j$ with $\mathcal{A}_j$ replaced by $f(\mathcal{A}_j)$. Hence

$$\begin{aligned}
p(\mathcal{A}_j) &= f(\mathcal{A}_j) \\
p(\mathcal{A}) &= f(\mathcal{A}_1) + \cdots + f(\mathcal{A}_k) \\
p(A) &= Pp(\mathcal{A})P^{-1} = Pf(\mathcal{A}_1)P^{-1} + \cdots + Pf(\mathcal{A}_k)P^{-1} \\
&\overset{\text{def}}{=} Pf(\mathcal{A})P^{-1}.
\end{aligned} \tag{4.7.12}$$

Since (cf. THEOREM **4.8.3**) there is a polynomial p behaving in the manner just described it thus appears that a reasonable *definition* of $f(A)$ is $p(A)$.

An alternative motivation for the definition of $f(A)$ as $p(A)$ is the following.

From the Taylor Formula with a Remainder one may conclude that there is a function g defined at and near each λ_i, i.e., there are positive r_i such that if $|\lambda - \lambda_i| < r_i$, $1 \leq i \leq k$ then $g(\lambda)$ is defined and for all such λ

$$p(\lambda) - f(\lambda) = m_A(\lambda)g(\lambda) \tag{4.7.13}$$

(cf. THEOREM **4.8.4**).

From (4.7.13) one may draw the following conclusion:

 Chapter 4. Useful forms for matrices

If $f(A)$ has any meaning consistent with algebraic and analytic manipulations then

$$p(A) - f(A) = m_A(A)g(A) = Og(A) = O,$$

i.e.,

$$p(A) = f(A) \qquad (4.7.14)$$

From this standpoint it appears that no matter how $g(A)$ is defined the only sensible definition for $f(A)$ is $p(A)$ — a *polynomial* function of the matrix A.

It is natural, in analogy with earlier definitions, to *define* $f(A)$ according to (4.7.14). Hence, although the Jordan normal form is used to motivate the development, the Jordan normal form is *not* needed for the calculation of $f(A)$. Any polynomial p that *agrees* with f on $\sigma(A)$, i.e., such that p and its derivatives agree with f and its derivatives on $\sigma(A)$ will serve for the calculations.

[**Note 4.7.1:** If A is diagonable then no *derivatives* occur in the "agreement" criterion for p. It suffices if A is diagonable for p and f to be equal at each point in the spectrum of A.]

Example 4.7.6. Let the 6×6 matrix A have the following Jordan normal form:

$$\mathcal{A} \stackrel{\text{def}}{=} \begin{pmatrix} 2 & 1 & & & & \\ & 2 & 1 & & & \\ & & 2 & & & \\ & & & 3 & 1 & \\ & & & & 3 & \\ & & & & & 4 \end{pmatrix}.$$

Thus, in the notation introduced earlier,

$$\mathcal{A}_1 = \begin{pmatrix} 2 & 1 & \\ & 2 & 1 \\ & & 2 \end{pmatrix}, \; \mathcal{A}_2 = \begin{pmatrix} 3 & 1 \\ & 3 \end{pmatrix}, \; \mathcal{A}_3 = \begin{pmatrix} 4 \end{pmatrix},$$

$\mathcal{A} = \mathcal{A}_1 + \mathcal{A}_2 + \mathcal{A}_3$, and $\sigma(A) =$ spectrum of $A = \{2, 3, 4\}$.

A direct check shows that the minimal polynomials for these matrices are

$$m_{\mathcal{A}_1}(\lambda) = (\lambda - 2)^3, \; m_{\mathcal{A}_2}(\lambda) = (\lambda - 3)^2, \; m_{\mathcal{A}_3}(\lambda) = \lambda - 4, \text{ and}$$
$$m_{\mathcal{A}}(\lambda) = (\lambda - 2)^3(\lambda - 3)^2(\lambda - 4) = m_{\mathcal{A}_1}(\lambda)m_{\mathcal{A}_2}(\lambda)m_{\mathcal{A}_3}(\lambda).$$

Hence if p is a polynomial and if

$$p(2) = \log 2, \ p'(2) = \frac{1}{2}, \ p''(2) = -\frac{1}{2^2}$$

$$p(3) = \log 3, \ p'(3) = \frac{1}{3}$$

$$p(4) = \log 4$$

then, by virtue of (4.7.12),

$$p(\mathcal{A}) = p(\mathcal{A}_1) + p(\mathcal{A}_2) + p(\mathcal{A}_3).$$

From (4.7.12) it follows that

$$p(\mathcal{A}_1) = \begin{pmatrix} \log 2 & \frac{1}{2} & -\frac{1}{8} \\ & \log 2 & \frac{1}{2} \\ & & \log 2 \end{pmatrix}, \ p(\mathcal{A}_2) = \begin{pmatrix} & & \\ \log 3 & \frac{1}{3} \\ & \log 3 \end{pmatrix},$$

$$\text{and } p(\mathcal{A}_3) = \begin{pmatrix} \\ \\ \log 4 \end{pmatrix}.$$

Hence,

$$\log(\mathcal{A}) = p(\mathcal{A}) = \begin{pmatrix} \log 2 & \frac{1}{2} & -\frac{1}{8} \\ & \log 2 & \frac{1}{2} \\ & & \log 2 \\ & & & \log 3 & \frac{1}{3} \\ & & & & \log 3 \\ & & & & & \log 4 \end{pmatrix}.$$

Exercise 4.7.5. Let A and B be two $n \times n$ matrices such that $AB = BA$. Use the functional calculus to show that

$$e^A e^B = e^{A+B}.$$

Exercise 4.7.6. Assume

$$A \overset{\text{def}}{=} \begin{pmatrix} 0 & 1 \\ 0 & 0 \end{pmatrix}$$

$$B \overset{\text{def}}{=} \begin{pmatrix} 0 & 0 \\ 1 & 0 \end{pmatrix}.$$

     Chapter 4. Useful forms for matrices

i. Show $AB \neq BA$.

ii. Show $e^A e^B \neq e^{A+B}$.

Exercise 4.7.7. Let A be a self-adjoint matrix. Show that e^{iA} is unitary.

Exercise 4.7.8. Let A be an $n \times n$ matrix with minimal polynomial m_A of degree μ_A. Show that if p is a polynomial there is a polynomial q such that:

i. the degree of q is no greater than $\mu_A - 1$;

ii. $q(A) = p(A)$.

(Hence, whenever the functional calculus is used and a polynomial p is such that for a given function f

$$p(A) = f(A)$$

then there is a polynomial q of degree not exceeding $\mu_A - 1$ and such that

$$q(A) = f(A).)$$

Exercise 4.7.9. Assume that the minimal polynomial m_A for an $n \times n$ matrix A has the form:

$$m_A(z) = z^k + z^{k-1} + \cdots + z + 1.$$

Show A is invertible.

[*Hint:* $z^{k+1} - 1 = (z - 1)(z^k + \cdots + 1)$.]

Exercise 4.7.10. Assume the $n \times n$ matrix A is nilpotent. Show $I - A$ is invertible.

[*Hint:* Jordan normal form (cf. **Exercise 4.5.12**).]

Exercise 4.7.11. Let D be the diagonal matrix consisting of the diagonal entries of the Jordan normal form $\mathcal{A}$ of a matrix A. Show:

i. D^k is the diagonal matrix consisting of the diagonal entries of the Jordan normal form of A^k;

ii. if p is a polynomial then $p(D)$ is the diagonal matrix consisting of the diagonal entries of the Jordan normal form of $p(A)$;

iii. if $\sigma(A)$ is the spectrum of A then

$$p(\sigma(A)) \overset{\text{def}}{=} \{ p(\lambda) \ : \ \lambda \in \sigma(A) \}$$

is the spectrum of $p(A)$, i.e., "$p(\sigma(A)) = \sigma(p(A))$."

Exercise 4.7.12. Let $A \stackrel{\text{def}}{=} (a_{ij})_{i,j=1}^{n,n}$ be an $n \times n$ matrix and assume

$$\chi_A(\lambda) = (-1)^n (\lambda - \lambda_1)^{s_1} \cdots (\lambda - \lambda_k)^{s_k}.$$

Show that A is diagonable iff

$$s_j = n - rank(A - \lambda_j I), \quad 1 \le j \le k.$$

Example 4.7.7. The functional calculus leads to an interesting analogy between complex numbers z such that $|z| = 1$ and unitary matrices. The equation $|z| = 1$ is equivalent to the equation $|z|^2 = 1$ which is equivalent to the equation $z\bar{z} = 1$. The analog, for a matrix U, of the complex conjugate $\bar{z}$ for a complex number z is U^*. The analog of the equation $z\bar{z} = 1$ for a complex number z is the equation $UU^* = I$ for a matrix U. In short,

$$z\bar{z} = 1 \leftrightarrow UU^* = I.$$

If, for some nonnegative r, z is written in the form $re^{i\theta}$ then $z\bar{z} = 1$ iff $r = 1$, i.e., iff $z = e^{i\theta}$. The number θ is *real*, i.e., $\theta = \bar{\theta}$, and in fact θ may be chosen so that $0 \le \theta < 2\pi$. Corresponding to the idea of a real number is the idea of a matrix H such that $H = H^*$, i.e., real numbers and self-adjoint matrices are analogous. The analogy is extended in that $UU^* = I$ iff for some self-adjoint matrix H

$$U = e^{iH}. \tag{4.7.15}$$

The proof of (4.7.15) runs as follows. Because $UU^* = I$ the matrix U is normal and hence for some unitary matrix P

$$P^*UP = \begin{pmatrix} \lambda_1 & & & \\ & \lambda_2 & & \\ & & \ddots & \\ & & & \lambda_n \end{pmatrix} \stackrel{\text{def}}{=} D$$

and furthermore for all k, $\lambda_k \overline{\lambda_k} = 1$. Hence for each λ_k there is a number θ_k such that $0 \le \theta < 2\pi$ and $\lambda_k = e^{i\theta_k}$. Let Δ be the diagonal matrix

$$\begin{pmatrix} \theta_1 & & & \\ & \theta_2 & & \\ & & \ddots & \\ & & & \theta_n \end{pmatrix}.$$

Then according to the functional calculus

$$e^{i\Delta} = \sum_{k=1}^{\infty} \frac{(i\Delta)^k}{n!}$$

$$= \begin{pmatrix} e^{i\theta_1} & & & \\ & e^{i\theta_2} & & \\ & & \ddots & \\ & & & e^{i\theta_n} \end{pmatrix} = D$$

$$Pe^{i\Delta}P^* = PDP^* = U = \sum_{k=1}^{\infty} \frac{(iP\Delta P^*)^k}{n!}.$$

If $H \overset{\text{def}}{=} P\Delta P^*$ it follows that $H^* = P\Delta^* P^* = P\Delta P^* = H^*$, i.e., $H = H^*$, H is self-adjoint and $U = e^{iH}$.

Example 4.7.8. Here is yet another extension of the analogy between complex numbers and matrices. The equation $z = |z|e^{i\theta}$ for some real θ may be rewritten $z = \sqrt{z\bar{z}}e^{i\theta}$. The corresponding situation for matrices is that there is a matrix S such that $S^2 = AA^*$ (whence S is denoted $\sqrt{AA^*}$) and a self-adjoint matrix H such that $A = \sqrt{AA^*}e^{iH}$. The fact that $\sqrt{z\bar{z}} \geq 0$ is reflected in the property that for every vector $\mathbf{x}$ $(S\mathbf{x}, \mathbf{x}) \geq 0$.

Here is the proof. The matrix $B \overset{\text{def}}{=} AA^*$ is self-adjoint and there is a unitary matrix P such that

$$P^*BP \overset{\text{def}}{=} D \overset{\text{def}}{=} \begin{pmatrix} \lambda_1 & & & \\ & \lambda_2 & & \\ & & \ddots & \\ & & & \lambda_n \end{pmatrix}$$

is a diagonal matrix. The numbers λ_k are the (not necessarily different) eigenvalues of AA^* and so if $\mathbf{x}_k$ are corresponding eigenvectors it follows that

$$(AA^*\mathbf{x}_k, \mathbf{x}_k) = \lambda_k \|\mathbf{x}_k\|^2$$
$$= \|A^*\mathbf{x}_k\|^2$$

whence $\lambda_k \geq 0$. Hence there are nonnegative numbers μ_k such that $\mu_k^2 = \lambda_k$ and if

$$M \overset{\text{def}}{=} \begin{pmatrix} \mu_1 & & & \\ & \mu_2 & & \\ & & \ddots & \\ & & & \mu_n \end{pmatrix} = M^*$$

then $M^2 = D$ and $(PMP^*)^2 = PDP^* = AA^* = B$. Thus $\sqrt{AA^*}$ may be taken to be PMP^*: $S \overset{\text{def}}{=} \sqrt{AA^*} \overset{\text{def}}{=} PMP^*$.

The remainder of the proof is in two parts, the first dealing with the case in which A is invertible and the second dealing with the case in which A is singular.

Case i. If A is invertible then so are AA^*, D, and M invertible. Consequently

$$(M^{-1}P^*AP)^*M^{-1}P^*AP = P^*A^*PM^{-1}M^{-1}P^*AP$$
$$= P^*A^*(PM^{-1}P^*)^2AP$$
$$= P^*A^*(AA^*)^{-1}AP = I,$$

i.e., $M^{-1}P^*AP$ is unitary. Since $S = PMP^* = \sqrt{AA^*}$ and since P is unitary it follows that $PM^{-1}P^*APP^* = S^{-1}AI = S^{-1}A \overset{\text{def}}{=} U$ is unitary, i.e.,

$$A = SU = \sqrt{AA^*}U.$$

According to **Example 4.7.7** there is a self-adjoint H such that $U = e^{iH}$ whence,

$$A = \sqrt{AA^*}e^{iH}.$$

Finally if $\mathbf{x}$ is any vector,

$$(S\mathbf{x},\mathbf{x}) = (MP^*\mathbf{x},P^*\mathbf{x}).$$

If

$$P^*\mathbf{x} \overset{\text{def}}{=} \mathbf{y} \overset{\text{def}}{=} \begin{pmatrix} y_1 \\ \vdots \\ y_n \end{pmatrix}$$

then

$$(S\mathbf{x},\mathbf{x}) = (M\mathbf{y},\mathbf{y}) = \sum_{i=1}^{n} \mu_i |y_i|^2 \geq 0,$$

as promised.

Case ii. If A is a singular $n \times n$ matrix and $rank(A) = m < n$ then also $rank(A^*) = m = dim(im(A^*))$. The relations derived in **Section 4.4** imply that there is in $\mathbf{C}^n$ an orthonormal basis

$$\{\mathbf{x}_1, \ldots, \mathbf{x}_m, \mathbf{x}_{m+1}, \ldots, \mathbf{x}_n\}$$

such that $\{\mathbf{x}_1, \ldots, \mathbf{x}_m\}$ is a basis for $im(A^*)$ and $\{\mathbf{x}_{m+1}, \ldots, \mathbf{x}_n\}$ is a basis for $ker(A)$.

Let W be the set of images of A confined to $im(A^*)$:

$$W \overset{\text{def}}{=} \{\, \mathbf{w} \ : \ \text{for some } \mathbf{y} \text{ in } im(A^*) \ \mathbf{w} = A\mathbf{y} \,\} = A(im(A^*)).$$

The set $\{\, A\mathbf{x}_j \ : \ 1 \leq j \leq m \,\}$ is linearly independent. Indeed, if

$$\mathbf{x} \overset{\text{def}}{=} \sum_{j=1}^{m} b_j \mathbf{x}_j$$

    Chapter 4. Useful forms for matrices

and

$$Ax = \sum_{j=1}^{m} b_j A\mathbf{x}_j = O$$

then, since $\mathbf{x} \in im(A^*)$, there is a $\mathbf{w}$ such that $\mathbf{x} = A^*\mathbf{w}$ and so

$$0 = (A\mathbf{x}, \mathbf{w}) = (AA^*\mathbf{w}, \mathbf{w}) = \|A^*\mathbf{w}\|^2,$$

i.e., $A^*\mathbf{w} = \mathbf{x} = O$. The vectors $\mathbf{x}_j$ are orthonormal, hence linearly independent and so each $b_j = 0$.

Since $span(\mathbf{x}_1, \ldots, \mathbf{x}_m) = im(A^*)$ it follows that $W = span(A\mathbf{x}_1, \ldots, A\mathbf{x}_m)$ and so $dim(W) = m$. Let the orthonormal basis $\{\mathbf{v}_1, \ldots, \mathbf{v}_n\}$ for $\mathbf{C}^n$ be such that $\{\mathbf{v}_1, \ldots, \mathbf{v}_m\}$ is an orthonormal basis for W.

Let V be the (unitary!) matrix in which the jth column is the column vector $\mathbf{v}_j$ and let X be the (unitary!) matrix in which the jth column is the column vector $\mathbf{x}_j$. Then $Y \overset{\text{def}}{=} XV^*$ is a unitary matrix. Furthermore $YV = X$, which is to say $Y\mathbf{v}_j = \mathbf{x}_j$, $1 \le j \le n$. Since $\{\mathbf{v}_1, \ldots, \mathbf{v}_m\}$ is a basis for W and $\{\mathbf{x}_{m+1}, \ldots, \mathbf{x}_n\}$ is a basis for $ker(A)$ it follows that

$$AY\mathbf{v}_j = A\mathbf{x}_j \in \begin{cases} W & \text{if } 1 \le j \le m \\ \{O\} & \text{if } m+1 \le j \le n. \end{cases}$$

Hence there is an $m \times m$ matrix $\widetilde{B} \overset{\text{def}}{=} (b_{jk})_{j,k=1}^{m,m}$ such that

$$AY\mathbf{v}_j = \sum_{k=1}^{m} b_{jk}\mathbf{v}_k, \ 1 \le j \le m,$$

$$AY\mathbf{v}_j = A\mathbf{x}_j = O, \ m+1 \le j \le n.$$

Let B be the block matrix

$$\begin{pmatrix} \widetilde{B} & O \\ O & O \end{pmatrix}.$$

In the notation and spirit of **Section 2.10** $AYV = BV$ and since V is invertible $AY = B$. Since Y is invertible, $rank(AY) = rank(A) = m$ and so $\widetilde{B}$ is an invertible $m \times m$ matrix.

According to the result in *Case i* there is an $m \times m$ matrix $\widetilde{S}$ and an $m \times m$ unitary matrix $\widetilde{U}$ such that

$$\widetilde{S}^2 = \widetilde{B}\widetilde{B}^*$$
$$\widetilde{B} = \widetilde{S}\widetilde{U}$$

and if $\mathbf{x} \in \mathbf{C}^m$

$$(\widetilde{S}\mathbf{x}, \mathbf{x}) \ge 0. \tag{4.7.16}$$

Let $\widetilde{Z}$ be any $(n-m) \times (n-m)$ unitary matrix and form the block matrix

$$\begin{pmatrix} \widetilde{U} & O \\ O & \widetilde{Z} \end{pmatrix}$$

---

which is an $n \times n$ unitary matrix P. If $PY^* \stackrel{\text{def}}{=} U$, then U is a unitary matrix. If S is the block matrix

$$\begin{pmatrix} \tilde{S} & O \\ O & O \end{pmatrix}$$

then $AY = SP$ and $A = SPY^* = SU$.

Since U is unitary there is a self-adjoint matrix H such that $U = e^{iH}$ and so

$$A = Se^{iH} = \sqrt{AA^*}\,e^{iH}.$$

The inequality (4.7.16) for any vector $\mathbf{x}$ in $\mathbf{C}^m$ implies that $(S\mathbf{y}, \mathbf{y}) \geq 0$ for any vector $\mathbf{y}$ in $\mathbf{C}^n$.

The expression Se^{iH} is called, in analogy with the polar form (polar coordinate representation) of a complex number, a *polar form* of A. Owing to the possibility of permuting the diagonal entries of the diagonal form of S the polar form of A need not be unique.

Exercise 4.7.13. Show that in **Example 4.7.8** *Case ii* $S^2 = AA^*$.

Exercise 4.7.14. Let U be the matrix

$$\begin{pmatrix} \cos\theta & \sin\theta \\ -\sin\theta & \cos\theta \end{pmatrix}.$$

Show that U is unitary and find a self-adjoint matrix H such that $U = e^{iH}$ and such that for all vectors $\mathbf{x}$: $(H\mathbf{x}, \mathbf{x}) \geq 0$.

Exercise 4.7.15. Let A be the matrix

$$\begin{pmatrix} 1 & 3 \\ 2 & 6 \end{pmatrix}.$$

Find a polar form of A.

4.8. Some theorems in algebra and analysis

Much of the development of this **Chapter** hinges on the existence of eigenvalues and eigenvectors for each $n \times n$ matrix A. The existence of eigenvalues in turn is certainly implied by the Fundamental Theorem of Algebra (FTA). For completeness there is given below a relatively simple proof of FTA. No matter what attempts at simplification are made, the proof appears to be unavoidably dependent on some aspect of the branch of mathematics called *analysis*, in which *limits, continuity, topology,* etc. are studied. In the proof below two theorems in analysis are cited and not proved. These are:

THEOREM **A.** IF p IS A POLYNOMIAL AND $R > 0$ THEN THERE IS A COMPLEX NUMBER z_0 SUCH THAT $|z_0| \leq R$ AND

$$|p(z_0)| = min\{\,|p(z)| \;:\; |z| \leq R\,\} \stackrel{\text{def}}{=} \mu(p, R).$$

THEOREM **A** says that in every closed disc

$$\Delta_R \overset{\text{def}}{=} \{\, z \;:\; |z| \le R \,\}$$

centered at the origin in **C** there is a point where the polynomial reaches its minimum absolute value on Δ_R.

THEOREM **B.** IF $z \in$ **C** AND $n \in$ **N** THERE IS IN **C** A w SUCH THAT $w^n = z$.

Of the two THEOREMS above **A** seems to be "deeper" than **B**. THEOREM **A** stems from the "completeness" of **R**, i.e., that **R**, in contrast with **Q**, the set of all rational numbers, has no "holes." For example, in **Q** there is no q such that $q^2 = 2$ while in **R** there is an r such that $r^2 = 2$.

The completeness of **R** is the consequence of an axiom system that declares **R** to be a complete ordered field or a consequence of a construction whereby **Q** is in some way "completed," e.g., by *Dedekind cuts*, by *Cantor's* technique of *Cauchy sequences*, by means of *continued fractions*, by means of *infinite decimals*, etc.

On the other hand THEOREM **B** may appear plausible to the reader who regards a nonzero complex number $z = x + iy$ as given by the equations

$$x + iy = |x + iy|(\cos\theta + i\sin\theta), \quad \text{(polar form)}$$

$$\cos\theta = \frac{x}{|x + iy|},$$

$$\sin\theta = \frac{y}{|x + iy|}.$$

Indeed, if $z = x + iy$ and if

$$w = |z|^{\frac{1}{n}}\left(\cos\frac{\theta}{n} + i\sin\frac{\theta}{n}\right)$$

then $w^n = z$ according to the theorem of de Moivre, e.g., via the addition formulae for the trigonometric functions.

Alternatively, according to Euler's formula:

$$\cos\theta + i\sin\theta = e^{i\theta}$$

and so

$$z = |z|e^{i\theta}.$$

If

$$w = |z|^{\frac{1}{n}} e^{i\frac{(\theta + 2k\pi)}{n}}, k = 0, \pm 1, \pm 2, \ldots$$

then

$$w^n = \left(|z|^{\frac{1}{n}} e^{i\frac{(\theta + 2k\pi i)}{n}}\right)^n$$

$$= |z|e^{i(\theta + 2k\pi)}$$

$$= |z|e^{i\theta}$$

$$= z.$$

Honesty requires a more rigorous development of the structure of $\mathbf{C}$ before the comments in the preceding paragraphs can withstand close scrutiny. Such rigorous development in turn leads back to the "completeness" of $\mathbf{R}$.

In recognition of the need to start somewhere, the approach in the next few paragraphs simply accepts THEOREMS $\mathbf{A}$ and $\mathbf{B}$ as valid and builds upon them.

THEOREM 4.8.1. (**The Fundamental Theorem of Algebra**) LET n BE IN $\mathbf{N}$ AND LET p BE THE POLYNOMIAL SUCH THAT

$$p(z) = z^n + \sum_{i=0}^{n-1} a_i z^i.$$

THEN THERE IS IN $\mathbf{C}$ A z_0 SUCH THAT $p(z_0) = 0$.

PROOF. Since, if $z \neq 0$,

$$p(z) = z^n \left(1 + \sum_{i=1}^{n-1} a_i z^{i-n}\right),$$

it follows that for some positive number R if $|z| \geq R$ then $|p(z)| \geq 1$.

According to $\mathbf{A}$ there is in $\mathbf{C}$ a z_0 such that $|z_0| \leq R$, and such that $|p(z_0)| = \mu(p, R)$. If $q(z) \overset{\text{def}}{=} p(z + z_0)$ then $|q(0)| = \mu(p, R)$. Thus if

$$q(z) = z^n + \sum_{i=0}^{n-1} b_i z^i$$

then $q(0) = b_0$ and it suffices to prove that $b_0 = 0$. For simplicity it may be assumed that for some k such that $0 < k \leq n - 1, b_1 = b_2 = \cdots = b_{k-1} = 0$ and $b_k \neq 0$. In other terms, for some polynomial r

$$q(z) = b_0 + b_k z^k + z^{k+1} r(z)$$

According to THEOREM $\mathbf{B}$ there is in $\mathbf{C}$ a w such that

$$w^k = \frac{-b_0}{b_k},$$

i.e.,

$$b_0 + b_k w^k = 0.$$

Then

$$q(tw) = (1 - t^k)b_0 + t^k(b_0 + b_k w^k) + (tw)^{k+1} r(tw)$$
$$= (1 - t^k)b_0 + (tw)^{k+1} r(tw)$$

      Chapter 4. Useful forms for matrices

and so if $0 < t < 1$,
$$|q(tw)| \le |b_0| - t^k(|b_0| - t|w^{k+1}r(tw)|).$$
If $b_0 \ne 0$ then for small enough t the second term above is negative whence
$$\mu(p, R) < |b_0| = |q(0)|,$$
a contradiction. [**Fe, Re, T**]

$$\Omega$$

Exercise 4.8.1. Show that the (implicit) assumption that p is a monic polynomial is admissible.

Exercise 4.8.2. Assume $p(z_0) = 0$. Show that there is a polynomial s such that
$$p(z) = (z - z_0)s(z)$$
(the The Factor Theorem).

[*Hint:* Note that $q(z - z_0) = p(z)$. Show that the constant term of q is $p(z_0)$ and thus that $(z - z_0)$ is a factor of $p(z)$. By mathematical induction prove that p is the product of n (not necessarily distinct) polynomials, each of degree 1.]

Exercise 4.8.3. Find all the zeros of the polynomial p such that
$$p(z) = z^n + z^{n-1} + \cdots + 1.$$

[*Hint:* Factor $z^{n+1} - 1$.]

THEOREM 4.8.2. LET p BE A POLYNOMIAL AND ASSUME THAT FOR SOME NUMBER a
$$p(a) = p'(a) = \cdots = p^{(r-1)}(a) = 0.$$
THEN THERE IS A POLYNOMIAL q SUCH THAT

$$p(z) = (z - a)^r q(z).$$

PROOF. According to the Taylor series expansion of p

$$p(z) = p(a) + p'(a)(z - a) + \cdots + \frac{p^{(r-1)}(a)}{(r-1)!}(z - a)^{r-1} + \frac{p^{(r)}(a)}{r!}(z - a)^r + \cdots$$

$$= (z - a)^r \left(\frac{p^{(r)}(a)}{r!} + \cdots \right)$$

$$\overset{\text{def}}{=} (z - a)^r q(z).$$

The function q is a polynomial since the Taylor expansion of a polynomial is a finite series.

$$\Omega$$

Example 4.8.1. Assume that a and b are two different numbers and that p is a polynomial such that

$$p(a) = p'(a) = \cdots = p^{(r-1)}(a) = 0$$
$$p(b) = p'(b) = \cdots = p^{(s-1)}(b) = 0.$$

Then there is a polynomial g such that

$$p(z) = (z - a)^r (z - b)^s g(z).$$

Indeed THEOREM **4.8.2** shows there is a polynomial q such that

$$p(z) = (z - a)^r q(z).$$

Thus $p(b) = 0 = (b - a)^r q(b)$ whence, since $a \neq b$, $q(b) = 0$. In like fashion

$$p'(z) = r(z - a)^{r-1} q(z) + (z - a)^r q'(z)$$
$$p'(b) = 0 = r(b - a)^{r-1} q(b) + (b - a)^r q'(b)$$
$$= (b - a)^r q'(b)$$

whence $q'(b) = 0$ and, by induction, $q^{(s-1)}(b) = 0$.

Hence there is a polynomial g such that $q(z) = (z - b)^s g(z)$, i.e., $p(z) = (z - a)^r (z - b)^s g(z)$.

Exercise 4.8.4. Show that if p is a polynomial and if $a_1, \ldots, a_k$ are k different numbers such that

$$p(a_1) = p'(a_1) = \cdots = p^{(r_1 - 1)}(a_1) = 0$$
$$p(a_2) = p'(a_2) = \cdots = p^{(r_2 - 1)}(a_2) = 0$$

$$\vdots$$

$$p(a_k) = p'(a_k) = \cdots = p^{(r_k - 1)}(a_k) = 0$$

then for some polynomial q

$$p(z) = (z - a)^{r_1} \cdots (z - a)^{r_k} q(z).$$

[*Hint:* See **Example 4.8.1** and then use mathematical induction.]

THEOREM **4.8.3.** LET $a_1 \ldots a_k$ BE k DIFFERENT NUMBERS AND LET f BE A FUNCTION THAT IS

$$r_1 - 1 \text{ TIMES DIFFERENTIABLE AT } a_1$$
$$r_2 - 1 \text{ TIMES DIFFERENTIABLE AT } a_2$$
$$\ldots$$
$$r_k - 1 \text{ TIMES DIFFERENTIABLE AT } a_k.$$

THEN THERE IS A POLYNOMIAL p OF DEGREE $n \overset{\text{def}}{=} r_1 + \cdots + r_k$ AND SUCH THAT

$$p(a_1) = f(a_1), p'(a_1) = f'(a_1), \ldots, p^{(r_1-1)}(a_1) = f^{(r_1-1)}(a_1)$$
$$\ldots$$
$$p(a_k) = f(a_k), p'(a_k) = f'(a_k), \ldots, p^{(r_k-1)}(a_k) = f^{(r_k-1)}(a_k).$$

PROOF. It suffices to show that there are $n + 1$ constants $b_0, \ldots b_n$ such that the polynomial p for which

$$p(z) = b_n z^n + b_{n-1} z^{n-1} + \cdots + b_0$$

behaves as asserted in the statement of the THEOREM.

In other words it suffices to show that there are constants b_i that are solutions of the following WIDE system of n linear equations in $n + 1$ unknowns (the b_i).

$$f(a_1) = b_n a_1^n + b_{n-1} a_1^{n-1} + \cdots + b_1 a_1 + b_0 1$$

$$f'(a_1) = b_n n a_1^{n-1} + \cdots + b_1 1 + b_0 0$$

$$\ldots$$

$$f^{(r_1-1)}(a_1) = b_n n(n-1) \cdots (n-r_1) a_1^{n-r_1+1} + \cdots$$
$$+ b_{r_1-1}(r_1 - 1)! + b_{r_1-2} 0 + \cdots + b_0 0$$

$$\ldots$$

$$f(a_k) = b_n n a_k^n + b_{n-1} a_k^{n-1} + \cdots + b_1 a_1 + b_0 1$$

$$f'(a_k) = b_n n a_k^{n-1} + \cdots + b_1 1 + b_0 0$$

$$\cdots$$

$$f^{(r_k-1)}(a_k) = b_n n(n-1)\cdots(n-r_k)a_k^{n-r_k+1} + \cdots$$
$$+ b_{r_k-1}(r_k-1)! + b_{r_k-2}0 + \cdots + b_0 0. \qquad (4.8.1)$$

In (4.8.1) there are $r_1 + \cdots + r_k\ (= n)$ equations in $n+1$ unknowns $b_0, \ldots, b_n$. The augmented matrix of the system is

$$\left(
\begin{array}{ccccc|c}
a_1^n & a_1^{n-1} & \cdots & a_1 & 1 & f(a_1) \\
na_1^{n-1} & (n-1)a_1^{n-2} & \cdots & 1 & 0 & f'(a_1) \\
\vdots & \vdots & & \vdots & \vdots & \vdots \\
a_k^n & a_k^{n-1} & \cdots & a_k & 1 & f(a_k) \\
na_k^{n-1} & (n-1)a_k^{n-2} & \cdots & 1 & 0 & f'(a_k) \\
\vdots & \vdots & & \vdots & \vdots & \vdots \\
\binom{n}{r_k}r_k! & \binom{n-1}{r_k}r_k! & & 0 & 0 & f^{(r_k-1)}(a_k)
\end{array}
\right). \qquad (4.8.2)$$

If it can be shown that the rank of the matrix to the *left* of the vertical line in the augmented matrix is n then, since there are only n rows in the augmented matrix, the rank of the augmented matrix is also n. Hence the system (4.8.1) has a solution (cf. THEOREM **2.9.2**).

It is shown next that the n columns numbered $2, \ldots, n+1$ just to the *left* of the vertical line in(4.8.2) are linearly independent. Indeed, if those columns are not linearly independent there is a polynomial

$$h(z) \overset{\text{def}}{=} c_{n-1}z^{n-1} + \cdots + c_1 z + c_0$$

such that

$$h(a_1) = h'(a_1) = \cdots = h^{(r_1-1)}(a_1) = 0$$
$$h(a_2) = h'(a_2) = \cdots = h^{(r_2-1)}(a_2) = 0$$
$$\cdots$$
$$h(a_k) = h'(a_k) = \cdots = h^{(r_k-1)}(a_k) = 0.$$

According to **Exercise 4.8.4** there is a polynomial q such that

$$h(z) = (z-a_1)^{r_1}\cdots(z-a_k)^{r_k}q(z).$$

Hence the degree of h is at least $r_1 + \cdots + r_k$, i.e., n. On the other hand the polynomial h *by definition* is a polynomial of degree $n-1$. The contradiction implies that the rank of (4.8.2) is n and so the system (4.8.1) has a solution, i.e., the polynomial p can be found.

$$\Omega$$

[**Remark 4.8.1:** THEOREM **4.8.3** is an example of an *application* of the part of linear algebra concerned with the solution of systems of linear equations.]

Exercise 4.8.5. Find a polynomial p such that

$$p(\log 2) = 2 \ (= e^{\log 2})$$
$$p(\log 3) = 3 \ (= e^{\log 3})$$
$$p'(\log 3) = 3 \ (= e^{\log 3}).$$

[*Hint:* The solution is a polynomial p that *agrees* with the exponential function exp at $\log 2$ and that together with its derivative agrees with exp and *its* derivative at $\log 3$.]

For the record one version of the theorem about the nth order TAYLOR FORMULA WITH A REMAINDER is quoted next (cf. [**O**]).

IF f IS A FUNCTION HAVING n CONTINUOUS DERIVATIVES ON THE CLOSED INTERVAL $[a, b]$, IF $f^{(n+1)}$ EXISTS ON THE OPEN INTERVAL (a, b), AND IF $a \leq x < y \leq b$ THEN FOR SOME ξ IN (x, y)

$$f(x) = f(y) + f'(y)(x - y) + \cdots + \frac{f^{(n)}(y)}{n!}(x - y)^n + \frac{f^{(n+1)}(\xi)}{(n + 1)!}(x - y)^{n+1}.$$

THEOREM **4.8.4.** LET $r_1, \ldots, r_k$ BE NATURAL NUMBERS, LET $a_1, \ldots, a_k$ BE k DIFFERENT NUMBERS, AND LET f BE A FUNCTION $m \overset{\text{def}}{=} r_1 + \cdots + r_k$ TIMES DIFFERENTIABLE AT AND NEAR EACH a_i SO THAT THE r_iTH ORDER TAYLOR FORMULA WITH A REMAINDER HOLDS AT AND NEAR a_i. IF

$$f^{(j)}(a_i) = 0, \ 1 \leq j \leq r_i - 1, \ 1 \leq i \leq k$$

THEN

$$\frac{f(z)}{(z - a_1)^{r_1} \cdots (z - a_k)^{r_k}}$$

IS A WELL-DEFINED FUNCTION $g(z)$ AT AND NEAR EACH a_i.

In other words there is a function g well-defined at and near each a_i and such that at and near each a_i

$$f(z) = (z - a_1)^{r_1} \cdots (z - a_k)^{r_k} g(z).$$

PROOF. According to the r_1st order Taylor Formula with a Remainder, at and near a_1 there is a function Q_1 such that $f(z) = (z - a_1)^{r_1} Q_1(z)$.

At and near a_2 the function Q_1 is the quotient

$$\frac{f(z)}{(z - a_1)^{r_1}}$$

of two $r_2 + \cdots r_k$ times differentiable functions. Using the product rule for differentiation one can establish successively that

$$Q_1(a_2) = 0, \ Q_1'(a_2) = 0, \ \ldots, \ \text{and} \ Q_1^{(r_2-1)}(a_2) = 0.$$

Hence there is a function Q_2 such that

$$Q_1(z) = (z - a_2)^{r_2} Q_2(z).$$

Mathematical induction implies that for some function g

$$f(z) = (z - a_1)^{r_1} \cdots (z - a_k)^{r_k} g(z).$$

$$\Omega$$

Exercise 4.8.6. Assume that f has $n + 1$ continuous derivatives at and near a. Let p be the polynomial such that

$$p(z) = f(a) + f'(a)(z - a) + \cdots + \frac{f^{(n)}(a)}{n!}(z - a)^n.$$

Show that

$$p^{(k)}(a) = f^{(k)}(a), \ 0 \le k \le n.$$

In other words the truncated Taylor series for f provides the polynomial that "agrees" with f at a.

PART II

PRACTICE

CHAPTER 5

APPLYING LINEAR ALGEBRA

5.1. Computational aspects of linear algebra

The applications of linear algebra occur in many fields of practical human endeavor. Each application brings with it the very real problem of implementing the theory relevant to the problem. It is reassuring to know that a problem *has* a solution. Without some well-defined procedure (algorithm) for finding (an approximation to) a solution, the fact that a solution exists is of very little practical consequence.

On the other hand, if an algorithm is offered for finding (an approximation to) a solution, the algorithm itself may be useless if its implementation is impossible in practical units of time, money, hardware, and software.

For example, in linear programming problems discussed later in this **Chapter**, there is an hierarchy of algorithms that *theoretically* lead to a solution of the problem.

i. One may test each of the finitely many vertices of a certain convex set in high-dimensional Euclidean space. Even though there are only finitely many vertices, their number can be so large, that a search among them is totally impractical.

ii. The so-called *simplex method* may be applied. Although experience suggests it is practical, theory *proves* that there are linear programming problems that are solvable practically by other means and that are out of reach practically via the simplex method.

iii. The Karmarkar technique is now available in software. The algorithm is theoretically valid and, to boot, it can be proved that the Karmarkar algorithm is, in a well-defined sense, faster than the simplex method.

In almost all practical operations with matrices and vectors the applications use *numbers* that are derived from the original data. Thus in the various forms of elimination, the $PROCESS$, the GEM, the GJM, etc., the matrices A_P, A_E, A_J, etc. are calculated and are used in back substitution. Alternatively, the decompositions LDU, $\Pi^{-1}LDU$, etc., are involved in solving systems of equations.

In subsequent work various approximative procedures for solving problems are used in place of the theoretically valid elimination-based methods. The Gram-Schmidt process and its associated least squares methods and the algorithms for the minimal polynomial of a SQUARE matrix, the finding of the eigenvalues of a SQUARE matrix, are other important procedures that are practically useful only when they yield numerical results that validly answer practical questions.

Theoretically all calculations can be carried out with precision. Actually some calculations can *not* be carried out and precision is often unattainable. The following items lie at the root of the difficulties.

Physical limitations. Computers are finite and the time allowable or affordable for a calculation is also finite. Thus the storage of data and of the (intermediate) results of computation is restricted. Storage questions aside, the time required for a calculation, e.g., of the 100! $(\approx 10^{150})$ terms in the classical formula for the determinant of a 100×100 matrix, is absurdly impractical.

Roundoff error. Sums and products of long sequences of m-digit numbers yield M-digit numbers and M can be arbitrarily greater than m. Since a computer or a piece of paper is finite, the time/space required for carrying out enormous calculations force the rounding off of numbers so that the less significant digits in a number are dropped and replaced by 0's, i.e., one uses *floating point* calculations with only a fixed number of retained (significant) digits. For example, if it is agreed ("programmed") that only five significant decimal digits are of interest, the product of 12,345 $(= 1.2345 \times 10^4)$ and 67,891 $(= 6.7891 \times 10^4)$ is taken to be 838,110,000 $(= 8.3811 \times 10^8)$ rather than 838,114,395 $(= 8.38114395 \times 10^8)$. If this kind of roundoff occurs 1,000,000 or 1,000,000,000 times in the course of finding numbers for the solution to a problem, the computed answer can deviate unacceptably from the true result. Hence there is a need to develop a set of guidelines, criteria, etc., that usefully predict the reliability of the results found in the face of inevitable roundoff.

In "five-figure" roundoff $0.00001 \neq 0$ whereas

$$(0.00001)^2 = 0.0000000001 \ "="\ 0.$$

Thus the matrix

$$A \stackrel{\text{def}}{=} \begin{pmatrix} 1 & 2 \\ 0 & 0.00001 \end{pmatrix}$$

is invertible whereas, $\doteq$ representing "almost equal,"

$$A^2 = \begin{pmatrix} 1 & 2.00002 \\ 0 & 0.0000000001 \end{pmatrix} \doteq \begin{pmatrix} 1 & 2.00002 \\ 0 & 0 \end{pmatrix}$$

is not. In five-figure roundoff the computer does not distinguish between $=$ and $\doteq$. In theory:

$$A \text{ is invertible iff } A^2 \text{ is invertible.}$$

In practice the displayed assertion is false in that it does not always hold. Similarly, $det(A) = 0.00001$, $[det(A)]^2 \doteq 0$. Many other rules of arithmetic and algebra are *invalid* when roundoff is in effect.

Instability. If the minimal polynomial m_B of a matrix B is such that $m_B(z) = z^2 + bz + c$ then the zeros of m_B are simple iff $b^2 - 4c \neq 0$. If the data giving rise to B are subject to error and $|b^2 - 4c|$ is small enough to be significantly affected by such error then the decision about the diagonability of B is also subject to error.

It is true that all matrices "close enough" to an invertible matrix are invertible. It is equally true that arbitrarily close to a singular matrix there are invertible matrices. [*Proof.* If A is singular and $\mathcal{A} \stackrel{\text{def}}{=} P^{-1}AP$ is its Jordan normal form, then some of the diagonal entries of $\mathcal{A}$ are 0's. If each such 0 is replaced by a nonzero small number ϵ to produce a matrix $\widetilde{\mathcal{A}}$ then $\widetilde{\mathcal{A}}$ is invertible as is $\widetilde{A} \stackrel{\text{def}}{=} P\widetilde{\mathcal{A}}P^{-1}$, which is also close to A if ϵ is small enough.] Hence, if the data, e.g., the entries of a matrix are subject to error, the invertibility of the matrix is subject to question.

The *PROCESS*, the GEM, and the GJM are, in the presence of roundoff, unstable. For example the systems

$$i. \quad \begin{array}{c} Mx + my = M \\ x + y = 2 \end{array} \quad \text{resp.} \quad ii. \quad \begin{array}{c} my + Mx = M \\ y + x = 2 \end{array}$$

are mathematically ("logically") equivalent. However if $\frac{m}{M} \doteq 0$, e.g., if five-figure roundoff is used and $|\frac{m}{M}| < 0.000001$, then these systems lead to the following solutions if the GEM is used:

$$sol(i.) \ y \doteq 1, x \doteq 1 \text{ resp. } sol(ii.) \ x \doteq 1, y \doteq 0.$$

If $M \neq m$ the precise solution to *both* systems is

$$y = \frac{M}{M - m}, x = \frac{M - 2m}{M - m}.$$

If Δq is the error arising in the computation of a nonzero quantity q then the *relative error* is taken to be

$$\frac{\Delta q}{q} \ (percent \ relative \ error = \frac{100\Delta q}{q}\%).$$

If m=1 and M=100,000 then the percent relative error for y in *sol(ii.)* is 100%.

The trouble with *sol(ii.)* stems from the use of the small pivot m as opposed to the large pivot M. If m and M represent data then in the result of one experiment $|\frac{m}{M}|$ can be very small and in the result of another experiment

---

$\frac{m}{M}$ can be quite large. The real situation underlying the experiment can well be the same. If system i is used without a sensible pivoting strategy, e.g., MAXSEARCH or PARTIALMAXSEARCH (cf. **Section 2.3**) the computed results can have no bearing on the real underlying situation.

Suppose that the system is

$$ax + by = p$$
$$cx + dy = q$$

and that $ad - cb \stackrel{\text{def}}{=} \delta$ is small in absolute value but not 0. Then, according to Cramer's rule

$$x = \frac{pd - qb}{\delta}, y = \frac{aq - cp}{\delta}.$$

One way to make δ small is to make a, b, c the same and d not much different from their common value: $a = b = c = 1$, $d = 1 + \delta$, and $|\delta|$ small but not 0. Then $ad - cb = \delta$. If $|\epsilon|$ is small and $q = p + \epsilon$ then

$$x = p - \frac{\epsilon}{\delta}, \; y = \frac{\epsilon}{\delta}.$$

If only ϵ is subject to small changes, e.g., if ϵ changes to $-\epsilon$ then both x and y change by $|2\frac{\epsilon}{\delta}|$. The relative changes for x resp. y are $|2\frac{\epsilon}{\delta p - \epsilon}|$ resp. 2. If $\delta = 10^{-6}$, $\epsilon = 10^{-5}$, and $p = 3$ then the true solutions change from

$$x = -7, y = 10 \;\; \text{to} \;\; x = 13, y = -10.$$

The relative percent changes for x resp. y are $\approx 314\%$ resp. 200% whereas the percent change in the data (the percent change in p) is a mere $\approx 0.0007\%$.

The phenomenon just examined illustrates the instability of *all* methods of exact solution of systems like

$$a_{11}x_1 + \cdots + a_{1n}x_n = b_1$$
$$\vdots \qquad \ddots \qquad \vdots \qquad \vdots \qquad\qquad\quad (5.1.1)$$
$$a_{m1}x_1 + \cdots + a_{mn}x_n = b_m.$$

This instability takes no account of roundoff, which simply exacerbates the difficulties.

Since a least squares solution of $A\mathbf{x} = \mathbf{b}$ is unique if A is invertible ($A^{-1} = A^{+}$) least squares solutions can be just as unstable as exact solutions.

The difficulty lies with the way in which a matrix A magnifies or "minifies" the length of a vector $\mathbf{x}$. If $\mathbf{x} \neq \mathbf{O}$ then

$$\frac{\|A\mathbf{x}\|}{\|\mathbf{x}\|}$$

is a measure of the degree of magnification or minification by A of $\|\mathbf{x}\|$. If $A \stackrel{\text{def}}{=} (a_{ij})_{i,j=1}^{m,n}$ and

$$\mathbf{x} \stackrel{\text{def}}{=} \begin{pmatrix} x_1 \\ \vdots \\ x_n \end{pmatrix}$$

then the Schwarz inequality leads to the following calculations:

$$\|A\mathbf{x}\|^2 = \sum_{i=1}^{m} |\sum_{j=i}^{n} a_{ij}x_j|^2$$

$$\leq [\sum_{i-1}^{m}(\sum_{j-1}^{n} |a_{ij}|^2)]\|\mathbf{x}\|^2 \stackrel{\text{def}}{=} M^2\|\mathbf{x}\|^2$$

$$\|A\mathbf{x}\| \leq M\|\mathbf{x}\|.$$

Hence

$$\sup_{\mathbf{x}\neq\mathbf{O}} \frac{\|A\mathbf{x}\|}{\|\mathbf{x}\|} \stackrel{\text{def}}{=} \|A\|_2 \leq M < \infty.$$

The numbers $\|A\|_2$ and M are called *norms* of A. To distinguish one from the other sometimes $\|A\|_2$ is called the *Euclidean norm* of A while M is denoted $\|A\|_F$ and sometimes is called the *Frobenius norm* of A:

$$\|A\|_F \stackrel{\text{def}}{=} \sqrt{(\sum_{i,j=1}^{m,n} |a_{ij}|^2)}.$$

[**Remark 5.1.1:** If A is written as a row vector having mn components:

$$\mathbf{A} \stackrel{\text{def}}{=} (a_{11}, \ldots, a_{mn}) \text{ in } \mathbf{C}_{mn}$$

then, for utter confusion, the Frobenius norm $\|A\|_F$ is actually the Euclidean length or norm of the *vector* $\mathbf{A}$. So much for terminology.]

If A is invertible then $\|A^{-1}\|_2$ is a measure of the magnification or minification by A^{-1} of $\|\mathbf{b}\|$ in the solution $\mathbf{x} = A^{-1}\mathbf{b}$ of the system $A\mathbf{x} = \mathbf{b}$.

If $\|A\|_2$ is large, small changes in the components x_i of $\mathbf{x}$ can be magnified and thereby give rise to large changes in $A\mathbf{x}$. Hence roundoff that leads to small errors in the "solution" $\mathbf{x}_0$ for $\mathbf{x}$ can cause $A\mathbf{x}_0$ to deviate greatly from $\mathbf{b}$. If $\|A^{-1}\|_2$ is large, small changes in $\mathbf{b}$ yield large changes in the exact solution $A^{-1}\mathbf{b}$ of $A\mathbf{x} = \mathbf{b}$.

If A^{-1} exists and $\mathbf{x} \neq \mathbf{O}$ then

$$\|\mathbf{x}\| = \|A^{-1}A\mathbf{x}\| \leq \|A^{-1}\|_2\|A\mathbf{x}\| \leq \|A^{-1}\|_2\|A\|_2\|\mathbf{x}\|$$

and so $1 \leq \|A^{-1}\|_2\|A\|_2 \stackrel{\text{def}}{=} \kappa(A)$, the *condition number* of the SQUARE matrix A, is sometimes used to adjudge how "stable" a matrix A is in the context of

<hr>

seeking solutions for $Ax = \mathbf{b}$. If $\kappa(A)$ is large then at least one of $\|A^{-1}\|_2$ and $\|A\|_2$ is large: one should be concerned about the possibility for trouble. The closer $\kappa(A)$ is to 1 the smaller is the distortion that A can inflict on a "solution" $\mathbf{x}_0$ for the system $Ax = \mathbf{b}$ or the smaller is the amplification that A^{-1} can cause in small variations of $\mathbf{b}$.

Exercise 5.1.1. Show that

$$\|A\|_2 = \sup_{\|\mathbf{x}\|=1} \|A\mathbf{x}\|.$$

[*Hint:* If $\mathbf{x} \neq \mathbf{0}$ and $\mathbf{y} \stackrel{\text{def}}{=} \frac{\mathbf{x}}{\|\mathbf{x}\|}$ then $\|\mathbf{y}\| = 1$ and $\frac{\|A\mathbf{x}\|}{\|\mathbf{x}\|} = \|A\mathbf{y}\|$.]

Exercise 5.1.2. Estimate closely the condition number of each of the following matrices:

i.

$$\begin{pmatrix} 1 & 1 \\ 1 & 1+\epsilon \end{pmatrix}, \ \epsilon = 0.1, 0.01, 0.001, 0.0001;$$

ii. a unitary matrix;

iii.

$$\begin{pmatrix} \lambda_1 & & & \\ & \lambda_2 & & \\ & & \ddots & \\ & & & \lambda_n \end{pmatrix};$$

iv. if $c_i \neq 0$ the $n \times n$ identity matrix in which column i is replaced by $(c_1, \ldots, c_n)^t$:

In each case calculate as well the determinant of the matrix.

Exercise 5.1.3. Let M be an arbitrary positive number. Find an invertible matrix A such that $M < \kappa(A)$.

Example 5.1.1. Assume $A \stackrel{\text{def}}{=} (a_{ij})_{i,j=1}^{n,n}$ is upper triangular and such that

$$a_{ij} = \begin{cases} 0 & \text{if } i > j, \\ 1 & \text{if } i = j, \\ -1 & \text{if } i < j. \end{cases}$$

Then $A^{-1} \stackrel{\text{def}}{=} B \stackrel{\text{def}}{=} (b_{ij})_{i,j=1}^{n,n}$ is such that

$$b_{ij} = \begin{cases} 0 & \text{if } i > j, \\ 1 & \text{if } i = j, \\ j - i & \text{if } i < j. \end{cases}$$

If $\mathbf{x} \stackrel{\text{def}}{=} (\frac{1}{\sqrt{n}}, \ldots, \frac{1}{\sqrt{n}})^t$ then

$$\|\mathbf{x}\| = 1$$

$$\|A\mathbf{x}\| \geq \frac{n-2}{\sqrt{n}}$$

$$\|B\mathbf{x}\| \geq \frac{(n-1)n(2n-1)}{6\sqrt{n}}$$

$$\|A\|_2 \cdot \|B\|_2 = \kappa(A) \geq \frac{(n-1)(n-2)(2n-1)}{6}.$$

For large n the value of $\kappa(A)$ is large, e.g., if $n = 100$ then $\kappa(A) \geq 321,783$, whereas $det(A) = 1$ for every n. Thus, in view of the results in **Exercise 5.1.1** and this **Example** there is no correlation between the condition number of a matrix and the determinant of a matrix. The size of one implies nothing about the size of the other.

[**Note 5.1.1:** A more refined estimate shows that for A as given above, $\kappa(A) \geq 2^{n-1}$, which dramatizes even more the lack of correlation between $det(A)$ and $\kappa(A)$.]

Exercise 5.1.4. For A as in **Example 5.1.1** and $n = 2, 3$ estimate closely $\kappa(A)$.

Exercise 5.1.5. In the items below the remarks in parentheses give the relevant descriptions of the symbols used. Prove each (in)equality.

   i. $\|pA\|_2 = |p|\|A\|_2$ ($p \in \mathbf{C}$, A an $m \times n$ matrix);
  ii. $\|A + B\|_2 \leq \|A\|_2 + \|B\|_2$ (A, B $m \times n$ matrices);
 iii. $\|AB\|_2 \leq \|A\|_2\|B\|_2$ (A, B $n \times n$ matrices);
  iv. $\|P^{-1}AP\|_2 \leq \kappa_P\|A\|_2$ (A, P $n \times n$ matrices, P invertible);
   v. $\|A\|_2 \leq \|A\|_F$;
  vi. $\|U\|_2 = 1$, $\|U\|_F = n$, $\kappa_U = 1$ (U a unitary $n \times n$ matrix);

 vii. $\|D\|_2 = \max_{1 \leq i \leq n} |d_{ii}|$ ($D \stackrel{\text{def}}{=} (d_{ij})_{i,j=1}^{n,n}$, D a diagonal matrix);
viii. $\|U^{-1}AU\|_2 = \|A\|_2$, $\kappa_{U^{-1}AU} = \kappa_A$ (A, U invertible $n \times n$ matrices, U unitary);
  ix. $\kappa_{AB} \leq \kappa_A\kappa_B$ (A, B invertible $n \times n$ matrices);
   x. $|\lambda| \leq \|A\|_2$ (λ an eigenvalue of A).

There is a connection between $\|A\|_2$ and the eigenvalues of the SQUARE matrix A. The exploration of this connection leads inexorably to some remarkable descriptions of a self-adjoint matrix A.

Recall that the spectrum $\sigma(A)$ of a SQUARE matrix A is the *finite* set of its different eigenvalues (cf. **Section 4.7**). The *spectral radius* ρ_A of A is the

---

radius of the smallest circle centered at the origin in $\mathbf{C}$ and containing $\sigma(A)$. Thus

$$\rho_A = \max\{\,|\lambda| \;:\; \lambda \text{ an eigenvalue of } A\,\}$$
$$= \max\left\{\,|\lambda| \;:\; \lambda \in \sigma(A) \overset{\text{def}}{=} spectrum(A)\,\right\}.$$

For any SQUARE matrix A if $\mathbf{x}$ is an eigenvector and $A\mathbf{x} = \lambda\mathbf{x}$ then

$$\|A\mathbf{x}\| = |\lambda| \cdot \|\mathbf{x}\| \le \rho_A \|\mathbf{x}\|$$

whence $\rho_A \le \|A\|_2$. However, is not necessarily true that $\rho_A = \|A\|_2$.

Example 5.1.2. If

$$A = \begin{pmatrix} 1 & 1 \\ 0 & 2 \end{pmatrix}$$

then the eigenvalues of A are 1 and 2. Furthermore

$$\|A\|_2 \ge \frac{\|A\mathbf{e}_2\|}{\|\mathbf{e}_2\|} = \sqrt{5} > \rho_A = 2.$$

Nevertheless, the following result is occasionally helpful.

THEOREM 5.1.1. IF A IS NORMAL $(AA^{*} = A^{*}A)$ THEN

$$\|A\|_2 = \rho_A = \sup_{\mathbf{x} \ne O} \frac{|(A\mathbf{x}, \mathbf{x})|}{\|\mathbf{x}\|^2}.$$

[**Remark 5.1.2:** The quantity

$$\frac{(A\mathbf{x}, \mathbf{x})}{\|\mathbf{x}\|^2}$$

is known as the *Rayleigh quotient* for A and $\mathbf{x}$. If A is self-adjoint (i.e., if $A = A^{*}$) then all the eigenvalues are real (cf. **Exercise 4.4.4**). If λ is the largest eigenvalue then $\rho_A = \lambda$.]

PROOF. Since A is normal there is for A an orthonormal eigenbasis $\{\mathbf{x}_i\}_{i=1}^n$. If $A\mathbf{x}_i = \lambda_i \mathbf{x}_i$ and

$$\mathbf{x} = \sum_{i=1}^{n} a_i \mathbf{x}_i$$

<hr>

then

$$Ax = \sum_{i=1}^{n} \lambda_i a_i \mathbf{x}_i$$

$$\|A\mathbf{x}\|^2 = \sum_{i=1}^{n} |\lambda_i a_i|^2 \le \rho_A^2 \sum_{i=1}^{n} |a_i|^2 = \rho_A^2 \|\mathbf{x}\|^2$$

$$\frac{\|A\mathbf{x}\|}{\|\mathbf{x}\|} \le \rho_A \text{ if } \mathbf{x} \ne \mathbf{O}$$

$$\rho_A \le \|A\|_2 \le \rho_A. \tag{5.1.2}$$

Similarly, if λ is an eigenvalue such that $|\lambda| = \rho_A$ and if $\mathbf{x}$ is a corresponding eigenvector then

$$|(A\mathbf{x}, \mathbf{x})| = \rho_A \|\mathbf{x}\|^2$$

$$\rho_A \le \sup_{\mathbf{x} \ne O} \frac{|(A\mathbf{x}, \mathbf{x})|}{\|\mathbf{x}\|^2} \le \frac{\|A\|_2 \|\mathbf{x}\|^2}{\|\mathbf{x}\|^2} = \|A\|_2 = \rho_A. \tag{5.1.3}$$

The conclusions follow from (5.1.2) and (5.1.3).

$$\Omega$$

Exercise 5.1.6. Use the Riesz representation theorem (cf. **Exercise 4.4.28**) and the Schwarz inequality (cf. (4.4.21)) to show that for any SQUARE matrix A

$$\|A\|_2 = \sup_{\mathbf{x}, \mathbf{y} \ne O} \frac{|(A\mathbf{x}, \mathbf{y})|}{\|\mathbf{x}\| \cdot \|\mathbf{y}\|}.$$

An important application of the Jordanizability of an arbitrary SQUARE matrix A is the following result.

THEOREM 5.1.2. LET A BE AN $n \times n$ MATRIX. THEN

$$\lim_{s \to \infty} \|A^s \mathbf{x}\| = 0 \text{ FOR EVERY VECTOR } \mathbf{x} \text{ IN } \mathbf{C}^n \text{ IFF } \rho_A < 1.$$

PROOF. Assume $|\rho_A| < 1$. Let P be an invertible matrix such that $P^{-1} A P \overset{\text{def}}{=} \mathcal{A}$ is in Jordan normal form.

Then, D denoting the diagonal matrix in which the diagonal entries are those of $\mathcal{A}$ and N denoting the nilpotent upper triangular matrix in which the entries are those above the diagonal of $\mathcal{A}$,

$$\mathcal{A} = D + N, \quad DN = ND.$$

Let $\lambda_1, \ldots, \lambda_m$ be the distinct eigenvalues of A, assume $|\lambda_1| \geq \cdots \geq |\lambda_m|$, and assume $N^k = O \neq N^{k-1}$. Then if $s \geq k$ and $\mathbf{x}$ is any vector in $\mathbf{C}^n$

$$\mathcal{A}^s \mathbf{x} = D^s \mathbf{x} + \cdots + \binom{s}{k-1} D^{s-k+1} N^{k-1} \mathbf{x}. \tag{5.1.4}$$

For any p in $\mathbf{N}$ if $0 < \rho$ then

$$\binom{s}{p} \rho^{s-p} = \frac{1}{p!} \rho^{-p} s(s-1) \cdots (s-p+1) \rho^s$$

$$\leq \frac{1}{p!} \rho^{-p} s^p \rho^s.$$

If $0 \leq \rho < 1$ L'Hôpital's rule shows that $\lim_{s \to \infty} s^p \rho^s = 0$. Hence for any vector

$$\mathbf{y} \overset{\text{def}}{=} \begin{pmatrix} y_1 \\ \vdots \\ y_n \end{pmatrix}$$

$$\|D^p \mathbf{y}\| \leq \rho_A^p \|\mathbf{y}\|$$

and so, since $\rho_A < 1$, as $s \to \infty$ the length of each of the finitely many vectors in the right member of (5.1.4) approaches 0, i.e.

$$\lim_{s \to \infty} \|\mathcal{A}^s \mathbf{x}\| = 0.$$

Hence

$$\lim_{s \to \infty} \|A^s \mathbf{x}\| \leq \|P\|_2 \|\mathcal{A}^s (P^{-1} \mathbf{x})\| = 0.$$

Conversely if $\|A^s \mathbf{x}\| \to 0$ as $s \to \infty$ for each $\mathbf{x}$ in $\mathbf{C}^n$ let $\mathbf{x}$ be an eigenvector corresponding to λ_1 (the eigenvalue of largest absolute value). Then

$$\|A^s \mathbf{x}\| = |\lambda_1|^s \|\mathbf{x}\| \to 0$$

as $s \to \infty$. Since $\|\mathbf{x}\| \neq 0$ it follows that

$$\lim_{s \to \infty} |\lambda_1|^s = 0,$$

which is true iff $|\lambda_1| < 1$.

$$\Omega$$

THEOREM **5.1.2** is useful in the discussion of various iterative procedures for solving a SQUARE system $Ax = \mathbf{b}$ and for finding the eigenvalues of a SQUARE matrix (cf. **Sections 5.2 and 5.3**).

Exercise **5.1.7.** Let A be a SQUARE matrix. Examine the proof of THEOREM **5.1.2.** Use the ideas there and **Exercise 5.1.5** to prove:

     Chapter 5. APPLYING LINEAR ALGEBRA

$$\|A^s\|_2 \to 0 \text{ as } s \to \infty \text{ iff } \rho_A < 1.$$

Give an example of a SQUARE matrix A such that $\|A\|_2 > 1$ and yet $\|A^s\|_2 \to 0$ as $s \to \infty$.

Exercise 5.1.8. Let A be a self-adjoint $n \times n$ matrix having (necessarily real) eigenvalues λ_i such that $\lambda_1 \le \lambda_2 \le \cdots \le \lambda_n$ and let $\{\mathbf{x}_1, \ldots, \mathbf{x}_n\}$ be an orthonormal eigenbasis such that $A\mathbf{x}_i = \lambda_i \mathbf{x}_i$.

i. Show that for any vector $\mathbf{x}$

$$\lambda_1 \|\mathbf{x}\|^2 \le |(A\mathbf{x}, \mathbf{x})| \le \lambda_n \|\mathbf{x}\|^2.$$

[*Hint:* Assume $\mathbf{x} = \sum_{i=1}^{n} a_i \mathbf{x}_i$ and use the right member to calculate $(A\mathbf{x}, \mathbf{x})$.]

ii. Use *i* to show that λ_1 is the minimum value of the Rayleigh quotient for A.

The next result, the *minmax principle* due to Hermann Weyl, (cf. **Exercise 4.4.52**) extends the conclusions in **Exercise 5.1.8**.

THEOREM 5.1.3. LET A BE A SELF-ADJOINT $n \times n$ MATRIX HAVING (NECESSARILY REAL) EIGENVALUES λ_i SUCH THAT $\lambda_1 \le \cdots \le \lambda_n$ AND LET $\{\mathbf{x}_1, \ldots, \mathbf{x}_n\}$ BE AN ORTHONORMAL EIGENBASIS SUCH THAT $A\mathbf{x}_j = \lambda_j \mathbf{x}_j$. THEN, S DENOTING A SUBSPACE OF $\mathbf{C}^n$,

$$\lambda_k = minimum_{dim(S)=k} \left[maximum_{S \ni \mathbf{x} \neq \mathbf{0}} \frac{(A\mathbf{x}, \mathbf{x})}{\|\mathbf{x}\|^2} \right].$$

PROOF. If $dim(S) = k$ there is in S an orthonormal basis $\{\mathbf{y}_1, \ldots, \mathbf{y}_k\}$ and there are numbers b_{ij} such that

$$\mathbf{y}_i = \sum_{j=1}^{n} b_{ij} \mathbf{x}_j.$$

Let $C \overset{\text{def}}{=} (c_{pq})_{p,q=1}^{k-1,k}$ be the $(k-1) \times k$ matrix in which

$$c_{pq} = b_{qp}, \ 1 \le p \le k-1, \ 1 \le q \le k.$$

As a WIDE system $C\mathbf{z} = \mathbf{O}$ has a nonzero solution and, since the system is homogeneous, there is a solution $\mathbf{a} \overset{\text{def}}{=} (a_1, \ldots, a_k)^t$ such that $\|\mathbf{a}\|^2 = 1$. If $\mathbf{x} \overset{\text{def}}{=} \sum_{i=1}^{k} a_i \mathbf{y}_i$ then, owing to the equation $C\mathbf{a} = \mathbf{O}$, it follows that

$$\|\mathbf{x}\|^2 = 1 = \sum_{j=1}^{n} |\sum_{i=1}^{k} a_i b_{ij}|^2 = \sum_{j=k}^{n} |\sum_{i=1}^{k} a_i b_{ij}|^2.$$

Hence

$$\frac{(A\mathbf{x},\mathbf{x})}{\|\mathbf{x}\|^2} = (A\mathbf{x},\mathbf{x}) = \sum_{j=1}^{n}\lambda_j\,|\sum_{i=1}^{k} a_i b_{ij}|^2$$

$$= \sum_{j=k}^{n}\lambda_j\,|\sum_{i=1}^{k} a_i b_{ij}|^2$$

$$\geq \lambda_k(\sum_{j=k}^{n}|\sum_{i=1}^{k} a_i b_{ij}|^2) = \lambda_k.$$

It follows that for S fixed and k-dimensional

$$maximum_{S\ni\mathbf{x}\neq O}\frac{(A\mathbf{x},\mathbf{x})}{\|\mathbf{x}\|^2} \geq \lambda_k$$

$$minimum_{dim(S)=k}\left[maximum_{S\ni\mathbf{x}\neq O}\frac{(A\mathbf{x},\mathbf{x})}{\|\mathbf{x}\|^2}\right] \geq \lambda_k.$$

If $S = span(\{\mathbf{x}_1,\ldots,\mathbf{x}_k\})$ then S is k-dimensional and if $\mathbf{x} = \mathbf{x}_k$ then

$$\frac{(A\mathbf{x},\mathbf{x})}{\|\mathbf{x}\|^2} = \lambda_k.$$

$$\Omega$$

Exercise 5.1.9. Show that if A is a self-adjoint matrix then its kth greatest eigenvalue is

$$minimum_{dim(S)=k}\left[maximum_{S\ni\mathbf{x},\|\mathbf{x}\|=1}(A\mathbf{x},\mathbf{x})\right].$$

Exercise 5.1.10. Let $A \stackrel{\text{def}}{=} (a_{ij})_{i,j=1}^{n,n}$ be a self-adjoint matrix with real entries. For

$$\mathbf{x} \stackrel{\text{def}}{=} \begin{pmatrix} x_1 \\ \vdots \\ x_n \end{pmatrix}$$

use **Exercise 5.1.9** and the method of Lagrange multipliers to set up the equations for finding

$$maximum_{\|\mathbf{x}\|=1}(A\mathbf{x},\mathbf{x}).$$

In **PART I** the principal concerns are:

i. for an arbitrary matrix A the analysis of the system $A\mathbf{x} = \mathbf{b}$;
ii. for a SQUARE matrix the eigenvalues of A and a useful form for A.

In the next three **Sections** there is a survey of the practical computational problems associated with analyzing $A\mathbf{x} = \mathbf{b}$ and with finding eigenvalues of a

---

    Chapter 5. APPLYING LINEAR ALGEBRA

SQUARE matrix. Once the eigenvalues are found the useful form for A is relatively easily computed. The interested reader is urged to consult [Fa, GvL, GW, KaR, Le, P, Wil, and WR] for more details about these and related questions. The literature on the subject is huge, is growing rapidly, and its expansion shows no signs of slowing. Detailed and final analyses of major problems in numerical analysis are not even in sight, much less in reach.

The aim of this **Chapter** is to indicate the kinds of problems to which linear algebra is applicable and to offer some of the theory behind the practice of applying linear algebra. The fascinating details of numerical analysis as they are relevant to linear algebra are treated in separate books cited throughout the **Chapter**. The sizes of those books suggest that their contents cannot be included in the pages of this one.

5.2. Decompositions and factorizations of matrices

The factorization arising from the equation

$$\Pi A = LDU$$

in **Section 2.7** is one instance of a useful decomposition of a matrix A into parts that are more easily handled than is the matrix A itself. In computational problems there are other useful decompositions of a matrix A. Some of these decompositions represent a matrix as a *product* of simpler and useful matrices and others represent a matrix as a *sum* of useful matrices.

Other examples of factorizations of SQUARE matrices are those that lead to upper triangular forms and, more particularly, to Jordan normal forms. Here is a THEOREM yielding a different factorization for an arbitrary matrix. The result gives rise to the *singular value decomposition* (SVD) of a matrix.

THEOREM 5.2.1. LET A BE AN $m \times n$ MATRIX. THERE ARE TWO UNITARY MATRICES, ONE IS AN $m \times m$ UNITARY MATRIX P AND THE OTHER IS AN $n \times n$ UNITARY MATRIX Q, AND PAQ IS AN $m \times n$ DIAGONAL MATRIX $\Sigma \overset{\text{def}}{=} (\sigma_{ij})_{i,j=1}^{m,n}$, i.e., $\sigma_{ij} = 0$ IF $i \neq j$. FURTHERMORE EACH NONZERO DIAGONAL ENTRY σ_{ii} IS POSITIVE AND THE NUMBER OF SUCH POSITIVE DIAGONAL ENTRIES IS THE RANK r OF A.

PROOF. The matrix $A^{\times}A \overset{\text{def}}{=} B$ is a self-adjoint $n \times n$ matrix for which there is an orthonormal eigenbasis $\{\mathbf{x}_j\}_{j=1}^{n}$.

Let Q be the $n \times n$ unitary matrix in which column j is the column vector $\mathbf{x}_j$. If λ is an eigenvalue of B and if $\mathbf{x}$ is a corresponding *nonzero* eigenvector then

$$(B\mathbf{x}, \mathbf{x}) = \lambda(\mathbf{x}, \mathbf{x})$$
$$= (A\mathbf{x}, A\mathbf{x}) \geq 0$$
$$\lambda = \frac{(A\mathbf{x}, A\mathbf{x})}{(\mathbf{x}, \mathbf{x})} \geq 0.$$

If $B\mathbf{x}_j = \lambda\mathbf{x}_j$ and $\lambda \neq 0$ let $\mathbf{y}_j$ be $\frac{A\mathbf{x}_j}{\sqrt{\lambda}}$. Then direct calculation shows $A\mathbf{x}_j \neq \mathbf{O}$ and indeed that $\|\mathbf{y}_j\|^2 = 1$. If also $j' \neq j$ and $B\mathbf{x}_{j'} = \lambda\mathbf{x}_{j'}$ then a similar calculation shows that for $\mathbf{y}_{j'} \overset{\text{def}}{=} \frac{A\mathbf{x}_{j'}}{\sqrt{\lambda}}$ it is true that $(\mathbf{y}_j, \mathbf{y}_{j'}) = 0$.

In short, it may be assumed that for B there are r nonzero eigenvalues λ_p, that $\lambda_1 \geq \cdots \geq \lambda_r > 0$, and that $\mathbf{x}_1, \ldots, \mathbf{x}_r$ are the eigenvectors corresponding to these nonzero eigenvalues. (Note that $r \leq n$ and that some of the λ_p may be equal since several of the $\mathbf{x}_j$ may correspond to the same eigenvalue.)

Thus if $p \leq r$ then $B\mathbf{x}_p = \lambda_p\mathbf{x}_p \neq \mathbf{O}$ and if

$$\mathbf{y}_p \overset{\text{def}}{=} \frac{A\mathbf{x}_p}{\sqrt{\lambda_p}}, \ 1 \leq p \leq r,$$

then $\{\mathbf{y}_1, \ldots \mathbf{y}_r\}$ is an orthonormal set of $m \times 1$ vectors. Thus $r \leq m$ and if $r < m$ the set $\{\mathbf{y}_1, \ldots \mathbf{y}_r\}$ may be filled out to a basis for $\mathbf{C}^m$ and that basis may be converted via the Gram-Schmidt process to an orthonormal basis $\{\mathbf{y}_1, \ldots, \mathbf{y}_r, \ldots, \mathbf{y}_n\}$. Let P be the unitary $m \times m$ matrix in which *row i* is $\mathbf{y}_i$.

The qth column of AQ is

$$\begin{cases} A\mathbf{x}_q & \text{if } q \leq r \\ \mathbf{O} & \text{if } q > r. \end{cases}$$

Hence the pq entry of PAQ is

$$\begin{cases} (\frac{A\mathbf{x}_p}{\sqrt{\lambda_p}}, A\mathbf{x}_q) = \sqrt{\lambda_p}\delta_{pq} & \text{if } 1 \leq p \leq r \\ 0 & \text{if } p > r. \end{cases}$$

Thus if $\sigma_p = \sqrt{\lambda_p}$ then

$$\sigma_1 \geq \cdots \geq \sigma_r \ (> \sigma_{r+1} = \cdots = \sigma_m = 0 \text{ if } r < m)$$

and PAQ is the $m \times n$ diagonal matrix

$$\Sigma \overset{\text{def}}{=} \begin{pmatrix} \sigma_1 & & & \\ & \sigma_2 & & \\ & & \ddots & \\ & & & \sigma_r \end{pmatrix}, \ r \leq minimum(m, n).$$

Hence $A = P^*\Sigma Q^*$ and A is factored into its *singular value decomposition* (SVD).

Note that if $A\mathbf{x} = \mathbf{O}$ then $A^*A\mathbf{x} = \mathbf{O}$ so that

$$ker(A) \subset ker(A^*A).$$

Conversely, if $A^*A\mathbf{x} = \mathbf{O}$ then $(A\mathbf{x}, A\mathbf{x}) = 0$ and hence

$$A\mathbf{x} = \mathbf{O}$$
$$ker(A^*A) \subset ker(A)$$
$$ker(A) = ker(A^*A)$$
$$dim[ker(A)] + dim[im(A)] = n = dim[dom(A)]$$
$$r = rank(A) = dim[im(A)] = dim[im(A^*A)] = rank(A^*A).$$

Since $rank(A^*A) = r$ it follows that the number of nonzero (diagonal) entries of PAQ is $rank(A)$.

$$\Omega$$

The numbers σ_i are the *singular values* of A.

Exercise 5.2.1. In the notation used above show that

$$Ax_i = \begin{cases} \sigma_i \mathbf{y}_i & \text{if } 1 \le i \le r \\ \mathbf{O} & \text{if } i > r \end{cases}$$
$$A^*\mathbf{y}_i = \begin{cases} \sigma_i \mathbf{x}_i & \text{if } 1 \le i \le r \\ \mathbf{O} & \text{if } i > r \end{cases}$$
$$ker(A) = span(\{\mathbf{x}_{r+1}, \ldots, \mathbf{x}_n\})$$
$$im(A) = span(\{\mathbf{y}_1, \ldots, \mathbf{y}_r\})$$
$$A = \sum_{i=1}^{r} \sigma_i \mathbf{y}_i \mathbf{x}_i^t.$$

Exercise 5.2.2. Show that if $A \overset{\text{def}}{=} (a_{ij})_{i,j=1}^{m,n}$ is an $m \times n$ *diagonal* matrix then its pseudo-inverse A^+ is an $n \times m$ diagonal matrix $(\alpha_{ij})_{i,j=1}^{m,n}$ such that

$$\alpha_{ii} = \begin{cases} \frac{1}{a_{ii}} & \text{if } a_{ii} \ne 0 \\ 0 & \text{if } a_{ii} = 0. \end{cases}$$

The SVD yields an explicit formula for A^+:

$$A^+ = Q\Sigma P. \tag{5.2.1}$$

The heart of the proof of (5.2.1) is the fact that unitary matrices preserve lengths of vectors. Thus to minimize $\|A\mathbf{x} - \mathbf{b}\|$ is to minimize

$$\|P^*\Sigma Q^*\mathbf{x} - \mathbf{b}\| = \|\Sigma Q^*\mathbf{x} - P\mathbf{b}\|.$$

If $Q^*\mathbf{x} \overset{\text{def}}{=} \mathbf{y}$ then $\overline{\mathbf{y}} \overset{\text{def}}{=} \Sigma P\mathbf{b}$ is the optimal (shortest) minimizing vector of $\|\Sigma\mathbf{y} - P\mathbf{b}\|$. Thus if

$$Q\overline{\mathbf{y}} = Q\Sigma P\mathbf{b} \overset{\text{def}}{=} \overline{\mathbf{x}}$$

---

then $\|\overline{\mathbf{x}}\| = \|\overline{\mathbf{y}}\|$ and so $\overline{\mathbf{x}}$ is the shortest minimizing vector of $\|A\mathbf{x} - \mathbf{b}\|$, i.e., (5.2.1).

If a SQUARE matrix A is written as the difference of two matrices, e.g., $A = P - Q$ then the system $A\mathbf{x} = \mathbf{b}$ is the same as the system $P\mathbf{x} = Q\mathbf{x} + \mathbf{b}$. If P is invertible then

$$\mathbf{x} = P^{-1}Q\mathbf{x} + P^{-1}\mathbf{b}. \tag{5.2.2}$$

The equation (5.2.2) suggests that the following procedure might be helpful in approximating a solution $\widetilde{\mathbf{x}}$ for $A\mathbf{x} = \mathbf{b}$ if a solution exists at all.

A vector $\mathbf{x}_0$ is chosen arbitrarily and then a sequence of vectors is defined according to the *recurrence scheme*

$$\mathbf{x}_{n+1} = P^{-1}Q\mathbf{x}_n + P^{-1}\mathbf{b}. \tag{5.2.3}$$

If the vectors $\mathbf{x}_n$ converge to a limiting vector $\widetilde{\mathbf{x}}$ then

$$\lim_{n \to \infty} \|\widetilde{\mathbf{x}} - \mathbf{x}_n\| = 0$$
$$\|P^{-1}Q\widetilde{\mathbf{x}} - P^{-1}Q\mathbf{x}_n\|$$
$$\leq \|P^{-1}Q\|_2 \|\widetilde{\mathbf{x}} - \mathbf{x}_n\| \to 0$$
$$\text{as } n \to \infty.$$

Owing to (5.2.3),

$$\widetilde{\mathbf{x}} = P^{-1}Q\widetilde{\mathbf{x}} + P^{-1}\mathbf{b}$$
$$P\widetilde{\mathbf{x}} = Q\widetilde{\mathbf{x}} + \mathbf{b}$$
$$(P - Q)\widetilde{\mathbf{x}} \equiv A\widetilde{\mathbf{x}} = \mathbf{b},$$

i.e., $\widetilde{\mathbf{x}}$ is a solution of the system $A\mathbf{x} = \mathbf{b}$.

If the system $A\mathbf{x} = \mathbf{b}$ has a solution $\widetilde{\mathbf{x}}$ the sequence $\{\mathbf{x}_n\}_{n=0}^{\infty}$ converges to $\widetilde{\mathbf{x}}$ iff the *error*

$$\mathbf{d}_n \overset{\text{def}}{=} \widetilde{\mathbf{x}} - \mathbf{x}_n$$
$$= P^{-1}Q(\widetilde{\mathbf{x}} - \mathbf{x}_{n-1})$$
$$= P^{-1}Q\mathbf{d}_{n-1}$$
$$= (P^{-1}Q)^n \mathbf{d}_0 \tag{5.2.4}$$

approaches $\mathbf{O}$ as $n \to \infty$.

If the spectral radius $\rho_{P^{-1}Q} < 1$ then, owing to THEOREM **5.1.2**,

$$\lim_{n \to \infty} (P^{-1}Q)^n \mathbf{d}_0 = \mathbf{O}$$

no matter what the choice of $\mathbf{x}_0$.

Hence:

     Chapter 5. APPLYING LINEAR ALGEBRA

If $\rho_{P^{-1}Q} < 1$ and if $A\mathbf{x} = \mathbf{b}$ has one solution then there is at most (hence only) one solution and the sequence defined by (5.2.3) converges to that solution no matter what the choice of $\mathbf{x}_0$.

Conversely, assume there is at most one solution for $A\mathbf{x} = \mathbf{b}$ and that $F(\mathbf{x}_0) \overset{\text{def}}{=} \lim_{n \to \infty} \mathbf{x}_n$ exists no matter what the choice of $\mathbf{x}_0$. Then $F(\mathbf{x}_0)$ is *the* solution $\tilde{\mathbf{x}}$ of $A\mathbf{x} = \mathbf{b}$, $(P^{-1}Q)^n \mathbf{d}_0$ converges to $\mathbf{O}$ no matter what the choice of $\mathbf{x}_0$, i.e., no matter what the vector $\mathbf{d}_0$ is, and so $\rho_{P^{-1}Q} < 1$.

In sum:

There is at most one solution for $A\mathbf{x} = \mathbf{b}$ and the sequence defined by (5.2.3) converges no matter what the choice of $\mathbf{x}_0$ iff $\rho_{P^{-1}Q} < 1$.

[**Note 5.2.1:** If the sequence $\{\mathbf{x}_n\}_{n=0}^{\infty}$ converges the vectors $\mathbf{x}_n$ are called *approximants* to the solution $F(\mathbf{x}_0)$.]

Thus the decomposition $A = P - Q$ is useful in a recurrence scheme if

i. P^{-1} exists;

ii. $\rho_{P^{-1}Q} < 1$.

The next **Example** not only illustrates the conclusion above but also illuminates its limitations.

Example 5.2.1. Let the system $A\mathbf{x} = \mathbf{b}$ be

$$\begin{pmatrix} 2 & -1 \\ 1 & -2 \end{pmatrix} \begin{pmatrix} x_1 \\ x_2 \end{pmatrix} = \begin{pmatrix} b_1 \\ b_2 \end{pmatrix}.$$

Then

$$A = \begin{pmatrix} 2 & -1 \\ 1 & -2 \end{pmatrix} = \begin{pmatrix} 2 & 0 \\ 1 & -2 \end{pmatrix} - \begin{pmatrix} 0 & 1 \\ 0 & 0 \end{pmatrix} \overset{\text{def}}{=} P - Q$$

and a direct calculation shows

$$P^{-1}Q = \begin{pmatrix} 0 & \frac{1}{2} \\ 0 & \frac{1}{4} \end{pmatrix}.$$

The eigen values of $P^{-1}Q$ are 0 and $\frac{1}{4}$ and $\rho_{P^{-1}Q} = \frac{1}{4} < 1$. If

$$\mathbf{x}_0 = \begin{pmatrix} a \\ b \end{pmatrix}$$

then

$$\mathbf{x}_{n+1} = (P^{-1}Q)^{n+1}\mathbf{x}_0 + \sum_{k=0}^{n} (P^{-1}Q)^k P^{-1}\mathbf{b}.$$

The identity $(I - B^{n+1}) = \sum_{k=0}^{n} B^k(I - B)$, stemming from the algebraic identity $1 - z^{n+1} = \sum_{k=0}^{n} z^k(1 - z)$, is valid for any SQUARE matrix B. Since $\rho_{P^{-1}Q} = \frac{1}{4}$

---

it follows that

$$I = \lim_{n \to \infty} \sum_{k=0}^{n} (P^{-1}Q)^k (I - P^{-1}Q)$$

$$\lim_{n \to \infty} \mathbf{x}_{n+1} = (I - P^{-1}Q)^{-1} P^{-1} \mathbf{b}$$

$$= \begin{pmatrix} 1 & \frac{2}{3} \\ 0 & \frac{4}{3} \end{pmatrix} \begin{pmatrix} \frac{1}{2} & 0 \\ \frac{1}{4} & -\frac{1}{2} \end{pmatrix} \begin{pmatrix} b_1 \\ b_2 \end{pmatrix}$$

$$= \begin{pmatrix} \frac{2}{3} & -\frac{1}{3} \\ \frac{1}{3} & -\frac{2}{3} \end{pmatrix} \begin{pmatrix} b_1 \\ b_2 \end{pmatrix}$$

$$= \begin{pmatrix} \frac{1}{3}(2b_1 - b_2) \\ \frac{1}{3}(b_1 - 2b_2) \end{pmatrix}. \tag{5.2.5}$$

A direct check shows that the vector in the right member of (5.2.5) is indeed the solution of $A\mathbf{x} = \mathbf{b}$.

On the other hand, the system may be rewritten $AE_{12}E_{12}\mathbf{x} = \mathbf{b}$, i.e., as follows:

$$\begin{pmatrix} -1 & 2 \\ -2 & 1 \end{pmatrix} \begin{pmatrix} x_2 \\ x_1 \end{pmatrix} = \begin{pmatrix} b_1 \\ b_2 \end{pmatrix}.$$

The matrix B of the system is AE_{12}, the unknown $\mathbf{y}$ of the system is $E_{12}\mathbf{x}$ and the right member of the system is unchanged: $B\mathbf{y} = \mathbf{b}$.

This time write

$$B = \begin{pmatrix} -1 & 0 \\ -2 & 1 \end{pmatrix} - \begin{pmatrix} 0 & -2 \\ 0 & 0 \end{pmatrix} \stackrel{\text{def}}{=} S - T.$$

Then

$$S^{-1}T = \begin{pmatrix} 0 & 2 \\ 0 & 4 \end{pmatrix}$$

for which the eigenvalues are 0 and 4 and for which $\rho_{S^{-1}T} = 4 > 1$. Furthermore, if

$$\mathbf{y}_0 \stackrel{\text{def}}{=} \begin{pmatrix} c \\ d \end{pmatrix}$$

then

$$(S^{-1}T)^n = \begin{pmatrix} 0 & 2 \cdot 4^{n-1} \\ 0 & 4^n \end{pmatrix}$$

$$\sum_{k=0}^{n} (S^{-1}T)^k = \begin{pmatrix} 1 & \frac{2}{3}(4^n - 1) \\ 0 & 1 + \frac{1}{3}(4^{n+1} - 1) \end{pmatrix}$$

$$\mathbf{y}_{n+1} = \begin{pmatrix} 2 \cdot 4^n d \\ 4^{n+1} d \end{pmatrix} + \begin{pmatrix} -b_1 + \frac{2}{3}(4^n - 1)(-2b_1 + b_2) \\ \frac{1}{3}(4^{n+1} - 1)(-2b_1 + b_2) \end{pmatrix}$$

$$= \begin{pmatrix} 4^n(2d + \frac{2}{3}(-2b_1 + b_2)) + \frac{1}{3}(b_1 - 2b_2) \\ 4^{n+1}(d + \frac{1}{3}(-2b_1 + b_2)) + \frac{1}{3}(2b_1 - b_2) \end{pmatrix}. \tag{5.2.6}$$

The sequence $\{y_n\}$ converges iff the coefficients of 4^n and 4^{n+1} in (5.2.6) are 0, i.e., iff

$$d = \tfrac{1}{3}(2b_1 - b_2).$$

In that case for all n

$$y_n = \begin{pmatrix} \tfrac{1}{3}(b_1 - 2b_2) \\ \tfrac{1}{3}(2b_1 - b_2) \end{pmatrix}$$

and the y_n converge (trivially) to the solution found before. Hence iff one uses for y_0 a vector in which the second component d is the very special number $\tfrac{1}{3}(2b_1 - b_2)$ does the sequence $\{y_n\}$ converge at all.

The particular decomposition of a SQUARE matrix $A \overset{\text{def}}{=} (a_{ij})_{i,j=1}^{n,n}$ as the difference of two matrices P and Q such that P is the lower triangular part of A and Q is simply $P - A$, the *negative* of the *strictly upper triangular* part of A should be called the *Gauss-Seidel* decomposition of A. If P is invertible and if $\rho_{P^{-1}Q} < 1$ the decomposition yields an iterative procedure or recurrence scheme for approximating the solution of $Ax = b$. Note that P in the Gauss-Seidel decomposition is invertible iff each diagonal entry a_{ii} of A is not 0.

The moral to be drawn from **Example 5.2.1** is that from the standpoint of numerical computation the Gauss-Seidel recursion scheme can fail badly even when the lower triangular part P of the decomposition is invertible. Logically valid rearrangements, i.e., interchanging rows and/or columns, of the system can give rise to new system matrices for which the Gauss-Seidel procedure succeeds while it fails for the original system. Such rearrangements correspond to the techniques of MAXSEARCH and PARTIALMAXSEARCH discussed in **Section 2.3**.

In numerical analysis of systems of linear equations there is nothing sacrosanct about the matrix A of a system $Ax = b$. Rearrangements of its columns or of its rows can produce order out of chaos. Such rearrangements correspond the MAXSEARCH or PARTIALMAXSEARCH. When the iterative schemes work they are used to solve SQUARE systems for which an approximate solution suffices and for which the execution of one of the elimination procedures is not cost-effective.

Exercise 5.2.3. For the first (convergent) Gauss-Seidel scheme in **Example 5.2.1** assume

$$b = \begin{pmatrix} 5 \\ -7 \end{pmatrix} \quad \text{and} \quad x_0 = \begin{pmatrix} 1 \\ 2 \end{pmatrix}.$$

Estimate the approximants x_n, $n = 1, 2, 3, 4, 5$ and compare them with the true solution.

Exercise 5.2.4. Let each of the matrices listed below be the matrix A of the system $Ax = b$. In each case determine whether the Gauss-Seidel decomposition

---

leads for all initial "guesses" $\mathbf{x}_0$ to a convergent sequence of approximants. If the decomposition fails determine whether a rearrangement leads to success.

$$i. \begin{pmatrix} 1 & 2 \\ 3 & 4 \end{pmatrix}; \quad ii. \begin{pmatrix} 1 & 2 \\ -1 & 1 \end{pmatrix};$$

$iii.$ a nonsingular $n \times n$ diagonal matrix

$iv.$ a nonsingular $n \times n$ matrix in Jordan normal form.

Exercise 5.2.5. Show that if $A \overset{\text{def}}{=} (a_{ij})_{i,j=1}^{n,n}$ then the Gauss-Seidel procedure applied to the system $A\mathbf{x} = \mathbf{b}$ may be described as follows:

$i.$ Let the ith component of the approximant $\mathbf{x}_n$ be x_{ni}.

$ii.$ Using the initial vector $\mathbf{x}_0$ in the first equation of the system define the first component x_{11} of the first approximant $\mathbf{x}_1$ as

$$\frac{1}{a_{11}}(b_1 - a_{12}x_{02} - \cdots - a_{1n}x_{0n}).$$

$iii.$ Using x_{11} and the last $n-2$ components of $\mathbf{x}_0$ in the second equation of the system define the second component x_{12} of the first approximant $\mathbf{x}_1$ as

$$\frac{1}{a_{22}}(b_2 - a_{21}x_{11} - a_{23}x_{03} - \cdots - a_{2n}x_{0n}).$$

$$\vdots$$

At each step the components numbered $1, 2, \ldots, k$ of the first approximant $\mathbf{x}_1$ together with the components numbered $k + 1, \ldots, n$ of $\mathbf{x}_0$ are used in the $(k + 1)$st equation to find $x_{1,k+1}$.

After all the components of $\mathbf{x}_1$ are found, they are used to cycle through the system as were the components of $\mathbf{x}_0$ to produce in n steps the components of $\mathbf{x}_2$.

$$\vdots$$

Example 5.2.2. For the system

$$3x_1 - 5x_2 = 2$$
$$-x_1 + 7x_2 = 1$$

$(A\mathbf{x} = \mathbf{b})$ the Gauss-Seidel decomposition of A shows that the eigenvalues of $P^{-1}Q$ are 0 and $\frac{5}{21}$

The cycling procedure shows that if the initial guess is $x_{10} = x_{20} = 0$ then the recursion formulae for $x_{1,k+1}$ and $x_{2,k+1}$ are

$$x_{1,k+1} = \frac{2}{3} + \frac{5}{3} \cdot \frac{1 + x_{1,k-1}}{7}$$

$$x_{2,k+1} = \frac{\frac{5}{3} + \frac{5}{3} \cdot x_{2k}}{7}. \tag{5.2.7}$$

These recursion formulae lead to the conclusions

$$\lim_{k \to \infty} x_{1,k+1} \stackrel{\text{def}}{=} x_1 = \frac{2}{3}\left(\sum_{m=0}^{\infty} \left(\frac{5}{21}\right)^m + \sum_{m=1}^{\infty} \left(\frac{5}{21}\right)^m \right)$$

$$= \frac{19}{16}$$

$$\lim_{k \to \infty} \stackrel{\text{def}}{=} x_2 = \sum_{m=1}^{\infty} \left(\frac{5}{21}\right)^m$$

$$= \frac{5}{16}$$

which are indeed the unique solutions for the given system.

Exercise 5.2.6. Rewrite the system in **Example 5.2.2** as follows:

$$-5x_2 + 3x_1 = 2$$
$$7x_2 - x_1 = 1.$$

Find the eigenvalues of the corresponding $P^{-1}Q$ for the rewritten system and set up the recursion formulae that are the analogues of (5.2.7). Do the new recursion formulae give rise to convergent geometric series? Correlate your answer with the eigenvalues of the new $P^{-1}Q$.

Exercise 5.2.7. Use the interpretation in **Exercise 5.2.5** to find the approximants for the systems in **Exercise 5.2.4**. For each system determine the stability of the approximation scheme under rearrangements, i.e., permutations, of the columns of A.

5.3. Finding eigenvalues

If A is a SQUARE $n \times n$ matrix $(a_{ij})_{i,j=1}^{n,n}$ the computational aspects of finding the set of eigenvalues of A are formidable. Various algorithms have been evolved for nailing down or at least approximating the eigenvalues of A. To the writer's knowledge, none of these is universally applicable. For each algorithm there is a matrix A such that roundoff, instability, or failure of convergence of the approximants afflicts the computation of the eigenvalues of A and no reliable result is obtained.

Nevertheless experience is less disappointing than theory and often the algorithms or procedures prove quite successful. In this **Section** there is a survey of a few of the currently available methods. Some are absurdly impractical except for small n. Others are popular and work well for large n if sufficient preparatory manipulation of A is performed before the application of the procedure itself. Some provide ballpark estimates for the eigenvalues. The applicability of others is theoretically questionable and yet empirically quite sound. Those that have been most satisfactory in practice are available as programs for computers.

Here is a short list of some of the methods:

i. THE CHARACTERISTIC POLYNOMIAL METHOD. Find the zeros of the characteristic polynomial χ_A. If $n > 4$ this approach is of questionable value, particularly since the first step, i.e., calculating χ_A is a minefield of lengthy and error-prone operations.

ii. THE MINIMAL POLYNOMIAL METHOD. Find the zeros of the minimal polynomial m_A. Since there are algorithms for finding A and computer programs for finding the zeros of a polynomial, this method can work for values of n such that n^4 is a practical number while $n!$ is not, e.g.,

$$20^4 = 160,000 \ll 20! \approx 2.43 \times 10^{18}.$$

Since small variations in the entries of A can change not merely the coefficients in m_A but even the *degree* μ_A of m_A this method offers its share of difficulty. On the other hand μ_A can be considerably smaller than n (whereas the degree of χ_A is always n). There is an added advantage to the use of the minimal polynomial if the difference between n and μ_A is large.

iii. THE POWER METHOD. For a fixed nonzero vector $\mathbf{x}$ calculate $\mathbf{y}_k \stackrel{\text{def}}{=} A^k\mathbf{x}$ for large values of k. In some circumstances this method works quite well in that via the *components* of the vectors $\mathbf{y}_k$ there can be calculated ratios that are approximations for the corresponding eigenvalue. Later in this **Section** the details of the favorable circumstances are taken up.

iv. THE QR ALGORITHM. This method is also described later. It is viewed with both awe and scepticism. Some use the word "magic" in connection with it. A serious look at the algorithm reveals a number of its flaws, both theoretical and computational. Before applying the QR algorithm one usually touches up A and then, with fingers crossed, hopes for convergence.

v. GERSCHGORIN'S METHOD. Approximate locations of the eigenvalues can be found by means of *Gerschgorin's theorem* discussed below.

Since the algorithm for $m_{A,\mathbf{x}}$ involves the calculation of $\mathbf{y}_k \stackrel{\text{def}}{=} A^k\mathbf{x}$ for a sequence of values of k there is a connection between *ii* and *iii* above.

The reason the power method can be helpful is found in the next result.

 Chapter 5. APPLYING LINEAR ALGEBRA

THEOREM **5.3.1.** LET A BE A DIAGONABLE $n \times n$ MATRIX. ASSUME THAT AMONG ITS DIFFERENT EIGENVALUES, SAY $\lambda_1, \ldots, \lambda_m$, ONE IS LARGER IN ABSOLUTE VALUE THAN ALL OTHERS, e.g., $|\lambda_1| > |\lambda_i|$, $2 \le i \le m$. ASSUME FURTHER THAT THE EIGENSPACE CORRESPONDING TO λ_1 IS 1-DIMENSIONAL:

$$dim[span(\{\, \mathbf{w} \ : \ \mathbf{w} \neq \mathbf{O}, A\mathbf{w} = \lambda_1 \mathbf{w} \,\})] = 1.$$

(THE GEOMETRIC MULTIPLICITY OF λ_1 IS 1.) ONE OF THE STANDARD BASIS VECTORS $\mathbf{e}_1, \ldots, \mathbf{e}_n$, SAY $\mathbf{y}$, HAS A NONZERO $\mathbf{x}_1$-COMPONENT a_1. IF

$$\mathbf{y}_k \overset{\text{def}}{=} A^k \mathbf{y} \overset{\text{def}}{=} \sum_{i=1}^{n} b_{ki} \mathbf{e}_i$$

THERE ARE NUMBERS j AND K, SUCH THAT IF $k > K$ THEN $b_{kj} \neq 0$ AND

$$\lim_{K < k \to \infty} \frac{b_{k+1,j}}{b_{kj}} = \lambda_1$$

$$\lim_{K < k \to \infty} \frac{\mathbf{y}_k}{\lambda_1^k} \overset{\text{def}}{=} \mathbf{v} \ \text{exists}$$

$$A\mathbf{v} = \lambda_1 \mathbf{v}.$$

PROOF. Since A is assumed to be diagonable there is for $\mathbf{C}^n$ a basis $\{\mathbf{x}_1, \ldots, \mathbf{x}_n\} \overset{\text{def}}{=} X$ consisting of A-eigenvectors each of which has length 1. It may be assumed that the notation is such that $A\mathbf{x}_i = r_i \mathbf{x}_i$, $r_1 = \lambda_1$, if $i > 1$ each r_i is one of $\lambda_2, \ldots, \lambda_m$, and $\|\mathbf{x}_i\| = 1$, $1 \le i \le n$.

Each $\mathbf{e}_i$ may be written in the form $\sum_{i=1}^{n} a_{ij} \mathbf{x}_j$. For one of the $\mathbf{e}_i$, say $\mathbf{y}$, the coefficient a_{i1} of $\mathbf{x}_1$ is not 0. (Otherwise X is not a basis.) For simplicity write $\mathbf{y} = \sum_{i=1}^{n} a_i \mathbf{x}_i$. Then:

$$\mathbf{y} = \sum_{i=1}^{n} a_i \mathbf{x}_i$$

$$\mathbf{y}_k \overset{\text{def}}{=} A^k \mathbf{x} = \sum_{i=1}^{n} a_i r_i^k \mathbf{x}_i$$

$$= \lambda_1^k \big(a_1 \mathbf{x}_1 + \sum_{i=2}^{n} a_i \big(\frac{r_i}{\lambda_1} \big)^k \mathbf{x}_i \big).$$

There is a number α such that if $i > 1$ then $\left| \frac{r_i}{\lambda_1} \right| \overset{\text{def}}{=} |\alpha_i| < \alpha < 1$. Hence if

---

$M \stackrel{\text{def}}{=} \sum_{i=2}^{n} |a_i|$ then

$$\| \sum_{i=2}^{n} a_i (\frac{r_i}{\lambda_1})^k \mathbf{x}_i \| \stackrel{\text{def}}{=} \|\mathbf{z}_k\| < \alpha^k M.$$

Hence

$$\lim_{k \to \infty} \frac{\mathbf{y}_k}{\lambda_1^k} = a_1 \mathbf{x}_1 \stackrel{\text{def}}{=} \mathbf{v}$$

$$\lim_{k \to \infty} \frac{A\mathbf{y}_k}{\lambda_1^k} = A\mathbf{v}$$

$$= \lambda_1 \lim_{k \to \infty} \frac{\mathbf{y}_{k+1}}{\lambda_1^{k+1}} = \lambda_1 \mathbf{v}.$$

Since $a_1 \neq 0$ the vector $\mathbf{v}$ is not $\mathbf{O}$ and so if $\mathbf{v} = \sum_{i=1}^{n} c_i \mathbf{e}_i$ one of the c_i, say c_j, is not 0. Hence if $\mathbf{y}_k \stackrel{\text{def}}{=} \sum_{i=1}^{n} b_{ki} \mathbf{e}_i$ then

$$\lim_{k \to \infty} \frac{b_{kj}}{\lambda_1^k} = c_j,$$

$b_{kj} \neq 0$ for all sufficiently large k, say if $k > K$, and

$$\lim_{K < k \to \infty} \frac{b_{k+1,j}}{b_{kj}} = \lim_{K < k \to \infty} \lambda_1 \frac{\frac{b_{k+1,j}}{\lambda_1^{k+1}}}{\frac{b_{kj}}{\lambda_1^k}} = \lambda_1 \frac{c_j}{c_j} = \lambda_1.$$

$$\Omega$$

Exercise 5.3.1. Let A be the matrix

$$\begin{pmatrix} 2 & 1 \\ 0 & 1.9 \end{pmatrix}.$$

The form of A reveals that its eigenvalues are 2 and 1.9. Ignore the obvious and find a general formula for b_{k1} if

$$A^k \mathbf{e}_1 = b_{k1} \mathbf{e}_1 + b_{k2} \mathbf{e}_2.$$

Show that

$$\lim_{k \to \infty} \frac{b_{k+1,1}}{b_{k1}} = 2.$$

[**Note 5.3.1:** Assume A is a SQUARE matrix satisfying the hypotheses of THEOREM **5.3.1**. Assume $A_\alpha \stackrel{\text{def}}{=} A - \alpha I$ satisfies the same hypotheses. For each such α let μ_α be the A_α-eigenvalue of greatest absolute value, let ν_α be one of the eigenvalues of next greatest absolute value, and let R_α be

$$\frac{\nu_\alpha}{\mu_\alpha}.$$

Sometimes one can choose α so that R_α is considerably smaller than R_0. In that event, the convergence of the power method for finding an eigenvalue of A_α is faster than the convergence of the power method applied to A ($= A_0$). Even then, although the eigenvalue μ_α emerges, $\alpha + \mu_\alpha$, the corresponding eigenvalue of A, need not be μ_0 ($= \lambda_1$).]

Exercise 5.3.2. Assume A is an $n \times n$ matrix having only three eigenvalues: $\lambda_1 = 1, \lambda_2 = .9, \lambda_3 = -.9$. Hence $R_0 = 0.9$. Show that if $\alpha \in \mathbf{C}$ then A_α satisfies the hypotheses of Theorem **5.3.1** and that for a suitable choice of α the value of R_α is $0.0270270270\ldots$ and that $\alpha + \mu_\alpha$ is not λ_1.

Exercise 5.3.3. Let U be a 3×3 unitary matrix in which the entries are real (in $\mathbf{R}$), let 1 be an eigenvalue of geometric multiplicity 1 (cf. **Section 4.3**), and let A be $U + .5I$. Show that A satisfies the hypotheses of Theorem **5.3.1** but that for some α the matrix A_α fails to satisfy those hypotheses.

Exercise 5.3.4. Let A be the unitary matrix

$$\begin{pmatrix} \cos\theta & \sin\theta \\ -\sin\theta & \cos\theta \end{pmatrix}.$$

i. Find the eigenvalues of A by finding the zeros of χ_A.
ii. Find a general formula for A^k.
iii. Show that if θ is not an integral multiple of 2π the power method for finding an eigenvalue of A fails.
iv. Show that for any θ there is an α such that the power method succeeds for A_α.

Exercise 5.3.5. Show that if A is a unitary matrix other than I and if one of the eigenvalues of A has geometric multiplicity 1 then the power method fails but that for a suitable α the power method succeeds for A_α.

Exercise 5.3.6. Let $A^{(n)}$ be the unitary matrices

$$\begin{pmatrix} 0 & (-1)^n \\ 1 & 0 \end{pmatrix}, \quad n = 1, 2.$$

Show that if $\Re\alpha \neq 0$ and $\Im\alpha = 0$, i.e., if α is real and not 0, then $A_\alpha^{(1)}$ fails to satisfy the hypotheses of Theorem **5.3.1**. Show that if $\Re\alpha = 0$ and $\Im\alpha \neq 0$, i.e., α is "purely imaginary" and not 0, then $A_\alpha^{(2)}$ fails to satisfy the hypotheses of Theorem **5.3.1**.

The reader may well wonder about the applicability of the power method. The **Exercises** above suggest that there are many situations in which the power method fails. Furthermore, the assurance provided by Theorem **5.3.1** is rather theoretical. If one has no inkling about the eigenvalues of A or whether, indeed, A is diagonable, it is impossible to confirm or deny that A falls in the domain of the THEOREM.

On the other hand, from a probabilistic standpoint, just as there is "no chance" that a matrix is singular or that a matrix is not diagonable, there is "no chance" that a matrix A is one for which the power method fails. In practice the eigenvalues of a matrix are of interest because of their relevance to the problem being treated. Familiarity with the practical aspects of the problem can often provide more than an inkling about the location of the eigenvalues and thus also about the applicability of the power method.

Even when the power method must succeed for a matrix A, computational considerations require that A be sparse, i.e., that most of the entries of A are 0's. The *Householder* matrices described next serve to "sparsify" a nonsparse matrix.

LEMMA **5.3.1.** If $\mathbf{u} \in \mathbf{C}^n$ AND $|\mathbf{u}| = 1$ THE MATRIX $H_{\mathbf{u}} \overset{\text{def}}{=} I - 2\mathbf{uu}^*$ IS A SELF-ADJOINT UNITARY MATRIX, i.e., IT IS ITS OWN INVERSE.

PROOF.
$$H_{\mathbf{u}}^* = I^* - 2(\mathbf{u}^*)^*\mathbf{u}^* = H_{\mathbf{u}}$$
$$H_{\mathbf{u}}H_{\mathbf{u}}^* = H_{\mathbf{u}}^2 = I - 4\mathbf{uu}^* + 4(\mathbf{uu}^*)\mathbf{uu}^*$$
$$= I - 4\mathbf{uu}^* + 4\mathbf{u}(\mathbf{u}^*\mathbf{u})\mathbf{u}^*$$
$$I - 4\mathbf{uu}^* + 4\mathbf{u} \cdot 1 \cdot \mathbf{u}^* = I.$$

$$\Omega$$

[**Note 5.3.2:** If $\|\mathbf{x}\| = \alpha \neq 0$ and $\mathbf{u} \overset{\text{def}}{=} \frac{\mathbf{x}}{\alpha}$ then $I - 2\mathbf{uu}^*$ is the *Householder matrix* denoted $H_{\mathbf{x}}$.]

LEMMA **5.3.2.** ASSUME

$$\mathbf{x} \overset{\text{def}}{=} \begin{pmatrix} x_1 \\ \vdots \\ x_n \end{pmatrix} \neq \mathbf{O}, \ \|\mathbf{x}\| = \alpha,$$

AND THAT $x_1 = re^{i\theta}$. THEN

$$H_{\mathbf{x}+\alpha e^{i\theta}\mathbf{e}_1}\mathbf{x} = -\alpha e^{i\theta}\mathbf{e}_1.$$

PROOF. - Since

$$\|\mathbf{x} + \alpha e^{i\theta}\mathbf{e}_1\|^2 = 2\alpha^2 + 2\alpha r = 2(\mathbf{x}^* + \alpha e^{-i\theta}\mathbf{e}_1)\mathbf{x}$$

it follows that

$$
\begin{aligned}
H_{\mathbf{x}+\alpha e^{i\theta}\mathbf{e}_1}\mathbf{x} &= \mathbf{x} - \frac{2}{\|\mathbf{x} + \alpha e^{i\theta}\mathbf{e}_1\|^2}(\mathbf{x} + \alpha e^{i\theta}\mathbf{e}_1)(\mathbf{x}^* + \alpha e^{-i\theta}\mathbf{e}_1)\mathbf{x} \\
&= \mathbf{x} - \frac{2}{2\alpha^2 + 2\alpha r}[(\mathbf{x}^* + \alpha e^{-i\theta}\mathbf{e}_i)\mathbf{x}] \cdot (\mathbf{x} + \alpha e^{i\theta}\mathbf{e}_1) \\
&= \mathbf{x} - (\mathbf{x} + \alpha e^{i\theta}\mathbf{e}_1) = -\alpha e^{i\theta}\mathbf{e}_1.
\end{aligned}
$$

$$\Omega$$

Call the matrix $H_{\mathbf{x}+\alpha e^{i\theta}\mathbf{e}_1}$ the *Householder matrix based on* $\mathbf{x}$. It rotates the vector $\mathbf{x}$ into the vector $\alpha e^{i\theta}\mathbf{e}_1$, a vector with only one nonzero component (and of the same length as the length of $\mathbf{x}$).

The power method for finding an eigenvalue of a SQUARE matrix A is speeded with the use of Householder matrices to sparsify A by producing a new matrix A_S *unitarily similar* to A and in which almost half the entries are 0's. If, to boot, $A = A^*$ then $A_S = A_S^*$ and it is in *tridiagonal* or *Hessenberg* form. Thus at worst A_S looks like this:

$$
\begin{pmatrix}
* & * & \cdots & * & * \\
* & * & \cdots & * & * \\
  & * & \cdots & * & * \\
  &   & \ddots & \vdots & \vdots \\
  &   &   & * & *
\end{pmatrix} \ ;
\tag{5.3.1}
$$

if $i - j > 1$ then the ij entry is 0.

If $A = A^*$ then A_S looks like this:

$$
\begin{pmatrix}
* & * & & & \\
* & * & * & & \\
  & \ddots & \ddots & \ddots & \\
  & & * & * & * \\
  & & & * & *
\end{pmatrix} \ ;
\tag{5.3.2}
$$

if $|i - j| > 1$ the ij entry is 0.

It is shown below that there is a unitary matrix U such that $U = U^*$ and such that $UAU^* = A_S$. The calculation of A^k is reduced to the calculation of $UA_S^k U$. The calculation of A_S^k is eased by virtue of the sparseness of A_S and so the power method moves more swiftly.

To find A_S first construct the $n - 1 \times n - 1$ Householder matrix $H^{(1)}$ based on the $n - 1 \times 1$ column vector $\mathbf{a}_1'$ consisting of the last $n - 1$ entries in the first column of A. Then let U_1 be the $n \times n$ block matrix

$$
\begin{pmatrix}
I & \\
& H^{(1)}
\end{pmatrix} .
$$

Owing to the form of U_1, the matrix $A_1 \stackrel{\text{def}}{=} U_1 A U_1$ is a block matrix of the form

$$
= \begin{pmatrix}
\begin{pmatrix} a_{11} I \\ -|\mathbf{a}_1'| \\ 0 \\ \vdots \\ 0 \end{pmatrix} & B_1 \\
& D_1
\end{pmatrix} .
$$

Treat the $n-1 \times n-1$ matrix D_1 with an appropriate $n-2 \times n-2$ Hessenberg matrix $H^{(2)}$. Then let U_2 be the block matrix

$$
\begin{pmatrix}
\begin{pmatrix} 1 & \\ & 1 \end{pmatrix} & \\
& H^{(2)}
\end{pmatrix}
$$

and set A_2 to be $U_2 A_1 U_2$. After $n-1$ performances of such operations there emerges a matrix A_S of the form (5.3.1), and, if $A = A^{\ast}$ then A_S has the form in (5.3.2).

> [**Remark 5.3.1:** If the procedure just described is applied by starting with the first column $\mathbf{a}_1$ of A (rather than with $\mathbf{a}_1'$) and if the corresponding operations are carried out n times rather than $n-1$ times then there emerges an upper triangular matrix R. The product Q of the unitary matrices used is such that $Q^{\ast} A \stackrel{\text{def}}{=} R$ is upper triangular and $QR \; (=A)$ is the QR factorization of A.]

The QR algorithm is so appealing because it is so easy to describe.

If A is an $n \times n$ matrix let $Q_0 R_0$ be a QR factorization of A and define A_1 by the formula $A_1 = R_0 Q_0$. Then $A_1 = Q_0^{-1} A Q_0$, i.e., A_1 is similar, indeed, unitarily equivalent, to A. In general, having found A_n let $Q_n R_n$ be a QR factorization of A_n and set A_{n+1} to be $R_n Q_n$. Then the sequence $A_1, A_2, \ldots$ is a sequence of matrices, each unitarily equivalent to A.

THEOREM 5.3.2. IF A IS AN $n \times n$ MATRIX AND IF THE UNITARY MATRICES

$$
\mathcal{Q}_n \stackrel{\text{def}}{=} \prod_{k=0}^{n} Q_k
$$

(THE PRODUCTS OF THE UNITARY MATRICES Q_k USED IN THE QR ALGORITHM) CONVERGE TO A LIMIT MATRIX $\mathcal{Q}$ THEN THE MATRICES $A_n \stackrel{\text{def}}{=} \mathcal{Q}_{n-1}^{\ast} A \mathcal{Q}_{n-1}$ CONVERGE TO AN UPPER TRIANGULAR MATRIX A_∞ IN WHICH THE DIAGONAL ENTRIES ARE THE EIGENVALUES OF A.

  Chapter 5. APPLYING LINEAR ALGEBRA

PROOF. Since

$$Q_n Q_{n+1} = Q_{n+1}$$
$$Q_{n+1} = Q_n^{\times} Q_{n+1}$$

it follows that $\lim_{n \to \infty} Q_n = Q^{\times} Q = I$. Furthermore

$$\lim_{n \to \infty} A_n = \lim_{n \to \infty} Q_{n-1}^{\times} A Q_{n-1} = Q^{\times} A Q \stackrel{\text{def}}{=} A_{\infty}.$$

Hence all the limits in the next display exist:

$$\lim_{n \to \infty} Q_n^{\times} A_n = \lim_{n \to \infty} R_n \stackrel{\text{def}}{=} R_{\infty} = A_{\infty}.$$

Thus $A_{\infty} = R_{\infty}$, A_{∞} is upper triangular, similar to A, and so the diagonal entries of A_{∞} are the eigenvalues of A.

$$\Omega$$

The charm of THEOREM **5.3.2** begins to fade after consideration of the following facts:

i. If all the entries of A are real but some of the eigenvalues of A are not real, e.g., if

$$A = \begin{pmatrix} 0 & -1 \\ 1 & 0 \end{pmatrix} \text{ (for which the eigenvalues are } \pm i),$$

then the QR algorithm applied to A must be such as to produce matrices A_n with complex diagonal entries for large n. If the entries of A are real the customary QR factorizations, e.g., via the Gram-Schmidt process, produce only real entries in the Q_n and the R_n In these circumstances the desired convergence cannot take place.

ii. If A is unitary, e.g., the A in i, and $A = QR$ via the customary QR factorization then $Q = A, R = I$, and the QR algorithm does not get off the ground. However if A is unitary its eigenvalues lie on the *unit circle* $\mathsf{T} \stackrel{\text{def}}{=} \{ z \ : \ z \in \mathsf{C}, |z| = 1 \}$. If A is not a already a diagonal matrix there is a nonzero α such that $0 < |\alpha| < 1$ and such that $A - \alpha I$ satisfies the hypotheses of THEOREM **5.3.1**.

iii. There are results other than THEOREM **5.3.2** that guarantee the success of the QR algorithm if certain conditions about A and the eigenvalues of A are known to obtain. For example, if

   a. A has n distinct eigenvalues λ_i;
   b. $|\lambda_1| > \cdots > |\lambda_n|$;
   c. there is an invertible matrix P such that $P^{-1}AP \stackrel{\text{def}}{=} D$ is diagonal and $P^{-1} = LU$ in which the diagonal entries of L are 1's;

then, although the matrices A_n need not converge, their diagonal entries do converge to the eigenvalues of A!

Not surprisingly, the proof of this result is not offered here.

Once the Householder matrices are used to produce A_S, the QR algorithm is applied to A_S rather than to A.

Exercise 5.3.7. Sparsify each of the matrices below by the use of Householder matrices:

$$
i. \quad \begin{pmatrix} 1 & 2 & 3 \\ 4 & 5 & 6 \\ 7 & 8 & 9 \end{pmatrix} ; \quad ii. \quad \begin{pmatrix} 1 & 2 & 3 \\ 2 & 4 & 5 \\ 3 & 5 & 6 \end{pmatrix} .
$$

(The result for ii should be in tridiagonal (Hessenberg) form.)

Exercise 5.3.8. Carry out the first two cycles of the QR algorithm for the sparsified versions of i and ii in **Exercise 5.3.7**.

Exercise 5.3.9. Assume $A \overset{\text{def}}{=} (a_{ij})_{i,j=1}^{2,2}$. Find A_S.

Exercise 5.3.10. Let A be the matrix in **Exercise 5.3.4**. Find A_S. Find an α such that $A_S - \alpha I$ can be successfully treated by the QR algorithm.

Gerschgorin's theorem may be stated as follows:

THEOREM **5.3.3.** IF $A \overset{\text{def}}{=} (a_{ij})_{i,j=1}^{n,n}$ AND IF r_i IS THE SUM OF THE AB-SOLUTE VALUES OF THE NONDIAGONAL ENTRIES IN THE iTH ROW OF A, i.e., IF

$$
r_i \overset{\text{def}}{=} \sum_{j \neq i} |a_{ij}|,
$$

THEN EVERY EIGENVALUE OF A LIES IN (SOME) ONE OF THE CLOSED DISCS

$$
D_i \overset{\text{def}}{=} \{ z \ : \ z \in \mathbf{C}, |z - a_{ii}| \leq r_i \}, \ 1 \leq i \leq n.
$$

PROOF. If

$$
\mathbf{x} \overset{\text{def}}{=} \begin{pmatrix} x_1 \\ \vdots \\ x_n \end{pmatrix} \neq \mathbf{O},
$$

$A\mathbf{x} = \lambda \mathbf{x}$, and $|x_i| \geq maximum_{j \neq i} |x_j|$ then

$$(\lambda - a_{ii})x_i = \sum_{j \neq i} a_{ij} x_j$$

$$\lambda - a_{ii} = \sum_{j \neq i} a_{ij} \frac{x_j}{x_i}$$

$$|\lambda - a_{ii}| \leq \sum_{j \neq i} |a_{ij}| = r_i.$$

$$\Omega$$

If the r_i are small then the search for the eigenvalues is confined to the discs D_i. In particular, if A is a diagonal matrix then each $r_i = 0$ and the diagonal elements are the eigenvalues of A.

Example 5.3.1. If A is a unitary matrix then each Gerschgorin radius $r_i \leq \sqrt{n}$ and each $|a_{ii}| \leq 1$ whence the D_i are discs of reasonable size and their centers are all inside the *standard unit disc*

$$\Delta_1 \stackrel{\text{def}}{=} \{ z \; : \; z \in \mathbf{C}, \; |z| \leq 1 \}.$$

(Note that all the eigenvalues of a unitary matrix lie on the circumference $\mathbf{T}$ of the disc Δ_1: {eigenvalues of A} $\subset \mathbf{T} \stackrel{\text{def}}{=} \partial \Delta_1$.

If A is upper or lower triangular then its diagonal entries are the eigenvalues of A but the Gerschgorin theorem might provide an absurd set of discs D_i to explore. (True, if A is upper triangular then $r_n = 0$ and if A is lower triangular then $r_1 = 0$. But then only one eigenvalue is narrowly, indeed, precisely, located by use of Gerschgorin's theorem.) More to the point is the next **Example**.

Example 5.3.2. Assume

$$A \stackrel{\text{def}}{=} \begin{pmatrix} a_{11} & M_1 & \\ & a_{22} & M_2 \\ & M_3 & a_{33} \end{pmatrix}$$

the Gerschgorin radii r_i are $|M_1|, |M_2|, |M_3|$ whereas

$$\chi_A(z) = (a_{11} - z)[(a_{22} - z)(a_{33} - z) - M_2 M_3]$$

One the eigenvalues is a_{11}. Knowing it one can find the other two as solutions of a quadratic equation. If the $|M_i|$ are large the Gerschgorin radii r_i are large and provide little help in locating any of the eigenvalues.

On the other hand if the diagonal elements are large in absolute value and the r_i are small the Gerschgorin radii provide some estimates having small relative errors.

Example 5.3.3. If all diagonal elements have absolute values that are at least 100 and if the Gerschgorin radii are all not more than 1, *any z in some*

Gerschgorin disc provides an approximant to an eigenvalue and the corresponding relative error is not more than 0.02.

Example 5.3.4. If

$$A(\epsilon) \overset{\text{def}}{=} \begin{pmatrix} 1 & 0 & M_1 \\ & 2 & M_2 \\ \epsilon & & 3 \end{pmatrix}$$

and then $\|A(\epsilon) - A(0)\|_2 = |\epsilon|$. If $M_1 = M_2 = 10^{200}$ and $\epsilon = 10^{-100}$ then the eigenvalues of $A(0)$ are 1,2, and 3. The eigenvalues of $A(\epsilon)$ are 2 and approximately $\pm 10^{50}$. The Gerschgorin disc centered at 3 is minuscule and yet contains *no* eigenvalues of $A(\epsilon)$. The other Gerschgorin radii are 10^{200} and provide little guidance toward estimating any of the eigenvalues. The matrices $A(\epsilon)$ demonstrate the instability of eigenvalues as functions of the entries. Although *sufficiently* small variations in the entries do not perturb the eigenvalues much, some *relatively* small variations produce gross perturbations in the eigenvalues. As functions of the entries some of the eigenvalues can have some partial derivatives of enormous absolute value.

If any inspiration is to be drawn from reading the **Sections 5.1 - 5.3** it may be directed to the improvement in computational techniques for applying linear algebra. A good first step is browsing in the vast literature, some of which is cited in **Section 5.1**.

5.4. Linear programming and game theory

In **Section 1.1, Examples 2.3.7** and **2.3.8**, and in **Exercises 2.3.11 - 2.3.14** the subject of linear programming is given a short introduction. At the root of the matter is the analysis of systems of linear *inequalities* rather than systems of linear *equations*. (To emphasize this distinction a few writers call inequalities *inequations*.) The simplest and most realistic context in which inequalities are studied is the set $\mathbf{R}$ of *real* numbers. The corresponding vector spaces are thus $\mathbf{R}^n$ and $\mathbf{R}_n$.

The system of linear equations

$$
\begin{aligned}
(1): &\quad x - 2y = 2 \ (L_1) \\
(2): &\quad x - y = 3 \ (L_2) \\
(3): &\quad 2x + y = 12 \ (L_3) \\
(4): &\quad -x + 2y = 9 \ (L_4) \\
(5): &\quad -x + y = 4 \ (L_5) \\
(6): &\quad x = 0 \ (L_6) \\
(7): &\quad y = 0 \ (L_7)
\end{aligned}
\qquad (5.4.1)
$$

represents a finite set of lines L_1 - L_7 in $\mathbf{R}^2$. The system of corresponding *inequalities*

$$
\begin{aligned}
(1): & \quad x - 2y \leq 2 \ (S_1) \\
(2): & \quad x - y \leq 3 \ (S_2) \\
(3): & \quad 2x + y \leq 12 \ (S_3) \\
(4): & \quad -x + 2y \leq 9 \ (S_4) \\
(5): & \quad -x + y \leq 4 \ (S_5) \\
(6): & \quad x \geq 0 \ (S_6) \\
(7): & \quad y \geq 0 \ (S_7)
\end{aligned}
\tag{5.4.2}
$$

taken from **Example 2.3.7** represents a finite set of *half-spaces* S_1 - S_7 in $\mathbf{R}^2$. Some lines and half-spaces are indicated in **Figure 5.4.1**.

The variously positioned $\triangle$s are used to indicate what the half-spaces are. A reversal ($\leq \rightarrow \geq$) of an inequality sign gives rise to a change of from one of the two half-spaces determined by the corresponding line to the other. The $\triangle$s are moved from one side of the corresponding line to the other.

The intersection Q of the half-spaces S_1 - S_7 varies accordingly.

Some of the different forms of Q are shown in **Figure 5.4.2** and **Figure 5.4.3**. In **Figure 5.4.2** Q is unbounded. In **Figure 5.4.3** Q is empty (L_1 and L_4 are parallel)!

If P and Q are two points in a half-space S then the line segment $\overline{PQ}$ is a subset of S. Hence if $\{S_\lambda\}_{\lambda \in \Lambda}$ is a set of half-spaces and if

$$
C \overset{\text{def}}{=} \bigcap_{\lambda \in \Lambda} S_\lambda
$$

is their intersection it follows that if P and Q are in C then:

   i. P and Q are in each S_λ;
  ii. the line segment $\overline{PQ}$ is contained in each S_λ;
 iii. the line segment $\overline{PQ}$ is contained in C.

Half-spaces and their intersections are special instances of *convex* sets, to be discussed following the **Figures**.

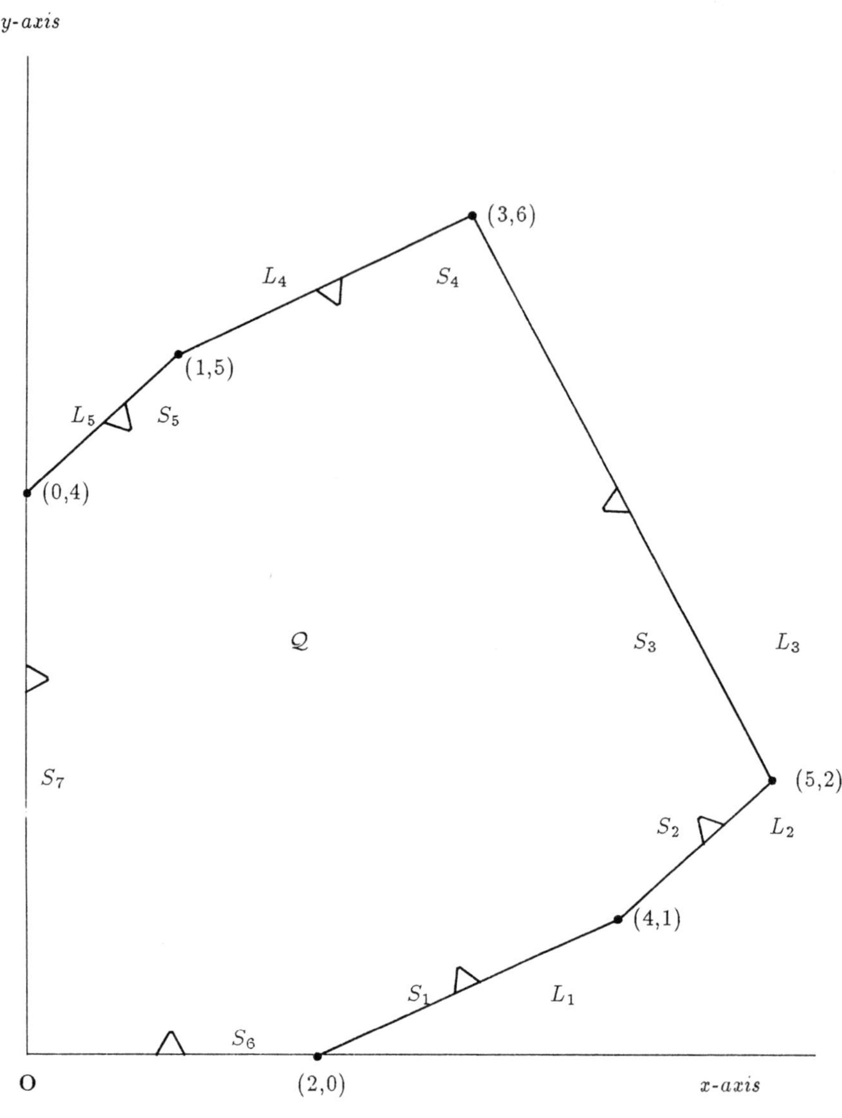

Figure 5.4.1.

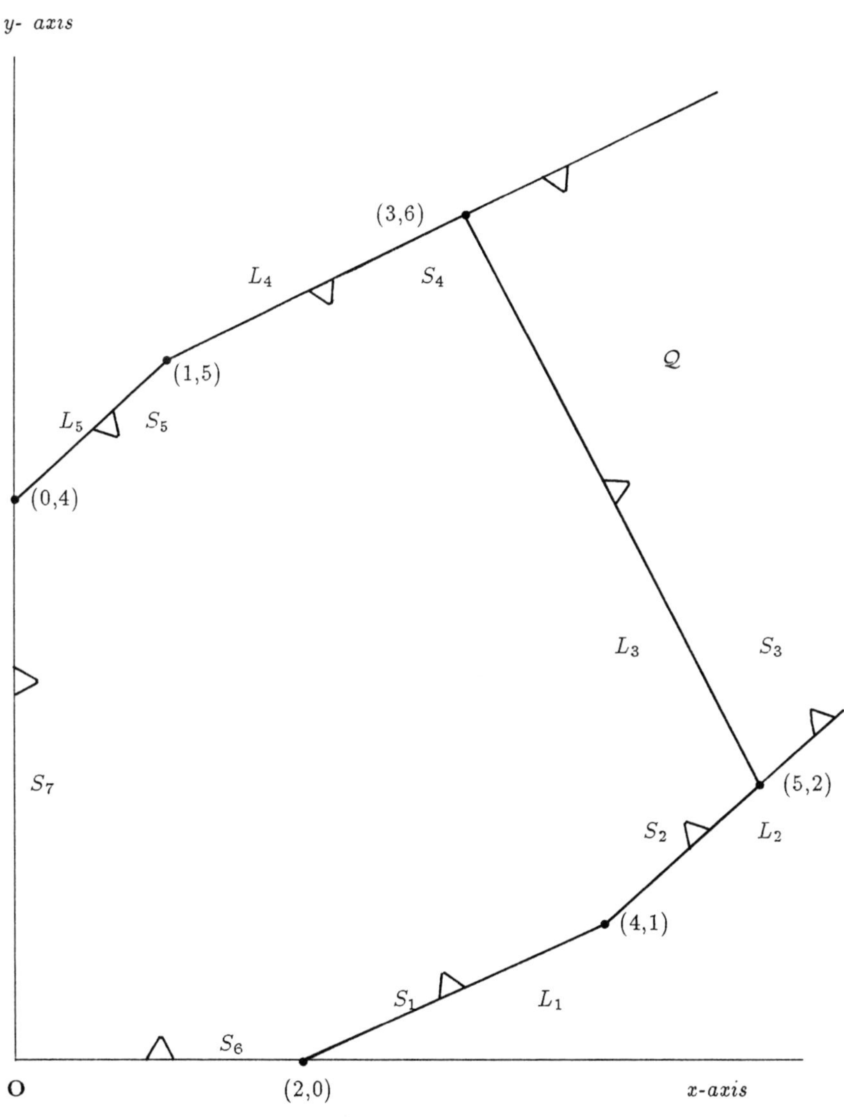

Figure 5.4.2.

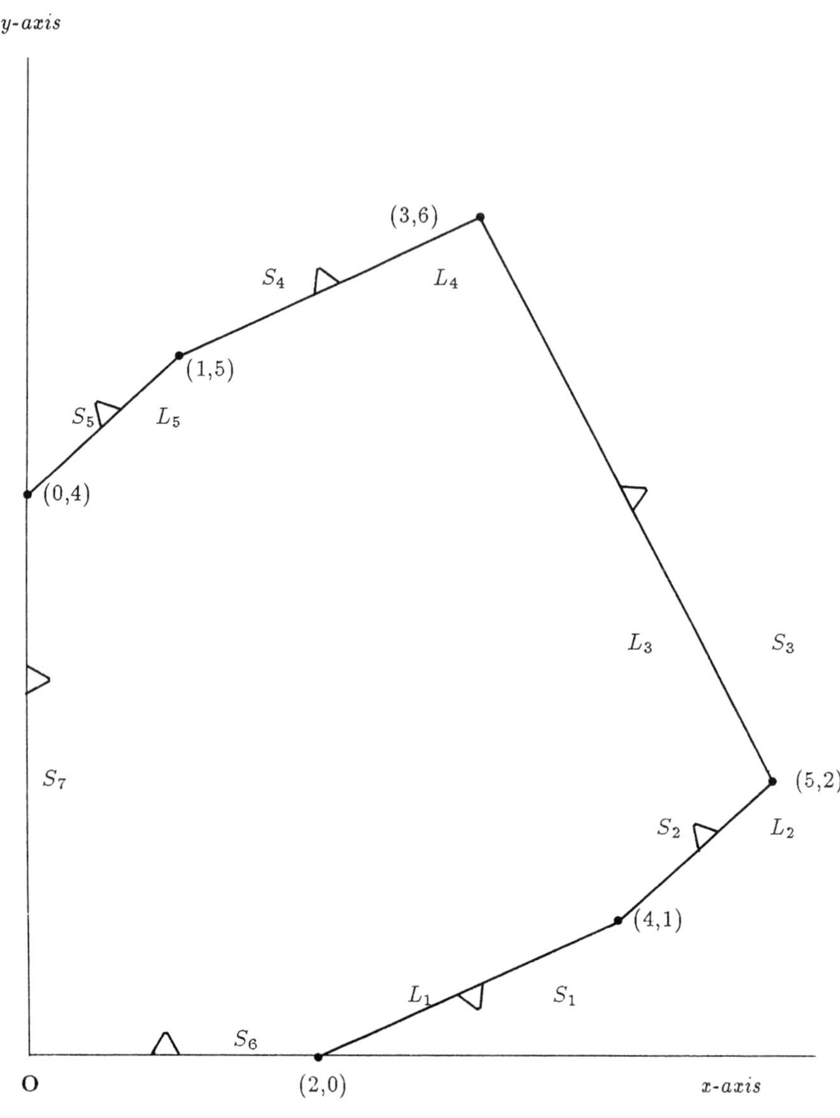

Figure 5.4.3.

In **Section 2.2** there is a passing reference to the convex set Q associated with the nutrition problem.

DEFINITION 5.4.1. A SET Q IN $\mathbf{C}^n$ IS CONVEX IFF WHENEVER x AND y ARE IN Q AND $0 \le t \le 1$ THEN $t\mathbf{x} + (1-t)\mathbf{y} \in Q$.

Example 5.4.1.

i. The entire vector space $\mathbf{C}^n$ is convex.

ii. A straight line or a straight line segment is convex.

iii. If

$$\mathbf{a} \overset{\text{def}}{=} \begin{pmatrix} a_1 \\ \vdots \\ a_n \end{pmatrix} \in \mathbf{R}^n,$$

and if $r \in \mathbf{R}$ then in $\mathbf{R}^n$ the set

$$S_{[\mathbf{a},r]} \overset{\text{def}}{=} \left\{ \mathbf{x} \ : \ \mathbf{x} \in \mathbf{R}^n, \ (\mathbf{a},\mathbf{x}) \ge r \right\}$$

is convex. The set $S_{[\mathbf{a},r]}$ is the *nonnegative half-space* determined by a and r. Since

$$S_{[\mathbf{a},r]} = \left\{ \mathbf{x} \ : \ \mathbf{x} \in \mathbf{R}^n, \ (\mathbf{a},\mathbf{x}) > r \right\} \cup \left\{ \mathbf{x} \ : \ \mathbf{x} \in \mathbf{R}^n, \ (\mathbf{a},\mathbf{x}) = r \right\}$$

$$\overset{\text{def}}{=} S^o_{[\mathbf{a},r]} \cup H_{[\mathbf{a},r]}$$

$S^o_{[\mathbf{a},r]}$ is called the *open half-space determined by* a *and* r and $H_{[\mathbf{a},r]}$ is called the *hyperplane determined by* a *and* r. The sets $S^o_{[\mathbf{a},r]}$ and $H_{[\mathbf{a},r]}$ are disjoint. Each is convex.

iv. The set consisting of the perimeter of a circle and its interior or of its interior alone is convex.

Exercise 5.4.1. Show that each of the sets described in **Example 5.4.1** is convex.

Exercise 5.4.2. Show that each of the following sets is convex:

i. the intersection of an arbitrary set of convex sets in $\mathbf{C}^n$;

ii. the set consisting of the perimeter and the interior of a triangle;

iii. the interior of a triangle;

iv. the complement $\mathbf{R}^n \setminus S_{[\mathbf{a},r]}$ in $\mathbf{R}^n$ of $S_{[\mathbf{a},r]}$.

Exercise 5.4.3. Show that each of the following sets is *not* convex:

i. the perimeter of a circle;

ii. the perimeter of a triangle;

iii. the complement in $\mathbf{R}^2$ of a straight line.

If $\{\mathbf{x}_1, \ldots, \mathbf{x}_k\} \subset \mathbf{C}^n$ and if

$$\alpha \stackrel{\text{def}}{=} (\alpha_1, \ldots, \alpha_k) \in \mathbf{R}_k, \ \alpha_i \geq 0, \ 1 \leq i \leq k, \ \text{and} \ \sum_{i=1}^{k} \alpha_i = 1 \qquad (5.4.3)$$

then

$$\sum_{i=1}^{k} \alpha_i \mathbf{x}_i,$$

which is a special kind of linear combination of the vectors $\mathbf{x}_i$, is called a *convex combination* of the $\mathbf{x}_i$. If S is a subset of $\mathbf{C}^n$ then $Conv(S)$, the *convex hull* of S is the set of all convex combinations of vectors in S. The set of all vectors α satisfying the conditions in (5.4.3) is

$$\Sigma_k \stackrel{\text{def}}{=} \left\{ (\alpha_1, \ldots, \alpha_k) \ : \ \alpha_i \geq 0, \ 1 \leq i \leq k, \ \sum_{i=1}^{k} \alpha_i = 1 \right\}$$

and is frequently termed the *basic simplex* in $\mathbf{R}_k$. Similarly $\Sigma^{(l)}$ is defined according to the equations

$$\Sigma^{(l)} \stackrel{\text{def}}{=} \left\{ \beta \ : \ \beta \in \mathbf{R}^l, \ \beta^t \in \Sigma_l \right\}$$

$$= \left\{ \begin{pmatrix} \beta_1 \\ \vdots \\ \beta_l \end{pmatrix} \ : \ \beta_j \geq 0, \ 1 \leq j \leq l, \sum_{j=1}^{l} \beta_j = 1 \right\}.$$

Example 5.4.2. If $S = \{\mathbf{e}_1, \ldots, \mathbf{e}_k\}$ is the set of standard basis vectors in $\mathbf{R}_k$ then Σ_k is the convex hull of S: $\Sigma_k = Conv(S)$.

Exercise 5.4.4.

i. Show that for any set S in $\mathbf{C}^n$ the convex hull $Conv(S)$ is convex.

[*Hint:* There is in Σ_k an α and in Σ_l a β and in S vectors $\mathbf{u}_1, \ldots, \mathbf{u}_k$ and $\mathbf{v}_1, \ldots, \mathbf{v}_l$, such that

$$\alpha_1 \mathbf{u}_1 + \cdots \alpha_k \mathbf{u}_k = \mathbf{x}$$
$$\beta_1 \mathbf{v}_1 + \cdots \beta_l \mathbf{v}_l = \mathbf{y}.$$

For t in $[0, 1]$ calculate $t\mathbf{x} + (1 - t)\mathbf{y}$.]

ii. Plot Σ_m in $\mathbf{R}_m$ for $1 \leq m \leq 3$. Give verbal descriptions of Σ_m, $1 \leq m \leq 3$.

     Chapter 5. APPLYING LINEAR ALGEBRA

Exercise 5.4.5. Show a set S in $\mathbf{C}^n$ is convex

iff $S = Conv(S)$

iff $S =$ the intersection of all convex sets containing S.

Exercise 5.4.6. Show that for any $\mathbf{y}$ in $\mathbf{R}^n$ and any r in $\mathbf{R}$ there is in $\mathbf{R}^n$ a $\mathbf{z}$ and in $\mathbf{R}$ an l such that

$$\{\,\mathbf{x}\ :\ (\mathbf{z},\mathbf{x}) \le l\,\} = S_{[\mathbf{y},r]}.$$

What are simple descriptions of $\mathbf{z}$ and l in terms of $\mathbf{y}$ and r?

If $\{g_1,\ldots,g_K\}$ is a finite set of real numbers and if $\mathbf{w} \overset{\text{def}}{=} (w_1,\ldots,w_K) \in \Sigma_K$ then $\overline{g}_{\mathbf{w}} \overset{\text{def}}{=} w_1 g_1 + \ldots w_K g_K$ is the *weighted average* of the g_k weighted by the w_k. The weighted average $\overline{g}_{\mathbf{w}}$ satisfies the inequalities

$$minimum_{1 \le k \le K}\ \{g_k\} \le \overline{g}_{\mathbf{w}} \le maximum_{1 \le k \le K}\ \{g_k\}. \qquad (5.4.4)$$

The relations in (5.4.4) constitute the *averaging principle*:

"worst grade $\le$ grade-point average $\le$ best grade."

[*Proof*:

$$
\begin{aligned}
minimum_{1 \le k \le K}\ \{g_k\} &= minimum_{1 \le k \le K}\ \{g_k(w_1 + \cdots + w_K)\} \\
&= w_1\ minimum_{1 \le k \le K}\ \{g_k\} + \cdots \\
&\quad + w_K\ minimum_{1 \le k \le K}\ \{g_k\} \\
&\le w_1 g_1 + \cdots + w_K g_K = \overline{g}_{\mathbf{w}} \\
&\le maximum_{1 \le k \le K}\ \{g_k(w_1 + \cdots + w_K)\} \\
&= maximum_{1 \le k \le K}\ \{g_k\}.]
\end{aligned}
$$

The essential fact about convex sets in $\mathbf{R}^n$ is given in the next result. The ideas are simple and intuitive and may be described as follows.

If Q is the convex hull of a finite set S of vectors and $\mathbf{O} \notin Q$ then there is a hyperplane $H_{[\mathbf{v},r]}$ that separates $\mathbf{O}$ from Q, i.e., $Q \subset S_{[\mathbf{v},r]}$ and $\mathbf{O}$ is on the "other" side of $H_{[\mathbf{v},r]}$: $\mathbf{O} \in \mathbf{R}^n \setminus S_{[\mathbf{v},r]}$.

In **Figure 5.4.4** there is an attempt to show schematically what happens when a convex set of the special (*polyhedral*) form of Q fails to include $\mathbf{O}$.

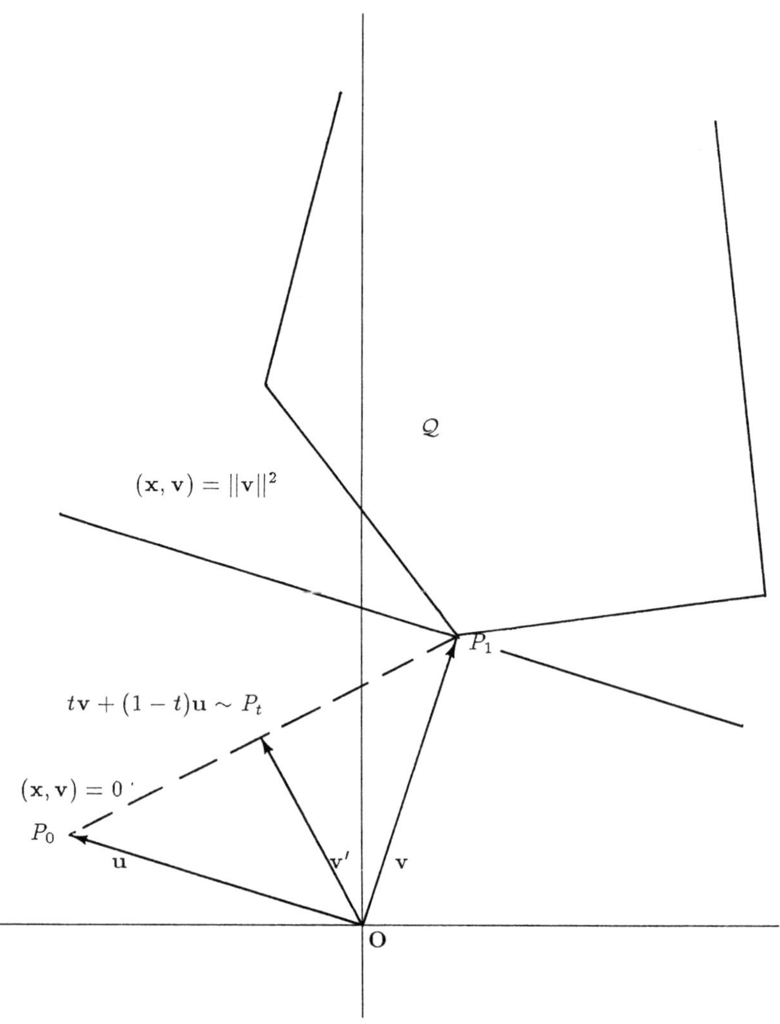

Figure 5.4.4.

What is intended is that there is in $\mathcal{Q}$ a vector $\mathbf{v}$ of minimal length, i.e., that there is in $\mathcal{Q}$ a point P_1 closest to the origin $\mathbf{O}$ and that the hyperplane through P_1 and perpendicular to the line containing $\mathbf{v}$ separates $\mathcal{Q}$ from $\mathbf{O}$. The equation of the hyperplane is $(\mathbf{v}, \mathbf{x}) = \|\mathbf{v}\|^2$. The hyperplane is thus $H_{[\mathbf{v}, \|\mathbf{v}\|^2]}$. It is parallel to the special hyperplane, indeed subspace, $\{\,\mathbf{y}\ :\ (\mathbf{v}, \mathbf{y}) = 0\,\} = H_{[\mathbf{v}, 0]}$, i.e., the set of all vectors perpendicular to $\mathbf{v}$.

It *seems* that the convex set $\mathcal{Q}$ is contained in the half-space $S_{[\mathbf{v},\|\mathbf{v}\|^2]}$. The intuition is valid and the precise statement is given next.

THEOREM 5.4.1. LET A BE A FINITE SET $\{\mathbf{a}_1, \ldots, \mathbf{a}_m\}$ OF VECTORS IN $\mathbf{R}^n$ AND LET $\mathcal{Q}$ BE THE CONVEX HULL OF A: $\mathcal{Q} \overset{\text{def}}{=} Conv(A)$. IF $\mathbf{O} \notin \mathcal{Q}$ THERE IS IN $\mathcal{Q}$ A VECTOR $\mathbf{v}$ SUCH THAT $\mathcal{Q} \subset S_{[\mathbf{v},\|\mathbf{v}\|^2]}$ AND $\mathbf{O} \notin S_{[\mathbf{v},\|\mathbf{v}\|^2]}$.

[**Remark 5.4.1:** Since $\mathbf{O} \notin \mathcal{Q}$ it follows that $\|\mathbf{v}\|^2 > 0$ and since $(\mathbf{v}, \mathbf{O}) = 0$ the vector $\mathbf{O}$ is on the "other" side of $H_{[\mathbf{v},\|\mathbf{v}\|^2]}$.]

PROOF. There is a nonnegative number d such that if $\mathbf{x} \in \mathcal{Q}$ then $\|\mathbf{x}\| \geq d$. Let D be the supremum of all such d. Hence there is in $\mathcal{Q}$ a sequence $\{\mathbf{v}_p\}_{p=1}^{\infty}$ such that $\lim_{p \to \infty} \|\mathbf{v}_p\| = D$. Because $\mathcal{Q}$ is convex it follows that for any p and q in $\mathbf{N}$

$$\frac{1}{2}(\mathbf{v}_p + \mathbf{v}_q) \in \mathcal{Q}. \tag{5.4.5}$$

The THEOREM *of Apollonius* to the effect that the sum of the squares of the lengths of the diagonals of a parallelogram is the same as the sum of the squares of the lengths of the sides of the parallelogram has a simple interpretation and proof derived from E1-E4:

$$\|\mathbf{x} + \mathbf{y}\|^2 + \|\mathbf{x} - \mathbf{y}\|^2 = 2(\|\mathbf{x}\|^2 + \|\mathbf{y}\|^2) \tag{5.4.6}$$

which follows from a careful calculation of the left member of (5.4.6).

Hence, by virtue of (5.4.6),

$$\|\frac{1}{2}(\mathbf{v}_p - \mathbf{v}_q)\|^2 = \frac{1}{2}\|\mathbf{v}_p\|^2 + \frac{1}{2}\|\mathbf{v}_q\|^2 - \|\frac{1}{2}(\mathbf{v}_p + \mathbf{v}_q)\|^2$$

$$\leq \frac{1}{2}(\|\mathbf{v}_p\|^2 + \|\mathbf{v}_q\|^2) - D^2. \tag{5.4.7}$$

As $p, q \to \infty$ each of the first two terms in the right member of (5.4.7) approaches $\frac{1}{2}D^2$ and thus the right member of (5.4.7) approaches 0 as $p, q \to \infty$. Thus the left member of (5.4.7) also approaches 0 as $p, \dot{q} \to \infty$.

If

$$(\alpha_{p1}, \ldots, \alpha_{pm}) \in \Sigma_m, \text{ and } \dot{\mathbf{v}}_p \overset{\text{def}}{=} \sum_{i=1}^{m} \alpha_{pi} \mathbf{a}_i,$$

it follows from the completeness of $\mathbf{R}$ that each sequence $\{\alpha_{pi}\}_{p=1}^{\infty}$ has a limit α_i as $p \to \infty$ and that

$$\alpha_i \geq 0, \sum_{i=1}^{m} \alpha_i = 1, \mathbf{v} = \sum_{i=1}^{m} \alpha_i \mathbf{a}_i,$$

---

i.e., the vector $\mathbf{v}$ is in $\mathcal{Q}$ and $||\mathbf{v}|| = D$. If also $\mathbf{u} \in \mathcal{Q}$ and $||\mathbf{u}|| = D$ then the convexity of $\mathcal{Q}$ implies that

$$\frac{1}{2}(\mathbf{u} + \mathbf{v}) \in \mathcal{Q}.$$

Since for vectors $\mathbf{x}$ and $\mathbf{y}$ in $\mathbf{R}^n$ it is true that $(\mathbf{x}, \mathbf{y}) = (\mathbf{y}, \mathbf{x})$, it follows that

$$\frac{1}{2}[D^2 + (\mathbf{u}, \mathbf{v})] = ||\frac{1}{2}(\mathbf{u} + \mathbf{v})||^2 \geq D^2$$

$$(\mathbf{u}, \mathbf{v}) \geq D^2$$

$$|(\mathbf{u}, \mathbf{v})| \leq ||\mathbf{u}|| \cdot ||\mathbf{v}|| = D^2 \text{ (Schwarz inequality)}$$

$$D^2 \geq |(\mathbf{u}, \mathbf{v})| \geq (\mathbf{u}, \mathbf{v}) \geq D^2$$

and thus that $\mathbf{u}$ and $\mathbf{v}$ are linearly dependent (cf. **Exercise 4.4.25**). Since neither $\mathbf{u}$ nor $\mathbf{v}$ is $\mathbf{O}$ (which is not in $\mathcal{Q}$), and since $||\mathbf{u}|| = ||\mathbf{v}|| = D$ it follows that $\mathbf{u} = \pm\mathbf{v}$. If $\mathbf{u} = -\mathbf{v}$ then $\mathcal{Q} \ni \frac{1}{2}(\mathbf{u} + \mathbf{v}) = \mathbf{O}$ a contradiction. Hence $\mathbf{u} = \mathbf{v}$ and so in $\mathcal{Q}$ no other vector $\mathbf{u}$ is as close to $\mathbf{O}$ as $\mathbf{v}$ is. The vector $\mathbf{v}$ is, among all vectors in $\mathcal{Q}$, the unique vector closest to $\mathbf{O}$:

$$||\mathbf{v}|| = minimum\,\{\,||\mathbf{x}||\ :\ \mathbf{x} \in \mathcal{Q}\,\}$$

and

$$\{\mathcal{Q} \ni \mathbf{u} \neq \mathbf{v}\} \Rightarrow \{||\mathbf{u}|| > ||\mathbf{v}||\}.$$

There remains the task of showing that $\mathcal{Q} \subset S_{[\mathbf{v}, ||\mathbf{v}||^2]}$.

Indeed, if $\mathbf{u} \in \mathcal{Q}$ then $||\mathbf{u}|| > 0$ and if also $\mathbf{u} \notin S_{[\mathbf{v}, ||\mathbf{v}||^2]}$, i.e., if also $(\mathbf{u}, \mathbf{v}) < ||\mathbf{v}||^2$, then $\mathbf{u} \neq \mathbf{v}$ and so

$$||\mathbf{v}||^2 - ||\mathbf{u}||^2 < 0 \text{ and } ||\mathbf{u} - \mathbf{v}||^2 > 0. \tag{$*$}$$

Define g according to the formula

$$g(t) \overset{\text{def}}{=} (t\mathbf{v} + (1 - t)\mathbf{u}, \mathbf{v} - \mathbf{u}).$$

Hence from $(*)$ it follows that

$$g(t) = ||\mathbf{u} - \mathbf{v}||^2\, t + (\mathbf{u}, \mathbf{v} - \mathbf{u})$$

$$= \begin{cases} (\mathbf{u}, \mathbf{v}) - ||\mathbf{u}||^2 < (\mathbf{u}, \mathbf{v}) - ||\mathbf{v}||^2 < 0 & \text{if } t = 0 \\ ||\mathbf{u}||^2 - (\mathbf{u}, \mathbf{v}) > 0 & \text{if } t = 1. \end{cases}$$

Since g is a linear function of t it follows that there is in $(0, 1)$ a number t_1 such that

$$g(t_1) = 0,$$

i.e., such that $\mathbf{v}_1 \overset{\text{def}}{=} t_1\mathbf{v} + (1 - t_1)\mathbf{u}$ is orthogonal to $\mathbf{v} - \mathbf{u}$. Because $0 < t_1 < 1$ it follows that $\mathbf{v}_1 \neq \mathbf{v}$. But then

$$(\mathbf{v} - \mathbf{v}_1, \mathbf{v}_1) = (1 - t_1)(\mathbf{v} - \mathbf{u}, \mathbf{v}_1) = 0$$

$$||\mathbf{v}||^2 = ||\mathbf{v} + \mathbf{v}_1 - \mathbf{v}_1||^2 = ||\mathbf{v} - \mathbf{v}_1||^2 + ||\mathbf{v}_1||^2$$

$$> ||\mathbf{v}_1||^2. \tag{5.4.8}$$

    Chapter 5. APPLYING LINEAR ALGEBRA

Since $\mathbf{u}, \mathbf{v} \in \mathcal{Q}$ and since $\mathcal{Q}$ is convex, it follows that $\mathbf{v}_1 \in \mathcal{Q}$ and (5.4.8) contradicts the minimality of the length of $\mathbf{v}$.

Thus $\mathcal{Q} \subset S_{[\mathbf{v}, \|\mathbf{v}\|^2]}$. Correspondingly,

$$Conv(A) \subset S_{[\mathbf{v}, \|\mathbf{v}\|^2]}$$
$$(\mathbf{v}, \mathbf{O}) = 0 < \|\mathbf{v}\|^2. \tag{5.4.9}$$

In (5.4.9) the inequality implies that $\mathbf{O}$ is on the "other" side of the hyperplane $H_{[\mathbf{v}, \|\mathbf{v}\|^2]}$, i.e., $\mathbf{O}$ is in the complement of $S_{[\mathbf{v}, \|\mathbf{v}\|^2]}$.

$$\Omega$$

[**Remark 5.4.2:**   The argument used above to prove the existence in the convex set $\mathcal{Q}$ of a vector $\mathbf{v}$ closest to $\mathbf{O}$ may be repeated with slight modification to show that the following is also true.

If W is a subspace of $\mathbf{C}^n$ and $\mathbf{x} \in \mathbf{C}^n$ then there is in (the convex set W) a vector $\mathbf{w}$ nearest to $\mathbf{x}$. Furthermore,

$$\mathbf{u} \overset{\text{def}}{=} \mathbf{x} - \mathbf{w} \in W^\perp$$

and the equation $\mathbf{x} = \mathbf{w} + \mathbf{u}$ displays the unique orthogonal decomposition of $\mathbf{x}$ into a sum of two vectors, one in W and one in $W^\perp$. The vector $\mathbf{w}$ is the *orthogonal projection* of $\mathbf{x}$ onto W (cf. **Section 4.4, E13**).]

A useful consequence of THEOREM **5.4.1** is the next result. In its statement the symbols $\succeq$, $\succ$, $\preceq$, $\prec$ (cf. **Section 1.1**) are rather helpful. They are used to describe in compact form relationships between vectors having real components. For the record, if $\mathbf{x}$ and $\mathbf{y}$ are vectors in $\mathbf{R}^n$ then:

$$\mathbf{x} \overset{\text{def}}{=} \begin{pmatrix} x_1 \\ \vdots \\ x_n \end{pmatrix} \succeq \begin{pmatrix} y_1 \\ \vdots \\ y_n \end{pmatrix} \overset{\text{def}}{=} \mathbf{y} \Leftrightarrow x_i \geq y_i, \ 1 \leq i \leq n$$

$$\mathbf{x} \overset{\text{def}}{=} \begin{pmatrix} x_1 \\ \vdots \\ x_n \end{pmatrix} \succ \begin{pmatrix} y_1 \\ \vdots \\ y_n \end{pmatrix} \overset{\text{def}}{=} \mathbf{y} \Leftrightarrow x_i > y_i, \ 1 \leq i \leq n$$

$$\mathbf{x} \overset{\text{def}}{=} \begin{pmatrix} x_1 \\ \vdots \\ x_n \end{pmatrix} \preceq \begin{pmatrix} y_1 \\ \vdots \\ y_n \end{pmatrix} \overset{\text{def}}{=} \mathbf{y} \Leftrightarrow x_i \leq y_i, \ 1 \leq i \leq n$$

$$\mathbf{x} \overset{\text{def}}{=} \begin{pmatrix} x_1 \\ \vdots \\ x_n \end{pmatrix} \prec \begin{pmatrix} y_1 \\ \vdots \\ y_n \end{pmatrix} \overset{\text{def}}{=} \mathbf{y} \Leftrightarrow x_i < y_i, \ 1 \leq i \leq n.$$

Exercise 5.4.7. Regard the vectors occurring below as matrices compatible for the operations performed. Show that for "$\succeq$" the following are true:

i. if $\mathbf{x} \succeq \mathbf{y}$ and $\mathbf{z} \succeq \mathbf{O}$ then $\mathbf{zx} \geq \mathbf{zy}$;
ii. if $\mathbf{x} \succeq \mathbf{y}$ and $\mathbf{x} \neq \mathbf{y}$ it is possible that $\mathbf{x} \not\succ \mathbf{y}$;
iii. if $\mathbf{x} \succeq \mathbf{y}$ and $t \geq 0$ then $t\mathbf{x} \succeq t\mathbf{y}$.

Replace "$\succeq$" resp. "$>$" in *i* - *iii* above by "$\succ$", "$\preceq$", and "$\prec$" resp. "$>$", "$\leq$", and "$<$" and prove or disprove each of the resulting statements.

THEOREM 5.4.2 below leads to THEOREM 5.4.3, the fundamental *minmax principle* of von Neumann [**vNM**].

THEOREM **5.4.2.** ASSUME $A \overset{\text{def}}{=} (a_{ij})_{i,j=1}^{m,n}$ IS A MATRIX IN WHICH ALL ENTRIES ARE REAL. THEN

EITHER

I. THERE IS IN Σ_m AN α_0 SUCH THAT

$$\alpha_0 A \preceq \mathbf{O} \tag{5.4.10}$$

OR

II. THERE IS IN $\Sigma^{(n)}$ A β_0 SUCH THAT

$$A\beta_0 \succ \mathbf{O}. \tag{5.4.11}$$

[**Remark 5.4.3:** Assume the *i*th component of α_0 is α_{0i} and that the *j*th component of β_0 is β_{0j}. Then

$$\alpha_0 A \preceq \mathbf{O} \text{ iff } \sum_{i=1}^{m} \alpha_{0i} a_{ij} \leq 0, \ 1 \leq j \leq n \tag{5.4.12}$$

$$A\beta_0 \succ \mathbf{O} \text{ iff } \sum_{j=1}^{n} a_{ij} \beta_{0j} > 0, \ 1 \leq i \leq m. \tag{5.4.13}$$

In (5.4.10) and in (5.4.12) the statements concern convex combinations of the *rows* of A. In (5.4.11) and in (5.4.13) the statements concern convex combinations of the *columns* of A.]

PROOF. Note first that **I** and **II** are mutually exclusive. Indeed, if **I** is true then for for any β in $\Sigma^{(n)}$

$$\alpha_0 A\beta \leq 0$$

and, in particular,

$$\alpha_0 A\beta_0 \leq 0. \tag{5.4.14}$$

However, if **II** is also true, (5.4.14) is a contradiction of the averaging principle (5.4.4) applied to the components of $A\beta_0$. Thus if **I** is true then **II** is false: **I** $\Rightarrow$ *not* **II**. Hence, according to the standard rules of logic: **II** $\Rightarrow$ *not* **I**. In sum, **I** and **II** are mutually exclusive.

Let $e_1, \ldots, e_m$ be the standard basis vectors in $\mathbf{R}_m$. It is shown next that

> **I** $\Leftrightarrow$ there is some nonnegative number P such that
> $$Conv(\mathbf{a}_1, \ldots, \mathbf{a}_m) \cap P \cdot Conv(-\mathbf{e}_1, \ldots, -\mathbf{e}_n) \neq \emptyset$$
> $$\Leftrightarrow Conv(\mathbf{a}_1, \ldots, \mathbf{a}_m, \mathbf{e}_1, \ldots, \mathbf{e}_n) \ni \mathbf{O}. \tag{5.4.15}$$

Indeed, **I** implies that

EITHER

> $\mathbf{O} \in Conv(\mathbf{a}_1, \ldots, \mathbf{a}_m)$, which is (5.4.15) when $P = 0$

OR

> there is some convex combination $\mathbf{x}$ of the rows of A that is also some convex combination of the vectors $-\mathbf{e}_1, \ldots, -\mathbf{e}_n$: for some α_0 in Σ_m
>
> $$\mathbf{x} \overset{\text{def}}{=} \sum_{i=1}^{m} \alpha_{0i} a_i$$

and, for nonnegative s_j, $1 \leq j \leq n$ such that $\sum_{j=1}^{n} s_j \overset{\text{def}}{=} P > 0$

$$\mathbf{O} \succeq \mathbf{x} = \sum_{j=1}^{n} s_j(-\mathbf{e}_j).$$

If $t_j \overset{\text{def}}{=} \frac{s_j}{P}$ then $t_j \geq 0$, $\sum_{j=1}^{n} t_j = 1$ and

$$\mathbf{x} = P \sum_{j=1}^{n} t_j(-\mathbf{e}_j), \text{ i.e.,}$$

$$\mathbf{x} \in Conv(\mathbf{a}_1, \ldots, \mathbf{a}_m) \cap P \cdot Conv(-\mathbf{e}_1, \ldots, -\mathbf{e}_n)$$

i.e., for some positive number P

$$P \cdot Conv(\mathbf{a}_1, \ldots, \mathbf{a}_m) \cap Conv(-\mathbf{e}_1, \ldots, -\mathbf{e}_n) \neq \emptyset.$$

Thus,

$$\mathbf{I} \Rightarrow \text{ there is a nonnegative number } P \text{ such that}$$
$$Conv(\mathbf{a}_1, \ldots, \mathbf{a}_m) \cap P \cdot Conv(-\mathbf{e}_1, \ldots, -\mathbf{e}_n) \neq \emptyset.$$

Hence there is in Σ_n a vector $(t_1, \ldots, t_n)$ such that

$$\sum_{i=1}^{m} \alpha_{0i} \mathbf{a}_i = -P \sum_{j=1}^{n} t_j \mathbf{e}_j$$

$$\sum_{i=1}^{m} \alpha_{0i} \mathbf{a}_i + P \sum_{j=1}^{n} t_j \mathbf{e}_j = \mathbf{O}$$

$$\sum_{i=1}^{m} \alpha_{0i} + P \sum_{j=1}^{n} t_j = 1 + P > 0$$

$$l_i \stackrel{\text{def}}{=} \frac{\alpha_{0i}}{1 + P}, \quad r_j \stackrel{\text{def}}{=} \frac{Pt_j}{1 + P} \geq 0$$

$$\sum_{i=1}^{m} l_i + \sum_{j=1}^{n} r_j = 1$$

$$\mathbf{O} = \sum_{i=1}^{m} l_i \mathbf{a}_i + \sum_{j=1}^{n} r_j \mathbf{e}_j \in Conv(\mathbf{a}_1, \ldots, \mathbf{a}_m, \mathbf{e}_1, \ldots, \mathbf{e}_n) \stackrel{\text{def}}{=} \mathcal{C},$$

i.e., if there is a nonnegative number P such that

$$Conv(\mathbf{a}_1, \ldots, \mathbf{a}_m) \cap P \cdot Conv(-\mathbf{e}_1, \ldots, -\mathbf{e}_n) \neq \emptyset$$

then $\mathbf{O} \in \mathcal{C}$.

The reversed implications (cf. **Exercise 5.4.8**) follow similarly and thus (5.4.15) is true.

Hence $\mathbf{I}$ is *not* true iff

$$\mathbf{O} \notin Conv(\mathbf{a}_1, \ldots, \mathbf{a}_m, \mathbf{e}_1, \ldots, \mathbf{e}_n).$$

Hence if $\mathbf{u} \in \mathcal{C}$ then $(\mathbf{u}, \mathbf{v}) > 0$. But $\mathbf{u} \in \mathcal{C}$ iff there are nonnegative numbers $l_1, \ldots, l_m, r_1, \ldots, r_n$ such that

$$\sum_{i=1}^{m} l_i + \sum_{j=1}^{n} r_j = 1 \tag{5.4.16}$$

$$\mathbf{u} = \sum_{i=1}^{m} l_i \mathbf{a}_i + \sum_{j=1}^{n} r_j \mathbf{e}_j. \tag{5.4.17}$$

Assume $\mathbf{v} \stackrel{\text{def}}{=} \sum_{j=1}^{n} v_j \mathbf{e}_j$. Then for $\mathbf{u}$ as in (5.4.17) it follows that

$$(\mathbf{u}, \mathbf{v}) = \sum_{i,j=1}^{m,n} v_j l_i a_{ij} + \sum_{j=1}^{n} v_j r_j > 0. \tag{5.4.18}$$

If each $l_i = 0$ and precisely one $r_j = 1$ then (5.4.18) implies each $v_j > 0$. If each $r_j = 0$ and precisely one $l_i = 1$ then (5.4.18) implies that

$$\sum_{j=1}^{n} v_j a_{ij} > 0, \ 1 \leq i \leq m.$$

Hence if

$$\beta_j \stackrel{\text{def}}{=} \frac{v_j}{\sum_{j=1}^{n} v_j}$$

then

$$\beta_j \geq 0, \ \sum_{j=1}^{n} \beta_j = 1$$

$$\sum_{j=1}^{n} \beta_j a_{ij} > 0, \ 1 \leq i \leq m,$$

i.e., if

$$\beta_0 \stackrel{\text{def}}{=} \begin{pmatrix} \beta_1 \\ \vdots \\ \beta_n \end{pmatrix} \ (\in \Sigma^{(n)})$$

then $A\beta_0 \succ \mathbf{O}$.

In sum,

if $\mathbf{I}$ is *false* there is in $\Sigma^{(n)}$ a β_0 such that $A\beta_0 \succ \mathbf{O}$.

If $\mathbf{I}$ is *true* then

$$\mathbf{O} \in Conv(\mathbf{a}_1, \ldots, \mathbf{a}_m, \mathbf{e}_1, \ldots, \mathbf{e}_n).$$

Hence there are nonnegative numbers $l_1, \ldots, l_m, r_1, \ldots, r_n$ such that

$$\sum_{i=1}^{m} l_i + \sum_{j=1}^{n} r_j = 1 \tag{5.4.19}$$

$$\sum_{i=1}^{m} l_i \mathbf{a}_i + \sum_{j=1}^{n} r_j \mathbf{e}_j = \mathbf{O} \tag{5.4.20}$$

$$\sum_{i=1}^{m} l_i a_{ij} = -r_j \leq 0, \ 1 \leq j \leq n. \tag{5.4.21}$$

If each $l_i = 0$ then, because the e_j are linearly independent, (5.4.20) implies that each $r_j = 0$ whence (5.4.19) is contradicted. Hence $\sum_{i=1}^{m} l_i > 0$ and so if

$$\alpha_i \stackrel{\text{def}}{=} \frac{l_i}{\sum_{i=1}^{m} l_i}, \ 1 \le i \le m$$

then $\sum_{i=1}^{m} \alpha_i = 1, \ \alpha_i \ge 0, \ 1 \le i \le m$, i.e., if

$$\boldsymbol{\alpha}_0 \stackrel{\text{def}}{=} (\alpha_1, \dots, \alpha_m)$$

then $\boldsymbol{\alpha}_0 \in \Sigma_m$ and (5.4.21) implies $\boldsymbol{\alpha}_0 A \preceq \mathbf{O}$. Since either $\mathbf{I}$ is true or $\mathbf{I}$ is false at long last it is established that:
EITHER

there is in Σ_m an $\boldsymbol{\alpha}_0$ such that

$$\boldsymbol{\alpha_0 A} \preceq \mathbf{O} \tag{5.4.22}$$

OR

there is in $\Sigma^{(n)}$ a β_0 such that

$$A\beta_0 \succ \mathbf{O}. \tag{5.4.23}$$

Whatever is not proved in the preceding lines is left for **Exercise 5.4.8.**

$$\Omega$$

Exercise 5.4.8. Show that the reversed implications

$$Conv(\mathbf{a}_1, \dots, \mathbf{a}_m, \mathbf{e}_1, \dots, \mathbf{e}_n) \ni \mathbf{O}$$
$$\Rightarrow \text{ there is a nonnegative number } P \text{ such that}$$
$$Conv(\mathbf{a}_1, \dots, \mathbf{a}_m) \cap P \cdot Conv(-\mathbf{e}_1, \dots, -\mathbf{e}_n) \ne \emptyset$$
$$\Rightarrow \mathbf{I}$$

are valid.

Exercise 5.4.9. From THEOREM **5.4.2** it follows in particular that:

$$\mathbf{II} \text{ is true} \Rightarrow \mathbf{I} \text{ is false.}$$

Derive the conclusion above without use of THEOREM **5.4.2**.

THEOREM **5.4.3.** LET $A \stackrel{\text{def}}{=} (a_{ij})_{i,j=1}^{m,n}$ BE AN $m \times n$ MATRIX IN WHICH EACH ENTRY IS REAL. THERE IS IN $\mathbf{R}$ A NUMBER $\mathbf{V}(A)$ SUCH THAT

$$\mathbf{V}(A) = maximum_{\beta \in \Sigma^{(n)}} minimum_{\alpha \in \Sigma_m} \alpha A \beta \stackrel{\text{def}}{=} \mathbf{m}(A)$$
$$= minimum_{\alpha \in \Sigma_m} maximum_{\beta \in \Sigma^{(n)}} \alpha A \beta \stackrel{\text{def}}{=} \mathbf{M}(A).$$

PROOF. For a given α the number

$$maximum_{\beta \in \Sigma^{(n)}} \alpha A \beta \stackrel{\text{def}}{=} \phi(\alpha)$$

depends on α. Similarly, for a given β the number

$$minimum_{\alpha \in \Sigma_m} \alpha A \beta \stackrel{\text{def}}{=} \psi(\beta)$$

depends on β.

First the simpler fact that $m(A) \leq M(A)$ is established in the next calculations.

i. For all α in Σ_m and β in $\Sigma^{(n)}$

$$\alpha A \beta \leq maximum_{\beta \in \Sigma^{(n)}} \alpha A \beta = \phi(\alpha)$$

ii. Hence for all α in Σ_m and β in $\Sigma^{(n)}$

$$\psi(\beta) = minimum_{\alpha \in \Sigma_m} \alpha A \beta \leq maximum_{\beta \in \Sigma^{(n)}} \alpha A \beta = \phi(\alpha)$$

and so

$$m(A) = maximum_{\beta \in \Sigma^{(n)}} \psi(\beta) = maximum_{\beta \in \Sigma^{(n)}} minimum_{\alpha \in \Sigma_m} \alpha A \beta$$

$$\leq maximum_{\beta \in \Sigma^{(n)}} \alpha A \beta = \phi(\alpha) \text{ (for all } \alpha \text{ in } \Sigma_m)$$

$$m(A) \leq minimum_{\alpha \in \Sigma_m} \phi(\alpha) = M(A),$$

i.e., ALWAYS "*maxmin* $\leq$ *minmax*."

On the other hand, if **I** in THEOREM **5.4.2** holds for A then the averaging principle implies that for any β in $\Sigma^{(n)}$

$$\alpha_0 A \beta \leq 0. \tag{5.4.24}$$

If **II** in THEOREM **5.4.2** holds for A then the averaging principle implies that for any α in Σ_m

$$\alpha A \beta_0 > 0. \tag{5.4.25}$$

Hence from (5.4.24) it follows that $m(A) \leq 0$ and from (5.4.25) it follows that $M(A) > 0$ and so

EITHER $m(A) \leq 0$ OR $M(A) > 0$,

i.e.,

NEVER $M(A) \leq 0 < m(A)$.

If $x \in \mathbf{R}$ and each entry a_{ij} in A is replaced by $a_{ij} - x$ there emerges a new matrix A_x. Owing to the fact that $\alpha \in \Sigma_m$ and $\beta \in \Sigma^{(n)}$ it follows that $m(A_x) = m(A) - x$ and $M(A_x) = M(A) - x$. Hence

---

for any real number x

$$\text{NEVER } \mathbf{M}(A) \leq x < \mathbf{m}(A). \tag{5.4.26}$$

In short, $\mathbf{M}(A) \geq \mathbf{m}(A)$ and so $\mathbf{m}(A) = \mathbf{M}(A)$. Their common value is denoted $\mathbf{V}(A)$.

$$\Omega$$

[**Note 5.4.1:** The functions ϕ and ψ, used in the proof of the inequality $\mathbf{m}(A) \leq \mathbf{M}(A)$, give rise to the sets

$$\widetilde{\Sigma}_m \overset{\text{def}}{=} \left\{ \widetilde{\alpha} \ : \ \alpha \in \Sigma_m, \ \phi(\widetilde{\alpha}) = minimum_{\alpha \in \Sigma_m} \phi(\alpha) \right\}$$
$$\widetilde{\Sigma}^{(n)} \overset{\text{def}}{=} \left\{ \widetilde{\beta} \ : \ \beta \in \Sigma^{(n)}, \ \psi(\widetilde{\beta}) = maximum_{\beta \in \Sigma^{(n)}} \psi(\beta) \right\}.$$

Owing to the completeness of $\mathbf{R}$ the sets $\widetilde{\Sigma}_m$ and $\widetilde{\Sigma}^{(n)}$ are nonempty. If $\widetilde{\alpha} \in \widetilde{\Sigma}_m$ and $\widetilde{\beta} \in \widetilde{\Sigma}^{(n)}$ then

$$\mathbf{V}(A) = \phi(\widetilde{\alpha}) = maximum_{\beta \in \Sigma^{(n)}} \widetilde{\alpha} A \beta \geq \widetilde{\alpha} A \widetilde{\beta}$$
$$\mathbf{V}(A) = \psi(\widetilde{\beta}) = minimum_{\alpha \in \Sigma_m} \alpha A \widetilde{\beta} \leq \widetilde{\alpha} A \widetilde{\beta}$$

whence $\mathbf{V}(A) = \widetilde{\alpha} A \widetilde{\beta}$.

Conversely, if $\widetilde{\alpha} \in \Sigma_m$, $\widetilde{\beta} \in \Sigma^{(n)}$, and $\widetilde{\alpha} A \widetilde{\beta} = \mathbf{V}(A)$ then, by definition, $\widetilde{\alpha} \in \widetilde{\Sigma}_m$ and $\widetilde{\beta} \in \widetilde{\Sigma}^{(n)}$.]

[**Remark 5.4.4:** THEOREM 5.4.3, one of many *"minmax"* theorems, is due to von Neumann, the founder of the theory of games. The matrix A is a matrix of payoffs to P_1, one of two players P_1 and P_2; $-A$ is a matrix of payoffs to P_2. What one wins the other loses: the game is a zero-sum two-person game. The number a_{ij} is what P_2 pays P_1 if P_1 chooses among the numbers $1, 2, \ldots, m$ the number i ("uses *pure strategy i*") while P_2 chooses among the numbers $1, 2, \ldots, n$ the number j ("uses pure strategy j"). If, in many plays of the game, P_1 uses his m pure strategies in the proportions $\alpha_1, \ldots, \alpha_m$, corresponding to the vector α in Σ_m, and P_2 uses his n pure strategies in the proportions $\beta_1, \ldots, \beta_n$, corresponding to the vector β in $\Sigma^{(n)}$, then the average or expected payoff to P_1 is $\alpha A \beta$. The vectors α resp. β are called *mixed strategies* for P_1 resp. P_2. The mixed strategies $\widetilde{\alpha}$ resp. $\widetilde{\beta}$ in $\widetilde{\Sigma}_m$ resp. $\widetilde{\Sigma}^{(n)}$ are called *optimal mixed strategies* for P_1 resp. P_2.

The number $\mathbf{V}(A)$, which is both a *minmax* and a *maxmin*, is called the *value of the game*. Since *minmax* is the average worst among the average best that P_1 can win, and *maxmin* is the average best among the average worst that P_1 can win, the burden of THEOREM 5.4.3 is that these two numbers are equal. If the word "average" is

---

omitted in the preceding statement, the statement can fail to hold. For example, if

$$A \overset{\text{def}}{=} \begin{pmatrix} 1 & -1 \\ -1 & 1 \end{pmatrix}$$

then the worst of the best is 1 and the best of the worst is -1:

$$maxmin = -1 < 1 = minmax.$$

In **Exercise 5.4.11** the reader is asked to calculate $\mathbf{V}(A)$ for some special matrices A.]

Example 5.4.3. Let $A \overset{\text{def}}{=} (a_{ij})_{i,j=1}^{2,n}$ be a $2 \times n$ matrix. For each β in Σ_2 one can plot the graph of the function $maximum_{\beta \in \Sigma^{(n)}} \alpha A \beta$ regarded as depending only on α_1 (since $\alpha_2 = 1 - \alpha_1$). The result is a broken line graph in which the ordinate of the lowest point is $\mathbf{V}(A)$. In **Figure 5.4.5** $n = 5$ and the graph is drawn for the matrix

$$\begin{pmatrix} 1 & -2 & 3 & -4 & 5 \\ -2 & 0 & 3 & 6 & -4 \end{pmatrix}.$$

The line labelled $\beta_i = 1$ corresponds to setting all β_j, $j \neq i$, to 0, setting β_i to 1, and then plotting $\alpha A \beta$ regarded as a function of α_1. For each value of α_1 the value of $maximum_{\beta \in \Sigma^{(5)}} \alpha A \beta$ is indicated as a point on the heavy line. The ordinate(s) of the lowest point(s) on the heavy line are all 3 which is the minimum of the maxima and is the value $\mathbf{V}(A)$ of the game. Thus

$$\widetilde{\Sigma_2} = \left\{ \alpha \ : \ \frac{3}{10} \leq \alpha_1 \leq \frac{7}{9}, \ \alpha_2 = 1 - \alpha_1 \right\}$$
$$\widetilde{\Sigma_5} = \{ \beta \ : \ \beta_3 = 1, \beta_i = 0, \ i \neq 3 \}.$$

Exercise 5.4.10. Plot in $\mathbf{R}_2$ the sets Σ_2 and $\widetilde{\Sigma_2}$. Give a simple verbal description of the geometric character of $\widetilde{\Sigma_5}$ (in $\mathbf{R}_5$).

Exercise 5.4.11. For each of the $2 \times n$ matrices A below use the method outlined in **Example 5.4.3** to find $\mathbf{V}(A)$:

$$i. \ \begin{pmatrix} 1 & -1 \\ -1 & 1 \end{pmatrix} ; \ \ ii. \ \begin{pmatrix} 1 & 2 & -3 \\ 0 & -1 & 4 \end{pmatrix} ; \ \ iii. \ \begin{pmatrix} 1 & 2 & 3 \\ 4 & 5 & 6 \end{pmatrix}.$$

Explain the (special) nature of the solution to iii.

Exercise 5.4.12. Assume that A is a matrix with real entries and that $A = -A^t$. Show that A is SQUARE and that $\mathbf{V}(A) = 0$.

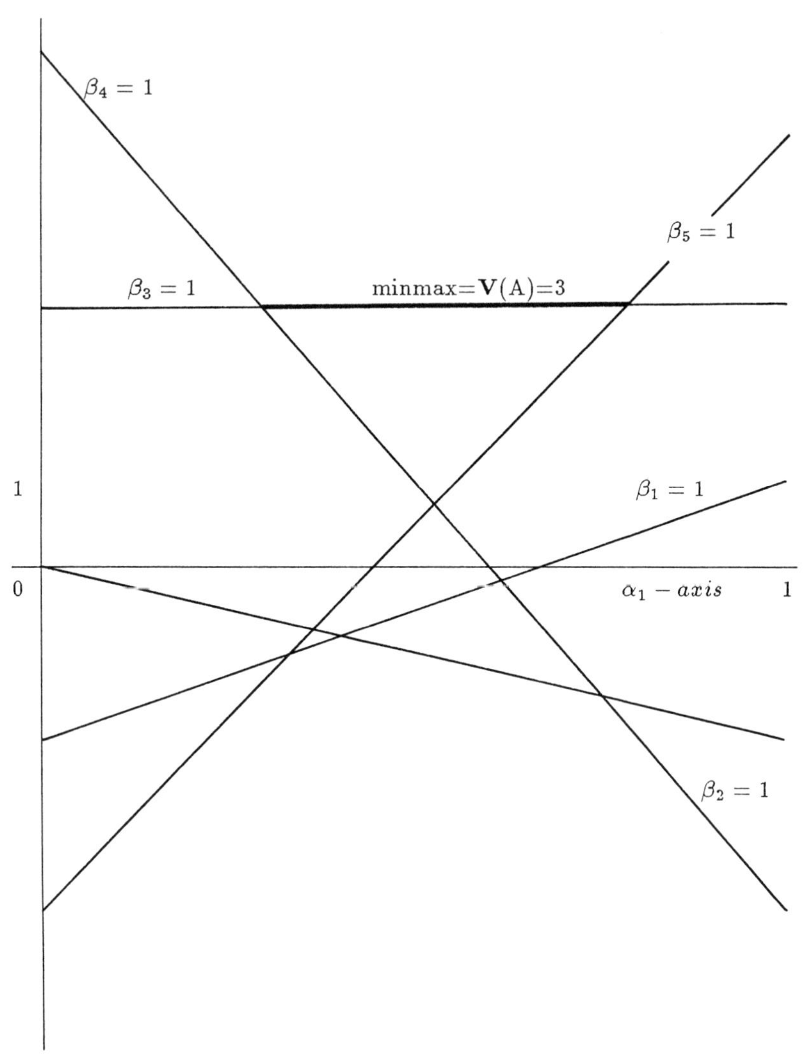

Figure 5.4.5.

The reader is urged to explore [**vNM**] for a detailed and illuminating discussion of the theory of games.

Exercise 5.4.13. Let A and B be two $m \times n$ matrices with real entries. Show that
EITHER
   there is in $\Sigma^{(n)}$ a β_0 such that $A\beta_0 \preceq B\beta_0$
OR
   there is in Σ_m an α_0 such that $\alpha_0 A \succ \alpha_0 B$.
   Show also that the two situations just described are mutually exclusive.

The relationship between the theory of games and the problem of linear programming is developed by studying not only systems of *equations* but systems of *inequalities*. The general problem of linear programming may be stated as follows.

The Primal Linear Programming Problem (PLPP)

i. There is given an $m \times n$ matrix

$$A \stackrel{\text{def}}{=} \begin{pmatrix} a_{11} & \cdots & a_{1n} \\ \vdots & \ddots & \vdots \\ a_{m1} & \cdots & a_{mn} \end{pmatrix}$$

in which each entry a_{ij} is real.

ii. There are given two vectors

$$\mathbf{p} \stackrel{\text{def}}{=} \begin{pmatrix} p_1 \\ \vdots \\ p_m \end{pmatrix}, \text{ resp. } \mathbf{c} \stackrel{\text{def}}{=} (c_1, \ldots, c_n) \neq \mathbf{O}.$$

in which each component p_i resp. c_j is real.

iii. It is required
EITHER
   to find an $n \times 1$ vector

$$\mathbf{x} \stackrel{\text{def}}{=} \begin{pmatrix} x_1 \\ \vdots \\ x_n \end{pmatrix}$$

in which each component x_j is nonnegative and such that the constraints represented by $A\mathbf{x} \succeq \mathbf{p}$ are satisfied and such that the number $\mathbf{cx}$ is minimal, i.e., that

$$\begin{pmatrix} a_{11}x_1 & +\cdots & +a_{1n}x_n \\ \vdots & \ddots & \vdots \\ a_{m1}x_1 & +\cdots & +a_{mn}x_n \end{pmatrix} \succeq \begin{pmatrix} p_1 \\ \vdots \\ p_m \end{pmatrix} \quad \text{(constraints)}$$

$$x_j \geq 0, \ 1 \leq j \leq n,$$

and

$$c_1 x_1 + \cdots + c_n x_n \text{ is minimal}$$

OR

to prove that no such vector $\mathbf{x}$ exists.

Associated with each primal linear programming problem **PLPP** is its *dual*.

The **D**ual **L**inear **P**rogramming **P**roblem (**DLPP**)

EITHER

find a $1 \times m$ vector $\mathbf{y}$ in which each component is nonnegative, such that the constraints represented by $\mathbf{y}A \preceq \mathbf{c}$ are satisfied, and such that the number $\mathbf{yp}$ is maximal, i.e., such that

$$\mathbf{y}A \stackrel{\text{def}}{=} \begin{pmatrix} y_1 a_{11} & + \cdots & + y_m a_{m1} \\ \vdots & \ddots & \vdots \\ y_1 a_{1n} & + \cdots & + y_m a_{mn} \end{pmatrix} \preceq \begin{pmatrix} c_1 \\ \vdots \\ c_n \end{pmatrix} \quad \text{(constraints)}$$

$$y_i \geq 0, \ 1 \leq i \leq m,$$

and

$$y_1 p_1 + \cdots + y_m p_m \text{ is maximal}$$

OR

show that no such $\mathbf{y}$ exists.

The functions $\mathbf{x} \mapsto \mathbf{cx}$ and $\mathbf{y} \mapsto \mathbf{yp}$ are called the *objective functions*.

Example 5.4.4. In the example of nutrition let AGRIBUSINESS be engaged in selling livestock and crops and let CHEMBUSINESS be engaged in selling acceptable man-made substitutes for PROTEINS, FATS, and CARBO-HYDRATES

Assume that CHEMBUSINESS can offer PROTEINS, FATS, and CARBO-HYDRATES in palatable form so that if \$1 of revenue comes in then \$$p_1$ comes from sales of PROTEINS, \$$p_2$ comes from sales of FATS, and \$$p_3$ comes from sales of CARBOHYDRATES. The price vector

$$\mathbf{p} \stackrel{\text{def}}{=} \begin{pmatrix} p_1 \\ p_2 \\ p_3 \end{pmatrix}$$

is such that $p_1 + p_2 + p_3 = 1$.

According to (1.1.3) one unit of cattle provides 24 units of PROTEINS, 18 units of FATS, and 20 units of CARBOHYDRATES. To be competitive with beef the prices p_1, p_2, p_3 must be such that the cost c_B of one unit of cattle cannot be less than the cost of the 24 units of PROTEINS, 18 units of FATS, and 20 units of CARBOHYDRATES that one unit of cattle can provide, i.e.,

$$24p_1 + 18p_2 + 20p_3 \leq c_B.$$

Similar inequalities must be satisfied if SHEEP, HOGS, etc., are to be replaced
by CHEMBUSINESS products. If the market is captured by CHEMBUSINESS
it provides 5 units of PROTEINS, 3 units of FATS, and 10 units of CARBOHY-
DRATES and the revenue R from the sales is then

$$5p_1 + 3p_2 + 10p_3.$$

Thus from the standpoint of CHEMBUSINESS there is a problem in linear
programming and it reads like this. Find nonnegative values p_1, p_2, p_3 so that:

$$24p_1 + 18p_2 + 20p_3 \leq 20$$
$$8p_1 + 5p_2 + 2p_3 \leq 15$$
$$21p_1 + 12p_2 + 0p_3 \leq 12$$
$$8p_1 + 0p_2 + 92p_3 \leq 5$$
$$12p_1 + 0p_2 + 138p_3 \leq 8$$
$$22p_1 + 2p_2 + 234p_3 \leq 6, \ \text{(constraints)} \tag{5.4.27}$$
$$\text{Average Revenue} \stackrel{\text{def}}{=} R \stackrel{\text{def}}{=} 5p_1 + 3p_2 + 10p_3 \ \text{is maximum.} \tag{5.4.28}$$

The following observations can be made upon comparison of (5.4.27) and
(5.4.28) with their counterparts (1.1.9) and (1.1.10):

   i. the matrices of the left members of the two systems (5.4.27) and (1.1.9) are
transposes of each other;

   ii. the right members of (1.1.9) are the *coefficients* in the right member of
(5.4.27);

   iii. the right members of (5.4.27) are the *coefficients* in the right member of
(1.1.10);

   iv. the inequalities in (1.1.9) and (5.4.27) are opposite to each other;

   v̄. T is to be minimized while R is to be maximized.

Because of *i - v* the linear programming problems of AGRIBUSINESS and
of CHEMBUSINESS are called *dual* to each other. The pair of dual problems
may be regarded as the ingredients of a game played between AGRIBUSINESS
and CHEMBUSINESS. Whatever AGRIBUSINESS earns through sales of its
products, CHEMBUSINESS loses through diminished sales of its products.

The data of a primal linear programming problem **PLPP** and of its dual
DLPP may be incorporated into a single matrix:

$$\mathcal{A} \stackrel{\text{def}}{=} \begin{array}{c} m \\ n \\ 1 \end{array} \begin{pmatrix} \overset{n}{O} & \overset{m}{A} & \overset{1}{-\mathbf{p}} \\ -A^t & O & \mathbf{c}^t \\ \mathbf{p}^t & -\mathbf{c} & O \end{pmatrix},$$

---

i.e.,

$$
\mathcal{A} \overset{\text{def}}{=} \begin{pmatrix}
0 & \cdots & 0 & a_{11} & \cdots & a_{1n} & -p_1 \\
\vdots & \ddots & \vdots & \vdots & \ddots & \vdots & \vdots \\
0 & \cdots & 0 & a_{m1} & \cdots & a_{mn} & -p_m \\
-a_{11} & \cdots & -a_{m1} & 0 & \cdots & 0 & c_1 \\
\vdots & \ddots & \vdots & \vdots & \ddots & \vdots & \vdots \\
-a_{1n} & \cdots & -a_{mn} & 0 & \cdots & 0 & c_n \\
p_1 & \cdots & p_m & -c_1 & \cdots & -c_n & 0
\end{pmatrix}
\tag{5.4.29}
$$

DEFINITION **5.4.2.** A NONNEGATIVE VECTOR $\mathbf{x}$ SUCH THAT $A\mathbf{x} \succ \mathbf{p}$ IS A *feasible* SOLUTION TO THE **PLPP**. A NONNEGATIVE VECTOR $\mathbf{y}$ SUCH THAT $\mathbf{y}A \prec \mathbf{c}$ IS A *feasible* SOLUTION TO THE **DLPP**.

The difference between a feasible solution and a solution for the **PLPP** resp. **DLPP** is that a feasible solution merely satisfies the constraints but is not necessarily a vector that minimizes resp. maximizes the objective function.

THEOREM **5.4.4.** ASSUME THAT $\{\mathbf{x}, \mathbf{y}\}$ IS A PAIR OF FEASIBLE SOLUTIONS FOR THE THE DUAL PAIR $\{\mathbf{PLPP}, \mathbf{DLPP}\}$. THEN $\mathbf{yp} \leq \mathbf{cx}$ AND IF $\mathbf{yp} = \mathbf{cx}$ THEN $\{\mathbf{x}, \mathbf{y}\}$ IS A PAIR OF SOLUTIONS FOR THE DUAL PAIR $\{\mathbf{PLPP}, \mathbf{DLPP}\}$.

[**Remark 5.4.5:** It is also true (cf. THEOREM **5.4.7**) that if $\{\mathbf{x}, \mathbf{y}\}$ is a pair of solutions for the dual pair $\{\mathbf{PLPP}, \mathbf{DLPP}\}$ then $\mathbf{yp} = \mathbf{cx}$.]

PROOF.

$$\mathbf{y}A \preceq \mathbf{c} \Rightarrow \mathbf{y}A\mathbf{x} \leq \mathbf{cx}$$
$$A\mathbf{x} \succeq \mathbf{p} \Rightarrow \mathbf{y}A\mathbf{x} \geq \mathbf{yp}$$

whence $\mathbf{yp} \leq \mathbf{cx}$.

If $\mathbf{yp} = \mathbf{cx}$ and if $\mathbf{x}$ is *not* a solution of the **PLPP** then for the **PLPP** there is a feasible solution $\mathbf{z}$ $(\succeq \mathbf{O})$ such that $\mathbf{cz} < \mathbf{cx}$. But then

$$\mathbf{yp} \leq \mathbf{y}(A\mathbf{z}) = (\mathbf{y}A)\mathbf{z} \leq \mathbf{cz} < \mathbf{cx},$$

in contradiction of the hypothesis: $\mathbf{yp} = \mathbf{cx}$.

A similar argument is valid if $\mathbf{y}$ is *not* a solution of the **DLPP**.

$$\Omega$$

The next result is crucial to the development of the the theory of linear programming.

THEOREM 5.4.5. LET $A \overset{\text{def}}{=} (a_{ij})_{i,j=1}^{m,n}$ BE A MATRIX IN WHICH EACH ENTRY IS REAL AND LET $\mathbf{b}$ BE A VECTOR IN $\mathbf{R}_n$. THEN
EITHER
THERE IS IN $\mathbf{R}_m$ A $\mathbf{y}$ SUCH THAT

$$\mathbf{y} \succeq \mathbf{O} \text{ AND } \mathbf{y}A = \mathbf{b} \tag{5.4.30}$$

OR
THERE IS IN $\mathbf{R}^n$ AN $\mathbf{x}$ SUCH THAT

$$A\mathbf{x} \succeq \mathbf{O} \text{ AND } \mathbf{b}\mathbf{x} < 0. \tag{5.4.31}$$

The proof of THEOREM **5.4.5** is given below after the next few paragraphs, which are aimed at clarifying its message and import.

In $\mathbf{R}_2$, as in **Figure 5.4.6**, the vector $\mathbf{b}$ is placed in three positions where it is denoted $\mathbf{b}_1$, $\mathbf{b}_2$, and $\mathbf{b}_3$. These illustrate essentially all interesting possibilities. The meaning of (5.4.30) is, since $\mathbf{y} \succeq \mathbf{O}$, that $\mathbf{b}$, denoted this time $\mathbf{b}_1$, is a linear combination *with nonnegative coefficients* of the rows $\mathbf{a}_i$ of A. In other words, $\mathbf{b}_1$ lies in the convex *cone* K spanned by the rows $\mathbf{a}_i$ of A. The cone K is shaded in **Figure 5.4.6**.

The precise definition of K is:

$$K \overset{\text{def}}{=} \{\, \mathbf{z} \; : \; \mathbf{z} = \mathbf{y}A, \; \mathbf{y} \succeq \mathbf{O} \,\}.$$

If
$$\mathbf{y} = (y_1, \ldots, y_m) \succeq \mathbf{O}$$
and $\mathbf{y} \neq \mathbf{O}$ let α be
$$\frac{1}{\sum_{i=1}^m y_i} \mathbf{y} \overset{\text{def}}{=} (\alpha_1, \ldots, \alpha_m).$$
Then $\alpha \in \Sigma_m$ and $\mathbf{y}A = (\sum_{i=1}^m y_i)\alpha A$. In other words

$$K = \{\, \mathbf{z} \; : \; \text{for some nonnegative number } P, \mathbf{z} \in P \cdot Conv(\mathbf{a}_1, \ldots \mathbf{a}_m) \,\}.$$

If (5.4.30) fails to hold, then
EITHER
$\mathbf{b}$, denoted this time $\mathbf{b}_2$, does *not* lie in K but makes an acute angle with the vector $\mathbf{a}$ that is in K and is closest to $\mathbf{b}_2$

OR there is in K no vector making an acute angle with $\mathbf{b}$, denoted $\mathbf{b}_3$. In this last circumstance the vector in K and closest to $\mathbf{b}$ is $\mathbf{O}$.

In either of the last two situations (5.4.31) means two things:

   i. that some $\mathbf{x}$ is such that $\mathbf{a}_i\mathbf{x} = (\mathbf{a}_i, \mathbf{x}^t) \geq 0$ for each row vector $\mathbf{a}_i$. This means that $\mathbf{x}^t$ makes an *acute* or a right angle with each row $\mathbf{a}_i$ of A;

  ii. at the same time $\mathbf{b}\mathbf{x} = (\mathbf{b}, \mathbf{x}^t) < 0$. This means that $\mathbf{x}^t$ makes an *obtuse* angle with $\mathbf{b}$. (It is shown below that if $\mathbf{b}$ is $\mathbf{b}_3$ then $\mathbf{x}^t = -\mathbf{b}_3 = -\mathbf{b}$.)

The interpretation of (5.4.30) in terms of angles stems from the formula in (4.4.3) relating the scalar product of two vectors with their lengths and the cosine of the angle between them. When neither vector is $\mathbf{O}$, their scalar product is nonnegative iff the cosine of the angle is nonnegative iff the angle is an acute or a right angle. The scalar product is negative iff the cosine of the angle is negative iff the angle is obtuse.

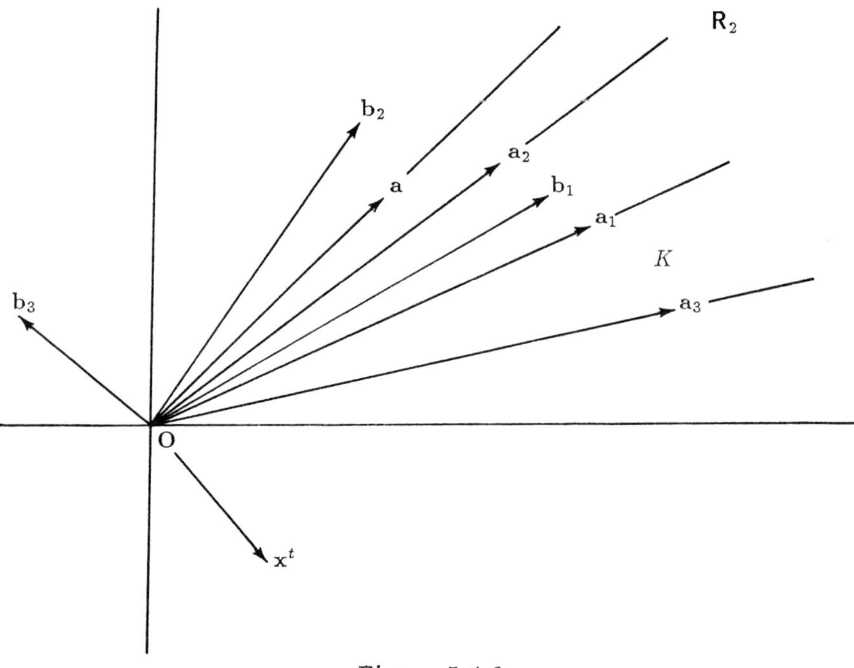

Figure 5.4.6.

Example 5.4.5. If

$$A \overset{\text{def}}{=} \begin{pmatrix} 1 & -1 \\ -1 & 1 \end{pmatrix}, \ \mathbf{y} \overset{\text{def}}{=} (y_1, y_2), \text{ and } \mathbf{b} \overset{\text{def}}{=} (1,1)$$

and if $\mathbf{y}$ is a solution of $\mathbf{y}A = \mathbf{b}$ then

$$y_1 = 1 + y_2 = 1 + (1 + y_1), \text{ i.e., } 0 = 2,$$

a contradiction. In particular there is no $\mathbf{y}$ such that $\mathbf{y} \succeq \mathbf{O}$ and $\mathbf{y}A = \mathbf{b}$. Hence, according to THEOREM **5.4.5**, there is an $\mathbf{x}$ such that

$$\mathbf{x} \stackrel{\text{def}}{=} \begin{pmatrix} x_1 \\ x_2 \end{pmatrix}, \; A\mathbf{x} \succeq \mathbf{O}, \text{ and } \mathbf{bx} < 0. \tag{$*$}$$

Indeed, if $(*)$ is true then

$$x_1 = x_2 \stackrel{\text{def}}{=} a, \; x_1 + x_2 = 2a < 0, \text{ e.g., } \mathbf{x} \stackrel{\text{def}}{=} \begin{pmatrix} -1 \\ -1 \end{pmatrix}.$$

The situation in **Figure 5.4.6** suggests a simple geometric argument for the proof, given next, of THEOREM **5.4.5**.

PROOF OF THEOREM **5.4.5**: Note first that the cone K is convex. Indeed, if $\mathbf{x}_i \in K$, $i = 1, 2$ and if $0 \le t \le 1$ then there are vectors $\mathbf{y}_i$ such that $\mathbf{y}_i \succeq \mathbf{O}$ and such that $\mathbf{x}_i = \mathbf{y}_i A$. Hence

$$\mathbf{y} \stackrel{\text{def}}{=} t\mathbf{y}_1 + (1-t)\mathbf{y}_2 \succeq \mathbf{O}$$
$$t\mathbf{x}_1 + (1-t)\mathbf{x}_2 = t\mathbf{y}_1 A + (1-t)\mathbf{y}_2 A \stackrel{\text{def}}{=} \mathbf{y}A \in K.$$

If there is no nonnegative $\mathbf{y}$ for which (5.4.30) holds then $\mathbf{b} \notin K$. Thus assume that (5.4.30) fails to hold for every nonnegative $\mathbf{y}$.

Since $\mathbf{b} \notin K$ it follows that $\mathbf{b} \ne \mathbf{O}$ and $\mathbf{O} \notin -\mathbf{b} + K$, which is also convex.

The "Apollonius-convexity" argument in the proof of THEOREM **5.4.1** yields the existence in K of a *unique* vector $\mathbf{a}$ such that $-\mathbf{b} + \mathbf{a}$ is closest, among all vectors in $-\mathbf{b} + K$, to $\mathbf{O}$.

$$\|\mathbf{a} - \mathbf{b}\| = minimum \, \{ \| - \mathbf{b} + \mathbf{c} \| \; : \; \mathbf{c} \in K \}$$

The argument in THEOREM **5.4.1** shows also that the convex set $-\mathbf{b} + K$ is contained in the half-space $S_{[\mathbf{a}-\mathbf{b}, \|\mathbf{a}-\mathbf{b}\|^2]}$.

If $\mathbf{a} = \mathbf{O}$ then for each $\mathbf{k}$ in K and each nonnegative t

$$\| - \mathbf{b} + t\mathbf{k}\|^2 \ge \|t\mathbf{k}\|^2$$
$$t^2(\mathbf{k}, \mathbf{k}) \ge 2t(\mathbf{b}, \mathbf{k})$$

whence $(\mathbf{b}, \mathbf{k}) \le 0$ for each $\mathbf{k}$ in K. Since each $\mathbf{a}_i$ is in K

$$A(-\mathbf{b})^t \succeq \mathbf{O} \text{ and } \mathbf{b}(-\mathbf{b})^t = -\|\mathbf{b}\|^2 < 0.$$

If $\mathbf{a} \neq \mathbf{O}$ it follows that $||\mathbf{a} - \mathbf{b}||^2 < ||\mathbf{b}||^2$. Upon simplification the last inequality yields $0 < ||\mathbf{a}||^2 < 2(\mathbf{a}, \mathbf{b})$ and, in particular, $(\mathbf{a}, \mathbf{b}) > 0$. (Hence the angle between $\mathbf{b}$ and $\mathbf{a}$ is acute.) Furthermore if $s \geq 0$ then $s\mathbf{a} \in K$ and

$$||\mathbf{b} - s\mathbf{a}||^2 = ||\mathbf{b}||^2 - 2s(\mathbf{b}, \mathbf{a}) + s^2||\mathbf{a}||^2 \qquad (5.4.32)$$

$$= ||\mathbf{a}||^2 \left[\left(s - \frac{(\mathbf{b}, \mathbf{a})}{||\mathbf{a}||^2} \right)^2 + \frac{||\mathbf{b}||^2}{||\mathbf{a}||^2} - \frac{(\mathbf{b}, \mathbf{a})^2}{||\mathbf{a}||^4} \right] \qquad (5.4.33)$$

the last quantity being least when

$$s = \frac{(\mathbf{b}, \mathbf{a})}{||\mathbf{a}||^2} \ \left(= \frac{(\mathbf{a}, \mathbf{b})}{||\mathbf{a}||^2} > 0 \right).$$

On the other hand, $||\mathbf{b} - s\mathbf{a}||^2$ is, by definition, least when $s = 1$. Thus

$$(\mathbf{a}, \mathbf{b}) = ||\mathbf{a}||^2. \qquad (5.4.34)$$

If $\mathbf{c} \in K$ then $-\mathbf{b} + \mathbf{c} \in -\mathbf{b} + K$. Since

$$-\mathbf{b} + K \subset S_{[\mathbf{a}-\mathbf{b}, ||\mathbf{a}-\mathbf{b}||^2]}$$

it follows that if $\mathbf{x}^t \overset{\text{def}}{=} \mathbf{a} - \mathbf{b}$ then

$$(-\mathbf{b} + \mathbf{c}, \mathbf{x}^t) \geq ||\mathbf{a} - \mathbf{b}||^2$$
$$\mathbf{c}\mathbf{x} = (\mathbf{c}, \mathbf{x}^t) \geq (\mathbf{b}, \mathbf{a} - \mathbf{b}) + ||\mathbf{a} - \mathbf{b}||^2$$
$$= (\mathbf{b}, \mathbf{a}) - ||\mathbf{b}||^2 + ||\mathbf{a}||^2 - 2(\mathbf{a}, \mathbf{b}) + ||\mathbf{b}||^2$$

and, by virtue of (5.4.34), the right member above is 0, whence

$$\mathbf{c}\mathbf{x} \geq 0.$$

Since $\mathbf{c}$ is an arbitrary vector in K, it follows that $\mathbf{y}A\mathbf{x} \geq 0$ for every nonnegative $\mathbf{y}$, in particular if $\mathbf{y}$ is any $\mathbf{e}_i$, $1 \leq i \leq m$. Hence

$$A\mathbf{x} \succeq \mathbf{O}. \qquad (5.4.35)$$

From (5.4.34) it follows that

$$(\mathbf{x}^t, \mathbf{a}) = 0$$
$$||\mathbf{x}^t||^2 + ||\mathbf{a}||^2 = ||\mathbf{b}||^2 > ||\mathbf{a}||^2$$
$$\mathbf{b}\mathbf{x} = (\mathbf{x}^t, \mathbf{b}) = (||\mathbf{a}|| - ||\mathbf{b}||)(||\mathbf{a}|| + ||\mathbf{b}||) < 0. \qquad (5.4.36)$$

Thus if $\mathbf{a} \neq \mathbf{O}$ (5.4.35) and (5.4.36) prove (5.4.31).

$$\Omega$$

[**Remark 5.4.6:** In [Ga] there is an alternative proof for THEOREM **5.4.5**.]

Exercise 5.4.14. Let $\{A; \mathrm{b}\}$ be any of the pairs listed below. Each pair consists of a matrix A and a (row) vector b. For each pair $\{A; \mathrm{b}\}$ determine which of the two alternatives in THEOREM **5.4.5** holds.

$$\left\{ \begin{pmatrix} 1 & 2 & -3 \\ -4 & 5 & 1 \end{pmatrix} ; (1,1,1) \right\} ; \left\{ \begin{pmatrix} -1 & 2 \\ 3 & -4 \\ -5 & 6 \end{pmatrix} ; (3,4) \right\} .$$

THEOREM **5.4.5** leads to the next result, which, in turn, opens the door to the fundamental facts in the theory of linear programming.

THEOREM **5.4.6.** IF A IS AN $m \times n$ MATRIX IN WHICH EACH ENTRY IS REAL AND IF $\mathrm{b} \in \mathbf{R}_n$ THEN

EITHER

I THERE IS IN $\mathbf{R}_m$ A y SUCH THAT $\mathrm{y} \succeq \mathbf{O}$ AND $\mathrm{y}A \preceq \mathrm{b}$

OR

II THERE IS IN $\mathbf{R}^n$ AN x SUCH THAT $\mathrm{x} \succeq \mathbf{O}$, $A\mathrm{x} \succeq \mathbf{O}$ AND $\mathrm{bx} < 0$. FURTHERMORE, **I** AND **II** ARE MUTUALLY EXCLUSIVE.

[**Remark 5.4.7:** Without the assertion $\mathrm{x} \succeq \mathbf{O}$ in **II** THEOREM **5.4.6** is a mere corollary to THEOREM **5.4.5**.]

PROOF. The mutual exclusivity of **I** and **II** follows from the next calculations:

$$\{\mathrm{x} \succeq \mathbf{O} \text{ and } \mathrm{y}A \preceq \mathrm{b}\} \Rightarrow \mathrm{y}A\mathrm{x} \preceq \mathrm{bx}$$
$$\mathrm{y}A\mathrm{x} \preceq \mathrm{bx} \text{ and } \mathrm{bx} < 0 \Rightarrow \mathrm{y}A\mathrm{x} < 0$$
$$\{\mathrm{y}A\mathrm{x} < 0 \text{ and } \mathrm{y} \succeq \mathbf{O}\} \Rightarrow A\mathrm{x} \not\succeq \mathbf{O}.$$

Hence **I** $\Rightarrow$ *not* **II** and so also **II** $\Rightarrow$ *not* **I**. If **I** is not true then there is in $\mathbf{R}_{m+n}$ no vector

$$\xi \stackrel{\text{def}}{=} (\xi_1, \ldots, \xi_m, \lambda_1, \ldots, \lambda_n)$$

such that

$$\xi \succeq \mathbf{O}$$
$$\sum_{i=1}^{m} \xi_i \mathrm{a}_i + \sum_{j=1}^{n} \lambda_j \mathrm{e}_j = \mathrm{b}. \tag{5.4.37}$$

If

$$\mathcal{A} \stackrel{\text{def}}{=} \begin{pmatrix} A \\ I \end{pmatrix}$$

then (5.4.37) may be rewritten

$$\xi \succeq \mathbf{O}$$
$$\xi \mathcal{A} = \mathbf{b}.$$

Hence, according to THEOREM **5.4.5**, there is in $\mathbf{R}^n$ an $\mathbf{x}$ such that

$$\mathcal{A}\mathbf{x} \succeq \mathbf{O} \text{ and } \mathbf{b}\mathbf{x} < 0,$$

i.e.,

$$\mathbf{a}_i \mathbf{x} \geq 0, \ \ 1 \leq i \leq m \tag{5.4.38}$$
$$\mathbf{e}_j \mathbf{x} \geq 0, \ \ 1 \leq j \leq n \tag{5.4.39}$$
$$\mathbf{b}\mathbf{x} < 0. \tag{5.4.40}$$

From (5.4.39) it follows that $\mathbf{x} \succeq \mathbf{O}$ and from (5.4.38) and (5.4.40) that $\mathcal{A}\mathbf{x} \succeq \mathbf{O}$ and $\mathbf{b}\mathbf{x} < 0$.

$$\Omega$$

THEOREM **5.4.7.** (the **Duality Theorem**)

i. IF

$$\{\mathbf{x}, \mathbf{y}\}$$

IS A PAIR OF FEASIBLE SOLUTIONS FOR THE DUAL PAIR

$$\{\mathbf{PLPP}, \mathbf{DLPP}\}$$

THEN THERE IS A PAIR

$$\{\overline{\mathbf{x}}, \overline{\mathbf{y}}\}$$

OF SOLUTIONS FOR THE DUAL PAIR $\{\mathbf{PLPP}, \mathbf{DLPP}\}$ AND

$$\overline{\mathbf{y}}\mathbf{p} = \mathbf{c}\overline{\mathbf{x}}.$$

ii. IF EITHER OF THE DUAL PAIR HAS A SOLUTION THEN SO DOES THE OTHER, i.e., EITHER *both* MEMBERS OF THE DUAL PAIR HAVE SOLUTIONS OR *neither* HAS A SOLUTION.

PROOF. ad*i*. Since THEOREM **5.4.6** is central to the argument, it is convenient to define the following block matrix and block vectors:

$$\widetilde{\mathcal{A}} \stackrel{\text{def}}{=} \begin{pmatrix} O & A & -\mathbf{p} \\ -A^t & O & \mathbf{c}^t \end{pmatrix}$$
$$\eta \stackrel{\text{def}}{=} (\mathbf{w}, \mathbf{z}^t)$$
$$\beta \stackrel{\text{def}}{=} (-\mathbf{p}^t, \mathbf{c}, 0).$$

   Chapter 5. APPLYING LINEAR ALGEBRA

If $\{x, y\}$ is a pair of feasible solutions for the dual pair $\{PLPP, DLPP\}$ and if the dual pair has no solution, then owing to THEOREM **5.4.4**, there is no nonnegative vector η that is a solution of the system

$$\eta \widetilde{A} \preceq \beta,$$

i.e.,

$$- z^t A^t \preceq -p^t$$
$$w A \preceq c$$
$$- wp + z^t c^t \leq 0. \tag{5.4.41}$$

Hence there is a nonnegative vector

$$\xi \stackrel{\text{def}}{=} \begin{pmatrix} u^t \\ v \\ s \end{pmatrix}$$

such that

$$\widetilde{A}\xi \succeq O \text{ and } \beta\xi < 0$$

i.e.,

$$Av \succeq s p \tag{5.4.42}$$
$$u A \preceq s c \tag{5.4.43}$$
$$cv < p^t u^t \ (= up). \tag{5.4.44}$$

The feasibility of $\{x, y\}$ and the nonnegativity of ξ imply that

$$Ax \succeq p$$
$$yA \preceq c$$
$$uAx \succeq up \tag{5.4.45}$$
$$yAv \preceq cv. \tag{5.4.46}$$

If $s = 0$ then (5.4.42), (5.4.43), (5.4.45), and (5.4.46) imply

$$up \leq uAx \leq 0 \leq yAv \leq cv,$$

contradicting (5.4.44). Hence $s > 0$. But then

$$\{\frac{v}{s}, \frac{u}{s}\}$$

is a feasible pair, whence, according to THEOREM **5.4.4**,

$$\frac{u}{s}p \leq c\frac{v}{s},$$

---

contradicting (5.4.44).

In sum, there is a pair $\{\widetilde{\mathbf{x}}, \widetilde{\mathbf{y}}\}$ of nonnegative vectors serving as a solution for the system (5.4.41) and THEOREM **5.4.4** implies that the pair $\{\widetilde{\mathbf{x}}, \widetilde{\mathbf{y}}\}$ is a solution for the dual pair $\{\mathbf{PLPP}, \mathbf{DLPP}\}$

ad*ii*. If the **DLPP** has no feasible solution then THEOREM **5.4.5** implies there is a nonnegative vector $\mathbf{x}_0$ such that

$$A\mathbf{x}_0 \succeq \mathbf{O}$$

$$\mathbf{c}\mathbf{x}_0 \overset{\text{def}}{=} q < 0.$$

If the **PLPP** has a feasible solution, say $\mathbf{w}$, then according to *i* the dual pair $\{\mathbf{PLPP}, \mathbf{DLPP}\}$ has a solution. On the other hand, for all nonnegative r

$$A(\mathbf{w} + r\mathbf{x}_0) \succeq \mathbf{p}$$

$$\mathbf{c}(\mathbf{w} + r\mathbf{x}_0) = \mathbf{c}\mathbf{w} + rq.$$

Since r may be arbitrarily large and positive and since $q < 0$ it follows that

$$infimum\,\{\,\mathbf{c}\mathbf{x}\ :\ A\mathbf{x} \succeq \mathbf{p}\,\} = -\infty,$$

in contradiction of the existence of a solution for the **PLPP**.

Since the dual of the **DLPP** is the **PLPP** it follows that if either of the dual pair $\{\mathbf{PLPP}, \mathbf{DLPP}\}$ fails to have a feasible solution then so does the other, i.e., if either has a feasible solution so does the other and the dual pair $\{\mathbf{PLPP}, \mathbf{DLPP}\}$ has a solution.

$$\Omega$$

Exercise 5.4.15. Associate with the dual pair $\{\mathbf{PLPP}, \mathbf{DLPP}\}$ the matrix and vector

$$\mathcal{A} \overset{\text{def}}{=} \begin{pmatrix} O & A & -\mathbf{p} \\ -A^t & O & \mathbf{c}^t \\ \mathbf{p}^t & -\mathbf{c} & 0 \end{pmatrix} \tag{5.4.47}$$

$$\xi \overset{\text{def}}{=} \begin{pmatrix} \mathbf{y}^t \\ \mathbf{x} \\ 1 \end{pmatrix}. \tag{5.4.48}$$

Show that the dual pair $\{\mathbf{PLPP}, \mathbf{DLPP}\}$ has a solution iff there is a nonnegative vector ξ that is a solution of the system

$$\mathcal{A}\xi \succeq \mathbf{O}. \tag{5.4.49}$$

The relationship between linear programming and game theory is revealed in the next THEOREMs. THEOREM **5.4.8** shows how and in what circumstances one can find a solution for a dual pair $\{\textbf{PLPP}, \textbf{DLPP}\}$ via the pair of optimal strategies for the game defined by the matrix $\mathcal{A}$ for the dual pair. Since a game always has a pair of optimal strategies whereas a dual pair may fail to have a solution it follows that the hypothesis for each THEOREM must include some hedge.

THEOREM **5.4.8.** LET $\mathcal{A}$ IN (5.4.47) BE THE MATRIX OF A GAME. LET Σ_{m+n+1} RESP. $\Sigma^{(m+n+1)}$, $\widetilde{\Sigma}_{m+n+1}$, RESP. $\widetilde{\Sigma}^{(m+n+1)}$, AND $\textbf{V}(\mathcal{A})$ HAVE THEIR USUAL MEANINGS. (IN PARTICULAR, IF $\widetilde{\alpha} \in \widetilde{\Sigma}_{m+n+1}$ AND $\widetilde{\beta} \in \widetilde{\Sigma}^{(m+n+1)}$ THEN $\textbf{V}(\mathcal{A}) = \widetilde{\alpha}\mathcal{A}\widetilde{\beta}$.) THEN THERE IS A NONNEGATIVE VECTOR

$$\begin{pmatrix} \mathbf{y}^t \\ \mathbf{x} \\ 1 \end{pmatrix}$$

THAT IS A SOLUTION FOR (5.4.49) IFF THERE IS SOME $\widetilde{\beta}$ SUCH THAT

$$\widetilde{\Sigma}^{(m+n+1)} \ni \widetilde{\beta} \overset{\text{def}}{=} \begin{pmatrix} \beta_1 \\ \vdots \\ \beta_{m+n+1} \end{pmatrix} \quad \text{AND (HEDGE)} \quad \beta_{m+n+1} > 0.$$

PROOF. Since $\mathcal{A}$ is antisymmetric ($\mathcal{A}^t = -\mathcal{A}$) it follows (cf. **Exercise 5.4.12**) that $\textbf{V}(\mathcal{A}) = 0$. If

$$\begin{pmatrix} \overline{\beta}_1 \\ \vdots \\ \overline{\beta}_{m+n+1} \end{pmatrix} \overset{\text{def}}{=} \overline{\beta} \in \Sigma^{(m+n+1)}$$

and if $\overline{\beta}_{m+n+1} \neq 0$ then for any α in Σ_{m+n+1}

$$\alpha\mathcal{A}\overline{\beta} \geq \textbf{V}(\mathcal{A}) = 0.$$

If $\mathcal{A}\overline{\beta} \not\succeq \textbf{O}$ then a suitable choice of α shows that

$$0 > \alpha\mathcal{A}\overline{\beta} \geq \textbf{V}(\mathcal{A}) = 0,$$

a contradiction. Hence $\mathcal{A}\overline{\beta} \succeq \mathbf{O}$ and so

$$
\begin{array}{llll}
a_{11}\overline{\beta}_{m+1} & +\cdots & +a_{1n}\overline{\beta}_{m+n} & -\overline{\beta}_{m+n+1}p_1 \geq 0 \\
\vdots & \ddots & \vdots & \vdots \\
a_{m1}\overline{\beta}_{m+1} & +\cdots & +a_{mn}\overline{\beta}_{m+n} & -\overline{\beta}_{m+n+1}p_m \geq 0 \\
-\overline{\beta}_1 a_{11} & -\cdots & -\overline{\beta}_m a_{m1} & +\overline{\beta}_{m+n+1}c_1 \geq 0 \\
\vdots & \ddots & \vdots & \vdots \\
-\overline{\beta}_1 a_{1n} & -\cdots & -\overline{\beta}_m a_{mn} & +\overline{\beta}_{m+n+1}c_n \geq 0
\end{array}
$$

$$
\overline{\beta}_i \geq 0,\ 1 \leq i \leq m+n+1,\ \sum_{k=1}^{m+n+1} \overline{\beta}_k = 1
$$

$$
p_1\overline{\beta}_1 + \cdots + p_m\overline{\beta}_m - c_1\overline{\beta}_{m+1} - \cdots - c_n\overline{\beta}_{m+n} \geq 0. \tag{5.4.50}
$$

If

$$
x_j \stackrel{\mathrm{def}}{=} \frac{\overline{\beta}_{m+j}}{\overline{\beta}_{m+n+1}},\ 1 \leq j \leq n
$$

$$
y_i \stackrel{\mathrm{def}}{=} \frac{\overline{\beta}_i}{\overline{\beta}_{m+n+1}},\ 1 \leq i \leq m
$$

and if

$$
\mathbf{x} \stackrel{\mathrm{def}}{=} \begin{pmatrix} x_1 \\ \vdots \\ x_n \end{pmatrix},\ \mathbf{y} \stackrel{\mathrm{def}}{=} (y_1, \ldots, y_m),\ \text{and}\ \xi \stackrel{\mathrm{def}}{=} \begin{pmatrix} \mathbf{y}^t \\ \mathbf{x} \\ 1 \end{pmatrix}
$$

then ξ is a nonnegative solution of (5.4.49).

Conversely let ξ as above be a nonnegative solution of (5.4.49) and let ζ be ξ^t.

A direct calculation shows since $\mathcal{A}$ is antisymmetric that

$$
\mathcal{A}\xi \succeq \mathbf{O} \text{ and } \zeta\mathcal{A} \preceq \mathbf{O}.
$$

Since $\xi \succeq \mathbf{O}$ and $\xi_{m+n+1} = 1$ it follows that

$$
\|\xi\|_1 \stackrel{\mathrm{def}}{=} \sum_{k=1}^{m+n+1} |\xi_k| = \sum_{k=1}^{m+n+1} \xi_k \geq 1.
$$

Hence

$$
\overline{\alpha} \stackrel{\mathrm{def}}{=} \frac{1}{\|\xi\|_1}\xi \in \Sigma^{(m+n+1)},\ \overline{\beta} \stackrel{\mathrm{def}}{=} \frac{1}{\|\zeta\|_1}\zeta \in \Sigma_{m+n+1}
$$

$$
\overline{\alpha}\mathcal{A}\overline{\beta} \geq 0 \geq \overline{\alpha}\mathcal{A}\overline{\beta}
$$

whence $\overline{\alpha}\mathcal{A}\overline{\beta} = 0 = \mathbf{V}(\mathcal{A})$. In particular, $\overline{\beta} \in \widetilde{\Sigma}^{(m+n+1)}$ and

$$
\overline{\beta}_{m+n+1} = \frac{1}{\|\zeta\|_1} \neq 0.
$$

$$\Omega$$

  Chapter 5. *APPLYING LINEAR ALGEBRA*

THEOREM 5.4.9. LET $A \stackrel{\text{def}}{=} (a_{ij})_{i,j=1}^{m,n}$ BE A MATRIX IN WHICH EACH ENTRY IS REAL AND LET $\tilde{\mathbf{V}}$ BE A NUMBER SO LARGE THAT $\mathbf{V}(A) + \tilde{\mathbf{V}} > 0$. CORRESPONDINGLY DEFINE b_{ij} TO BE $a_{ij} + \tilde{\mathbf{V}}$. THUS

$$B \stackrel{\text{def}}{=} (b_{ij})_{i,j=1}^{m,n} = A + \sum_{i,j=1}^{m,n} \tilde{\mathbf{V}} U_{ij}$$

(CF. **Exercise 1.1.15** WHERE THE *matrix units* U_{ij} ARE DEFINED). FINDING OPTIMAL STRATEGIES $\tilde{\alpha}$ AND $\tilde{\beta}$ FOR P_1 AND P_2 IN THE GAME REPRESENTED BY A IS EQUIVALENT TO FINDING A SOLUTION PAIR $\{\overline{\mathbf{x}}, \overline{\mathbf{y}}\}$ FOR $\{\mathbf{PLPP}, \mathbf{DLPP}\}$ ASSOCIATED WITH THE MATRIX (HEDGE)

$$\tilde{\mathcal{B}} \stackrel{\text{def}}{=} \begin{pmatrix} 0 & \cdots & 0 & b_{11} & \cdots & b_{1n} & -1 \\ \vdots & \ddots & \vdots & \vdots & \ddots & \vdots & \vdots \\ 0 & \cdots & 0 & b_{m1} & \cdots & b_{mn} & -1 \\ -b_{11} & \cdots & -b_{m1} & 0 & \cdots & 0 & 1 \\ \vdots & \ddots & \vdots & \vdots & \ddots & \vdots & \vdots \\ -b_{1n} & \cdots & -b_{mn} & 0 & \cdots & 0 & 1 \\ 1 & \cdots & 1 & -1 & \cdots & -1 & 0 \end{pmatrix}. \tag{5.4.51}$$

PROOF. If $\tilde{A}$ is the matrix resulting from adding to each entry of A the number $\tilde{\mathbf{V}}$, i.e., if

$$\tilde{A} = A + \tilde{\mathbf{V}} \sum_{i,j=1}^{m,n} U_{ij},$$

then $\mathbf{V}(\tilde{A}) = \mathbf{V}(A) + \tilde{\mathbf{V}}$ and a pair $\{\tilde{\alpha}, \tilde{\beta}\}$ of strategies is optimal for A iff it is optimal for $\tilde{A}$. Thus it may be assumed at the outset that $\mathbf{V}(A) > 0$, i.e., that $A = \tilde{A}$.

The strategy $\tilde{\alpha}$ is optimal for P_1, i.e., $\tilde{\alpha} \in \tilde{\Sigma}_m$, iff for all strategies β in $\Sigma^{(n)}$

$$\tilde{\alpha} A \beta \leq \mathbf{V}(A).$$

Similarly the strategy $\tilde{\beta}$ is optimal for P_2, i.e., $\tilde{\beta} \in \tilde{\Sigma}^{(n)}$, iff for all strategies α in Σ_m

$$\alpha A \tilde{\beta} \geq \mathbf{V}(A).$$

In particular, if $\beta = \mathbf{e}_j \in \mathbf{R}^m$ and if $\alpha = \mathbf{e}_i \in \mathbf{R}_m$ it follows that

$$A\tilde{\beta} \succeq \begin{pmatrix} \mathbf{V}(A) \\ \vdots \\ \mathbf{V}(A) \end{pmatrix} \tag{5.4.52}$$

$$\tilde{\alpha} A \preceq (\mathbf{V}(A), \ldots, \mathbf{V}(A)).$$

[**Note 5.4.2:** From (5.4.52) it follows that every entry of A is positive.]
In other words, if

$$\mathbf{E}^{(m)} \overset{\text{def}}{=} \begin{pmatrix} 1 \\ \vdots \\ 1 \end{pmatrix} = \sum_{i=1}^{m} \mathbf{e}_i$$

$$\mathbf{E}_n \overset{\text{def}}{=} (1, \ldots, 1) = \sum_{j=1}^{n} \mathbf{e}_j$$

then

$$A\widetilde{\beta} \succeq \mathbf{V}(A)\mathbf{E}^{(m)}$$
$$\widetilde{\alpha}A \preceq \mathbf{V}(A)\mathbf{E}_n$$
$$\widetilde{\alpha}\mathbf{V}(A)\mathbf{E}^{(m)} = \mathbf{V}(A)\mathbf{E}_n\widetilde{\beta} = \mathbf{V}(A).$$

Hence if

$$\overline{\mathbf{x}} = \frac{1}{\mathbf{V}(A)}\widetilde{\beta} \text{ and } \overline{\mathbf{y}} = \frac{1}{\mathbf{V}(A)}\widetilde{\alpha}$$

then

$$A\overline{\mathbf{x}} \geq \mathbf{E}^{(m)}$$
$$\overline{\mathbf{y}}A \leq \mathbf{E}_n$$
$$\mathbf{O} \neq \overline{\mathbf{x}} \succeq \mathbf{O}, \ \mathbf{O} \neq \overline{\mathbf{y}} \succeq \mathbf{O}$$
$$\overline{\mathbf{y}}\mathbf{E}^{(m)} \geq \mathbf{E}_n\overline{\mathbf{x}}$$

whence $\{\overline{\mathbf{x}}, \overline{\mathbf{y}}\}$ is a solution pair for the dual pair $\{\mathbf{PLPP}, \mathbf{DLPP}\}$ of linear programming problems associated with the matrix $\widetilde{\mathcal{A}}$.

Conversely, if $\{\overline{\mathbf{x}}, \overline{\mathbf{y}}\}$ is a solution pair of nonnegative vectors associated with (5.4.51) (which, by assumption, is

$$\begin{pmatrix} 0 & \ldots & 0 & a_{11} & \ldots & a_{1n} & -1 \\ \vdots & \ddots & \vdots & \vdots & \ddots & \vdots & \vdots \\ 0 & \ldots & 0 & a_{m1} & \ldots & a_{mn} & -1 \\ -a_{11} & \ldots & -a_{m1} & 0 & \ldots & 0 & 1 \\ \vdots & \ddots & \vdots & \vdots & \ddots & \vdots & \vdots \\ -a_{1n} & \ldots & -a_{mn} & 0 & \ldots & 0 & 1 \\ 1 & \ldots & 1 & -1 & \ldots & -1 & 0 \end{pmatrix})$$

then

$$A\overline{\mathbf{x}} \succeq \mathbf{E}^{(m)}, \ \mathbf{y}A \preceq \mathbf{E}_n,$$

whence $\overline{\mathbf{x}}, \overline{\mathbf{y}} \neq \mathbf{O}$.

Thus at least one of the components $\overline{x}_j$ of $\overline{\mathbf{x}}$ is not 0 (and hence is positive) and at least one of the components $\overline{y}_i$ of $\overline{\mathbf{y}}$ is not 0 (and hence is positive). It follows that s

$$\sum_{j=1}^{n} \overline{x}_j = \sum_{i=1}^{m} \overline{y}_i \overset{\text{def}}{=} \mathbf{V} > 0.$$

Hence if

$$\widetilde{\alpha} \stackrel{\text{def}}{=} \frac{1}{\mathbf{V}}\overline{\mathbf{y}} \quad \text{and} \quad \widetilde{\beta} \stackrel{\text{def}}{=} \frac{1}{\mathbf{V}}\overline{\mathbf{x}}$$

then $\widetilde{\alpha} \in \Sigma_m$ and $\widetilde{\beta} \in \Sigma^{(n)}$. Furthermore

$$A\widetilde{\beta} \succeq \mathbf{V}\mathbf{E}^{(m)}$$
$$\widetilde{\alpha}A \preceq \mathbf{V}\mathbf{E}_n.$$

Hence for all α in Σ_m and all β in $\Sigma^{(n)}$

$$\alpha A\widetilde{\beta} \geq \mathbf{V} \geq \widetilde{\alpha}A\beta.$$

In particular,

$$\widetilde{\alpha}A\widetilde{\beta} \geq \mathbf{V} \geq \widetilde{\alpha}A\widetilde{\beta}. \tag{5.4.53}$$

On the other hand,

$$minimum_{\alpha \in \Sigma_m} \alpha A\widetilde{\beta} \geq \mathbf{V}$$
$$\mathbf{V}(A) = maximum_{\beta \in \Sigma^{(n)}} minimum_{\alpha \in \Sigma_m} \alpha A\beta \geq \mathbf{V}$$
$$\mathbf{V} \geq maximum_{\beta \in \Sigma^{(n)}} \widetilde{\alpha}A\beta \tag{5.4.54}$$
$$\mathbf{V} \geq minimum_{\alpha \in \Sigma_m} maximum_{\beta \in \Sigma^{(n)}} \alpha A\beta = \mathbf{V}(A). \tag{5.4.55}$$

It follows from (5.4.53) that $\mathbf{V} = \widetilde{\alpha}A\widetilde{\beta}$ and from (5.4.54) and (5.4.55) that $\mathbf{V} = \mathbf{V}(A)$, whence

$$\mathbf{V}(A) = \widetilde{\alpha}A\widetilde{\beta}.$$

$$\Omega$$

Exercise 5.4.16.

i. Show that if the objective function is unbounded below in the **PLPP** resp. unbounded above in the **DLPP** then Q in the **DLPP** resp. the **PLPP** is empty.

ii. Show that if Q for both the **PLPP** and the **DLPP** are not empty then both the **PLPP** and the **DLPP** have solutions.

iii. Give an example of a **PLPP** for which Q for *both* the **PLPP** and the **DLPP** are empty.

The conclusions above suggest that finding the optimal mixed strategies for a game might be approached by finding the solutions of an associated dual pair of linear programming problems and that finding solutions, if they exist, of a linear programming problem might be approached by finding a pair of optimal strategies for an associated game.

There are iterative methods for finding a pair of optimal mixed strategies for a game corresponding to an $m \times n$ matrix $A \stackrel{\text{def}}{=} (a_{ij})_{i,j=1}^{m,n}$.

In effect Robinson [**Ro**] shows that if the game is played R times and:

i. P_2 uses $\mathbf{e}_1$ k_1 times, ..., $\mathbf{e}_n$ k_n times ($k_1 + \cdots + k_n = R$);
ii. P_1 uses $\mathbf{e}_1$ l_1 times, ..., $\mathbf{e}_m$ l_m times ($l_1 + \cdots + l_m = R$);
iii.

$$\boldsymbol{\alpha}_R \stackrel{\text{def}}{=} \frac{1}{R} \sum_{i=1}^{m} l_i \mathbf{e}_i;$$

iv.

$$\boldsymbol{\beta}_R \stackrel{\text{def}}{=} \frac{1}{R} \sum_{j=1}^{n} k_j \mathbf{e}_j;$$

v. in the $(R+1)$st play of the game P_1 chooses in Σ_m the vector $\mathbf{e}_{i_0}$ such that

$$\mathbf{e}_{i_0} A \boldsymbol{\beta}_R = maximum_{1 \leq i \leq n} \mathbf{e}_i A \boldsymbol{\beta}_R;$$

vi. in the $(R+1)$st play of the game P_2 chooses in $\Sigma^{(n)}$ the vector $\mathbf{e}_{j_0}$ such that

$$\boldsymbol{\alpha}_R A \mathbf{e}_{j_0} = minimum_{1 \leq j \leq n} \boldsymbol{\alpha}_R A \mathbf{e}_j;$$

then

$$\lim_{R \to \infty} \boldsymbol{\alpha}_R \stackrel{\text{def}}{=} \alpha \text{ resp. } \lim_{R \to \infty} \boldsymbol{\beta}_R \stackrel{\text{def}}{=} \beta$$

exist and are in $\widetilde{\Sigma}_m$ resp. $\widetilde{\Sigma}^{(n)}$.

Hence by starting with an arbitrary pair $\mathbf{e}_i$ and $\mathbf{e}_j$ of (pure) strategies and using the iterative procedure just described, one can study the game and find approximately optimal mixed strategies for it. Computers can quickly carry out sufficiently many steps of such an iterative procedure and thereby find satisfactory (nearly optimal) strategies for games and, in some instances, solutions for linear programming problems.

If the task is to solve a linear programming problem and if the averaging approximants for the associated game conform to the conditions of THEOREMs **5.4.8** and **5.4.9** then there are corresponding approximate solutions to the the linear programming problem.

If a **PLLP** has a solution it can always be found by the so-called *simplex method*, a device that has proven extremely useful in practice. The method is an outgrowth of work done by several authors although the systematization is properly attributed to Dantzig ([**Da**]) and, for a special facet ("cycling"), to Charnes ([**Ch**]).

There are several computer programs that run the Fourier/Hitchcock/Dantzig/Charnes/*et al* algorithm. These programs are used widely and effectively in commerce, industry, and engineering.

Several years after the development of the simplex method Khachiyan ([**Ha, Kh**]) discovered the *"ellipsoid"* algorithm to be applied to the **PLPP**. Finally

---

([**Karm**]) developed a different algorithm for solving the **PLPP**. His algorithm is, in worst cases, faster even than the Khachiyan algorithm. Furthermore, the Karmarkar algorithm has proven to be economically programmable. In fact, at this writing, software exists for the implementation of the Karmarkar algorithm and some now regard the Dantzig/Charnes/Bland simplex method as destined for replacement by the Karmarkar algorithm.

The data of the **PLPP** are the entries in A and c. There are constants $\mathcal{D}$, $\mathcal{K}_\infty$, and $\mathcal{K}_\in$ depending only on the data of the **PLPP** but *not* on the matrix size $m \times n$. In a worst case situation the Dantzig algorithm can require as many as $\mathcal{D}(m + n)^{\frac{n}{2}}$ computations (cf. [**Kl**]), the Khachiyan algorithm can require as many as $\mathcal{K}_\infty n^6$ computations, and the Karmarkar algorithm can require as many $\mathcal{K}_\in n^{3.5}$ computations. For the Khachiyan and Karmarkar algorithms the number of computations is bounded by a function that grows no faster than a *polynomial* function of n. For the Dantzig algorithm the number of computations can grow as rapidly as an *exponential* function of n. For large n the difference is decisive. For example if $n = 50$ the Dantzig bound is at least 3×10^{42} whereas the Karmarkar bound is about 9×10^5.

On the other hand, Borgwardt [**Bor**] and Smale [**Sm1**], [**Sm2**] show that, in a well-defined sense, the simplex method algorithm "on the average" grows *polynomially*. In other words, the chance of running into a **PLPP** for which the computations are inordinately lengthy is small.

In view of the developments noted and of the highly technical nature of both the Dantzig and Karmarkar algorithms they are treated in conceptual terms in the following **Sections**.

5.5. The simplex method: Theory (and some practice)

Let $A \stackrel{\text{def}}{=} (a_{ij})_{i,j=1}^{m,n}$ be a matrix with real entries and let p resp. c be vectors in $\mathbf{R}^m$ resp. $\mathbf{R}_n$. The primal problem **PLPP** is:

Find in $\mathbf{R}^n$ a vector $\mathbf{x}_{Opt}$ such that

$$A\mathbf{x}_{Opt} \succeq \mathbf{p} \tag{5.5.1}$$

$$\mathbf{x}_{Opt} \succeq \mathbf{O} \tag{5.5.2}$$

$$\mathbf{c}\mathbf{x}_{Opt} \leq \mathbf{c}\mathbf{x} \text{ for all } \mathbf{x} \text{ such that } \mathbf{x} \succeq \mathbf{O} \text{ and } A\mathbf{x} \succeq \mathbf{p} \tag{5.5.3}$$

$$= minimum\{\,\mathbf{c}\mathbf{x} \; : \; \mathbf{x} \succeq \mathbf{O}, \; A\mathbf{x} \succeq \mathbf{p}\,\}.$$

Any vector x such that $\mathbf{x} \succeq \mathbf{O}$ and $A\mathbf{x} \succeq \mathbf{O}$ is a *feasible solution*. Thus x is a feasible solution iff $\mathbf{x} \in \mathcal{Q}$. Any vector $\mathbf{x}_{Opt}$ satisfying (5.5.1) - (5.5.3) is an *optimal solution* or, more simply, a *solution of the* **PLPP**.

The function $z : \mathbf{x} \mapsto \mathbf{c}\mathbf{x}$ is the *objective function* and its *value* cx is the *Cost* at x: $Cost = \mathbf{c}\mathbf{x} = z(\mathbf{x})$.

The conditions (5.5.1) resp. (5.5.3) are alternatives to

$$A\mathbf{x}_{Opt} \preceq \mathbf{p}$$

$$A\mathbf{x}_{Opt} = \mathbf{p}$$

resp.

$$\mathbf{cx}_{Opt} \geq \mathbf{cx} \text{ for all } \mathbf{x} \text{ such that } \mathbf{x} \succeq \mathbf{O} \text{ and } A\mathbf{x} \succeq \mathbf{p}$$
$$\mathbf{cx}_{Opt} \leq \mathbf{cx} \text{ for all } \mathbf{x} \text{ such that } \mathbf{x} \succeq \mathbf{O} \text{ and } A\mathbf{x} \succeq \mathbf{p}$$
$$\mathbf{cx}_{Opt} \leq \mathbf{cx} \text{ for all } \mathbf{x} \text{ such that } \mathbf{x} \succeq \mathbf{O} \text{ and } A\mathbf{x} \preceq \mathbf{p}.$$

etc.

All told, the original, the alternatives, and mixes thereof give rise to a large number of different formulations of the **PLPP**.

To be useful the simplex method is best developed in the context of a standard format into which all linear programming problems can be thrown. Some of the adjustments useful in achieving the standard format are given next.

  i. Every maximization problem can be replaced by a minimization problem via appropriate sign reversals:

$$maximum\left\{ \mathbf{cx} \,:\, \mathbf{x} \succeq \mathbf{O}, A\mathbf{x} \succeq \mathbf{p} \right\} = -minimum\left\{ -\mathbf{cx} \,:\, \mathbf{x} \succeq \mathbf{O}, A\mathbf{x} \succeq \mathbf{p} \right\}.$$

Hence the standard format of the **PLPP** *always* asks for a minimum.

  ii. Some linear programming problems, e.g., the **DLPP**, involve constraints of the "$\leq$" kind and in fact there is no reason to exclude constraints of the "$=$" kind. However, each constraint drawn from a system such as $A\mathbf{x} \preceq \mathbf{p}$ or a system such as $A\mathbf{x} = \mathbf{p}$ or from a mixture of such systems can be brought into a standard *equational* format $A\mathbf{X} = \mathbf{q}$ coupled with a single type of constraint of the type $\mathbf{X} \succeq \mathbf{O}$, signifying only the NONNEGATIVITY of the variables. The trick is to modify inequalities or to introduce first more inequalities and then introduce more variables.

For example:

  i. Changing signs on both side of a "$\geq$" constraint produces a "$\leq$" constraint and vice versa;
  ii. Replacing an "$=$" constraint by a pair of constraints, one a "$\leq$" constraint and the other a "$\geq$" constraint, then reversing signs on the "$\leq$" constraint leads ultimately to the replacement of an "$=$" constraint by two "$\geq$" constraints. Thus the **PLPP** is to minimize *Cost* while $\mathbf{x}$ is constrained by the conditions of the form: $\mathbf{x} \succeq \mathbf{O}$ and $A\mathbf{x} \succeq \mathbf{p}$.
  iii. Since $A\mathbf{x} \succeq \mathbf{p}$ iff there is some vector $\mathbf{w}$ such that

$$\mathbf{w} \succeq \mathbf{O} \text{ and } A\mathbf{x} - \mathbf{w} = \mathbf{p},$$

introducing as NONNEGATIVE *slack* variables the m components of $\mathbf{w}$ converts each constraint in $A\mathbf{x} \succeq \mathbf{p}$ into an equational constraint.

Thus define the vector $\mathbf{X}$ in $\mathbf{R}^{n+m}$ to be

$$\mathbf{X} \overset{\text{def}}{=} \begin{matrix} 1 \\ n \\ m \end{matrix} \begin{pmatrix} \mathbf{x} \\ \mathbf{w} \end{pmatrix} \overset{\text{def}}{=} \begin{pmatrix} x_1 \\ \vdots \\ x_n \\ w_1 \\ \vdots \\ w_m \end{pmatrix} \overset{\text{def}}{=} \begin{pmatrix} X_1 \\ \vdots \\ X_{n+m} \end{pmatrix}.$$

Correspondingly define the matrix $m \times (n + m)$ matrix $\widetilde{A}$ to be

$$\widetilde{A} \overset{\text{def}}{=} m \;\;\overset{\displaystyle n \;\;\; m}{\left(A \;\; -I \right)}.$$

Finally, define $\mathbf{C}$ be the $1 \times (n + m)$ vector to be

$$\mathbf{C} \overset{\text{def}}{=} 1 \;\;\overset{\displaystyle n \;\;\; m}{\left(\mathbf{c} \;\; \mathbf{O} \right)}.$$

Then the **PLPP** is equivalent to seeking an $\mathbf{X}_{Opt}$ such that

$$\widetilde{A}\mathbf{X}_{Opt} = \mathbf{p} \tag{5.5.4}$$

$$\mathbf{X}_{Opt} \succeq \mathbf{O} \tag{5.5.5}$$

$$\mathbf{C}\mathbf{X}_{Opt} \leq \mathbf{C}\mathbf{X} \text{ for all } \mathbf{X} \text{ such that } \mathbf{X} \succeq \mathbf{O} \text{ and } \widetilde{A}\mathbf{X} = \mathbf{p} \tag{5.5.6}$$

$$= minimum \left\{ Cost \overset{\text{def}}{=} \mathbf{C}\mathbf{X} \;:\; \mathbf{X} \succeq \mathbf{O}, \; \widetilde{A}\mathbf{X} = \mathbf{p} \right\}.$$

If $\mathbf{p} \succeq \mathbf{O}$ the system above is in standard format. However if some components of $\mathbf{p}$ are negative then a change of all signs in both members of the corresponding equations of (5.5.4) gives rise to a matrix A replacing $\widetilde{A}$ and a vector $\mathbf{q}$ replacing $\mathbf{p}$.

Indeed, let $\widetilde{\mathbf{a}_i}$ resp. $\mathbf{a}_i$ be the ith row of $\widetilde{A}$ resp. A and let q_i be the ith component of $\mathbf{q}$. Then:

$$\mathbf{a}_i \overset{\text{def}}{=} \begin{cases} \widetilde{\mathbf{a}}_i & \text{if } p_i \geq 0 \\ -\widetilde{\mathbf{a}}_i & \text{if } p_i < 0 \end{cases}$$

$$A \overset{\text{def}}{=} \begin{pmatrix} \mathbf{a}_{11} & \cdots & \mathbf{a}_{1n} \\ \vdots & \ddots & \vdots \\ \mathbf{a}_{m1} & \cdots & \mathbf{a}_{mn} \end{pmatrix}$$

$$q_i \overset{\text{def}}{=} \begin{cases} p_i & \text{if } p_i \geq 0 \\ -p_i & \text{if } p_i < 0 \end{cases}$$

$$\mathbf{q} \overset{\text{def}}{=} \begin{pmatrix} q_1 \\ \vdots \\ q_m \end{pmatrix}.$$

The standard formulation of the **PLPP** is thus to seek an $\mathbf{X}_{Opt}$ such that

$$A\mathbf{X}_{Opt} = \mathbf{q} \tag{5.5.7}$$

$$\mathbf{X}_{Opt} \succeq \mathbf{O}, \; \mathbf{q} \succeq \mathbf{O} \tag{5.5.8}$$

$$\mathbf{C}\mathbf{X}_{Opt} \leq \mathbf{C}\mathbf{X} \text{ for all } \mathbf{X} \text{ such that } \mathbf{X} \succeq \mathbf{O} \text{ and } A\mathbf{X} = \mathbf{q} \tag{5.5.9}$$

$$= minimum \left\{ \mathbf{C}\mathbf{X} \;:\; \mathbf{X} \succeq \mathbf{O}, \; A\mathbf{X} = \mathbf{q} \right\}.$$

The condition $\mathbf{X} \succeq \mathbf{O}$ means that each component X_j is nonnegative:

$$X_j \geq 0, \ 1 \leq j \leq n + m.$$

Such a vector is *nonnegative*. The set of all nonnegative vectors in any $\mathbf{R}^k$ is the nonnegative *orthant* of $\mathbf{R}^k$ and is denoted $\mathbf{R}^{(k,+)}$. Similarly a vector $\mathbf{X}$ is *positive* iff $\mathbf{X} \succ \mathbf{O}$. The algebraic treatment of the **PLPP** is more easily discussed in terms of orthants and of nonnegative and positive vectors.

Example 5.5.1. The nonnegative orthant $\mathbf{R}^{(2,+)}$ of $\mathbf{R}^2$ is the *first quadrant*; in $\mathbf{R}^3$ the nonnegative orthant $\mathbf{R}^{(3,+)}$ is the *first octant*.

The simplex method, for the problem as described by (5.5.1) - (5.5.3) has a simple geometric description in the context of $\mathbf{R}^{(n,+)}$. The experience in the context of $\mathbf{R}^{(2,+)}$ as exemplified in **Section 5.4** and earlier in **Section 2.3** offers the motivation and guidance. Here are the two basic steps:

i. Seek a vertex $\mathbf{x}_0$ in the convex set $\mathcal{Q}$. If no vertex can be found STOP.

ii. If a vertex $\mathbf{x}_0$ is found choose an *optimal edge* E_0. In other words, find at $\mathbf{x}_0$ an edge E_0 along which *Cost* decreases most rapidly. If, perchance, $\mathbf{x}_0$ is optimal then there is no edge along which *Cost* decreases.

iii. If *Cost* decreases without bound on E_0, there is no solution: STOP.

iv. If on E_0 there is a minimum value of *Cost* that minimum value occurs at a vertex, say $\mathbf{x}_1$. If $\mathbf{x}_1 = \mathbf{x}_0$ then $\mathbf{x}_0$ is an optimal vertex and is a solution of the **PLPP**. If $\mathbf{x}_1 \neq \mathbf{x}_0$ find an optimal edge at $\mathbf{x}_1$.

The idea is that after finitely many repetitions of *i* - *iv* there *should* emerge a vertex that is optimal and solves the **PLPP**. If there is an optimal vertex of $\mathcal{Q}$, the simplex method as described in geometric terms above does yield an optimal vertex in finitely many repetitions of *i* - *iv*.

The *geometric* description above needs an *algebraic* interpretation so that *vertex, edge, neighboring vertex,* etc., are terms that are meaningful to a computer. Although one picture (geometric description) is worth a thousand words (algebraic interpretation), computers need verbal/numerical input, i.e., an algebraic interpretation.

Just as analytic geometry "algebraicizes" Euclidean geometry, so the simplex algorithm algebraicizes the geometrical version of the **PLPP** and makes possible its solution in concrete numerical decisions, e.g., reveals how much of MEAT PRODUCTS, of DAIRY PRODUCTS, etc., to buy, etc.

[**Remark 5.5.1:** In the usual algebraic/numerical interpretation and implementation of the algorithm described geometrically above there *can* arise "*degeneracies*" that sometimes cause repetition of algorithm steps that fail to move from one vertex of $\mathcal{Q}$ to another one. (In effect, several neighboring vertices have "degenerated" to one vertex.) This kind of repetition can give rise to *cycling*:

---

at some nonoptimal vertex the algorithm keeps grinding away
forever without causing a move away from the nonoptimal
vertex to a better one.

Instances of this behavior are given in **Section 5.6**.

In practice the simplex method just plain works and although degeneracies *do* occur, cycling does *not*. Experience with literally millions
of " real world" applications of the simplex method turns up few, if any,
examples of cycling. Thus in practice many of the computer programs
for the simplex method ignore the remote threat of cycling.

Nevertheless, there are examples of of **PLPP**s in which degeneracy
and cycling do occur. Some are given in and discussed in **Section
5.6**. To solve the cycling problem, there are variants of the Dantzig
algorithm. These are due to Bland and Charnes (cf. **[Bl]**, **[Ch]**, **[Mu]**).
The modified Dantzig algorithms theoretically dispose of the cycling
problem. However they are computationally expensive.

It should be noted again that there is now an algorithm due to Karmarkar (**[Karm]**). In worst-case situations the Karmarkar algorithm is
much faster than the simplex algorithm, which may soon give way to its
faster competitor. However, the insights provided by the study of the
simplex method are helpful in understanding the Karmarkar algorithm,
which is treated in **Section 5.7**.]

Note that (5.5.7) is a system of constraints that are *equations* rather than
inequalities like (5.5.1). The only constraints that are inequalities are those
specifying the NONNEGATIVITY of the variables. Since the last m components
of $\mathbf{C}$ are 0's the minimization problems (5.5.3) and (5.5.9) are the same. For each
$\mathbf{X}$ in $\mathbf{R}^{(n+m,+)}$ there is in $\mathbf{R}^{(n,+)}$ a vector $\mathbf{x}$ — the *associate* of $\mathbf{X}$ — consisting
of the first n components of $\mathbf{X}$.

[**Note 5.5.1:** If $\mathbf{x} \in \mathcal{Q}$ there is in $\mathbf{R}^{(n+m,+)}$ at least one $\mathbf{X}$ such that
$A\mathbf{X} = \mathbf{q}$ and of which $\mathbf{x}$ is the associate. Indeed, if

$$\mathbf{w} \stackrel{\text{def}}{=} A\mathbf{x} - \mathbf{p} \ (\in \mathbf{R}^{(m,+)})$$

and if

$$\mathbf{X} \stackrel{\text{def}}{=} \begin{pmatrix} \mathbf{x} \\ \mathbf{w} \end{pmatrix} \ (\in \mathbf{R}^{(n+m,+)})$$

then $\mathbf{x}$ is the associate of $\mathbf{X}$. The definition given for $\mathbf{X}$ implies that
$A\mathbf{X} = \mathbf{q}$. Furthermore $\mathbf{c}\mathbf{x} = \mathbf{C}\mathbf{X}$ and so if $\mathbf{x}$ is optimal so is $\mathbf{X}$.]

Exercise 5.5.1. Write each of the linear programming problems in **Examples 2.3.7** and **2.3.8** and in **Exercise 2.3.12** in the format exemplified by
(5.5.7) - (5.5.9).

To go from geometry to algebra it is helpful to rephrase the **PLPP** in a way that relates the **PLPP** to the methods used to solve systems: $A\mathbf{X} = \mathbf{q}$. Properly rephrased the **PLPP** is essentially reduced to two problems, listed next.

   i. For the system $A\mathbf{X} = \mathbf{q}$ seek a solution $\mathbf{X}$ *under the constraint that* $\mathbf{X} \succeq \mathbf{O}$. Such a vector $\mathbf{X}$ is a *feasible* solution for the system $A\mathbf{X} = \mathbf{q}$. Its associate $\mathbf{x}$ is necessarily a feasible solution for the system $A\mathbf{x} \succeq \mathbf{p}$. Under the heading SEEKING A STARTING VERTEX below it is shown that feasible solutions need not exist: Q might be empty.

   ii. For a given vector $\mathbf{C}$ (in which the last m components are 0's) seek a feasible solution $\mathbf{X}_{Opt}$ that also minimizes $\mathbf{C}\mathbf{X}$. Since $\mathbf{C}\mathbf{X} = \mathbf{c}\mathbf{x}$ and since $\mathbf{X} \succeq \mathbf{O}$ it follows that the associate $\mathbf{x}_{Opt}$ of $\mathbf{X}_{Opt}$ is an *optimal* solution of the **PLPP**. It is shown by illustration below that even when Q is not empty there may be no $\mathbf{X}_{Opt}$ (hence no $\mathbf{x}_{Opt}$).

How are the geometric description of the **PLPP** in $\mathbf{R}^n$ and the algebraic description of the same **PLPP** in $\mathbf{R}^{(n+m)}$ related?

Consider the $n + m$ equations in **Table 5.5.1**.

$$
\begin{array}{cc}
\underline{\text{Equation}} & \underline{\text{Hyperplane in } \mathbf{R}^n} \\[4pt]
\sum_{j=1}^{n} a_{1j} x_j = p_1 & H_1 \\[4pt]
\vdots & \vdots \\[4pt]
\sum_{j=1}^{n} a_{mj} x_j = p_m & H_m \\
x_1 = 0 & H_{m+1} \\
\vdots & \vdots \\
x_n = 0 & H_{m+n}
\end{array}
$$

Table 5.5.1

The intersection of each H_k with Q is a *face* F_k, $1 \leq k \leq n + m$, of Q. To describe a *vertex* of Q it is helpful to introduce the $(n + m) \times n$ matrix

$$
\widetilde{A} \stackrel{\text{def}}{=} \begin{matrix} n \\ m \\ n \end{matrix} \begin{pmatrix} A \\ I \end{pmatrix} \stackrel{\text{def}}{=}
\begin{pmatrix}
a_{11} & \cdots & a_{1n} \\
\vdots & \ddots & \vdots \\
a_{m1} & \cdots & a_{mn} \\
1 & \cdots & 0 \\
\vdots & \ddots & \vdots \\
0 & \cdots & 1
\end{pmatrix}
$$

---

and the $(n+m) \times 1$ column vector

$$\widetilde{\mathbf{p}} \stackrel{\text{def}}{=} \begin{matrix} 1 \\ m \\ n \end{matrix} \begin{pmatrix} \mathbf{p} \\ \mathbf{O} \end{pmatrix} \stackrel{\text{def}}{=} \begin{pmatrix} p_1 \\ \vdots \\ p_m \\ 0 \\ \vdots \\ 0 \end{pmatrix} \stackrel{\text{def}}{=} \begin{pmatrix} \widetilde{p}_1 \\ \vdots \\ \widetilde{p}_{n+m} \end{pmatrix}.$$

Row k of $\widetilde{A}$ corresponds to the left member of the kth equation describing H_k. The system (5.5.1) - (5.5.2) may be written $\widetilde{A}\mathbf{x} \succeq \widetilde{\mathbf{p}}$.

If n rows of $\widetilde{A}$ are linearly independent then the intersection of the corresponding hyperplanes (by abuse of language, the *linearly independent hyperplanes*) is nonempty and, indeed, a single vector $\mathbf{x}$ in $\mathbf{R}^n$. If, to boot, $\mathbf{x} \in \mathcal{Q}$ then the vector $\mathbf{x}$ is a *vertex* of $\mathcal{Q}$.

DEFINITION 5.5.1. A VERTEX $\mathbf{x}$ OF $\mathcal{Q}$ IS A VECTOR BELONGING TO $\mathcal{Q}$ AND TO n LINEARLY INDEPENDENT HYPERPLANES H_k.

Example 5.5.2. The n hyperplanes $H_{m+1}, \ldots, H_{m+n}$ correspond to the last n rows of $\widetilde{A}$ and these rows are linearly independent. The corresponding system of equations is homogeneous and hence consistent. The vector $\mathbf{x}$ they determine is the origin $\mathbf{O}$ of $\mathbf{R}^n$. If $\mathbf{O} \in \mathcal{Q}$ then $\mathbf{O}$ is a vertex of $\mathcal{Q}$.

If $\mathbf{x}$ is a vertex of $\mathcal{Q}$ and if $\mathbf{x}$ is the intersection of n linearly independent hyperplanes $H_{k_1}, \ldots, H_{k_n}$ then the intersection of $n-1$ of those hyperplanes is a line L. The intersection of L with $\mathcal{Q}$ is an *edge* of $\mathcal{Q}$. Thus there are at least $\binom{n}{n-1}$, i.e., n, edges containing a vertex $\mathbf{x}$. (Since $\mathbf{x}$ might be definable by several different sets of n linearly independent hyperplanes the number of edges containing $\mathbf{x}$ can be greater than n.)

Example 5.5.3. There are situations in which $\mathcal{Q}$ has a special form.

i. There is only one vector $\mathbf{x}$ in $\mathcal{Q}$. Then $\mathbf{x}$ is a vertex and, indeed, an optimal vertex of $\mathcal{Q}$.

ii. The dimension μ of $\mathcal{Q}$ is less than n. In other words, $\mathcal{Q}$ lies in some μ-dimensional hyperplane and for some vectors $\mathbf{x}_0, \mathbf{x}_1, \ldots, \mathbf{x}_\mu$ in $\mathcal{Q}$ every vector in $\mathcal{Q}$ may be written in the form $\mathbf{x}_0 + \sum_{i=1}^{\mu} a_i(\mathbf{x}_i - \mathbf{x}_0)$.

iii. The **PLPP** has more than one solution. This situation can arise if:

    a. $m < n$;

    b. an optimal vertex $\mathbf{x}$ is in the intersection of hyperplanes $H_{i_1}, \ldots, H_{i_m}$

$$\mathbf{x} \in \bigcap_{j=1}^{m} H_{i_j} \stackrel{\text{def}}{=} \mathcal{Q}';$$

c. c is in the span of the rows numbered $i_1, \ldots, i_m$ in A.

$iv.$ There is more than one vector in $\mathcal{Q}$ and every vector in $\mathcal{Q}$ is a solution of the **PLPP**, e.g., if, in iii, $\mathcal{Q}' = \mathcal{Q}$.

Exercise 5.5.2. Show that if $\mathbf{O} \in \mathcal{Q}$ and $\mathbf{c} \in \mathbf{R}_{(n,+)}$ (the nonnegative orthant of $\mathbf{R}_n$) then $\mathbf{O}$ is an optimal vertex of $\mathcal{Q}$.

Exercise 5.5.3. Give an example of a **PLPP** for which more than one set of n linearly independent hyperplanes H_k determine a vertex of $\mathcal{Q}$.

DEFINITION 5.5.2. LET $\mathbf{x}$ BE A VERTEX OF $\mathcal{Q}$: THERE ARE n LINEARLY INDEPENDENT HYPERPLANES H_{k_i}, $1 \leq i \leq n$ SUCH THAT

$$\mathbf{x} \stackrel{\text{def}}{=} \bigcap_{i=1}^{n} H_{k_i}.$$

A VERTEX $\mathbf{x}'$ IS A *neighboring* VERTEX OR SIMPLY A *neighbor* OF $\mathbf{x}$ IFF

$i.$ $\mathbf{x}' \in \mathcal{Q}$;

$ii.$ $\mathbf{x}' \neq \mathbf{x}$;

$iii.$ $\mathbf{x}'$ BELONGS TO AN EDGE THROUGH $\mathbf{x}$.

Since A has m rows and since its last m columns are linearly independent it follows that $rank(A) = m$. Hence simple elimination via the GEM without ROW EXCHANGE is applied to the system $A\mathbf{X} = \mathbf{q}$ and leads to a solution

$$X_{b_1} = q_1 - \sum_{j=1}^{n} \alpha_{1j} X_{f_j}$$

$$\vdots \qquad \vdots$$

$$X_{b_m} = q_m - \sum_{j=1}^{n} \alpha_{mj} X_{f_j} \tag{5.5.10}$$

in which the *free variables* X_{f_j}, $1 \leq j \leq n$ may be assigned arbitrary values. The variables X_{b_i}, $1 \leq i \leq m$ are *basic variables*. The free variables and the basic variables give rise to two vectors:

$$\text{the } \textit{free vector:} \ \ \mathbf{X}_F \stackrel{\text{def}}{=} \begin{pmatrix} X_{f_1} \\ \vdots \\ X_{f_n} \end{pmatrix}, \text{ of size } n \times 1,$$

$$\text{the } \textit{basic vector:} \ \ \mathbf{X}_B \stackrel{\text{def}}{=} \begin{pmatrix} X_{b_1} \\ \vdots \\ X_{b_m} \end{pmatrix}, \text{ of size } m \times 1.$$

Since the GEM without ROW EXCHANGE is used, the indices $b_1, \ldots, b_m$ as listed are not necessarily in their natural order. For example if $m = 4, n = 3$ then $b_1 = 6, b_2 = 3, b_3 = 4, b_4 = 1$ is a possible listing. (In this connection cf. **Section 2.4(V)**, in particular, the paragraph following **Table 2.4.2**.)

To achieve (5.5.10) the $m \times m$ matrix A_B formed from the columns numbered

$$b_1, \ldots, b_m$$

in A must be invertible. Then

$$\begin{pmatrix} q_1^{(1)} \\ \vdots \\ q_m^{(1)} \end{pmatrix} \overset{\text{def}}{=} \mathbf{q}^{(1)} = A_B^{-1} \mathbf{q}.$$

If $\mathbf{q}^{(1)} \succeq \mathbf{O}$ the matrix A_B is a *basic matrix*. Thus a basic submatrix of A is an invertible $m \times m$ matrix A_B such that $A_B^{-1} \mathbf{q} \succeq \mathbf{O}$. A *free matrix* A_F consists of the columns remaining after removal from A of the columns of a basic matrix A_B. Although invertible $m \times m$ submatrices of A must exist since $rank(A) = m$, there may be no basic matrix in A, e.g., as is shown below, if $\mathcal{Q}$ is empty.

In light of the definitions above, if A_B is basic and if

$$A_B^{-1} A_F \overset{\text{def}}{=} \begin{pmatrix} \alpha_{11} & \cdots & \alpha_{1n} \\ \vdots & \ddots & \vdots \\ \alpha_{m1} & \cdots & \alpha_{mn} \end{pmatrix}$$

then (5.5.10) may be rewritten in compact form as

$$\mathbf{X}_B = A_B^{-1} \mathbf{q} - A_B^{-1} A_F \mathbf{X}_F$$

If A_B is basic then setting each free variable at 0, i.e., setting $\mathbf{X}_F$ at $\mathbf{O}$, determines a *nonnegative* solution of $A\mathbf{X} = \mathbf{q}$:

$$\mathbf{X} \overset{\text{def}}{=} \begin{matrix} m \\ n \end{matrix} \begin{pmatrix} A_B^{-1} \mathbf{q} \\ \mathbf{O} \end{pmatrix}.$$

For the associate $\mathbf{x}$ of $\mathbf{X}$, it follows that $A\mathbf{x} \succeq \mathbf{p}$ and $\mathbf{x} \succeq \mathbf{O}$, i.e., $\mathbf{x} \in \mathcal{Q}$. In particular, if there is a basic matrix then $\mathcal{Q}$ is nonempty. The next result shows that even more is true.

THEOREM 5.5.1. IF $\mathbf{X}$ IS A NONNEGATIVE SOLUTION OF $A\mathbf{X} = \mathbf{q}$ THEN THE ROWS NUMBERED $f_1, \ldots, f_n$ OF $\tilde{A}$ ARE LINEARLY INDEPENDENT. FURTHERMORE THE ASSOCIATE $\mathbf{x}$ OF $\mathbf{X}$ IS A VERTEX OF $\mathcal{Q}$ IFF $\mathbf{X}_F = \mathbf{O}$.

[**Remark 5.5.2:** The linear independence of the rows numbered $f_1, \ldots, f_n$ in $\widetilde{A}$ means that the corresponding hyperplanes are linearly independent and hence that their intersection is a vertex of Q if their intersection is in $\mathcal{Q}$.]

PROOF. Let the rows of $\widetilde{A}$ be the row vectors denoted $\widetilde{\mathbf{a}}_1, \ldots, \widetilde{\mathbf{a}}_{m+n}$. Thus

$$\widetilde{A} = \begin{pmatrix} \widetilde{\mathbf{a}}_1 \\ \vdots \\ \widetilde{\mathbf{a}}_{m+n} \end{pmatrix}$$

$$\widetilde{\mathbf{a}}_s = \begin{cases} \mathbf{a}_s \stackrel{\text{def}}{=} (a_{s1} \ldots, a_{sn}) & \text{if } 1 \le s \le m \\ \mathbf{e}_{s-m} \stackrel{\text{def}}{=} (0, \; \ldots, \; \underset{\underset{s-m}{\uparrow}}{1}, \; \ldots, \; 0) & \text{if } m+1 \le s \le m+n. \end{cases}$$

For each f_j define ϕ_j according to the following rule:

$$\phi_j = \begin{cases} f_j - n & \text{if } n+1 \le f_j \le n+m \\ m + f_j & \text{if } 1 \le f_j \le n. \end{cases}$$

Then, generally, if $A\mathbf{Y} = \mathbf{q}$ it follows that for the associate $\mathbf{y}$ of $\mathbf{Y}$

$$\widetilde{\mathbf{a}}_{\phi_j}\mathbf{y} - Y_{\phi_j} = \widetilde{q}_{\phi_j}, \quad 1 \le j \le n.$$

According to DEFINITION **5.5.1** the vector $\mathbf{y}$ is a vertex of Q iff $\mathbf{y}$ is in Q and the rows $\widetilde{\mathbf{a}}_{\phi_1}, \ldots, \widetilde{\mathbf{a}}_{\phi_n}$ are linearly independent.

If the rows $\widetilde{\mathbf{a}}_{\phi_1}, \ldots, \widetilde{\mathbf{a}}_{\phi_n}$ are linearly dependent, i.e., if

$$\sum_{j=1}^{n} \alpha_j \widetilde{\mathbf{a}}_{\phi_j} = \mathbf{O}$$

then

$$(\sum_{j=1}^{n} \alpha_j \widetilde{\mathbf{a}}_{\phi_j})\mathbf{y} - \sum_{j=1}^{n} \alpha_j Y_{f_j} = \sum_{j=1}^{n} \alpha_j \widetilde{p}_{\phi_j} \stackrel{\text{def}}{=} P$$

$$0 - \sum_{j=1}^{n} \alpha_j Y_{f_j} = P.$$

If some $\alpha_{j'} \ne 0$ then since each Y_{f_j} is free one may assume

$$Y_{f_j} = \begin{cases} \frac{(1-P)}{\alpha_{j'}} & \text{if } j = j' \\ 0 & \text{otherwise.} \end{cases}$$

Substituting the chosen values for the Y_{f_j} leads to the contradiction: $0 = 1$. Hence the rows numbered $\phi_1, \ldots, \phi_n$ are linearly independent and so the hyperplanes $H_{\phi_1}, \ldots, H_{\phi_n}$ are linearly independent.

     Chapter 5. APPLYING LINEAR ALGEBRA

Finally $\mathbf{x}$ is in each of the hyperplanes H_{ϕ_j}, $1 \leq j \leq n$, iff

$$X_{f_j} = 0, \ 1 \leq j \leq n,$$

i.e., iff $\mathbf{X}_F = \mathbf{O}$.

$$\Omega$$

[**Note 5.5.2:**   If $A\mathbf{X} = \mathbf{q}$ then setting $\mathbf{X}_F$ at $\mathbf{O}$ leads to a vertex of $\mathcal{Q}$.]

Now assume one of the components, say $X_{f_{j'}}$, of $\mathbf{X}_F$ is positive while the other components of $\mathbf{X}_F$ are 0's. If each X_{b_i} is nonnegative, the resulting vector $\mathbf{X}'$ has an associate $\mathbf{x}'$ on an *edge E* of $\mathcal{Q}$. The edge E starts at the vertex $\mathbf{x}$. If $\alpha_{ij'} > 0$ in the ith equation of (5.5.10) then $X_{f_j'}$ cannot exceed the nonnegative quantity

$$\frac{q_i}{\alpha_{ij'}}$$

if X_{b_i} is to remain nonnegative. Hence

$$X_{f_{j'}} \leq \mu_{j'} \overset{\text{def}}{=} minimum \left\{ \frac{q_i}{\alpha_{ij'}} \ : \ \alpha_{ij'} > 0 \right\} \overset{\text{def}}{=} \frac{q_{i'}}{\alpha_{i'j'}}.$$

Since $\alpha_{i'j'} > 0$ one may solve for $X_{f_{j'}}$ in the equation

$$X_{b_{i'}} = q_{i'} - \sum_{j=1}^{n} \alpha_{i'j} X_{f_j}$$

and substitute the result in all the other equations of the system

$$\mathbf{X}_B = A_B^{-1}\mathbf{q} - A_B^{-1} A_F \mathbf{X}_F.$$

Consequently, there is a SWITCH

$$X_{b_{i'}} \leftrightarrow X_{f_{j'}}.$$

It gives rise to a system of equations in which the basic variables are all the original basic variables, $X_{b_{i'}}$ excepted and replaced by $X_{f_{j'}}$. The new free variables are all the original free variables, $X_{f_{j'}}$ excepted and replaced by $X_{b_{i'}}$ (cf. **Example 5.5.4** below).

When each new free variable is set at 0 then the value of the new basic variable $X_{f_{j'}}$ is μ'_j. At the same time $X_{b_{i'}}$ and all other X_{f_j} are 0's. The corresponding $\mathbf{x}'$ is a neighboring vertex of $\mathbf{x}$.

Moving from vertex to vertex is equivalent to a sequence of such SWITCHes:

$$\text{basic variable} \leftrightarrow \text{free variable}.$$

Owing to the close relationship between the underlying steps of the simplex algorithm and the various forms of the simple elimination algorithm, the column

---

numbered j' in A is the *simplex pivot column* and the row numbered i' in A is the *simplex pivot row*. The $i'j'$ entry of A is the *simplex pivot entry*.

If the **PLPP** has a solution, such SWITCHes, properly chosen, lead to a solution.

If, for all i the coefficient $\alpha_{ij} \leq 0$ then X_{f_j} may be arbitrarily large and E is a half-line starting at $\mathbf{x}$.

The *equational version* of the simplex algorithm as discussed above may be summarized as follows:

If the system $A\mathbf{X} = \mathbf{q}$ gives rise via simple elimination to an equivalent system

$$\mathbf{X}_B = A_B^{-1}\mathbf{q} - A_B^{-1}A_F\mathbf{X}_F$$

in which $A_B^{-1}\mathbf{q} \succeq \mathbf{O}$ then the vertices and edges of $\mathcal{Q}$ can be found by choosing appropriate nonnegative values for the free variables and making appropriate SWITCHes between basic and free variables.

A crucial step in the simplex algorithm is the decision about which free variable is to serve as one half of a SWITCH. This decision is made via the objective function written in terms of the free variables. Let $\mathbf{C}_B$ resp. $\mathbf{C}_F$ be the $1 \times m$ resp. $1 \times n$ vector having the components $C_{b_1}, \ldots, C_{b_m}$ resp. $C_{f_1}, \ldots, C_{f_n}$. Then

$$\begin{aligned}
\mathbf{CX} &= \mathbf{C}_B\mathbf{X}_B + \mathbf{C}_F\mathbf{X}_F \\
&= \mathbf{C}_B(A_B^{-1}\mathbf{q} - A_B^{-1}A_F\mathbf{X}_F) + \mathbf{C}_F\mathbf{X}_F \\
&= \mathbf{C}_B A_B^{-1}\mathbf{q} + (\mathbf{C}_F - \mathbf{C}_B A_B^{-1}A_F)\mathbf{X}_F.
\end{aligned} \qquad (5.5.11)$$

Since the free vector $\mathbf{X}_F$ is the vector of free variables and must remain nonnegative, it follows that $Cost \stackrel{\text{def}}{=} \mathbf{CX}$ cannot be decreased if

$$(\mathbf{C}_F - \mathbf{C}_B A_B^{-1}A_F) \succeq \mathbf{O}.$$

In that event $\mathbf{X} = \mathbf{X}_{Opt}$ and

$$Cost_{minimum} \stackrel{\text{def}}{=} \mathbf{C}_B A_B^{-1}\mathbf{q}.$$

On the other hand if

$$(\mathbf{C}_F - \mathbf{C}_B A_B^{-1}A_F) \not\succeq \mathbf{O}$$

there can be a possibility of decreasing $Cost$. The next **Example** shows what is going on.

Example 5.5.4. The development thus far may be applied to the data in **Example 2.3.7**. In standard format the **PLPP** reads as follows:

$$A = \begin{pmatrix} 1 & -2 & 1 & 0 & 0 & 0 & 0 \\ 1 & -1 & 0 & 1 & 0 & 0 & 0 \\ 2 & 1 & 0 & 0 & 1 & 0 & 0 \\ -1 & 2 & 0 & 0 & 0 & 1 & 0 \\ -1 & 1 & 0 & 0 & 0 & 0 & 1 \end{pmatrix}, \ \mathbf{q} = \begin{pmatrix} 2 \\ 3 \\ 12 \\ 9 \\ 4 \end{pmatrix}, \ \mathbf{c} = (-5, -2).$$

Then a suitable basic matrix A_B is the 5×5 identity matrix consisting of columns 3–7 of A:

$$A_B = I$$
$$\mathbf{X}_B = (X_3, X_4, X_5, X_6, X_7)^t$$
$$\mathbf{X}_F = (X_1, X_2)^t$$
$$\mathbf{C} = (-5, -2, 0, 0, 0, 0, 0)$$
$$\mathbf{C}_B = (0, 0, 0, 0, 0), \quad \mathbf{C}_F = (-5, -2)$$
$$\mathbf{q} = (2, 3, 12, 9, 4)^t$$

and

$$X_3 = 2 - X_1 + 2X_2 \ (*) \tag{5.5.12}$$
$$X_4 = 3 - X_1 + X_2 \tag{5.5.13}$$
$$X_5 = 12 - 2X_1 - X_2 \tag{5.5.14}$$
$$X_6 = 9 + X_1 - 2X_2 \tag{5.5.15}$$
$$X_7 = 4 + X_1 - X_2 \tag{5.5.16}$$
$$\mathbf{CX} = -5X_1 - 2X_2. \tag{5.5.17}$$

Thus $b_1 = 3, \ldots, b_5 = 7$; $f_1 = 1, f_2 = 2$. If $\mathbf{X}_F \overset{\text{def}}{=} (X_1, X_2) = (0, 0)$ then
$$Cost = \mathbf{C}_B A_B^{-1} \mathbf{q} = \mathbf{O}(2, 3, 12, 9, 4)^t = 0$$
$$\mathbf{C}_F - \mathbf{C}_B A_B^{-1} A_F = (-5, -2).$$

Since the associate $\mathbf{x}$ of $\mathbf{X}$ is $(X_1, X_2)^t$, from (5.5.17) it follows that $Cost = 0$ at the vertex $\mathbf{x}_1 \overset{\text{def}}{=} (0, 0)^t$ corresponding to setting X_1 and X_2 at 0. $Cost$ can be reduced by increasing either X_1 or X_2 because each has a negative coefficient in the formula for $\mathbf{CX}$. To pass to a better and neighboring vertex one of the two free variables may be increased while the other is held at 0 and some basic variable is reduced to 0. In that way $\mathbf{X}$ continues to be a solution of $A\mathbf{X} = \mathbf{q}$ and at least 2 components of $\mathbf{X}$ are 0's.

From (5.5.12)-(5.5.16) it follows that if X_2 is held at 0 and X_1 is increased then
$$f_{j'} = 1, \ j' = 1, \ i' = 1, \ \mu_{j'} = \mu_1 = 2$$

and so X_1 should be increased to 2. The determining equation is flagged $(*)$. The move is made along the edge corresponding to setting X_2 at 0 and allowing X_1 to increase to 2. In the notation used above:

the SWITCH is $X_3 \leftrightarrow X_1$

$$A_B^{-1} = I, \quad A_F = \begin{pmatrix} 1 & -2 \\ 1 & -1 \\ 2 & 1 \\ -1 & 2 \\ -1 & 1 \end{pmatrix}.$$

When the SWITCH $X_3 \leftrightarrow X_1$ is complete the new system is

$$X_1 = 2 + 2X_2 - X_3$$
$$X_4 = 1 - X_2 + X_3 \; (*)$$
$$X_5 = 8 + 2X_2 - 5X_3$$
$$X_6 = 11 - X_3$$
$$X_7 = 6 + X_2 - X_3$$
$$B_B = (X_1, X_4, X_5, X_6, X_7)^t, \; X_F = (X_2, X_3)^t$$
$$\mathbf{C}_B = (-5, 0, 0, 0, 0), \; \mathbf{C}_F = (-2, 0)$$
$$\mathbf{q} = (2, 1, 8, 1, 6)^t$$
$$\mathbf{CX} = -5X_1 - 2X_2 = -10 - 10X_2 + 5X_3$$
$$Cost = \mathbf{C}_B(2, 1, 8, 11, 6) = -10.$$

At the vertex corresponding to setting $\mathbf{X}_F$ at $\mathbf{O}$ the value ($Cost$) of Z is -10. $Cost$ can be diminished by increasing X_2. According to the $(*)$-flagged equation X_2 may not exceed 1. Setting the new $\mathbf{X}_F$ at $\mathbf{O}$ leads from vertex $\mathbf{x}_1 = (0, 0)^t$ to the vertex $\mathbf{x}_2 \overset{\text{def}}{=} (2, 0)^t$. The SWITCH $X_4 \leftrightarrow X_2$ is in order.

Without further comment there are presented below the successive systems in which the equations admitting SWITCHes are flagged $(*)$'s and the $Cost$s are calculated.

$$X_4 \leftrightarrow X_2$$

$$X_1 = 4 + X_3 - 2X_4$$
$$X_2 = 1 + X_3 - X_4$$
$$X_5 = 3 - X_3 + 5X_4 \; (*)$$
$$X_6 = 11 - X_3$$
$$X_7 = 7 - X_4$$
$$X_B = (X_1, X_2, X_5, X_6, X_7)^t, \; X_F = (X_3, X_4)^t$$
$$\mathbf{C}_B = (-5, -2, 0, 0, 0), \mathbf{C}_F = (0, 0)$$
$$\mathbf{CX} = -22 - 7X_3 + 12X_4$$
$$\mathbf{q} = (4, 1, 3, 11, 7)^t$$
$$Cost = \mathbf{C}_B(4, 1, 3, 11, 7) = -22.$$

$$X_5 \leftrightarrow X_3$$

$$X_1 = 5 - \frac{1}{3}X_4 - \frac{1}{3}X_5$$

$$X_2 = 2 + \frac{2}{3}X_4 - \frac{1}{3}X_5$$

$$X_3 = 1 + \frac{5}{3}X_4 - \frac{1}{3}X_5$$

$$X_6 = 10 - \frac{5}{3}X_4 + \frac{1}{3}X_5$$

$$X_7 = 7 - X_4$$

$$X_B = (X_1, X_2, X_3, X_6, X_7)^t, \quad X_F = (X_4, X_5)^t$$

$$\mathbf{C}_B = (-5, -2, 0, 0, 0), \quad \mathbf{C}_F = (0, 0)$$

$$\mathbf{q} = (5, 2, 1, 10, 7)^t$$

$$\mathbf{CX} = -29 + \frac{1}{3}X_4 + \frac{7}{3}X_5$$

$$Cost = \mathbf{C}_B(5, 2, 1, 10, 7) = -29.$$

In the last system no equation is flagged because there is no possibility for diminishing *Cost*: *all the coefficients of the free variables in the formula for* $\mathbf{CX}$ *are nonnegative*. The **PLPP** is solved. The optimal vertex is $\mathbf{x}_{Opt} = (5, 2)^t$ where the minimal *Cost* is -29.

Exercise 5.5.4.

i. In **Figure 5.4.1** trace out the edges along which the simplex algorithm as applied above moves in going from $\mathbf{x}_1$ to $\mathbf{x}_{Opt}$.

ii. In the first round let X_2 increase and hold X_1 at 0 (instead of the other way round). Complete the execution of the algorithm and identify all the SWITCHes.

iii. Redo **Example 5.5.4** but show that the matrix

$$\begin{pmatrix} 1 & -2 & 1 & 0 & 0 \\ 1 & -1 & 0 & 1 & 0 \\ 2 & 1 & 0 & 0 & 1 \\ -1 & 2 & 0 & 0 & 0 \\ -1 & 1 & 0 & 0 & 0 \end{pmatrix}$$

is also basic and then use it to find minimum *Cost*. Identify all sets of basic variables, all sets of free variables, and all SWITCHes.

iv. Repeat *i* for the operations in *ii* and *iii*..

Exercise 5.5.5. Use the equational version of the simplex method to solve the **PLPP** of **Example 2.3.8**.

Here are some noteworthy features of the operation:

i. If

$$A' \stackrel{\text{def}}{=} A_B^{-1} A,$$
$$A'_F \stackrel{\text{def}}{=} A_B^{-1} A_F = (\beta_{ij})_{i,j=1}^{m,n},$$
$$\mathbf{q}' \stackrel{\text{def}}{=} A_B^{-1} \mathbf{q},$$

then the new system $\mathbf{X}_B = \mathbf{q}' - A'_F \mathbf{X}_F$ may be rewritten in the form $A'\mathbf{X} = \mathbf{q}'$ or $A'\mathbf{X} - \mathbf{q}' = 0$. The matrix A' contains an embedded $m \times m$ identity matrix I. The column vector $\mathbf{e}_i$ of I is in column b_i, $1 \le i \le m$ of A'. Hence I is strewn throughout A' but the columns of I do not necessarily appear in their natural order.

ii. Once a basic matrix is found the new system $A'\mathbf{X} - \mathbf{q}' = 0$ is equivalent to $A\mathbf{X} = \mathbf{q}$. The vector $\mathbf{C}$ remains unchanged. There is an important distinction between the old and the new systems: I is embedded in A' while only by happenstance, as in **Example 5.5.4**, is I embedded in A. The new system may be viewed as representing in standard format the original **PLPP**.

Solving for the basic variables can be achieved via row-reduction of A within the augmented matrix $A \mid - \mathbf{q}$. Hence it suffices for the analysis of the **PLPP** to operate with the augmented matrix $A \mid - \mathbf{q}$ and to dispense with the many systems of equations like those appearing in the discussion of **Example 5.5.4**. The operations with $A \mid - \mathbf{q}$ are very much like row-reduction as it appears in refinements of the *PROCESS*. However there are differences because in the simplex algorithm one chooses pivots with greater care and with less freedom.

A sequence $X_{b_{i'}} \leftrightarrow X_{f_{j'}}$, etc., of SWITCHes gives rise to a sequence $A'' \mid - \mathbf{q}''$, etc., of augmented matrices. Each augmented matrix is the result of row-reducing its predecessor in one column that is *not* the last column nor one of the columns of the embedded identity matrix. The row-reduction introduces a new column of a new embedded identity matrix and alters one of the columns of the old embedded identity matrix. The vector $\mathbf{C}$ remains unchanged throughout.

iii. Let A be $A^{(0)}$, $\mathbf{q}$ be $\mathbf{q}^{(0)}$, etc. If the $m \times m$ identity matrix I is embedded in A^0 let $A^{(r)}$, $\mathbf{q}^{(r)}$, etc., be the counterparts of $A^{(0)}$, $\mathbf{q}^{(0)}$, etc., after r row reductions are performed on $A^{(0)} \mid - \mathbf{q}^{(0)}$.

If $r \ge 1$ then I is embedded in $A^{(r)}$. Furthermore, if $r \ge 1$ the passage from $A^{(r)} \mid - \mathbf{q}^{(r)}$ to $A^{(r+1)} \mid - \mathbf{q}^{(r+1)}$ is brought about by multiplication with a very simple matrix. Indeed, let the SWITCH $X_{b_i} \leftrightarrow X_{f_j}$ be in order and let $A_F^{(r)} \stackrel{\text{def}}{=} (\alpha_{ij}^{(r)})_{i,j=1}^{m,m}$ be the corresponding free matrix. If $L_{ij}^{(r)}$ is the identity matrix in which column i is replaced by

$$\begin{pmatrix} -\dfrac{\alpha_{1j}^{(r)}}{\alpha_{ij}^{(r)}} \\ \vdots \\ \dfrac{1}{\alpha_{ij}^{(r)}} \\ \vdots \\ -\dfrac{\alpha_{mj}^{(r)}}{\alpha_{ij}^{(r)}} \end{pmatrix} \quad \leftarrow \; i\text{th row} \quad ,$$

i.e., if

$$L_{ij}^{(r)} = \;\; i\text{th row} \;\rightarrow\; \begin{pmatrix} 1 & \cdots & & -\dfrac{\alpha_{1j}^{(r)}}{\alpha_{ij}^{(r)}} & & \cdots & 0 \\ \vdots & \ddots & & \vdots & & \ddots & \vdots \\ 0 & \cdots & & \dfrac{1}{\alpha_{ij}^{(r)}} & & \cdots & 0 \\ \vdots & & \ddots & \vdots & & \ddots & \vdots \\ 0 & \cdots & & -\dfrac{\alpha_{mj}^{(r)}}{\alpha_{ij}^{(r)}} & & \cdots & 1 \end{pmatrix} ,$$

with the ith column indicated above the matrix.

then

$$A^{(r+1)} \mid - \mathbf{q}^{(r+1)} = L_{ij}^{(r)} \left(A^{(r)} \mid - \mathbf{q}^{(r)} \right) .$$

iv. Although $\mathbf{C}$ remains the same throughout, the vectors $\mathbf{X}_B^{(r)}$, $\mathbf{X}_F^{(r)}$, $\mathbf{C}_B^{(r)}$, and $\mathbf{C}_F^{(r)}$ are not necessarily the same as their counterparts at the beginning: if $b_1^{(r)}, \ldots, b_m^{(r)}$ resp. $f_1^{(r)}, \ldots, f_n^{(r)}$ are the indices of the basic variables resp. free variables then

$$\mathbf{C}_B^{(r)} = (C_{b_1^{(r)}}, \ldots, C_{b_m^{(r)}})$$
$$\mathbf{C}_F^{(r)} = (C_{f_1^{(r)}}, \ldots, C_{f_n^{(r)}}).$$

v. A free variable $X_{f_j^{(r)}}$ is *eligible* for SWITCH iff

    a. some $C_{f_j^{(r)}} < 0$

and

    b. some coefficient is positive in column $f_j^{(r)}$ of $A^{(r)} \mid - \mathbf{q}^{(r)}$.

If an eligible free variable X_{f_j} increases then *Cost* decreases *so long as all variables remain nonnegative*. What free variable is to be SWITCHed at the rth stage is the eligible free variable with the *smallest* index, say $f_{j'}^{(r)}$. (This is one of two *Bland selection rules*. It determines the chosen eligible free variable

---

uniquely.) The free variable so chosen is the *entering free variable*. Then column $f_{j'}^{(r)}$ of $A^{(r)}| - \mathbf{q}^{(r)}$ is the *simplex pivot column*.

vii. A basic variable $X_{b_{i'}^{(r)}}$ is *eligible* for SWITCH with $X_{f_{j'}^{(r)}}$ iff

$$\alpha_{i'j'}^{(r)} > 0$$

$$\frac{q_{i'}^{(r)}}{\alpha_{i'j'}^{(r)}} = minimum \left\{ \frac{q_i^{(r)}}{\alpha_{ij'}^{(r)}} \; : \; \alpha_{ij'}^{(r)} > 0 \right\}$$

$$= \mu_{j'}.$$

What basic variable is to be SWITCHed is the *eligible* basic variable with the *smallest* index, say $b_{i'}^{(r)}$. (This is the other Bland selection rule. It determines the chosen basic variable uniquely.) The eligible basic variable so chosen is the *leaving* basic variable. The row in which the leaving basic variable is found is the *simplex pivot row*. The coefficient $\alpha_{i'j'}^{(r)}$ is the *simplex pivot entry*.

The simplex algorithm as just described is the *augmented matrix version*. It is illustrated in the next **Example**.

Example 5.5.5. Assume the **PLPP** is given by the the following data:

$$A \stackrel{\text{def}}{=} \begin{pmatrix} 1 & 2 & -3 & 0 \\ 1 & -1 & -1 & 2 \\ 2 & 1 & 1 & 6 \end{pmatrix}, \quad \mathbf{p} \stackrel{\text{def}}{=} \begin{pmatrix} p_1 \\ p_2 \\ p_3 \end{pmatrix}, \quad \mathbf{c} \stackrel{\text{def}}{=} (c_1, c_2, c_3, c_4)$$

Then $m = 3$, $n = 4$ and if $\mathbf{p} \succeq \mathbf{O}$

$$A^{(0)} = \begin{pmatrix} 1 & 2 & -3 & 0 & -1 & & \\ 1 & -1 & -1 & 2 & & -1 & \\ 2 & 1 & 1 & 6 & & & -1 \end{pmatrix},$$

$$\mathbf{q}^{(0} = \mathbf{p} = \begin{pmatrix} p_1 \\ p_2 \\ p_3 \end{pmatrix},$$

$$A^{(0}| - \mathbf{q}^{(0)} = \begin{pmatrix} 1 & 2 & -3 & 0 & -1 & & & \\ 1 & -1 & -1 & 2 & & -1 & & \\ 2 & 1 & 1 & 6 & & & -1 & \end{pmatrix} \left| \begin{matrix} -p_1 \\ -p_2 \\ -p_3 \end{matrix} \right)$$

$$\mathbf{C} = (c_1, c_2, c_3, c_4, 0, 0, 0)$$

$$\stackrel{\text{def}}{=} (C_1, C_2, C_3, C_4, C_5, C_6, C_7).$$

A row-reduction of $A^{(0)}| - \mathbf{q}^{(0)}$ leads to

$$A^{(1)}| - \mathbf{q}^{(1)} \stackrel{\text{def}}{=} \begin{pmatrix} 1 & 0 & 0 & \frac{8}{3} & 0 & -\frac{1}{3} & -\frac{1}{3} & \\ 0 & 1 & 0 & -\frac{2}{15} & -\frac{1}{5} & \frac{7}{15} & -\frac{3}{15} & \\ 0 & 0 & 1 & \frac{4}{5} & \frac{1}{5} & \frac{1}{5} & -\frac{1}{5} & \end{pmatrix} \left| \begin{matrix} -\frac{p_2+p_3}{3} \\ -\frac{3p_1-7p_2+2p_3}{15} \\ -\frac{-p_1-p_2+p_3}{5} \end{matrix} \right)$$

in which the identity matrix is embedded. Furthermore

$$A_B^{(0)} = \begin{pmatrix} 1 & 2 & -3 \\ 1 & -1 & -1 \\ 2 & 1 & 1 \end{pmatrix}$$

$$\left(A_B^{(0)}\right)^{-1} = \begin{pmatrix} 0 & \frac{1}{3} & \frac{1}{3} \\ \frac{1}{5} & -\frac{7}{15} & \frac{2}{15} \\ -\frac{1}{5} & -\frac{1}{5} & \frac{1}{5} \end{pmatrix}$$

$$\mathbf{X}_B^{(1)} = (X_1, X_2, X_3), \quad \mathbf{X}_F^{(1)} = (X_4, X_5, X_6, X_7)$$

$$\mathbf{C}_B^{(1)} = (c_1, c_2, c_3), \quad \mathbf{C}_F^{(1)} = (c_4, c_5, c_6, c_7)$$

and according to (5.5.11)

$$
\begin{aligned}
Cost^{(1)} &= \mathbf{C}_B^{(1)}\left(A_B^{(0)}\right)^{-1}\mathbf{q}^{(0} + \left(\mathbf{C}_F^{(1)} - \mathbf{C}_B^{(1)}\left(A_B^{(0)}\right)^{-1}\right)\mathbf{X}_F^{(1)} \\
&= \frac{(3c_2 - 3c_3)p_1 + (5c_1 - 7c_2 - 3c_3)p_2 + (5c_1 + 2c_2 + 3c_3)p_3}{15} \\
&\quad + \frac{(-40c_1 + 2c_2 - 12c_3 + 15c_4)X_4}{15} + \frac{(3c_2 - 3c_3)X_5}{15} \\
&\quad + \frac{(5c_1 - 7c_2 - 3c_3)X_6}{15} + \frac{(5c_1 + 2c_2 + 3c_3)X_7}{15}.
\end{aligned}
$$

At this stage $b_i = i$, $1 \le i \le 3$, $f_j = j$, $4 \le j \le 7$.

If $\mathbf{X}_F^{(1)} = \mathbf{O}$ then

$$Cost^{(1)} = \frac{(3c_2 - 3c_3)p_1 + (5c_1 - 7c_2 - 3c_3)p_2 + (5c_1 + 2c_2 + 3c_3)p_3}{15}$$

the associate of $\mathbf{X}$ is the vector $\mathbf{x} \overset{\text{def}}{=} \begin{pmatrix} \frac{p_2 + p_3}{3} \\ \frac{3p_1 - 7p_2 + 2p_3}{15} \\ \frac{-p_1 - p_2 + p_3}{5} \\ 0 \end{pmatrix}$

arising when $X^{(1)} = \mathbf{O}$. If $\mathbf{x} \succeq \mathbf{O}$ then $\mathbf{x}$ is a vertex of $\mathcal{Q}$.
Consider two cases.

Case 1. Assume

$$\mathbf{p} = \begin{pmatrix} 3 \\ 2 \\ 5 \end{pmatrix} \quad \text{and} \quad \mathbf{c} = (1, 1, 1, -1).$$

Then

$$A^{(1)}| - \mathbf{q}^{(1)} = \begin{pmatrix} 1 & 0 & 0 & \frac{8}{3} & 0 & -\frac{1}{3} & -\frac{1}{3} & \Big| & -\frac{7}{3} \\ 0 & 1 & 0 & -\frac{2}{15} & -\frac{1}{5} & \frac{7}{15} & -\frac{2}{15} & \Big| & -\frac{1}{3} \\ 0 & 0 & 1 & \frac{4}{5} & \frac{1}{5} & \frac{1}{5} & -\frac{1}{5} & \Big| & 0 \end{pmatrix}$$

$$\mathbf{X}_B^{(1)} = (X_1, X_2, X_3), \quad \mathbf{X}_F^{(1)} = (X_4, X_5, X_6, X_7)$$

$$\mathbf{C}_B^{(1)} = (1, 1, 1), \quad \mathbf{C}_F^{(1)} = (-1, 0, 0, 0)$$

$$Cost^{(1)} = \frac{8}{3} - \frac{13}{3}X_4 + 0X_5 - \frac{1}{3}X_6 + \frac{2}{3}X_7.$$

$Cost^{(1)}$ is $\frac{8}{3}$ at the vertex

$$\mathbf{x}_1 \stackrel{\text{def}}{=} \begin{pmatrix} \frac{7}{3} \\ \frac{1}{3} \\ 0 \\ 0 \end{pmatrix},$$

corresponding to setting $\mathbf{X}_F$ at $\mathbf{O}$.

At this point the eligible free variables are X_4 and X_6; hence $f_{j'}^{(1)} = 4$. Furthermore $i' = b_3 = 3$ and $\mu_{j'}^{(1)} = 0$. Thus the SWITCH $X_3 \leftrightarrow X_4$ is in order. Since $\mu_{j'}^{(1)} = 0$ the SWITCH gives rise to no diminution of $Cost^{(1)}$ and no change in vertex.

[**Remark 5.5.3:** This situation is an instance of a *degeneracy*, occasioned by the fact that the third component of $\mathbf{q}^{(1)}$ is 0.]

To effect the SWITCH $X_3 \leftrightarrow X_4$ one multiplies the matrix $A^{(1)}| - \mathbf{q}^{(1)}$ by $L_{24}^{(1)}$. In this case

$$L_{24}^{(1)} = \begin{pmatrix} 1 & 0 & \frac{-\frac{8}{3}}{\frac{4}{5}} \\ 0 & 1 & \frac{\frac{2}{15}}{\frac{4}{5}} \\ 0 & 0 & \frac{\frac{1}{4}}{\frac{4}{5}} \end{pmatrix} = \begin{pmatrix} 1 & 0 & -\frac{10}{3} \\ 0 & 1 & \frac{1}{6} \\ 0 & 0 & \frac{5}{4} \end{pmatrix}$$

$$L_{24}^{(1)} A^{(1)}| - \mathbf{q}^{(1)} = \begin{pmatrix} 1 & 0 & -\frac{1}{10} & 0 & -\frac{2}{3} & -1 & \frac{1}{3} & \bigg| & -\frac{7}{3} \\ 0 & 1 & \frac{1}{6} & 0 & -\frac{1}{6} & \frac{1}{2} & \frac{1}{6} & \bigg| & -\frac{1}{3} \\ 0 & 0 & \frac{5}{4} & 1 & \frac{1}{4} & \frac{1}{4} & -\frac{1}{4} & \bigg| & 0 \end{pmatrix}$$
$$= A^{(2)}| - \mathbf{q}^{(2)}.$$

Hence

$$\mathbf{X}_B^{(2)} = (X_1, X_2, X_4), \quad \mathbf{X}_F^{(2)} = (X_3, X_5, X_6, X_7)$$
$$\mathbf{C}^{(2)} = (1, 1, -1), \quad \mathbf{C}_F^{(2)} = (1, 0, 0, 0)$$
$$Cost^{(2)} = \frac{8}{3} + \frac{65}{12}X_3 + \frac{13}{4}X_5 + \frac{3}{4}X_6 - \frac{5}{12}X_7.$$

The associate of $\mathbf{X}^{(2)}$ when $\mathbf{X}_F^{(2)} = \mathbf{O}$ is again $(\frac{7}{3}, \frac{1}{3}, 0, 0)^t$. This time the only eligible free variable is X_7, $f_{j'}^{(2)} = 7$, the only eligible basic variable is X_1, $\mu_{j'}^{(2)} = 7$, and the SWITCH $X_1 \leftrightarrow X_7$ is in order.

Hence

$$L_{17}^{(2)} = \begin{pmatrix} 3 & 0 & 0 \\ \frac{1}{2} & 1 & 0 \\ \frac{3}{4} & 0 & 1 \end{pmatrix}$$

$$L_{17}^{(2)} A^{(2)}| - \mathbf{q}^{(2)} = \begin{pmatrix} 3 & 0 & -10 & 0 & -2 & -3 & 1 & | & -7 \\ \frac{1}{2} & 1 & -\frac{3}{2} & 0 & -\frac{1}{2} & 0 & 0 & | & -\frac{3}{2} \\ \frac{3}{4} & 0 & -\frac{5}{4} & 1 & -\frac{1}{4} & -\frac{1}{2} & 0 & | & -\frac{7}{4} \end{pmatrix}$$

$$= A^{(3)}| - \mathbf{q}^{(3)}$$

$$\mathbf{X}_B^{(3)} = (X_2, X_4, X_7), \quad \mathbf{X}_F^{(3)} = (X_1, X_3, X_5, X_6,)$$

$$\mathbf{C}_B^{(3)} = (1, -1, 0), \quad \mathbf{C}_F^{(3)} = (1, 0, 0, 0)$$

$$Cost^{(3)} = -\frac{1}{4} + \frac{5}{4}X_1 + \frac{5}{4}X_3 + \frac{1}{4}X_5 - \frac{1}{2}X_6.$$

The only potentially eligible free variable is X_6. Since there is no positive entry in the the simplex pivot column (column 6) there is no simplex pivot row and no simplex pivot entry. It follows that X_6 may be arbitrarily large and that $Cost$ may be driven arbitrarily low: there is no solution.

Case 2. Assume $\mathbf{p}$ is as in *Case 1* but that

$$\mathbf{c} = (-1, 1, 1, 1).$$

Then, because $\mathbf{p}$ is unchanged, $A| - \mathbf{q}$ and $A^{(1)}| - \mathbf{q}^{(1)}$ are the same as they are in *Case 1*. Since $\mathbf{c}$ is different

$$Cost^{(1)} = -2 + 2X_4 + 0X_5 - 3X_6 + 0X_7.$$

Only X_6 is an eligible free variable, $j' = 6$, $\mu_{j'}^{(1)} = \mu_6^{(1)} = 0$, and the Bland rule dictates X_3 as the choice among the eligible basic variables X_2 and X_3. The SWITCH $X_3 \leftrightarrow X_6$ is in order. Thus

$$L_{36}^{(1)} = \begin{pmatrix} 1 & 0 & \frac{5}{3} \\ 0 & 1 & -\frac{7}{3} \\ 0 & 0 & 5 \end{pmatrix}$$

$$L_{36}^{(1)} A^{(1)}| - \mathbf{q}^{(1)} = \begin{pmatrix} 1 & 0 & \frac{5}{3} & 4 & \frac{1}{3} & 0 & -\frac{2}{3} & | & -\frac{7}{3} \\ 0 & 1 & -\frac{7}{3} & -2 & -\frac{2}{3} & 0 & \frac{1}{3} & | & -\frac{1}{3} \\ 0 & 0 & 5 & 4 & 1 & 1 & -1 & | & 0 \end{pmatrix}$$

$$\mathbf{X}_B^{(2)} = (X_1, X_2, X_6), \quad \mathbf{X}_F^{(2)} = (X_3, X_4, X_5, X_7)$$

$$\mathbf{C}_B^{(2)} = (-1, 1, 0), \quad \mathbf{C}_F^{(2)} = (1, 1, 0, 0)$$

$$Cost^{(2)} = -2 + 5X_3 + 7X_4 + \frac{1}{3}X_7.$$

There is no eligible free variable. *Cost* cannot be driven below -2. An optimal vertex is

$$\mathbf{x}_{Opt} = \begin{pmatrix} \frac{7}{3} \\ \frac{1}{3} \\ 0 \\ 0 \end{pmatrix}.$$

Exercise 5.5.6. For *Case 1* in **Example 5.5.5** describe the edge along which the simplex method moves the vector in going from the x_1 to x_2 and then along the half-line emanating from x_2 and along which *Cost* may driven down without bound.

Exercise 5.5.7. By choosing different vectors p and c in **Example 5.5.5** explore the basic **PLPP** given by A. Construct at least one instance in which there is no solution because $\mathcal{Q}$ is empty.

It is time to study in more detail the geometry of $\mathcal{Q}$.

THEOREM 5.5.2. IF THE **PLPP** HAS A FEASIBLE SOLUTION THEN THERE IS AT LEAST ONE VERTEX IN $\mathcal{Q}$.

[**Remark 5.5.4:** Thus if the convex set $\mathcal{Q}$ is not empty then $\mathcal{Q}$ contains a vertex.]

PROOF. Let x be any feasible solution, i.e., $x \in \mathcal{Q}$. Then x is the associate of some $\mathbf{X}$ (cf. **Note 5.5.1**) and $A\mathbf{X} = \mathbf{q}$. If the number K of 0's among the components of $\mathbf{X}$ is at least n then x is a vertex.

Assume that $K < n$. By an appropriate change of notation it may be assumed that

$$X_1, \ldots, X_{n+m-K} > 0 = X_{n+m-K+1}, \ldots, X_{n+m}.$$

Let $\mathbf{A}^1, \ldots, \mathbf{A}^{n+m}$ be the $n + m$ columns of A. Each column of A is an $m \times 1$ column vector. Since $K < n$ it follows that $n + m - K > m$ and, as remarked earlier, $rank(A) = m$, whence $\mathbf{A}^1, \ldots \mathbf{A}^{n+m-K}$ are linearly dependent. Hence there are numbers

$$\gamma_1, \ldots, \gamma_{n+m-K},$$

at least one of which is *positive*, and numbers

$$\gamma_{n+m-K+1}, \ldots, \gamma_{n+m},$$

each equal to 0, and such that

$$\sum_{k=1}^{n+m} \gamma_k \mathbf{A}^k = \mathbf{O}.$$

In other words, if

$$\mathbf{\Gamma} \stackrel{\text{def}}{=} \begin{pmatrix} \gamma_1 \\ \vdots \\ \gamma_{n+m} \end{pmatrix}$$

then $A\mathbf{\Gamma} = \mathbf{O}$.

If

$$\mathbf{Y}(t) \stackrel{\text{def}}{=} \mathbf{X} + t\mathbf{\Gamma} \stackrel{\text{def}}{=} \begin{pmatrix} X_1 \\ \vdots \\ X_{n+m} \end{pmatrix} + t \begin{pmatrix} \gamma_1 \\ \vdots \\ \gamma_{n+m} \end{pmatrix} \succeq \mathbf{O}$$

then, since $A\mathbf{\Gamma} = \mathbf{O}$, for any t

$$AY(t) = A\mathbf{X} + tA\mathbf{\Gamma} = \mathbf{q} + \mathbf{O} = \mathbf{q}.$$

For any t the last K components of $\mathbf{Y}(t)$ are 0's. Thus for any t if the vector $\mathbf{Y}(t) \in \mathbf{R}^{(n,+)}$ then the associate $y(t)$ of $\mathbf{Y}(t)$ is in $\mathcal{Q}$.

   i. If $\mathbf{\Gamma} \succeq \mathbf{O}$ then $\mathbf{Y}(t) \succeq \mathbf{O}$ for all nonnegative t. Furthermore if

$$t_1 \stackrel{\text{def}}{=} maximum \left\{ -\frac{X_i}{\gamma_i} \; : \; \gamma_i > 0 \right\} \stackrel{\text{def}}{=} -\frac{X_{i_1}}{\gamma_{i_1}}$$

then $i_i \leq n + m - K$, $t_1 < 0$, and $\mathbf{Y}(t) \succeq \mathbf{O}$ iff $t_1 \leq t < \infty$. The i_1st component of $\mathbf{Y}(t_1)$ is 0. Because $i_i \leq n + m - K$, the number of 0's among the components of $\mathbf{Y}(t_1)$ is at least $K + 1$.

  ii. If $\mathbf{\Gamma} \not\succeq \mathbf{O}$ some component of $\mathbf{\Gamma}$ is negative and some other component of $\mathbf{\Gamma}$ is positive. If

$$t_1 \stackrel{\text{def}}{=} maximum \left\{ -\frac{X_k}{\gamma_k} \; : \; \gamma_k > 0 \right\} \stackrel{\text{def}}{=} -\frac{X_{i_1}}{\gamma_{i_1}}$$

$$t_2 \stackrel{\text{def}}{=} minimum \left\{ -\frac{X_k}{\gamma_k} \; : \; \gamma_k < 0 \right\} \stackrel{\text{def}}{=} -\frac{X_{i_2}}{\gamma_{i_2}}$$

then, $i_1, i_2 \leq n + m - K$, $t_1 < 0 < t_2$, and $\mathbf{Y}(t) \succeq \mathbf{O}$ iff $t_1 \leq t \leq t_2$. The i_1st component of $\mathbf{Y}(t_1)$ is 0 and the i_2nd component of $\mathbf{Y}(t_2)$ is 0. Because $i_1, i_2 \leq n + m - K$, the number of 0's among the components of $\mathbf{Y}(t_1)$ and the number of 0's among the components of $\mathbf{Y}(t_2)$ is at least $K + 1$.

In each of the situations *i* - *ii* the vector there is in $\mathcal{Q}$ a vector $\mathbf{Y}(t)$ having at least $K + 1$ components that are 0's. Repeated application of the procedure just described yields ultimately a vector $\mathbf{Y}$ having at least n components that are 0's. Then the associate y of $\mathbf{Y}$ is a vertex of $\mathcal{Q}$.

$$\Omega$$

Exercise 5.5.8. Use the notation and the ideas of the proof above to show that there is a positive ϵ such that if $-\epsilon \leq t \leq \epsilon$ then $\mathbf{Y}(t) \succeq \mathbf{O}$.

Exercise 5.5.9. Show that if $\mathbf{Y}(t) \succeq \mathbf{O}$ then $\mathbf{CY}(t) = \mathbf{CX} + t\mathbf{C\Gamma}$ and thus that for a given $\mathbf{X}$ there is a $\mathbf{Y}(t)$ such that the associate y of $\mathbf{Y}(t)$ is a vertex of $\mathcal{Q}$ and $\mathbf{CY}(t) \leq \mathbf{CX}$.

Section 5.5. The simplex method: Theory (and some practice)

[*Hint:* See **Exercise 5.5.8.**]

The next result shows why vertices of Q are so important in the analysis of the **PLPP**.

THEOREM **5.5.3.** IF THE **PLPP** HAS AN OPTIMAL SOLUTION THEN AT LEAST ONE OF THE VERTICES OF Q IS OPTIMAL.

PROOF. Let $\mathbf{x}$ be a solution of the **PLPP** and let $\mathbf{X}$ correspond in $\mathbf{R}^{(n+m,+)}$ to $\mathbf{x}$ in $\mathbf{R}^{(n,+)}$. The argument used in the proof of THEOREM **5.5.2** shows that if $\mathbf{x}$ is not a vertex, i.e., $\mathbf{X}$ has fewer than n 0's among its components, there is an algorithm that leads to a new solution corresponding to a $\mathbf{Y}(t)$ having more 0's among its components. In not more than n steps $\mathbf{x}$ can be replaced, say by a **PLPP** solution $\mathbf{y}$ that *is* a vertex. The next few lines show that $\mathbf{y}$ is not merely a vertex but an optimal solution.

The idea is to show that, since $\mathbf{X}$ corresponds to an optimal solution $\mathbf{x}$,

$$\mathbf{C\Gamma} = 0.$$

According to **Exercises 5.5.8** and **5.5.9** if $\mathbf{C\Gamma} > 0$ and $-\epsilon \leq t < 0$ then $\mathbf{Y}(t) \succeq \mathbf{O}$ and $\mathbf{CY}(t) < \mathbf{CX}$, whence $\mathbf{x}$ is not optimal, a contradiction. Similarly, if $\mathbf{C\Gamma} < 0$ and $\epsilon \geq t > 0$ then $\mathbf{Y}(t) \succeq \mathbf{O}$ and $\mathbf{CY}(t) < \mathbf{CX}$, whence $\mathbf{x}$ is not optimal, a contradiction. It follows that $\mathbf{C\Gamma} = 0$. Hence $\mathbf{CY}(t) = \mathbf{CX}$, and so $\mathbf{Y}(t)$ corresponds not only to a vertex $\mathbf{y}$ of Q but to an optimal vertex as well.

Ω

[Remark **5.5.5**:   From THEOREMs **5.5.1**, **5.5.2**, and **5.5.3** it follows that one way to attack the **PLPP** is to seek among the solutions $\mathbf{X}$ of (5.5.7) - (5.5.9) those in which at least n of the $n + m$ components of $\mathbf{X}$ are 0's. In essence, the simplex method translates this idea into algebraic/numerical operations that seek a solution.

The parallel between the simplex method and the various forms of elimination is worth noting. In each case the methods employed provide

EITHER

a solution to the problem

OR

a demonstration that the problem has no solution.

Thus there is good reason to seek an optimal solution by the simplex method. It starts at any vertex of Q, moves to a neighboring and better vertex, and, if cycling does not occur, after finitely many such moves arrives at an optimal vertex.]

[**Remark 5.5.6:** Suppose that $\mathbf{x}$ is a feasible solution for the **PLPP**. The algorithm for proceeding from $\mathbf{x}$ to a vertex $\mathbf{v}$ of Q is such that $\mathbf{v}$ is optimal if $\mathbf{x}$ is optimal. Hence if $\mathbf{cv} < \mathbf{cx}$ then $\mathbf{x}$ is *not* optimal and so if the algorithm leads from $\mathbf{x}$ to some vertex $\mathbf{v}$ where $\mathbf{cv} \neq \mathbf{cx}$ then $\mathbf{x}$ is *not* optimal.]

What is of crucial importance is the question:

When $\mathbf{x}_0$ is a nonoptimal vertex of Q, is there is an edge E starting at $\mathbf{x}_0$ and along which the value of the objective function *decreases*?

Happily, the answer is, "YES." The following development provides the proof that at any nonoptimal vertex there is an edge along which the objective function decreases.

Let $H_1, \ldots, H_n$ be n linearly independent hyperplanes in $\mathbf{R}^n$, and let $\mathcal{K}$ be the (nonempty!) convex set that is the intersection of the nonnegative half-spaces $S_1, \ldots, S_n$ defined by the H_i, $1 \le i \le n$. Let $E_1, \ldots, E_n$ be the n different edges determined in $\mathcal{K}$ by the H_i, $1 \le i \le n$. Thus there are row vectors $\mathbf{h}_i \overset{\text{def}}{=} (h_{i1}, \ldots, h_{in})$, $1 \le i \le n$, there is a column vector $\mathbf{p} \overset{\text{def}}{=} (p_1, \ldots, p_n)^t$ and

$$H_i \overset{\text{def}}{=} \{\, \mathbf{x} \;:\; \mathbf{h}_i \mathbf{x} = p_i \,\}$$
$$S_i \overset{\text{def}}{=} \{\, \mathbf{x} \;:\; \mathbf{h}_i \mathbf{x} \ge p_i \,\}$$
$$\mathcal{K} \overset{\text{def}}{=} \bigcap_{i=1}^{n} S_i$$
$$E_i \overset{\text{def}}{=} \mathcal{K} \cap \left(\bigcap_{j \neq i} H_j \right).$$

Let $\mathbf{x}_0$ be the unique vector in the intersection of all the hyperplanes H_i. Then, as the simple elimination shows, each edge E_i is a half-line. Thus there is on each edge E_i a vector $\mathbf{x}_i$ different from $\mathbf{x}_0$.

THEOREM 5.5.4. THE SET $\{\mathbf{x}_1 - \mathbf{x}_0, \ldots, \mathbf{x}_n - \mathbf{x}_0\} \overset{\text{def}}{=} B$ IS LINEARLY INDEPENDENT AND IF $\mathbf{w} \in \mathcal{K}$ THERE ARE UNIQUE NONNEGATIVE NUMBERS $t_1, \ldots, t_n$ SUCH THAT

$$\mathbf{w} = \sum_{i=1}^{n} t_i (\mathbf{x}_i - \mathbf{x}_0) + \mathbf{x}_0.$$

PROOF. If $\sum_{i=1}^{n} q_i(\mathbf{x}_i - \mathbf{x}_0) = \mathbf{O}$ then

$$\sum_{i=1}^{n} q_i \mathbf{h}_k(\mathbf{x}_i - \mathbf{x}_0) = 0.$$

However

$$\mathbf{x}_i \in \bigcap_{j \neq i} H_j \text{ and } \mathbf{x}_0 = \bigcap_{i=1}^{n} H_j,$$

i.e., each $\mathbf{x}_i$ is in $\mathcal{K}$ and also on precisely $n - 1$ of the hyperplanes while $\mathbf{x}_0$ is in $\mathcal{K}$ and on all n hyperplanes. It follows that there are positive numbers δ_i, $1 \leq i \leq n$, such that

$$\mathbf{h}_k \mathbf{x}_i \stackrel{\text{def}}{=} \eta_{ki} = \begin{cases} p_k & \text{if } k \neq i \\ p_i + \delta_i > p_i & \text{if } k = i \end{cases}$$

$$\mathbf{h}_k(\mathbf{x}_i - \mathbf{x}_0) = \begin{cases} p_k - p_k = 0 & \text{if } k \neq i \\ \eta_{ii} - p_i > 0 & \text{if } k = i. \end{cases}$$

Thus $q_i \eta_{ii} = 0$, hence $q_i = 0$, $1 \leq i \leq n$, and so B is linearly independent.

Since $\mathbf{w} - \mathbf{x}_0 \in \mathbf{R}^n$ it follows that there are unique numbers t_i such that

$$\mathbf{w} - \mathbf{x}_0 = \sum_{i=1}^{n} t_i(\mathbf{x}_i - \mathbf{x}_0).$$

Since $\mathbf{w} \in \mathcal{K}$ it follows that $\mathbf{h}_k \mathbf{w} \stackrel{\text{def}}{=} w_k \geq p_k$ whence

$$\mathbf{h}_k(\mathbf{w} - \mathbf{x}_0) = t_k(\eta_{kk} - p_k)$$
$$= w_k - p_k \geq 0$$

and so $t_k \geq 0$.

$$\Omega$$

THEOREM 5.5.5. IF $\mathbf{x}_0$ IS A VERTEX OF $\mathcal{Q}$ THEN THERE ARE ONLY THREE MUTUALLY EXCLUSIVE POSSIBILITIES:

i. $\mathbf{x}_0$ IS AN OPTIMAL VERTEX;

ii. THERE IS ON SOME EDGE STARTING AT $\mathbf{x}_0$ A VECTOR $\mathbf{x}$ SUCH THAT $\mathbf{cx} < \mathbf{cx}_0$;

iii. THE **PLPP** HAS NO SOLUTION.

PROOF. If neither *i* nor *iii* is true then there is in $\mathcal{Q}$ an optimal vertex $\mathbf{x}_{Opt}$:

$$\mathbf{cx}_{Opt} < \mathbf{cx}_0. \tag{5.5.18}$$

If *ii* is also false then at every $\mathbf{x}$ on any edge starting at $\mathbf{x}_0$

$$\mathbf{cx} \geq \mathbf{cx}_0 > \mathbf{cx}_{Opt}. \tag{5.5.19}$$

As the next lines show, a contradiction is at hand.

Indeed, the vertex $\mathbf{x}_0$ is the intersection of n linearly independent hyperplanes H_{k_i}, $1 \leq i \leq n$, among the hyperplanes H_k, $1 \leq k \leq n+m$:

$$\mathbf{x}_0 = \bigcap_{i=1}^{n} H_{k_i}.$$

Let $\mathcal{Q}'$ be the intersection of the nonnegative half-spaces determined by the H_{k_i}, $1 \leq i \leq n$:

$$\mathcal{Q}' \stackrel{\text{def}}{=} \bigcap_{i=1}^{n} \{\, \mathbf{z} \; : \; \mathbf{a}_{k_i}\mathbf{z} \geq p_{k_i} \,\} \stackrel{\text{def}}{=} \bigcap_{i=1}^{n} S_i.$$

Because

$$\{H_{k_1}, \ldots, H_{k_n}\} \subset \{H_1, \ldots, H_{n+m}\}$$

it follows that $\mathcal{Q}' \supset \mathcal{Q}$.

On the other hand according to THEOREM **5.5.4** if $1 \leq i \leq n$ there is in $\mathcal{Q}'$ an edge E_i and on E_i a vector $\mathbf{x}_i$ different from $\mathbf{x}_0$. Furthermore if $\mathbf{w} \in \mathcal{Q}'$ there are unique nonnegative numbers $t_1, \ldots, t_n$ such that

$$\mathbf{w} - \mathbf{x}_0 = \sum_{i=1}^{n} t_i(\mathbf{x}_i - \mathbf{x}_0)$$

Since $\mathbf{x}_{Opt} \in \mathcal{Q} \subset \mathcal{Q}'$ and since $\mathbf{x}_0$ is not optimal

$$0 > \mathbf{c}(\mathbf{x}_{Opt} - \mathbf{x}_0) = \sum_{i=1}^{n} t_i(\mathbf{c}\mathbf{x}_i - \mathbf{c}\mathbf{x}_0)$$

$$\geq 0,$$

a contradiction.

$$\Omega$$

Exercise 5.5.10. Show that if $\mathbf{x}_0$ is a vertex of $\mathcal{Q}$ and if the **PLPP** has a solution, then for some optimal vertex $\mathbf{x}_{Opt}$ there is a finite set $\{E_0, \ldots, E_P\}$ of edges such that

$$\mathbf{x}_0 \in E_0$$
$$E_p \cap E_{p+1} \neq \emptyset, \; 1 \leq p \leq P$$
$$\mathbf{x}_{Opt} \in E_P.$$

In other words, show that for some optimal vertex $\mathbf{x}_{Opt}$ there is an *"edge-path"* leading from $\mathbf{x}_0$ to $\mathbf{x}_{Opt}$.

[*Hint:* Use THEOREM **5.5.5** and the fact that there are only finitely many vertices in $\mathcal{Q}$.]

[**Note 5.5.3:** THEOREM **5.5.5** and **Exercise 5.5.10** establish the very important fact that if the **PLPP** has a solution then for any vertex

$\mathbf{x}_0$ of $\mathcal{Q}$ there is a sequence of edges leading from $\mathbf{x}_0$ to an optimal vertex $\mathbf{x}_{Opt}$. Thus the simplex method, which provides a systematic way of passing from a vertex to a better neighboring vertex, is an appropriate mechanism for solving the **PLPP** if it has a solution.]

Exercise 5.5.11. Show that if $\mathcal{Q}$ is not empty but the **PLPP** has no solution then one of the edges of $\mathcal{Q}$ is a half-line.

Exercise 5.5.12. Assume

$$A \overset{\text{def}}{=} \begin{pmatrix} 2 & 1 & 0 \\ -3 & 1 & 0 \\ 1 & 1 & 1 \end{pmatrix}, \; \mathbf{p} \overset{\text{def}}{=} \begin{pmatrix} 0 \\ 0 \\ -1 \end{pmatrix}, \; \mathbf{c} = (1, 2, -1).$$

Show that the **PLPP** in which a vector $\mathbf{x}_0$ is sought such that $\mathbf{x}_0 \succeq \mathbf{O}$ and

$$\mathbf{c}\mathbf{x}_0 = minimum\,\{\,\mathbf{c}\mathbf{x} \; : \; A\mathbf{x} \succeq \mathbf{p}, \; \mathbf{x} \succeq \mathbf{O}\,\}$$

has no solution although $\mathcal{Q}$ is not empty. Seek in $\mathcal{Q}$ an edge that is a half-line.

Exercise 5.5.13. Assume

$$A \overset{\text{def}}{=} \begin{pmatrix} 2 & -1 \\ -3 & 1 \end{pmatrix}, \; \mathbf{p} \overset{\text{def}}{=} \begin{pmatrix} -3 \\ 1 \end{pmatrix}, \; \mathbf{c} = (1, 2).$$

Show that the **PLPP** in which a vector $\mathbf{x}_{Opt}$ is sought so that $\mathbf{x}_{Opt} \succeq \mathbf{O}$ and

$$\mathbf{c}\mathbf{x}_{Opt} = minimum\,\{\,\mathbf{c}\mathbf{x} \; : \; A\mathbf{x} \succeq \mathbf{p}, \; \mathbf{x} \succeq \mathbf{O}\,\}$$

has a solution and that $\mathcal{Q}$ contains an edge that is a half-line. Hence the **PLPP** can have a solution even when $\mathcal{Q}$ contains an edge that is a half-line.

The following THEOREMs imply that the set of edges of $\mathcal{Q}$ is *connected*. The author is indebted to William Zame for suggesting the method of proof for the second result. The basic idea is that some of the *topological properties* of $\mathcal{Q}$ can be explored by means of an appropriately chosen **PLPP**. From the standpoint of linear programming the result what follows is embroidery – but rather beautiful embroidery.

As the intersection of nonnegative half-spaces determined by the hyperplanes H_k the set $\mathcal{Q}$ is convex. Hence the line segment containing any two vectors of $\mathcal{Q}$ lies entirely in $\mathcal{Q}$. A vector $\mathbf{x}$ of $\mathcal{Q}$ is called an *extreme point* of $\mathcal{Q}$ iff $\mathbf{x}$ is *not inside* any line segment containing two *different* vectors of $\mathcal{Q}$.

DEFINITION 5.5.3. A VECTOR $\mathbf{x}$ IS AN EXTREME POINT OF $\mathcal{Q}$ IFF IT IS IMPOSSIBLE TO FIND IN $\mathcal{Q}$ TWO DIFFERENT VECTORS $\mathbf{x}_1$ AND $\mathbf{x}_2$ AND A NUMBER t STRICTLY BETWEEN 0 AND 1 $(0 < t < 1)$ SO THAT $\mathbf{x} = t\mathbf{x}_1 + (1 - t)\mathbf{x}_2$.

Exercise 5.5.14. Show, using the notation just introduced for the definition of an extreme point, that neither x_1 nor x_2 is x.

THEOREM 5.5.6. A VECTOR x IS A VERTEX OF $\mathcal{Q}$ IFF x IS AN EXTREME POINT OF $\mathcal{Q}$.

[**Remark 5.5.7:** Thus in geometric terms a vertex x of $\mathcal{Q}$ is the same as an extreme point of $\mathcal{Q}$.]

PROOF. Let the rows of the matrix $\tilde{A}$ be the $1 \times n$ vectors $a_1, \ldots, a_{n+m}$.

i. Assume x_0 is a vertex of $\mathcal{Q}$. Thus in $\tilde{A}$ there are n linearly independent rows, $a_{k_1}, \ldots, a_{k_n}$ such that x_0 is the *unique* vector solution of the system

$$a_{k_1} x = \tilde{p}_{k_1}$$
$$\vdots \qquad \vdots$$
$$a_{k_n} x = \tilde{p}_{k_n}$$

which may be written $Mx = \tilde{p}$.

If x_1 and x_2 are two different vectors in $\mathcal{Q}$, if $0 < t < 1$, and if $x_0 = tx_1 + (1-t)x_2$ then, because $x_0 \neq x_1, x_2$ and because the system $Mx = \tilde{p}$ has the *unique* solution x_0, it follows that $Mx_1, Mx_2 \neq \tilde{p}$. Because $x_1, x_2 \in \mathcal{Q}$ it follows that

$$Mx_1, \ Mx_2 \succeq \tilde{q},$$

whence for some j among the numbers $1, \ldots, n$,

$$a_{k_j} x_1 > \tilde{p}_{k_j}, \ a_{k_j} x_2 \geq \tilde{p}_{k_j}.$$

Consequently

$$a_{k_j} x_0 = t a_{k_j} x_1 + (1-t) a_{k_j} x_2$$
$$> t \tilde{p}_{k_j} + (1-t) \tilde{p}_{k_j} = \tilde{p}_{k_j}.$$

It follows that $Mx_0 \neq \tilde{p}$, a contradiction. Hence the vertex x_0 is an extreme point.

ii. Conversely, if x_0 is an extreme point of $\mathcal{Q}$ then x_0, as a vector in $\mathcal{Q}$, satisfies the whole system $\tilde{A}x \succeq p$. If $\tilde{A}x_0 \succ \tilde{p}$ then there is a sufficiently small positive number ϵ such that if $0 < y_1, \ldots, y_n < \epsilon$ and if

$$y \stackrel{\text{def}}{=} \begin{pmatrix} y_1 \\ \vdots \\ y_n \end{pmatrix}$$

then $x_0 \pm y \stackrel{\text{def}}{=} z_\pm \in \mathcal{Q}$. Furthermore $z_+ \neq z_-$ and $x_0 = \frac{1}{2}z_+ + (1 - \frac{1}{2})z_-$, a contradiction of the assumption that x_0 is an extreme point of $\mathcal{Q}$.

Hence some inequality ("$\geq$") in the system $\tilde{A}\mathbf{x}_0 \succeq \mathbf{p}$ is an equation ("="). Let the number of equations be s and let the rank of the matrix of coefficients in these s equations be r. Then $1 \leq r \leq s \leq n + m$ and $n + m - s$ of the inequalities in the system $\tilde{A}\mathbf{x}_0 \succeq \tilde{\mathbf{p}}$ are strict inequalities (">" rather than "$\geq$"). It is sufficient to show that in any event $r = n$.

If $\mathbf{x}_0 = \mathbf{O}$ then $\mathbf{x}_0$ is a vertex because

$$\mathbf{x}_0 = \mathbf{O} \in \bigcap_{k=m+1}^{n+m} H_k$$

and the matrix of coefficients corresponding to $H_{m+1}, \ldots, H_{n+m}$ is the $n \times n$ matrix I. Hence also $s \geq n$ and $r = n$.

Thus it may be assumed that some component of $\mathbf{x}_0$ is positive.

    a. If $r < n$ and *all* components of $\mathbf{x}_0$ are positive, i.e., $\mathbf{x}_0 \succ \mathbf{O}$, apply the GJM to the system of s equations for which the corresponding matrix consists of s rows of A. An appropriate change in notation permits the assumption that there are numbers

$$b_1, \ldots, b_r, \; b_{ij}, \; 1 \leq i \leq r, \; r + 1 \leq j \leq n$$

such that the row-reduction leads to the system

$$x_1 = b_1 + \sum_{j=r+1}^{n} \alpha_{1j} x_j$$

$$\vdots \qquad \vdots$$

$$x_r = b_r + \sum_{j=r+1}^{n} \alpha_{rj} x_j. \tag{5.5.20}$$

Then $\mathbf{x}_0$ is a solution of (5.5.20) and sufficiently small perturbations in the last $n - r$ components of $\mathbf{x}_0$ give rise via (5.5.20) to *positive* vectors close to $\mathbf{x}_0$. Furthermore if the perturbations are sufficiently small the resulting vectors also satisfy the strict inequalities (">") in the system $A\mathbf{x} \succeq \mathbf{p}$. More precisely, if

$$\mathbf{x}_0 \stackrel{\text{def}}{=} \begin{pmatrix} x_{01} \\ \vdots \\ x_{0,n+1} \end{pmatrix}$$

there is a positive ϵ such that if $0 \leq |x_{0j} - y_j| < \epsilon$, $r + 1 \leq j \leq n$, then

---

each $y_j > 0$. Furthermore if

$$y_1 \overset{\text{def}}{=} b_1 + \sum_{j=r+1}^{n} \alpha_{1j} y_j$$

$$\vdots \qquad \vdots$$

$$y_r \overset{\text{def}}{=} b_r + \sum_{j=r+1}^{n} \alpha_{rj} y_j$$

$$\mathbf{y} \overset{\text{def}}{=} \begin{pmatrix} y_1 \\ \vdots \\ y_n \end{pmatrix}$$

then $\mathbf{y} \in \mathcal{Q}$. In particular, if,

$$y_j \overset{\text{def}}{=} x_j + \frac{1}{2}\epsilon \text{ and } z_j \overset{\text{def}}{=} x_j - \frac{1}{2}\epsilon, \ r+1 \leq j \leq n$$

and if

$$z_1 \overset{\text{def}}{=} b_1 + \sum_{j=r+1}^{n} \alpha_{1j} z_j$$

$$\vdots \qquad \vdots$$

$$z_r \overset{\text{def}}{=} b_r + \sum_{j=r+1}^{n} \alpha_{rj} z_j$$

$$\mathbf{z} \overset{\text{def}}{=} \begin{pmatrix} z_1 \\ \vdots \\ z_n \end{pmatrix}$$

then $\mathbf{y}$ and $\mathbf{z}$ are two different vectors in $\mathcal{Q}$. At the same time

$$\mathbf{x} = \frac{1}{2}\mathbf{y} + \frac{1}{2}\mathbf{z}$$

in denial of the fact that $\mathbf{x}$ is an extreme point of $\mathcal{Q}$. Hence in this case $r = n$.

b. If $r < n$ and some components of $\mathbf{x}_0$ are 0's choose the notation so that $x_{01} = \cdots = x_{0\nu} = 0 < x_{0\nu+1}, \ldots, x_{0n}$. Then $r \geq \nu$. After application

---

of the GJM to the system of s equations they look like this:

$$x_1 = 0$$
$$x_2 = 0$$
$$\vdots \qquad \vdots$$
$$x_\nu = 0$$
$$x_{\nu+1} = \sum_{j=r+1}^{n} \alpha_{\nu j} x_j$$
$$\vdots \qquad \vdots$$
$$x_r = \sum_{j=r+1}^{n} \alpha_{r j} x_j.$$

Again, sufficiently small perturbations of $x_{0r+1}, \ldots, x_{0n}$ give rise in $\mathcal{Q}$ to two different vectors $\mathbf{y}$ and $\mathbf{z}$ such that
$$\mathbf{x} = \frac{1}{2}\mathbf{y} + \frac{1}{2}\mathbf{z},$$
again in denial of the fact that $\mathbf{x}$ is an extreme point of $\mathcal{Q}$. Hence in all instances $r = n$.

$$\Omega$$

THEOREM 5.5.7. ASSUME THAT $\mathcal{K}$ IS THE INTERSECTION OF THE NON-NEGATIVE HALF-SPACES DETERMINED BY A FINITE SET $\{H_1, \ldots, H_N\}$ OF HYPERPLANES IN $\mathbf{R}^n$ AND THAT $\mathcal{K}$ CONTAINS AT LEAST ONE VERTEX $\mathbf{v}$. IF $\mathbf{x}$ AND $\mathbf{y}$ ARE VECTORS LYING ON EDGES OF $\mathcal{Q}$, THEN THERE IS IN $\mathcal{Q}$ A SEQUENCE

$$\{E_i\}_{i=1}^{N}$$

OF EDGES SUCH THAT

$$E_i \cap E_{i+1} \neq \emptyset, \ 1 \leq i \leq N - 1$$
$$\mathbf{x} \in E_1 \text{ and } \mathbf{y} \in E_N.$$

PROOF. Assume $\mathbf{x}$ lies on an edge denoted E_1, the intersection of $n-1$ linearly independent hyperplanes. There is another hyperplane that, forms with those determining E_1 a linearly independent set. Then E_1 contains the vertex $\mathbf{v}$ determined by those n linearly independent hyperplanes. Let $\mathcal{Q}'$ be the convex hull of the *finite* set $\{\mathbf{v}_1, \ldots \mathbf{v}_K\}$ of all vertices, other than $\mathbf{v}$ of $\mathcal{Q}$:

$$\mathcal{Q}' \overset{\text{def}}{=} Conv(\{\mathbf{v}_1, \ldots, \mathbf{v}_K\}).$$

The next few lines show that $\mathbf{v}$ is not in $\mathcal{Q}'$. Indeed, if $\mathbf{v} \in \mathcal{Q}'$ then there is in Σ_K a vector $(\alpha_1, \ldots, \alpha_K)$ such that

$$\mathbf{v} = \sum_{k=1}^{K} \alpha_k \mathbf{v}_k.$$

Since $\mathbf{v} \notin \{\mathbf{v}_1, \ldots, \mathbf{v}_K\}$ it follows that some α_k, say α_1, is strictly between 0 and 1. Thus if $\beta \overset{\text{def}}{=} \sum_{k=2}^{K} \alpha_k$ then $\alpha_1 + \beta = 1$, $\alpha_1, \beta > 0$

$$\mathbf{v} = \alpha_1 \mathbf{v}_1 + \beta\left(\sum_{k=2}^{K} \frac{\alpha_k}{\beta} \mathbf{v}_k\right) \overset{\text{def}}{=} \alpha_1 \mathbf{v}_1 + \beta \mathbf{w}.$$

Since $\sum_{k=2}^{K} \frac{\alpha_k}{\beta} = 1$ it follows that $\mathbf{w} \in \mathcal{Q}$ and that the vertex $\mathbf{v}$ is *not* an extreme point of $\mathcal{Q}$, a contradiction.

An argument like that used in the PROOF of THEOREM **5.4.1** shows that there is in $\mathcal{Q}'$ a unique vector z that is nearest to $\mathbf{v}$ and that the hyperplane H defined as

$$\{\mathbf{u} \; : \; (\mathbf{u} - \mathbf{y}, \mathbf{z} - \mathbf{v}) = 0\}$$

separates $\mathbf{v}$ from $\mathcal{Q}'$:

$$(\mathbf{v} - \mathbf{z}, \mathbf{z} - \mathbf{v}) = -\|\mathbf{v} - \mathbf{z}\|^2 < 0$$
$$\mathbf{w} \in \mathcal{Q}' \Rightarrow (\mathbf{w} - \mathbf{z}, \mathbf{z} - \mathbf{v}) \geq 0.$$

Hence if a **PLPP** is defined by the vector $\mathbf{c}^t \overset{\text{def}}{=} (\mathbf{z} - \mathbf{v})^t$, i.e., the **PLPP** is to minimize $\mathbf{c}\mathbf{x}$ among all $\mathbf{x}$ in $\mathcal{Q}$ then the unique optimal vertex in $\mathcal{Q}$ is $\mathbf{v}$.

The proof that there is on the edge E_1 a vertex of $\mathcal{Q}$ may repeated to show that there is in $\mathcal{Q}$ an edge, say E_N, containing a vertex $\mathbf{u}$ of $\mathcal{Q}$. If $\mathbf{u}=\mathbf{v}$ then $N = 2$ and the job is done. If $\mathbf{u} \neq \mathbf{v}$ then according to THEOREM **5.5.5** and **Exercise 5.5.10**, since $\mathbf{v}$ is the only optimal vertex, there is a sequence of edges leading from $\mathbf{u}$ to $\mathbf{v}$. Thus starting at $\mathbf{x}$ one can move along the edge E_1 to $\mathbf{v}$, then along edges to $E_2, \ldots, E_{N-1}$ to $\mathbf{u}$, and finally along E_N to $\mathbf{y}$.

$$\Omega$$

The question addressed next is that of systematically and economically seeking a starting vertex. A random row-reduction of the system $A\mathbf{X} = \mathbf{q}$ might produce some negative $q_i^{(1)}$. Then setting each free variable at 0 leads to a $\mathbf{X}$ *not* in the nonnegative orthant.

Without some well-defined algorithm the search for a starting vertex can be very long. The number of sets of linearly independent rows in the $(n + m) \times n$ matrix $\widetilde{A}$ can be as large as $\binom{n+m}{n}$. If $m = 10$ and $n = 15$ then

$$\binom{n + m}{n} = \binom{25}{15} = 69,920.$$

If $m = 60$ and $n = 40$ then

$$\binom{n + m}{n} = \binom{100}{40} \approx 1.7 \times 10^{29}.$$

The next part of this **Section** is devoted to a reasonable method for seeking a vertex of $\mathcal{Q}$. The method is such that it finds a vertex or shows that $\mathcal{Q}$ is empty.

SEEKING A STARTING VERTEX

In some instances a vertex of $\mathcal{Q}$ is easy to find.

i.. If each entry of A is nonnegative and each row of A contains at least one positive entry then a positive vector in which each entry is sufficiently large serves as a feasible solution. Thus if $A \stackrel{\text{def}}{=} (a_{ij})_{i,j=1}^{m,n}$ and if

$$a_{ij_i} > 0, \ 1 \le i \le m,$$

let $\mathbf{x}$ be the vector

$$\sum_{i=1}^{m} \frac{|p_i|}{a_{ij_i}} \mathbf{e}_{j_i}.$$

Then $\mathbf{x}$ is nonnegative, has precisely m nonzero components, and $A\mathbf{x} \succeq \mathbf{p}$: $\mathbf{x}$ is a feasible solution.

Since in the nutrition problem and in many other practical linear programming problems all the entries of A are nonnegative and each row has at least one positive entry each such problem has a feasible solution.

ii. If each component of $\mathbf{p}$ is 0 or a negative number then $\mathbf{O}$ is a feasible solution.

iii. If A contains an embedded identity matrix I or if there is a clearly and easily identifiable basic matrix A_B then a vertex is at hand.

If A does not contain an embedded identity matrix and there is no obvious basic matrix A_B some extra analysis is required. One defines an **Auxiliary Linear Programming Problem (ALPP)** as follows.

i. Adjoin to A the $m \times m$ identity matrix I and thereby produce the matrix A':

$$A' \stackrel{\text{def}}{=} m \ \begin{pmatrix} \overset{n+m}{A} & \overset{m}{I} \end{pmatrix}.$$

---

ii. Introduce a new $1 \times (2n + m)$ vector $\mathbf{C}'$ in which the first $n + m$ components are 0's and the last m components are 1's:

$$\mathbf{C}' \stackrel{\text{def}}{=} 1 \begin{array}{cc} \overset{n+m}{} & \overset{m}{} \\ \left(\mathbf{O} \right. & \left. \textstyle\sum_{i=1}^{m} \mathbf{e}_i \right). \end{array}$$

iii. Introduce m more new variables

$$X_{n+m+1}, \ldots, X_{n+2m}$$

and thereby a new $1 \times (2n + m)$ vector $\mathbf{X}'$:

$$\mathbf{X}' \stackrel{\text{def}}{=} \begin{pmatrix} X_1 \\ \vdots \\ X_{n+m} \\ X_{n+m+1} \\ \vdots \\ X_{n+2m} \end{pmatrix}.$$

The new variables $X_{n+m+1}, \ldots, X_{n+2m}$ are called *artificial* variables as opposed to the *slack* variables $X_{n+1}, \ldots, X_{n+m}$.

The **ALPP** is that of seeking an $\mathbf{X}'_{Opt}$ such that:

$$\mathbf{X}'_{Opt} \succeq \mathbf{O}$$
$$A'\mathbf{X}'Opt = \mathbf{q} \; (\succeq \mathbf{O})$$
$$\mathbf{C}'\mathbf{X}'_{Opt} = minimum \left\{ \mathbf{C}'\mathbf{X}' \; : \; \mathbf{X}' \in \mathsf{R}_{2n+m}, \; \mathbf{X}' \succeq \mathbf{O} \right\}.$$

Since, by definition, $\mathbf{q} \succeq \mathbf{O}$ the vector

$$\mathbf{X}'_0 \stackrel{\text{def}}{=} \begin{array}{c} \\ n+m \\ m \end{array} \begin{pmatrix} \overset{1}{\mathbf{O}} \\ \mathbf{q} \end{pmatrix}$$

is a feasible solution of the **ALPP**. Since $\mathbf{C}' \succeq \mathbf{O}$ it follows that $\mathbf{C}'\mathbf{X}'$ for the **ALPP** is bounded below by 0. It follows that the **ALPP** has a solution $\mathbf{X}'_{Opt}$.

Then, *Cost'* denoting the value of the objective function for the **ALPP**, if $Cost' = \mathbf{C}'\mathbf{X}'_{Opt} = 0$ the last m components of $\mathbf{X}'_{Opt}$ are 0's. Hence, if $\mathbf{X}_0$ is the $1 \times (n + m)$ vector that results from dropping the last m components of $\mathbf{X}'_{Opt}$, then

$$\mathbf{X}_0 \succeq \mathbf{O}, \; A\mathbf{X}_0 = \mathbf{q}.$$

In other words, the associate x_0 of $\mathbf{X}_0$ is in $\mathcal{Q}$. The method described in the proof of THEOREM **5.5.3** and discussed further in **Remark 5.5.6** leads from

---

x_0 to a vertex in $\mathcal{Q}$. The stage is set for applying the simplex method to the **PLPP**.

On the other hand, if $Cost' = \mathbf{C}'\mathbf{X}'_{Opt} > 0$ at least one of the last m components of $\mathbf{X}'_{Opt}$ is not 0. In this case the **PLPP** has no solution because $\mathcal{Q} = \emptyset$. Indeed, if $\mathcal{Q} \neq \emptyset$ and if $x_0 \in \mathcal{Q}$ there is in $\mathbf{R}^{(n+m,+)}$ a vector X_0 such that $A\mathbf{X}_0 = \mathbf{q}$ and for which the associate is x_0 (cf. **Note 5.5.1**). Assume

$$\mathbf{Y}'_0 \overset{\text{def}}{=} \begin{matrix} \\ n+m \\ m \end{matrix} \overset{\displaystyle 1}{\left(\begin{matrix} \mathbf{X}_0 \\ \mathbf{O} \end{matrix} \right)}$$

Then $\mathbf{Y}'_0 \in \mathcal{Q}'$ and $\mathbf{C}'\mathbf{Y}'_0 = 0 < \mathbf{C}'\mathbf{X}'_{Opt}$ in denial of the optimality of $\mathbf{X}'_{Opt}$ for the **ALPP**.

Thus:

   i. $\mathcal{Q}'$ is *never* empty and the **ALPP** *always* has a solution;

   ii. $\mathcal{Q}$ is not empty iff the minimum value of $Cost'$ for the **ALPP** is zero iff the last m components of the optimal solution $\mathbf{X}'_{Opt}$ of the **ALPP** are 0's;

   iii. if the last m components of the *optimal* solution $\mathbf{X}'_{Opt}$ of the **ALPP** are 0's then the first $n+m$ components of $\mathbf{X}'_{Opt}$ constitute a $(n+m) \times 1$ vector $\mathbf{X}_0$ for which the associate x_0 is a *feasible* solution of the **PLPP**.

5.6. Computational methods in linear programming

In this **Section** there is given an organized approach to the practical application of the simplex algorithm.

In light of the developments above it can be said that the technique of moving from vertex to vertex of $\mathcal{Q}'$ and then of $\mathcal{Q}$ is the essence of the simplex method. That vertex-to-vertex technique can be systematized by stripping away the place-holding symbols X_i from the formal presentation of either the **ALPP** or the **PLPP** and reducing the whole operation to manipulations of the entries in a sequence of arrays. Each array is called a *tableau* (plural: tableaux). Among the ingredients of each tableau are those in the following list:

   i. $A^{(r)}$, in which the identity matrix is embedded;

   ii. $\mathbf{q}^{(r)}$;

   iii. $\mathbf{X}$;

   iv. $\mathbf{X}_B^{(r)}$;

   v. $Cost^{(r)}$;

   vi. $\mathbf{C}_B^{(r)}$;

   vii. $C - \mathbf{C}_B^{(r)}A \overset{\text{def}}{=} C^{(r)} \overset{\text{def}}{=} (C_1^{(r)}, \ldots, C_{n+m}^{(r)})$.

Tableaux are used *only when a vertex of $\mathcal{Q}$ is known.* Thus tableaux are used at once in the **ALPP** and in the **PLPP** if A contains an embedded identity matrix. If A has no embedded identity matrix then tableaux are used only if a

     Chapter 5. APPLYING LINEAR ALGEBRA

vertex of $\mathcal{Q}$ is somehow found, e.g., either by finding a basic matrix A_B or via the **ALPP**.

Tableaux are more elaborate than matrices because the simplex algorithm requires more than mere simple elimination. Attention must be paid to the nonnegativity of $\mathbf{X}$, to the components of $\mathbf{C}_b^{(r)}$, to the components of $\mathcal{C}^{(r)}$, to the SWITCH decisions, and to $Cost^{(r)}$. The model for $\mathcal{T}^{(r)}$, the tableau at stage r, is the following.

$\mathbf{X}_B^{(r)}$	X_1	$\cdots$	$X_{b_i^{(r)}}$	$\cdots$	$-\mathbf{q}^{(r)}$	$\mathbf{C}_B^{(r)}$
$X_{b_1^{(r)}}$	$a_{11}^{(r)}$	$\cdots$	0	$\cdots$	$-q_1^{(r)}$	$C_{b_1^{(r)}}$
$\vdots$	$\vdots$	$\ddots$	$\vdots$	$\ddots$	$\vdots$	$\vdots$
$X_{b_i^{(r)}}$	$a_{i1}^{(r)}$	$\cdots$	1	$\cdots$	$-q_i^{(r)}$	$C_{b_i^{(r)}}$
$\vdots$	$\vdots$	$\ddots$	$\vdots$	$\ddots$	$\vdots$	$\vdots$
$X_{b_m^{(r)}}$	$a_{m1}^{(r)}$	$\cdots$	0	$\cdots$	$-q_m^{(r)}$	$C_{b_m^{(r)}}$
$\mathcal{C}^{(r)}$	$c_1^{(r)}$	$\cdots$	0	$\cdots$	$\mathbf{C}_B^{(r)}\mathbf{q}^{(r)}$	

Figure 5.6.1. Model of $\mathcal{T}^{(r)}$

The tableau consists of several sections.

i. The top box in which various ingredients are merely labelled. No algebraic manipulations apply to the entries in the top box, although as a result of algebraic manipulations, the entries bearing a superscript $^{(r)}$ change.

ii. In the central box:
- a. below $X_B^{(r)}$ are listed its components;
- b. below each named variable are listed its coefficients in the m equations of the system $A^{(r)}\mathbf{X} - \mathbf{q}^{(R)} = \mathbf{O}$; under each basic variable $X_{b_i^{(r)}}$ is the column vector $\mathbf{e}_i$, $1 \le i \le m$; together these standard basis vectors make up the embedded identity matrix I;
- c. below $-\mathbf{q}^{(r)}$ are listed its components;
- d. below $\mathbf{C}_B^{(r)}$ are listed its components.

iii. In the bottom box there is the label $\mathcal{C}^{(r)}$ of the vector

$$\mathbf{C} - \mathbf{C}_B^{(r)} A^{(r)}$$

and to the right of that label are its components followed by

$$\mathbf{C}_B^{(r)}\mathbf{q}^{(r)}, \text{ i.e., } 0 - \mathbf{C}_B^{(r)}(-\mathbf{q}^{(r)}).$$

All algebraic manipulations bear on the entries between the interior vertical lines and in the central and bottom boxes . These entries constitute the matrix

$$\mathcal{A}^{(r)} \stackrel{\text{def}}{=} \left(\begin{array}{ccc|c} a_{11}^{(r)} & \cdots & a_{1,n+m}^{(r)} & -q_1^{(r)} \\ \vdots & \ddots & \vdots & \vdots \\ a_{m1}^{(r)} & \cdots & a_{m,n+m}^{(r)} & -q_m^{(r)} \\ \hline C_1^{(r)} & \cdots & C_{n+m}^{(r)} & \mathbf{C}_B^{(r)}\mathbf{q}^{(r)} \end{array} \right) . \tag{5.6.1}$$

The entries in the bottom box and between the interior vertical lines of the tableau may be regarded as the components of

$$\begin{array}{c} n+m+1 \\ 1 \ (\quad \mathbf{C} \quad 0\ ) - \mathbf{C}_B^{(r)}(A^{(r)}|-\mathbf{q}^{(r)}). \end{array}$$

The vector $\mathcal{C}^{(r)}$ in the last row of $\mathcal{A}^{(r)}$ contains the information leading to the choice of the next pivot entry. The entries in that last row are 0's at the components numbered $b_1^{(r)},\ldots,b_m^{(r)}$; the entries at the components numbered $f_1^{(r)},\ldots,f_n^{(r)}$ are the components of the vector $\mathbf{X}_F^{(r)}$. If q is the smallest value of j for which $C_j < 0$ and if p is the smallest value of i' for which

$$\frac{q_{i'}^{(r)}}{a_{i'q}^{(r)}} = minimum \left\{ \frac{q_i^{(r)}}{a_{iq}^{(r)}} \ : \ a_{iq}^{(r)} > 0 \right\}$$

then SWITCH $X_p \leftrightarrow X_q$ is dictated. It amounts to converting the qth column of $\mathcal{A}^{(r)}$ to the $(m+1) \times 1$ vector $\mathbf{e}_p$ (in $\mathbf{R}^{m+1}$).

At the start $A^{(0)}$ is assumed to contain an identity matrix embedded in columns $b_1^{(0)},\ldots,b_m^{(0}$. At the corresponding vertex of $\mathcal{Q}$ *Cost* is $\mathbf{C}_B^{(0)}\mathbf{q}$. Consequently $\mathcal{A}^{(1)}$ is the block matrix arising from the complete row-reduction of

$$\left(\begin{array}{c|c} A^{(0)} & -\mathbf{q}^{(0)} \\ \hline \mathbf{C} & 0 \end{array} \right).$$

Hence

$$\mathcal{A}^{(1)} = \left(\begin{array}{c|c} A^{(0)} & -\mathbf{q}^{(0)} \\ \hline \mathbf{C} - \mathbf{C}_B^{(0)}A^{(0)} & \mathbf{C}_B^{(0)}\mathbf{q}^{(0)} \end{array} \right).$$

THEOREM 5.6.1. IF $\mathcal{T}^{(r)}$ DICTATES THE SWITCH

$$X_{b_i^{(r)}} \leftrightarrow X_{f_j^{(r)}}$$

THEN $\mathcal{T}^{(r+1)}$ IS THE RESULT OF ROW-REDUCTION OF $\mathcal{A}^{(r)}$ VIA THE PIVOT $a_{if_j^{(r)}}^{(r)}$.

     Chapter 5. APPLYING LINEAR ALGEBRA

[**Remark 5.6.1:**   In other words, if $\mathcal{L}^{(r)}_{if_j^{(r)}}$ is the $(m+1) \times (m+1)$ identity matrix in which the ith column is replaced by:

$$
\begin{array}{c}
-\dfrac{a^{(r)}_{1f_j^{(r)}}}{a^{(r)}_{if_j^{(r)}}} \\[2ex]
\vdots \\[1ex]
\dfrac{1}{a^{(r)}_{if_j^{(r)}}} \qquad \leftarrow i\text{th row} \\[2ex]
\vdots \\[1ex]
-\dfrac{a^{(r)}_{mf_j^{(r)}}}{a^{(r)}_{if_j^{(r)}}} \\[2ex]
-\dfrac{c_{f_j^{(r)}}}{a^{(r)}_{if_j^{(r)}}}
\end{array}
$$

then

$$
\mathcal{A}^{(r+1)} = \mathcal{L}^{(r)}_{if_j^{(r)}} \mathcal{A}^{(r)}.]
$$

PROOF.   Row-reduction pivoted on $a^{(r)}_{if_j^{(r)}}$ and applied to the first m rows of $\mathcal{A}^{(r)}$ is exactly the operation that effects the SWITCH between the chosen basic and free variables. Such row-reduction enables one to "solve," in the ith equation $A^{(r)}\mathbf{X} = \mathbf{q}^{(r)}$, for the free variable $X_{f_j^{(r)}}$ in terms of the basic variable $X_{b_i^{(r)}}$ and then to substitute the result in the other equations of the system.

The principal result to be established is that the same row-reduction applied to the bottom row of $\mathcal{A}^{(r)}$ yields the proper bottom row of $\mathcal{A}^{(r+1)}$.

For simplicity of notation the superscripts $^{(r)}$ resp. $^{(r+1)}$ are dropped in favor of no superscript at all resp. $'$: for any symbol $\star$: $\star^{(r)} = \star$ and $\star^{(r+1)} = \star'$. Furthermore it is assumed that $b_i^{(r)} \overset{\text{def}}{=} b_i = i$, $1 \le i \le m$, and that

$$
f_j^{(r)} = m + j \overset{\text{def}}{=} k, \ 1 \le j \le n.
$$

In other words, at the start the basic variables are $X_1, \ldots, X_m$ and the free variables are $X_{m+1}, \ldots, X_{m+n}$. The SWITCH is $X_i \leftrightarrow X_k$ and the pivot for the SWITCH is $a^{(r)}_{b_i^{(r)} f_j^{(r)}} \overset{\text{def}}{=} a_{ik} \ne 0$. Finally $\mathcal{C}^{(r)} \overset{\text{def}}{=} \mathcal{C} \overset{\text{def}}{=} (\mathcal{C}_1, \ldots, \mathcal{C}_{n+m})$ and $\mathbf{C}_B^{(r)} \overset{\text{def}}{=} \mathbf{C}_B = (\mathcal{C}_1, \ldots, \mathcal{C}_m)$.

Let $\mathbf{a}_\mu$ resp. $\mathbf{a}'_\mu$ denote the μth row of A resp. A'; let $\mathbf{a}^\nu$ resp. $\mathbf{a}'^\nu$ denote the νth column of A resp. A'; (subscripts μ for rows, superscripts ν for columns).

The next equations:

$$\mathbf{C}'_B = \mathbf{C}_B + (C_k - C_i)\mathbf{e}_i$$
$$\mathcal{C} \overset{\text{def}}{=} \mathbf{C} - \mathbf{C}_B A$$
$$\mathcal{C}_k = C_k - \mathbf{C}_B a^k \tag{5.6.2}$$
$$\mathbf{q}' = \mathcal{L}_{ik}\mathbf{q} = \mathbf{q} - (I - \mathcal{L}_{ik})\mathbf{q}$$
$$A' = \mathcal{L}_{ik}A = A - (I - \mathcal{L}_{ik})A$$

$$
I - \mathcal{L}_{ik} =
\begin{pmatrix}
0 & \cdots & \frac{a_{1k}}{a_{ik}} & \cdots & 0 \\
\vdots & \ddots & \vdots & \ddots & \vdots \\
0 & \cdots & 1 - \frac{1}{a_{ik}} & \cdots & 0 \\
\vdots & \ddots & \vdots & \ddots & \vdots \\
0 & \cdots & \frac{a_{mk}}{a_{ik}} & \cdots & 0
\end{pmatrix},
$$

which are valid almost by definition, constitute the hinge of the calculation. Note that the only nonzero column of $I - \mathcal{L}_{ik}$ is the ith and that

$$1 - \frac{1}{a_{ik}}$$

is the ii (diagonal) entry of $I - \mathcal{L}_{ik}$.

Indeed, from those equations it follows that

$$\mathbf{C} - \mathbf{C}'_B A' = \mathbf{C} - \mathbf{C}_B A + \mathbf{C}(I - \mathcal{L}_{ik})A - (C_k - C_i)\mathbf{e}_i \mathcal{L}_{ik} A$$
$$\mathbf{C}'_B \mathbf{q}' = \mathbf{C}_B \mathbf{c} - \mathbf{C}_B(I - \mathcal{L}_{ik})\mathbf{q} + (C_k - C_i)\mathbf{e}_i \mathcal{L}_{ik}\mathbf{q}$$
$$\mathbf{C}_B(I - \mathcal{L}_{ik}) = \begin{pmatrix} \mathbf{O} & \mathbf{C}_B a^k - C_i & \mathbf{O} \end{pmatrix}$$
$$\underset{i}{\uparrow}$$
$$\mathbf{C}_B(I - \mathcal{L}_{ik})A = \frac{\mathbf{C}_B a^k - C_i}{a_{ik}} a_i$$
$$(C_k - C_i)\mathbf{e}_i \mathcal{L}_{ik} A = \frac{(C_k - C_i)}{a_{ik}} a_i$$
$$\mathbf{C} - \mathbf{C}'_B A' = \mathbf{C} - \mathbf{C}_B A + \frac{\mathbf{C}_B a^k - C_k}{A_{ik}} a_i$$
$$\mathbf{C}'_q \mathbf{q}' = \mathbf{C}_q \mathbf{q} - \frac{\mathbf{C}_B a^k - C_k}{a_{ik}} b_i.$$

In view of (5.6.2) and the definition of $\mathcal{C}'$ the last two equations above yield the desired result.

$$\Omega$$

Example 5.6.1. Let the **PLPP** be that in **Example 5.5.4.** The first tableau $\mathcal{T}^{(0)}$ arising from row-reduction operations is:

     Chapter 5. APPLYING LINEAR ALGEBRA

$\mathbf{X}_B^{(0)}$	X_1	X_2	X_3	X_4	X_5	X_6	X_7	$-\mathbf{q}^{(0)}$	$\mathbf{C}_B^{(0)}$
$X_3^{(0)}$	1^*	-2	1	0	0	0	0	-2	0
$X_4^{(0)}$	1	-1	0	1	0	0	0	-3	0
$X_5^{(0)}$	2	1	0	0	1	0	0	-12	0
$X_6^{(0)}$	-1	2	0	0	0	1	0	-9	0
$X_7^{(0)}$	-1	1	0	0	0	0	1	-4	0
$\mathcal{C}^{(0)}$	-5	-2	0	0	0	0	0	0	

Figure 5.6.2. Example 5.5.4 $\mathcal{T}^{(0)}$

By happenstance, in the current **Example** I is embedded in $\widetilde{A}$. Hence $\widetilde{A} = A^{(0)}$. Nevertheless in some circumstances I can fail to be embedded in $\widetilde{A}$.

The first negative entry in the bottom row is -5. Since there are positive entries in column 1 the simplex pivot column is column 1. The relevant ratios associated with column 1 are:

$$\frac{2}{1} = 2 \ (\text{first row})$$

$$\frac{3}{1} = 3 \ (\text{second row})$$

$$\frac{12}{2} = 6 \ (\text{third row}).$$

Hence $j' = 1$, $\mu_{j'}^{(0)} = 2$, and the the simplex pivot entry is 1 (flagged $*$). The SWITCH $X_3 \leftrightarrow X_1$ is in order.

This SWITCH amounts to choosing as the basic variables

$$X_1, X_4, X_5, X_6, \text{ and } X_7$$

and choosing as the free variables X_2 and X_3. The basic variable X_3 is the *leaving variable* and the free variable X_1 is the *entering variable* in the SWITCH $X_3 \leftrightarrow X_1$. Thus the columns numbered $1, 4, 5, 6, 7$ in $A^{(1)}$ are those among which the columns of I are found. In effect this means that ROW SCALING is not needed because the simplex pivot entry is 1. The simplex pivot column 1 of $\mathcal{A}^{(0)}$ must be converted to e_1 by means of row-reduction (ROW COMBINATION). In the process the entries under $\mathbf{q}^{(0)}$ are affected. The result, after the recording of $Cost^{(1)}$ is $\mathcal{T}^{(1)}$:

$$
\begin{array}{c|ccccccc c|c}
\mathbf{X}_B^{(1)} & X_1 & X_2 & X_3 & X_4 & X_5 & X_6 & X_7 & -\mathbf{q}^{(1)} & \mathbf{C}_B^{(1)} \\
\hline
X_1^{(1)} & 1 & -2 & 1 & 0 & 0 & 0 & 0 & -2 & -5 \\
X_4^{(1)} & 0 & 1^* & -1 & 1 & 0 & 0 & 0 & -1 & 0 \\
X_5^{(1)} & 0 & 5 & -2 & 0 & 1 & 0 & 0 & -8 & 0 \\
X_6^{(1)} & 0 & 0 & 1 & 0 & 0 & 1 & 0 & -11 & 0 \\
X_7^{(1)} & 0 & -1 & 1 & 0 & 0 & 0 & 1 & -6 & 0 \\
\hline
\mathcal{C}^{(1)} & 0 & -12 & 5 & 0 & 0 & 0 & 0 & -10 & \\
\end{array}
$$

Figure 5.6.3. Example 5.5.4 $\mathcal{T}^{(1)}$

In $\mathcal{T}^{(1)}$ the bottom row shows that the simplex pivot column is column 2. The relevant ratios are $\frac{1}{1}$ (second row) and $\frac{8}{5}$ (third row). Hence $j' = 2$, $\mu_{j'}^{(1)} = 1$, the simplex pivot row is row 2, and the simplex pivot entry is 1 (flagged $*$). The SWITCH $X_4 \leftrightarrow X_2$ is in order. Since I is embedded in columns 1,4,5,6,7 it follows that the $\mathbf{C}$ entries from those columns are used for $\mathbf{C}_B^{(1)}$: $\mathbf{C}_B^{(1)} = (-5, 0, 0, 0, 0)$. Hence $Cost^{(1)} = \mathbf{C}^{(1)}\mathbf{q}^{(1)} = -10$.

The successive tableaux are given below without further comment.

$$
\begin{array}{c|ccccccc c|c}
\mathbf{X}_B^{(2)} & X_1 & X_2 & X_3 & X_4 & X_5 & X_6 & X_7 & -\mathbf{q}^{(2)} & \mathbf{C}_B^{(2)} \\
\hline
X_1^{(2)} & 1 & 0 & -1 & 2 & 0 & 0 & 0 & -4 & -5 \\
X_2^{(2)} & 0 & 1 & -1 & 1 & 0 & 0 & 0 & -1 & -2 \\
X_5^{(2)} & 0 & 0 & 3^* & -5 & 1 & 0 & 0 & -3 & 0 \\
X_6^{(2)} & 0 & 0 & 1 & 0 & 0 & 1 & 0 & -11 & 0 \\
X_7^{(2)} & 0 & 0 & 0 & 1 & 0 & 0 & 1 & -7 & 0 \\
\hline
\mathcal{C}^{(2)} & 0 & 0 & -7 & 12 & 0 & 0 & 0 & -22 & \\
\end{array}
$$

Figure 5.6.4. Example 5.5.4 $\mathcal{T}^{(2)}$

$X_5 \leftrightarrow X_3$

$\mathbf{X}_B^{(3)}$	X_1	X_2	X_3	X_4	X_5	X_6	X_7	$-\mathbf{q}^{(3)}$	$\mathbf{C}_B^{(3)}$
$X_1^{(3)}$	1	0	0	$\frac{1}{3}$	$\frac{1}{3}$	0	0	-5	-5
$X_2^{(3)}$	0	1	0	$-\frac{2}{3}$	$\frac{1}{3}$	0	0	-2	-2
$X_3^{(3)}$	0	0	1	$-\frac{5}{3}$	$\frac{1}{3}$	0	0	-1	0
$X_6^{(3)}$	0	0	0	$\frac{5}{3}$	$-\frac{1}{3}$	1	0	-10	0
$X_7^{(3)}$	0	0	0	1	0	0	1	-7	0
$\mathcal{C}^{(3)}$	0	0	0	$\frac{1}{3}$	$\frac{7}{3}$	0	0	-29	

Figure 5.6.5. Example 5.5.4 $\mathcal{T}^{(3)}$

In $\mathcal{T}^{(3)}$ all entries for $\mathcal{C}^{(3)}$ are nonnegative. Hence there is no simplex pivot column, no simplex pivot row, and no flagged simplex pivot entry. Thus the operation of the algorithm stops and the optimal $Cost$ is -29. To find the optimal vertex set the free variables X_4 and X_5 at 0 and calculate the corresponding values of the basic variables:

$$X_1 = 5, \ X_2 = 2. \ X_3 = 1, \ X_6 = 8, \ X_7 = 7.$$

The associated optimal vertex of $\mathcal{Q}$ is thus $(5, 2)^t$, as found earlier.

Exercise 5.6.1. Apply the simplex method to the duals of the problems in Example 5.5.4.

[**Remark 5.6.2:** If, among the free variables, there are several for which the coefficients are the same and also most negative there is a "tie" in the contest for pivot column. Similarly if, among the rows of the underlying matrix, there are several for which, in the given pivot column, $\mu_{j'}$ is achieved there is a "tie" in the contest for pivot row. The choices of "first" or "least index" free variable in ii and of "first" or "least index" basic variable in iii constitute, as noted earlier, the *Bland selection rules*. They obviate indecision when a "tie" for choice of entering and/or leaving variables presents itself. Furthermore, it can be shown that these rules obviate cycling. The proof that they do is deferred to the end of this **Section**.

What appears to be a more promising procedure is the replacement of iib by the alternative *"steepest descent"* rules

iib'. Choose in the formula for $\mathbf{CX}$ *any* free variable having the most negative coefficient.

and the replacement of $iiib$ by the alternative rule

---

iiib'. Choose *any* basic variable for which $\mu_{j'}$ is achieved.

The rule *iib'* appears to be attractive because it provides for the quickest decrease in the value of *Cost*. The rule *iiib'* is attractive because it dispenses with searching for i'. In other words, the "firsts" or the "least indices" need not be sought.]

THEOREM **5.6.2**. IF, IN ROUND r OF THE SIMPLEX ALGORITHM,

i. SOME COMPONENT OF $C^{(r)}$ IS NEGATIVE

AND

ii. EVERY COMPONENT OF $\mathbf{q}^{(r)}$ IS POSITIVE,

THEN

EITHER

THERE IS AN ELIGIBLE FREE VARIABLE AND A MOVE TO A BETTER VERTEX IS POSSIBLE

OR

THERE IS NO ELIGIBLE BASIC VARIABLE AND *Cost* CAN BE DECREASED WITHOUT BOUND (*every* ENTRY IN *every* POTENTIAL SIMPLEX PIVOT COLUMN IS NONPOSITIVE).

PROOF. If there is a positive entry in some simplex pivot column then *ii* insures that some $\mu_{j'}^{(r)} > 0$ and hence setting $X_{f_{j'}}^{(r)}$ at $\mu_{j'}^{(r)}$ decreases *Cost*.

If every entry in every simplex pivot column is nonpositive then every $X_{f_j}^{(r)}$ may be increased without bound and *Cost* may be driven negative without bound.

$$\Omega$$

However, if some component of $\mathbf{q}^{(r)}$ is 0 (no component can be negative) then a *degeneracy* is at hand. Necessarily $\mu_{j'}^{(r)} = 0$ and a basic variable $X_{b_i^{(r)}}^{(r)}$ is eligible only if $q_i^{(r)} = 0$. If there are several 0's among the components of $\mathbf{q}^{(r)}$ there is a "tie" among the eligible leaving variables.

As **Examples 5.6.2** and **5.6.3** show, if there is no rule for breaking the "tie," e.g., as can occur if the *"steepest descent"* approach is used, degeneracy can lead to cycling: after a finite number of steps a set of basic variables found at one stage comes up again *before* either an optimal vertex is reached or it is discovered that there is no solution.

In summary, the following is the state of affairs in the matter of degeneracy/cycling:

i. if there is no degeneracy there can be no cycling;

ii. the Charnes technique obviates degeneracy and therefore cycling;

iii. the Bland rules do not obviate degeneracy but *do* obviate cycling.

The validity of *i* is established above. After the next **Examples** the truth of *ii* and *iii* are demonstrated.

The **PLPP**s given below provide instances in which cycling arises unless some device such as the Bland rules or the Charnes technique is employed to obviate cycling. The first is due to E. M. L. Beale ([**Bea**]) and the second to K. T. Marshall and J. W. Suurballe ([**MaS**]). It should be stressed that the **Examples** are ad hoc — designed to show the possibility of cycling — and are not representative of what proves to be daily experience in linear programming.

Example 5.6.0.2. Assume

$$
A \stackrel{\text{def}}{=} \begin{pmatrix} 1 & 0 & 0 & 1 & 1 & 1 & 1 \\ 0 & 1 & 0 & 0.5 & -5.5 & -2.5 & 9 \\ 0 & 0 & 1 & 0.5 & -1.5 & -0.5 & 1 \end{pmatrix}
$$

$$
\mathbf{p} \stackrel{\text{def}}{=} \begin{pmatrix} 1 \\ 0 \\ 0 \end{pmatrix} \ (= \mathbf{e}_1)
$$

$$
\mathbf{c} \stackrel{\text{def}}{=} \begin{pmatrix} 0 & 0 & 0 & -1 & 7 & 1 & 2 \end{pmatrix}.
$$

Find in $\mathbf{R}^{(7,+)}$ an $\mathbf{x}_{Opt}$ such that

$$
A\mathbf{x}_{Opt} = \mathbf{p}
$$
$$
\mathbf{c}\mathbf{x}_{Opt} = minimum\{\, \mathbf{c}\mathbf{x} \ : \ \mathbf{x} \succeq \mathbf{O}, \ A\mathbf{x} = \mathbf{p} \,\}.
$$

Here there is no need for slack variables: $A = A$. The first set of basic variables is $\{X_1, X_2, X_3\}$, corresponding to the embedded identity matrix in the first three columns of A. The SWITCHes

$$
\begin{aligned}
X_2 &\leftrightarrow X_4 \\
X_3 &\leftrightarrow X_5 \\
X_4 &\leftrightarrow X_6 \\
X_5 &\leftrightarrow X_7 \\
X_6 &\leftrightarrow X_2 \\
X_7 &\leftrightarrow X_3
\end{aligned}
$$

lead to a cycle.

Example 5.6.3. Assume

$$A \overset{\text{def}}{=} \begin{pmatrix} -\frac{1}{4} & 8 & 1 & -9 \\ -\frac{1}{2} & 12 & \frac{1}{2} & -3 \\ 0 & 0 & -1 & 0 \end{pmatrix}$$

$$\mathbf{c} \overset{\text{def}}{=} \begin{pmatrix} -\frac{3}{4} & 20 & -\frac{1}{2} & 6 \end{pmatrix}$$

$$\mathbf{p} \overset{\text{def}}{=} \begin{pmatrix} 0 \\ 0 \\ -1 \end{pmatrix}$$

Find in $\mathbf{R}^{(4,+)}$ an $\mathbf{x}_{Opt}$ such that

$$A\mathbf{x}_{Opt} \succeq \mathbf{p}$$
$$\mathbf{c}\mathbf{x}_{Opt} = minimum\{\,\mathbf{c}\mathbf{x} \,:\, \mathbf{x} \succeq \mathbf{O},\ A\mathbf{x} \succeq \mathbf{p}\,\}.$$

After the introduction of slack variables there are seven variables in the standard format. The relevant augmented matrix is

$$A \overset{\text{def}}{=} \left(\begin{array}{ccccccc|c} \frac{1}{4} & -8 & -1 & 9 & 1 & 0 & 0 & 0 \\ \frac{1}{2}^{*} & -12 & -\frac{1}{2} & 3 & 0 & 1 & 0 & 0 \\ 0 & 0 & 1 & 0 & 0 & 0 & 1 & 1 \end{array} \right).$$

Note that sign changes have resulted in the presence of an embedded identity matrix I.

The following sequence of sets of basic variables and SWITCHes constitutes a cycle that, in consonance with the simplex algorithm, is repeated endlessly.

$$\begin{aligned}
&\{X_5, X_6, X_7\} \\
&\{X_1, X_6, X_7\} \quad (X_5 \leftrightarrow X_1); \\
&\{X_1, X_2, X_7\} \quad (X_6 \leftrightarrow X_2); \\
&\{X_3, X_2, X_7\} \quad (X_1 \leftrightarrow X_3); \\
&\{X_3, X_4, X_7\} \quad (X_2 \leftrightarrow X_4); \\
&\{X_5, X_4, X_7\} \quad (X_3 \leftrightarrow X_5); \\
&\{X_5, X_6, X_7\} \quad (X_4 \leftrightarrow X_6).
\end{aligned}$$

Exercise 5.6.2. Explain how in each of the **Examples** above the Bland selection rules are violated at some point.

Example 5.6.4. The following sequence of tableaux demonstrates the simplex algorithm applied together with the Bland rules to the **PLPP** in **Example 5.6.2**.

In the early stages the pivot entry is flagged with an asterisk (*). The pivot column is found by SEARCHing in the bottom row for the negative entry with

---

the least ("first") index. The pivot row (and thus the pivot entry) is found by SEARCHing in the pivot column for entry for which $\mu_{j'}$ is achieved and for which j' is least ("first"). Note that cycling does not occur even though there is degeneracy. The initial tableau for **Example 5.6.2** is the one given next.

The simplex pivot entry is flagged *.

$\mathbf{X}_B^{(0)}$	X_1	X_2	X_3	X_4	X_5	X_6	X_7	$-\mathbf{q}^{(0)}$	$\mathbf{C}_B^{(0)}$
$X_1^{(0)}$	1	0	0	1	1	1	1	-1	0
$X_2^{(0)}$	0	1	0	$\frac{1}{2}^*$	$-\frac{11}{2}$	$-\frac{5}{2}$	9	0	0
$X_3^{(0)}$	0	0	1	$\frac{1}{2}$	$-\frac{3}{2}$	$-\frac{1}{2}$	1	0	0
$\mathcal{C}^{(0)}$	0	0	0	-1	7	1	2	0	

Figure 5.6.6. Example 5.6.2 $\mathcal{T}^{(0)}$

Thus $j' = 4$, $\mu_4^{(0)} = 0$, and the simplex pivot entry is $\frac{1}{2}$ (flagged *).

After the SWITCH $X_2 \leftrightarrow X_4$ the $\mathcal{T}^{(1)}$ is constructed.

$\mathbf{X}_B^{(1)}$	X_1	X_2	X_3	X_4	X_5	X_6	X_7	$-\mathbf{q}^{(1)}$	$\mathbf{C}_B^{(1)}$
$X_1^{(1)}$	1	-2	0	0	12	6	-17	-1	0
$X_4^{(1)}$	0	2	0	1	-11	-5	18	0	-1
$X_3^{(1)}$	0	-1	1	0	4^*	2	-8	0	0
$\mathcal{C}^{(1)}$	0	2	0	0	-4	-4	20	0	

Figure 5.6.7. Example 5.6.2 $\mathcal{T}^{(1)}$

The SWITCH $X_3 \leftrightarrow X_5$ is in order. Without further comment the subsequent tableaux dictated by the Bland rules are given below. Simplex pivot entries are flagged and SWITCHes are indicated.

$\mathbf{X}_B^{(2)}$	X_1	X_2	X_3	X_4	X_5	X_6	X_7	$-\mathbf{q}^{(2)}$	$\mathbf{C}_B^{(2)}$
$X_1^{(2)}$	1	1	-3	0	0	0	7	-1	0
$X_4^{(2)}$	0	$-\frac{3}{4}$	$\frac{11}{4}$	1	0	$\frac{1}{2}^*$	-4	0	-1
$X_5^{(2)}$	0	$-\frac{1}{4}$	$\frac{1}{4}$	0	1	$\frac{1}{2}$	-2	0	7
$\mathcal{C}^{(2)}$	0	1	1	0	0	-2	12	0	

Figure 5.6.8. Example 5.6.2 $\mathcal{T}^{(2)}$

$X_4 \leftrightarrow X_6$

$\mathbf{X}_B^{(3)}$	X_1	X_2	X_3	X_4	X_5	X_6	X_7	$-\mathbf{q}^{(3)}$	$\mathbf{C}_B^{(3)}$
$X_1^{(3)}$	1	1	-3	0	0	0	7	-1	0
$X_6^{(3)}$	0	$-\frac{3}{2}$	$\frac{11}{2}$	2	0	1	-8	0	1
$X_5^{(3)}$	0	$\frac{1}{2}^*$	$-\frac{5}{2}$	-1	1	0	2	0	7
$\mathcal{C}^{(3)}$	0	-2	12	4	0	0	-4	0	

Figure 5.6.9. Example 5.6.2 $\mathcal{T}^{(3)}$

$X_5 \leftrightarrow X_2$

$\mathbf{X}_B^{(4)}$	X_1	X_2	X_3	X_4	X_5	X_6	X_7	$-\mathbf{q}^{(4)}$	$\mathbf{C}_B^{(4)}$
$X_1^{(4)}$	1	0	2	2	-2	0	3	-1	0
$X_6^{(4)}$	0	0	-2	-1	3	1	-2	0	1
$X_2^{(4)}$	0	1	-5	-2	2	0	4	0	0
$\mathcal{C}^{(4)}$	0	0	2	0	4	0	4	0	

Figure 5.6.10. Example 5.6.2 $\mathcal{T}^{(4)}$

Since all the entries in the bottom row are nonnegative an optimal vertex is at hand. Since the last entry in the bottom row is 0 it follows that the minimum *Cost* is 0.

[**Remark 5.6.3:** In carrying out the SWITCH/row-reduction operation, first the bottom row of the tableau is reduced. If all its entries are nonnegative, only the reduction of the "q" column is required to compute the minimum *Cost*. Only if some entry in the bottom row is negative is there a need to SEARCH in a potential pivot column. *That*

SEARCH is carried out after *all* the rows of the tableau are reduced.]

Exercise 5.6.3. Apply the simplex algorithm and the Bland rules to the **PLPP** of Example 5.6.3.

THE CHARNES TECHNIQUE

Assume that at some stage of the simplex algorithm, the constant term in the expression of each basic variable is positive. Then the next round of the simplex algorithm leads to a strictly better vertex since the entering free variable may be given some positive value. Hence cycling can occur only if the constant term in the expression for some basic variable is 0.

In rough form the argument below runs as follows:

A perturbation $\mathbf{q}(\epsilon)$ of $\mathbf{q}$ gives rise to a $\mathcal{Q}(\epsilon)$, a perturbed version of $\mathcal{Q}$. If $\mathbf{x}(\epsilon)$ is an optimal vertex of $\mathcal{Q}(\epsilon)$ then there is in $\mathcal{Q}$ a vertex $\mathbf{x}$ nearest to $\mathbf{x}(\epsilon)$. Since the *coefficients* in the objective function are unchanged $\mathbf{x}$ is an optimal vertex of $\mathcal{Q}$.

THEOREM 5.6.3. LET $A, \mathbf{q}, \mathbf{C}$ BE THE DATA OF A **PLPP** IN EQUATIONAL FORMAT. THERE IS A POSITIVE NUMBER ϵ_0 SUCH THAT IF $0 < \epsilon < \epsilon_0$ AND IF

$$\mathbf{q}(\epsilon) \stackrel{\mathrm{def}}{=} \mathbf{q} + \begin{pmatrix} \epsilon \\ \epsilon^2 \\ \vdots \\ \epsilon^m \end{pmatrix}$$

THEN $A, \mathbf{q}(\epsilon), \mathbf{C}$ ARE DATA FOR A **PLPP**(ϵ) IN WHICH DEGENERACY IN THE SIMPLEX ALGORITHM IS IMPOSSIBLE. FURTHERMORE, IF $\mathbf{x}(\epsilon)$ IS AN OPTIMAL VERTEX FOR **PLPP**(ϵ) THEN AS $\epsilon \to 0$ THE OPTIMAL VERTEX $\mathbf{x}(\epsilon)$ CONVERGES TO A VERTEX $\mathbf{x}$ THAT IS AN OPTIMAL VERTEX OF $\mathcal{Q}$.

PROOF. After the first round of the simplex algorithm the vector $\mathbf{q}^{(0)}$ is replaced by a vector $\mathbf{q}^{(1)}$ that is the result of multiplying $\mathbf{q}^{(0)}$ by the inverse A_B^{-1} of a basic matrix, one of the finitely many invertible $m \times m$ submatrices of A. Simultaneously A is changed to $A_B^{-1}A$. The next round applies the simplex algorithm to the **PLPP** in which the data are, for some vector $\dot{\mathbf{C}}^{(1)}$,

$$A_B^{-1}A \stackrel{\mathrm{def}}{=} A^{(1)}, \ A_B^{-1}\mathbf{q} \stackrel{\mathrm{def}}{=} \mathbf{q}^{(1)}, \ \mathbf{C}^{(1)}.$$

There are only finitely many basic matrices in each $A^{(r)}$ Hence, even if cycling occurs, there are only finitely many different vectors $\mathbf{q}^{(r)}$, $1 \le r \le R$, that can be produced in the application of the simplex algorithm to **PLPP**(0), the original **PLPP**.

Correspondingly there are only finitely many different vectors

$$\mathbf{q}^{(r)}(\epsilon), \ 1 \leq r \leq R,$$

that can be produced by the application of the simplex algorithm to $\mathbf{PLPP}(\epsilon)$. Each component of any $\mathbf{q}^{(r)}(\epsilon)$ is a polynomial function of ϵ. Since each polynomial has only finitely many zeros it follows that for *some* positive ϵ_0 *no* component of *any* $\mathbf{q}^{(r)}(\epsilon)$, $1 \leq r \leq R$, is 0 if $0 < \epsilon < \epsilon_0$. Hence if $0 < \epsilon < \epsilon_0$

EITHER

the simplex algorithm leads to an optimal vertex $\mathbf{x}_{Opt}(\epsilon)$ in $\mathcal{Q}(\epsilon)$

OR

there is a sequence $\{\epsilon_n\}_{n=1}^{\infty}$ such that $\epsilon_n \downarrow 0$ and the simplex algorithm shows that each $Cost(\epsilon_n)$ for $\mathcal{Q}(\epsilon_n)$ is unbounded below.

As $\epsilon \to 0$ the vertices of $\mathcal{Q}(\epsilon)$ approach vertices of $\mathcal{Q}$ and every vertex of $\mathcal{Q}$ is approached by at least one vertex in $\mathcal{Q}(\epsilon)$. If ϵ is small $Cost$ at a vertex of $\mathcal{Q}$ and $Cost(\epsilon)$ at some vertex of $\mathcal{Q}(\epsilon)$ are near each other. Thus if there are arbitrarily small values of ϵ for which $Cost(\epsilon)$ is unbounded below, then $Cost$ is also unbounded below. A determination for the original $\mathbf{PLPP}$ is at hand: there is no solution.

Alternatively, there is an ϵ_1 such that $0 < \epsilon_1 \leq \epsilon_0$ and there is a finite number C such that $C \leq Cost(\epsilon)$ for all ϵ in $(0, \epsilon_1)$. In that case $\mathbf{PLPP}(\epsilon)$ has a solution. For each small ϵ there is an optimal vertex $\mathbf{x}(\epsilon)$. Since $Cost$ is near $Cost(\epsilon)$, $Cost$ itself is also bounded below. As $\epsilon \to 0$ the optimal vertices $\mathbf{x}(\epsilon)$ converge to a vertex $\mathbf{x}$ of $\mathcal{Q}$.

If $\mathbf{x}$ is not optimal there is in $\mathcal{Q}$ another vertex $\mathbf{y}$ where $Cost$ is below $Cost$ at $\mathbf{x}$. If $\mathbf{y}(\epsilon)$ are vertices of $\mathcal{Q}(\epsilon)$ that approach $\mathbf{y}$ then, for very small ϵ, $Cost(\epsilon)$ at $\mathbf{y}(\epsilon)$ is so near to $Cost$ at $\mathbf{y}$ that $Cost(\epsilon)$ at $\mathbf{y}(\epsilon)$ is below $Cost(\epsilon)$ at the optimal vertex $\mathbf{x}(\epsilon)$, a contradiction. Thus $\mathbf{x}$ is an optimal vertex of $\mathcal{Q}$.

$$\Omega$$

In effect the result above, which is due essentially to Charnes, says that by making a very small perturbation in $\mathbf{q}$, one can rescue a $\mathbf{PLPP}$ from degeneracy and hence from cycling. The probable reason that cycling is not been experienced in practice is that roundoff errors provide enough perturbation to obviate degeneracy. Note too that the Charnes perturbation need not be applied until a $\mathbf{q}^{(r)}$ having at least one zero component arises.

Example 5.6.5. If the Charnes technique is applied to the $\mathbf{PLPP}$ in Example 5.6.3 the sequence of tableaux is the following. SWITCHes resp. simplex pivot entries are noted resp. flagged.

$\mathbf{X}_B^{(0)}$	X_1	X_2	X_3	X_4	X_5	X_6	X_7	$-\mathbf{q}^{(0)}$	$\mathbf{C}_B^{(0)}$
$X_5^{(0)}$	$\frac{1}{4}$	-8	-1	9	1	0	0	$-\epsilon$	0
$X_6^{(0)}$	$\frac{1}{2}^*$	-12	$-\frac{1}{2}$	3	0	1	0	$-\epsilon^2$	0
$X_7^{(0)}$	0	0	1	0	0	0	1	$-1-\epsilon^3$	0
$\mathcal{C}^{(0)}$	$-\frac{3}{4}$	20	$-\frac{1}{2}$	6	0	0	0	0	

Figure 5.6.11. Example 5.6.3 $\mathcal{T}^{(0)}$

Here steepest descent is worth using since no degeneracy is possible. Hence $j' = 1$. Since $\frac{\epsilon^2}{2} < \frac{\epsilon}{4}$ so long as $0 < \epsilon < 2$ the SWITCH $X_6 \leftrightarrow X_1$ corresponding to the flagged entry is in order. Here are the next tableaux, flagged entries, and SWITCHes.

$\mathbf{X}_B^{(1)}$	X_1	X_2	X_3	X_4	X_5	X_6	X_7	$-\mathbf{q}^{(1)}$	$\mathbf{C}_B^{(1)}$
$X_5^{(1)}$	0	-2	$-\frac{3}{4}$	$\frac{15}{2}$	1	$-\frac{1}{2}$	0	$-\epsilon+\frac{\epsilon^2}{2}$	0
$X_1^{(1)}$	1	-24	-1	6	0	2	0	$-2\epsilon^2$	$-\frac{3}{4}$
$X_7^{(1)}$	0	0	1^*	0	0	0	1	$-1-\epsilon^3$	0
$\mathcal{C}^{(1)}$	0	2	$-\frac{5}{4}$	$\frac{21}{2}$	0	$\frac{3}{2}$	0	$-\frac{3\epsilon^2}{2}$	

Figure 5.6.12. Example 5.6.3 $\mathcal{T}^{(1)}$

$X_7 \leftrightarrow X_3$

$\mathbf{X}_B^{(2)}$	X_1	X_2	X_3	X_4	X_5	X_6	X_7	$-\mathbf{q}^{(2)}$	$\mathbf{C}_B^{(2)}$
$X_5^{(2)}$	0	-2	0	$\frac{15}{2}$	1	$-\frac{1}{2}$	$\frac{3}{4}$	$-\frac{3}{4}-\epsilon+\frac{\epsilon^2}{2}-\frac{3\epsilon^2}{4}$	0
$X_1^{(2)}$	1	-24	0	6	0	2	1	$-1-2\epsilon^2-\epsilon^3$	$-\frac{3}{4}$
$X_3^{(2)}$	0	0	1	0	0	0	1	$-1-\epsilon^3$	$-\frac{1}{2}$
$\mathcal{C}^{(2)}$	0	2	0	$\frac{21}{2}$	0	$\frac{3}{2}$	$\frac{5}{4}$	$-\frac{5}{4}-\frac{3\epsilon^2}{2}-\frac{5\epsilon^3}{2}$	

Figure 5.6.13. Example 5.6.3 $\mathcal{T}^{(2)}$

Because all entries for $\mathcal{C}^{(2)}$ are nonnegative a solution is at hand and if $\epsilon \to 0$ minimum $Cost$ is seen to be $-\frac{5}{4}$.

Exercise 5.6.4.

i. Apply the simplex method using steepest descent and the Charnes technique

---

to the **PLPP** in **Example 5.6.2**. Compare the sequences of SWITCHes found in the two methods.

ii. Apply the simplex method and the Bland rules to the **PLPP** in **Example 5.6.3**. Again compare the sequences of SWITCHes found in the two methods.

Exercise 5.6.5. Find in this **Section** the numerical **PLPP Exercises** and **Examples** in which there is no embedded identity matrix in the equational format. For each use the method of artificial variables (**ALPP**) to find a starting vertex. If cycling occurs apply the Bland rules. Then go back and apply the Charnes technique.

Exercise 5.6.6. Explain how the final Bland and Charnes tableaux can be different and yet do provide the same minimal *Cost*.

THE JUSTIFICATION OF THE BLAND RULES

Assume that the Bland rules are used in the simplex algorithm applied to a **PLPP** in which the basic data are $A, \mathbf{p}, \mathbf{c}$. Assume further that cycling occurs in either the **ALPP** or the **PLPP**. The argument that follows reveals a contradiction. The notations introduced and used above are used in the discussion.

To say that cycling occurs is to say that at some stage r there appears for the first time a set $\{b_1^{(r)}, \ldots, b_m^{(r)}\}$ of basic variable indices and that at some later stage $r + s \stackrel{\text{def}}{=} t$ the same set of basic variable indices appears for a second time:

$$b_i^{(t)} = b_i^{(r)}, \ 1 \le i \le m.$$

In passing from stage r to t some of the basic variables *leave* in the course of SWITCHes and others do not. All the basic variables that leave in the course of the SWITCHes eventually enter again and thereby give rise to the cycle. Let Σ be the set of indices of basic variables that leave and re-enter.

Hence:

If k is an index not in Σ then k is the index of a variable that remains basic during the cycle or of a variable that remains free during the cycle.

Otherwise put, the preceding statement is:

$k \notin \Sigma \Rightarrow$ during the cycle X_k is always basic or X_k is always free.

Finally let σ be the largest number in Σ: $\sigma = maximum \{ s \ : \ s \in \Sigma \}$ and, suppressing superscripts, let $\mathcal{T}$ be a tableau in which X_σ is found, according to the Bland rules, to be the new entering free variable: if $i < \sigma$ then $C_i \ge 0$. In the course of the cycle there is in Σ a τ different from σ and there appears a tableau $\mathcal{T}'$ in which X_σ is the leaving basic variable for the SWITCH: $\tau < \sigma$ and $X_\sigma \hookleftarrow X_\tau$. It may be assumed that the indices of the basic variables in $\mathcal{T}'$ are

<hr>

$1, 2, \ldots, m$: the identity matrix I embedded in A' occupies the first m columns of A'.

Introduce the matrices

$$M \overset{\text{def}}{=} \begin{array}{c} \\ m \\ 1 \end{array}\!\!\begin{pmatrix} \overset{n+m}{A} & \overset{1}{\mathbf{O}} \\ \mathbf{O} & 1 \end{pmatrix} \text{ and } M' \overset{\text{def}}{=} \begin{array}{c} \\ m \\ 1 \end{array}\!\!\begin{pmatrix} \overset{n+m}{A'} & \overset{1}{\mathbf{O}} \\ \mathbf{O} & 1 \end{pmatrix}$$

and the $(n + m + 1) \times 1$ vector

$$\mathbf{y} \overset{\text{def}}{=} \begin{pmatrix} y_1 \\ \vdots \\ y_{n+m+1} \end{pmatrix} \overset{\text{def}}{=} \begin{pmatrix} a'_{1\tau} \\ \vdots \\ a'_{m\tau} \\ 0 \\ \vdots \\ -1 \\ 0 \\ \vdots \\ 0 \\ c'_\tau \end{pmatrix} \begin{array}{l} \\ \\ \\ \\ \\ \leftarrow \text{ row } \tau \\ \\ \\ \\ \leftarrow \text{ row } n+m+1 \end{array}$$

The argument that follows is given in a "Statement-Reason" format similar to that used traditionally in proofs of theorems in Euclidean geometry. Each "Statement" is followed by its "Reason" in parentheses.

i. $M'\mathbf{y} = \mathbf{O}$. (Cycling, hence degeneracy is assumed to take place.)

ii. $M\mathbf{y} = \mathbf{O}$. (There is a matrix P, a product of invertible "$\mathcal{L}$"-matrices and such that $P^{-1}A' = A$.)

iii. $\mathcal{C}\mathbf{y} = \sum_{i=1}^{n+m} \mathcal{C}_i y_i + 1\dot{\mathcal{C}}'_\tau = 0$. (See *ii.*)

iv. $\mathcal{C}'_\tau < 0$. (The SWITCH is $X_\sigma \leftrightarrow X_\tau$.)

v. There is a u such that $1 \le u \le n + m$ and such that $\mathcal{C}_u y_u > 0$. (See *iii* and *iv.*)

vi. $\mathcal{C}_u, \; y_u \ne 0$. (See *v.*)

vii. $u \in \{1, \ldots, m, \tau\}$. (See *iv* and the definition of $\mathbf{y}$.)

viii. X_u is not a basic variable in $\mathcal{T}$. (In any tableau each $\mathcal{C}$-component corresponding to a basic variable is 0. See *v.*)

ix. $u \in S$. (If $u \notin \Sigma$ then, during the cycle, X_u is always basic or always free. See *viii.*)

x. $u \le \sigma$. (See *ix* and the definition of σ.)

xi. $u < \sigma$. (If $u = \sigma$ then $\tau < \sigma = u$ whence $\sigma = u \le m < \tau < \sigma$, a contradiction. See *vi*, *vii*, and the definition of $\mathbf{y}$.)

xii. $\mathcal{C}_u, \; y_u > 0$. (See *v*, *vi*, *xi*, and definition of σ.)

xiii. $y_u = a'_{u\tau}$. (See definition of $\mathbf{y}$.)

xiv. $q'_u = 0$. (The value of each SWITCHed basic variable is 0 in each tableau of the cycle.)

---

xv. $\sigma \leq u$. (Throughout the cycle the minimum ratio is 0 for the decision about the SWITCH. In row u of T' the ratio is 0. Hence the Bland rules force the SWITCH row index σ in T' to be not more than u.)

xvi. Contradiction. ($\sigma \leq u \leq m < \tau < \sigma$. See *xi.*)

$$\Omega$$

5.7. The Karmarkar algorithm

The practical experience with the simplex algorithm has been excellent. Although calculations prove [**Kl**] that for "worst-case" scenarios the simplex algorithm (even in raw form, without modifications such as the Bland rules or the Charnes technique) is *im*practical nevertheless the "worst cases," like instances of cycling, do not crop up in practice.

The simplex algorithm is "on the average" ([**Bor**], [**Sm1**]) reasonably fast and experience suggests that the average is the rule. On the other hand, an algorithm that is practical not merely on the average but in all instances is clearly desirable. The Karmarkar algorithm [**Karm**], discussed below is just such an algorithm. It has the further advantage that it avoids entirely the possibility of cycling encountered in the simplex algorithm.

The exposition that follows presents a version of the ideas put forth in [**Karm**]. The outline of the Karmarkar idea is the following:

i. In view of the discussion in **Section 5.4** it suffices to regard the search for a solution of the **PLPP** as the search for a solution of

$$
\begin{aligned}
A\mathbf{x} &\succeq \mathbf{q} \\
\mathbf{u}A &\preceq \mathbf{c} \\
\mathbf{c}\mathbf{x} &= \mathbf{u}\mathbf{q} \\
\mathbf{x} &\succeq \mathbf{O} \\
\mathbf{u} &\succeq \mathbf{O}.
\end{aligned}
\tag{5.7.1}
$$

ii. A reformulation of (5.7.1) leads to a very succinct system, readily described in terms of the following notations.

Once and for all

$$
\mathbf{e} \stackrel{\text{def}}{=} \begin{pmatrix} 1 \\ \vdots \\ 1 \end{pmatrix}
$$

and

$$
\mathbf{a}_0 \stackrel{\text{def}}{=} \frac{\mathbf{e}}{n},
$$

the centroid of $\Sigma^{(n)}$.

The **H**omogeneous **P**rimal **L**inear **P**rogramming **P**roblem (**HPLPP**)

There is a matrix, again denoted A, and a vector, again denoted $\mathbf{c}$, such that $A\mathbf{a}_0 = \mathbf{O}$ and such that the original **PLPP** is equivalent to the following problem (**HPLPP**):

$$\text{minimize } \mathbf{cx}$$
$$\text{subject to } \mathbf{x} \in \Sigma^{(n)} \cap ker(A). \tag{5.7.2}$$

Note that the matrix A and the vector $\mathbf{c}$ in (5.7.2) are *not* necessarily the same as the matrix A and the vector $\mathbf{c}$ of the original **PLPP**. The same symbols are used to simplify the exposition. In the reformulation the minimal value of $\mathbf{cx}$ under the constraints is nonnegative. The original **PLPP** has a solution iff in the **HPLPP** the minimum of $\mathbf{cx}$ is 0.

iii. Starting with $\mathbf{a}_0$ and by means of the Karmarkar algorithm one defines a sequence $\mathbf{x}_1, \mathbf{x}_2, \ldots$ of vectors such that

EITHER

$$\mathbf{cx}_k > \mathbf{cx}_{k+1} \downarrow 0, \text{ as } k \to \infty \tag{5.7.3}$$

OR

for some k the algorithm signals a halt and, as well, the fact that

$$minimum\left\{ \mathbf{cx} \ : \ \mathbf{x} \in ker(A) \cap \Sigma^{(n)} \right\} > 0. \tag{5.7.4}$$

If the latter alternative occurs the original **PLPP** has no solution. (The situation is akin to the **PLPP/ALPP** situation. There the **ALPP** always has a solution that corresponds to a solution of the **PLPP** iff the minimal value of the objective function for the **ALPP** is 0.) In any event there is a constant K such that for each natural number q if no halt is signalled before the algorithm goes through k rounds and if $k \geq Kn(q + \ln n)$ then

$$\frac{\mathbf{cx}_k}{\mathbf{ca}_0} \leq 2^{-q}.$$

[**Remark 5.7.1:** According to **Exercise 5.5.9** there is a subalgorithm that carries one from an interior point $\mathbf{x}$ of $\Sigma^{(n)}$ to a nearby vertex $\mathbf{y}$ where $\mathbf{cy} \leq \mathbf{cx}$. Hence, at any stage before there is a signal to stop the Karmarkar algorithm (because the **PLPP** has no solution) the vector $\mathbf{x}_k$ gives rise to a corresponding vertex $\mathbf{y}_k$, which can be

---

tested to see whether $\mathbf{cy}_k = 0$. If this testing is done periodically and if the **PLPP** has a solution then the Karmarkar algorithm leads to an optimal vertex.

It should be noted that roundoff makes the use of the subalgorithm somewhat extravagant since for large enough q

$$2^{-q}\mathbf{ca}_0$$

is, within roundoff, in fact 0. Karmarkar suggests that the subalgorithm be used whenever the time spent on the main algorithm since the last use of the subalgorithm exceeds the time required to carry out the subalgorithm again.]

The heart of the algorithm lies in the observation that it is simple to describe a subroutine or subalgorithm that minimizes $\mathbf{cx}$ if $\mathbf{x}$ is confined to lie in a ball

$$B(\mathbf{a},\rho) \stackrel{\text{def}}{=} \left\{ \mathbf{x} \; : \; \|\mathbf{x} - \mathbf{a}\| \leq \rho \; (> 0) \right\}.$$

Indeed, each value t of $\mathbf{cx}$ corresponds to a hyperplane

$$H_t \stackrel{\text{def}}{=} \left\{ \mathbf{x} \; : \; \mathbf{cx} = t \right\}.$$

These hyperplanes are such that precisely two of them H_{t_1} resp. H_{t_2} meet $B(\mathbf{a},\rho)$ in just one point (of tangency) $\mathbf{x}_1$ resp. $\mathbf{x}_2$. If $t_1 < t_2$ then $\mathbf{cx}_1$ resp. $\mathbf{cx}_2$ is the minimum resp. the maximum of $\mathbf{cx}$ on the ball $B(\mathbf{a},\rho)$. Furthermore, if $\mathbf{u}$ is the unit vector defined by $\mathbf{c}$, i.e., if

$$\mathbf{u} \stackrel{\text{def}}{=} \frac{\mathbf{c}}{\|\mathbf{c}\|},$$

then

$$\mathbf{x}_1 = \mathbf{a} - \rho\mathbf{u} \text{ resp. } \mathbf{x}_2 = \mathbf{a} + \rho\mathbf{u}.$$

(See THEOREM 5.7.1.)

The remainder of this **Section** is devoted to describing in detail and with **Examples** and **Exercises** just what goes on in the Karmarkar algorithm.

From PLPP to HPLPP.

i. Introduce slack variables $\mathbf{y}$ and $\mathbf{v}$ into the Primal **D**ual Linear Programming Problem (**PDLPP**):

$$A\mathbf{x} - \mathbf{y} = \mathbf{q} \tag{5.7.5}$$

$$\mathbf{u}^t A + \mathbf{v}^t = \mathbf{c} \tag{5.7.6}$$

$$\mathbf{cx} = \mathbf{u}^t\mathbf{q} \tag{5.7.7}$$

$$\mathbf{x} \succeq \mathbf{O},\, \mathbf{y} \succeq \mathbf{O}$$

$$\mathbf{u} \succeq \mathbf{O},\, \mathbf{v} \succeq \mathbf{O}.$$

ii. Introduce *one* artificial variable: λ (a kind of "Lagrange multiplier"). Let $\mathbf{x}_0$, $\mathbf{v}_0$ resp. $\mathbf{y}_0$, $\mathbf{u}_0$, be positive vectors in the orthants $\mathbf{R}^{(n,+)}$ resp. $\mathbf{R}^{(m,+)}$:

$$\mathbf{x}_0 \succ \mathbf{O}, \ \mathbf{y}_0 \succ \mathbf{0}, \ \mathbf{u}_0 \succ \mathbf{O}, \ \text{and} \ \mathbf{v}_0 \succ \mathbf{O}.$$

Then the problem

$$\text{minimize } \lambda$$
$$\text{subject to } \begin{cases} A\mathbf{x} - \mathbf{y} + (\mathbf{q} - A\mathbf{x}_0 + \mathbf{y}_0)\lambda = \mathbf{q} \\ \mathbf{u}^t A + \mathbf{v}^t + (\mathbf{c} - \mathbf{u}_0^t A + \mathbf{y}_0^t)\lambda = \mathbf{c} \\ \mathbf{c}\mathbf{x} - \mathbf{u}^t\mathbf{q} + (-\mathbf{c}\mathbf{x}_0 + \mathbf{u}_0^t\mathbf{q})\lambda = 0 \\ \mathbf{x}, \mathbf{y}, \mathbf{u}, \mathbf{v} \succeq \mathbf{O}, \ \lambda \geq 0 \end{cases} \qquad (5.7.8)$$

is such that if the minimal admissible value of λ is 0 then the corresponding $\mathbf{x}$, $\mathbf{y}$, $\mathbf{u}$, $\mathbf{v}$ are solutions of (5.7.5)-(5.7.7) whence $\mathbf{x}$ is a solution of the **PLPP**.

Conversely, if there is a solution of the (original) **PLPP** then its dual, the **DLPP**, has a solution. In that case 0 is an admissible value for a feasible solution for the **PDLPP** and hence 0 is the minimal admissible value of λ for (5.7.8):

the **PLPP** has a solution

iff

minimal admissible value of λ is 0.

In other words, the target value of the objective function for the **PDLPP** is 0.

The virtue of the **PDLPP** as stated in (5.7.8) is that for *it* a feasible solution is

$$\mathbf{x} = \mathbf{x}_0, \ \mathbf{y} = \mathbf{y}_0, \ \mathbf{u} = \mathbf{u}_0, \ \mathbf{v} = \mathbf{v}_0, \text{and } \lambda = 1;$$

furthermore the **PDLPP** always has a solution, say

$$\mathbf{x}_1, \mathbf{y}_1, \mathbf{u}_1, \mathbf{v}_1, \lambda_1.$$

(See the discussion of the **ALPP** in **Section 5.6**.)

The **PLPP** has a solution iff for some solution for the **PDLPP** $\lambda_1 = 0$. If $\lambda_1 = 0$ for some solution for the **PDLPP** the solution for the **PLPP** can be read off from $\mathbf{x}_1, \mathbf{y}_1$. Furthermore the **PDLPP** permits *one* application of the simplex algorithm (or any other algorithm) to serve rather than two separate applications as in the procedure used for the the **ALPP/PLPP** pair when a feasible solution of the **PLPP** is not easily discerned.

Exercise 5.7.1. Show that if

$$
A' \stackrel{\text{def}}{=} \begin{array}{c} \\ m \\ n \\ 1 \end{array}
\begin{array}{cccccc}
n & m & n & m & 1 \\
\left(\begin{array}{ccccc}
A & -I & O & O & q - A\mathbf{x}_0 + \mathbf{y}_0 \\
O & O & I & A^t & \mathbf{c}^t - A^t\mathbf{u}_0 - \mathbf{v}_0 \\
\mathbf{c} & O & O & -\mathbf{q}^t & \mathbf{c}\mathbf{x}_0 + \mathbf{q}^t\mathbf{u}_0
\end{array} \right)
\end{array}
$$

$$
\mathbf{x}' \stackrel{\text{def}}{=} \begin{array}{c} \\ n \\ m \\ n \\ m \\ 1 \end{array}
\begin{array}{c}
1 \\
\left(\begin{array}{c}
\mathbf{x} \\
\mathbf{y} \\
\mathbf{v} \\
\mathbf{u} \\
\lambda
\end{array} \right)
\end{array}
$$

$$
\mathbf{q}' \stackrel{\text{def}}{=} \begin{array}{c} \\ m \\ n \\ 1 \end{array}
\begin{array}{c}
1 \\
\left(\begin{array}{c}
\mathbf{q} \\
\mathbf{c}^t \\
0
\end{array} \right)
\end{array}
$$

$$
\mathbf{c}' \stackrel{\text{def}}{=} \begin{array}{c} 1 \end{array}
\begin{array}{ccccc}
n & m & n & m & 1 \\
\left(\begin{array}{ccccc}
O & O & O & O & 1
\end{array} \right)
\end{array}
$$

then (5.7.8) may be written as

$$
\text{minimize } \mathbf{c}'\mathbf{x}'
$$
$$
\text{subject to } A'\mathbf{x}' = \mathbf{q}' \text{ and } \mathbf{x}' \succeq O.
$$

The last part of the reformulation is homogenization.

iii. Modify the data $A', \mathbf{q}', \mathbf{c}'$ in **Exercise 5.7.1** to achieve the homogeneous form (5.7.2). Since the **PDLPP** has a clearly visible feasible solution, say

$$
\begin{array}{c} n \\ m \\ n \\ m \\ 1 \end{array}
\begin{array}{c}
1 \\
\left(\begin{array}{c}
\mathbf{x}_0 \\
\mathbf{y}_0 \\
\mathbf{v}_0 \\
\mathbf{u}_0 \\
1
\end{array} \right)
\end{array}
\stackrel{\text{def}}{=} \mathbf{s} \stackrel{\text{def}}{=}
\left(\begin{array}{c}
s_1 \\
\vdots \\
s_{2(m+n)+1}
\end{array} \right),
$$

let N be $2(m + n) + 1$ and for $\mathbf{x}'$ in $\mathbf{R}^{(N,+)}$ define in $\mathbf{R}^{(N+1,+)}$ the vector

$$
\mathbf{x}'' \stackrel{\text{def}}{=} \left(\begin{array}{c}
x_1'' \\
\vdots \\
x_{N+1}''
\end{array} \right)
$$

---

according to the formulae:

$$x_j'' \overset{\text{def}}{=} \frac{x_j'/s_j}{1 + \sum_{j=1}^{N} x_j'/s_j}, \ 1 \le j \le N$$

$$x_{N+1}'' \overset{\text{def}}{=} 1 - \sum_{j=1}^{N} x_j'' \ (> 0).$$

According to these formulae there is defined a map

$$\widetilde{T}_{\mathbf{s}} : \mathbf{R}^{(N,+)} \ni \mathbf{x}' \overset{\text{def}}{=} (x_1',\dots,x_N')^t \mapsto \mathbf{x}'' \overset{\text{def}}{=} (x_1'',\dots,x_{N+1}'')^t \overset{\text{def}}{=} \widetilde{T}_{\mathbf{s}}\mathbf{x}' \in \mathbf{R}^{(N+1,+)}.$$

Then, $\Sigma^{(N+1)}$ denoting as in **Section 5.4** the basic simplex in $\mathbf{R}^{N+1}$, the map $\widetilde{T}_{\mathbf{s}}$ enjoys the following properties:

$$\widetilde{T}_{\mathbf{s}}(\mathbf{R}^{(n,+)}) = \left\{ \mathbf{x}'' \ : \ \mathbf{x}'' \overset{\text{def}}{=} \begin{pmatrix} x_1'' \\ \vdots \\ x_{N+1}'' \end{pmatrix} \in \Sigma^{N+1}, x_{N+1}'' \ne 0 \right\} \overset{\text{def}}{=} \widetilde{\Sigma}^{N+1}$$

$$\widetilde{T}_{\mathbf{s}}\mathbf{s} = \frac{\sum_{j=1}^{N+1} \mathbf{e}_j}{N+1} \overset{\text{def}}{=} \mathbf{a}_0$$

$\widetilde{T}_{\mathbf{s}}$ is one-one, i.e.,

$$\widetilde{T}_{\mathbf{s}}\mathbf{x}' = \widetilde{T}_{\mathbf{s}}\mathbf{y}' \Rightarrow \mathbf{x}' = \mathbf{y}',$$

$$\mathbf{x}' \succ \mathbf{O} \Rightarrow \widetilde{T}_{\mathbf{s}}\mathbf{x}' \overset{\text{def}}{=} \mathbf{x}'' \succ \mathbf{O}.$$

If $\mathbf{x}'' \overset{\text{def}}{=} (x_1'',\dots,x_{N+1}'')^t \in \mathbf{R}^{(N+1,+)}$ and $x_{N+1}'' \ne 0$ then

$$\widetilde{T}_{\mathbf{s}}^{-1}\mathbf{x}'' = \mathbf{x}' \overset{\text{def}}{=} \frac{1}{x_{N+1}''} \begin{pmatrix} s_1 x_1'' \\ \vdots \\ s_N x_N'' \end{pmatrix}.$$

Furthermore, A'^j denoting column j of A',

$$\sum_{j=1}^{N} s_j x_j'' A'^j - x_{N+1}'' \mathbf{q} = \mathbf{O}.$$

Hence if

$$A''^j \overset{\text{def}}{=} s_j A'^j, \ 1 \le j \le N, \ A''^{N+1} \overset{\text{def}}{=} -\mathbf{q}$$

then, A'' denoting the matrix in which column j is A''^j, $1 \le j \le N+1$,

$$A''\mathbf{x}'' = \mathbf{O}.$$

The first part of the homogenization is achieved.

Since the target value of the objective function for the **PDLPP** is 0, the homogenization described next, meshes with the homogenization above.

Let $\mathbf{c}''$ be $(c_1'', \ldots, c_{N+1}'')$, defined according to the formulae

$$c_j'' \overset{\text{def}}{=} s_j c_j', \ 1 \le j \le N, \ c_{N+1}'' = 0.$$

Then $\mathbf{c}''\mathbf{x}'' = 0 \Leftrightarrow \mathbf{c}'\mathbf{x}_0' = 0$. Hence if a solution $\mathbf{x}_0''$ is found for

$$\text{minimize } \mathbf{c}''\mathbf{x}''$$
$$\text{subject to} \begin{cases} A''\mathbf{x}'' = \mathbf{O} \\ \mathbf{x}'' \in \Sigma^{N+1} \end{cases}$$

and $\mathbf{c}''\mathbf{x}_0'' = 0$ then $\mathbf{c}'\mathbf{x}' = 0$ which leads to the solution of the original **PLPP**. Thus after

$$\mathbf{c}'' \text{ is replaced by } \mathbf{c}$$
$$\mathbf{x}'' \text{ is replaced by } \mathbf{x}$$
$$A'' \text{ is replaced by } A$$
$$N + 1 \text{ is replaced by } n$$

the homogeneous **PDLPP** may be renamed the **HPLPP** ("H" for "Homogeneous") and may be stated as follows:

$$\text{minimize } \mathbf{cx}$$
$$\text{subject to } \mathbf{x} \in \Sigma^{(n)} \cap ker(A) \tag{5.7.9}$$

which is the form in (5.7.2).

> [**Note 5.7.1:** In the homogeneous form achieved there is always available the feasible solution $\mathbf{a}_0$. Since $\mathbf{c}''\mathbf{x}'' = 0 \Leftrightarrow \mathbf{c}'\mathbf{x}' = 0$, since $\mathbf{c}'\mathbf{s} = 1 > 0$, and since $\widetilde{T}_s\mathbf{s} = \mathbf{a}_0$ it follows that $\mathbf{c}''\mathbf{a}_0 > 0$. In the notation achieved by dropping each superscript $''$: $\mathbf{ca}_0 > 0$. Furthermore, the target minimum value for $\mathbf{cx}$ is 0 and thus at any point in the execution of the algorithm it may be assumed that $\mathbf{cx} \ge 0$ and, if the execution of the algorithm is to be continued, that $\mathbf{cx} > 0$.]

Exercise 5.7.2. Reformulate each of the numerically given **PLPP**s in this **Chapter** and achieve for each the (equivalent) format (5.7.9).

Minimization of cx on a ball

For the **HPLPP** the ball over which **cx** is to be minimized is not n-dimensional since the constraint

$$\mathbf{x} \in ker(A) \cap \Sigma^{(n)} \overset{\text{def}}{=} \mathcal{D}$$

must be observed as well. However, since $\mathbf{a}_0 \in \mathcal{D}$ it follows that

$$B(\mathbf{a}_0, \rho) \cap \mathcal{D}$$

is simply a lower-dimensional ball.

[**Note 5.7.2:** Since $\mathbf{a}_0 \in \mathcal{D}$ it follows that the radius of the lower-dimensional ball $B(\mathbf{a}_0, \rho) \cap \mathcal{D}$ is again ρ.]

Exercise 5.7.3. Let Z be the subspace

$$\{\, \mathbf{x} \ : \ (\mathbf{x}, \mathbf{e}) = 0 \,\}.$$

i. Show that if A in the **HPLPP** is an $m \times n$ matrix then $rank(A) = m$.

ii. Show $dim[ker(A) \cap Z] \overset{\text{def}}{=} \nu \leq n - m$.

iii. Let $\mathbf{a}_0 + Z$ be the the hyperplane

$$\{\, \mathbf{x} \ : \ \mathbf{x} - \mathbf{a}_0 \in Z \,\},$$

or *translate* by $\mathbf{a}_0$ of Z. Show

$$\mathcal{D} = (\mathbf{a}_0 + Z) \cap ker(A) \cap \mathbf{R}^{(n,+)}.$$

iv. Show that $\mathbf{x} \in \mathbf{a}_0 + Z$ iff $(\mathbf{x}, \mathbf{e}) = 1$.

(Thus $B(\mathbf{a}_0, \rho) \cap \mathcal{D}$ is, for some ν not exceeding $n - m$, the intersection of $B(\mathbf{a}_0, \rho)$ and the ν-dimensional *subspace translate* $\mathbf{a}_0 + (ker(A) \cap Z)$ passing through $\mathbf{a}_0$.)

In **Figure 5.7.1** there are pictured in $\mathbf{R}^2$ the subspace W, its *translate*

$$\mathbf{a} + W \overset{\text{def}}{=} \{\, \mathbf{x} \ : \ \mathbf{x} - \mathbf{a} \in W \,\}$$

of W and the procedure for minimizing **cx** when **x** is constrained to lie in the set

$$B(\mathbf{a}, \rho) \cap (\mathbf{a} + W).$$

The vector $\mathbf{c}_W$ is the orthogonal projection of $\mathbf{c}^t$ into W and is assumed to be nonzero; **u** is the corresponding *unit vector*

$$\frac{\mathbf{c}_W}{\|\mathbf{c}_W\|}.$$

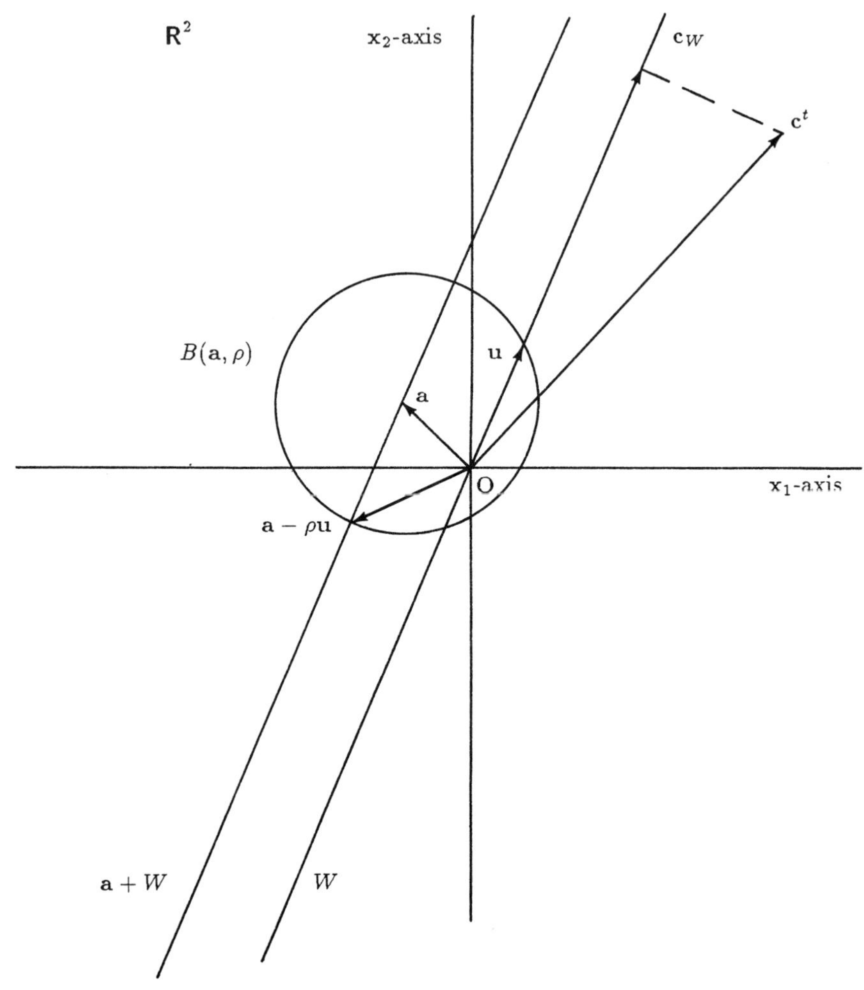

Figure 5.7.1. Minimization over a ball

The vector $\mathbf{a} - \rho\mathbf{u}$ is the point where the minimum of $\mathbf{cx}$ is achieved.

In verbal terms, the minimization procedure in $\mathbf{R}^n$ is to project $\mathbf{c}^t$ orthogonally into W and then to move:

i. in $B(\mathbf{a}, \rho) \cap (\mathbf{a} + W)$ from the center $\mathbf{a}$;

ii. a distance ρ in the direction $-\mathbf{u}$ to the

$$\text{boundary of } B(\mathbf{a}, \rho) \stackrel{\text{def}}{=} \partial B(\mathbf{a}, \rho) \stackrel{\text{def}}{=} \{\, \mathbf{x} \ : \ \|\mathbf{x} - \mathbf{a}\| = \rho \,\}.$$

The reason for moving in the direction $-\mathbf{u}$ rather than in the direction $-\mathbf{c}$ is that one needs the point of hyperplane/ball tangency to be in W. If $\mathbf{c}^t$ is not projected into W the family of hyperplanes orthogonal to $\mathbf{c}^t$ (rather than the family of hyperplanes orthogonal to $\mathbf{u}$) yields two hyperplanes tangent to the ball $B(\mathbf{a}, \rho)$ at points not necessarily in W.

Exercise 5.7.4. Assume that

$$W \stackrel{\text{def}}{=} \{\, \mathbf{x} \ : \ (\mathbf{x}, \mathbf{b}) = 0 \,\}.$$

Use each set of data given below to calculate the corresponding $\mathbf{a} - \rho\mathbf{u}$.

$$(i.) \quad \mathbf{b} = \begin{pmatrix} -1 \\ 1 \end{pmatrix}, \quad \mathbf{c} = (2, 3), \quad \rho = 4, \quad \mathbf{a} = \begin{pmatrix} 1 \\ -2 \end{pmatrix};$$

$$(ii.) \quad \mathbf{b} = \begin{pmatrix} 0 \\ 1 \end{pmatrix}, \quad \mathbf{c} = (c_1, c_2), \quad \rho = \rho, \quad \mathbf{a} = \begin{pmatrix} a_1 \\ a_2 \end{pmatrix};$$

$$(iii.) \quad \mathbf{b} = \begin{pmatrix} b_1 \\ \vdots \\ b_n \end{pmatrix}, \quad \mathbf{c} = (c_1, \ldots, c_n), \quad \rho = \rho, \quad \mathbf{a} = \begin{pmatrix} a_1 \\ \vdots \\ a_n \end{pmatrix}.$$

Exercise 5.7.5. For each of the data sets above what is the equation of the hyperplane H that passes through $\mathbf{a} - \rho\mathbf{u}$ and is tangent to $B(\mathbf{a}, \rho)$?

[**Remark 5.7.2:** If the projection $\mathbf{c}_W$ of $\mathbf{c}^t$ into W is $\mathbf{O}$ the vector $\mathbf{u}$ is not defined. See **Exercise 5.7.9** below for what conclusions can be drawn.]

The Karmarkar algorithm uses *projective* transformations $T_{\mathbf{a}}$ of $\Sigma^{(n)}$ into itself. Each such a transformation is determined by a *positive* vector $\mathbf{a}$ in $\Sigma^{(n)}$:

$$\Sigma^{(n)} \ni \mathbf{a} \succ \mathbf{O},$$

i.e.,

$$(a_1, \ldots, a_n) \stackrel{\text{def}}{=} \mathbf{a} \in \Sigma^{(n)\circ} \stackrel{\text{def}}{=} \left\{\, \mathbf{x} \ : \ \mathbf{x} \in \Sigma^{(n)} \text{ and } \mathbf{x} \succ \mathbf{O} \,\right\}$$

$$\stackrel{\text{def}}{=} \text{ the } \textit{interior} \text{ of } \Sigma^{(n)}.$$

The projective transformation $T_{\mathbf{a}}$ determined by $\mathbf{a}$ is given in terms of the diagonal matrix

$$D_{\mathbf{a}} \stackrel{\text{def}}{=} \begin{pmatrix} a_1 & & & \\ & a_2 & & \\ & & \ddots & \\ & & & a_n \end{pmatrix}$$

---

and the projective transformation itself is

$$T_{\mathbf{a}} : \Sigma^{(n)} \ni \mathbf{x} \mapsto \frac{D_{\mathbf{a}}^{-1}\mathbf{x}}{e^t D_{\mathbf{a}}^{-1}\mathbf{x}} \stackrel{\text{def}}{=} \mathbf{x}'_{\mathbf{a}} \in \Sigma^{(n)},$$

i.e.,

$$T_{\mathbf{a}}(x_1,\ldots,x_n)^t = \frac{1}{\left(\sum_{j=1}^{n} x_i/a_i\right)}(x_1/a_1,\ldots,x_n/a_n)^t.$$

Owing to the assumption that $\mathbf{a}$ is *positive* and that $\mathbf{x} \in \Sigma^{(n)}$ it follows that $e^t D_{\mathbf{a}}^{-1}\mathbf{x} > 0$. A direct calculation shows that the inverse of $T_{\mathbf{a}}$ is given by the formula

$$T_{\mathbf{a}}^{-1} : \Sigma^{(n)} \ni \mathbf{x}'_{\mathbf{a}} \mapsto \frac{D_{\mathbf{a}}\mathbf{x}'_{\mathbf{a}}}{e^t D_{\mathbf{a}}\mathbf{x}'_{\mathbf{a}}} \in \Sigma^{(n)}.$$

Exercise 5.7.6. Verify the formula for the inverse $T_{\mathbf{a}}^{-1}$ of $T_{\mathbf{a}}$. Show:

  i. $T_{\mathbf{a}}\mathbf{a} = \mathbf{a}_0$ $(\mathbf{a}'_{\mathbf{a}} = \mathbf{a}_0)$;
  ii. $(\mathbf{x}'_{\mathbf{a}} = \mathbf{y}'_{\mathbf{a}} \Leftrightarrow \mathbf{x} = \mathbf{y})$, i.e., $T_{\mathbf{a}}$ is one-one;
  iii. $T_{\mathbf{a}}(\Sigma^{(n)}) = \Sigma^{(n)}$ $(\Sigma_{\mathbf{a}}^{(n)'} = \Sigma^{(n)})$;
  iv. $\mathbf{x}'_{\mathbf{a}} \succ \mathbf{O} \Leftrightarrow \mathbf{x} \succ \mathbf{O}$, i.e., $T_{\mathbf{a}}\Sigma^{(n)^{\circ}} = \Sigma^{(n)^{\circ}}$;
  v. that if $\mathbf{c}'_{\mathbf{a}} \stackrel{\text{def}}{=} \mathbf{c}D_{\mathbf{a}}$ then $\mathbf{c}'_{\mathbf{a}}\mathbf{x}'_{\mathbf{a}} = 0 \Leftrightarrow \mathbf{c}\mathbf{x} = 0$.

[**Remark 5.7.3:** If $\mathbf{x} \in \Sigma^{(n)}$ and $t \neq 0, 1$ the formula for $T_{\mathbf{a}}\mathbf{x}$ may be used for $T_{\mathbf{a}}t\mathbf{x}$. However then

$$T_{\mathbf{a}}t\mathbf{x} = T_{\mathbf{a}}\mathbf{x} \neq tT_{\mathbf{a}}\mathbf{x},$$

which suggests that $T_{\mathbf{a}}$ can*not* be extended so as to be a linear transformation of $\mathbf{R}^n$ into itself.]

The vectors

$$(x'_{\mathbf{a}1},\ldots,x'_{\mathbf{a}n})^t \stackrel{\text{def}}{=} \mathbf{x}'_{\mathbf{a}} = T_{\mathbf{a}}\mathbf{x} \stackrel{\text{def}}{=} T_{\mathbf{a}}(x_1,\ldots,x_n)^t$$

$$\mathbf{c}'_{\mathbf{a}} \stackrel{\text{def}}{=} \mathbf{c}D_{\mathbf{a}}$$

serve to define functions f and f' according to the formulae

$$f(\mathbf{x}) \stackrel{\text{def}}{=} \sum_{j=1}^{n} \ln\left[\frac{\mathbf{c}\mathbf{x}}{x_i}\right]$$

$$f'(\mathbf{x}'_{\mathbf{a}}) \stackrel{\text{def}}{=} f(T_{\mathbf{a}}^{-1}\mathbf{x}'_{\mathbf{a}}) = f(\mathbf{x}).$$

(Note that for consistency of notation f' denotes *not* the derivative of f but a function related to f.) Then direct calculation shows

$$infimum\{\, f(\mathbf{x}) \ : \ \mathbf{x} \in \mathcal{D} \,\} = -\infty \Leftrightarrow infimum\{\, \mathbf{c}\mathbf{x} \ : \ \mathbf{x} \in \mathcal{D} \,\} = 0$$

$$f'(\mathbf{x}'_{\mathbf{a}}) = \sum_{j=1}^{n} \ln\left[\frac{\mathbf{c}'_{\mathbf{a}}\mathbf{x}'_{\mathbf{a}}}{x'_{\mathbf{a}j}}\right] - \sum_{j=1}^{n} \ln a_j$$

$$f(\mathbf{p}) - f(\mathbf{q}) = f'(\mathbf{p}'_{\mathbf{a}}) - f'(\mathbf{q}'_{\mathbf{a}}) \text{ (by definition).} \tag{5.7.10}$$

Since the original **PLPP** has a solution iff, in the notation of the **HPLPP**,

$$minimum_{\mathbf{x} \in \mathcal{D}} \, \mathbf{cx} = 0$$

the solution of the original **PLPP** is achievable iff

$$infimum_{\mathbf{x} \in \mathcal{D}} \, f(\mathbf{x}) = -\infty$$

i.e., iff there is in $\mathcal{D}$ a sequence $\{\mathbf{x}_k\}_{k=0}^{\infty}$ such that

$$f(\mathbf{x}_k) \to -\infty \text{ as } k \to \infty.$$

Karmarkar's development shows that if the **PLPP** has a solution there is in $\mathcal{D}$ a sequence $\{\mathbf{x}_k\}_{k=0}^{\infty}$ such that if $0 < \alpha < 1$ then

$$f(\mathbf{x}_{k+1}) \leq f(\mathbf{x}_k) - \ln(1 + \alpha), \ 0 \leq m < \infty. \tag{5.7.11}$$

The following list summarizes notations used thus far or yet to be used to describe the algorithm:

$$ker(A) \overset{\text{def}}{=} \Omega$$

$$ker(A) \cap \Sigma^{(n)} \overset{\text{def}}{=} \mathcal{D}$$

$$ker(AD_{\mathbf{a}}) \overset{\text{def}}{=} \Omega'_{\mathbf{a}}$$

$$M_{\mathbf{a}} \overset{\text{def}}{=} \begin{pmatrix} AD_{\mathbf{a}} \\ \mathbf{e}^t \end{pmatrix}$$

$$Z \overset{\text{def}}{=} \{\, \mathbf{x} \ : \ (\mathbf{x}, \mathbf{e}) = 0 \,\}$$

$$Z_1 \overset{\text{def}}{=} \mathbf{a}_0 + Z \ (= \{\, \mathbf{x} \ : \ (\mathbf{x}, \mathbf{e}) = 1 \text{ i.e., } \mathbf{e}^t \mathbf{x} = 1 \,\})$$

$$\Omega'_{\mathbf{a}} \cap Z_1 \overset{\text{def}}{=} \Omega''_{\mathbf{a}}$$

$$B(\mathbf{a}_0, \rho) \overset{\text{def}}{=} \{\, \mathbf{x} \ : \ \|\mathbf{x} - \mathbf{a}_0\| \leq \rho \,\}$$

$$\mathbf{c}'_{\mathbf{a}} \overset{\text{def}}{=} \mathbf{c} D_{\mathbf{a}}$$

$$\alpha \overset{\text{def}}{=} \text{ a number in } (0,1).$$

Exercise 5.7.7. Show that if $\mathbf{x} \succ \mathbf{O}$ then

$$\mathbf{x} \in \Omega \Leftrightarrow \mathbf{x}'_{\mathbf{a}} \in ker(AD_{\mathbf{a}})$$
$$\mathbf{x} \in \Omega \Leftrightarrow \mathbf{x}'_{\mathbf{a}} \in \Omega'_{\mathbf{a}}$$
$$\mathbf{x} \in \Omega''_{\mathbf{a}} \Leftrightarrow \mathbf{x} - \mathbf{a}_0 \in ker(M_{\mathbf{a}})$$
$$\Leftrightarrow \mathbf{x} \in \mathbf{a}_0 + ker(M_{\mathbf{a}}).$$

[**Note 5.7.3:**   The results in **Exercise 5.7.7** are used repeatedly and without further comment in what follows.]

In the notation of **Exercise 5.7.1** there is a feasible solution

$$\mathbf{s} \overset{\text{def}}{=} (\mathbf{x}_0, \mathbf{y}_0, \mathbf{u}_0, \mathbf{v}_0, 1)^t$$

for the **PDLPP** modified by the artificial variable ("Lagrange multiplier") λ. That solution is sent by $\widetilde{T}_{\mathbf{s}}$ into the vector $\mathbf{a}_0$ that is in $\mathcal{D}$ and is a feasible solution of the **HPLPP**. In particular $A\mathbf{a}_0 = \mathbf{O}$, i.e., $\mathbf{a}_0 \in \Omega$. Since $\Sigma_{\mathbf{a}}^{(n)} = \Sigma^{(n)}$ it follows that $\mathbf{a}_0 \in \Omega'_{\mathbf{a}}$ and since $\mathbf{a}_0 \in \mathcal{D}$ it follows that $\mathbf{a}_0 \in \Omega''_{\mathbf{a}}$.

Exercise 5.7.8. Let $B(\mathbf{a}_0, R)$ resp. $B(\mathbf{a}_0, r)$ be the smallest resp. largest ball centered at $\mathbf{a}_0$ and circumscribed about resp. inscribed in $\Sigma^{(n)}$. Show that

$$R = \sqrt{\frac{n-1}{n}} \text{ and } r = \sqrt{\frac{1}{n(n-1)}},$$

(whence $\frac{R}{r} = n - 1$).

> [*Hint:* To find R calculate the distance from $\mathbf{a}_0$ to a *vertex* of $\Sigma^{(n)}$; to find r calculate the distance of $\mathbf{a}_0$ to the centroid of a *face* of $\Sigma^{(n)}$, e.g., to the centroid of $\Sigma^{(n)} \cap \{ \mathbf{x} \ : \ (\mathbf{x}, \mathbf{e}_1) = 0 \}$.]

The key principles in the Karmarkar algorithm may be described in the following terms.

   i. If $\mathbf{a} \in \Sigma^{(n)}$ and if the value of $\mathbf{ca}$ is to be diminished, map $\mathbf{a}$ by $T_{\mathbf{a}}$ into $\mathbf{a}_0$ and operate in the context as modified by $T_{\mathbf{a}}$. The modified context involves $\mathbf{c}'_{\mathbf{a}}$ and $\Omega''_{\mathbf{a}}$.

   ii. If $0 < \alpha < 1$ and if the original **PLPP** has a solution there is in

$$B(\mathbf{a}_0, \alpha r) \cap \Omega''_{\mathbf{a}}$$

a vector $\mathbf{b}'$ such that $f'(\mathbf{b}') \leq f(\mathbf{a}_0) - \ln(1 + \alpha)$.

   iii. There is in $B(\mathbf{a}_0, \alpha r) \cap \Omega''_{\mathbf{a}}$ a vector $\mathbf{b}''$ that minimizes $\mathbf{c}'_{\mathbf{a}}\mathbf{x}$ on

$$B(\mathbf{a}_0, \alpha r) \cap \Omega''_{\mathbf{a}},$$

and there is a positive number $\delta(\alpha, n)$ such that

$$f'(\mathbf{b}'') \leq f'(\mathbf{a}_0) - \delta(\alpha, n).$$

Hence if $\mathbf{b} \overset{\text{def}}{=} T_{\mathbf{a}}^{-1}\mathbf{b}''$ then $f(\mathbf{b}) \leq f(\mathbf{a}) - \delta(\alpha, n)$.

The important difference between *ii* and *iii* above is that *ii* merely asserts the *existence* of a $\mathbf{b}'$ such that $f'(\mathbf{b}') \leq f(\mathbf{a}_0) - \ln(1 + \alpha)$, while *iii* uses *ii*

---

to provide an explicit algorithm for finding a $\mathbf{b}''$, a $\delta(\alpha, n)$, and a $\mathbf{b}$ for which $f(\mathbf{b}) \leq f(\mathbf{a}) - \delta(\alpha, n)$.

Since $\delta(\alpha, n)$ is independent of $\mathbf{a}$, if the **PLPP** has a solution, repeated application of the procedure starting with $\mathbf{a} = \mathbf{a}_0 \overset{\text{def}}{=} \mathbf{x}_0$, gives rise to a sequence $\{\mathbf{x}_k\}_{k=0}^{\infty}$ such that

$f(\mathbf{x}_k) \leq f(\mathbf{x}_{k-1}) - k\delta(\alpha, n)$, $1 \leq k < \infty$. Finally, at any stage of the algorithm, the inequality

$$f(\mathbf{x}_k) > f(\mathbf{x}_{k-1}) - \delta(\alpha, n)$$

is a *signal* that the **PLPP** has no solution and that the operation of the algorithm should stop. (The **HPLPP** always has a solution but the minimal value of $\mathbf{cx}$ can be positive rather than 0.)

The details are taken up below.

The formal description of the algorithm uses a subalgorithm or subroutine for minimizing $\mathbf{c}'_{\mathbf{a}_k}\mathbf{x}$ on $B(\mathbf{a}_0, \alpha r) \cap \Omega''_{\mathbf{a}_k}$. This latter set is itself a ball of radius αr in $\mathcal{D}$. Since $dim(Z) < n$ movement within $B(\mathbf{a}_0, \alpha r) \cap \Omega''_{\mathbf{x}_k}$ must be restricted, in particular, to movement within $\Omega''_{\mathbf{x}_k}$. Hence movement is not necessarily in the direction of $(\mathbf{c}'_{\mathbf{x}_k})^t$ itself since it may not lie in Z. Instead movement is in the direction of the *orthogonal projection of* $(\mathbf{c}'_{\mathbf{x}_k})^t$ *into* Z.

Hence movement is:

i. from $\mathbf{a}_0$;

ii. in $B(\mathbf{a}_0, \alpha r) \cap \Omega''_{\mathbf{x}_k}$;

iii. in the direction of decreasing $\mathbf{c}'_{\mathbf{x}_k}\mathbf{x}$ along the orthogonal projection of $(\mathbf{c}'_{\mathbf{x}_k})^t$ into $ker(AD_{\mathbf{x}_k})$;

iv. to $\mathbf{b}''_{k+1}$ on the boundary $\partial B(\mathbf{a}_0, \alpha r) \cap \Omega''_{\mathbf{x}_k}$ of the ball $B(\mathbf{a}_0, \alpha r) \cap \Omega''_{\mathbf{x}_k}$.

Why does the Karmarkar algorithm work? The answer is based on the analysis that follows.

It is assumed that the **PLPP** has a solution, i.e., that the minimal value of $\mathbf{cx}$ on $\mathcal{D}$ is 0 rather than a positive number. It is shown that there is a sequence $\{\mathbf{x}_k\}_{k=0}^{\infty}$ such that

$$f(\mathbf{x}_{k+1}) \leq f(\mathbf{x}_k) - \delta(\alpha, n) \leq \cdots \leq f(\mathbf{a}_0) - k\delta(\alpha, n).$$

Hence if the **PLPP** has a solution $f(\mathbf{x}_k)$ does not merely decrease as $k+1$ replaces k but decreases by at least $\delta(\alpha, n)$ and so $f(\mathbf{x}_k) \to -\infty$ as $k \to \infty$.

Since each component of each $\mathbf{x}_k$ is positive and less than 1

$$f(\mathbf{x}) \geq \ln(\mathbf{cx})$$

whence it follows that $\mathbf{cx}_k \to 0$ as $k \to \infty$. In fact, there is a constant K such that if $k \geq Kn(q + \ln n)$ and k rounds of the algorithm are performed before there is the signal to stop the operation of the algorithm then

$$\mathbf{cx}_k \leq 2^{-q}\mathbf{ca}_0$$

(cf. **Remark 5.7.1**). If the signal to stop the operation is never given then

$$infimum\,\{\,\mathbf{cx}\ :\ \mathbf{x} \in \mathcal{D}\,\} = 0.$$

The next part of this **Section** consists of formal proofs that validate the claims made above for the Karmarkar algorithm. In many instances the proofs are more readable when the vectors are not encumbered by subscripts, super-scripts, etc. For the most part below such appendices are dropped.

THEOREM 5.7.1. LET:

i. W BE A SUBSPACE OF $\mathbf{R}^n$;

ii. $\mathbf{a}$ AND $\mathbf{c}^t$ BE VECTORS IN $\mathbf{R}^n$;

iii. $B(\mathbf{a},\rho)$ BE A BALL OF (POSITIVE) RADIUS ρ;

iv. $H_{\mathbf{a}}$ BE THE HYPERPLANE: $\{\,\mathbf{x}\ :\ \mathbf{x} - \mathbf{a} \in W\,\} \overset{\text{def}}{=} \mathbf{a} + W$, THE *translate by* $\mathbf{a}$ OF W;

v. $\mathbf{c}_W$ BE THE ORTHOGONAL PROJECTION OF $\mathbf{c}^t$ INTO W.

IF $\mathbf{c}_W \neq \mathbf{O}$, AND IF

$$\mathbf{u} \overset{\text{def}}{=} \frac{\mathbf{c}_W}{\|\mathbf{c}_W\|}$$

THEN THE VECTOR $\mathbf{a} - \rho\mathbf{u} \overset{\text{def}}{=} \mathbf{y}$ IS THE VECTOR THAT MINIMIZES $\mathbf{cx}$ ON THE SET $B(\mathbf{a},\rho) \cap (\mathbf{a} + W) \overset{\text{def}}{=} B_{H_{\mathbf{a}}}(\mathbf{a},\rho)$.

PROOF. Since $\mathbf{a} \in H_{\mathbf{a}}$ and $\mathbf{u} \in W$ it follows that $\mathbf{y} \in H_{\mathbf{a}}$. Since $\|\mathbf{u}\| = 1$ it follows that $\mathbf{y} \in B_{H_{\mathbf{a}}}(\mathbf{a},\rho)$ and so $\mathbf{y} \in B_{H_{\mathbf{a}}}(\mathbf{a},\rho)$.

If $\mathbf{z} \in B_{H_{\mathbf{a}}}(\mathbf{a},\rho)$ then $\mathbf{y} - \mathbf{z} \in W$ and hence $(\mathbf{c} - \mathbf{c}_W) \perp (\mathbf{y} - \mathbf{z})$. The following equations and inequalities yield the conclusion:

$$\mathbf{c}(\mathbf{y} - \mathbf{z}) = \mathbf{c}_W(\mathbf{y} - \mathbf{z}) = (\mathbf{c}_W, \mathbf{a} - \rho\mathbf{u} - \mathbf{z})$$
$$= \|\mathbf{c}_W\|(\mathbf{u}, \mathbf{a} - \mathbf{z} - \rho\mathbf{u})$$
$$(\mathbf{u}, \mathbf{a} - \mathbf{z}) \leq \|\mathbf{a} - \mathbf{z}\| \leq \rho \ \text{(Schwarz inequality)}$$
$$\mathbf{cy} \leq \mathbf{cz}.$$

Ω

[**Note 5.7.4:** Let P be the orthogonal projection (idempotent) matrix such that $P\mathbf{R}^n = W$, in particular $P\mathbf{c}^t = \mathbf{c}_W$. Then since P is not only idempotent but also self-adjoint it follows that if $\mathbf{c}_W \neq \mathbf{O}$ then

$$\mathbf{cu} = (\mathbf{c}^t, \mathbf{u}) = \frac{1}{\|\mathbf{c}_W\|}(\mathbf{c}^t, P\mathbf{c}^t)$$
$$= \frac{1}{\|\mathbf{c}_W\|}(\mathbf{c}^t, P^2\mathbf{c}^t) = \frac{1}{\|\mathbf{c}_W\|}\|P\mathbf{c}^t\|^2$$
$$= \frac{1}{\|\mathbf{c}_W\|}\|\mathbf{c}_W\|^2 = \|\mathbf{c}_W\| > 0.$$

Hence if furthermore $0 < \alpha_1 < \alpha_2 < 1$ and $\mathbf{y}_i = \mathbf{a}_0 - \alpha_i \rho \mathbf{u}$, $i = 1, 2$, then, since $\mathbf{c}_W \neq \mathbf{O}$ it follows that $\mathbf{cu} > 0$ whence

$$\mathbf{cy}_1 - \mathbf{cy}_2 = (\alpha_2 - \alpha_1)\rho\mathbf{cu} > 0,$$

i.e., $\mathbf{cy}_2 < \mathbf{cy}_1$. In other words, the larger the ball centered on $\mathbf{a}$ and contained in $\mathbf{a} + W$ the smaller is the minimum of $\mathbf{cx}$ on the intersection of that ball with $\mathbf{a} + W$.]

Exercise 5.7.9. Show that if, in the context of THEOREM **5.7.1**, $\mathbf{c}_W = \mathbf{O}$ then $minimum\{\,\mathbf{cx} \ : \ \mathbf{x} \in \mathbf{a} + W\,\} = \mathbf{ca}$.

Exercise 5.7.10. The following data provide an opportunity to apply THE-OREM **5.7.1**:

$$\mathbf{c} \overset{\text{def}}{=} (1, -2, 3)$$
$$\mathbf{a}^t \overset{\text{def}}{=} (-1, 3, 2)$$
$$r \overset{\text{def}}{=} 2$$
$$A \overset{\text{def}}{=} \begin{pmatrix} 1 & 2 & 3 \\ 4 & 5 & 6 \end{pmatrix}$$
$$W \overset{\text{def}}{=} ker(A).$$

Use the procedure described above to find

$$minimum\{\,\mathbf{cx} \ : \ \mathbf{x} \in B(\mathbf{a}, r) \cap (\mathbf{a} + W)\,\}.$$

Exercise 5.7.11. Use the (calculus) method of Lagrange multipliers to find

$$minimum\{\,\mathbf{cx} \ : \ \mathbf{x} \in B(\mathbf{a}, r) \cap (\mathbf{a} + W)\,\}$$

for the data of **Exercise 5.7.10**.

THEOREM **5.7.1** applied to the **HPLPP** leads to the next result, the fundamental reduction possibility for $f(\mathbf{x})$ if the **PLPP** has a solution.

THEOREM 5.7.2. ASSUME:

i. THE ORIGINAL **PLPP** HAS A SOLUTION;
ii. $0 < \alpha < 1$;
iii. $\mathbf{a} \in \mathcal{D} \cap \Sigma^{(n)^{\circ}}$ (HENCE $\mathbf{a} \succ \mathbf{O}$);

THEN THERE IS IN $B(\mathbf{a}_0, \alpha r) \cap \Omega_{\mathbf{a}}''$ A VECTOR $\mathbf{b}'$ SUCH THAT

$$f'(\mathbf{b}') \leq f'(\mathbf{a}_0) - \ln(1 + \alpha).$$

PROOF.   Since the **PLPP** has a solution there is for the **HPLPP** a vector z that minimizes $\mathbf{cx}$ on $\mathcal{D}$. Thus $\mathbf{z}$ is such that $\mathbf{cz} = 0$ hence

$$\mathbf{c}_{\mathbf{a}}' \mathbf{z}_{\mathbf{a}}' = 0 \text{ (cf. \textbf{Exercise 5.7.6}}v)$$
$$\mathbf{z}_{\mathbf{a}}' \in B(\mathbf{a}_0, R) \cap \Omega_{\mathbf{a}}''$$
$$\|\mathbf{z}_{\mathbf{a}}' - \mathbf{a}_0\| \leq R.$$

Since $\mathbf{ca}_0 > 0$ (cf. **Note 5.7.1**) it follows from **Exercise 5.7.9** that the projection of $\mathbf{c}_{\mathbf{a}}'$ into $ker(AD_{\mathbf{a}})$ is not $\mathbf{O}$. Since $r > \alpha r$, it follows from **Note 5.7.4** that

$$minimum\{\, \mathbf{c}_{\mathbf{a}}'\mathbf{x} \;:\; \mathbf{x} \in B(\mathbf{a}_0, r) \cap \Omega_{\mathbf{a}}''\,\} < minimum\{\, \mathbf{c}_{\mathbf{a}}'\mathbf{x} \;:\; \mathbf{x} \in B(\mathbf{a}_0, \alpha r) \cap \Omega_{\mathbf{a}}''\,\}$$
$$minimum\{\, \mathbf{c}_{\mathbf{a}}'\mathbf{x} \;:\; \mathbf{x} \in \mathcal{D}\,\} < minimum\{\, \mathbf{c}_{\mathbf{a}}'\mathbf{x} \;:\; \mathbf{x} \in B(\mathbf{a}_0, \alpha r) \cap \Omega_{\mathbf{a}}''\,\}$$

and so $\mathbf{z}_{\mathbf{a}}' \notin B(\mathbf{a}_0, \alpha r)$.

Let $\mathbf{z}_{\mathbf{a}}'$ be $(z_{\mathbf{a}1}', \ldots, z_{\mathbf{a}n}')^t$ and let $\mathbf{b}'$ be $(b_1', \ldots, b_n')^t$, the vector where the line segment connecting $\mathbf{z}_{\mathbf{a}}'$ to $\mathbf{a}_0$ meets the boundary

$$\partial B(\mathbf{a}_0, \alpha r) \overset{\text{def}}{=} \{\, \mathbf{x} \;:\; \|\mathbf{x} - \mathbf{a}_0\| = \alpha r\,\}$$

of $B(\mathbf{a}_0, \alpha r)$. Thus for some λ in $(0, 1)$: $\mathbf{b}' = (1 - \lambda)\mathbf{a}_0 + \lambda \mathbf{z}_{\mathbf{a}}'$. Then, since $\mathbf{z}_{\mathbf{a}}'$ and $\mathbf{a}_0$ are in the convex set $\Omega_{\mathbf{a}}''$ it follows that

$$\mathbf{b}' \in B(\mathbf{a}_0, \alpha r) \cap \Omega_{\mathbf{a}}''$$
$$\mathbf{c}_{\mathbf{a}}' \mathbf{z}_{\mathbf{a}}' = 0$$
$$\mathbf{c}_{\mathbf{a}}' \mathbf{b}' = (1 - \lambda)\mathbf{c}_{\mathbf{a}}' \mathbf{a}_0 + \lambda \mathbf{c}_{\mathbf{a}}' \mathbf{z}_{\mathbf{a}}'$$
$$= (1 - \lambda)\mathbf{c}_{\mathbf{a}}' \mathbf{a}_0$$
$$\frac{\mathbf{c}_{\mathbf{a}}' \mathbf{a}_0}{\mathbf{c}_{\mathbf{a}}' \mathbf{b}'} = \frac{1}{1 - \lambda}$$
$$b_j' = \frac{(1 - \lambda)}{n} + \lambda \mathbf{z}_{\mathbf{a}}'.$$

Hence

$$f'(\mathbf{a}_0) - f'(\mathbf{b}') = \sum_{j=1}^{n} \ln \left[n \frac{\mathbf{c}'_\mathbf{a} \mathbf{a}_0 b'_j}{\mathbf{c}'_\mathbf{a} \mathbf{b}'} \right]$$

$$= \sum_{j=1}^{n} \ln \left[1 + \frac{\lambda n}{1 - \lambda} z'_{\mathbf{a}j} \right].$$

Since

$$\ln \left[1 + \frac{\lambda n}{1 - \lambda} z'_{\mathbf{a}j} \right] \geq 0$$

since $\mathbf{e}^t \mathbf{z}'_\mathbf{a} = 1$, and since $P_j \geq 0$ it follows that

$$\prod_{j=1}^{n} (1 + P_j) \geq 1 + \sum_{j=1}^{n} P_j.$$

Consequently

$$f'(\mathbf{a}_0) - f'(\mathbf{b}') \geq \ln \left[1 + \frac{n\lambda}{1 - \lambda} \mathbf{e}^t \mathbf{z}'_\mathbf{a} \right]$$

$$f'(\mathbf{a}_0) - f'(\mathbf{b}') \geq \ln \left[1 + \frac{n\lambda}{1 - \lambda} \right]$$

$$\mathbf{b}' - \mathbf{a}_0 = \lambda(\mathbf{z}'_\mathbf{a} - \mathbf{a}_0)$$

$$\alpha r = \|\mathbf{b}' - \mathbf{a}_0\| = \lambda \|\mathbf{z}'_\mathbf{a} - \mathbf{a}_0\| \leq \lambda R$$

$$\lambda \geq \alpha \frac{r}{R} = \frac{\alpha}{n - 1} \quad (\text{cf. Exercise 5.7.8})$$

$$1 + \frac{\lambda n}{1 - \lambda} \geq 1 + \alpha$$

$$f'(\mathbf{a}_0) - f'(\mathbf{b}') \geq \ln(1 + \alpha)$$

$$f'(\mathbf{b}') \leq f'(\mathbf{a}_0) - \ln(1 + \alpha). \tag{5.7.12}$$

$$\Omega$$

[**Note 5.7.5:** Even if the **PLPP** has no solution the vector a, via the projective transformation $T_\mathbf{a}$, determines the vector $\mathbf{b}'$ in the manner described in the PROOF of THEOREM **5.7.2**.

If, to boot, the **PLPP** has a solution one may conclude as well that

$$f'(\mathbf{b}') \leq f(\mathbf{a}_0) - \ln(1 + \alpha).$$

The inverse $T_\mathbf{a}^{-1}$ applied to b' yields a vector $\mathbf{b} \stackrel{\text{def}}{=} \phi(\mathbf{a})$. Thus if $\mathbf{a} = \mathbf{a}_0$ and $\mathbf{b}' \stackrel{\text{def}}{=} \mathbf{a}'_1$, then $\mathbf{a}'_1$ determines a vector $\mathbf{a}'_2 \stackrel{\text{def}}{=} \phi(\mathbf{a}'_1)$ and

$$f'(\mathbf{a}'_2) \leq f'(\mathbf{a}'_1) - \ln(1 + \alpha) \leq f'(\mathbf{a}_0) - 2\ln(1 + \alpha),$$

Thus by induction there can be defined in $\Sigma^{(n)^\circ}$ a sequence $\{a'_k\}_{k=0}^\infty$ such that

$$f'(a'_k) \leq f'(a'_{k-1}) - k\ln(1+\alpha), \ 1 \leq k < \infty.$$

If $a_k \overset{\text{def}}{=} T^{-1}_{a'_{k-1}} a'_k$ and if the **PLPP** has a solution then

$$f(a_k) \leq f(a_{k-1}) - k\ln(1+\alpha).$$

The *positive* number $\ln(1+\alpha)$ depends only on α and hence *not* on any a_k. Whence if the **PLPP** has a solution then $f(a_k) \downarrow -\infty$ as $k \to \infty$.

If, for any k, $f(a_k) > f(a_{k-1}) - \ln(1+\alpha)$ the **PLPP** has no solution.]

The next part of Karmarkar's development is to provide a simple procedure for finding for each vector a'_k a nearby vector a''_k and a constant $\delta(\alpha, n)$, dependent only on α and n so that if the **PLPP** has a solution and if $x_k \overset{\text{def}}{=} T^{-1}_{a_{k-1}} a''_k$ then

$$f(x_k) \leq f(x_{k-1}) - k\delta(\alpha, n). \tag{5.7.13}$$

Again, it is to emphasized that the sequence $\{x_k\}_{k=0}^\infty$ may be constructed even if the **PLPP** has no solution. However the relation (5.7.13) is derivable on the assumption that the **PLPP** has a solution. Hence, if for some k (5.7.13) fails to hold the **PLPP** has no solution and the operation halts.

The notations used are the following:

$$b' \overset{\text{def}}{=} \text{the vector described in Theorem 5.7.2}$$

$$b'' \overset{\text{def}}{=} \text{the vector that minimizes } c'_a x \text{ on } B(a_0, \alpha r) \cap \Omega''_a$$

$$W \overset{\text{def}}{=} ker(M_a)$$

$$c_W \overset{\text{def}}{=} \text{the orthogonal projection of } (c'_a)^t \text{ into } W$$

$$u \overset{\text{def}}{=} \frac{c_W}{\|c_W\|} \ (\text{iff } c_W \neq O)$$

$$\tilde{f}(x) \overset{\text{def}}{=} n\ln\left[\frac{c'_a x}{c'_a a_0}\right].$$

Theorem 5.7.3. There is a constant $\delta(\alpha, n)$ such that if

$$f'(b') \leq f'(a_0) - \ln(1+\alpha) \tag{5.7.14}$$

then $f'(b'') \leq f'(a_0) - \delta(\alpha, n)$. Furthermore, if $\alpha = 0.25$ then $\delta(\alpha, n) \geq 0.125$.

   Chapter 5. APPLYING LINEAR ALGEBRA

PROOF. Let $\mathbf{b}_m$ be the vector that minimizes $f'(\mathbf{y})$ over $B(\mathbf{a}_0, \alpha r) \cap \Omega''_{\mathbf{a}}$. Then

$$f'(\mathbf{a}_0) - f'(\mathbf{b}') = f'(\mathbf{a}_0) - f'(\mathbf{b}_m) + f'(\mathbf{b}_m) - f'(\mathbf{b}')$$
$$= [f'(\mathbf{a}_0) - f'(\mathbf{b}_m)]$$
$$+ [f'(\mathbf{b}_m) - \left(f'(\mathbf{a}_0) + \widetilde{f}(\mathbf{b}_m)\right)]$$
$$- [f'(\mathbf{b}') - \left(f'(\mathbf{a}_0) + \widetilde{f}(\mathbf{b}_m)\right)]$$
$$+ [\widetilde{f}(\mathbf{b}_m) - \widetilde{f}(\mathbf{b}')].$$

If $\mathbf{x} \in B(\mathbf{a}_0, \alpha r) \cap \Omega''_{\mathbf{a}}$ then by direct calculation

$$f'(\mathbf{x}) - \left(f'(\mathbf{a}_0) + \widetilde{f}(\mathbf{x})\right) = -\sum_{j=1}^{n} \ln n x_j.$$

At this point two inequalities derivable via the methods of the calculus are useful. The proofs of the inequalities are deferred (cf. LEMMAS 5.7.1, 5.7.2) until the other parts of the proof of THEOREM 5.7.3 are given. The inequalities are stated in terms of the parameter

$$\beta \stackrel{\text{def}}{=} \alpha \sqrt{\frac{n}{n-1}}.$$

If α is sufficiently small and positive then $0 < \beta < 1$ for all n in $\mathbb{N} \setminus 1$. The inequalities for t in $[0, \beta)$ and $\mathbf{x}$ in $B(\mathbf{a}_0, \alpha r)$ are:

$$|\ln(1 + t) - t| \leq \frac{t^2}{2(1 - \beta)}$$

$$\left| \sum_{j=1}^{n} \ln n x_j \right| \leq \frac{\beta^2}{2(1 - \beta)}.$$

Hence

$$\left| f'(\mathbf{x}) - \left(f'(\mathbf{a}_0) + \widetilde{f}(\mathbf{x})\right) \right| \leq \frac{\beta^2}{2(1 - \beta)}.$$

The formula for $\widetilde{f}(\mathbf{x})$ shows that it and $\mathbf{c}'_{\mathbf{a}}\mathbf{x}$ achieve their minima at the same vector, namely $\mathbf{b}''$. Hence

$$\widetilde{f}(\mathbf{b}_m) \geq \widetilde{f}(\mathbf{b}')$$

$$f'(\mathbf{a}_0) - f'(\mathbf{b}') \geq \ln(1 + \alpha) - \frac{\beta^2}{(1 - \beta)}.$$

If $|t| < 1$ the Maclaurin series

$$\ln(1 + t) = t - \frac{t^2}{2} + \cdots + (-1)^{m+1} \frac{t^m}{m} + \cdots$$

is valid and so, since $0 < \alpha < 1$,

$$\ln(1+\alpha) = \alpha - \frac{\alpha^2}{2} + \alpha^3[(\frac{1}{3} - \frac{\alpha}{4}) + \alpha^2(\frac{1}{5} - \frac{\alpha}{6}) + \cdots]$$

$$\geq \alpha - \frac{\alpha^2}{2}.$$

Hence

$$f'(\mathbf{a}_0) - f'(\mathbf{b}') \geq \alpha - \frac{\alpha^2}{2} - \frac{\alpha^2 n^2}{n(n-1)\left[1 - \alpha\sqrt{\frac{n}{n-1}}\right]} \stackrel{\text{def}}{=} \delta(\alpha, n)$$

$$f'(\mathbf{b}') \leq f'(\mathbf{a}_0) - \delta(\alpha, n).$$

If $\alpha = 0.25$ then $\lim_{n \to \infty} \delta(\alpha, n) \geq 0.125$.

$$\Omega$$

LEMMA 5.7.1. IF $|t| \leq \beta < 1$ THEN

$$|\ln(1+t) - t| \leq \frac{t^2}{2(1-\beta)}.$$

PROOF. The Maclaurin series for $\ln(1+t)$ shows that if $|t| \leq \beta < 1$ then

$$|\ln(1+t) - t| \leq \frac{\beta^2}{2}(\sum_{m=0}^{\infty} \beta^m) = \frac{\beta^2}{2(1-\beta)}.$$

$$\Omega$$

LEMMA 5.7.2. IF

$$\beta \stackrel{\text{def}}{=} \alpha\sqrt{\frac{n}{n-1}} < 1 \text{ AND } \mathbf{x} \in B(\mathbf{a}_0, \alpha r)$$

THEN

$$\left|\sum_{j=1}^{n} \ln n x_j\right| \leq \frac{\beta^2}{2(1-\beta)}.$$

    Chapter 5. APPLYING LINEAR ALGEBRA

PROOF.

$$\sum_{j=1}^{n}(nx_j - 1)^2 \le n^2\alpha^2 r^2 = \beta^2$$

$$|nx_j - 1| \le \beta, \ 1 \le j \le n$$

$$|\ln(1 + nx_j - 1) - nx_j - 1| \le \frac{1}{2(1-\beta)}(nx_j - 1)^2$$

$$\sum_{j=1}^{n}\ln(nx_j) = \sum_{j=1}^{n}\ln(1 + nx_j - 1)$$

$$\sum_{j=1}^{n}|\ln(nx_j) - (nx_j - 1)| \le \frac{1}{2(1-\beta)}\sum_{j=1}^{n}(nx_j - 1)^2 \le \frac{\beta^2}{2(1-\beta)}$$

$$\sum_{j=1}^{n}(nx_j - 1) = n\sum_{j=1}^{n}x_j - \sum_{j=1}^{n}1 = n - n = 0$$

$$\left|\sum_{j=1}^{n}\ln(nx_j) - \sum_{j=1}^{n}(nx_j - 1)\right| \le \sum_{j=1}^{n}|\ln(nx_j) - (nx_j - 1)|$$

$$\le \frac{\beta^2}{2(1-\beta)}.$$

$$\Omega$$

Exercise 5.7.12. Show that if, in the context of THEOREM 5.7.3, $c_W = O$ then $b_m = a_0$.

The formula for c_W is, according to **Exercise 4.4.43**,

$$c_W = (I - M_\mathbf{a}^\times M_\mathbf{a}^{\times +})c_\mathbf{a}'.$$

Since the context is $\mathbf{R}^n$ the formula is

$$c_W = (I - M_\mathbf{a}^t M_\mathbf{a}^{t+})c_\mathbf{a}'. \tag{5.7.15}$$

If the columns of $M_\mathbf{a}^t$, i.e., if the rows of $M_\mathbf{a}$ are linearly independent, then

$$c_W = [I - M_\mathbf{a}^t(M_\mathbf{a}M_\mathbf{a}^t)^{-1}M_\mathbf{a}]c_\mathbf{a}'. \tag{5.7.16}$$

[**Remark 5.7.4:** It is automatic that the rows of A are linearly independent, i.e., that if $\mathbf{x} \in \mathbf{R}_m$ and $\mathbf{x}A = O$ then $\mathbf{x} = O$. Since $D_\mathbf{a}$ is invertible if $\mathbf{x}AD_\mathbf{a} = O$ then $\mathbf{x}A = O$, i.e., the rows of $AD_\mathbf{a}$ are linearly independent.

However, the rows of $M_\mathbf{a}$ might well fail to be linearly independent, e.g.,

$$AD_\mathbf{a} \overset{\text{def}}{=} \begin{pmatrix} 1 & 0 & 1 & 1 & 0 & 0 \\ 0 & 1 & 0 & 0 & 1 & 1 \end{pmatrix}$$

$$M_\mathbf{a} = \begin{pmatrix} 1 & 0 & 1 & 1 & 0 & 0 \\ 0 & 1 & 0 & 0 & 1 & 1 \\ 1 & 1 & 1 & 1 & 1 & 1 \end{pmatrix}.$$

If C is an $m \times n$ matrix then the matrix CC^t is invertible iff there is an $m \times m$ matrix R such that $CC^t R = I$. If CC^t is invertible and if there is in $\mathbf{R}_m$ an $\mathbf{x}$ such that $\mathbf{x}C = \mathbf{O}$ then $\mathbf{x}CC^t R = \mathbf{x} = \mathbf{O}$, i.e., CC^t is invertible iff the rows of C are linearly independent. Hence (5.7.15) is always a valid formula and (5.7.16) is valid iff the rows of $M_\mathbf{a}$ are linearly independent.]

The Karmarkar algorithm, instead of sticking with the given $\mathbf{c}$, finds a sequence $\mathbf{c}_{\mathbf{x}_1}, \mathbf{c}_{\mathbf{x}_2}, \ldots$ of vectors and for each $\mathbf{c}_{\mathbf{x}_k}$ finds a vector $\mathbf{b}_k''$ that minimizes $\mathbf{c}_{\mathbf{x}_k}\mathbf{x}$ on

$$B(\mathbf{a}_0, \rho) \cap \Omega_{\mathbf{x}_k}'' \cap \Sigma^{(n)},$$

a (lower-dimensional) ball contained in $\Sigma^{(n)^\circ}$. Then from the sequence

$$\mathbf{b}_1'', \mathbf{b}_2'', \ldots$$

the algorithm creates by *projective* transformations $T_{\mathbf{x}_k}^{-1}$ the vectors

$$\mathbf{x}_1, \mathbf{x}_2, \ldots$$

that validate the alternatives (5.7.3)/(5.7.4).

Thus, x_{kj} denoting the jth component of $\mathbf{x}_k$, the inequality

$$f(\mathbf{x}_k) \leq f(\mathbf{x}_{k-1}) - \delta(\alpha, n)$$

implies

$$n \ln \frac{\mathbf{c}\mathbf{x}_k}{\mathbf{c}\mathbf{a}_0} \leq n \ln n \left[\frac{\sum_{j=1}^n c_j x_{kj}}{\sum_{j=1}^n c_j} \right] - k\delta(\alpha, n).$$

Since $0 < x_{kj} \leq 1$ and $\mathbf{c}\mathbf{a}_0 > 0$

$$\sum_{j=1}^n \left[\frac{c_j}{\sum_{i=1}^n c_i} \right] = 1$$

whence

$$\left[\frac{\sum_{j=1}^n c_j x_{kj}}{\sum_{j=1}^n c_j} \right] \leq 1$$

$$n \ln \frac{\mathbf{c}\mathbf{x}_k}{\mathbf{c}\mathbf{a}_0} \leq n - \frac{k}{n}\delta(\alpha, n) \leq n \ln n - k\delta(\alpha, n).$$

Thus for some constant K if the algorithm operates through k rounds then

$$k \geq K n(q + \ln n) \Rightarrow \frac{\mathbf{c}\mathbf{x}_k}{\mathbf{c}\mathbf{a}_0} \leq 2^{-q}.$$

There is the following interpretation of the Karmarkar algorithm:

An *ellipsoid* is defined by means of a self-adjoint matrix H and a vector $\mathbf{b}_0$. It is assumed that for all $\mathbf{x}$ $(H\mathbf{x}, \mathbf{x}) \geq 0$.

An ellipsoid $\mathcal{E} \stackrel{\text{def}}{=} \{\, \mathbf{x} \;:\; (H(\mathbf{x} - \mathbf{b}_0), \mathbf{x} - \mathbf{b}_0) \leq 1 \,\}$ is centered at $\mathbf{b}_0$ and its boundary is $\partial\mathcal{E} \stackrel{\text{def}}{=} \{\, \mathbf{x} \;:\; (H(\mathbf{x} - \mathbf{b}_0), \mathbf{x} - \mathbf{b}_0) = 1 \,\}$.

Exercise 5.7.13. Show that if $\mathbf{a} \in \mathcal{D}$ and $\mathbf{a} \succ \mathbf{O}$ then

$$T_{\mathbf{a}}^{-1} B(\mathbf{a}_0, \alpha r)$$

is an *ellipsoid* $\mathcal{E}$ contained in $\mathcal{D}$. Show also that if $\mathbf{b}'$ is the minimizing vector for $\mathbf{c}_{\mathbf{a}}'\mathbf{x}$ on $B(\mathbf{a}_0, \alpha r) \cap \Omega_{\mathbf{a}}''$ then $T_{\mathbf{a}}^{-1}\mathbf{b}' \in \partial\mathcal{E}$.

For details and refinements associated with the Karmarkar algorithm see [**Karm**].

5.8. Some special linear programming problems

In this book the subject of linear programming arises in a version of the *diet problem*: provide an adequate diet at least cost. In disguised form the diet problem occurs in other contexts.

i. Produce a petroleum-based fuel blend with required properties and at least cost.
ii. Produce a stock portfolio that maximizes return within the constraint of a given risk level.

In **Section 5.4** the relation of linear programming to the optimization of strategies for games of chance is discussed.

Other forms of **PLPP**s deserve notice:

i. the *transportation problem*;
ii. the *network flow problem*;
iii. the *travelling salesman problem*.

These are treated below.

THE TRANSPORTATION PROBLEM

Assume there are s sources of a product that is to be shipped to m markets. Let the unit shipping cost from source i to market j be c_{ij}, $1 \leq i \leq s$, $1 \leq j \leq m$. Finally assume that source i can supply no more than S_i and that market j

---

demands at least D_j. If source i supplies x_{ij} units to market j there arises the following **PLPP**:

$$\text{minimize} \sum_{i=1,j=1}^{s,m} c_{ij} x_{ij}$$

$$\text{subject to} \begin{cases} \sum_{i=1}^{s} x_{ij} \geq D_j \\ \sum_{j=1}^{m} x_{ij} \leq S_i \\ x_{ij} \geq 0. \end{cases}$$

Exercise 5.8.1. Rewrite the transportation problem in the form:

$$\text{minimize } \mathbf{cx}$$

$$\text{subject to} \begin{cases} A\mathbf{x} = \mathbf{q} \\ \mathbf{x} \succeq \mathbf{O}. \end{cases}$$

THE NETWORK FLOW PROBLEM

Assume there is a source from which boundless quantities of a substance can flow and a sink into which boundless quantities of the substance can flow. Assume furthermore that there is a network of *links* (pipes if the substance is a fluid, electric cables if the substance consists of units of electric current, railroad lines or roads if the substance is merchandise (or even people) transported by rail, bus, or truck, etc.) through which the source supplies the sink. The network of links has *junctions* or *nodes* where some links meet or are joined to other links. These junctions are pumping stations on pipeline networks, junction boxes or substations in electrical networks, railheads, railroad stations, bus stations, or distribution warehouses in rail/bus/truck systems, etc. In particular, the source and the sink are junctions.

Some junctions are connected by links to the source, some to the sink, some to neither but to other junctions. It is assumed that each junction j is connected by links through a sequence of junctions leading from j back to the source and by links through a sequence of junctions leading from j forward to the sink. The link between one junction and another is assumed to have a (finite) carrying capacity. The problem is then to maximize the (size of the) flow from source to sink and to obey the constraints imposed by the carrying capacities of the links. In fluid or electrical networks the calculation of the maximal flow is useful in determining the size of the source. For example, there is no point in building a 10-megawatt electric generator complex (source) to feed a city (sink) if the transmission network between generator complex and city can carry only 5 megawatts.

Let the junctions be numbered $1, \ldots, N$. (The number of the source junction So is 1 and the number of the sink junction Si is N). Assume that if junction i and junction j are linked then:

i. the nonnegative flow between junction i and junction j is x_{ij};
ii. the *nonnegative* capacity of the link between junction i and j is c_{ij}.

The mathematical formulation of the problem is:

$$\text{maximize} \sum_{j=2}^{N} x_{1j} \tag{5.8.1}$$

$$\text{subject to} \begin{cases} 0 \le x_{ij} \le c_{ij}, \ 0 \le i \le N-1, \ 1 \le j \le N \\ \sum_{i=1}^{N-1} x_{ij} - \sum_{k=2}^{N} x_{jk} = 0, \ 0 \le i \le N-1, \ 1 \le j \le N. \end{cases} \tag{5.8.2}$$

Condition (5.8.2) says that no amount of the substance accumulates at any junction other than junction N. Hence everything leaving junction 1 gets to junction N. Hence the requirement (5.8.1) is the mathematical form of the requirement to maximize the flow through the network.

In **Figure 5.8.1** there is depicted a schematic form of a simple network.

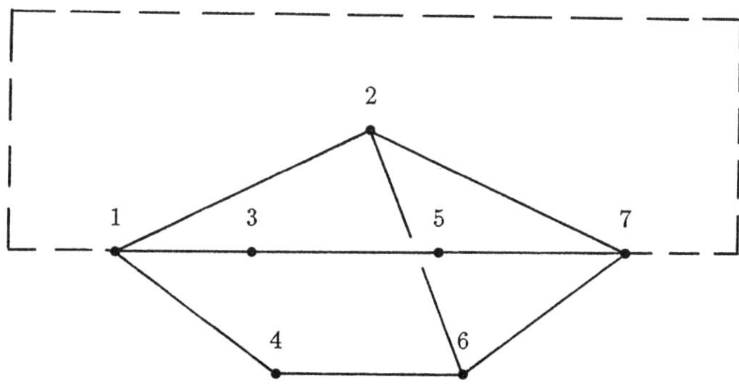

Figure 5.8.1. A simple network

If the network is augmented by a fictitious link (e.g., the line of dashes in the **Figure 5.8.1** above) from junction N to junction 1 and if the fictitious link is assumed to have unlimited capacity the network flow problem can be given a simple description:

$$\text{maximize } x_{N1}$$

$$\text{subject to} \begin{cases} 0 \le x_{ij} \le c_{ij}, \ 1 \le i \le N, \ 1 \le j \le N \\ \sum_{i=1}^{N} x_{ij} - \sum_{k=1}^{N} x_{jk} = 0, \ 1 \le j \le N. \end{cases} \tag{5.8.3}$$

[Associated with the network flow problem (with fictitious link) is an example of an *incidence matrix*. Each row corresponds to one of the N junctions and each column corresponds to one of the links (a pair ij) through which a flow can take place. Each entry 0, 1, or -1 expresses whether (and how) a junction and a link are incident, i.e., are in some way related.

  i. If the entry is 0 then the junction is not part of the link.
  ii. If the entry is 1 then there is a flow *out* of the junction and through the link.
  iii. If the entry is -1 there is a flow *into* the junction from the link.
  iv. In each column there is just one 1 and just one -1.]

Example 5.8.1. The incidence matrix associated with the network in **Figure 5.8.1** is:

$$
\begin{array}{cccccccccc}
(12) & (13) & (14) & (26) & (27) & (35) & (46) & (57) & (67) & (71)
\end{array}
$$

$$
\begin{pmatrix}
1 & 1 & 1 & & & & & & & -1 \\
-1 & & & 1 & 1 & & & & & \\
& -1 & & & & 1 & & & & \\
& & -1 & & & & 1 & & & \\
& & & & & -1 & -1 & 1 & & \\
& & & 1 & & & & & 1 & \\
& & & & -1 & & & -1 & -1 & 1
\end{pmatrix}.
$$

(Entries not printed are 0's.)

Exercise 5.8.2. For the network flow problem formulated as in (5.8.3) there is the standard **PLPP** format:

$$
\text{minimize } \mathbf{cx}
$$

$$
\text{subject to } \begin{cases} A\mathbf{x} \succeq \mathbf{q} \\ \mathbf{x} \succeq \mathbf{O}. \end{cases}
$$

Describe $\mathbf{c}$, $\mathbf{x}$, $\mathbf{q}$, and A.

Exercise 5.8.3. Formulate the **DLPP** corresponding to the **PLPP** that describes the network flow problem.

An important fact about the network flow problem is another "minmax" statement. It can be phrased in terms of the notion of a *cut* in the set of junctions of a network.

DEFINITION 5.8.1. LET $\mathcal{J}$ BE THE SET OF JUNCTIONS OF A NETWORK. LET $C \overset{\text{def}}{=} (\mathcal{J}_1, \mathcal{J}_2)$ BE A PAIR OF SUBSETS OF $\mathcal{J}$. THEN C IS A *cut* IFF

$$
\begin{aligned}
& 1 \in \mathcal{J}_1, \ N \in \mathcal{J}_2 \\
& \mathcal{J}_1 \cap \mathcal{J}_2 = \emptyset \\
& \mathcal{J}_1 \cup \mathcal{J}_2 = \mathcal{J}.
\end{aligned}
$$

In other words, a cut C is a pair of complementary nonempty subsets of $\mathcal{J}$; the source is in one and the sink is in the other. It follows that:

If F is a flow through the network then F is also the flow across any cut, i.e., for some cut $(\mathcal{J}_1, \mathcal{J}_2)$, a flow from $\mathcal{J}_1$ to $\mathcal{J}_2$.
Let

$\mathcal{C}$ be the set of all cuts C in the network,
$\mathcal{F}$ be the set of all flows, F in the network,
$\mathcal{F}(C)$ be the set of all flows F across the cut C,
F_{max} be the maximal flow through the network:

$$F_{max} \overset{\text{def}}{=} maximum\{ F \ : \ F \in \mathcal{F} \}.$$

The *cut capacity* of a cut $C \overset{\text{def}}{=} (\mathcal{J}_1, \mathcal{J}_2)$ is the sum of all capacities of links joining junctions of $\mathcal{J}_1$ to junctions of $\mathcal{J}_2$. In other words, if the restrictions imposed by what can flow through each of $\mathcal{J}_1$ and $\mathcal{J}_2$ are ignored, the cut capacity $cap(C)$ of a cut $C \overset{\text{def}}{=} (\mathcal{J}_1, \mathcal{J}_2)$ is the *maximal* flow from one side $\mathcal{J}_1$ of C to the other side $\mathcal{J}_2$:

$$cap(C) \overset{\text{def}}{=} maximum_{F \in \mathcal{F}(C)} F.$$

THEOREM 5.8.1. (The minmax cut theorem) THE MAXIMAL NETWORK FLOW IS THE MINIMUM OF ALL CUT CAPACITIES:

$$F_{max} = minimum_{C \in \mathcal{C}} cap(C)$$
$$= minimum_{C \in \mathcal{C}} maximum_{F \in \mathcal{F}(C)} F.$$

PROOF. Let C_{min} be the minimum of all cut capacities:

$$C_{min} \overset{\text{def}}{=} minimum_{C \in \mathcal{C}} cap(C).$$

Then $C_{min} \geq F_{max}$ and in particular F_{max} is finite.

The idea of the proof is to find a cut C across which the flow is the capacity $cap(C)$ of C. Then $C_{min} \geq F_{max} = cap(C) \geq C_{min}$, i.e., $F_{max} = C_{min}$.

If each link is fully used then for every cut C: $F_{max} = cap(C) = C_{min}$.

If some links are under-used for the flow F_{max} let $\mathcal{J}_1$ consist of the source and of all other junctions connected to the source (junction 1) by under-used links. Let $\mathcal{J}_2$ consist of all other junctions. By definition, the source (junction 1) is in $\mathcal{J}_1$. More to the point is the fact that the sink (junction N) is in $\mathcal{J}_2$.

(Otherwise the source and the sink are connected by a link that is under-used and thus the flow is not maximal.) Hence the pair $(\mathcal{J}_1, \mathcal{J}_2)$ is a cut C.

All links connecting a junction in $\mathcal{J}_1$ to a junction in $\mathcal{J}_2$ are fully used. (Otherwise some junctions in $\mathcal{J}_2$ should be in $\mathcal{J}_1$.) Hence the flow across C is maximal, i.e., is $cap(C)$. The idea of the proof is realized.

$$\Omega$$

Exercise 5.8.4. How many cuts are possible for the network depicted in **Figure 5.8.1**? How many cuts are possible for the general network with junctions numbered $1, \ldots, N$?

Example 5.8.2. Let a network consist of a source labelled So, a sink labelled Si, m junctions labelled $X_1, \ldots, X_m$, and n other junctions labelled $Y_1, \ldots Y_n$. Assume

   i. there is a link of capacity 1 from So to each of the junctions $X_1, \ldots, X_m$;

   ii. there is a link of capacity 1 from each junction $Y_1, \ldots, , Y_n$ to the sink Si;

   iii. there is at least one link from some X_i to some Y_j and the capacity of such a link is m;

   iv. there is no link other than those described in *i-iii*.

Let $\mathcal{J}$ be the set of all junctions. The cut $(So, \mathcal{J} \setminus Si)$ has capacity m. If the maximal flow F_{max} is less than m there is a cut C for which $cap(C) < m$. After appropriate relabelling the cut C may be described as the pair in which $\{So; X_1, \ldots, X_p; Y_1, \ldots, Y_q\}$ is the half containing So.

The capacity of the cut $(So, \mathcal{J} \setminus So)$ is m whence $1 \leq p$.

The capacity of the cut one half of which contains $\{So, X_1, \ldots X_m\}$ is at least m whence $p < m$

The capacity of the cut one half of which contains $\{Si, Y_1, \ldots, Y_n\}$ is at least m whence $1 \leq q$.

Hence the capacity of C is the sum of $m - p$ (the capacity of the links from So to the junctions $X_{p+1}, \ldots, X_m$) and q (the capacity of the links from $Y_1, \ldots, Y_q$ to Si): $cap(C) = m - p + q < m$. Thus $q < p$.

Since $cap(C) < m$ there is no link between any of the p junctions $X_1, \ldots, X_p$ and any of the junctions $Y_{q+1}, \ldots, Y_n$. Hence the only links between the p junctions $X_1, \ldots, X_p$ and the junctions $Y_1, \ldots, Y_n$ can be with some of the $Y_1, \ldots, Y_q$. Since $p < q$ there can be no one-one pairing off or *assignment* of each X_i to some Y_j.

Let each X_i represent a girl, let each Y_j represent a boy, and let the existence of a link between a girl and a boy indicate their mutual compatibility. Then there is a one-one correspondence $i \mapsto j(i)$ such that $1 \leq i \leq m$ and each $Y_{j(i)}$ and X_i are compatible iff the maximal flow is m. The maximal flow is m iff the minimal cut capacity is m iff there is no set of p girls for which there are fewer than p compatible boys.

The minmax network flow result may be rephrased as:

    Chapter 5. APPLYING LINEAR ALGEBRA

THE MARRIAGE THEOREM: THERE IS A MATCHING UP OF EACH GIRL
TO SOME BOY WITH WHOM SHE IS COMPATIBLE IFF FOR EACH SET OF
p GIRLS THERE ARE AT LEAST p COMPATIBLE BOYS.

Let each X_i represent a person, each Y_j represent a job, and let the existence
of a link between a person and a job indicate that the person is qualified for the
job. Then there is an *assignment* giving each person a job for which that person
is qualified iff the maximal flow is m iff there is no set of p persons qualified for
fewer than p jobs.

The minmax network flow result may be rephrased also as:

THE ASSIGNMENT THEOREM: THERE IS A COMPLETE ASSIGNMENT
OF EACH PERSON TO A JOB FOR WHICH THE PERSON IS QUALIFIED
IFF FOR EACH SET OF p PERSONS THERE ARE AT LEAST p JOBS FOR
WHICH THEY ARE QUALIFIED.

An interesting application of the MARRIAGE/ASSIGNMENT THEOREM is the
proof of the result of Frobenius (cf. **Exercise 3.3.10**).

THE FROBENIUS THEOREM: LET $A \stackrel{\text{def}}{=} (a_{ij})_{i,j=1}^{n,n}$ BE AN $n \times n$ MATRIX. IN
THE CLASSICAL FORMULA FOR $det(A)$, i.e.,

$$
det\left[\begin{pmatrix} a_{11} & \ldots & a_{1n} \\ \vdots & \ddots & \vdots \\ a_{n1} & \ldots & a_{nn} \end{pmatrix}\right] = \sum_{\pi \in S_{1,2,\ldots,n}} sgn(\pi) a_{1\pi(1)} a_{2\pi(2)} \cdots a_{n\pi(n)}, \qquad (5.8.4)
$$

(cf. (3.2.4)) EVERY TERM IS 0 IFF THERE IS A k SUCH THAT $1 \leq k \leq n$ AND
THERE IS IN A A $k \times n - k + 1$ SUBMATRIX $O_{k \times n-k+1}$ (IN WHICH EACH ENTRY
IS 0).

The proof that each term in (5.8.4) is 0 if there is a submatrix $O_{k \times n-k+1}$ is
the burden of **Exercise 3.3.10**. The converse is proved as follows.

A set T of n entries in a matrix $A \stackrel{\text{def}}{=} (a_{ij})_{i,j=1}^{n,n}$ is a *transversal* iff there is in
T an entry from each row and from each column of A. Thus the right member of
the formula (5.8.4) may be described as a sum of $n!$ terms, each a signed product
of the entries in a transversal of A.

For each matrix $A \stackrel{\text{def}}{=} (a_{ij})_{i,j=1}^{n,n}$ there is an associated marriage
problem:

Let there be n girls and n boys such that girl i is compatible with boy
j iff $a_{ij} \neq 0$. Is there a matching up of girls with compatible boys so
that each girl has a husband?

THEOREM 5.8.2. THE MARRIAGE PROBLEM FOR THE n GIRLS AND n BOYS
HAS A SOLUTION IFF THERE IS IN A A TRANSVERSAL CONSISTING ENTIRELY OF
NONZERO ENTRIES.

PROOF. If there is a one-one matching up of compatible boys and girls, say
girl i with boy $j(i)$, then the set $T \stackrel{\text{def}}{=} \{a_{1j(1)}, \ldots, a_{nj(n)}\}$ is a transversal of M.

Conversely, if T above is a transversal then the matching up $i \mapsto j(i)$ provides a solution to the marriage problem.

$$\Omega$$

Consequently, the marriage problem has *no* solution iff every transversal in A contains at least one 0, i.e., iff each term of (5.8.4) is 0. But the marriage problem has no solution iff there is a set of k girls for which there are fewer than k compatible boys. Thus it may be assumed that the k girls are numbered $1, \ldots, k$. If the set of boys compatible with the k girls are numbered $1, \ldots, q$ then $q \le k - 1$ and the $k \times n - k + 1$ submatrix $O_{k \times n - k + 1}$ is in the northeast corner of A. The assumption that the k girls and q boys are numbered in the manner assumed is equivalent to the statement that $O_{k \times n - k + 1}$ is *some* submatrix of A, i.e., that the rows and columns of A can be permuted to place $O_{k \times n - k + 1}$ in the northeast corner, i.e., that the result of Frobenius is valid.

Exercise 5.8.5. Assume $A \overset{\text{def}}{=} (a_{ij})_{i,j=1}^{m,n}$. Formulate a useful definition of a transversal in A when $m \ne n$. Then formulate a corresponding Marriage/Assignment problem in terms of transversals in A. Discuss the situations: $m < n$ and $m > n$.

THE TRAVELLING SALESMAN PROBLEM

Assume that there are cities numbered $1, \ldots, N$, that any pair ij of cities is connected by a straight road of length d_{ij}, and that a travelling salesman starting at city 1 must visit each city and then return home. What is the shortest trip he can make?

Exercise 5.8.6. Formulate the travelling salesman problem as a **PLPP**. Then formulate the **DLPP**.

Exercise 5.8.7. How many different circuits are available for the travelling salesman? (The answer suggests why a solution by exhaustion, i.e., by listing all circuits and choosing the shortest one, is impractical.)

A problem related to the travelling salesman problem is the *shortest distance problem*:

Assume only that some pairs (not necessarily all pairs) of cities are connected by straight roads but that each pair is connected through the network of roads. For each pair ij of cities what is the shortest road distance between them?

Exercise 5.8.8. Formulate the shortest distance problem as a **PLPP**. Formulate the **DLPP**.

For the transportation problem, the network flow problem, and the shortest distance problem the simplex or Karmarkar algorithms applied to their associated **PLPP**s prove to be cumbersome and impractical. In [**ZA**] there are descriptions of algorithms that deal more efficiently with these problems. The travelling salesman problem is more intractable. The writer is unaware of any algorithm, save exhaustion, which is impractical, for solving it.

5.9. Markov chains

In early discussions of probability theory the context is that of a finite *sample space* $S \stackrel{\text{def}}{=} \{s_1, \ldots, s_n\}$ consisting of *states* or *events*. Associated with each state s_i is a nonnegative number p_i, the *probability* of (being in) state s_i. It is assumed that $\sum_{i=1}^{n} p_i = 1$, whence $0 \leq p_i \leq 1$. The $n \times 1$ vector

$$\mathbf{p} \stackrel{\text{def}}{=} \begin{pmatrix} p_1 \\ \vdots \\ p_n \end{pmatrix}$$

is the *state vector*.

Example 5.9.1. In the study of weather in a given city S might well consist of four states:

$$\{s_1 \stackrel{\text{def}}{=} \text{clear}, s_2 \stackrel{\text{def}}{=} \text{cloudy}, s_3 \stackrel{\text{def}}{=} \text{rainy}, s_4 \stackrel{\text{def}}{=} \text{snowy}\}.$$

Then at any time and in the absence of other information about weather at previous times, p_1 is the probability of clear weather, $\ldots$, p_4 is the probability of snowy weather.

A more refined approach to weather study takes into account the weather today in predicting the weather tomorrow. Hence to each pair ij there is attached a *transition probability* p_{ij}, $0 \leq p_{ij} \leq 1$, the probability that if the state (weather) today is s_i then the state (weather) tomorrow will be s_j. Hence for weather there is a 4×4 *transition matrix* $P \stackrel{\text{def}}{=} (p_{ij})_{i,j=1}^{4,4}$. Furthermore since certainly *some* weather will be observed tomorrow no matter what the weather is today it follows that

$$\sum_{j=1}^{4} p_{ij} = 1, \ 1 \leq i \leq 4.$$

More generally if $S \stackrel{\text{def}}{=} \{s_1, \ldots, s_n\}$ there is a *transition matrix* or *stochastic matrix* $P \stackrel{\text{def}}{=} (p_{ij})_{i,j=1}^{n,n}$ such that

$$0 \leq p_{ij} \leq 1, \ \sum_{j=1}^{n} p_{ij} = 1, \ 1 \leq i \leq n.$$

A simplifying assumption about the weather may be phrased as follows:

MARKOV ASSUMPTION: If $\mathbf{p}$ is the state vector today, $\mathbf{p}^t P$ is the state vector for tomorrow, $\ldots$, $\mathbf{p}^t P^m$ is the state vector for the day m days from today.

Example 5.9.2. If

$$\mathbf{p}^t = (.25, .5, .125, .125)$$

then one-quarter of the time it is clear, one-half the time it is cloudy, one-eighth of the time it is rainy, and one-eighth of the time it is snowy. Under the MARKOV ASSUMPTION the chance of rain tomorrow is:

$$\mathbf{p}^t \text{ (third column of } P) = .25p_{13} + .5p_{23} + .125p33 + .125p_{43}.$$

The formula above corresponds to the *assumption* that the chance of rain tomorrow when it is clear today is $.25p_{13}$, $\ldots$, the chance of rain tomorrow when it is snowy today is $.125p_{43}$. Hence regardless of the weather today the chance of rain tomorrow is the sum of the four chances just calculated.

The calculation is based on the intuitively appealing "multiplication rule for probabilities." Although the rule seems to be reasonable it does not always apply.

For example, one might argue that after 30 successive days of rain the chance of a change in weather is greater than it is after only one day of rain. The MARKOV ASSUMPTION is equivalent to the assumption that only today's weather has anything to do with tomorrow's.

A more persuasive argument against the indiscriminate use of the multiplication rule is the following. Assume there are four states, $s_1, \ldots, s_4$ with respective probabilities $.5, .25, .125, .125$. The probability of (being in) s_1 *or* s_2 is $.75 = .5 + .25$. The probability of (being in) s_2 *or* s_3 is $.375 = .25 + .125$. The probability of (being in) s_1 *or* s_2 *and* of (being in) s_2 *or* s_3 is just $.25$, i.e., the probability of (being in) s_2. On the other hand,

$$.25 < (.75) \times (.375) = .28125.$$

Thus the MARKOV ASSUMPTION that the weather on day $m + 1$ depends only on the state vector on day 1 and on the mth power P^m of the transition matrix P is quite restrictive.

     Chapter 5. APPLYING LINEAR ALGEBRA

A Markov chain in simplest terms defined by a state vector p, a transition matrix P and the sequence

$$\mathbf{p}^t P, \mathbf{p}^t P^2, \ldots$$

of row vectors.

Example 5.9.3. There is a Markovian aspect to **Example 1.1.2.** Suppose that $x_1(n), \ldots x_4(n)$ represent the relative *proportions* of plant matter, $\ldots$, water at the beginning of the nth period. Hence it may be assumed that

$$x_1(n) + \cdots + x_4(n) = 1.$$

Similarly L_{ij} may be regarded as the fractional part of matter labelled j that is converted to matter labelled i. Then

$$\sum_{i=1}^{4} L_{ij} = 1, \ 1 \le j \le 4.$$

In that case $\mathcal{L}^t$ is a transition matrix and

$$X_{n+1}^t = X_n^t \mathcal{L}^t = X_1^t (\mathcal{L}^t)^n.$$

THEOREM 5.9.1. ASSUME P IS A TRANSITION MATRIX. LET e BE $\sum_{i=1}^{n} \mathbf{e}_i$ (THE COLUMN VECTOR IN WHICH EACH COMPONENT IS 1). THEN:

  i. $P\mathbf{e} = \mathbf{e}$;
  ii. THE NUMBER 1 IS AN EIGENVALUE OF P;
  iii. IF λ IS AN EIGENVALUE OF P THEN $1 \ge |\lambda|$.

PROOF. The condition $\sum_{j=1}^{n} p_{ij} = 1$ implies *i* and *ii*.

ad*iii.* If λ is an eigenvalue and if $\mathbf{x} \stackrel{\text{def}}{=} (x_1, \ldots, x_n)^t$ is a corresponding eigenvector then, since each $p_{ij} \ge 0$,

$$\sum_{j=1}^{n} p_{ij} x_j = \lambda x_i$$

$$\sum_{j=1}^{n} p_{ij} |x_j| \ge \left| \sum_{j=1}^{n} p_{ij} x_j \right| = |\lambda| \cdot |x_i|. \tag{5.9.1}$$

If

$$|x_M| \stackrel{\text{def}}{=} maximum_{1 \leq i \leq n} |x_i|$$

then $|x_M| > 0$ since, as an eigenvector, $\mathbf{x} \neq \mathbf{O}$. The averaging principle and (5.9.1) imply

$$|x_M| \geq \sum_{j=1}^{n} p_{Mj} |x_j| \geq |\lambda| |x_M|$$

whence $|\lambda| \leq 1$.

$$\Omega$$

Exercise 5.9.1. Show that if P is a transition matrix and $k \in \mathbf{N}$ then P^k is also a transition matrix.

Exercise 5.9.2. Show that if every entry of an $n \times n$ matrix M is positive and if P is an $n \times n$ transition matrix then every entry of PM is also positive (and hence if $k \in \mathbf{N}$ then every entry of $P^k M$ is positive).

To motivate the next and fundamental result about Markov chains consider again the weather model above.

One might the guess that even if one does not know $\mathbf{p}$, knowing P one might calculate a limiting state vector $\mathbf{p}_\infty$. In other words, starting with any state vector $\mathbf{p}$, one might be assured that if m is large then $\mathbf{p}^t P^m$ is close to $\mathbf{p}_\infty$. One may view $\mathbf{p}_\infty$ as a state vector toward which all other state vectors converge when they are subjected to repeated applications of P.

Hence if one studies a locale where the state vector $\mathbf{p}$ is given and if one knows the transition matrix P one imagines that the long-term state vector $\mathbf{p}_\infty$ can be calculated.

The state vector $\mathbf{p}$ represents the average weather as observed in the past. The transition matrix P tells how the average weather is changing. In some circumstances the successive state vectors $\mathbf{p}^t P^m$ approach a final state vector $\mathbf{p}_\infty$, and $\mathbf{p}_\infty$ *is independent of* $\mathbf{p}$. The transition matrix P is a settling influence on the weather: knowing P one can bet on the weather and know the odds.

Example 5.9.4. If there are only two kinds of weather, clear or rainy and if the transition matrix is

$$P \stackrel{\text{def}}{=} \begin{pmatrix} \frac{1}{2} & \frac{1}{2} \\ \frac{1}{2} & \frac{1}{2} \end{pmatrix}$$

one might well guess that regardless of today's weather the chance for clear weather and the chance for rainy weather tomorrow are forever both $\frac{1}{2}$: $p_1 = p_2 = \frac{1}{2}$.

On the other hand if the transition matrix is

$$Q \stackrel{\text{def}}{=} \begin{pmatrix} 0 & 1 \\ 1 & 0 \end{pmatrix}$$

---

the chance for clear weather and the chance for rainy weather tomorrow are both $\frac{1}{2}$ if nothing is known about the current weather. But if it is clear today it will (certainly, i.e., with probability 1) rain tomorrow and if it is rainy today it will (certainly, i.e., with probability 1) be clear tomorrow. This time if the state vector

$$\mathbf{p} \stackrel{\text{def}}{=} \begin{pmatrix} \frac{1}{3} \\ \frac{2}{3} \end{pmatrix} \text{ then } \mathbf{p}^t Q^m = \begin{cases} \left(\frac{2}{3}, \frac{1}{3}\right) & \text{if } m \text{ is odd} \\ \left(\frac{1}{3}, \frac{2}{3}\right) & \text{if } m \text{ is even.} \end{cases}$$

More generally, if the state vector is

$$\mathbf{p} \stackrel{\text{def}}{=} \begin{pmatrix} p_1 \\ p_2 \end{pmatrix} \text{ then } \mathbf{p}^t Q^m = \begin{cases} (p_2, p_1) & \text{if } m \text{ is odd} \\ (p_1, p_2) & \text{if } m \text{ is even.} \end{cases}$$

There are no fixed odds against which to bet.

What is "wrong" with the transition matrix Q? The minimum value of all entries in all powers Q^k of Q is 0. DEFINITION **5.9.1** and THEOREM **5.9.2** show how the absence of a 0 entry in some power of a transition matrix stabilizes matters.

DEFINITION **5.9.1**. A TRANSITION MATRIX P IS *regular* IFF EACH ENTRY OF P IS POSITIVE.

THEOREM **5.9.2**. LET P BE AN $n \times n$ TRANSITION MATRIX SUCH THAT FOR SOME NATURAL NUMBER K EACH ENTRY IN P^K IS *positive*, i.e., P^K IS REGULAR.. LET P^k BE $(p_{ij}^{(k)})_{i,j=1}^{n,n}$, $1 \leq k < \infty$. THEN:

i. IF $k \geq K$ EVERY ENTRY IN P^k IS POSITIVE;

ii. THERE IS A TRANSITION MATRIX $P^{(\infty)} \stackrel{\text{def}}{=} (p_{ij}^{(\infty)})_{i,j=1}^{n,n}$ SUCH THAT

$$\lim_{k \to \infty} p_{ij}^{(k)} = p_{ij}^{(\infty)}, \ 1 \leq i, j \leq n, \ \text{i.e.,} \ \lim_{k \to \infty} P^k = P^{(\infty)};$$

iii. ALL ROWS OF $P^{(\infty)}$ ARE THE SAME AND EACH ENTRY IS POSITIVE, i.e., e DENOTING THE COLUMN VECTOR IN WHICH EVERY COMPONENT IS 1, THERE IS A UNIQUE ROW VECTOR $\pi \stackrel{\text{def}}{=} (\pi_1, \ldots, \pi_n)$ SUCH THAT

$$P^{(\infty)} = \mathbf{e}\pi = \begin{pmatrix} \pi_1 & \cdots & \pi_n \\ \vdots & \ddots & \vdots \\ \pi_1 & \cdots & \pi_n \end{pmatrix},$$

i.e., $p_{ij}^{(\infty)} = \pi_j$, $1 \leq i, j, \leq n, \pi \succ \mathbf{O}$;

iv. FOR ANY STATE VECTOR $\mathbf{q} \stackrel{\text{def}}{=} (q_1, \ldots, q_n)^t \lim_{k \to \infty} \mathbf{q}^t P^k = \pi.$

[**Note 5.9.1:** The limit $P^{(\infty)}$ can well exist even if no P^k is regular. For example if $P = I$ then $P^k = I$, $k \in \mathbf{N}$, no P^k is regular , and yet $P^{(\infty)}$ exists and is P $(= I)$. Furthermore, although when some P^k is regular the rank of $P^{(\infty)}$ is 1, the rank of $I^{(\infty)}$ is full.]

PROOF. ad$i$. See **Exercise 5.9.2.**
adii. If $\mathbf{x} \stackrel{\text{def}}{=} (x_1, \ldots, x_n)^t \in \mathbf{R}^n$ and if

$$m \stackrel{\text{def}}{=} minimum_{1 \le i \le n} x_i, \quad M \stackrel{\text{def}}{=} maximum_{1 \le i \le n} x_i$$

$$P^k \mathbf{x} \stackrel{\text{def}}{=} \begin{pmatrix} x_1^{(k)} \\ \vdots \\ x_n^{(k)} \end{pmatrix}$$

$$m^{(k)} \stackrel{\text{def}}{=} minimum_{1 \le i \le n} x_i^{(k)}$$

$$M^{(k)} \stackrel{\text{def}}{=} maximum_{1 \le i \le n} x_i^{(k)}$$

then by the *averaging principle* explained in **Section 5.4**

$$m \le m^{(1)} \le \cdots \le m^{(k)} \le M^{(k)} \le \cdots \le M^{(1)} \le M$$
$$M^{(k+1)} - m^{(k+1)} \le M^{(k)} - m^{(k)}, \; k \in \mathbf{N}.$$

Let μ_i be the minimum entry in row i of P^K and let μ be the minimum of all μ_i:

$$\mu_i \stackrel{\text{def}}{=} minimum_{1 \le j \le n} p_{ij}^{(K)} \stackrel{\text{def}}{=} p_{ij_0}^{(K)}, \; 1 \le i \le n$$

$$\mu \stackrel{\text{def}}{=} minimum_{1 \le i \le n} \mu_i.$$

Then the hypothesis implies that $\mu > 0$. Hence

$$x_i^{(K)} \le (1 - p_{ij_0}^{(K)})M + p_{ij_0}^{(K)}m, \; 1 \le i \le n$$

$$maximum_{1 \le i \le n} x_i^{(K)} = M^{(K)}$$

$$\le (1 - p_{ij_0}^{(K)})M + p_{ij_0}^{(K)}m, \; 1 \le i \le n$$

Since $p_{ij_0}^{(K)} \ge \mu$ it follows that

$$M^{(K)} \le M - \mu(M - m). \tag{5.9.2}$$

If $\mathbf{x}$ is replaced by $\mathbf{y} \stackrel{\text{def}}{=} -\mathbf{x}$ the rôles of *maximum* resp. *minimum* are taken by $-minimum$ and $-maximum$ and (5.9.2) becomes

$$-m^{(K)} \le -m - \mu(-m + M)$$

whence $M^{(K)} - m^{(K)} \le (1 - 2\mu)(M - m)$.

     Chapter 5. APPLYING LINEAR ALGEBRA

For any natural number N it follows that

$$M^{(NK)} - m^{(NK)} \leq (1 - 2\mu)^N (M - m).$$

Each P^k is a transition matrix and $\mu > 0$ whence $0 < \mu < 1$ and so $|1 - 2\mu| < 1$. Since

$$M^{(k+1)} - m^{(k+1)} \leq M^{(k)} - m^{(k)}, \ k \in \mathbf{N}$$

it follows that $\lim_{k \to \infty} M^{(k)} - m^{(k)} = 0$.

If $\mathbf{x} = \mathbf{e}_j$ then $M = 1$ and $m = 0$. Furthermore $P^k \mathbf{e}_j$ is the jth column of P^k. Thus the maximum and the minimum of the entries in the jth column of $P^k \mathbf{e}_j$ approach a common limit π_j as $k \to \infty$. In other words

$$\lim_{k \to \infty} P^k \mathbf{e}_j = \pi_j \mathbf{e}$$

$$\lim_{k \to \infty} p_{ij}^{(k)} = \pi_j, \ 1 \leq i \leq n$$

$$p_{ij}^{(\infty)} = \pi_j, \ 1 \leq i \leq n$$

$$P^{(\infty)} = \begin{pmatrix} p_{11}^{(\infty)} & \cdots & p_{1n}^{(\infty)} \\ \vdots & \ddots & \vdots \\ p_{n1}^{(\infty)} & \cdots & p_{nn}^{(\infty)} \end{pmatrix}.$$

Since

$$\sum_{j=1}^{n} p_{ij} = 1, \ 1 \leq i \leq 1$$

it follows that $P^{(\infty)}$ is a transition matrix and thus that

$$\sum_{j=1}^{n} \pi_j = 1.$$

adiii. Since $0 < \mu$ each component of $P^K \mathbf{e}_j$ is at least μ and the averaging principle implies that for any l in $\mathbf{N}$ each component of $P^{K+l} \mathbf{e}_j$ is at least μ and so $\pi_j \geq \mu$: all entries in $P^{(\infty)}$ are positive.

iv. If $\mathbf{q}$ is a state vector then $\lim_{k \to \infty} \mathbf{q}^t P^k = \mathbf{q}^t P^{(\infty)}$. The jth component of the $1 \times n$ vector $\mathbf{q}^t P^{(\infty)}$ is simply

$$\mathbf{q}^t \begin{pmatrix} \pi_j \\ \vdots \\ \pi_j \end{pmatrix}.$$

According to the averaging principle the jth component is therefore π_j: $\lim_{k \to \infty} \mathbf{q}^t P^k = (\pi_1, \ldots, \pi_n) = \pi$. In particular, $\pi P^{(\infty)} = \pi$. If $\mathbf{q}$ is a state vector and $\mathbf{q}^t P^{(\infty)} = \mathbf{q}^t$ then $\lim_{k \to \infty} \mathbf{q}^t P^k = \mathbf{q}^t P^{(\infty)} = \pi$ whence $\mathbf{q}^t = \pi$ and so the uniqueness assertion is proved.

$$\Omega$$

THEOREM **5.9.2** may be interpreted for **Example 5.9.3** as saying that if each L_{ij} is positive then no matter what the proportions of plant matter, …, water at any given time, those proportions tend to limiting proportions, i.e., stable proportions, say $\pi_1, \ldots, \pi_4$, as life goes on.

Exercise 5.9.3. Show that if $P^{(\infty)}$ exists and $k \in \mathbf{N}$ then

$$P^k P^{(\infty)} = P^{(\infty)} P^k = P^{(\infty)}.$$

Exercise 5.9.4. Let P be the transition matrix

$$\begin{pmatrix} \frac{1}{3} & \frac{2}{3} \\ \frac{3}{4} & \frac{1}{4} \end{pmatrix}.$$

i. Find the eigenvalues of P and a Jordan normal form $\mathcal{P}$ for P.

ii. Find an invertible matrix M such that

$$M^{-1} P M = \mathcal{P}.$$

iii. Use the Jordan normal form to find $P^{(\infty)}$.

Exercise 5.9.5. Show that if P is a regular transition matrix and if J is a Jordan block for the eigenvalue 1 in a Jordan normal form $\mathcal{P}$ of P then $J = I$.

What follows sharpens the statement in **Exercise 5.9.5**.

Example 5.9.5. Every permutation matrix Π is a transition matrix. If $\Pi \overset{\text{def}}{=} (p_{ij})_{i,j=1}^{n,n}$ looks like this:

$$\begin{pmatrix} 0 & 1 & 0 & \\ 0 & 0 & 1 & \\ \vdots & \vdots & \vdots & \\ 1 & 0 & & \end{pmatrix}$$

then Π is a *circulant matrix*. It corresponds to a Markov chain in which there is certainty of changing from state s_m to state s_{m+1} if $0 \le m \le n-1$ and of changing from state s_n to state s_1: there is circulation

$$s_1 \longrightarrow s_2 \longrightarrow \cdots \longrightarrow s_{n-1} \longrightarrow s_n \longrightarrow s_1$$

among the states. If $n > 2$ then Π is not symmetric but, as a special kind of permutation matrix, $\Pi^k = I$ for some k. In fact, $m_\Pi(z) = z^n - 1$ and so the eigenvalues of Π are

$$e^{i\frac{2\pi}{n}m}, \ 0 \le m \le n-1.$$

    Chapter 5. APPLYING LINEAR ALGEBRA

If $n > 2$ a circulant matrix is an example of a transition matrix for which some eigenvalues are not real.

THEOREM 5.9.3. LET P BE AN $n \times n$ TRANSITION MATRIX AND LET THE NOT NECESSARILY DISTINCT ZEROS OF χ_P BE $\lambda_1, \ldots, \lambda_n$ ARRANGED SO THAT

$$|\lambda_1| \geq |\lambda_2| \geq \cdots \geq |\lambda_n|.$$

IF SOME POWER P^k IS REGULAR THEN $\lambda_1 = 1 > |\lambda_2|$, i.e., IF SOME POWER P^k IS REGULAR THEN THE EIGENVALUE 1 IS A SIMPLE ZERO OF χ_P AND ANY EIGENVALUE λ OTHER THAN 1 HAS ABSOLUTE VALUE LESS THAN 1: $|\lambda| < 1$.

The object of **Exercises 5.9.6-5.9.12** is to provide the basic ingredients of a proof of THEOREM **5.9.3**.

Part of the idea in the proof is that if λ is an eigenvalue, $\lambda \neq 1$, and $|\lambda| = 1$ then for one reason or another

$$\lim_{k \to \infty} P^k$$

does not exist. If $\mathcal{P}$ is a Jordan normal form for P it may be assumed that there is a natural number k, a $k \times k$ identity matrix I_k, and an $(n-k) \times (n-k)$ diagonal block matrix B consisting of Jordan blocks in each of which the diagonal entries are equal to some eigenvalue other than 1, hence an eigenvalue having absolute value not more than 1.

$$\mathcal{P} = \begin{array}{c} \\ k \\ n-k \end{array} \overset{\begin{array}{cc} k & n-k \end{array}}{\begin{pmatrix} I_k & O \\ O & B \end{pmatrix}}.$$

In **Exercises 5.9.6-5.9.12** P is a transition matrix such that for some natural number K, P^K is regular.

Exercise 5.9.6. If $|\lambda| = 1$ there is a real number θ such that

$$0 \leq \theta < 1, \text{ and } \lambda = e^{i2\pi\theta}.$$

Show that if $\theta \neq 0$ then every Jordan block for which each diagonal entry is λ is a diagonal matrix.

[*Hint:* Use THEOREM **5.9.1** and the discussion in and following **Example 4.7.2**. Examine the $1n$ entry of the powers J^m of J.]

Exercise 5.9.7. Show that if θ is rational ($\theta \in \mathbf{Q}$) then $\theta = 0$, i.e., $\lambda = 1$. [*Hint:* Examine the diagonal entries of $\mathcal{P}$. Use THEOREM **5.9.1**.]

If x is any real number then $[x]$ is by definition the largest integer not greater than x, e.g., $[3] = 3, [2.5] = 2, [\pi] = 3, [-.5] = -1$. Correspondingly,

---

$$\{x\} \overset{\text{def}}{=} x - [x], \text{ e.g.,}$$

$$\{3\} = 0, \{2.5\} = .5, \{\pi\} = .141597\ldots, \{-.5\} = .5.$$

Thus $0 \leq \{x\} < 1$ and if $\phi \in \mathbf{R}$ then

$$e^{i2\pi\phi} = \begin{cases} e^{i2\pi\{\phi\}} & \text{if } \phi \geq 0 \\ e^{i2\pi(1-\{\phi\})} & \text{if } \phi < 0. \end{cases}$$

Exercise 5.9.8. Show that if $0 < \theta < 1$ and $\theta \notin \mathbf{Q}$, i.e., θ is *irrational*, then for any t in the interval $[0, 1]$ there is a sequence $\{n_k : 1 \leq k < \infty\}$ of natural numbers such that

$$\lim_{k \to \infty} \{n_k \theta\} = t.$$

[*Hint:*

i. There is a least natural number n such that:

$$\frac{1}{n} < \theta < \frac{2}{n},$$

and then $\{n\theta\} < \theta$.

ii. For n in $\mathbf{N}$ let δ_n denote $\{n\theta\}$ and let δ be the greatest lower bound of the set Δ of all numbers δ_n:

$$\delta = g.l.b. \{ \delta_n : \delta_n \in \Delta \} = infimum \{ \{n\theta\} : n \in \mathbf{N} \}.$$

If $q \in \mathbf{N}$ then $\{q\delta_n\} \in \Delta$. If $0 < \delta$ and $\delta \in \mathbf{Q}$ then for p and q in $\mathbf{N}$

$$\delta = \frac{p}{q}, \quad \{q\delta\} = 0,$$

and there is a sequence $\{\delta_{n_k}\}_{k=1}^{\infty}$ such that

$$\delta_{n_1} \geq \delta_{n_2} \geq \cdots, \quad \delta_{n_k} \downarrow \delta.$$

Then

$$q\delta_{n_k} \downarrow q\delta, \quad \{q\delta_{n_k}\} \downarrow \{q\delta\} = 0, \quad \delta = 0,$$

a contradiction.

If $0 < \delta$ and $\delta \notin \mathbf{Q}$ then (see i) there is in $\mathbf{N}$ an N such that

$$\frac{1}{N} < \delta < \frac{2}{N}, \quad 1 < N\delta < 2, \quad \{N\delta\} < \delta,$$

a contradiction. Ineluctably $\delta = 0$.

---

iii. If $t = 0$ the previous result shows there is a sequence $\{\delta_{n_k}\}_{k=1}^{\infty}$ such that $\delta_{n_k} \downarrow t$. If $0 < t \leq 1$ and if $k \in \mathbb{N}$ there is in Δ a δ_{m_k} such that $\delta_{m_k} < \frac{1}{k}$ and then one, say δ_k', of the numbers

$$\delta_{m_k}, 2\delta_{m_k}, \ldots,$$

is in $[0, 1] \cap \Delta$ and $|t - \delta_k'| < \frac{1}{k}$:

$$\lim_{k \to \infty} \delta_k' = t.]$$

[**Remark 5.9.1:** The following valid group theoretic statements together constitute an alternative argument for **Exercise 5.9.8.**

i. The set $\{\, \{n\theta\} \ : \ n \in \mathbb{N} \,\} \overset{\text{def}}{=} G$ is an infinite subgroup of $[0, 1)$ regarded as a group **I** with respect to "addition modulo 1."

ii. Every infinite subgroup of **I** is dense in **I**.

Actually if θ is irrational a stronger result is valid: if $0 \leq \alpha < \beta \leq 1$ and if $\nu_n([\alpha, \beta])$ is the number of numbers

$$\{\theta\}, \ldots, \{n\theta\}$$

in $[\alpha, \beta]$ then

$$\lim_{n \to \infty} \frac{\nu_n([\alpha, \beta])}{n} = \beta - \alpha.$$

This and similar results about the *equidistribution* of sequences

$$\{x_n\}_{n=1}^{\infty}$$

lying in $[0, 1]$ are due to theorems of Hermann Weyl. More about this subject can be found in the first volume of [**N**].]

Exercise 5.9.9. Show that if λ is an eigenvalue of P and $|\lambda| = 1$ then $\lambda = 1$.

Exercise 5.9.10. Show that if λ is an eigenvalue, $|\lambda| < 1$, and J is a corresponding Jordan block then

$$\lim_{k \to \infty} J^k = O,$$

i.e., if $J^k \overset{\text{def}}{=} (\alpha_{ij}^{(k)})_{i,j=1}^{n,n}$ then

$$\lim_{k \to \infty} \alpha_{ij}^{(k)} = 0, \ 1 \leq i, j \leq n.$$

Exercise 5.9.11. Use the results of **Exercises 5.9.6-5.9.10** and the existence of a Jordan normal form $\mathcal{P}$ for P to provide another proof of THEOREM **5.9.2**.

Exercise 5.9.12. Show that there is an invertible matrix M such that $M^{-1}PM \overset{\text{def}}{=} \mathcal{P}$ is a Jordan form of P and

$$
M^{-1}P^{(\infty)}M = \begin{matrix} & & 1 & n-1 \\ & 1 & \\ & n-1 \end{matrix} \begin{pmatrix} 1 & \\ & O \end{pmatrix}.
$$

[*Hint:* See THEOREM **5.9.2**iii and consider the rank of $P^{(\infty)}$.]

At this point the conclusions in **Exercises 5.9.6-5.9.12** constitute a proof of THEOREM **5.9.3**.

Exercise 5.9.13. Show that if

$$
P = \begin{pmatrix} .5 & .5 & 0 \\ .5 & 0 & .5 \\ 0 & .5 & .5 \end{pmatrix}
$$

then P^K is regular for some natural number K. Find the least K for which P^K is regular. Use the Jordan normal form $\mathcal{P}$ of P to find $P^{(\infty)}$.

Exercise 5.9.14. If the states $s_1, \ldots, s_n$ are imagined to be points with coordinates $1, \ldots n$ in $\mathbf{R}$ then p_{ij} may be viewed as the probability of moving from s_i to s_j and P describes a *random walk* on $\mathbf{R}$. Assume that for some p in $(0,1)$

$$
p_{ij} = \begin{cases} p & \text{if } 1 < i < n \text{ and } j = i+1 \\ 1 - p \overset{\text{def}}{=} q & \text{if } 1 < i < n \text{ and } j = i-1 \\ 0 & \text{if } i = 1 \text{ and } j > 1 \text{ or } i = n \text{ and } j < n. \end{cases}
$$

Show the general form of P and give a verbal description of what goes on in the random walk. The condition

$$
p_{12} = p_{n,n-1} = 0 \text{ if } i = 1 \text{ and } j > 1 \text{ or } i = n \text{ and } j < n
$$

is interpreted as describing the states s_1 and s_n as *absorbing barriers* for the random walk. Explain this terminology. Assume $n = 3$. Is P^K regular for some natural number K? Does

$$
\lim_{k \to \infty} P^k
$$

exist? If it exists what is it?

Exercise 5.9.15. If, for a random walk, the hypothesis is

$$p_{ij} = \begin{cases} p > 0 & \text{if } 1 \le i < n \text{ and } j = i+1 \\ 1 - p \overset{\text{def}}{=} q > 0 & \text{if } 1 < i \le n \text{ and } j = i-1 \end{cases}$$

the random walk is described as having reflective barriers. Explain the terminology. Assume $n = 3$. Is P^K regular for some natural number K? Does

$$\lim_{k \to \infty} P^k$$

exist? If it exists what is it?

Exercise 5.9.16. Define p_{ij} so that for the corresponding random walk the boundary s_1 is reflecting and the boundary s_n is absorbing. Assume $n = 3$. Is P^K regular for some natural number K? Does

$$\lim_{k \to \infty} P^k$$

exist? If it exists what is it?

It appears that if, for a transition matrix P, some of the statements listed below obtain then they are relevant to the existence of $P^{(\infty)}$:

i. there is in $\mathbf{N}$ some K such that P^K is regular;
ii. if $\lambda_1, \ldots, \lambda_n$ are the not necessarily distinct eigenvalues of P and they are listed so that $|\lambda_1| \ge |\lambda_2| \ge \cdots \ge |\lambda_n|$ then $|\lambda_1| > |\lambda_2|$ (hence, in particular, $\lambda_1 = 1$);

Another important property of *any* matrix A in which each entry is non-negative is *decomposability* (or its absence, *indecomposability*).

DEFINITION 5.9.2. IF A IS A SQUARE MATRIX A IS *decomposable* IFF FOR SOME PERMUTATION MATRIX Π THERE ARE SQUARE MATRICES A_{11} AND A_{22} SUCH THAT

$$\Pi A \Pi^t = \begin{pmatrix} A_{11} & A_{12} \\ O & A_{22} \end{pmatrix} \overset{\text{def}}{=} A_\Pi. \tag{5.9.3}$$

IN OTHER WORDS, BY PERMUTING ROWS AND COLUMNS OF A *in the same way* ONE CAN PRODUCE THE MATRIX A_Π. A MATRIX THAT IS NOT DECOMPOSABLE IS *indecomposable*.

Example 5.9.6. An upper triangular matrix is decomposable. On the other hand the circulant matrix

$$Q \overset{\text{def}}{=} \begin{pmatrix} 0 & 1 \\ 1 & 0 \end{pmatrix}$$

---

is indecomposable.

Exercise 5.9.17. Show that if Π_n is the circulant $n \times n$ matrix then

$$\Pi_n = E_{1n} E_{n,n-1} \cdots E_{32}$$

and Π_n is indecomposable.

[*Hint:* Show that if Π_n is decomposable then $m_{\Pi_n} = (z^k - 1)(z^l - 1)$ for some k and l in $\mathbf{N}$.]

If a transition matrix P is decomposable then for no k in $\mathbf{N}$ is P^k regular, i.e., for each k some entry in P^k is 0. Indeed, if P is decomposable then there is a permutation matrix Π such that $P_\Pi \overset{\text{def}}{=} \Pi P \Pi^t$ has the form in (5.9.3). Since Π is a product of ROW EXCHANGE matrices E_{ij} and since $E_{ij} = E_{ij}$ it follows that $\Pi^t = \Pi^{-1}$. Hence block multiplication implies that for some matrix A'_{12} (depending on k)

$$\Pi P^k \Pi = \begin{pmatrix} A_{11}^k & A'_{12} \\ O & A_{22}^k \end{pmatrix}.$$

Hence for no k is P^k regular.

Thus if P is a transition matrix such that P^K is regular for some K then P is indecomposable.

Exercise 5.9.18. Determine whether

$$B \overset{\text{def}}{=} \begin{pmatrix} 0 & 0 & 2 \\ 0 & 2 & 0 \\ 2 & 0 & 0 \end{pmatrix}$$

is indecomposable. If there is in $\mathbf{N}$ a k such that A^k is decomposable is A itself necessarily decomposable?

The conclusions of **Exercises 5.9.19-5.9.23** below constitute a proof of THEOREM **5.9.4** given after them. The goal is to relate indecomposability and ultimate regularity for a transition matrix. The latter implies the former and a little bit more. That little bit more and indecomposability imply ultimate regularity.

Throughout P is a transition matrix and its not necessarily distinct eigenvalues $\lambda_1, \ldots, \lambda_n$ are listed so that $\lambda_1 = 1 \geq |\lambda_2| \geq \cdots \geq |\lambda_n|$.

Exercise 5.9.19. Show that if Π is a permutation matrix and $R \overset{\text{def}}{=} \Pi P \Pi^t$ then

$$R^{(\infty)} \text{ exists iff } P^{(\infty)} \text{ exists}$$

$$\text{if } R^{(\infty)} \text{ exists then } R^{(\infty)} = \Pi P^{(\infty)} \Pi^t$$

$$R \text{ is indecomposable iff } P \text{ is indecomposable.}$$

Exercise 5.9.20. Show that If A is indecomposable then the column vector **O** is not one of the columns of A and the row vector **O** is not one of the rows of A.

Exercise 5.9.21. Show that if $1 > |\lambda_2|$ then $P^{(\infty)}$ exists.

[*Hint:* Consider the Jordan normal form of P.]

Exercise 5.9.22. Show that if P is *not* ultimately regular and $1 > |\lambda_2|$ then at least one column of $P^{(\infty)}$ is **O**.

Exercise 5.9.23. Show that if at least one column of $P^{(\infty)}$ is **O** there is a permutation matrix Π such that the first column of $\Pi P^{(\infty)} \Pi$ is **O**. Show further that if $R \overset{\mathrm{def}}{=} \Pi P \Pi$ then the first column of $R^{(\infty)}$ is **O**.

[*Hint:* Show that $P^{(\infty)} \neq O$. Let m be the number of columns in $P^{(\infty)}$ that are **O**'s. Show that there is a permutation matrix Π such that $R \overset{\mathrm{def}}{=} \Pi P \Pi$ has the block form

$$
\begin{array}{c}
\\
m \\
n - m
\end{array}
\begin{array}{c}
\begin{array}{cc} m & n-m \end{array} \\
\begin{pmatrix} O & A_{12} \\ O & A_{22} \end{pmatrix}
\end{array} .]
$$

THEOREM **5.9.4.** LET P BE A TRANSITION MATRIX. ASSUME THAT THE NOT NECESSARILY DISTINCT EIGENVALUES OF P ARE $\lambda_1, \ldots, \lambda_n$ LISTED SO THAT

$$
1 = \lambda_1 \geq |\lambda_2| \geq \cdots \geq |\lambda_n|.
$$

THEN:

P IS ULTIMATELY REGULAR IFF P IS INDECOMPOSABLE AND $1 > |\lambda_2|$.

PROOF. See THEOREM **5.9.3** and **Exercises 5.9.19-5.9.23**.

Exercise 5.9.24. Give an example of a transition matrix P that is indecomposable but not ultimately regular.

Exercise 5.9.25. Show that every SQUARE matrix is similar to a decomposable matrix.

For a more detailed discussion of Markov chains see [**KeS**], [**Ros**].

It should be noted that some of the theory of indecomposable transition matrices can be extended to arbitrary indecomposable matrices in which all entries are nonnegative. The work goes back to Perron, Frobenius, and, somewhat later, Wielandt [**Wi**]. In [**Ros**] there is an exposition of much of the Wielandt approach.

Exercise 5.9.26. Use the averaging principle to show that if $\mathbf{O} \neq \mathbf{x} \succeq \mathbf{O}$ and if $\frac{a}{0}$ is interpreted as ∞ whenever $a > 0$ then for the transition matrix $P \overset{\text{def}}{=} (p_{ij})_{i,j=1}^{n,n}$

$$1 = maximum_{\mathbf{O} \neq \mathbf{x} \succeq \mathbf{O}} \, minimum_{1 \leq i \leq n} \frac{\sum_{j=1}^{n} p_{ij} x_j}{x_i}$$

$$= maximum_{\mathbf{x} \succ \mathbf{O}} \, minimum_{1 \leq i \leq n} \frac{\sum_{j=1}^{n} p_{ij} x_j}{x_i}.$$

[**Note 5.9.2:** The P-eigenvalue of largest absolute value is always 1. Part of the Wielandt argument contains the assertion that if $A \overset{\text{def}}{=} (a_{ij})_{i,j=1}^{n,n}$ is an indecomposable matrix with nonnegative entries then

$$\lambda_1 \overset{\text{def}}{=} maximum_{\mathbf{O} \neq \mathbf{x} \succeq \mathbf{O}} \, minimum_{1 \leq i \leq n} \frac{\sum_{j=1}^{n} a_{ij} x_j}{x_i}.$$

(a *maxmin*) is a finite positive number and is in fact the A-eigenvalue of largest absolute value. Furthermore:

i. λ_1 is the only eigenvalue for which a corresponding eigenvector is nonnegative;
ii. λ_1 is a simple zero of the polynomial χ_A;
iii. the eigenspace corresponding to λ_1 is one-dimensional;
iv. if $\lambda_1, \ldots, \lambda_n$ are the not necessarily distinct eigenvalues of A and if they are listed so that $\lambda_1 \geq |\lambda_2| \geq \cdots \geq |\lambda_n|$ then A is ultimately regular iff $\lambda_1 > |\lambda_2|$.

Each of *i-iv* above is a generalization to indecomposable nonnegative matrices of results derived earlier for indecomposable transition matrices.

A circulant matrix is an example of an indecomposable matrix such that each of its eigenvalues has absolute value 1. For every decomposable matrix A with nonnegative entries if λ_m, $2 \leq m \leq k$ are the eigenvalues such that $|\lambda_m| = \lambda_1$ they may be enumerated so that $\lambda_m = \lambda_1 e^{i \frac{2\pi}{k}(m-1)}$, $2 \leq m \leq k$. Then A is not ultimately regular and indeed, after a suitable permutation of both its rows and columns, A is

---

a ("circulant") block matrix of the form

$$
\begin{pmatrix}
O & A_{12} & O & \\
O & O & A_{23} & O \\
\vdots & \vdots & \vdots & \\
A_{n1} & O & &
\end{pmatrix}
$$

in which the diagonal blocks are SQUARE.]

PART III

THEORY

CHAPTER 6

GENERAL VECTOR SPACES

6.1. Axioms and definitions

In Section 1.1 among **THE LAWS OF MATRIX ALGEBRA** are the following:

$A + B = B + A$ (addition is commutative);

$A + (B + C) = (A + B) + C$ (addition is associative);

$O + A = A + O = A$;

$r(A + B) = rA + rB$;

$r(sA) = (rs)A$;

$(r + s)A = rA + sA$.

If column vectors or row vectors are regarded as special kinds of matrices the laws above apply to them. These laws lead to an axiomatic approach to vector spaces defined over *fields*. Fields as well as far more general algebraic entities are defined in terms of the rules for combining elements of these entities with other elements of the same entities or of other entities. For example, in a vector space there are rules for vector addition, involving the combination of two vectors and rules for multiplication by scalars (scalar multiplication), involving the combination of a number and a vector. It is helpful to formulate such rules in terms of *Cartesian products* as defined next.

DEFINITION **6.1.1.** IF S AND T ARE SETS THEIR CARTESIAN PRODUCT, DENOTED $S \times T$, IS THE SET OF ALL PAIRS $\{s,t\}$ CONSISTING OF AN s IN S AND A t IN T:

$$S \times T \stackrel{\text{def}}{=} \{ \{s,t\} \ : \ s \in S, \ t \in T \}.$$

Definition **6.1.2.** A *field* is a set **K** in which there are defined two operations:

$$+ : \mathbf{K} \times \mathbf{K} \ni \{a, b\} \mapsto a + b \in \mathbf{K} \text{ ADDITION}$$
$$: \mathbf{K} \times \mathbf{K} \ni \{a, b\} \mapsto ab \in \mathbf{K} \text{ MULTIPLICATION.}$$

Addition and multiplication are commutative and associative operations related by the *distributive law*

$$a(b + c) = ab + ac.$$

There are in **K** two special elements denoted 0 and 1 and such that for each a in **K**: $1a = a$ and $0 + a = a$. Finally:

for each a in **K** there is in **K** a b such that $a + b = 0$;
if $a \neq 0$ there is in **K** a b such that $ab = 1$.

The sets **R**, **C**, **Q** are fields. Other examples of fields are:

i. the set of all rational functions (quotients of polynomials with coefficients in a field **K**);
ii. the set $\{0, 1\}$ in which

$$0 + 0 = 0, 0 + 1 = 1, 1 + 1 = 0;$$
$$0 \cdot 0 = 0, 0 \cdot 1 = 0, 1 \cdot 1 = 1.$$

iii. for a prime number p the set $\{0, 1, \ldots, p - 1\}$ in which, by ambiguous abuse of "+",

$$a + b = \begin{cases} a + b & \text{if } a + b \leq p - 1 \\ a + b - p & \text{if } a + b > p - 1. \end{cases}$$
$$ab = \begin{cases} ab & \text{if } ab \leq p - 1 \\ r & \text{if } ab = qp + r, \ q \in \mathbf{N}, \ 0 \leq r < p. \end{cases}$$

This field is frequently denoted $\mathbf{Z}/(p)$ or $\mathbf{Z}$ mod p. The field in *ii* is $\mathbf{Z}/(2)$.

Exercise 6.1.1. Show that in a field **K**

i. if $a + b = a$ for all a then $b = 0$; if $ba = a$ for all a then $b = 1$;
ii. for each a in **K** there is only one b such that $a + b = 0$ (the unique *additive inverse* for a is denoted $-a$: $a + (-a) = 0$;)
iii. if $a \neq 0$ there is only one b such that $ba = 1$ (if $a \neq 0$ the unique *multiplicative inverse* for a is denoted a^{-1} or $1/a$: $aa^{-1} = a(1/a) = 1$;)
iv. $(-1)a = -a$;
v. $0a = 0$.

DEFINITION 6.1.3. LET **K** BE A FIELD AND LET

$$V \overset{\text{def}}{=} \{\mathbf{x}, \mathbf{y}, \ldots\}$$

BE A SET. THEN V IS A *vector space over* **K** IF THERE ARE DEFINED TWO MAPS

$$+ : V \times V \ni \{\mathbf{x}, \mathbf{y}\} \mapsto \mathbf{x} + \mathbf{y} \in V \text{ ADDITION}$$
$$: \mathbf{K} \times V \ni \{a, \mathbf{x}\} \mapsto a\mathbf{x} \in V \text{ SCALAR MULTIPLICATION.}$$

THE RULES FOR ADDITION ARE:

$\mathbf{x} + \mathbf{y} = \mathbf{y} + \mathbf{x}$ (ADDITION IS COMMUTATIVE);
$\mathbf{x} + (\mathbf{y} + \mathbf{z}) = (\mathbf{x} + \mathbf{y}) + \mathbf{z}$ (ADDITION IS ASSOCIATIVE).

THERE IS IN V AN ELEMENT **O** SUCH THAT FOR EVERY **v** IN V

$$\mathbf{O} + \mathbf{v} = \mathbf{v} \ (= \mathbf{v} + \mathbf{O}).$$

FOR EACH **v** IN V THERE IS IN V AN ELEMENT **w** SUCH THAT

$$\mathbf{v} + \mathbf{w} = \mathbf{O}.$$

THE ELEMENTS OF **K** ARE COMBINED WITH THE ELEMENTS OF V ACCORDING TO THE FOLLOWING RULES:

$r(\mathbf{x} + \mathbf{y}) = r\mathbf{x} + r\mathbf{y};$
$r(s\mathbf{x}) = (rs)\mathbf{x};$
$(r + s)\mathbf{x} = r\mathbf{x} + s\mathbf{x};$
$1\mathbf{x} = \mathbf{x}.$

THE ELEMENTS OF A VECTOR SPACE ARE CALLED *vectors*.

Exercise 6.1.2. Show that in a vector space V

i. if $\mathbf{x} + \mathbf{v} = \mathbf{v}$ for every **v** in V then $\mathbf{x} = \mathbf{O}$;
ii. if $\mathbf{v} + \mathbf{w} = \mathbf{v} + \mathbf{w}'$ then $\mathbf{w} = \mathbf{w}'$ and hence that the additive inverse of each **v** is unique (the unique additive inverse of **v** is denoted $-\mathbf{v}$);
iii. $(-1)\mathbf{v} = -\mathbf{v}$;
iv. $0\mathbf{v} = \mathbf{O}$.

Examples of vector spaces abound and some are given next. In each instance the operations of vector addition and multiplication of vectors by numbers are the standard and familiar operations associated with the example in question.

i. $\mathbf{R}^n$ (a vector space over **R**);
ii. $\mathbf{C}^n$ (a vector space over **C** and also over **R**);

---

iii. the set of all polynomials having coefficients in $\mathbf{C}$ (a vector space over $\mathbf{C}$ and also over $\mathbf{R}$);

iv. the set of all polynomials having coefficients in $\mathbf{C}$ and of degree not exceeding n (a vector space over $\mathbf{C}$ and also over $\mathbf{R}$);

v. the set of all $\mathbf{C}$-valued functions defined on $\mathbf{R}$ (a vector space over $\mathbf{C}$ and also over $\mathbf{R}$);

vi. the set of all $\mathbf{R}$-valued *Riemann integrable* functions defined on $[0,1]$ (a vector space over $\mathbf{R}$);

vii. the set
$$\left\{ \sum_{k=0}^{n}(a_k \cos kx + b_k \sin kx) \ : \ a_k, b_k \in \mathbf{C}, \ n \in \mathbf{N} \right\};$$

of all *trigonometric polynomials* (a vector space over $\mathbf{C}$ and also over $\mathbf{R}$);

viii. the set Mat_{mn} of all $m \times n$ matrices; (If the entries in the matrices are all in some field $\mathbf{K}$ then Mat_{mn} is a vector space over $\mathbf{K}$.)

ix. the set of all "arrows" having their tails at the origin in $\mathbf{R}^3$ and combined according "vector addition" via the "head to tail" (parallelogram) rule and scalar multiplication via appropriate stretching, shrinking, and direction reversal (a vector space over $\mathbf{R}$).

Assume that $\mathbf{K}$ is a finite field. There are only finitely many field elements in the set

$$1, 1+1, 1+1+1, \ldots .$$

Let m denote $\sum_{i=1}^{m} 1$ viewed as a element in $\mathbf{N}$. Hence there are in $\mathbf{N}$ two numbers k and l such that:

i. $k < l$ in $\mathbf{N}$;

ii. $k = l$ in $\mathbf{K}$;

iii with respect to i and ii k and l are minimal.

Then $l - k = 0$ in $\mathbf{K}$. If the natural number $l - k$ has two prime factors, say p and q, then in $\mathbf{N}$: $l - k = pqr$ and $1 < p, q, \ 1 \le r < l - k$. Since $pqr = 0$ in $\mathbf{K}$ it follows that in $\mathbf{K}$ $p = 0$, $q = 0$, or $r = 0$.

If $p = 0$ in $\mathbf{K}$ then $k - p < l$ in $\mathbf{N}$ and $k - p = k$ in $\mathbf{K}$, a contradiction of the minimality of k.

If $q = 0$ in $\mathbf{K}$ then $k - q = k$ in $\mathbf{K}$, a contradiction of the minimality of both k and l.

If $r = 0$ in $\mathbf{K}$ then in $\mathbf{N}$ $r = l - k - s$ for some s in $\mathbf{N}$ whence $k = l - s$ in $\mathbf{K}$ a contradiction of the minimality of l.

Hence $l - k$ is a prime, say p. Thus $\mathbf{K} \supset \mathbf{Z}/(p)$ and $\mathbf{K}$ is a vector space over $\mathbf{Z}/(p)$. Since $\mathbf{K}$ is finite there is in $\mathbf{K}$ a maximal linearly independent set $X \stackrel{\text{def}}{=} \{\mathbf{x}_1, \ldots, \mathbf{x}_s\}$, i.e., a basis for $\mathbf{K}$.

Let $X \overset{\text{def}}{=} \{\mathbf{x}_1, \ldots, \mathbf{x}_s\}$ be a basis for $\mathbf{K}$. Hence every element of $\mathbf{K}$ is a unique linear combination

$$\sum_{i=1}^{s} a_i \mathbf{x}_i, \ a_i \in \mathbf{Z}/(p), \ 1 \leq i \leq s$$

and so the number of elements in $\mathbf{K}$ is $p^s \overset{\text{def}}{=} q$.

Exercise 6.1.3. Show that in the preceding context if $x \in \mathbf{K}$ then:

i. there is a natural number m such that $m \leq q$ and $x^m = x$. If $x \neq 0$ then $x^{m-1} = 1$, i.e., every nonzero element of $\mathbf{K}$ is a root of unity;

ii. if $x \neq 0$ and $\{y, z\} \subset \mathbf{K} \setminus \{0\} \overset{\text{def}}{=} G$ the sets

$$\{y, yx, \ldots, yx^{m-2}\} \text{ and } \{z, zx, \ldots, zx^{m-2}\}$$

are either the same or disjoint;

iii. if $x \neq 0$ then

$$G = \bigcup_{y \in G} \{y, yx^2, \ldots, yx^{m-2}\};$$

iv. the number $q - 1$ of elements in G is divisible by $m - 1$;

v. if $x \neq 0$ then $x^{q-1} = 1$.

[**Remark 6.1.1:** The set G above is the *multiplicative group* of nonzero elements of $\mathbf{K}$. The statement

$$q - 1 \text{ is divisible by } m - 1$$

is a special case of a classical theorem in the theory of groups:

If F is a finite group and N is a normal subgroup of F then the number $\#(F)$ of elements in F is divisible by the number $\#(N)$ of elements in N. In the special case above $\{1, x, \ldots, x^{m-2}\} \overset{\text{def}}{=} N$ is a normal subgroup of the group G.]

If V is a vector space and if $W \subset V$ then W is a *subspace* iff W is also a vector space with respect to the vector operations of V.

Exercise 6.1.4. Refer to **Remark 4.3.1**. Assume that V is a vector space over a finite field $\mathbf{K}$, e.g., for p a prime, $\mathbf{Z}/(p)$. Let k be the number of elements in $\mathbf{K}$: $k = \#(\mathbf{K})$. Let $W_1, \ldots W_J$ be subspaces of V and assume that:

i. no W_j is contained in the union of the other $W_{j'}$:

$$W_j \not\subset \bigcup_{j' \neq j} W_{j'}, \ 1 \leq j \leq J;$$

ii. $k - 2 > J$.

Show that if

$$V = \bigcup_{j=1}^{J} W_j$$

then one of the W_j is V.

6.2. Linear independence, bases

The notions of *linear combination, linear dependence, linear independence,* and *basis,* as described in **Section 2.8** make sense in any vector space.

Whether a given vector space *has* a basis is a subtle question. It seems to go to the very foundations of mathematics and to be intimately tied to such questions as the acceptability of the *Axiom of Choice* and its numerous equivalents such as the *Well-ordering Principle, Zorn's Lemma,* etc., (cf. [**Me**]). This text uses an equivalent formulation developed next.

DEFINITION 6.2.1. IF S IS A SET IT IS *ordered* IFF THERE IS DEFINED IN $S \times S$ A *binary relation* OR *order* (DENOTED $<$) SUCH THAT:

i. IF $a, b \in S$ THEN $a = b$, $a < b$, OR $b < a$ AND THE THREE POSSIBILITIES ARE MUTUALLY EXCLUSIVE;

ii. IF $a < b$ AND $b < c$ THEN $a < c$ (THE ORDER IS *transitive*).

DEFINITION 6.2.2. A SET S IS *specially ordered* IFF S IS ORDERED AND

i. IF $T \subset S$ THERE IS IN T A *first* ELEMENT, i.e., THERE IS IN T AN ELEMENT t_0 SUCH THAT FOR EVERY t IN T: $t_0 = t$ OR $t_0 < t$; ("Every subset of S has a first element.")

ii. IF S IS NOT FINITE, i.e., IF S CAN*not* BE PUT INTO ONE-ONE CORRESPONDENCE WITH SOME SET OF THE FORM $\{1, \ldots, n\}$, AND IF $s \in S$ THERE IS IN S AN s' SUCH THAT $s < s'$; ("If S is not finite there is no 'last' element in S.")

iii. IF $s \in S$ THEN S AND $S' \stackrel{\text{def}}{=} \{ s' \ : \ s' < s \}$ CANNOT BE PUT IN ONE-ONE CORRESPONDENCE. ("$\#(S') < \#(S)$.")

Example 6.2.1.

i. The finite set $\{1, \ldots, n\}$ with its usual order is specially ordered.

ii. The set $\mathbf{N}$ of natural numbers with its usual order is specially ordered.

iii. The set $\mathbf{Z}$ of all integers with its usual order is *not* specially ordered ($\mathbf{Z}$ itself has no first element).

iv. The subset $(0, 1)$ of $\mathbf{R}$ with its usual order is *not* specially ordered.

v. The set
$$\left\{1, \frac{1}{2}, \frac{1}{3}, \ldots, \frac{1}{n}, \ldots\right\}$$

with its order derived from the usual order of **R** is *not* specially ordered while in the order presented the set *is* specially ordered.

vi. The set
$$\left\{0, \frac{1}{2}, \frac{3}{4}, \frac{7}{8}, \ldots, (1 - 2^{-n}), \ldots\right\}$$

is specially ordered with its order derived from the usual order of **R**. That order is the same as the order in which the set is presented.

vii. The set
$$\left\{1, 0, \frac{1}{2}, \frac{3}{4}, \frac{7}{8}, \ldots, (1 - 2^{-n}), \ldots\right\}$$

is *not* specially ordered with its order derived from that in **R** but *is* specially ordered as presented.

One form of the Well-ordering Principle (Axiom) is the **Special Ordering Principle (SOP)**:

Every set S can be specially ordered.

Thus the Well-ordering Principle implies that even though in their usual orders neither **Z** nor **R** is specially ordered there is for each of them *some* order with respect to which it is specially ordered. For example, the presentation

$$\{0, 1, -1, 2, -2, \ldots\}$$

of **Z** induces an "unnatural" yet special order among the elements of **Z**. Indeed, any set that is one-one correspondence with **N** may be specially ordered via the one-one correspondence with **N**. There is no known special ordering of **R** even though the **SOP** implies that such a special ordering of **R** exists.

THEOREM 6.2.1. IF V IS A VECTOR SPACE THEN THERE IS IN V A SPE-CIALLY ORDERED BASIS.

PROOF.   Let S be $V \setminus O$. Then there is for S a binary relation $<_1$ with respect to which S is specially ordered. Thus there is a specially ordered set Λ and a one-one correspondence

$$\Lambda \leftrightarrow S$$

between the elements of Λ and the vectors in S. In other words, the vectors in S may be *indexed* by the elements in Λ:

$$S = \{\mathbf{x}_\lambda \,:\, \lambda \in \Lambda\}.$$

The indexing is such that $\lambda < \lambda'$ iff $\mathbf{x}_\lambda <_1 \mathbf{x}_{\lambda'}$.

By *transfinite induction* define as follows a set $\{B_\lambda\}$ of subsets of S.

---

i. Let $\mathbf{x}_1$ be the first element of S and let B_1 be the singleton set $\{\mathbf{x}_1\}$. If $\mathbf{x}_2$ is the least element of $S \setminus B_1$ let B_2 be

$$\begin{cases} B_1 & \text{if } \mathbf{x}_2 \in span(B_1) \\ \{\mathbf{x}_2\} \cup B_1 & \text{otherwise.} \end{cases}$$

ii. Having defined B_λ for all λ such that $\mathbf{x}_\lambda <_1 \mathbf{x}_{\lambda'}$ define $B_{\lambda'}$ as

$$\bigcup_{\mathbf{x}_\lambda <_1 \mathbf{x}_{\lambda'}} B_\lambda \text{ if } \mathbf{x}_{\lambda'} \in span\left(\bigcup_{\mathbf{x}_\lambda <_1 \mathbf{x}_{\lambda'}} B_\lambda\right)$$

$$\{\mathbf{x}_{\lambda'}\} \cup \left(\bigcup_{\mathbf{x}_\lambda <_1 \mathbf{x}_{\lambda'}} B_\lambda\right) \text{ if } \mathbf{x}_{\lambda'} \notin span\left(\bigcup_{\mathbf{x}_\lambda <_1 \mathbf{x}_{\lambda'}} B_\lambda\right).$$

(Hence $\mathbf{x}_\lambda$ is ALWAYS in $span(B_\lambda)$.)

iii. Define B as

$$\bigcup_{\mathbf{x}_\lambda \in S} B_\lambda.$$

The set B is a basis for V in the sense that every vector $\mathbf{x}$ in V is a *unique finite linear combination of vectors in B*. Here is the reasoning.

iv. If $\mathbf{x}_\lambda <_1 \mathbf{x}_{\lambda'}$ then $B_\lambda \subset B_{\lambda'}$. [*Proof:* See *ii.*]

v. If $\{\mathbf{x}_{\lambda_1}, \ldots, \mathbf{x}_{\lambda_m}\} \subset B$ it may be assumed that $\mathbf{x}_{\lambda_1} <_1 \cdots <_1 \mathbf{x}_{\lambda_m}$ and then

$$\mathbf{x}_{\lambda_m} \notin span\left(\bigcup_{\mathbf{x}_\lambda <_1 \mathbf{x}_{\lambda_m}} B_\lambda\right).$$

[*Proof.* See *ii.*] Hence B is linearly independent.

vi. If $span(B) \neq V$ let $\mathbf{x}_{\lambda'}$ be the first vector not in $span(B)$ Then

$$\mathbf{x}_{\lambda'} \in span(B_{\lambda'}) \subset span(B),$$

a contradiction.

From *v* and *vi* it follows that every vector in V is a unique linear combination of the vectors in B, i.e., B is a basis.

$$\Omega$$

Exercise 6.2.1. Show that for the sets B_λ in the argument above if $\lambda \neq \lambda'$ then $B_\lambda \subset B_{\lambda'}$ or $B_{\lambda'} \subset B_\lambda$.

THEOREM 6.2.2. IF V IS A VECTOR SPACE AND IF X AND Y ARE BASES FOR V THEN THERE IS A ONE-ONE CORRESPONDENCE BETWEEN THE VECTORS IN X AND THE VECTORS IN Y.

[**Remark 6.2.1:**   Theorem **6.2.2** is a generalization of Theorem **2.9.1**. As the next lines show, the same technique is used to prove both results. Since there are no finiteness restrictions in Theorem **6.2.2** the **SOP** and its consequence Theorem **6.2.1** (sometimes called *Hamel's Theorem*) are invoked. The basis in Hamel's Theorem is a *Hamel basis*.]

Proof.   The **SOP** implies that both X and Y can be specially ordered. Let X resp. Y be specially ordered with respect to $<_X$ resp. $<_Y$ and let X resp. Y be

$$\{\, \mathbf{x}_\lambda \; : \; \lambda \in \Lambda \,\}$$

resp.

$$\{\, \mathbf{y}_\phi \; : \; \phi \in \Phi \,\}$$

Let $\mathbf{x}_1$ and $\mathbf{y}_1$ be the first vector in X resp. Y and define $\mathbf{y}_1$ to be $F(\mathbf{x}_1)$. If $F(\mathbf{x}_\lambda)$ is defined for all λ preceding λ' let $F(\mathbf{x}_{\lambda'})$ be

(a) undefined if $Y_{\phi(\lambda')} \overset{\text{def}}{=} \{\, F(\mathbf{x}_\lambda) \; : \; \mathbf{x}_\lambda <_X \mathbf{x}_{\lambda'} \,\} = Y$

(b) the first vector in $Y \setminus Y_{\phi(\lambda')}$ otherwise.

On the set (in X) where the function F is defined it is one-one:

$$F(\mathbf{x}_\lambda) = F(\mathbf{x}_{\lambda'}) \text{ iff } \mathbf{x}_\lambda = \mathbf{x}_{\lambda'}.$$

If (a) above occurs for some λ' then F sets up a one-one correspondence between Y and a proper subset of X. If (a) above never occurs then F maps X in a one-one manner into or onto Y.

Assume (a) occurs for some λ'. Write each vector $\mathbf{x}_\lambda$ in X as a linear combination of finitely many vectors in Y:

$$\mathbf{x}_\lambda = \sum_{\phi \in \Phi} a_{\lambda\phi} \mathbf{y}_\phi, \ \lambda \in \Lambda. \tag{6.2.1}$$

Note that for each λ the sum in (6.2.1) is really a finite sum (all but finitely many of the $a_{\lambda\phi}$ are 0's).

Owing to the fact that X is specially ordered, a modification of the *PROCESS* may be applied to the system (6.2.1). The infinite matrix of numerical entries is

$$A \overset{\text{def}}{=} (a_{\lambda\phi})_{\lambda \in \Lambda, \phi \in \Phi} = \begin{matrix} & \begin{matrix} 1 & \cdots & \phi & \cdots \end{matrix} \\ \begin{matrix} 1 \\ \vdots \\ \lambda \\ \vdots \end{matrix} & \begin{pmatrix} a_{11} & \cdots & a_{1\lambda} & \cdots \\ \vdots & \ddots & \vdots & \ddots \\ a_{\lambda 1} & \cdots & a_{\lambda\phi} & \cdots \\ \vdots & \ddots & \vdots & \ddots \end{pmatrix} \end{matrix}.$$

If

$$\mathcal{X} \stackrel{\text{def}}{=} \begin{pmatrix} \mathbf{x}_1 \\ \vdots \\ \mathbf{x}_\lambda \\ \vdots \end{pmatrix},$$

a vectorial column vector in which the components are the vectors of the basis X and if

$$\mathcal{Y} \stackrel{\text{def}}{=} \begin{pmatrix} \mathbf{y}_1 \\ \vdots \\ \mathbf{y}_\phi \\ \vdots \end{pmatrix},$$

a vectorial vector in which the components are the vectors of the basis Y then the system (6.2.1) may be written

$$A\mathcal{Y} = \mathcal{X}.$$

Correspondingly the augmented matrix for the system (6.2.1) is $A_{aug} \stackrel{\text{def}}{=} A|\mathcal{X}$. In the first column of A_{aug} let λ_1 be the first index such that $a_{\lambda_1 1} \neq 0$. If $\lambda_1 < \lambda < \lambda'$ either $a_{\lambda 1} = 0$ or one can reduce $a_{\lambda 1}$ to 0 by ROW COMBINATION wherein row λ_1 is the active row and row λ is the affected row. Thus row λ_1 is the marked row and all rows below row λ_1 are unchanged or affected. (SEARCH/ELIMINATION is carried out on column 1 of A_{aug}.)

Assume $\phi' \in \Phi$ and that row-marking and corresponding SEARCH/ELIMINATION (S/E) are performed on all columns indexed by a ϕ preceding ϕ'. Perform S/E on column ϕ' itself. Since each row of A_{aug} contains only finitely many nonzero numerical entries each row of A_{aug} is affected only finitely many times. Consequently, when A_{aug} is PROCESSed to yield, say $A_E|\mathcal{X}_E$, each $\mathbf{x}_\lambda$ in $\mathcal{X}$ remains unchanged or is replaced by a finite linear combination of other vectors in X.

On the other hand, since (a) is assumed to occur, after all columns of A are handled row λ' of A_E is a row of 0's. Consequently the vector in row λ' of $\mathcal{X}_E$ is via the GEM a linear combination with nonzero coefficients

$$\sum_{i=1}^{m} a_i \mathbf{x}_{\lambda_i} \ \lambda_i < \lambda', \ 1 \leq i \leq m,$$

of vectors in the basis X and on the other hand is $\mathbf{O}$, in denial of the linear independence of X.

In sum:

If X and Y are two bases for V the situation (a) can NEVER occur.

Hence (b) occurs. If $F(X) \subsetneq Y$ there is definable a map $G : Y \mapsto X$ and a map $\gamma : \Phi \mapsto \Lambda$ such that

$$G(\mathbf{y}_{\phi'}) = \begin{cases} \text{(c) undefined if } X_{\gamma(\phi')} \stackrel{\text{def}}{=} \{ G(\mathbf{y}_\phi) \ : \ \mathbf{y}_\phi <_Y \mathbf{y}_{\phi'} \} = X \\ \text{(d) the first vector in } X \setminus X_{\gamma(\phi')} \text{ otherwise} \end{cases}$$

---

and if $\mathbf{y} = F(\mathbf{x})$ then $G(\mathbf{y}) = \mathbf{x}$.

Since (c) cannot occur for G it follows that the one-one map F is such that $F(X) = Y$.

$$\Omega$$

[**Remark 6.2.2:** What goes on in the discussions above is a disguised version of *transfinite induction* (cf. [**Me**]). A simplified way of describing how to demonstrate the existence of a basis in a vector space is to say:

> Run through the specially ordered set $V \setminus \mathbf{O}$ and for each vector $\mathbf{x}$ cast it out if it is in the span of all the vectors already chosen. Otherwise add $\mathbf{x}$ to the set of all vectors already chosen. Transfinite induction rigorizes, with the help of the **SOP**, the simplified description above.

It should be noted as well that the argument proving THEOREM **6.2.2** uses the transfinite version of the argument in the proof of THEOREM **2.9.1**: if (a) oçurs then A is transfinitely HIGH.

If an abstract vector space V has a finite basis X, i.e., a basis that is in one-one correspondence with some set $\{1, 2, \ldots, n\}$ of natural numbers, then V is *finite-dimensional*. Otherwise V is *infinite-dimensional*.]

The preceding discussion leads one to describe a matrix more generally as a doubly indexed set of numbers or of vectors etc. If Λ and Φ are two index sets and if for each pair λ, ϕ in $\Phi \times \Lambda$ there is defined a number $a_{\lambda\phi}$ then the matrix

$$A \overset{\text{def}}{=} (a_{\lambda\phi})_{\lambda \in \Lambda, \phi \in \Phi}$$

is well-defined. If there is a one-one correspondence between Λ and Φ then Λ and Φ may be taken as the same:

$$A = (a_{\lambda\lambda'})_{\lambda, \lambda' \in \Lambda}$$

and A is "SQUARE." If each row of A contains only finitely many nonzero entries (A is *row-finite*) and if

$$B \overset{\text{def}}{=} (b_{\lambda, \lambda'})_{\lambda, \lambda' \in \Lambda}$$

is any other matrix then the product C of A and B may be defined according to the formula

$$C \overset{\text{def}}{=} (c_{\rho\sigma})_{\rho, \sigma \in \Lambda} \overset{\text{def}}{=} \left(\sum_{\lambda \in \Lambda} a_{\rho\lambda} b_{\lambda\sigma} \right)_{\rho, \sigma \in \Lambda}$$

However BA is not readily definable unless B is also row-finite. In particular, the row-finiteness of A permits the definition of powers A^k, $1 \le k < \infty$, of A and polynomials $p(A)$, etc.

Exercise 6.2.2. Show that if A is row-finite then in accordance with the definition of matrix product polynomials $p(A)$ are well-defined.

Exercise 6.2.3. Interpret the statement: A matrix is a function of two variables.

Exercise 6.2.4. Let V be a vector space and let W be a subspace of V. Let $X \stackrel{\text{def}}{=} \{\, x_\lambda \ : \ \lambda \in \Lambda \,\}$ be a Hamel basis for W. Use the proof of THEOREM **6.2.1** as a model for a procedure for filling X out to a basis Y for V.

6.3. Linear transformations and matrices

If V and W are vector spaces over a field $\mathbf{K}$ and $T : V \ni \mathbf{v} \mapsto T\mathbf{v} \stackrel{\text{def}}{=} \mathbf{w} \in W$ is a map then T is *linear* iff for α_1 and α_2 in $\mathbf{K}$ and $\mathbf{v}_1$ and $\mathbf{v}_2$ in V

$$T(\alpha_1 \mathbf{v}_1 + \alpha_2 \mathbf{v}_2) = \alpha_1 T\mathbf{v}_1 + \alpha_2 T\mathbf{v}_2.$$

Such a map is a *linear transformation* and

$$T(\mathbf{O}) = \mathbf{O}, \ T(-\mathbf{v}) = -T(\mathbf{v}), \ \text{etc.}$$

The zero map

$$O : V \ni \mathbf{x} \mapsto \mathbf{O} \in W$$

is linear. If $V = W$ the identity map

$$I : V \ni \mathbf{x} \mapsto \mathbf{x} \in V$$

is also linear.

The set of all linear maps from V to W is denoted $\mathcal{L}(V, W)$. If $V = W$ then $\mathcal{L}(V, W) = \mathcal{L}(V, V)$ and is denoted $\mathcal{L}(V)$.

A linear map $T : V \mapsto W$ is *invertible* iff there is a linear map $S : W \mapsto V$ such that $ST : V \mapsto V$ and $TS : W \mapsto W$ are the identity maps.

If $R : V \mapsto W$ and $T : V \mapsto W$ are linear maps and $a, b \in \mathbf{K}$ then $aR + bT$ is a linear transformation according to the natural rule

$$(aR + bT)\mathbf{x} = aR\mathbf{x} + bT\mathbf{x}.$$

If U, V, and W are vector spaces over $\mathbf{K}$ and if $T : U \mapsto V$ and $S : V \mapsto W$ are linear maps then their composition

$$R \stackrel{\text{def}}{=} S \circ T : U \ni \mathbf{x} \mapsto S(T\mathbf{x}) \in W$$

is a linear map. It is usually denoted simply ST.

If $U = V = W$ then the set $\mathcal{L}(V)$ of all linear maps of V into itself has the structure of an *algebra* over $\mathbf{K}$. In other words there are three maps

$$+ : \mathcal{L}(V) \times \mathcal{L}(V) \ni \{S, T\} \mapsto S + T \in \mathcal{L}(V) \text{ addition}$$
$$: \mathcal{L}(V) \times \mathcal{L}(V) \ni \{S, T\} \mapsto ST \in \mathcal{L}(V) \text{ composition}$$
$$: \mathbf{K} \times \mathcal{L}(V) \ni \{a, T\} \mapsto aT \in \mathcal{L}(V) \text{ scalar multiplication}$$

that obey the following rules:

i. addition is commutative: $S + T = T + S$;
ii. the distributive law: $R(S + T) = RS + RT$;
iii. addition and composition are associative:

$$R + (S + T) = (R + S) + T, \ \ R(ST) = (RS)T;$$

iv. $O + T = T$, $OT = O$, $IT = T$;
v. $(ab)T = a(bT)$, $S(aT) = a(ST)$, $(a + b)T = aT + bT$;
vi. for each T there is an additive inverse $-T$ such that $T + (-T) = O$.

Exercise 6.3.1. Show that $(-1)T = -T$ and $0T = O$.

If $T : V \mapsto W$ is a linear map of vector spaces the notions *kernel, image, eigenvalue, eigenvector*, etc., for T are what they are for matrices. For example,

$$ker(T) \stackrel{\text{def}}{=} \{\, \mathbf{x} \ : \ T\mathbf{x} = \mathbf{O} \,\} \stackrel{\text{def}}{=} T^{-1}(\{\mathbf{O}\});$$
$$im(T) \stackrel{\text{def}}{=} \{\, \mathbf{y} \ : \ \mathbf{y} = T\mathbf{x} \text{ for some } \mathbf{x} \,\} \stackrel{\text{def}}{=} T(V).$$

An eigenvalue is a number λ such that for some nonzero vector $\mathbf{x}$: $T\mathbf{x} = \lambda\mathbf{x}$ and $\mathbf{x}$ is the corresponding eigenvector.

If $X \stackrel{\text{def}}{=} (\mathbf{x}_\lambda)_{\lambda \in \Lambda}$ resp. $Y \stackrel{\text{def}}{=} (\mathbf{y}_\phi)_{\phi \in \Phi}$ are bases for V resp. W then there is a unique row-finite matrix $T_{XY} \stackrel{\text{def}}{=} (t_{\lambda\phi})_{\lambda \in \Lambda, \phi \in \Phi}$ such that

$$T\mathbf{x}_\lambda = \sum_{\phi \in \Phi} t_{\lambda\phi}\mathbf{y}_\phi.$$

Conversely, if a row-finite matrix $A \stackrel{\text{def}}{=} (a_{\lambda\phi})_{\lambda \in \Lambda, \phi \in \Phi}$ is given and if

$$\mathbf{v} \stackrel{\text{def}}{=} \sum_{\lambda \in \Lambda} v_\lambda \mathbf{x}_\lambda$$

(a finite sum) one may define a linear mapping

$$S : V \ni \mathbf{v} \mapsto \sum_{\lambda \in \Lambda} v_\lambda \left(\sum_{\phi \in \Phi} a_{\lambda\phi}\mathbf{y}_\phi \right) \stackrel{\text{def}}{=} \mathbf{w} \in W.$$

Furthermore the procedure whereby T engenders the matrix T_{XY} engenders for S a matrix $S_{XY} \overset{\text{def}}{=} (s_{\lambda\phi})_{\lambda\in\Lambda,\phi\in\Phi}$. A direct calculation shows that $S_{XY} = A$.

On the other hand, the existence of the bases X resp. Y suggests that every vector in $\mathbf{v}$ resp. $\mathbf{w}$ in V resp. W may be identified with the finite set of coefficients used to express $\mathbf{v}$ resp. $\mathbf{w}$ in terms of the vectors in X resp. Y. In this way vectors in V resp. W are identified with certain functions defined on the index sets Λ resp. Φ. These functions are restricted in only one way: each is 0 at all but a finite number of indices. Thus there are the correspondences

$$V \ni \mathbf{v} \leftrightarrow \begin{pmatrix} \vdots \\ v_\lambda \\ \vdots \end{pmatrix} \overset{\text{def}}{=} \mathbf{v}_X$$

$$W \ni \mathbf{w} \leftrightarrow \begin{pmatrix} \vdots \\ w_\phi \\ \vdots \end{pmatrix} \overset{\text{def}}{=} \mathbf{w}_Y$$

between abstract vectors and *transfinite* numerical column vectors each of which contains only finitely many nonzero components. As in **Section 2.10** when V resp. W are construed as sets of *column-finite* transfinite column vectors there obtain relationships associated with matrix transposes:

$$T\mathbf{v} \overset{\text{def}}{=} \mathbf{w}$$
$$T_{XY}^t \mathbf{v}_X = (T\mathbf{v})_Y = \mathbf{w}_Y.$$

The transposes T_{XY}^t are *column-finite* matrices that are not necessarily row-finite. Since the column vectors are, by abuse of language, column- finite, there are no algebraic difficulties in forming $T_{XY}^t \mathbf{v}_X$.

[**Remark 6.3.1:** As shown above, set theoretic considerations lead to the existence of bases for arbitrary (abstract) vector spaces. Bases lead to the identification of abstract vector spaces with special kinds of *function spaces*. From that point it is a small step to looking at other special kinds of function spaces such as:

    i. the set $\mathbf{C}([0,1])$ of all continuous $\mathbf{C}$-valued functions defined on $[0,1]$;

    ii. the set l^2 of all $\mathbf{C}$-valued sequences $\{a_n\}_{n=1}^\infty$ such that

$$\sum_{n=1}^\infty |a_n|^2 < \infty;$$

    iii. the set $\widetilde{L}^2([0,1])$ of all $\mathbf{C}$-valued continuous functions f defined on $\mathbf{R}$ and such that

$$\int_{\mathbf{R}} |f(x)|^2 \, dx < \infty.$$

Much more on this topic can be found in [**Ber**].]

Exercise 6.3.2. Show that with respect to the natural definitions of addition and multiplication by numbers l^2 and $\widetilde{L}^2(\mathbf{R})$ are vector spaces.

If $X \stackrel{\text{def}}{=} \{\mathbf{x}_\lambda\}_{\lambda \in \Lambda}$ resp. $Y \stackrel{\text{def}}{=} \{\mathbf{y}_\mu\}_{\mu \in M}$ resp. $Z \stackrel{\text{def}}{=} \{\mathbf{z}_\nu\}_{\nu \in N}$ are bases for the vector spaces U resp. V resp. W there emerge according to the equations

$$T\mathbf{x}_\lambda = \sum_{\mu \in M} t_{\lambda\mu}\mathbf{y}_\mu, \ \lambda \in \Lambda$$

$$S\mathbf{y}_\mu = \sum_{\nu \in N} s_{\mu\nu}\mathbf{z}_\nu, \ \mu \in M$$

$$R\mathbf{x}_\lambda = \sum_{\nu \in N} r_{\lambda\nu}\mathbf{z}_\nu, \ \lambda \in \Lambda$$

the row-finite matrices

$$T_{XY} \stackrel{\text{def}}{=} \{t_{\lambda\mu}\}_{\lambda \in \Lambda, \mu \in M}, \ S_{YZ} \stackrel{\text{def}}{=} \{s_{\mu\nu}\}_{\mu \in M, \nu \in N}, \ \text{and} \ R_{XZ} \stackrel{\text{def}}{=} \{r_{\lambda\nu}\}_{\lambda \in \Lambda, \nu \in N}.$$

Exercise 6.3.3. Show that $R_{XZ} \stackrel{\text{def}}{=} (ST)_{XZ} = T_{XY}S_{YZ}$. (Note the match-up of indices and the reversal of matrices.)

If V is a finite-dimensional vector space, if $T : V \mapsto V$ is a linear map, and if X is a T-eigenbasis for V then T_{XX} is diagonal. Hence the search for a diagonal matrix that represents T is really a search for an eigenbasis for V.

More generally, a Jordanizing basis X for V is "nearly" an eigenbasis in the sense that for each $\mathbf{x}$ in X either the vector $T\mathbf{x}$ is a multiple of $\mathbf{x}$ ($\mathbf{x}$ is an eigenvector) or $T\mathbf{x}$ is a linear combination of $\mathbf{x}$ and one other vector in X. Hence the search for the Jordan block matrix that represents T is really a search for a Jordanizing basis for V.

There is a lexicon of terminology that has gained increasing acceptance in the study of algebra. Here are the terms frequently used and their meanings:

▷ *homomorphism*: a map $H : L \mapsto M$ that "respects" the algebraic relations in L by communicating them to M. Thus if, e.g., L and M are fields then

$$H(ab) = H(a)H(b), \ H(a + b) = H(a) + H(b).$$

▷ *monomorphism*: a homomorphism $H : L \mapsto M$ that is one-one.
▷ *epimorphism*: a homomorphism $H : L \mapsto M$ that is "onto," i.e., $H(L) = M$.
▷ *isomorphism*: a homomorphism $H : L \mapsto M$ that is both a monomorphism and an epimorphism.

$\triangleright$ *endomorphism*: a homomorphism $H : L \mapsto L$.

$\triangleright$ *automorphism*: an endomorphism that is an isomorphism.

Exercise 6.3.4. Show that if V and W are finite-dimensional vector spaces and $T : V \mapsto W$ is a map then T is:

   i. a homomorphism iff T is linear;

  ii. an isomorphism iff T is an epimorphism and $dim(V) = dim(W)$;

 iii. an automorphism iff $V = W$ and T is invertible;

 iv. a monomorphism iff $ker(T) = \{\mathbf{O}\}$;

  v. an epimorphism iff $im(T) = W$;

 vi. *not* an epimorphism if $dim(V) < dim(W)$.

Exercise 6.3.5. Let V be an n-dimensional vector space over $\mathbf{K}$. Show that each basis
$$X \stackrel{\text{def}}{=} \{\mathbf{x}_1, \ldots, \mathbf{x}_n\}$$
for V engenders an isomorphism

$$\mathcal{I}_X : V \ni \mathbf{x} \stackrel{\text{def}}{=} \sum_{i=1}^{n} a_i \mathbf{x}_i \mapsto \begin{pmatrix} a_1 \\ \vdots \\ a_n \end{pmatrix} \in \mathbf{K}^n.$$

The set of all homomorphisms $H : L \mapsto M$ is denoted $[L, M]$. If L and M are vector spaces then $[L, M] = \mathcal{L}(L, M)$.

Hence in the study of finite-dimensional vector spaces one may, confine one's attention to studying the "concrete" vector spaces $\mathbf{K}^n$. In **Parts I** and **II** of this book $\mathbf{K}$ is usually $\mathbf{C}$ and occasionally $\mathbf{R}$.

Exercise 6.3.6. Let V resp. W be n- resp. m-dimensional vector spaces with bases X resp. Y. Show that $[V, W]$ is a vector space and that there is an isomorphism
$$\mathcal{J}_{XY} : [V, W] \mapsto Mat_{mn}$$
between the vector spaces $[V, W]$ and Mat_{mn}.

[*Hint:* See **Exercise 6.3.3.**]

Exercise 6.3.7. Let V be a vector space and let X be a basis for V. Assume furthermore that the vectors in X are indexed by the specially ordered set $\Lambda \stackrel{\text{def}}{=} \{\lambda\}$. Let $Mat_{\Lambda \times \Lambda}$ be the algebra of all row-finite matrices with entries indexed by $\Lambda \times \Lambda$ Show that the map

$$\mathcal{M} : [V, V] \ni T \mapsto T_{XX}^t \in Mat_{\Lambda \times \Lambda}$$

is an isomorphism of the *algebras* $[V, V]$ and $Mat_{\Lambda \times \Lambda}$. Generalize the result for two vector spaces V resp. W, bases X resp. Y and $[V, W]$.

[*Hint:* When $V \neq W$ is composition defined among the elements of $[V, W]$? Is there a homomorphism of *algebras* or of *vector spaces*? Is there an isomorphism of *algebras* or of *vector spaces*?]

6.4. Linear transformations in Euclidean vector spaces

The eigenvalue problem in nonfinite-dimensional vector spaces is quite different from the eigenvalue problem in finite-dimensional vector spaces.

Example 6.4.0.1. Let V be the set of all *two-sided sequences*

$$\mathbf{a} \stackrel{\text{def}}{=} \{a_n\}_{-\infty < n < \infty}$$

of complex numbers of which only finitely many are nonzero. Let T be defined according to the rule:

$$\text{if } T\mathbf{a} \stackrel{\text{def}}{=} \mathbf{b} \stackrel{\text{def}}{=} \{b_n\}_{-\infty < n < \infty} \text{ then } b_n = a_{n+1}, \ -\infty < n < \infty.$$

If λ is an eigenvalue of T and $\mathbf{x} \stackrel{\text{def}}{=} \{x_n\}_{-\infty < n < \infty}$ is a corresponding eigenvector then

$$x_{n+1} = \lambda x_n, \ -\infty < n < \infty.$$

There are integers K and L such that

$$x_n \begin{cases} = 0 & \text{if } n < K \text{ or } n > L \\ \neq 0 & \text{if } n = K \text{ or } n = L. \end{cases}$$

Hence $x_{L+1} = 0 = \lambda x_L$ whence $\lambda = 0$. Hence $x_L = \lambda x_{L-1} = 0$ a contradiction. Thus T has *no* eigenvalues.

However, in a finite-dimensional vector space an eigenvalue of a linear transformation or matrix T is a number λ such that $T - \lambda I$ is not invertible. In that context the set of eigenvalues is the spectrum $\sigma(T)$ of T. Thus let the *spectrum* of T be re-interpreted as the set of λ such that $T - \lambda I$ is not invertible.

DEFINITION **6.4.0.1.** IF T IS A LINEAR MAPPING OF A VECTOR SPACE V INTO ITSELF THEN THE SPECTRUM $\sigma(T)$ IS

$$\{\lambda \ : \ T - \lambda I \text{ IS NOT INVERTIBLE}\}.$$

Example 6.4.2. Let T be as in **Example 6.4.1**. It is shown next that if $\lambda \neq 0$ then $T - \lambda I$ is not invertible. Indeed if $T - \lambda I$ is invertible let $\mathbf{x}$ be the vector such that

$$x_n = \begin{cases} 1 & \text{if } n = 0 \\ 0 & \text{otherwise.} \end{cases}$$

If $(T - \lambda I)^{-1}\mathbf{x} \overset{\text{def}}{=} \mathbf{y}$ then $\mathbf{y} \neq O$. There is in $\mathbf{N}$ an L such that

$$y_n \begin{cases} = 0 & \text{if } n > L \\ \neq 0 & \text{if } n = L. \end{cases}$$

If $\lambda \neq 0$ then $x_L = y_{L+1} - \lambda y_L$ whence $L = 0$. But then

$$0 \neq y_0 = \lambda^k y_{-k}, \ k \in \mathbf{N}$$

whence $y_{-k} \neq 0$, $k \in \mathbf{N}$, a contradiction. Hence $\lambda = 0$ and $\sigma(T) = \mathbf{C} \setminus \{0\}$. By contrast, if $n \in \mathbf{N}$ and T is a linear transformation of an n-dimensional vector space V into itself then $\sigma(T)$ is the set of eigenvalues of T and is necessarily finite $(\#[\sigma(T)] \leq n)$.

Exercise 6.4.1. Show that if V is a finite-dimensional vector space over $\mathbf{K}$ then for some n in $\mathbf{N}$ there is a one-one mapping

$$\mathcal{I} : V \to \mathbf{K}^n$$

such that
$$\mathcal{I}(a\mathbf{x} + b\mathbf{y}) = a\mathcal{I}\mathbf{x} + b\mathcal{I}\mathbf{y} \ (\mathcal{I} \text{ is a linear mapping})$$
$$\mathcal{I}(V) = \mathbf{K}^n$$

(cf. **Exercise 6.3.5**).

Exercise 6.4.2. Let V be the vector space of all polynomials with coefficients in $\mathbf{C}$.

  i. Show that V is not finite-dimensional.
  ii. Show that
$$X \overset{\text{def}}{=} \{1, x, x^2, \ldots, x^n, \ldots\}$$

    as presented is a specially ordered Hamel basis for V.
  iii. Show that if $P \overset{\text{def}}{=} \{p_0, p_1, \ldots, p_n, \ldots\}$ is a sequence of polynomials such that $deg(p_n) = n$ then P as presented is a specially ordered basis for V.
  iv. Assume that if $p \in V$ then $T(p)(x) = \int_0^x p(t)\, dt$, i.e.,

$$T : V \ni p \mapsto \int_0^x p(t)\, dt.$$

a. Show that there is a linear mapping

$$S : V \mapsto V$$
$$ST = I$$
$$TS \neq I.$$

(Hence T is NOT invertible. In contrast to the situation in Mat_{nn} the map S is a *left inverse* but *not* a *right inverse* of T.)

b. What is S? Show that 0 is its only eigenvalue.

c. Show that T has no eigenvalues.

d. Show that $\sigma(T) = \mathbf{C}$ and that $\sigma(S) = 0$.

e. If X is the basis in *ii* what is T_{XX}?

f. If $p_n(x) = x^n + x^{n-1} + \cdots + x + 1$ and P is the basis in *iii* what is T_{PP}?

v. Assume $q \in V$ and that

$$R : V \ni p \mapsto qp \in V.$$

a. Show R is a linear mapping.

b. If $q(x) = a_0 x^n + \cdots + a_n$ and if X is the basis in *ii* what is R_{XX}?

b. If $p_n(x) = x^n + x^{n-1} + \cdots + x + 1$ and P is the basis in *iii* what is R_{PP}?

Exercise 6.4.0.3. Let V be the set of all rational functions

$$f : \mathbf{C} \ni z \mapsto f(z) \in \mathbf{C}.$$

Each f in V is the quotient of two polynomials p and q: $f = p/q$. Let k be a fixed polynomial of positive degree and let T be defined by the equation

$$T : V \ni f \mapsto T(f) \overset{\text{def}}{=} kf \ (= \frac{kp}{q}) \in V.$$

Show that with respect to the natural definitions of addition and scalar multiplication V is a vector space over $\mathbf{C}$ and that T is a linear transformation of V into itself. Show that the spectrum $\sigma(T)$ is empty: $\sigma(T) = \emptyset$. Again there is a contrast with what happens if T is a linear transformation of $\mathbf{C}^n$ into itself. For such a T the spectrum $\sigma(T)$ is never empty.

In light of the **Examples** above there arises the question of what kinds of useful forms for linear transformations can be found. The full answer to the question are not known at the time of this writing. In particular, there is no satisfactory analog of the Jordan normal form valid for all infinite-dimensional spaces. However von Neumann, in pursuit of his interests in quantum mechanics, took some giant steps beyond the accomplishments of Hilbert in analyzing the structure of a linear transformation T operating in a special kind of infinite-dimensional vector space: *Hilbert space* ([**Ber**], [**vN1**], [**vN2**]).

Von Neumann's development began with the formulation of the axioms E1-E4 (cf. **Section 4.4**). In a vector space V endowed with an inner product that gives V the structure of a Euclidean space the notion of a *normal linear transformation* can be introduced without reference to matrices.

Exercise 6.4.0.4. Show that if V is a Euclidean vector space and $T \in [V, V]$ then there is in $[V, V]$ at most one S such that $(T\mathbf{x}, \mathbf{y}) = (\mathbf{x}, S\mathbf{y})$.

DEFINITION **6.4.0.2.** LET V BE A EUCLIDEAN VECTOR SPACE (V IS ENDOWED WITH AN INNER PRODUCT). IF $T \in [V, V]$ AND IF THERE IS IN $[V, V]$ AN S SUCH THAT FOR ALL VECTORS $\mathbf{x}$, $\mathbf{y}$ IN V

$$(T\mathbf{x}, \mathbf{y}) = (\mathbf{x}, S\mathbf{y})$$

THEN S IS THE *adjoint* OF T AND IS DENOTED T^*. A T IN $[V, V]$ IS *self-adjoint* IFF FOR ALL $\mathbf{x}$ AND $\mathbf{y}$ IN V

$$(T\mathbf{x}, \mathbf{y}) = (\mathbf{x}, T\mathbf{y})$$

(WHENCE T^* EXISTS AND $T - T^*$). A T IN $[V, V]$ IS *normal* IFF T^* EXISTS AND $TT^* = T^*T$. A U IN $[V, V]$ IS *unitary* IFF U^* EXISTS AND $UU^* = U^*U = I$.

Exercise 6.4.0.5. Show that if V is a Euclidean vector space and $U \in [V, V]$ then U is unitary iff: $im(U) = V$ and $\|U\mathbf{x}\| = \|\mathbf{x}\|$ for all $\mathbf{x}$ in V.

If V is a finite-dimensional Euclidean vector space the matrix representation of a T in $[V, V]$ permits the definition of T^* in matrix terms:

If V is finite-dimensional, if $T \in [V, V]$, if X is an orthonormal basis for V, and if T_{XX} is the corresponding matrix representation of T, then corresponding to the matrix $(T_{XX})^*$ there is in $[V, V]$ an S and $S = T^*$.

Exercise 6.4.0.6. Show that if V is a finite-dimensional Euclidean space and if $T : V \mapsto V$ is a linear map then T^* exists and T is normal iff for each orthonormal basis X for V the matrix T_{XX} is a normal matrix:

$$T_{XX}(T_{XX})^* = (T_{XX})^*T_{XX}.$$

If V is not finite-dimensional just as the question of the existence of T^{-1} is subtle (cf. **Exercise 6.4.2**) so also the question of the existence of T^* has no simple answer.

Example 6.4.3. Let $\mathcal{P}$ be the set of all polynomials

$$p : [0,1] \ni x \mapsto p(x) \stackrel{\text{def}}{=} \sum_{k=1}^{n} a_k x^k, \ a_k \in \mathbf{C}, \ n \in \mathbf{N}.$$

Then $\mathcal{P}$ is infinite-dimensional. Introduce into $\mathcal{P}$ the inner product

$$(\, , \,) : \mathcal{P} \times \mathcal{P} \ni \{p, q\} \mapsto \int_0^1 p(x)\overline{q(x)}\, dx \stackrel{\text{def}}{=} (p, q).$$

Then $\mathcal{P}$ is a Euclidean vector space and

$$T : V \ni p \mapsto \frac{dp}{dx}$$

is a linear transformation. However T has no adjoint. Indeed if T^* is the adjoint of T, if $p_n(x) \stackrel{\text{def}}{=} x^n$, $0 \le n < \infty$, then

$$(p_n, T^* p_0) = \int_0^1 x^n \overline{T^* p_0(x)}\, dx = \int_0^1 T(x^n) p_0\, dx = \int_0^1 n x^{n-1}\, dx = 1.$$

Let p be the polynomial $T^* p_0$. Then according to the Schwarz inequality (4.4.21)

$$|(p_n, p)| = 1 \le ||p_n|| \cdot ||p|| = \sqrt{\frac{1}{2n+1}} \cdot ||p||.$$

If n is such that $||p|| < \sqrt{2n+1}$ there emerges the contradiction: $1 < 1$.

In l^2 let U be defined as follows:

$$U : l^2 \ni \mathbf{a} \stackrel{\text{def}}{=} (a_1, a_2, \ldots) \mapsto U\mathbf{a} \stackrel{\text{def}}{=} (0, a_1, a_2, \ldots) \in l^2.$$

If S is defined as follows:

$$S : l^2 \ni \mathbf{a} \stackrel{\text{def}}{=} (a_1, a_2, \ldots) \mapsto S\mathbf{a} \stackrel{\text{def}}{=} (a_2, a_3, \ldots) \in l^2$$

then $S = U^*$ and $U^* U = I$ but $UU^* \ne I$.

The difficulty in the first part of **Example 6.4.3** is in some measure the fact that

$$||p_n|| \to 0 \text{ as } n \to \infty$$
$$||Tp_n|| = \frac{n}{\sqrt{2n-1}} \to \infty \text{ as } n \to \infty,$$

i.e., that T is not continuous. The difficulty in the second part lies in the fact that $U(l^2) \subsetneq l^2$.

One way to eliminate the apparent chaos exemplified above in infinite-dimensional vector spaces is to introduce topologies. In Euclidean vector spaces a topology is at hand via the norm $||\ ||$: $||\mathbf{x}|| \stackrel{\text{def}}{=} \sqrt{(\mathbf{x}, \mathbf{x})}$.

DEFINITION **6.4.0.3.** IN A EUCLIDEAN VECTOR SPACE V THE *distance* BETWEEN TWO VECTORS $\mathbf{x}$ AND $\mathbf{y}$ IS $\|\mathbf{x} - \mathbf{y}\|$. IF V AND W ARE EUCLIDEAN VECTOR SPACES THE LINEAR TRANSFORMATION $T : V \mapsto W$ IS CONTINUOUS IFF $\|T\mathbf{x}_n - T\mathbf{x}\| \to 0$ WHENEVER $\|\mathbf{x}_n - \mathbf{x}\| \to 0$ AS $n \to \infty$.

In analogy with the completeness of $\mathbf{R}$ and $\mathbf{C}$ an arbitrary Euclidean vector space V is *complete* iff

E5. Whenever

$$\|\mathbf{x}_n - \mathbf{x}_m\| \to 0 \text{ as } n, m \to \infty$$

there is in V an $\mathbf{x}$ such that

$$\|\mathbf{x}_n - \mathbf{x}\| \to 0 \text{ as } n \to \infty.$$

In a finite-dimensional Euclidean vector space V the axiom E5 is automatically true because:

i. there is a finite orthonormal X basis in V;
ii. V, via the basis X, is, for some n in $\mathbf{N}$, effectively $\mathbf{C}^n$;
iii. $\mathbf{C}^n$ is complete.

Example 6.4.0.4. If V is an arbitrary vector space and if X is a Hamel basis for V it is possible to define an inner product via the basis X: if

$$\mathbf{x} = \sum_{\lambda \in \Lambda} a_\lambda \mathbf{x}_\lambda$$

$$\mathbf{y} = \sum_{\lambda \in \Lambda} b_\lambda \mathbf{x}_\lambda$$

then

$$(\mathbf{x}, \mathbf{y}) \stackrel{\text{def}}{=} \sum_{\lambda \in \Lambda} a_\lambda \overline{b_\lambda}.$$

All sums above are finite, E1-E4 obtain for (,), and X is an orthonormal set in V. However if V is not finite-dimensional and is endowed with the Euclidean structure just described then V is *not* complete. Indeed, if $\{\mathbf{x}_{\lambda_i}\}_{i=1}^\infty$ is an infinite subsequence of X and if

$$\mathbf{y}_n \stackrel{\text{def}}{=} \sum_{k=1}^n 2^{-k} \mathbf{x}_{\lambda_k}$$

then

$$\|\mathbf{y}_n - \mathbf{y}_m\| = \sum_{k=n+1}^m 2^{-2k} \to 0, \text{ as } n, m \to \infty.$$

There is in V no $\mathbf{y}$ such that $\|\mathbf{y}_n - \mathbf{y}\| \to 0$ as $n \to \infty$ and the axiom E5 fails to hold in V.

  Chapter 6. GENERAL VECTOR SPACES

Exercise 6.4.0.7. Prove the last assertion in **Example 6.4.4**.

[*Hint:* Write $\mathbf{y}$ in terms of X: $\mathbf{y} = \sum_{\lambda \in \Lambda} c_\lambda \mathbf{x}_\lambda.$]

Example 6.4.0.5. Let X be the Hamel basis in **Example 6.4.4** and let $\mathcal{H}_V$ be the set of all column vectors

$$\boldsymbol{\alpha} \stackrel{\mathrm{def}}{=} \begin{pmatrix} \vdots \\ \alpha_\lambda \\ \vdots \end{pmatrix} \ \text{ such that } \ \sum_{\lambda \in \Lambda} |\alpha_\lambda|^2 < \infty. \qquad (6.4.0.1)$$

The Minkowski inequality (4.4.22) implies that $\mathcal{H}_V$ is indeed a vector space. The Schwarz inequality (4.4.21) implies that if

$$\beta \stackrel{\mathrm{def}}{=} \begin{pmatrix} \vdots \\ \beta_\lambda \\ \vdots \end{pmatrix} \in \mathcal{H}_V$$

then

$$\sum_{\lambda \in \Lambda} \alpha_\lambda \overline{\beta_\lambda} \qquad (*)$$

converges. By definition $(*)$ is $(\boldsymbol{\alpha}, \beta)$. Define the map

$$\mathcal{J} : V \ni \sum_{\lambda \in \Lambda} \alpha_\lambda \mathbf{x}_\lambda \mapsto \boldsymbol{\alpha} \stackrel{\mathrm{def}}{=} \begin{pmatrix} \vdots \\ \alpha_\lambda \\ \vdots \end{pmatrix} \in \mathcal{H}_V.$$

Then $\mathcal{J}$ is one-one and by abuse of notation $V \subsetneq \mathcal{H}_V$ and for $\mathcal{H}_V$ not only do E1-E4 hold but E5 holds as well.

[**Note 6.4.0.1:** The vector $\boldsymbol{\alpha}$ in (6.4.1) need not be column-finite. On the other hand at most *countably* many components are not 0's. In other words, there is a subsequence $\{\lambda_k\}_{k=1}^\infty$ such that $\alpha_\lambda = 0$ if $\lambda \notin \{\lambda_k\}_{k=1}^\infty$.

In **Example 6.4.3** the set $X \stackrel{\mathrm{def}}{=} \{p_n\}_{n=0}^\infty$ is a Hamel basis for $\mathcal{P}$. The Gram-Schmidt process applied to X yields an orthonormal Hamel basis Y for $\mathcal{P}$. The procedure described above gives rise to $\mathcal{H}_\mathcal{P}$ and the corresponding $\mathcal{J}$ is now a one-one map from $\mathcal{J} : \mathcal{P} \mapsto l^2$ (cf. **Remark 6.3.1**ii).

If V is finite-dimensional then X with respect to $(\, , \,)$ is a standard orthonormal basis for V and with respect to $\| \ \|$ V *is* complete. In any event $\mathcal{H}_V$ is the *completion* of V.]

Exercise 6.4.0.8. Show that $\mathcal{H}_V$ as defined in **Example 6.4.5** is a vector space, that it is complete, and that $(*)$ converges.

[*Hint:* Show that if $S \overset{\text{def}}{=} \{\mathbf{x}_n\}_{n=1}^{\infty}$ is a sequence of vectors in $\mathcal{H}_V$ then the set of all coefficients used to represent the vectors in S is a countable set (in one-one correspondence with $\mathbf{N}$).]

Exercise 6.4.0.9. Let V be $\widetilde{L}^2([0,1])$ as defined in **Remark 6.3.1***iii*. Show that if $f, g \in V$ and $s, t \in \mathbf{C}$ then $sf + tg \in V$ and

$$(f, g) \overset{\text{def}}{=} \int_{\mathbf{R}} f(x)\overline{g(x)}\, dx$$

exists. Show also that E1-E4 hold in V but that E5 fails to hold in V.

[*Hint:* Show that for every positive number R

$$(f, g)_R \overset{\text{def}}{=} \int_{-R}^{R} f(x)\overline{g(x)}\, dx$$

exists and that for $(\ ,\ )_R$ V is a vector space in which E1-E4 hold.]

DEFINITION **6.4.0.4.** LET V BE A EUCLIDEAN VECTOR SPACE AND LET T BE A LINEAR TRANSFORMATION OF V INTO V. IF $(T\mathbf{x}, \mathbf{x}) \geq 0$ FOR EVERY VECTOR $\mathbf{x}$ THEN T IS *positive semidefinite*. IF WHENEVER $\mathbf{x} \neq \mathbf{O}$ $(T\mathbf{x}, \mathbf{x}) > 0$ THEN T IS *positive definite*.

Exercise 6.4.0.10. Show that if V is a finite-dimensional Euclidean vector space and if T is a positive semidefinite linear transformation of V into V then T is self-adjoint (and thus diagonable).

The next paragraphs are devoted to proving that if $T : \mathcal{H} \mapsto \mathcal{H}$ is a continuous linear transformation of a Hilbert space $\mathcal{H}$ (for which E1-E5 hold) then the adjoint T^* exists. A Euclidean vector space in which E1-E5 hold is a *Hilbert space*.

DEFINITION **6.4.0.5.** IF $\mathcal{H}$ IS A HILBERT SPACE A SUBSET S OF $\mathcal{H}$ IS *closed* IFF WHENEVER

$$\{\mathbf{x}_n\}_{n=1}^{\infty} \subset S$$

AND

$$\|\mathbf{x}_n - \mathbf{x}\| \to 0 \text{ AS } n \to \infty$$

THEN $\mathbf{x} \in S$.

Exercise 6.4.0.11. Show that:

i. $\mathcal{H}$ and $\emptyset$ are closed;

ii. if $\{F_\lambda\}_{\lambda \in \Lambda}$ is a set of closed sets in $\mathcal{H}$ then

$$\bigcap_{\lambda \in \Lambda} F_\lambda$$

is closed;

iii. if $F_1, \ldots, F_k$ are closed then

$$\bigcup_{j=1}^{k} F_j$$

is closed.

THEOREM 6.4.0.1. IF $\mathcal{Q}$ IS A CONVEX AND CLOSED SUBSET OF A HILBERT SPACE $\mathcal{H}$ AND IF $\mathbf{x} \in \mathcal{H}$ THERE IS IN $\mathcal{Q}$ A VECTOR $\mathbf{v}$ SUCH THAT

$$\|\mathbf{x} - \mathbf{v}\| = infimum\,\{\,\|\mathbf{x} - \mathbf{w}\| \ : \ \mathbf{w} \in \mathcal{Q}\,\} \overset{\text{def}}{=} d.$$

PROOF. Assume $\{\mathbf{v}_n\|_{n=1}^{\infty}\} \subset \mathcal{Q}$ and that

$$\|\mathbf{x} - \mathbf{v}_n\| \to d, \ \text{as } n \to \infty.$$

Since $\mathcal{Q}$ is convex it follows that for any n, m

$$\frac{\mathbf{v}_n + \mathbf{v}_m}{2} \in \mathcal{Q}.$$

As in the proof of THEOREM 5.4.1 the THEOREM of Apollonius (5.4.6) implies that $\|\mathbf{v}_n - \mathbf{v}_m\| \to 0$ as $n, m \to \infty$. Then E5 implies there is a $\mathbf{v}$ such that

$$\|\mathbf{v}_n - \mathbf{v}\| \to d \text{ as } n \to \infty.$$

Since $\mathcal{Q}$ is closed it follows that $\mathbf{v} \in \mathcal{Q}$. Since $\mathbf{x} \in \mathcal{H}$ it follows that $\mathbf{x} \neq \mathbf{v}$ and so $\|\mathbf{x} - \mathbf{v}\| \overset{\text{def}}{=} d \ (\geq 0)$.

$$\Omega$$

Exercise 6.4.0.12. Show:

i. there is in $\mathcal{Q}$ only one vector $\mathbf{v}$ "nearest" $\mathbf{x}$ (in particular, $\mathbf{v}=\mathbf{x}$ if $\mathbf{x} \in \mathcal{Q}$);

ii. that if $\mathbf{z} \in \mathcal{Q}$ then $\Re(\mathbf{z} - \mathbf{v}, \mathbf{x} - \mathbf{v}) \leq 0$.

Every subspace M of $\mathcal{H}$ is convex and if M is closed as well then for each $\mathbf{x}$ in $\mathcal{H}$ there is in M a unique $\mathbf{v}$ closest to $\mathbf{x}$. That vector $\mathbf{v}$ is the *orthogonal projection of* $\mathbf{x}$ into M. The map

$$P_M : \mathcal{H} \ni \mathbf{x} \mapsto P_M\mathbf{x} \overset{\mathrm{def}}{=} \mathbf{v} \in M$$

is the *orthogonal projection* onto M.

[**Note 6.4.0.2:** If $\mathbf{x} \in M$ then $P_M\mathbf{x} = \mathbf{x}$.]

THEOREM **6.4.0.2.** IF M IS A CLOSED SUBSPACE OF $\mathcal{H}$ THEN P_M IS A LINEAR, CONTINUOUS, SELF-ADJOINT IDEMPOTENT $(P_M^2 = P_M)$ AND

$$||P_M\mathbf{x}|| \le ||\mathbf{x}||.$$

PROOF. The "least squares" character of $\mathbf{v}$ implies that

$$\mathbf{x} - \mathbf{v} \in M^{\perp}.$$

Indeed, if $\mathbf{z} \in M$, if $(\mathbf{x} - \mathbf{v}, \mathbf{z}) = re^{i\theta}$, and if $t \in \mathbf{R}$ then $\mathbf{v} + te^{i\theta}\mathbf{z} \in M$ and so

$$||\mathbf{x} - (\mathbf{v} + te^{i\theta}\mathbf{z})||^2 = ||\mathbf{x} - \mathbf{v}||^2 + t^2||\mathbf{z}||^2 - 2tr \ge ||\mathbf{x} - \mathbf{v}||^2$$
$$t(t||\mathbf{z}||^2 - 2tr) \ge 0 \qquad\qquad (6.4.0.2)$$

which is true for all real t iff $r = 0$, i.e., iff $\mathbf{x} - \mathbf{v} \perp \mathbf{z}$, i.e., iff $\mathbf{x} - \mathbf{v} \in M^{\perp}$.

It follows that for every $\mathbf{x}$ in $\mathcal{H}$ there is in M a unique $\mathbf{v}$ and in $M^{\perp}$ a unique $\mathbf{u}$ $(= \mathbf{x} - \mathbf{v})$ such that

$$\mathbf{x} = \mathbf{v} + \mathbf{u}. \qquad\qquad (6.4.0.3)$$

The linearity and idempotence of P_M follow from the uniqueness of the representation (6.4.3). Furthermore $||P_M\mathbf{x}|| \le ||\mathbf{x}||$ whence P_M is continuous. Finally if $\mathbf{x}' \in \mathcal{H}$ and $\mathbf{x}' = \mathbf{v}' + \mathbf{u}'$ is the corresponding representation for $\mathbf{x}'$ then, owing to the orthogonality of vectors in M to vectors in $M^{\perp}$,

$$(P_M\mathbf{x}, \mathbf{x}') = (\mathbf{v}, \mathbf{v}') + (\mathbf{u}, \mathbf{u}') = (\mathbf{x}, P_M\mathbf{x}')$$

and so P_M^* exists and $P_M = P_M^*$.

$$\Omega$$

[**Remark 6.4.0.1:** A continuous, linear, self-adjoint idempotent map P is called an *orthogonal idempotent* or *orthogonal projection* to distinguish it from a mere idempotent or (oblique) projection. For example, the matrix

$$Q \overset{\mathrm{def}}{=} \begin{pmatrix} 1 & 0 \\ 2 & 0 \end{pmatrix}$$

is idempotent ($Q^2 = Q$) but it is not self-adjoint ($Q^* \neq Q$). It is an oblique projection because

$$Q\mathbf{x} \overset{\text{def}}{=} Q \begin{pmatrix} s \\ t \end{pmatrix} = \begin{pmatrix} s \\ 2s \end{pmatrix},$$

i.e., Q projects $\mathbf{x}$ obliquely onto the line

$$L \overset{\text{def}}{=} \{\, (s,t) \; : \; t = 2s \,\}$$

whereas P_L projects $\mathbf{x}$ orthogonally onto the line L.

In $\mathbf{R}^2$ the projection Q is such that for any vector $\mathbf{x}$ the vector $(Q - I)\mathbf{x}$ lies on the vertical axis (the first component of $(Q - I)\mathbf{x}$ is 0) as in **Figure 6.4.1.**]

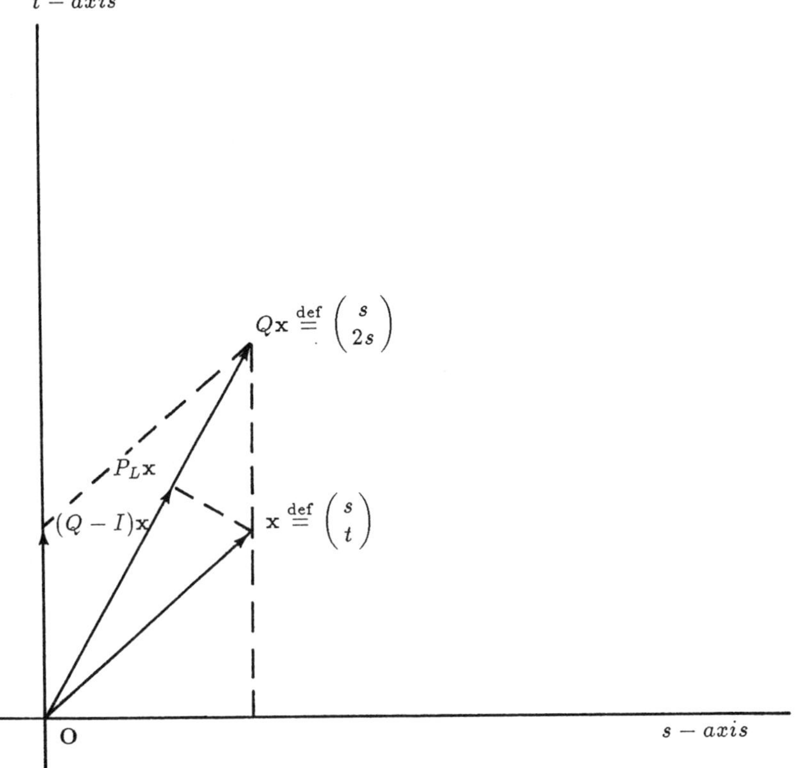

Figure 6.4.0.1.

Note that it is *not* true that for every $\mathbf{x}$ and every $\mathbf{y}$

$$(Q\mathbf{x}, (Q - I)\mathbf{y}) = 0.$$

For example,

$$
\begin{aligned}
Q\mathbf{e}_1 &= \mathbf{e}_1 + 2\mathbf{e}_2 \\
(Q - I)\mathbf{e}_1 &= \mathbf{e}_1 - (\mathbf{e}_1 + 2\mathbf{e}_2) = -2\mathbf{e}_2 \\
(Q\mathbf{e}_1, (Q - I)\mathbf{e}_1) &= (\mathbf{e}_1 + 2\mathbf{e}_2, -2\mathbf{e}_2) = -4 \neq 0.
\end{aligned}
$$

An idempotent P is an orthogonal idempotent, i.e., is self-adjoint, iff

$$im(P) \perp im(I - P),$$

i.e., iff for every $\mathbf{x}$ and every $\mathbf{y}$

$$(P\mathbf{x}, (I - P)\mathbf{y}) = 0.$$

Indeed, if $P = P^*$ then

$$(P\mathbf{x}, (I - P)\mathbf{y}) = ((I - P)P\mathbf{x}, \mathbf{y}) = (\mathbf{O}, \mathbf{y}) = 0.$$

Conversely, if $im(P) \perp im(I - P)$ then for every $\mathbf{x}$

$$(P\mathbf{x}, (I - P)(I - P^*)P\mathbf{x}) = \|(I - P^*)P\mathbf{x}\|^2 = 0$$

whence

$$
\begin{aligned}
(I - P^*)P = O &= O^* \\
&= ((I - P^*)P)^* = P^*(I - P)
\end{aligned}
$$

and so $P = P^*P = P^*$, i.e., P is self-adjoint.

Exercise 6.4.0.13. Show that the representation (6.4.3) is unique.

Exercise 6.4.0.14. Show that $\|\mathbf{x}\|^2 = \|P_M\mathbf{x}\|^2 + \|(I - P_M)\mathbf{x}\|^2$.

Exercise 6.4.0.15. Explain why the representation (6.4.3) is an *orthogonal decomposition* (with respect to M and $M^\perp$) of $\mathbf{x}$.

THEOREM 6.4.0.3. IF f IS A CONTINUOUS LINEAR TRANSFORMATION MAPPING A HILBERT SPACE $\mathcal{H}$ INTO $\mathbf{C}$ THERE IS IN $\mathcal{H}$ A VECTOR $\mathbf{w}_f$ SUCH THAT FOR EVERY $\mathbf{y}$ IN $\mathcal{H}$

$$f(\mathbf{y}) = (\mathbf{y}, \mathbf{w}_f).$$

[**Remark 6.4.2:** A linear transformation $f : \mathcal{H} \mapsto \mathbf{C}$ is a *linear functional*. THEOREM **6.4.3** is the Riesz representation theorem (cf. **Remark 4.4.9**) for Hilbert space.]

PROOF. Since f is continuous, $ker(f) \stackrel{\mathrm{def}}{=} N$ is closed. If $f(\mathbf{x}) \equiv 0$ then $\mathbf{w}_f = \mathbf{O}$. If $f(\mathbf{x}) \neq 0$ for some $\mathbf{x}$ in $\mathcal{H}$, then $N \subsetneq \mathcal{H}$. Hence

$$\mathbf{w} \stackrel{\mathrm{def}}{=} \frac{(I - P_N)\mathbf{x}}{f(\mathbf{x})} \ (\in N^\perp)$$

is well-defined and since $P_N \mathbf{x} \in N = ker(f)$ it follows that $f(\mathbf{w}) = 1$. If $\mathbf{y} \in \mathcal{H}$ and $f(\mathbf{y}) = a$ then $\mathbf{y} - a\mathbf{w} \in N$ and

$$\mathbf{y} = (\mathbf{y} - a\mathbf{w}) + a\mathbf{w}$$

is the orthogonal decomposition of $\mathbf{y}$ with respect to N and $N^\perp$. Furthermore

$$(\mathbf{y}, \mathbf{w}) = a\|\mathbf{w}\|^2$$

and so if

$$\mathbf{w}_f \stackrel{\mathrm{def}}{=} \frac{\mathbf{w}}{\|\mathbf{w}\|^2}$$

then for any $\mathbf{y}$

$$(\mathbf{y}, \mathbf{w}_f) = a = f(\mathbf{y}).$$

The uniqueness of $\mathbf{w}_f$ follows because if $(\mathbf{y}, \mathbf{z}) = f(\mathbf{y})$ for all $\mathbf{y}$ then

$$(\mathbf{y}, \mathbf{z} - \mathbf{w}_f) = 0$$

for all $\mathbf{y}$ and thus $\mathbf{z} = \mathbf{w}_f$.

$$\Omega$$

Exercise 6.4.16. Show that

$$\|\mathbf{w}_f\| = supremum\{\, |f(\mathbf{x})| \ : \ \|\mathbf{x}\| = 1 \,\}.$$

[*Hint:* Show first that if $\mathbf{x} \in \mathcal{H}$ then

$$\|\mathbf{x}\| = supremum\{\, |(\mathbf{x}, \mathbf{y})| \ : \ \|\mathbf{y}\| = 1 \,\}.$$

Then use the Schwarz inequality (4.4.21).]

THEOREM **6.4.3** leads to the proof that if $\mathcal{H}$ is a Hilbert space and if $T : \mathcal{H} \mapsto \mathcal{H}$ is a *continuous* linear transformation then T^* exists and is itself a continuous linear map.

Indeed, the Schwarz inequality implies that for $\mathbf{y}$ fixed the map $\mathcal{H} \ni \mathbf{x} \mapsto (T\mathbf{x}, \mathbf{y}) \overset{\text{def}}{=} f_{\mathbf{y}}(\mathbf{x})$ is a continuous linear functional and so there is in $\mathcal{H}$ a $\mathbf{w}_{f_{\mathbf{y}}}$ such that

$$(T\mathbf{x}, \mathbf{y}) = (\mathbf{x}, \mathbf{w}_{f_{\mathbf{y}}})$$

The map $S : \mathcal{H} \ni \mathbf{y} \mapsto \mathbf{w}_{f_{\mathbf{y}}}$ is linear. Furthermore

$$(T\mathbf{x}, \mathbf{y}) = (\mathbf{x}, S\mathbf{y})$$
$$\|S\mathbf{y}\| = \|\mathbf{w}_{f_{\mathbf{y}}}\| = supremum \left\{ |(\mathbf{w}_{f_{\mathbf{y}}}, \mathbf{x})| \ : \ |\mathbf{x}| = 1 \right\}$$
$$= supremum \left\{ |(T\mathbf{x}, \mathbf{y})| \ : \ \|\mathbf{x}\| = 1 \right\}$$
$$\leq \|T\| \cdot \|\mathbf{y}\| \ \text{(Schwarz inequality)}$$

whence S is continuous and is denoted T^*. Furthermore $\|T^*\| \leq \|T\|$.

Exercise 6.4.0.17. By showing that $(T^*)^* \overset{\text{def}}{=} T^{**}$ exists, is continuous, and is T show that $\|T\| = \|T^*\|$.

Exercise 6.4.0.18. Show that if V is a finite-dimensional Euclidean space and $T : V \mapsto V$ is a linear map then for every ON basis X for V

$$T^*_{XX} = (T_{XX})^*.$$

Show that T is normal iff T_{XX} is normal.

It is helpful to introduce a *partial order* "$<$" in $\mathbf{C}$ according to the rule:
If $x \overset{\text{def}}{=} a + ib$ and $y \overset{\text{def}}{=} c + id$ then $x < y$ iff $a < c$ and $b < d$. Thus $x < y$ iff x lies in the open quadrant Q_y that is "southwest" of y. The statement $x \leq y$ requires careful definition:

$$x \leq y \ \text{iff} \ a \leq c \ \text{and} \ b \leq d.$$

For any λ in $\mathbf{C}$ the set

$$\mathcal{Q}_\lambda \overset{\text{def}}{=} \left\{ \lambda' \ : \ \lambda' \leq \lambda \right\}$$

is the quadrant "southwest" of λ. The "north" and "east" edges of that quadrant are subsets of $\mathcal{Q}_\lambda$.

Let $T : V \mapsto V$ be a normal linear map of the finite-dimensional Euclidean space V and let $\{\lambda_1, \ldots \lambda_k\}$ be the set of different eigenvalues of T. If $\lambda \in \mathbf{C}$ let M_λ be the span of all eigenvectors corresponding to eigenvalues λ_i such that $\lambda_i \leq \lambda$, i.e., such that $\lambda_i \in \mathcal{Q}_\lambda$. (If λ lies southwest of all λ_i then $M_\lambda \overset{\text{def}}{=} \{\mathbf{O}\}$.) Let P_λ be the orthogonal projection map corresponding to the orthogonal decomposition with respect to M_λ and $M_\lambda^\perp$: $P_\lambda : V \mapsto M_\lambda$.

     Chapter 6. GENERAL VECTOR SPACES

Exercise 6.4.0.19. Show that there is in $\mathbf{C}$ a λ_{max} such that $P_\lambda = I$ if $\lambda_{max} \le \lambda$.

If $x \overset{\text{def}}{=} a + ib < c + id \overset{\text{def}}{=} y$ and $u \overset{\text{def}}{=} a + id < v \overset{\text{def}}{=} c + ib$ then y, u, x, and v are the corners, enumerated counterclockwise and starting from y, of a rectangle $\mathcal{R}$. Note that

$$\mathcal{R} = (\mathcal{Q}_y \setminus \mathcal{Q}_u) \setminus (\mathcal{Q}_x \setminus \mathcal{Q}_v)$$
$$y \in \mathcal{R}, \ u \notin \mathcal{R}, \ x \notin \mathcal{R}, \ v \notin \mathcal{R}$$
$$(u, y] \cup (v, y] \subset \mathcal{R}$$
$$[x, u) \cup [x, v] \subset \mathbf{C} \setminus \mathcal{R}.$$

The relations above are indicated in **Figure 6.4.2**.

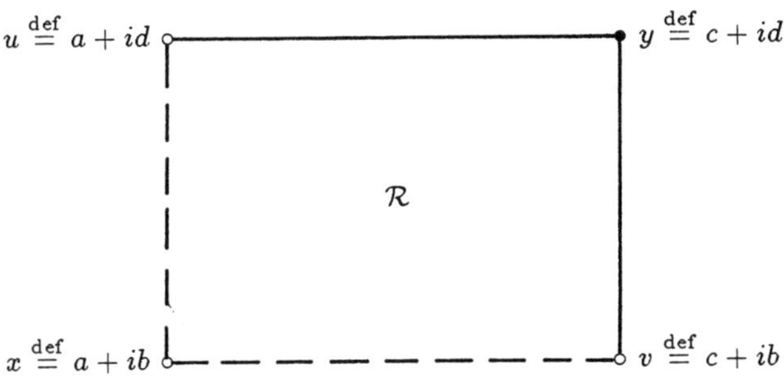

Figure 6.4.0.2.

Define $P_{\mathcal{R}}$ to be

$$P_y - P_u + P_x - P_v.$$

Then a direct calculation shows that $P_{\mathcal{R}}$ is itself an orthogonal idempotent.

Exercise 6.4.0.20. Show that if $\mathcal{R}$ is a rectangle to which precisely one eigenvalue λ belongs then $im(P_{\mathcal{R}})$ is the span of the set of all eigenvectors corresponding to λ:

$$im(P_{\mathcal{R}}) \overset{\text{def}}{=} W = span(\{\, \mathbf{x} \ : \ \mathbf{x} \ne \mathbf{O}, \ T\mathbf{x} = \lambda\mathbf{x} \,\}).$$

(Thus $im(P_{\mathcal{R}})$ is the eigenspace corresponding to λ.)

Exercise 6.4.0.21. Show that if $\mathcal{R}_1$ and $\mathcal{R}_2$ are rectangles then

$$\mathcal{R}_1 \cap \mathcal{R}_2 = \emptyset \Rightarrow P_{\mathcal{R}_1} P_{\mathcal{R}_2} = P_{\mathcal{R}_2} P_{\mathcal{R}_1} = O.$$

Exercise 6.4.0.22. Show that if $\{\mathcal{R}_1, \ldots \mathcal{R}_m\}$ is a set of rectangles no two of which intersect then

$$P_{\mathcal{R}_1} + \cdots + P_{\mathcal{R}_m}$$

is an orthogonal idempotent.

Exercise 6.4.0.23. Show that if $\mathcal{R}_1$ and $\mathcal{R}_2$ are rectangles then

$$\mathcal{R}_1 \subset \mathcal{R}_2 \Rightarrow P_{\mathcal{R}_1} P_{\mathcal{R}_2} = P_{\mathcal{R}_2} P_{\mathcal{R}_1} = P_{\mathcal{R}_1}.$$

Exercise 6.4.0.24. Show that if $\mathcal{R}_1$ and $\mathcal{R}_2$ are arbitrary rectangles then

$$P_{\mathcal{R}_1} P_{\mathcal{R}_2} = P_{\mathcal{R}_2} P_{\mathcal{R}_1}.$$

Exercise 6.4.0.25. Show that if $\{\mathcal{R}_1, \ldots \mathcal{R}_k\}$ is a set of rectangles no two of which intersect and if λ_i is in the interior of $\mathcal{R}_i$, $1 \leq i \leq k$ then

$$I = \sum_{j=1}^{k} P_{\mathcal{R}_j}. \tag{6.4.0.4}$$

Show that for any $\mathbf{x}$ in V

$$T\mathbf{x} = \sum_{j=1}^{k} \lambda_j P_{\mathcal{R}_j} \mathbf{x} \tag{6.4.0.5}$$

and any polynomial p

$$p(T)\mathbf{x} = \sum_{j=1}^{k} p(\lambda_j) P_{\mathcal{R}_j}. \tag{6.4.0.6}$$

[*Hint:* For each j find a basis for the eigenspace W_j (span of the eigenvectors) corresponding to λ_j. Write $\mathbf{x}$ and $T\mathbf{x}$ in terms of the basis made up of all the basis vectors in the pairwise orthogonal eigenspaces W_j, $1 \leq j \leq k$.]

The formulae (6.4.4) and (6.4.5) may be written symbolically

$$I = \int_C dP_\lambda \tag{6.4.0.7}$$

$$T = \int_C \lambda \, dP_\lambda \tag{6.4.0.8}$$

and from these equations von Neumann ([**Be, vN1, vN2**]) drew the inspiration
for writing any continuous, normal, linear map $T : \mathcal{H} \mapsto \mathcal{H}$ of an arbitrary
Hilbert space $\mathcal{H}$ in the form (6.4.8).

THE SPECTRAL THEOREM: IF $T : \mathcal{H} \mapsto \mathcal{H}$ IS A CONTINUOUS, NORMAL,
LINEAR MAP OF A HILBERT SPACE $\mathcal{H}$ INTO ITSELF THEN THERE IS A SET

$$\{\, P_\lambda \; : \; \lambda \in \mathbf{C} \,\} \qquad\qquad (6.4.0.9)$$

OF ORTHOGONAL PROJECTIONS SUCH THAT

$$I = \int_{\mathbf{C}} dP_\lambda \qquad\qquad (6.4.0.10)$$

$$T = \int_{\mathbf{C}} \lambda \, dP_\lambda \qquad\qquad (6.4.0.11)$$

AND FOR ANY POLYNOMIAL p

$$p(T) = \int_{\mathbf{C}} p(\lambda) \, dP_\lambda. \qquad\qquad (6.4.0.12)$$

Note that:

i. (6.4.10)-(6.4.12) are valid even if T has no eigenvalues;
ii. (6.4.10) says that (6.4.9) is a *resolution of the identity* (cf. (4.3.9) and
(4.6.1));
iii. (6.4.12) is the genesis of a *functional calculus* for continuous, normal, linear
maps T.

[**Remark 6.4.0.3:** Von Neumann's principal concern was with self-
adjoint linear transformations. In his axiomatization of quantum mechanics self-adjoint linear transformations correspond to "observables"
in physics. Many regard von Neumann's mathematical formulation of
quantum mechanics as the most successful rigorous approach to the
subject (cf. [**Ma**]).]

If V is a finite-dimensional Euclidean space and H is self-adjoint then
$\sigma(H) \subset \mathbf{R}$. If an eigenvalue λ lies in a rectangle $\mathcal{R}$ then

$$\mathcal{R} \cap \mathbf{R} \neq \emptyset.$$

Since there are only finitely many eigenvalues λ_j there are finitely many pairwise
disjoint rectangles $\mathcal{R}_j$ such that:

i. $\mathcal{R}_j \cap \mathbf{R} \overset{\text{def}}{=} \mathcal{I}_j$ is a half-open interval on $\mathbf{R}$;
ii. if $S_{\mathcal{I}_j} \overset{\text{def}}{=} P_{\mathcal{R}_j}$ and p is a polynomial then

$$I = \sum_{j=1}^{k} S_{\mathcal{I}_j} \qquad\qquad (6.4.0.13)$$

$$Hx = \sum_{j=1}^{k} \lambda_j S_{\mathcal{I}_j} x \qquad (6.4.0.14)$$

$$p(H)x = \sum_{j=1}^{k} p(\lambda_j) S_{\mathcal{I}_j} x. \qquad (6.4.0.15)$$

If V is a finite-dimensional Euclidean space and U is unitary then

$$\sigma(U) \subset \{ z \; : \; z \in \mathbf{C}, \; |z| = 1 \} \overset{\text{def}}{=} \mathbf{T},$$

i.e.,

$$\sigma(U) \subset \{ z \; : \; \text{for some real number } \theta \; z = e^{i\theta} \} .$$

This time if an eigenvalue λ lies in a rectangle $\mathcal{R}$ then

$$\mathcal{R} \cap \mathbf{T} \neq \emptyset.$$

Since there are only finitely many eigenvalues $\lambda_j \overset{\text{def}}{=} e^{i\theta_j}$ there are finitely many pairwise disjoint rectangles $\mathcal{R}_j$ such that:

i. $\mathcal{R}_j \cap \mathbf{T} \overset{\text{def}}{=} \mathcal{A}_j$ is a closed subarc of $\mathbf{T}$;

ii. if $Q_{\mathcal{A}_j} \overset{\text{def}}{=} P_{\mathcal{R}_j}$ and p is a polynomial then

$$I = \sum_{j=1}^{k} Q_{\mathcal{A}_j} \qquad (6.4.0.16)$$

$$Ux = \sum_{j=1}^{k} e^{i\theta_j} Q_{\mathcal{A}_j} x \qquad (6.4.0.17)$$

$$p(U) = \sum_{j=1}^{k} p(e^{i\theta_j}) Q_{\mathcal{A}_j} x. \qquad (6.4.0.18)$$

Exercise 6.4.0.26. For an (infinite-dimensional) Hilbert space write the integral formulae that are the analogs of (6.4.13) - (6.4.18).

In [vN1], [vN2] there is an extensive motivation and discussion of the theory just outlined. In [Ta] there is a complete derivation of the results above.

Example 6.4.0.6. Refer to **Sections 4.3** and **4.4**. One proof that a normal matrix is diagonable is based on the diagonability criterion:

a SQUARE matrix A is diagonable iff the zeros of its minimal polynomial m_A are simple.

Associated with m_A are the idempotents $F_1, \ldots, F_k$. When A is normal the idempotents F_i are self-adjoint, i.e., orthogonal idempotents. The proof follows.

i. Since $F_i = \pi_i(A)$ it follows as in the proof of Theorem **4.4.2** that

$$ker(F_i) = ker(F_i^*).$$

Hence

$$im(F_i) = [ker(F_i^*)]^\perp = [ker(F_i)]^\perp = im(F_i^*) = [ker(F_i)]^\perp. \qquad (6.4.0.19)$$

ii. Since F_i and F_i^* are idempotents it follows from (6.4.19) that

$$F_i F_i^* = F_i^* \text{ and } F_i^* F_i = F_i.$$

Hence

$$F_i = (F_i^*)^* = (F_i F_i^*)^* = F_i F_i^* = F_i^*$$

and each F_i is self-adjoint.

Exercise 6.4.0.27. Let M be a closed subspace of a Hilbert space $\mathcal{H}$. Show that P_M is unique, i.e., show that if $P : \mathcal{H} \mapsto M$ is an orthogonal projection then $P = P_M$.

It follows that if $\lambda_i \in \mathcal{R}_i$ and λ_i is the only eigenvalue in $\mathcal{R}_i$ then $F_i = P_{\mathcal{R}_i}$. Hence if $\mathbf{x}_i \in im(F_i)$ and $i \neq j$ then $(\mathbf{x}_i, \mathbf{x}_j) = 0$:

Eigenvectors corresponding to different eigenvalues of a normal matrix A (a normal linear map T) are orthogonal.

[**Remark 6.4.0.4:** The last assertion may be derived directly as follows:
$$\begin{aligned}
(\mathbf{x}_i, \mathbf{x}_j) &= (F_i \mathbf{x}_i, F_j \mathbf{x}_j) \\
&= (F_j^* F_i \mathbf{x}_i, \mathbf{x}_j) \\
&= (F_j F_i \mathbf{x}_i, \mathbf{x}_j) = (O \mathbf{x}_i, \mathbf{x}_j) = 0.]
\end{aligned}$$

Definition **6.4.0.6.** If S is a subset of a Hilbert space $\mathcal{H}$ then the *closure* $\overline{S}$ of S is the intersection of all closed sets F such that $S \subset F$.

In other words, if $\mathcal{F}_S$ is the set of closed sets containing S then

$$\overline{S} \stackrel{\text{def}}{=} \bigcap_{F \in \mathcal{F}_S} F.$$

Exercise 6.4.0.28. Show that if S is a subset of a Hilbert space $\mathcal{H}$ then:

i. $\mathcal{F}_S$ is not empty;

ii. $S \subset \overline{S}$;

iii. if $S \subset R \subset \mathcal{H}$ then $\overline{S} \subset \overline{R}$;

iv. $\overline{S}$ is closed;

v. $\overline{\overline{S}} = \overline{S}$;

vi. $S = \overline{S}$ iff S is closed;

vii. if M is a subspace of $\mathcal{H}$ then $\overline{M}$ is a (closed) subspace of $\mathcal{H}$.

THEOREM 6.4.0.4. LET $\mathcal{H}$ BE A HILBERT SPACE AND LET

$$Z \overset{\text{def}}{=} \left\{ \mathbf{z}_\lambda \; : \; \lambda \in \Lambda \right\}$$

BE AN ORTHONORMAL SUBSET OF $\mathcal{H}$. IF $\mathbf{x} \in \mathcal{H}$ THEN

$$\|\mathbf{x}\|^2 \geq \sum_{\lambda \in \Lambda} |(\mathbf{x}, \mathbf{z}_\lambda)|^2 \text{ (BESSEL'S INEQUALITY)} \tag{6.4.0.20}$$

AND EQUALITY OBTAINS IN (6.4.20) IFF $\mathbf{x} \in \overline{span(Z)}$ (PARSEVAL'S THEOREM).

PROOF. Bessel's inequality (cf. **Section 4.4**) for *finite* orthonormal sets implies that for every finite subset Λ_f of Λ

$$\|\mathbf{x}\|^2 \geq \sum_{\lambda \in \Lambda_f} |(\mathbf{x}, \mathbf{z}_\lambda)|^2$$

whence (6.4.20) is true. In particular,

$$\left\{ \lambda \; : \; \lambda \in \Lambda, \text{ and } (\mathbf{x}, \mathbf{z}_\lambda) \neq 0 \right\}$$

is countable.

If $\mathbf{x} \in \overline{span(Z)}$ there is in $span(Z)$ a sequence

$$\{\mathbf{x}_k\}_{k=1}^{\infty}$$

such that

$$\|\mathbf{x}_k - \mathbf{x}\|^2 \to 0 \text{ as } k \to \infty.$$

Hence if $a_{k\lambda} \overset{\text{def}}{=} (\mathbf{x}_k, \mathbf{z}_\lambda)$ then for each k there is in Λ a finite subset

$$\{\lambda_{k1}, \ldots, \lambda_{kn_k}\}$$

such that

$$\mathbf{x}_k = \sum_{j=1}^{n_k} a_{k\lambda_{kj}} \mathbf{z}_{\lambda_{kj}}.$$

If the countable set of all $z_{\lambda_{k_j}}$ is enumerated, say as z_{λ_r}, $1 \leq r < \infty$, then, since $\|\mathbf{x}_k - \mathbf{x}\| \to 0$ as $k \to \infty$,

$$a_{k\lambda_r} \to (\mathbf{x}, z_{k_{\lambda_r}}) \text{ as } k \to \infty.$$

If $\epsilon > 0$ there is in $\mathbf{N}$ a k_ϵ such that $\|\mathbf{x} - \mathbf{x}_k\| < \epsilon$ if $k > k_\epsilon$ and for each such k there is an R such that $\mathbf{x}_k = \sum_{r=1}^{R}(\mathbf{x}_k, z_{\lambda_r})z_{\lambda_r}$. Hence if $k \geq k_\epsilon$ and R is large enough Bessel's inequality implies

$$\left\| \mathbf{x} - \sum_{r=1}^{R}(\mathbf{x}, z_{\lambda_r})z_{\lambda_r} \right\|$$

$$\leq \|\mathbf{x} - \mathbf{x}_k\| + \left\| \mathbf{x}_k - \sum_{r=1}^{R}(\mathbf{x}_k, z_{\lambda_r})z_{\lambda_r} \right\|$$

$$+ \left\| \sum_{r=1}^{R}(\mathbf{x}_k - \mathbf{x}, z_{\lambda_r})z_{\lambda_r} \right\|$$

$$\leq \|\mathbf{x} - \mathbf{x}_k\| + 0 + \|\mathbf{x}_k - \mathbf{x}\| \leq 2\epsilon.$$

Since ϵ is an arbitrary positive number it follows that

$$\|\mathbf{x}\|^2 - \sum_{r=1}^{R}|(\mathbf{x}, z_{\lambda_r})|^2 \; \left(= \left\| \mathbf{x} - \sum_{r=1}^{R}(\mathbf{x}, z_{\lambda_r})z_{\lambda_r} \right\|^2 \right) \tag{6.4.0.21}$$

approaches 0 as $R \to \infty$:

$$\mathbf{x} = \sum_{\lambda \in \Lambda}(\mathbf{x}, z_\lambda)z_\lambda.$$

Conversely (6.4.21) implies that if equality obtains in the Bessel inequality then $\mathbf{x} \in \overline{span(Z)}$.

$$\Omega$$

For a Hilbert space $\mathcal{H}$ the analog of a Hamel basis for a vector space V is an orthonormal set $\mathcal{F}$ such that

$$\mathcal{H} = \overline{span(\mathcal{F})}.$$

THEOREM 6.4.0.5. IF $\mathcal{H}$ IS A HILBERT SPACE THERE IS IN $\mathcal{H}$ AN OR-THONORMAL SET $\mathcal{F}$ SUCH THAT $\mathcal{H} = \overline{span(\mathcal{F})}$.

PROOF. The procedure used to establish the existence of $\mathcal{F}$ is an imitation of the procedure used to prove the existence of a Hamel basis in a vector space V.

Let Λ, "$<$", "$<_1$" be as in the PROOF of THEOREM **6.2.1** let S be $\mathcal{H} \setminus \mathbf{O}$. By transfinite induction define a set $\{\mathcal{F}_\lambda\}$ of orthonormal subsets of S as follows.

i. Let $\mathbf{x}_1$ be the first element of S and let $\mathcal{F}_1$ be the singleton set

$$\left\{ \frac{\mathbf{x}_1}{\|\mathbf{x}_1\|} \right\} \overset{\text{def}}{=} \{\mathbf{u}_1\}.$$

ii. Having defined $\mathcal{F}_\lambda$ for all λ such that $\lambda < \lambda'$ define $\mathcal{F}_{\lambda'}$ as

$$\begin{cases} \bigcup_{\lambda < \lambda'} \mathcal{F}_l \quad \text{if } \mathbf{x}_{\lambda'} \in \overline{span\left(\bigcup_{\lambda < \lambda'} \mathcal{F}_\lambda\right)} \overset{\text{def}}{=} M_{\lambda'} \\ \left\{ \frac{\mathbf{x}_{\lambda'} - P_{M'_\lambda}\mathbf{x}_{\lambda'}}{\|\mathbf{x}_{\lambda'} - P_{M'_\lambda}\mathbf{x}_{\lambda'}\|} \right\} \cup \left(\bigcup_{\lambda < \lambda'} \mathcal{F}_\lambda\right) \overset{\text{def}}{=} \{\mathbf{u}_{\lambda'}\} \cup \left(\bigcup_{\lambda < \lambda'} \mathcal{F}_\lambda\right) \quad \text{if } \mathbf{x}_{\lambda'} \notin M_{\lambda'}. \end{cases}$$

iii. Define $\mathcal{F}$ to be

$$\bigcup_{\lambda \in \Lambda} \mathcal{F}_\lambda.$$

Exercise 6.4.0.29. Show that $\mathcal{F}$ as constructed above is an orthonormal set such that

$$\mathcal{H} = \overline{span(\mathcal{F})}. \tag{6.4.0.22}$$

$$\Omega \cdot$$

An orthonormal set $\mathcal{F}$ for which (6.4.22) holds is a *complete orthonormal set* (CON) for $\mathcal{H}$. Thus THEOREM **6.4.4** may be interpreted as saying that an orthonormal (ON) set $\{\mathbf{z}_\lambda\}_{\lambda \in \Lambda}$ is a CON iff for every $\mathbf{x}$ in $\mathcal{H}$ there obtains

$$\|\mathbf{x}\|^2 = \sum_{\lambda \in \Lambda} |(\mathbf{x}, \mathbf{z}_\lambda)|^2 \ (< \infty)$$

$$\mathbf{x} = \sum_{\lambda \in \Lambda} (\mathbf{x}, \mathbf{z}_\lambda)\mathbf{z}_\lambda, \text{ a convergent series,}$$

(the *Parseval theorem*). Furthermore if

$$\sum_{\lambda \in \Lambda} |a_\lambda|^2 < \infty$$

then there is a unique $\mathbf{x}$ such that

$$(\mathbf{x}, \mathbf{z}_\lambda) = a_\lambda$$

$$\mathbf{x} = \sum_{\lambda \in \Lambda} a_\lambda \mathbf{z}_\lambda, \text{ a convergent series.}$$

Exercise 6.4.0.30. Show that the ON set $\{\mathbf{z}_\lambda\}_{\lambda \in \Lambda}$ is a CON iff: $(\mathbf{x}, \mathbf{z}_\lambda) = 0$ for all $\lambda \Rightarrow \mathbf{x} = \mathbf{O}$.

Exercise 6.4.0.31. Show that U in $[\mathcal{H}, \mathcal{H}]$ is unitary iff $im(U) = \mathcal{H}$ and for all $\mathbf{x}$, $\mathbf{y}$ in $\mathcal{H}$

$$(U\mathbf{x}, U\mathbf{y}) = (\mathbf{x}, \mathbf{y}).$$

Exercise 6.4.0.32.

i. Show that U in $[\mathcal{H}, \mathcal{H}]$ is unitary iff for some CON set $\{\, \mathbf{z}_\lambda \ : \ \lambda \in \mathbf{\Lambda} \,\}$ the set $\{\, U\mathbf{z}_\lambda \ : \ \lambda \in \mathbf{\Lambda} \,\}$ is also a CON set.

ii. Show that if

$$\{\, \mathbf{z}_\lambda \ : \ \lambda \in \mathbf{\Lambda} \,\} \text{ and } \{\, \mathbf{w}_\mu \ : \ \mu \in \mathbf{\Lambda} \,\}$$

are two CON sets and if

$$U : \mathbf{x} \overset{\text{def}}{=} \sum_{\lambda \in \mathbf{\Lambda}} a_\lambda \mathbf{z}_\lambda \mapsto U\mathbf{x} \overset{\text{def}}{=} \sum_{\lambda \in \mathbf{\Lambda}} a_\lambda \mathbf{w}_\lambda$$

then U is unitary.

[**Remark 6.4.0.5:** It is also true that if

$$\{\, \mathbf{z}_\lambda \ : \ \lambda \in \mathbf{\Lambda} \,\} \text{ and } \{\, \mathbf{w}_\mu \ : \ \mu \in \mathbf{M} \,\}$$

are CON sets for a Hilbert space $\mathcal{H}$ then $\mathbf{\Lambda}$ and $\mathbf{M}$ are in one-one correspondence ([**Gem**]). Hence in **Exercise 6.4.32** there it is permissible to use the same indexing set $\mathbf{\Lambda}$ for the two CON sets.]

CHAPTER 7

QUADRATIC FORMS

7.1. Quadratic forms and linear algebra

If $A \overset{\text{def}}{=} (a_{ij})_{i,j=1}^{n,n}$ is a self-adjoint matrix and if

$$\mathbf{x} \overset{\text{def}}{=} \begin{pmatrix} x_1 \\ \vdots \\ x_n \end{pmatrix}, \ \mathbf{y} \overset{\text{def}}{=} \begin{pmatrix} y_1 \\ \vdots \\ y_n \end{pmatrix}$$

then

$$B : \mathbf{C}^n \times \mathbf{C}^n \ni \{\mathbf{x}, \mathbf{y}\} \mapsto B(\mathbf{x}, \mathbf{y}) \overset{\text{def}}{=} (A\mathbf{x}, \mathbf{y}) = \sum_{i,j=1}^{n,n} a_{ij} x_i \overline{y_j}$$

is a *bilinear form* or *bilinear functional* defined on $\mathbf{C}^n \times \mathbf{C}^n$.

Example 7.1.1. If $A = I$ then $B = (\ ,\ )$, the standard inner product.

If $\mathbf{x} = \mathbf{y}$ then

$$B(\mathbf{x}, \mathbf{x}) \overset{\text{def}}{=} Q(\mathbf{x}) = \sum_{i,j=1}^{n,n} a_{ij} x_i \overline{x_j}$$

is a *quadratic form*. Because A is self-adjoint its eigenvalues $\lambda_1, \ldots, \lambda_n$ are real and there is a unitary matrix U such that

$$U^* A U = \begin{pmatrix} \lambda_1 & & & \\ & \lambda_2 & & \\ & & \ddots & \\ & & & \lambda_n \end{pmatrix}.$$

For each $\mathbf{x}$ let $U^* \mathbf{x}$ be $\mathbf{y}$. Then

$$Q(\mathbf{x}) = Q(U\mathbf{y}) = (AU\mathbf{y}, U\mathbf{y}) = (U^* A U \mathbf{y}, \mathbf{y})$$

$$= \sum_{i=1}^{n} \lambda_i |y_i|^2. \tag{7.1.1}$$

If

$$D \stackrel{\text{def}}{=} \begin{pmatrix} d_1 & & & \\ & d_2 & & \\ & & \ddots & \\ & & & d_n \end{pmatrix}$$

and D is invertible then

$$Q(DU\mathbf{y}) = \sum_{i=1}^{n} |d_i|^2 \lambda_i |y_i|^2 \tag{7.1.2}$$

and thus there is no unique linear combination of "squares" to which $Q(\mathbf{x})$ can be reduced. On the other hand, since $|d_i|^2 > 0$ the number of positive coefficients in (7.1.1) is the same as the number of positive coefficients in (7.1.2). In fact, this state of affairs is generally true according to THEOREM **7.1.1** given after LEMMA **7.1.1** following.

LEMMA **7.1.1.** IF V AND W ARE SUBSPACES OF $\mathbf{C}^n$ THEN

$$dim(V) + dim(W) - dim(V \cap W) \le n. \tag{7.1.3}$$

PROOF. Let X be a basis for $V \cap W$ and fill out X to a basis Y resp. Z for V resp. W:

$$X \stackrel{\text{def}}{=} \{\mathbf{x}_1, \dots, \mathbf{x}_k\}$$
$$Y \stackrel{\text{def}}{=} \{\mathbf{x}_1, \dots, \mathbf{x}_k\} \cup \{\mathbf{y}_1, \dots, \mathbf{y}_l\} \stackrel{\text{def}}{=} X \cup Y_1, X \cap Y_1 = \emptyset$$
$$Z \stackrel{\text{def}}{=} \{\mathbf{x}_1, \dots, \mathbf{x}_k\} \cup \{\mathbf{z}_1, \dots, \mathbf{z}_m\} \stackrel{\text{def}}{=} X \cup Z_1, X \cap Z_1 = \emptyset.$$

It suffices to prove that $X \cup Y_1 \cup Z_1$ is linearly independent since then $k+l+m \le n$ and

$$dim(V) + dim(W) - dim(V \cap W) = (k + l) + (k + m) - k$$
$$= k + l + m \le n.$$

If

$$\sum_{p=1}^{k} a_p \mathbf{x}_p + \sum_{q=1}^{l} b_q \mathbf{y}_q + \sum_{r=1}^{m} c_r \mathbf{z}_r \stackrel{\text{def}}{=} \mathbf{x} + \mathbf{y} + \mathbf{z} = \mathbf{O}$$

then $V \ni \mathbf{y} = -\mathbf{x} - \mathbf{z} \in W$, i.e., $\mathbf{y} \in V \cap W$ whence for some coefficients d_s

$$\mathbf{y} \stackrel{\text{def}}{=} \sum_{q=1}^{l} b_q \mathbf{y}_q = \sum_{s=1}^{k} d_s \mathbf{x}_s.$$

Since $X \cup Y_1$ (like $X \cup Z_1$) is a basis it is linearly independent and so $b_q = 0$, $1 \le q \le l$, $d_s = 0$, $1 \le s \le k$. Similarly it follows that $c_r = 0$, $1 \le r \le m$, and then that $a_p = 0$, $1 \le p \le k$.

$$\Omega$$

THEOREM 7.1.1. (SYLVESTER'S LAW OF INERTIA) IF Q IS A QUADRATIC FORM CORRESPONDING TO THE SELF-ADJOINT MATRIX A, IF T AND S ARE INVERTIBLE MATRICES, AND IF

$$T^{-1}\mathbf{x} \overset{\text{def}}{=} \mathbf{y} \overset{\text{def}}{=} \begin{pmatrix} y_1 \\ \vdots \\ y_n \end{pmatrix}$$

$$S^{-1}\mathbf{x} \overset{\text{def}}{=} \mathbf{z} \overset{\text{def}}{=} \begin{pmatrix} z_1 \\ \vdots \\ z_n \end{pmatrix}$$

$$Q(\mathbf{x}) = Q(T\mathbf{y}) \overset{\text{def}}{=} \sum_{i=1}^{n} t_i |y_i|^2$$

$$= Q(S\mathbf{z}) \overset{\text{def}}{=} \sum_{i=1}^{n} s_i |z_i|^2$$

THEN THE t_i RESP. THE s_i ARE REAL AND THE NUMBER τ OF POSITIVE t_i IS THE SAME AS THE NUMBER σ OF POSITIVE s_i.

[**Remark 7.1.1:** Hence consideration of $-Q$ shows that the number of negative t_i is the same as the number of negative s_i and hence the number of t_i that are 0's is the same as the number of s_i that are 0's.]

PROOF. Setting y_i resp. z_i at 1 and the other y_j resp. z_j at 0 shows that t_i resp. s_i is real.

It may be assumed that there is a positive τ such that $t_1, \ldots, t_\tau > 0$ and all other τ_j are nonpositive and that there are some positive s_i, namely $s_1, \ldots, s_\sigma$ and $\sigma < \tau$. (If there is no positive s_i and if $T\mathbf{e}_1 \overset{\text{def}}{=} \mathbf{x}$ then

$$\mathbf{x} \neq \mathbf{O}$$

$$S^{-1}\mathbf{x} \overset{\text{def}}{=} \mathbf{z} \neq \mathbf{O}$$

$$Q(T\mathbf{e}_1) = t_1 = Q(\mathbf{x}) > 0$$

$$Q(S\mathbf{z}) = Q(\mathbf{x}) \le 0,$$

a contradiction. This line of argument is the basis of the proof that follows.)

Let W_τ be $T[span(\{\mathbf{e}_1, \ldots, \mathbf{e}_\tau\})]$. Define $V_{n-\sigma}$ by the formula

$$V_{n-\sigma} \overset{\text{def}}{=} \begin{cases} S[span(\{\mathbf{e}_{\sigma+1}, \ldots, \mathbf{e}_n\})] & \text{if there are some negative } s_i \\ \{\mathbf{O}\} & \text{otherwise.} \end{cases}$$

Then since T and S are invertible

$$dim(W_\tau) + dim(V_{n-\sigma}) = \tau + (n - \sigma) = n + (\tau - \sigma) > n.$$

Hence, by virtue of **Lemma 7.1.1**, $dim(V_{n-\sigma} \cap W_\tau) > 0$ and so $\{\mathbf{O}\} \subsetneq M \overset{\text{def}}{=} V_{n-\sigma} \cap W_\tau$.

Thus there is in $span(\{\mathbf{e}_1, \ldots, \mathbf{e}_\tau\})$ resp. $span(\{\mathbf{e}_{\sigma+1}, \ldots, \mathbf{e}_n\})$ a vector $\mathbf{y} \overset{\text{def}}{=} (y_1, \ldots, y_n)^t$ resp. a $\mathbf{z} \overset{\text{def}}{=} (z_1, \ldots, z_n)^t$ such that $T\mathbf{y} = S\mathbf{z} \overset{\text{def}}{=} \mathbf{x} \neq \mathbf{O}$. Hence

$$Q(\mathbf{x}) = Q(T\mathbf{y}) = \sum_{i=1}^{\tau} t_i |y_i|^2 > 0$$

$$= Q(S\mathbf{z}) = \sum_{j=\sigma+1}^{n} s_j |z_j|^2 \leq 0,$$

a contradiction.

$$\Omega$$

Exercise 7.1.1. In the context and notation of the Weyl minmax principle (cf. **Exercise 4.4.53**), which is a theorem about quadratic forms, construct the alternative proof sketched below.

i. Construct an ON basis $Y_l \overset{\text{def}}{=} \{\mathbf{y}_1, \ldots, \mathbf{y}_l\}$ for $span(V_{k-1})$ and fill Y_l out to an ON basis $Y \overset{\text{def}}{=} \{\mathbf{y}_i\}_{i=1}^n$ for $\mathbf{C}^n$.

ii. Show $l \leq k - 1$.

iii. The vectors $\mathbf{z}_i$, $1 \leq i \leq n$ serving as an orthonormal eigenbasis, define a map

$$\widetilde{U} : \mathbf{C}^n \ni \mathbf{x} \overset{\text{def}}{=} \sum_{i=1}^{n} a_i \mathbf{y}_i \mapsto \widetilde{U}(\mathbf{x}) \overset{\text{def}}{=} \mathbf{z} = \sum_{i=1}^{n} a_i \mathbf{z}_i \in \mathbf{C}^n.$$

Show $\widetilde{U}$ is unitary.

iii. Show

$$(\widetilde{U}^* A \widetilde{U} \mathbf{x}, \mathbf{x}) = (A\mathbf{z}, \mathbf{z}).$$

iv. Show $\mathbf{x} \in W$ iff $\widetilde{U}\mathbf{x} \in span(\{\mathbf{z}_{l+1}, \ldots, \mathbf{z}_n\})$.

v. Show

$$M(V_{k-1}) \geq \lambda_k$$

$$\lambda_k = minimum_L\, maximum \left\{ (A\mathbf{x}, \mathbf{x}) \; : \; \begin{cases} \|\mathbf{x}\| = 1 \\ L \text{ a subspace} \\ dim(M) \leq k - 1 \\ \mathbf{x} \in L^\perp \end{cases} \right\}$$

$$= minimum_L\, maximum \left\{ (A\mathbf{v}, \mathbf{v}) \; : \; \begin{cases} \|\mathbf{v}\| = 1 \\ L \text{ a subspace} \\ dim(L) \geq n - k + 1 \\ \mathbf{v} \in L. \end{cases} \right\}$$

Exercise 7.1.2. Validate the following interpretation of the min-max theorem.

If $\mathbf{x}$ is confined to the unit ball in an $n - k + 1$-dimensional subspace then each $\mathbf{x}$ may be specified by $n - k + 1$ components and $(A\mathbf{x}, \mathbf{x})$ may be regarded as a function of those $n - k + 1$ components:

$$(A\mathbf{x}, \mathbf{x}) \stackrel{\text{def}}{=} \sum_{i,j=1}^{n-k+1} c_{ij} \xi_i \overline{\xi}_j.$$

The matrix $C \stackrel{\text{def}}{=} (c_{ij})_{i,j=1}^{n-k+1,n-k+1}$ is self-adjoint and its (necessarily real) eigenvalues $\{\mu_i\}_{i=1}^{n-k+1}$ may be arranged so that

$$\mu_1 \geq \cdots \geq \mu_{n-k+1}.$$

If $1 \leq h \leq n - k + 1$ then $\lambda_h \geq \mu_h \geq \lambda_{h+k}$.

In many applications the eigenvalues are related to the resonant frequencies of an oscillatory (vibratory) system — the larger the eigenvalue the lower the associated resonant frequencies. Hence the interpretation above says that the more one restricts oscillation the higher are the resonant frequencies. A strong and rigid system (a "stiff" system) resonates at very high frequencies. With *no* restrictions (in a "flabby" system) there are no oscillations at all. To increase the pitch of violin string, stretch it.

7.2. Positive definite and positive semidefinite quadratic forms

DEFINITION 7.2.1. IF V IS A EUCLIDEAN VECTOR SPACE AND $B \in [V]$ THEN B IS *positive definite* IFF

$$\mathbf{x} \neq \mathbf{O} \Rightarrow (B\mathbf{x}, \mathbf{x}) > 0.$$

IF (MERELY) $(B\mathbf{x}, \mathbf{x}) \geq 0$ THEN B IS *positive semidefinite*.

[**Note 7.2.1:** A positive definite linear transformation is positive semidefinite. The converse is false, e.g., if $n > 1$ each diagonal $n \times n$ matrix unit U_{ii} corresponds to a positive semidefinite linear transformation that is not positive definite.]

Example 7.2.1. The identity I is positive definite. If A is invertible then AA^* is positive definite. In any event AA^* is positive semidefinite.

Exercise 7.2.1. Show that if B is positive semidefinite then B is self-adjoint.

[*Hint:* Prove and use the identity

$$(B\mathbf{x}, \mathbf{y}) = \frac{1}{4}\{(B\mathbf{x} + \mathbf{y}, B\mathbf{x} + \mathbf{y}) - (B\mathbf{x} - \mathbf{y}, B\mathbf{x} - \mathbf{y})$$
$$+ i(B\mathbf{x} + i\mathbf{y}, \mathbf{x} + i\mathbf{y}) - i(B\mathbf{x} - i\mathbf{y}, \mathbf{x} - i\mathbf{y})\}.]$$

Exercise 7.2.2. Show that if V is finite-dimensional then a unitarily diagonable B in $[V]$ is positive definite resp. semidefinite iff:

i. each of its eigenvalues is positive resp. nonnegative;
ii. there is an invertible A resp. an A such that $B = AA^*$.

[**Note 7.2.2:** From i above it follows that if B is positive definite resp. semidefinite then $det(B) > 0$ resp. $det(B) \geq 0.$]

If the context is $\mathbf{C}^n$ and $\mathbf{x}$ is confined to lie in $span(\{\mathbf{e}_1, \ldots, \mathbf{e}_k\}) \overset{\text{def}}{=} W_k$ then $(B\mathbf{x}, \mathbf{x})$ may be regarded as a function of k variables. If $B \overset{\text{def}}{=} (b_{ij})_{i,j=1}^{n,n}$, $B_k \overset{\text{def}}{=} (b_{ij})_{i,j=1}^{k,k}$, and if B is positive definite resp. semidefinite so is B_k. Hence if B is positive definite then

$$det(B_k) > 0, \ 1 \leq k \leq n. \tag{7.2.1}$$

If $\mathbf{x}$ is confined to $span(\{\mathbf{e}_{i_1}, \ldots, \mathbf{e}_{i_k}\})$ there is a $k \times k$ matrix $B_{i_1, \ldots, i_k}$, a *principal minor*, such that if $\mathbf{x} \overset{\text{def}}{=} \sum_{\alpha=1}^{k} x_\alpha \mathbf{e}_{i_\alpha}$ then

$$(B\mathbf{x}, \mathbf{x}) = \sum_{\alpha, \beta=1}^{k} b_{i_\alpha j_\beta} x_\alpha \overline{x}_\beta.$$

If B is positive semidefinite so is each $B_{i_1, \ldots, i_k}$ and so

$$det(B_{i_1, \ldots, i_k}) \geq 0, \ 1 \leq i_1 < i_2 < \cdots < i_k \leq n. \tag{7.2.2}$$

Exercise 7.2.3 below is helpful in the next development.

Exercise 7.2.3. Show that if $\mathcal{E}$ is the product of lower triangular ROW COMBINATION matrices $E_{ij}(t_{ij})$ $(i < j)$ and A is a nonsingular lower triangular matrix A then the diagonal entries of A and of $\mathcal{E}A$ are the same.

THEOREM 7.2.1. IF B IS A SELF-ADJOINT MATRIX THEN B IS POSITIVE DEFINITE RESP. SEMIDEFINITE IFF (7.2.1) RESP. (7.2.2) IS TRUE.

PROOF. The necessity of the conditions is proved above.

Case i. Assume (7.2.1) is true. If the the diagonal entries of the echelon form $B_E \overset{\text{def}}{=} \mathcal{E}B$ of B are $\beta_1, \ldots, \beta_n$ then (7.2.1) implies that no ROW EXCHANGEs are performed when the GEM is applied to B. Hence

$$det(B_k) = \prod_{i=1}^{k} \beta_{ii}$$

and so each $\beta_{ii} > 0$. Since $B = B^*$ it follows that

$$\mathcal{E}B = \mathcal{E}B^* = B_E$$
$$B\mathcal{E}^* = B_E^*$$
$$\mathcal{E}B\mathcal{E}^* = B_E\mathcal{E}^* = \mathcal{E}B_E^*.$$

Since $\mathcal{E}B_E^*$ is lower triangular and $B_E\mathcal{E}^*$ is upper triangular each is diagonal. According to **Exercise 7.2.3** the diagonal entries of B_E^* and $\mathcal{E}B_E^*$ are the same: they are the β_{ii} and so

$$\mathcal{E}B\mathcal{E}^* = \begin{pmatrix} \beta_{11} & & \\ & \ddots & \\ & & \beta_{nn} \end{pmatrix}$$

If $\mathbf{x} \neq \mathbf{O}$ there is a a nonzero $\mathbf{y} \overset{\text{def}}{=} (y_1, \ldots, y_n)^t$ such that $\mathbf{x} = \mathcal{E}^*\mathbf{y}$ and thus

$$(B\mathbf{x}, \mathbf{x}) = (B\mathcal{E}^*\mathbf{y}, \mathcal{E}^*\mathbf{y}) = \sum_{i=1}^{n} \beta_{ii}|y_i|^2 > 0.$$

Hence (7.2.1) implies that B is positive definite.

Case ii. Assume (7.2.2) is true. Thus each diagonal entry of B is nonnegative. If each diagonal entry of B is 0 then (7.2.2) and the self-adjointness of B imply first that each supradiagonal entry $b_{i,i+1} = 0$ and each subdiagonal entry $b_{j+1,j} = 0$. Continued argument by mathematical induction shows $B = O$.

That simple instance aside, some diagonal entry of B is positive. For a suitable permutation matrix Π and for some k in $\mathbf{N}$ the first k diagonal entries of $\Pi B\Pi \overset{\text{def}}{=} C \overset{\text{def}}{=} (c_{ij})_{i,j=1}^{n,n}$ are positive and if $k < n$ the others are 0's. If $\mathcal{E}C \overset{\text{def}}{=} C_E$ the argument given in *Case i.* shows that $\mathcal{E}C\mathcal{E}^*$ is a diagonal matrix in which the diagonal entries are those of C. Then for each $\mathbf{x}$ there is a $\mathbf{y}$ such that $\mathbf{x} = \Pi\mathbf{y}$ and if $\mathcal{E}^*\mathbf{y} \overset{\text{def}}{=} \mathbf{z}$

$$(B\mathbf{x}, \mathbf{x}) = (C\mathbf{y}, \mathbf{y}) = (C_E\mathbf{z}, \mathbf{z}) \geq 0.$$

$$\Omega$$

[**Remark 7.2.1:** The arguments given above show also that a self-adjoint matrix B is positive definite resp. semidefinite iff each of its pivots is positive resp. nonnegative.]

     Chapter 7. Quadratic forms

Example 7.2.2. In the study of a twice continuously differentiable function f of n real variables there is the Taylor Formula with Remainder (cf. THEOREM 4.8.4): For $\mathbf{x} \overset{\text{def}}{=} (x_1, \ldots, x_n)^t$, and $\mathbf{a} \overset{\text{def}}{=} (a_1, \ldots, a_n)^t$ there is a vector

$$\theta \overset{\text{def}}{=} \begin{pmatrix} \theta_1 \\ \vdots \\ \theta_n \end{pmatrix}$$

such that θ_i lies between x_i and a_i, $1 \le i \le n$, and

$$f(x_1, \ldots, x_n) = f(a_1, \ldots, a_n) + \sum_{i=1}^{n} \frac{\partial f}{\partial x_i}(a_1, \ldots, a_n)(x_i - a_i)$$

$$\frac{1}{2!} \sum_{i,j=1}^{n} \frac{\partial^2 f}{\partial x_i \partial x_j}(\theta_1, \ldots, \theta_n)(x_i - a_i)(x_j - a_j). \quad (7.2.3)$$

If

$$\nabla f(\mathbf{a}) \overset{\text{def}}{=} \left(\frac{\partial f}{\partial x_1}(a_1, \ldots, a_n), \ldots, \frac{\partial f}{\partial x_n}(a_1, \ldots, a_n) \right) \in \mathbf{C}^n$$

$$H f(\mathbf{a}) \overset{\text{def}}{=} \left\{ \frac{\partial^2 f}{\partial x_i \partial x_j}(a_1, \ldots, a_n) \right\}_{i,j=1}^{n} \in Mat_{nn}$$

then (7.2.3) may be written

$$f(\mathbf{x}) = f(\mathbf{a}) + \nabla f(\mathbf{a})(\mathbf{x} - \mathbf{a}) + \frac{1}{2!}(H f(\theta)(\mathbf{x} - \mathbf{a}), \mathbf{x} - \mathbf{a}).$$

The function f assumes an extreme value at a vector $\mathbf{a}$ in the interior of the domain of f iff $\nabla f(\mathbf{a}) = \mathbf{O}$. If the entries in $H f(\mathbf{x})$ are continuous at and near $\mathbf{a}$ the number $f(\mathbf{a})$ is

$$\begin{cases} \text{locally maximal} & \text{if } -H f(\mathbf{a}) \text{ is positive definite} \\ \text{locally minimal} & \text{if } H f(\mathbf{a}) \text{ is positive definite.} \end{cases} \quad (7.2.4)$$

When $f(\mathbf{a})$ is locally minimal resp. locally maximal $\mathbf{a}$ is by abuse of language a *local minimum* resp. *local maximum*.

Any of the criteria for positive definiteness may be applied to determine whether $-H f(\mathbf{a})$ resp. $H f(\mathbf{a})$ is positive definite. In particular, if $n = 2$ and

$$\frac{\partial f}{\partial x_i}(\mathbf{a}) \overset{\text{def}}{=} f_i, \quad \frac{\partial^2 f}{\partial \mathbf{x}_i \partial \mathbf{x}_j}(\mathbf{a}) \overset{\text{def}}{=} f_{ij}, \quad 1 = 1, 2,$$

the criterion (7.2.1) is the familiar rule:

$$\begin{cases} \text{if } f_1 > 0, \ \Delta \overset{\text{def}}{=} (f_{11} f_{22} - f_{21} f_{12}) > 0 & \mathbf{a} \text{ is a local minimum} \\ \text{if } -f_1 > 0, \ \Delta > 0 & \mathbf{a} \text{ is a local maximum.} \end{cases}$$

Exercise 7.2.4. Explain why the condition $\Delta > 0$ appears *both* in the rule for a local minimum *and* in the rule for a local maximum.

[**Remark 7.2.2:** The matrix $Hf(\mathbf{a})$ is the *Hessian,* whence the "H." The conditions in (7.2.4) are sufficient but *not* necessary for the two kinds of extrema.]

Exercise 7.2.5. A self-adjoint matrix B is negative definite resp. semidefinite iff $-B$ is positive definite resp. semidefinite. For negative definiteness resp. semidefiniteness formulate and prove analogs of all criteria given above for positive definiteness resp. semidefiniteness.

If B is a positive definite $n \times n$ matrix then the function

$$[\,,\,] : \{\mathbf{x}, \mathbf{y}\} \mapsto (B\mathbf{x}, \mathbf{y}) \overset{\text{def}}{=} [\mathbf{x}, \mathbf{y}]$$

is an inner product satisfying E1-E4. If A is a self-adjoint matrix then $B^{-1}A$ is *B-self-adjoint*, i.e., self-adjoint with respect to $[\,,\,]$:

$$[B^{-1}A\mathbf{x}, \mathbf{y}] = (BB^{-1}A\mathbf{x}, \mathbf{y}) = (A\mathbf{x}, \mathbf{y}) = (\mathbf{x}, A\mathbf{y})$$
$$[\mathbf{x}, B^{-1}A\mathbf{y}] = ((B\mathbf{x}, B^{-1}A\mathbf{y})$$
$$= ((B^{-1})^* B\mathbf{x}, A\mathbf{y}) = (B^{-1}B\mathbf{x}, A\mathbf{y}) = (\mathbf{x}, A\mathbf{y}).$$

Hence there is for $\mathbf{C}^n$ a $B^{-1}A$-eigenbasis $X \overset{\text{def}}{=} \{\mathbf{x}_1, \ldots, \mathbf{x}_n\}$ such that for eigenvalues $\lambda_1, \ldots, \lambda_n$

$$B^{-1}A\mathbf{x}_i = \lambda_i \mathbf{x}_i$$
$$[\mathbf{x}_i, \mathbf{x}_j] = \delta_{ij}.$$

If $\mathbf{x} \overset{\text{def}}{=} \sum_{i=1}^n x_i \mathbf{x}_i$ then

$$(B\mathbf{x}, \mathbf{x}) = [\mathbf{x}, \mathbf{x}] = \sum_{i=1}^n |x_i|^2$$
$$(A\mathbf{x}, \mathbf{x}) = [B^{-1}A\mathbf{x}, \mathbf{x}] = \sum_{i=1}^n \lambda_i |x_i|^2.$$

The last conclusion is known as the SIMULTANEOUS REDUCTION THEOREM FOR QUADRATIC FORMS. It is a consequence of the unitary diagonability of a self-adjoint matrix if "unitary" and "self-adjoint" are interpreted in the context of $[\,,\,]$.

Exercise 7.2.6. Show that the eigenvalues $\lambda_1, \ldots, \lambda_n$ above are the zeros of the polynomial $det(A - \lambda B)$.

[**Note 7.2.3:** A distinction should be drawn between the simultaneous diagonability of a finite set of pairwise commuting diagonable matrices (cf. **Exercise 4.3.17**) and the simultaneous reduction of two quadratic forms $(A\mathbf{x}, \mathbf{x})$ and $(B\mathbf{x}, \mathbf{x})$. The matrices A and B need not commute for the simultaneous reduction of the quadratic forms.

The reduction of the quadratic form $(A\mathbf{x}, \mathbf{x})$ may be achieved in various ways by a substitution of the form $A \mapsto T^*AT$ for some non-singular T. Diagonalization is achieved by a substitution of the form $A \mapsto P^{-1}AP$ for some nonsingular P. For the functional calculus of matrices the substitution $A \mapsto T^*AT$ is not necessarily adequate: if p is a polynomial $p(T^*AT)$ and $T^*p(A)T$ are not necessarily equal whereas $p(P^{-1}AP) = P^{-1}p(A)P$.]

Exercise 7.2.7. Assume

$$B \stackrel{\text{def}}{=} \begin{pmatrix} 1 & 2+i \\ 2-i & 6 \end{pmatrix}$$

$$A \stackrel{\text{def}}{=} \begin{pmatrix} -1 & 1-3i \\ 1+3i & 0 \end{pmatrix}.$$

Show B is positive definite and that $AB \neq BA$. Simultaneously reduce the quadratic forms $(B\mathbf{x}, \mathbf{x})$ and $(A\mathbf{x}, \mathbf{x})$. Show that there is no invertible matrix P (unitary or not) such that both $P^{-1}AP$ and $P^{-1}BP$ are diagonal matrices.

Exercise 7.2.8. Show that two $n \times n$ diagonable matrices A and B are simultaneously diagonable iff $AB = BA$ (cf. **Exercise 4.3.17**).

Exercise 7.2.9. Generalize and prove the Weyl minmax principle (cf. **Exercise 4.4.53**) in the context of simultaneous reduction of two quadratic forms as follows.

$i.$ Show that the eigenvalues λ_i are real and thus that they may be ordered so that $\lambda_1 \geq \lambda_2 \geq \cdots \geq \lambda_n$.

$ii.$ Show that

$$\lambda_k = minimum_M\, maximum \left\{ [B^{-1}A\mathbf{x}, \mathbf{x}] : \begin{cases} [\mathbf{x}, \mathbf{x}] = 1 \\ M \text{ a subspace} \\ dim(M) \leq k-1 \\ \mathbf{x} \in M^{\perp} \end{cases} \right\}$$

$$= minimum_L\, maximum \left\{ [B^{-1}A\mathbf{v}, \mathbf{v}] : \begin{cases} [\mathbf{v}, \mathbf{v}] = 1 \\ L \text{ a subspace} \\ dim(L) \geq n-k+1 \\ \mathbf{v} \in L \end{cases} \right\}.$$

The Hadamard product of $A \stackrel{\text{def}}{=} (a_{ij})_{i,j=1}^{n,n}$ and $B \stackrel{\text{def}}{=} (b_{ij})_{i,j=1}^{n,n}$ is

$$A \circ B \overset{\text{def}}{=} \{a_{ij}b_{ij}\}_{i,j=1}^{n} \overset{\text{def}}{=} (c_{ij})_{i,j=1}^{n,n} \overset{\text{def}}{=} C$$

(cf. **Exercise 3.3.14**).

THEOREM **7.2.2**. IF A AND B ARE POSITIVE DEFINITE SO IS $A \circ B$. IF A AND B ARE POSITIVE SEMIDEFINITE SO IS $A \circ B$.

PROOF. Assume A and B are positive definite. Let $U \overset{\text{def}}{=} (u_{ij})_{i,j=1}^{n,n}$ be a unitary matrix such that

$$U^{*}BU \overset{\text{def}}{=} D \overset{\text{def}}{=} \begin{pmatrix} \lambda_1 & & & \\ & \lambda_2 & & \\ & & \ddots & \\ & & & \lambda_n \end{pmatrix}.$$

Since B is positive definite each λ_i is positive. If $\mathbf{x} \overset{\text{def}}{=} (x_1, \ldots, x_n)^t$ then

$$b_{ij} = \sum_{k=1}^{n} \lambda_k u_{ik} \overline{u}_{jk}$$

$$(A \circ B\mathbf{x}, \mathbf{x}) = \sum_{i,j=1}^{n} a_{ij} b_{ij} x_i \overline{x}_j$$

$$= \sum_{i,j,k=1}^{n} a_{ij} \lambda_k u_{ik} \overline{u}_{jk} x_i \overline{x}_j$$

and if $Q(x_1, \ldots, x_n) \overset{\text{def}}{=} (A\mathbf{x}, \mathbf{x})$ then

$$(A \circ B\mathbf{x}, \mathbf{x}) = \sum_{k=1}^{n} \lambda_k Q(u_{1k}x_1, \ldots, u_{nk}x_n). \tag{7.2.5}$$

Since A is positive definite and U is nonsingular it follows that $(A \circ B\mathbf{x}, \mathbf{x}) > 0$ if $\mathbf{x} \neq \mathbf{O}$.

A similar argument shows that $A \circ B$ is positive semidefinite if both A and B are positive semidefinite.

$$\Omega$$

THEOREM **7.2.2** is due to I. Schur.

Exercise 7.2.10. Verify that if

$$A \overset{\text{def}}{=} \begin{pmatrix} 1 & 2+i \\ 2-i & 6 \end{pmatrix}, \; B \overset{\text{def}}{=} \begin{pmatrix} 2 & 1-3i \\ 1+3i & 6 \end{pmatrix}$$

then A, B, and $A \circ B$ are positive definite.

Exercise 7.2.11. If $t > 0$ there is in $\mathbf{R}$ an s such that $t = e^s$. Prove the analog for positive definite matrices: if B is a positive definite matrix there is a self-adjoint matrix A such that $B = e^A$.

Exercise 7.2.12. If $t \geq 0$ then $1 + t \neq 0$ and $(1 + t)^{-1}$ exists. Prove the analog for positive semidefinite matrices: if A is a positive semidefinite matrix then $I + A$ is nonsingular.

CHAPTER 8

MISCELLANY

8.1. Duality

If V is a vector space and if $X \overset{\text{def}}{=} \{\, \mathbf{x}_\lambda \ : \ \lambda \in \Lambda \,\}$ is a Hamel basis for V then each $\mathbf{x}$ in V may be written

$$\mathbf{x} - \sum_{\lambda \in \Lambda} a_\lambda \mathbf{x}_\lambda$$

in which only finitely many coefficients a_λ are not 0's. For each λ the *coefficient functional*

$$\mathbf{x}'_\lambda : V \ni \mathbf{x} \mapsto \mathbf{x}'_\lambda(\mathbf{x}) \overset{\text{def}}{=} a_\lambda$$

is a linear transformation, a *linear functional*, i.e., an element in $\mathcal{L}(V, \mathbf{K}) = [V, \mathbf{K}]$, denoted V', the *dual* of V. With respect to the natural definitions of addition and multiplication by scalars V' is also a vector space. Each $\mathbf{x}'_\lambda$ is a special kind of linear functional in that

$$\mathbf{x}'_\mu(\mathbf{x}_\lambda) = \delta_{\mu\lambda} \overset{\text{def}}{=} \begin{cases} 1 & \text{if } \mu = \lambda \\ 0 & \text{otherwise,} \end{cases} \tag{8.1.1}$$

i.e., each $\mathbf{x}'_\mu$ takes the value 0 off a finite set, namely the singleton set $\{\mathbf{x}_\mu\}$. The set $X' \overset{\text{def}}{=} \{\mathbf{x}'_\lambda\}_{\lambda \in \Lambda}$ is the set of coefficient functionals. Owing to (8.1.1) the pair $\{X, X'\}$ is a *biorthonormal pair*.

DEFINITION 8.1.1. THE *weak dual* $\widetilde{V}'$ OF V IS THE SET OF LINEAR FUNCTIONALS EACH OF WHICH IS 0 OFF A FINITE SUBSET OF X.

THEOREM 8.1.1. IF V IS A VECTOR SPACE AND $X \overset{\text{def}}{=} \{\, \mathbf{x}_\lambda \ : \ \lambda \in \Lambda \,\}$ IS A HAMEL BASIS FOR V THEN THE SET $X' \overset{\text{def}}{=} \{\, \mathbf{x}'_\lambda \ : \ \lambda \in \Lambda \,\}$ OF COEFFICIENT FUNCTIONALS IS A HAMEL BASIS FOR $\widetilde{V}'$.

PROOF. If $\mathbf{x}' \in \widetilde{V}'$ and $\mathbf{x} \in V$ then

$$\mathbf{x}'(\mathbf{x}) = \sum_{\lambda \in \Lambda} a_\lambda \mathbf{x}'(\mathbf{x}_\lambda)$$

$$= \sum_{\lambda \in \Lambda} \mathbf{x}'_\lambda(\mathbf{x})\mathbf{x}'(\mathbf{x}_\lambda)$$

$$= [\sum_{\lambda \in \Lambda} \mathbf{x}'(\mathbf{x}_\lambda)](\mathbf{x}).$$

Since $\sum_{\lambda \in \Lambda} \mathbf{x}'(\mathbf{x}_\lambda)$ is a finite sum it follows that

$$\mathbf{x}' = \sum_{\lambda \in \Lambda} \mathbf{x}'(\mathbf{x}_\lambda).$$

Hence $\widetilde{V}' = span(X')$. However, just as an orthonormal set in a Euclidean vector space is linearly independent, so, owing to the biorthonormality of the pair $\{X, X'\}$, each of X and X' is linearly independent.

Indeed, $\mathbf{O}'$ denoting the linear functional identically 0 ($\mathbf{O}'(\mathbf{x}) = 0$ for each $\mathbf{x}$) if

$$\sum_{\lambda \in \Lambda} b_\lambda \mathbf{x}'_\lambda = \mathbf{O}'$$

then

$$[\sum_{\lambda \in \Lambda} b_\lambda \mathbf{x}'_\lambda](\mathbf{x}_\mu) = b_\mu = 0, \ \mu \in \Lambda.$$

Thus X' is a spanning linearly independent set for $\widetilde{V}'$.

$$\Omega$$

Exercise 8.1.1. Let X be a linearly independent set in a vector space V. Define a set X' in $\widetilde{V}'$ so that $\{X, X'\}$ is a biorthonormal pair. Show that:

 i. if X is a Hamel basis for V there is in V' one and only one set X' such that $\{X, X'\}$ is a biorthonormal pair for the dual pair $\{V, V'\}$;

 ii. if X is a Hamel basis for V then X' is a Hamel basis for $\widetilde{V}'$.

Example 8.1.1. If V is the vector space of all polynomials with coefficients in $\mathbf{C}$ the functional

$$F : V \ni \sum_{k=0}^{n} a_k x^k \mapsto \sum_{k=1}^{n} k a_k$$

is in $V' \setminus \widetilde{V}'$. To see this let $P \overset{\text{def}}{=} \{p_0, p_1, \ldots\}$ be the sequence of polynomials such that $p_n(x) = x^n$. Then (cf. **Exercise 6.4.2**) P is a Hamel basis. If P' is such that $\{P, P'\}$ is the unique biorthonormal pair associated with P then P' is a Hamel basis for $\widetilde{V}'$ but, as shown next, *not* for V'.

Indeed, if F defined above is in $\widetilde{V}'$ and if $P' \overset{\text{def}}{=} \{q_n\}_{n=0}^{\infty}$ then

$$F = \sum_{n=0}^{\infty} b_n q_n, \qquad\qquad (8.1.2)$$

a sum in which only finitely many summands are not 0's. Hence

$$m = F(p_m) = \sum_{n=0}^{\infty} b_n q_n (p_m) = b_m$$

whence the sum (8.1.2) is *not* finite, a contradiction.

THEOREM **8.1.2** below justifies the terminology *dual pair*.

THEOREM **8.1.2**. IF V IS A VECTOR SPACE THEN V IS THE WEAK DUAL OF $\widetilde{V}'$: $V = \big(\widetilde{V'}\big)'$.

PROOF. If $\mathbf{x} \in V$ then the map

$$\mathbf{x}'' : \widetilde{V}' \ni \mathbf{x}' \mapsto \mathbf{x}'(\mathbf{x}) \overset{\text{def}}{=} \mathbf{x}''(\mathbf{x}') \in \mathbf{K}$$

is in $\big(\widetilde{V'}\big)'$. The map

$$\Phi : V \ni \mathbf{x} \mapsto \Phi(\mathbf{x}) \overset{\text{def}}{=} \mathbf{x}'' \in \big(\widetilde{V'}\big)'$$

is a monomorphism from V to $\big(\widetilde{V'}\big)'$: if $\Phi(\mathbf{x}) = \Phi(\mathbf{y})$ then, for any $\mathbf{x}'$ in $\widetilde{V}'$, $\mathbf{x}'(\mathbf{x} - \mathbf{y}) = 0$. If $\mathbf{z} \overset{\text{def}}{=} \mathbf{x} - \mathbf{y} \neq \mathbf{O}$ and if X is a Hamel basis for V then some coefficient functional $\mathbf{x}'_\lambda$ is such that $\mathbf{x}'_\lambda(\mathbf{z}) \neq 0$, a contradiction.

Let $\{X, X'\}$ be a biorthonormal pair of Hamel bases for the weak dual pair $\{V, \widetilde{V}'\}$. If $\phi \in \big(\widetilde{V'}\big)'$ it suffices to show that there is in V an $\mathbf{x}$ such that for any $\mathbf{x}'$ in $\widetilde{V}'$

$$\Phi(\mathbf{x})(\mathbf{x}') = \phi(\mathbf{x}').$$

However since $\phi \in \big(\widetilde{V'}\big)'$ there are only finitely many nonzero numbers in

$$\{\, \phi(\mathbf{x}'_\lambda) \ : \ \mathbf{x}'_\lambda \in X' \,\}.$$

If

$$\mathbf{x} \overset{\text{def}}{=} \sum_{\lambda \in \Lambda} \phi(\mathbf{x}'_\lambda) \mathbf{x}_\lambda$$

$$\mathbf{x}' \overset{\text{def}}{=} \sum_{\mu \in \Lambda} b_\mu \mathbf{x}'_\mu \ \ (\text{in } \widetilde{V}')$$

---

then

$$\Phi(\mathbf{x})(\mathbf{x}') = \mathbf{x}''(\mathbf{x}') = \mathbf{x}'\Big(\sum_{\lambda \in \Lambda} \phi(\mathbf{x}'_\lambda)\mathbf{x}_\lambda\Big)$$

$$= \Big[\sum_{\mu \in \Lambda} b_\mu \mathbf{x}'_\mu\Big]\Big(\sum_{\lambda \in \Lambda} \phi(\mathbf{x}'_\lambda)\mathbf{x}_\lambda\Big)$$

$$= \sum_{\mu \in \Lambda} b_\mu \phi(\mathbf{x}'_\mu) = \phi(\mathbf{x}').$$

Hence $\Phi(V) = (\widetilde{V'})'$.

$$\Omega$$

Exercise 8.1.2. Show that if V is a vector space, W is a subspace of V, and $\mathbf{x} \notin W$ then there is in $\widetilde{V}'$ an ϕ such that

$$\phi(\mathbf{x}) = 1, \text{ and } \phi(\mathbf{w}) = 0 \text{ if } \mathbf{w} \in W.$$

(Thus $\phi \in W^\perp \subset \widetilde{V}'$.)

8.2. Multilinear functionals (forms)

If V is a vector space let $V^{k\times}$ denote the k-fold Cartesian product of V with itself:

$$V^{k\times} \stackrel{\text{def}}{=} \Big\{ \mathbf{X} \stackrel{\text{def}}{=} (\mathbf{x}_1, \ldots, \mathbf{x}_k) \; : \; \mathbf{x}_j \in V, \; 1 \le j \le k \Big\}.$$

[**Remark 8.2.1:** Although $V^{k\times}$ is a vector space with respect to "component-wise" addition and multiplication by numbers:

$$\mathbf{X} + \mathbf{Y} \stackrel{\text{def}}{=} (\mathbf{x}_1 + \mathbf{y}_1, \ldots, \mathbf{x}_k + \mathbf{y}_k)$$

$$t\mathbf{X} \stackrel{\text{def}}{=} (t\mathbf{x}_1, \ldots, t\mathbf{x}_k)$$

this aspect of $V^{k\times}$ is ignored for the moment. When, as above, $V^{k\times}$ is structured as a vector space it is denoted $\bigoplus^k V$, the *k-fold direct sum* of V with itself.]

A functional $f : V^{k\times} \mapsto \mathbf{C}$ is *multilinear* or a *k-form* iff

$$f[(\mathbf{x}_1, \ldots, \mathbf{x}_j + \mathbf{y}_j, \ldots, \mathbf{x}_k)] = f[(\mathbf{x}_1, \ldots, \mathbf{x}_j, \ldots, \mathbf{x}_k)] + f[(\mathbf{x}_1, \ldots, \mathbf{y}_j, \ldots, \mathbf{x}_k)]$$

$$f[(\mathbf{x}_1, \ldots, t\mathbf{x}_j, \ldots, \mathbf{x}_k)] = tf[(\mathbf{x}_1, \ldots, \mathbf{x}_j, \ldots, \mathbf{x}_k)]$$

$$1 \le j \le k,$$

in other words, $f(\mathbf{X})$ depends linearly on each vector component of $\mathbf{X}$.

A k-form ϕ is *alternating* iff $\phi(\mathbf{X}) = 0$ whenever two of the vector components of $\mathbf{X}$ are the same. If $\mathbf{X} \in V^{k\times}$ and $\pi \in S_{1,2,\ldots,k}$ (the set of all permutations of the numbers $1, \ldots, k$) then

$$\pi\mathbf{X} \stackrel{\text{def}}{=} (\mathbf{x}_{\pi(1)}, \ldots, \mathbf{x}_{\pi(k)}).$$

[**Remark 8.2.2:** The applications of k-forms are widespread, particularly in the study of differential geometry.]

Example 8.2.1. If Mat_{nn} is regarded as $\mathbf{C}_n{}^{n\times}$ then (cf. **Section 3.1**) det is a nonzero alternating n-form on $\mathbf{C}_n{}^{n\times}$.

Exercise 8.2.1. Show that if ϕ is a k-form and $\pi \in S_{1,2,\ldots,k}$ then ϕ is an alternating k-form iff for each $\mathbf{X}$ in $V^{k\times}$: $\phi(\pi\mathbf{X}) \stackrel{\mathrm{def}}{=} \pi\phi(\mathbf{X}) = sgn(\pi)\phi(\mathbf{X})$.

Exercise 8.2.2. Show that if the vector components of $\mathbf{X}$ are linearly dependent and ϕ is an alternating k-form then $\phi(\mathbf{X}) = 0$.

Exercise 8.2.3. Show that if $dim(V) < k$ and ϕ is an alternating k-form defined on $V^{k\times}$ then $\phi(V^{k\times}) = \{0\}$ ($\phi \equiv$ the zero functional $\stackrel{\mathrm{def}}{=} O_k$).

Exercise 8.2.4. Show that if $dim(V) = k$, if the components of

$$X \stackrel{\mathrm{def}}{=} \{\mathbf{x}_1, \ldots, \mathbf{x}_k\}$$

are linearly independent, and ϕ is a nonzero alternating k-form defined on $V^{k\times}$ then $\phi(\mathbf{X}) \neq 0$.

[**Note 8.2.1:** If the field in which an alternating k-form takes its values is not $\mathbf{C}$ but some field $\mathbf{K}$ in which $1 + 1 = 0$ then *every* function h taking values in $\mathbf{K}$ is such that $h = -h$. Yet if $V = \mathbf{K}^2$ and

$$h[(\mathbf{x}, \mathbf{y})] \stackrel{\mathrm{def}}{=} x_1 y_1 + x_2 y_2$$

then $h[(\mathbf{e}_1, \mathbf{e}_1)] = 1 \neq 0$. In this instance the conclusion of **Exercise 8.2.1** is false.]

The development in **Section 3.1** demonstrates that up to a constant factor there is only one nonzero alternating n-form, namely det, defined on $\mathbf{C}_n{}^{n\times}$.

It is shown below, and without reference to determinants, that there is only one nonzero alternating n-form defined on $\mathbf{C}_n{}^{n\times}$. The proof is simplified by the notational convenience:

$$\pi_{ij}(l) \stackrel{\mathrm{def}}{=} \begin{cases} l & \text{if } l \neq i, j \\ j & \text{if } l = i \\ i & \text{if } l = j. \end{cases}$$

Thus π_{ij} interchanges i and j and leaves any other l unchanged.

THEOREM 8.2.1. IF THE VECTOR SPACE V IS n-DIMENSIONAL THEN:

i. IF $k \leq n$ THERE IS A NONZERO ALTERNATING k-FORM DEFINED ON $V^{k\times}$;

ii. IF $k = n$ THERE IS ESSENTIALLY ONE AND ONLY ONE NONZERO ALTERNATING n-FORM DEFINED ON $V^{n\times}$.

PROOF. If $n = 1$ and $\mathbf{x}$ is a basis vector for V the functional ϕ defined by the equation

$$\phi(a\mathbf{x}) = a$$

is essentially the only 1-form (which is automatically alternating) defined on $V^{1\times}\ (= V)$.

The next part of the argument proceeds via mathematical induction. If the result is false there is a natural number n, $n > 1$, and in $\{1, \ldots, n\}$ there is a k for which the following are true.

 i. There is an n-dimensional vector space V such that:
  a. there is no nonzero alternating k-form defined on $V^{k\times}$;
  b. if $1 \le dim(W) < n$ and $1 \le l \le dim(W)$ there is a nonzero alternating l-form defined on $W^{l\times}$.

Let $X \overset{\text{def}}{=} \{\mathbf{x}_1, \ldots, \mathbf{x}_k\}$ be a linearly independent set in V and let W be $span(\{\mathbf{x}_1, \ldots, \mathbf{x}_{k-1}\})$. By inductive hypothesis there is a nonzero alternating $k - 1$-form γ defined on $W^{k-1\times}$.

Let η in V' be such that $\eta \in W^{\perp}$ and $\eta(\mathbf{x}_k) = 1$ (cf. **Exercise 8.1.2**). The functional

$$\Gamma : (\mathbf{z}_1, \ldots, \mathbf{z}_k) \mapsto \gamma[(\mathbf{z}_1, \ldots, \mathbf{z}_{k-1})]\eta(\mathbf{z}_k)$$

is a k-form.

Define ϕ by the formula

$$\phi[(\mathbf{z}_1, \ldots, \mathbf{z}_k)] = \sum_{i=1}^{k-1} \pi_{ik}\Gamma[(\mathbf{z}_1, \ldots, \mathbf{z}_k)] - \Gamma[(\mathbf{z}_1, \ldots, \mathbf{z}_k)]. \tag{8.2.1}$$

(If $k = 3$ then (8.2.1) reads as follows:

$$\phi[(\mathbf{z}_1, \mathbf{z}_2, \mathbf{z}_3)] =$$
$$\gamma[(\mathbf{z}_3, \mathbf{z}_2)]\eta(\mathbf{z}_1) + \gamma[(\mathbf{z}_1, \mathbf{z}_3)]\eta(\mathbf{z}_2) - \gamma[(\mathbf{z}_1, \mathbf{z}_2)]\eta(\mathbf{z}_3).)$$

Since X is linearly independent, $\eta \in W^{\perp}$, and $\eta(\mathbf{x}_k) = 1$ it follows that

$$\phi[(\mathbf{x}_1, \ldots, \mathbf{x}_k)] = -\gamma[(\mathbf{x}_1, \ldots, \mathbf{x}_{k-1})] \ne 0.$$

Assume $\mathbf{Z} \overset{\text{def}}{=} \{\mathbf{z}_1, \ldots, \mathbf{z}_k\}$ is such that $\mathbf{z}_p = \mathbf{z}_q \overset{\text{def}}{=} \mathbf{a}$.

 i. If $1 \le p < q < k$ then the last term in (8.2.1) is 0 because γ is an alternating $k - 1$-form.

If $i \ne p, q$ the ith summand in (8.2.1) is 0 because γ is an alternating k-form. On the other hand,

$$\pi_{pk}\Gamma(\mathbf{Z}) = -\pi_{qk}\Gamma(\mathbf{Z})$$

because in each member above the argument of η is $\mathbf{a}$ and the positions of $\mathbf{z}_k$ and $\mathbf{a}$ are the reverses of each other in the two members above. Hence in this case $\phi(\mathbf{Z}) = 0$.

---

ii. If $1 \leq p < q = k$ then every summand except the pth in

$$\sum_{i=1}^{k-1} \pi_{ik} \Gamma[(\mathbf{z}_1, \ldots, \mathbf{z}_k)]$$

is 0 because γ is an alternating $k - 1$-form. Furthermore

$$\pi_{pk} \Gamma(\mathbf{Z}) = \Gamma(\mathbf{Z})$$

and so again $\phi(\mathbf{Z}) = 0$. Hence ϕ is an alternating k-form. The induction is complete.

To see that there is essentially one and only one nonzero alternating n-form use the existence of some nonzero alternating n-form, say ϕ, and the "break-out" operation (cf. (3.1.6)) to determine that if $\mathbf{E} \overset{\text{def}}{=} (\mathbf{e}_1, \ldots, \mathbf{e}_n)$ then

$$\phi[(\sum_{i=1}^{n} a_{1i}\mathbf{e}_i, \ldots, \sum_{i=1}^{n} a_{ni}\mathbf{e}_i)] =$$
$$sgn(\pi) \sum_{\pi \in S_{1,2,\ldots,n}} a_{1\pi(1)} \cdots a_{n\pi(n)} \phi(\mathbf{E}).$$

If ψ is an alternating n-form then

$$\psi = \frac{\psi(E)}{\phi(E)} \phi.$$

$$\Omega$$

[**Note 8.2.2:** The alternating n-form ψ such that $\psi(E) = 1$ is *det*.]

8.3. Tensor products

If $\{V_1, \ldots, V_k\}$ is a finite set of vector spaces let

$$L(V_1 \times V_2 \times \cdots \times V_k) \overset{\text{def}}{=} \mathbf{V}_k \overset{\text{def}}{=} L$$

denote the set of all **K**-valued *finitely supported* functions f, i.e., associated with each f there is in $\mathbf{V}_k$ a finite subset $supp(f)$, the *support of f*, and f is 0 off $supp(f)$: if $\mathbf{X} \overset{\text{def}}{=} (\mathbf{x}_1, \ldots, \mathbf{x}_k) \notin supp(f)$ then $f(\mathbf{X}) = 0$.

In $L(\mathbf{V}_k)$ are two special kinds of functions:

> $i.$ if $supp(f) = \{(\mathbf{x}_1,\ldots,\mathbf{x}_k),(\mathbf{x}_1,\ldots,t\mathbf{x}_j,\ldots,\mathbf{x}_k)\}$
> and if $t \in \mathbf{K}$
> then $f[(\mathbf{x}_1,\ldots,t\mathbf{x}_j,\ldots,\mathbf{x}_k)] = -tf[(\mathbf{x}_1,\ldots,\mathbf{x}_k)]$;
>
> $ii.$ if $supp(f) = \{(\mathbf{x}_1,\ldots,\mathbf{x}_j,\ldots\mathbf{x}_k),(\mathbf{x}_1,\ldots,\mathbf{y}_j,\ldots,\mathbf{x}_k),$
> $(\mathbf{x}_1,\ldots,\mathbf{x}_j+\mathbf{y}_j,\ldots,\mathbf{x}_k)\}$
> then $f[(\mathbf{x}_1,\ldots,\mathbf{x}_j+\mathbf{y}_j,\ldots,\mathbf{x}_k)] =$
> $$- f[(\mathbf{x}_1,\ldots,\mathbf{x}_j,\ldots\mathbf{x}_k)]$$
> $$- f[(\mathbf{x}_1,\ldots,\mathbf{y}_j,\ldots,\mathbf{x}_k)].$$

The span of all functions of the kinds described in i and ii is a subspace W of the vector space $L(\mathbf{V}_k)$. Two functions that differ by a function in W are regarded as equal. In other words, there is an *equivalence relation* $\sim$ among the functions in L: $f \sim g$ iff $f - g \in W$. The relation $\sim$, like similarity among matrices, is

$i.$ reflexive $(f \sim f)$;
$ii.$ symmetric $(f \sim g$ iff $g \sim f)$;
$iii.$ transitive (if $f \sim g$ and $g \sim h$ then $f \sim h$).

The relation $\sim$ decomposes L into pairwise disjoint *equivalence classes*. Each f in L is in one and only one equivalence class and g belongs to the equivalence class containing f iff $f \sim g$.

If $supp(f) = (\mathbf{x}_1,\ldots,\mathbf{x}_k) \stackrel{\text{def}}{=} \mathbf{X}$ and $f(\mathbf{X}) = 1$ then the equivalence class containing f is denoted

$$\mathbf{x}_1 \otimes \cdots \otimes \mathbf{x}_k. \tag{8.3.1}$$

If

$$supp(g_j) = \mathbf{X}_j \stackrel{\text{def}}{=} (\mathbf{x}_{j1},\ldots,\mathbf{x}_{jk})$$
$$g_j(\mathbf{X}_j) = 1$$
$$a_j \in \mathbf{K}, \ 1 \leq j \leq n$$

then the equivalence class of $a_1 g_1 + \cdots + a_n g_n$ is denoted

$$\sum_{j=1}^{n} a_j \mathbf{x}_{j1} \otimes \cdots \otimes \mathbf{x}_{jk} \tag{8.3.2}$$

The tensor product $\bigotimes_{i=1}^{k} V_i$ is the set of all sums of the form given in (8.3.2). If the natural laws of function addition and multiplication by scalars are observed then $\bigotimes_{i=1}^{k} V_i$ is a vector space, the *tensor product* of the vector spaces V_i, $1 \leq i \leq k$. The elements of $\bigotimes_{i=1}^{k} V_i$ are *tensors*.

[**Note 8.3.1:** In group theoretic terms the tensor product is the *quotient space* of L by W:

$$\bigotimes_{i=1}^{k} V_i = L/W.$$

The equivalence classes that are the elements of the tensor product are the *cosets* of W.]

Example 8.3.1. Let V_1 be the set $C([0,1])$ of continuous $\mathbf{C}$-valued functions defined on $[0,1]$. Let V_2 be $\mathbf{C}^m$. Then, $C([0,1], \mathbf{C}^m)$ denoting the set of continuous $\mathbf{C}^m$-valued functions defined on $[0,1]$, there is a natural identification

$$V_1 \otimes V_2 \leftrightarrow C([0,1], \mathbf{C}^m).$$

Indeed, if $F \in C([0,1], \mathbf{C}^m)$ then there are in $C([0,1])$ functions

$$a_1(t), \ldots, a_m(t)$$

such that

$$F(t) = \sum_{j=1}^{m} a_j(t)\mathbf{e}_j.$$

Then F may be identified with

$$\sum_{j=1}^{m} a_j(t) \otimes \mathbf{e}_j \ \ (\text{in } V_1 \otimes V_2).$$

Conversely, if

$$\sum_{p=1}^{n} b_p(t) \otimes \mathbf{x}_p \in V_1 \otimes V_2 \tag{8.3.3}$$

then each $\mathbf{x}_p$ may be written as a linear combination of the $\mathbf{e}_j$:

$$\mathbf{x}_p \overset{\text{def}}{=} \sum_{j=1}^{m} c_{pj}\mathbf{e}_j, \ \ 1 \le p \le n. \tag{8.3.4}$$

According to the rules for combining tensors,

$$\sum_{p=1}^{n} b_p(t) \otimes \mathbf{x}_p = \sum_{j=1}^{m}(\sum_{p=1}^{n} b_p(t)c_{pj}) \otimes \mathbf{e}_j$$

$$\overset{\text{def}}{=} \sum_{j=1}^{m} d_j(t) \otimes \mathbf{e}_j$$

---

which may be identified with

$$G(t) \stackrel{\text{def}}{=} \sum_{j=1}^{m} d_j(t)\mathbf{e}_j \ (\text{in } C([0,1], \mathbf{C}^m)).$$

Exercise 8.3.1. Show that the identifications established in **Example 8.3.1** are one-one and inverses of each other, so that there is an isomorphism between $C([0,1], \mathbf{C}^m)$ and $V_1 \otimes V_2$.

THEOREM 8.3.1. THERE IS AN ISOMORPHISM BETWEEN THE TENSOR PRODUCT $\bigotimes_{i=1}^{k} V_i$ AND THE WEAK DUAL $\mathcal{D} \stackrel{\text{def}}{=} \bigoplus_{i=1}^{k} \widetilde{V_i}'$ OF THE *direct sum* OF THE WEAK DUALS $\widetilde{V_i}'$ OF THE V_i.

PROOF. If $\mathbf{x}'' \in \mathcal{D}$ then the maps

$$f_i : \widetilde{V_i}' \ni \mathbf{x}' \mapsto \mathbf{x}''[(0,\ldots,\mathbf{x}',\ldots)] \in \mathbf{K}, \ 1 \le i \le k$$

are in $((\widetilde{V_i}'))'$, whence, via duality, in V_i. Hence identify $\mathbf{x}''$ with $\tau \stackrel{\text{def}}{=} f_1 \otimes \cdot \otimes f_k$ in $\bigotimes_{i=1}^{k} V_i$.

Conversely if

$$\tau \stackrel{\text{def}}{=} \sum_{i=1}^{n} a_i \mathbf{x}_{i1} \otimes \mathbf{x}_{ik} \in \bigotimes_{i=1}^{k} V_i$$

$$\mathbf{X} \stackrel{\text{def}}{=} (\mathbf{x}_1', \ldots, \mathbf{x}_k') \in \bigoplus_{i=1}^{k} \widetilde{V_i}'$$

then the formula

$$\tau(\mathbf{X}) \stackrel{\text{def}}{=} \sum_{j=1}^{k}[\sum_{i=1}^{n} a_i \mathbf{x}_i'(\mathbf{x}_{ij})] \in \mathbf{K}$$

defines τ as an element $\mathbf{y}''$ of $\mathcal{D}$. A direct calculation shows that, according to the rules above,

$$\mathbf{x}'' \mapsto \tau \text{ and } \tau \mapsto \mathbf{y}'' \Rightarrow \mathbf{x}'' = \mathbf{y}''.$$

Similarly, according to the rules above,

$$\tau \mapsto \mathbf{y}'' \text{ and } \mathbf{y}'' \mapsto \sigma \Rightarrow \tau = \sigma.$$

$$\Omega$$

THEOREM 8.3.1 is an emphatic expression of the multilinear character of the construction of the tensor product. By forming L/W one "divides out" the *non*multilinear functions in L.

BIBLIOGRAPHY

Bea — Beale, E. M. L., *Cycling in the dual simplex algorithm*, Naval Research Logistics Quarterly 2(1955), 269–276.

Ber — Berberian, S. K., *Lectures in functional analysis and operator theory*, Springer-Verlag, New York, 1974.

Bl — Bland, R. G., *New finite pivoting rules for the simplex method*, Mathematics of Operations Research, 2 (May,1977), 103–107.

Bô — Bôcher, M., *Introduction to higher algebra*, Macmillan, New York, 1907; Dover, New York, 1964.

Bor — Borgwardt, K. H., *The average number of pivot steps required by the simplex method is polynomial*, Zeitschrift für Operations Research, Serie A-B, 26 (1982) no. 5, A157– A177.

Ch — Charnes, A., *Optimality and degeneracy in linear programming*, Econometrica, 20, (1952), 160–170.

Co — Coury, J. E., *On the measure of zeros of coordinate functions*, Proceedings of the American Mathematical Society, 25 (1970), 16–20.

Da — Dantzig, G. B., *Maximization of linear functions of variables subject to linear inequalities*, cf. [**Ko**], 339–347.

Fa — Faddeev, D. K. and Faddeev, V. N., *Computational methods of linear algebra*, Translated by Robert C. Williams, W. H. Freeman, San Francisco, 1963.

Fe — Fefferman, C., *An easy proof of the fundamental theorem of algebra*, Amer. Math. Monthly, 74 (1967), 854–5.

Fi — Filippov, A. F., *A short proof of the theorem on reduction of a matrix to Jordan form*, (Russian), Vestnik Moskov. Univ., Ser. I Math. Meh. 2 (1971), 18–9.

Ga — Gale, D., *The theory of economic models*, Mc Graw-Hill Book Company, New York, 1960.

GaW — Galperin, A. and Waksman, Z., *An elementary approach to Jordan theory*, Amer. Math. Monthly, 87 (1980), 728–32.

GeA — Gelbaum, B. R. , *On von Neumann's theorem on Abelian families of operators*, Proceedings of the American Mathematical Society, 15 (1964), 391–2.

GeL — Gelbaum, B. R. , *On relatively free subsets of Lie Groups*, Proceedings of the American Mathematical Society, 58 (1976), 301–5.

Gem — Gelbaum, B. R. , *Problems in analysis*, Springer-Verlag, New York, 1982..

Gep — Gelbaum, B. R. , *An algorithm for the minimal polynomial of a matrix*, Amer. Math. Monthly, 90 (1983), 43–4.

GeS — Gelbaum, B. R. and Schanuel, S., *A characterization of the semigroup of matrix units*, J. Austral. Math. Soc. (Series A) **29** (1980), 291-6.

GL — Glazman, I. M. and Liubich, Iu. I., *Finite-dimensional linear analysis via problems*, (Russian), Moskva, 1969.

GvL — Golub, G. H. and van Loan, C. F., *Matrix computations*, The Johns Hopkins University Press, Baltimore, 1983.

GW — Golub, G. H. and Wilkinson, J. H., *Ill-conditioned eigensystems and the computation of the Jordan canonical form*, SIAM Rev. 18 (1976), 578–619.

GR — Gunning, R. and Rossi, H., *Analytic functions of several complex variables*, Prentice-Hall, Englewood Cliffs, N. J. , 1965.

Ha — Hačijan, L. C., *A polynomial algorithm in linear programming*, Soviet Math-Doklady 20, 1 (1979), 191–194.

KaR — Kagstrom, B. and Ruhe, A., *An algorithm for the numerical computation of the Jordan form of a complex matrix*, ACM Trans. Math. Software, 6 (1980), 398–419.

Karm — Karmarkar, N., *A new polynomial-time algorithm for linear programming*, Combinatorica, 4 (4), (1984), 373–395.

KeS — Kemeny, J. G. and Snell, J. L., *Finite Markov chains*, D. Van Nostrand Company, Inc., Princeton, 1960.

Kh — Khachiyan, L. G., *Polynomial algorithms in linear programming*, Zhurnal Vichislitel'noi Matematiki i Matematicheskoi Fiziki (Russian) 20, 1 (1980), 51–68.

Kl — Klee, V. and Minty, G. L., *How good is the simplex algorithm?*, Proceedings of the Third Symposium, UCLA, 1969, 159–175; Inequalities III, Academic Press, New York, 1972.

Kle — Klein, E., *A comprehensive etymological dictionary of the English language*, Elsevier Science Publishing Company, Inc. New York, 1971.

Ko — Koopmans, T. C., *Activity analysis of production and allocation*, Cowles Commission for Research in Economics, Monograph 13, John Wiley & Sons, Inc., New York, 1951.

L — Lang, S., *Algebra*, Addison-Wesley Publishing Company, Inc., Palo-Alto, 1965.

Le — Leach, R. J., *On Gelbaum's algorithm for computing the minimal polynomial of a matrix*, Amer. Math. Monthly, 92 (1985), 208–9.

Ma — Mackey, G. W., *Mathematical foundations of quantum mechanics*, W. A. Benjamin, Inc., New York, 1963.

MaS — Marshall, K. T. and Suurballe, J. W., *A note on cycling in the simplex method*, Naval Research Logistics Quarterly, 16 (1969), 121–137.

Me — Mendelson, E., *Introduction to mathematical logic*, D. van Nostrand Company, Inc., New York, 1964.

MiW — Miller, W. and Wrathall, C., *Software for roundoff analysis of matrix algorithms*, Academic Press, New York, 1980.

Miy — Mirsky, L., *An introduction to linear algebra*, Oxford University Press, London, 1955.

Mo — Morris, William (Editor), *The American Heritage dictionary of the English language*, American Heritage Publishing Company, Inc., New York, 1973.

Mu — Murty, K. G., *Linear programming*, John Wiley & Sons, New York, 1983.

O — Olmsted, J. M. H., *Calculus with analytic geometry*, Appleton-Century-Crofts, New York, 1966.

P — Pan, V., *How to multiply matrices faster*, Springer-Verlag, New York, 1984.

Re — Redheffer, R. M., *What! Another note just on the fundamental theorem of algebra,* Amer. Math. Monthly, 71 (1964), 180–5.

Ri — Rice, N. M., *An algorithmic derivation of the Jordan canonical form,* preprint.

Ro — Robinson, J., *An iterative method of solving a game*, Ann. Math., (2), 54,(1951), 296–301.

Ros — Rosenblatt, M., *Random processes*, Oxford University Press, New York, 1962.

Rud — Rudin, W., *Real and complex analysis*, Second edition, McGraw-Hill, Inc., New York, 1974.

Sm1 — Smale, S., *On the average speed of the simplex method of linear programming*, Technical report, Department of Mathematics, University of California, Berkeley, 1982.

Sm2 — *ibid., On the average number of steps of the simplex method of linear programming*, Math. Programming, 27 (1983), no. 3, 241–262.

Ta — Taylor, A. E., *Introduction to functional analysis*, John Wiley & Sons, Inc., New York, 1958.

T — Terkelsen, F., *The fundamental theorem of algebra*, Amer. Math. Monthly, 83 (1976), 646–7.

vN1 — von Neumann, J., *Mathematische Grundlagen der Quantenmechanik*, Springer-Verlag, Berlin, 1932..

vN2 — von Neumann, J., *Mathematical foundations of quantum mechanics*, Princeton University Press, Princeton, 1955..

vNM — von Neumann, J. and Morgenstern, O., *Theory of games and economic behavior, Second Edition*, Princeton University Press, 1947.

Wi — Wielandt, H., *Unzerlegbare, nichtnegative Matrizen*, Math. Z. 52 (1950), 642-8.

Wil — Wilkinson, J. M., *The algebraic eigenvalue problem*, Oxford University Press, London, 1965.

WR — Wilkinson, J. M. and Reinsch, C., *Handbook for automatic computation, linear algebra*, Springer-Verlag, New York, 1971.

ZA — Zukhovitskiy, S. I. and Avdeyeva, L. I., *Linear and convex programming*, W. B. Saunders Company, Philadelphia, 1966.

SYMBOL LIST

The notation $x.y.z$ indicates page z in **Section x.y.**

$(a_{ij})_{i,j=1}^{m,n}$ 2.1.31
$(A|\mathbf{b})$ 2.3.36
A^k 1.1.16
A^t 1.1.11
A^{-1} 2.8.101
A_E 2.4.61
A_J 2.4.62
A_P 2.4.63
$A_P|\mathbf{c}$ 2.3.40, 2.4.63
A^+ 4.4.321
A_R 2.7.99
A_S 5.3.407
$A^{\times}$ 2.9.146
$\overline{A}$ 2.9.146
$\mathbf{C}$ 1.1.10
$\mathbf{C}^n$ 2.8.121
$\mathbf{C}_n$ 2.8.121
CON 6.4.592
$Conv(S)$ 5.4.418
$deg(p)$ 4.2.237
$det(A)$ 3.1.173
$dim(W)$ 2.9.133
$dom(A)$ 2.9.134
d_x 4.5.338
e 5.9.539
dim 2.9.133
$\mathcal{E}$ 2.10.154
E_{ij} 2.6.70
$E_i(s)$ 2.6.72
$E_{ij}(r)$ 2.6.70
FTA 4.8.370
$\mathcal{G}$ 4.4.318
GCD 4.3.272
$H_{[\mathbf{a},r]}$ 5.4.417
$H_{\mathbf{x}}$ 5.3.406
$im(A)$ 2.9.131
$\Im(z)$ 2.9.150
$ker(A)$ 2.9.131

$\mathbf{K}$ 6.1.556
$\mathbf{K}^n$ 6.3.570
LCM 4.4.282
$\mathcal{L}(V)$ 6.3.566
$\mathcal{L}(V,W)$ 6.3.566
$m \times n$ 1.1.5
$\mathbf{m}(A)$ 5.4.428
$\mathbf{M}(A)$ 5.4.428
Mat_{mn} 3.1.179
$Mat_{\Lambda \times \Lambda}$ 6.3.570
m_A 4.3.258
$m_{A,\mathbf{x}}$ 4.3.255
$\mathbf{N}$ 1.1.19
$\mathcal{N}_A$ 2.9.131
$\mathbf{O}$ 2.3.46
O' 8.1.607
$\mathbf{O}^c$ 2.8.110
$\mathbf{O}_r$ 2.8.110
$\mathbf{Q}$ 4.8.371
$\mathbf{R}$ 1.1.10
$\mathbf{R}^n$ 2.3.52
$\mathcal{R}_A$ 2.9.131
$\Re(z)$ 2.9.150
S/E 2.3.40
$S_{[\mathbf{v},r]}$ 5.4.417
$S^o_{[\mathbf{a},r]}$ 5.4.417
$sgn(\pi)$ 3.2.192
SOP 6.2.561
$S^{\perp}$ 4.4.286
$\#(S)$ 3.2.193
$supp(f)$ 8.3.612
SVD 5.2.393
$tr(A)$ 4.1.230
$T^+(\mathbf{b})$ 4.4.320
U_{ij} 1.1.17
$\{u\}^{\perp}$ 4.4.297
$[V,W]$ 6.3.570
V' 8.1.606

$\widetilde{V}'$ 8.1.606
$\mathbf{V}(A)$ 5.4.430
$(\mathbf{x},\mathbf{y})$ 4.4.285
$[\mathbf{x},\mathbf{y}]$ 7.2.602
$\mathbf{x} \times \mathbf{a}$ 4.4.317
X-components 2.10.156
$[x]$ 4.5.338
$\{x\}$ 5.9.544
$\mathbf{Z}$ 3.2.204
$\mathbf{Z}/(p)$ 6.1.556
$\mathbf{Z}\mathrm{mod}\,(p)$ 6.1.556

Γ 4.4.308
δ_{ij} 2.9.145
$\delta_{\mu\lambda}$ 8.1.606
Δ_R 4.8.371
∂ 5.3.411
$\kappa(A)$ 5.1.385
μ_A 4.3.258
$\mu_{A,\mathbf{x}}$ 4.3.255
π 3.2.192
π^{-1} 3.2.193
$\pi\mathbf{X}$ 8.2.609
ρ_A 5.1.387
$\Sigma^{(l)}$ 5.4.418
Σ_k 5.4.418
$\sigma(A)$ 4.7.361
$\perp$ 4.4.286
$\oplus$ 2.10.171
$\otimes$ 8.3.613
$\prec$ 1.1.20
$\preceq$ 1.1.20
$\succ$ 1.1.20
$\succeq$ 1.1.20
$\Rightarrow$ 2.8.106
$\Leftrightarrow$ 2.6.83

INDEX

The notation $x.y.z$ indicates page z in **Section x.y.**